Methods in Enzymology

Volume 85
STRUCTURAL AND CONTRACTILE PROTEINS
Part B
The Contractile Apparatus and the Cytoskeleton

METHODS IN ENZYMOLOGY

EDITORS-IN-CHIEF

Sidney P. Colowick   Nathan O. Kaplan

Methods in Enzymology

Volume 85

Structural and Contractile Proteins

Part B

The Contractile Apparatus and the Cytoskeleton

EDITED BY

Dixie W. Frederiksen

DEPARTMENT OF BIOCHEMISTRY
VANDERBILT UNIVERSITY
NASHVILLE, TENNESSEE

Leon W. Cunningham

DEPARTMENT OF BIOCHEMISTRY
VANDERBILT UNIVERSITY
NASHVILLE, TENNESSEE

1982

ACADEMIC PRESS

A Subsidiary of Harcourt Brace Jovanovich, Publishers

New York London
Paris San Diego San Francisco São Paulo Sydney Tokyo Toronto

ACADEMIC PRESS, INC.
111 Fifth Avenue, New York, New York 10003

United Kingdom Edition published by
ACADEMIC PRESS, INC. (LONDON) LTD.
24/28 Oval Road, London NW1 7DX

Library of Congress Cataloging in Publication Data
Main entry under title:

Structural and contractile proteins.

    (Methods in enzymology ; v. 85,)
    Includes bibliographical references and index.
    Contents: pt. A. Extracellular matrix -- pt. B.
The Contractile apparatus and the cytoskeleton.
    1. Proteins. I. Colowick, Sidney P. II. Series.
QP601.M49 vol. 85, etc.   574.19'25s   81-20622
ISBN 0-12-181985-X   [574.19'245]   AACR2

PRINTED IN THE UNITED STATES OF AMERICA

82 83 84 85   9 8 7 6 5 4 3 2 1

Table of Contents

Section I. Methods for Striated Muscle Chemistry

Section II. Methods for Smooth Muscle Chemistry

Section III. Methods for Study of the Motility Apparatus in Nonmuscle Cells

Section IV. Special Techniques for the Study of the Contractile Protein Complex and the Cytoskeleton

Contributors to Volume 85

Article numbers are in parentheses following the names of contributors.
Affiliations listed are current.

ROBERT S. ADELSTEIN (27, 28), *Laboratory of Molecular Cardiology, National Heart, Lung, and Blood Institute, National Institutes of Health, Bethesda, Maryland 20205*

JOHN F. ASH (48), *Department of Anatomy, The University of Utah College of Medicine, Salt Lake City, Utah 84132*

MICHAEL BÁRÁNY (53), *Department of Biological Chemistry, University of Illinois at the Medical Center, Chicago, Illinois 60612*

CHRISTOPHER W. BELL (43), *Pacific Biomedical Research Center, University of Hawaii, Honolulu, Hawaii 96822*

PAULINE M. BENNETT (47), *Medical Research Council, Cell Biophysics Unit, London WC2B 5RL, England*

FRANCIS D. CARLSON (49), *The Thomas C. Jenkins Department of Biophysics, The Johns Hopkins University, Baltimore, Maryland 21218*

ISAAC COHEN (35), *Atherosclerosis Program, Department of Molecular Biology, Northwestern University Medical School, Chicago, Illinois 60611*

JIMMY H. COLLINS (32), *Laboratory of Cell Biology, National Heart, Lung, and Blood Institute, National Institutes of Health, Bethesda, Maryland 20205*

PETER COOKE (24), *Department of Physiology, University of Connecticut School of Medicine, Farmington, Connecticut 06032*

ROGER COOKE (51), *Department of Biochemistry/Biophysics, and The Cardiovascular Research Institute, University of California, San Francisco, California 94143*

JOHN A. COOPER (19, 20, 29), *Department of Cell Biology and Anatomy, The Johns Hopkins University School of Medicine, Baltimore, Maryland 21205*

SUSAN W. CRAIG (29), *Department of Physiological Chemistry, The Johns Hopkins University School of Medicine, Baltimore, Maryland 21205*

PETER J. A. DAVIES (30), *Department of Pharmacology, University of Texas Medical School at Houston, Houston, Texas 77025*

EVAN EISENBERG (56), *Section on Cellular Physiology, Laboratory of Cell Biology, National Heart, Lung, and Blood Institute, National Institutes of Health, Bethesda, Maryland 20205*

MYRA J. ELFVIN (16), *The Medical College of Pennsylvania, Philadelphia, Pennsylvania 19129*

HANS M. EPPENBERGER (15), *Department of Cell Biology, Swiss Federal Institute of Technology (ETH), 8093 Zürich, Switzerland*

KEVIN W. FARRELL (37), *Department of Biological Sciences, University of California, Santa Barbara, California 93106*

F. S. FAY (25), *Department of Physiology, University of Massachusetts Medical School, Worcester, Massachusetts 01605*

JAMES R. FERAMISCO (48), *Cold Spring Harbor Laboratory, Cold Spring Harbor, New York 11724*

CLARENCE L. FRASER (43), *Pacific Biomedical Research Center, University of Hawaii, Honolulu, Hawaii 96822*

DIXIE W. FREDERIKSEN (26, 34), *Department of Biochemistry, Vanderbilt University, Nashville, Tennessee 37232*

FELICIA GASKIN (41), *Departments of Pathology and Physiology and Biophysics, Albert Einstein College of Medicine, Bronx, New York 10461*

JOHN GERGELY (1), *Department of Muscle Research, Boston Biomedical Research Institute, Boston, Massachusetts 02114*

I. R. GIBBONS (43), *Pacific Biomedical Research Center, University of Hawaii, Honolulu, Hawaii 96822*

THOMAS GLONEK (53), *Nuclear Magnetic Resonance Laboratory, Chicago College of Osteopathic Medicine, Chicago, Illinois 60615*

CHARLES C. GOODNO (12), *Department of Physiology 206H, University of North Carolina School of Medicine, Chapel Hill, North Carolina 27514*

B. L. GRANGER (46), *Division of Biology, California Institute of Technology, Pasadena, California 91125*

W. B. GRATZER (44), *Medical Research Council, Cell Biophysics Unit, King's College, London WC2B 5RL, England*

LOIS E. GREENE (56), *Section on Cellular Physiology, Laboratory of Cell Biology, National Heart, Lung, and Blood Institute, National Institutes of Health, Bethesda, Maryland 20205*

JOHN H. HARTWIG (45), *Hematology-Oncology Unit, Massachusetts General Hospital, Boston, Massachusetts 02114*

R. HOFFMANN (25), *Department of Physiology, University of Massachusetts Medical School, Worcester, Massachusetts 01605*

TIMOTHY L. KARR (40, 42), *Department of Biochemistry and Biophysics, University of California School of Medicine, San Francisco, California 94143*

CYRIL M. KAY (54), *Medical Research Council Group in Protein Structure and Function, Department of Biochemistry, University of Alberta, Edmonton, Alberta T6G 2H7, Canada*

JOHN KENDRICK-JONES (33), *MRC Laboratory of Molecular Biology, Cambridge CB2 2QH, England*

CLAUDE B. KLEE (27), *Laboratory of Biochemistry, National Cancer Institute, National Institutes of Health, Bethesda, Maryland 20205*

PETER J. KNIGHT (2, 3), *Muscle Biology Division, Agricultural Research Council,*

Meat Research Institute, Langford, Bristol BS18 7DY, England

JANE F. KORETZ (6), *Department of Biology, Rensselaer Polytechnic Institute, Troy, New York 12181*

EDWARD D. KORN (32), *Laboratory of Cell Biology, National Heart, Lung, and Blood Institute, National Institutes of Health, Bethesda, Maryland 20205*

SONJA KRAUSE (17), *Department of Chemistry, Rensselaer Polytechnic Institute, Troy, New York 12181*

DAVID KRISTOFFERSON (42), *Department of Chemistry, University of California, Santa Barbara, California 93106*

CHRISTINE L. LANCASHIRE (29), *Department of Physiological Chemistry, The Johns Hopkins University School of Medicine, Baltimore, Maryland 21205*

E. LAZARIDES (46), *Division of Biology, California Institute of Technology, Pasadena, California 91125*

S. LECLAIR (25), *Department of Physiology, University of Massachusetts Medical School, Worcester, Massachusetts 01605*

JAMES C. LEE (36), *Edward Doisy Department of Biochemistry, St. Louis University School of Medicine, St. Louis, Missouri 63104*

RHEA J. C. LEVINE (16), *Department of Anatomy, The Medical College of Pennsylvania, Philadelphia, Pennsylvania 19129*

STEPHEN J. LOVELL (3), *Department of Experimental Pathology, The John Curtin School of Medical Research, The Australian National University, Canberra City ACT 2061, Australia*

SUSAN LOWEY (7), *The Rosenstiel Basic Medical Sciences Research Center, Brandeis University, Waltham, Massachusetts 02254*

WILLIAM D. McCUBBIN (54), *Medical Research Council Group in Protein Structure and Function, Department of Biochemistry, University of Alberta, Edmonton, Alberta T6G 2H7, Canada*

ROBERT K. MACNEAL (40), *Department of Chemistry, University of California, Santa Barbara, California 93106*

SARKIS S. MARGOSSIAN (7), *Department of Biochemistry and Medicine, Albert Einstein College of Medicine, Montefiore Hospital and Medical Center, Bronx, New York 10467*

HIROSHI MARUTA (32), *Laboratory of Cell Biology, National Heart, Lung, and Blood Institute, National Institutes of Health, Bethesda, Maryland 20205*

P. MERRIAM (25), *Department of Physiology, University of Massachusetts Medical School, Worcester, Massachusetts 01605*

CHARLES MONTAGUE (49), *The Thomas C. Jenkins Department of Biophysics, The Johns Hopkins University, Baltimore, Maryland 21218*

NANCY LETKO MUNSON (17), *Department of Chemistry, Rensselaer Polytechnic Institute, Troy, New York 12181*

JOHN M. MURRAY (4), *Department of Structural Biology, School of Medicine, Stanford University, Stanford, California 94305*

GEORGE C. NA (38), *Eastern Regional Research Center, United States Department of Agriculture, Philadelphia, Pennsylvania 19118*

REGINA NIEBIESKI (9), *Biology Department, Brandeis University, Waltham, Massachusetts 02254*

GERALD OFFER (14), *Muscle Biology Division, Agricultural Research Council, Meat Research Institute, Langford, Bristol BS18 7DY, England*

JOEL D. PARDEE (18), *Department of Structural Biology, Fairchild Center, Stanford University, Stanford, California 94305*

IRA PASTAN (30), *Laboratory of Molecular Biology, National Cancer Institute, National Institutes of Health, Bethesda, Maryland 20205*

MARY D. PATO (28), *Laboratory of Molecular Cardiology, National Heart, Lung, and Blood Institute, National Institutes of Health, Bethesda, Maryland 20205*

THOMAS D. POLLARD (13, 19, 20, 31), *Department of Cell Biology and Anatomy, The Johns Hopkins University School of Medicine, Baltimore, Maryland 21205*

JAMES D. POTTER (22), *Section of Contractile Proteins, Department of Pharmacology and Cell Biophysics, University of Cincinnati College of Medicine, Cincinnati, Ohio 45267*

DANIEL L. PURICH (40, 42), *Department of Chemistry, University of California, Santa Barbara, California 93106*

DIANNE D. REES (26), *Department of Physiology, University of Massachusetts Medical School, Worcester, Massachusetts 01605*

EMIL REISLER (10), *Department of Chemistry and Biochemistry and the Molecular Biology Institute, University of California, Los Angeles, California 90024*

JOEL L. ROSENBAUM (39), *Department of Biology, The Kline Biology Tower, Yale University, New Haven, Connecticut 06520*

WINFIELD S. SALE (43), *Department of Anatomy, Emory University, Atlanta, Georgia 30322*

VITALY SAWYNA (16), *Department of Surgery, Allentown Sacred Heart Hospital, Allentown, Pennsylvania 18102*

J. M. SCHOLEY (33), *MRC Laboratory of Molecular Biology, Cambridge CB2 2QH, England*

JOHN C. SEIDEL (52), *Department of Muscle Research, Boston Biomedical Research Institute, Boston, Massachusetts 02114*

YUTAKA SHIZUTA (30), *Department of Biochemistry, Kochi Medical School, Kohasu, Okocho, Nangoku City, Kochi, Japan*

ROGER D. SLOBODA (39), *Department of Biological Sciences, Dartmouth College, Hanover, New Hampshire 03755*

L. B. SMILLIE (21), *Medical Research Council of Canada, Group in Protein Structure and Function, Department of Biochemistry, University of Alberta, Edmonton, Alberta T6G 2H7 Canada*

JAMES A. SPUDICH (18), *Department of Structural Biology, Fairchild Center, Stanford University, Stanford, California 94305*

WALTER F. STAFFORD III (50), *Department of Muscle Research, Boston Biomedical Research Institute, Boston, Massachusetts 02114, and Department of Neurology, Harvard Medical School, Boston, Massachusetts 02114*

ROGER STARR (14), *Muscle Biology Division, Agricultural Research Council, Meat Research Institute, Langford, Bristol BS18 7DY, England*

THOMAS P. STOSSEL (45), *Hematology-Oncology Unit, Massachusetts General Hospital, Boston, Massachusetts 02114*

EMANUEL E. STREHLER (15), *Department of Cell Biology, Swiss Federal Institute of Technology (ETH), 8093 Zürich, Switzerland*

ANDREW G. SZENT-GYÖRGYI (9), *Biology Department, Brandeis University, Waltham, Massachusetts 02254*

WEN-JING Y. TANG (43), *Pacific Biomedical Research Center, University of Hawaii, Honolulu, Hawaii 96822*

K. A. TAYLOR (33), *Department of Anatomy, Duke University Medical Center, Durham, North Carolina 27710*

BRIAN J. TERRY (40), *Department of Biochemistry, Duke University Medical Center, Durham, North Carolina 27710*

SERGE N. TIMASHEFF (38), *Graduate Department of Biochemistry, Brandeis University, Waltham, Massachusetts 02254*

JOHN A. TRINICK (2, 5), *Muscle Biology Division, Agricultural Research Council, Meat Research Institute, Langford, Bristol BS18 7DY, England*

PAUL D. WAGNER (8), *School of Medicine, University of California, San Francisco, San Francisco, California 94143*

KUAN WANG (23, 48), *Department of Chemistry and Clayton Foundation Biochemical Institute, University of Texas at Austin, Austin, Texas 78712*

JERRY P. WEIR (34), *Laboratory of Biology of Viruses, National Institute of Allergy and Infectious Diseases, National Institutes of Health, Bethesda, Maryland 20205*

JAMES A. WELLS (11), *Department of Biochemistry, Stanford University School of Medicine, Stanford, California 94305*

HOWARD D. WHITE (55), *Department of Biochemistry, University of Arizona, Tucson, Arizona 85721*

ROBLEY C. WILLIAMS, JR. (36), *Department of Molecular Biology, Vanderbilt University, Nashville, Tennessee 37232*

RALPH G. YOUNT (11), *Biochemistry/Biophysics Program, Department of Chemistry and the Institute of Biological Chemistry, Washington State University, Pullman, Washington 99164*

Preface

The techniques and preparative procedures of the contractile protein chemist are becoming increasingly important to investigators whose interests do not involve the molecular basis for contraction in striated muscle. The interest in smooth muscle contraction has grown tremendously over the past ten years. In addition, actin, myosin, and an array of other contractile–cytoskeletal proteins have been discovered in a large number of nonmuscle cell types. The study of these proteins has provided important new insight into the molecular biology of cell growth, maintenance, and function.

The need for a coherent treatment of the experimental methods used for the investigation of the contractile and cytoskeletal proteins and of their interactions has been evident for some time. This volume is designed to fill that need.

In some cases the state of the art is just not developed to the extent that explicit procedures are established. Nevertheless, this volume does provide a unified representation of the current techniques used by contractile–cytoskeletal protein chemists. We hope that this work will provide a better basis for the integration of data from various laboratories and will, as well, prove useful to many other enzymologists, chemists, and biologists.

The cooperation, suggestions, and contributions of the individual authors have been most helpful. The staff of Academic Press has provided support and encouragement. We are grateful for this assistance. Very special thanks are due Dr. Sidney P. Colowick for his continual wisdom and insight.

DIXIE W. FREDERIKSEN
LEON W. CUNNINGHAM

METHODS IN ENZYMOLOGY

EDITED BY

Sidney P. Colowick and Nathan O. Kaplan

VANDERBILT UNIVERSITY
SCHOOL OF MEDICINE
NASHVILLE, TENNESSEE

DEPARTMENT OF CHEMISTRY
UNIVERSITY OF CALIFORNIA
AT SAN DIEGO
LA JOLLA, CALIFORNIA

METHODS IN ENZYMOLOGY

EDITORS-IN-CHIEF

Sidney P. Colowick   Nathan O. Kaplan

[1] Introduction

By JOHN GERGELY

This chapter considers a number of current problems in contractility and relates them to the techniques described in this volume. References are rather sparingly used, since they are largely to be found in other chapters of this volume. A few that include useful summaries of related topics not covered are included. Reference is also made to the various contributions in this volume (numbers in brackets designate the chapters referred to).

It is now well established that the contraction of muscle cells—skeletal, cardiac, and smooth—depends on the interaction of actin and myosin, in a process that involves the hydrolysis of ATP. This ATP–protein interaction results in the transduction of chemical energy into mechanical work. Many types of motions in nonmuscle cells also are based on actin–myosin interactions, but some motions, such as those of cilia, involve tubulin and dynein (see [43]). An important difference between muscle and nonmuscle systems is found in the organization of the proteins that form the contractile machinery. In muscle cells myosin and actin are laid down in permanent structures, thick and thin filaments, respectively, sliding past each other in the course of contraction. In striated muscle the filaments are part of the structure of the hierarchy of the sarcomere. Nonmuscle cells are characterized by a more fleeting organization. In muscle cells the contraction process is regulated through the control of the interaction of thick- and thin-filament components, notably the myosin heads and actin. In nonmuscle systems, not only is there control of actin-myosin interaction, but the supramolecular structures made up of actin and myosin are in a dynamic state, and their assembly and disassembly assumes a regulatory role.

While the constituents of the contractile machinery are built on the same pattern in striated and smooth muscle, the control system shows a difference in the two types. Striated muscles operate with a control system located in the thin filaments. This consists of tropomyosin [21]—containing a pair of identical or nonidentical subunits—and the three-subunit-complex of troponin [22]. In smooth muscle, according to the current, generally held opinion (a somewhat different view is held by Ebashi and his colleagues[1]), the key process is the phosphorylation of a subunit of myosin by a specific kinase. Both striated and smooth muscle

---

[1] S. Ebashi, *Proc. R. Soc. London Ser. B* **207**, 259 (1980).

resemble each other in that the activation of muscle is triggered by an increase in the ionized calcium level, and it is in the final pathway of the activation process that the difference lies. In striated muscle calcium binds to troponin and, in an as yet not fully understood way, releases the inhibition of the interaction of actin and myosin. In smooth muscle, calcium ions, according to the generally accepted view, are necessary for the activation of the myosin light-chain kinase, and once phosphorylation has taken place the muscle is active. Inactivation in striated muscle occurs upon sequestration of calcium by the sarcoplasmic reticulum[2] through the dissociation of calcium from troponin; in smooth muscle, lowering of the calcium level through uptake into sarcoplasmic reticulum and possibly by other mechanisms results in an inactivation of the light-chain kinase [27] and a shift to dephosphorylation through the activity of a phosphatase [28].

These processes can be studied on various levels, and one of the simplest and most complete systems is represented by myofibrils whose preparation is described in [2]. Although myofibrils contain all the components of the contractile and regulatory apparatus, a more detailed analysis can be carried out by investigation of the thin filaments native or reconstituted from their constituents, as well as by the use of reconstituted and native thick filaments [3–6]. Analogous systems exist for smooth muscle [24–26]. In view of the fact that different muscles contain different components, it has been of considerable interest to prepare hybrid systems [4,8] containing individual components that come from different sources; these permit the pinpointing of what is essential and what is not essential in the function of the individual components.

The chief component of the thick filaments, myosin, is itself a microcosm. It consists of six subunits, two heavy and four light. It is a two-headed structure with the two heavy chains forming a coiled coil in the C-terminal half, whereas the N-terminal ends form two distinct globular heads. It is with each of these heads that a pair of light chains is associated. Myosin can be studied by investigating the subunits and the combination of the heavy chains of one type with the homologous light chains or with light chains from a different type of muscle [8]. Thus hybrids can be formed between skeletal and cardiac subunits or among skeletal subunits from fast-twitch and slow-twitch muscles. The use of sodium dodecyl sulfate gel electrophoresis and, more recently, of electrophoresis under nondissociating conditions utilizing a pyrophosphate medium has been of great help. In addition to the dissociation of myosin into its subunits, its finer structure can be analyzed by investigation of proteolytic fragments that can be identified in terms of their location in the intact

---

[2] M. Tada, T. Yamamoto, and Y. Tonomura, *Physiol. Rev.* **58**, 1 (1978).

structure [7]. The head regions give rise to subfragment-1, while light meromyosin and the larger rod portion correspond to the core of the thick filaments. The connection between core and head is made through that portion of myosin which is identified as subfragment-2. The two-headed fragment containing the heads and subfragment-2 is heavy meromyosin.

In addition to the regulatory mechanism represented by the striated muscle of vertebrates, in which the troponin–tropomyosin system plays a key role, and the phosphorylation-dependent system of smooth muscle, a third system has been discovered in a number of invertebrate muscles and in this volume is represented by scallop myosin [9]. In this type of system, calcium combines directly with myosin in the presence of a so-called regulatory light chain. This light chain is related to the phosphorylatable light chain of smooth muscle and to one of the light chains in skeletal muscle, which is also phosphorylatable, although in the latter case the role of phosphorylation has not been established. It should be noted that protein sequence studies have revealed extensive homologies among the various myosin light chains, troponin C, and calmodulin. The latter is a ubiquitous calcium-binding protein, which is also the calcium-binding subunit of the light-chain kinases of smooth and striated muscle.

Considerable progress was made during the 1970s in arriving at a detailed description of the elementary steps of the ATPase reaction catalyzed by myosin and actomyosin. These developments have greatly benefited from the use of rapid kinetic techniques, such as stop flow and rapid quenching [55,56], and have been facilitated by the existence of sulfhydryl groups in myosin whose modification affects the ATPase activity [10]. It is also possible to attach to these groups probes that then respond to the combination of myosin with ATP or its analogs and exhibit various changes as the myosin molecule goes through the ATPase cycle. The crosslinking of two thiol groups found in the myosin head makes it possible to trap the nucleotide at its site [11]; another kind of trapping at the active site has been possible by replacing phosphate with vanadate [12]. The fuller implications of these findings are yet to be explored.

Probes attached to the thiol groups, and indeed to other groups, on myosin may exhibit distinct optical absorption features, may upon illumination emit fluorescent light, or have magnetic properties to be studied by electron spin resonance techniques [6]. These techniques serve a dual function. They can be used to detect changes in the environment of the probe, or they can serve to follow motions, if the probe is rigidly attached, of certain segments of the molecule. The use of probes is not restricted to myosin. In the case of myosin, probes serve as monitors of the motion of the myosin heads; in the case of actin or tropomyosin, they can detect dynamic changes in the conformation of the corresponding subunits. Monitoring of myosin head motions is of particular interest since, according to currently accepted views, myosin heads undergo a cyclic process of attachment to actin, change in orientation, and eventual detachment that involves the hydrolysis of 1 ATP molecule.

Before turning to myosin's partner in the contractile process, actin, it should be pointed out that thick filaments contain various other proteins, such as C-protein, H-protein, X-protein, M-line proteins, creatine kinase, and that appropriate techniques for isolation and purification are important [14,15]. In some specialized systems in lower organisms, myosin filaments are formed by association with paramyosin [16,17], the latter forming the core of the filaments. Some muscles, such as those of *Limulus*, that contain paramyosin show a length change in the thick filaments on shortening that is in contrast to the constant-length behavior of thick filaments in rabbits, say. The involvement of phosphorylation in this process increases its interest from a comparative point of view.

The main constituent of thin filaments of muscle is actin. Actin is isolated in globular form and one of the remarkable properties of its proteins is its ability to undergo polymerization known as the G–F transformation accompanied by the hydrolysis of ATP. Actin filaments together with tubulin have been found to undergo treadmilling (see [42]); that is, preferential polymerization at one end and depolymerization at the other. This constitutes an energy-requiring steady-state process. Over the years appropriate preparative and analytical techniques have evolved [18,19], as well as methods for characterizing networks that result from association of F-actin filaments [20]. Actin is one of the most sociable proteins in that each globular unit contains tightly bound metal ions, a nucleotide, and interacts within the thin filament with as many as four other actins and with the regulatory proteins, tropomyosin and troponin. In a contracting muscle, actin also undergoes cyclic interactions with the myosin heads. The thin filaments are attached to the Z disks in whose structure α-actinin as well as other proteins, including desmin, are involved. Most proteins of the thin filament have been sequenced (for methods, see footnote 3); the light chains' sequences are also known, but the complete sequencing of the heavy chains (much progress has already been made, particularly by M. Elzinga and his colleagues) presents a more difficult task. Progress is likely to be accelerated by current developments in sequencing the corresponding genes. Knowledge of the primary structure of the myofibrillar proteins will be particularly useful for obtaining their three-dimensional structure when crystals for X-ray studies become available. Up to the present, X-ray studies[4] have played an important role—supplemented by

---

[3] This series, Vol. 57, Section VIII.
[4] J. S. Wray and K. C. Holmes, *Annu. Rev. Physiol.* **43**, 553 (1981).

electron microscopic and optical diffraction techniques [47]—in the analysis of the disposition of various constituents within the filaments and of the structure of the complex of myosin, or its heads, with actin and the thin filaments.

The physicochemical methods described in this volume (see Section IV) are important in studies of the properties of the individual components and the interaction of the high-molecular-weight constituents of the various muscle filaments in solution, which in many instances permit inferences concerning the interaction in more organized systems. Dynamic studies utilizing fluorescence and magnetic probes on supramolecular structures, in conjunction with X-ray techniques utilizing time resolution on a millisecond scale,[5] together with physiological investigations of rapid transients[6] of contracting muscle, promise a more detailed, truly molecular elucidation of the contraction process. Nuclear magnetic resonance [53] is coming into its own in the study of muscle, and probably of other contractile systems, as a monitor of molecular dynamics, but also as a method for *in situ* metabolic studies of fundamental and clinical significance.

As pointed out above, nonmuscle systems have been shown to contain essentially all the components found in the contractile apparatus of the muscle cells with the possible exception of troponin. These nonmuscle proteins (see Section III) exhibit considerable similarities to their skeletal counterparts, but also differ in many aspects of their submolecular structure. Thus the interesting myosin from *Acanthamoeba castellanii* [32] undergoes phosphorylation of its heavy chain. Phosphorylation of other nonmuscle myosins play a role in the stabilization of the thick-filament structure [33]. Nonmuscle cytoplasmic actin *in vitro* undergoes polymerization as does its skeletal counterpart, and this process, as well as depolymerization, may play important roles *in vivo* in nonmuscle systems. They are probably regulated by a number of proteins [31], some of them similar to filamins [30], found in smooth muscles, in their actin-binding properties.

Tubulin and its associated proteins [36–42] are the building blocks of the microtubules that are found in a variety of tissues and are particularly prominent in nervous tissue. These microtubules share with actin the ability to undergo treadmilling in an end-selective polymerization and depolymerization process [42]. Tubulin-containing microtubules are usually considered as cytoskeletal elements, but their potential to be involved in motile processes is exemplified by their interaction with dynein [43],

[5] H. E. Huxley, R. M. Simmons, A. R. Faruqi, M. Kress, J. Bordas, and M. H. J. Koch, *Proc. Natl. Acad. Sci. U.S.A.* **78**, 2297 (1981).
[6] A. F. Huxley and R. M. Simmons, *Nature (London)* **233**, 533 (1971).

which constitutes another ATP-linked mechanochemical transduction system in cilia. Proteins present in the ubiquitous intermediate filaments (10-nm filaments) carry a variety of names depending on the tissue in which they are found. Thus desmin and vimentin are different, but similar, constituents of intermediate filaments [46], presumably playing slightly different roles.

The red cell membrane affords an interesting system in which the interaction of cytoskeletal elements, membrane constituents, and actin takes place [44]. Spectrin in some sense may be an analog of myosin: it interacts with actin and its chain molecular weight is very close to that of the heavy chain of myosin. Phosphorylation of spectrin has been demonstrated, although its precise functional significance in determining cell shape or such processes as endocytosis remains to be fully established.

This volume clearly portrays the complexity of motile and cytoskeletal systems, which in turn require a complex armamentarium for their study. The methods described bring these systems closer to investigators new to the field and will put powerful means into the hands of those already well versed in this fascinating area of study.

Acknowledgment

The author's work in this field has been supported by grants from the National Institutes of Health, National Heart, Lung and Blood Institute (HL-05949), the National Science Foundation, and the Muscular Dystrophy Association.

Section I

Methods for Striated Muscle Chemistry

[2] Preparation of Myofibrils

By PETER J. KNIGHT and JOHN A. TRINICK

Isolated myofibrils are fully competent fragments of the contractile apparatus of muscle, capable of calcium-dependent shortening in the presence of ATP. Their major difference from myofibrils *in vivo* is that they are broken, usually at or near the Z line, into comparatively short lengths, generally less than 20 sarcomeres.

Myofibrils are an important tool in the study of contraction because they bridge the gap between work on whole muscle, on the one hand, and studies of purified proteins on the other. Since they can be pipetted in precise amounts, they are valuable not only for studies of their structure and composition, but for enzyme assays of the myosin ATPase. Because myofibrils retain much of the structural organization of whole muscle, yet are sufficiently small that diffusion in and out of them occurs very quickly, they offer one of the few means by which the effects of immobilization of the myosin ATPase in the filament lattice can be assessed.

Preparative Procedures

A variety of procedures are available for the preparation of myofibrils, and we detail four of these here. Their applicability depends on such considerations as the type and quantity of myofibrils needed, how accurately sarcomere length is to be controlled, and whether the preparation is to be carried through to completion or stored until a later date. The methods that offer the greatest control over the myofibrils during their preparation are better suited to production of small (<1 g) quantities of material. This is because they necessitate maintaining muscle strips at fixed length until the intracellular calcium or ATP is reduced to levels below which contraction cannot occur. Homogenization is then performed in a "rigor buffer" so that as many cross-bridges as possible are attached to thin filaments, thereby forming a rigid protective lattice. Much larger quantities of myofibrils can be produced by methods that involve mincing or blending of relaxed muscle immediately postmortem, but here there is less control over the transition into rigor, and the resulting preparations tend to have less clearly defined band patterns.

Myofibrils from Fresh Rabbit Psoas Muscle. The psoas muscles are excised, teased into thin strips 3–5 mm in diameter, and attached by thin string to Perspex sticks. They are then depleted of calcium by incubation at 0° in an EGTA–Ringer's solution (100 mM NaCl, 2 mM KCl, 2 mM MgCl$_2$, 6 mM potassium phosphate, 1 mM EGTA, 0.1% glucose, pH 7.0 at 0°). (See also this volume [5] on the preparation of thick filaments.) At the stage of attachment to the sticks, the muscle strips are quite relaxed and, if required, can be stretched before tying to obtain myofibrils with a long sarcomere length; alternatively, they can be electrically stimulated to shorten the sarcomeres. After incubation in Ringer's solution for at least 12 hr, the muscle is chopped into short segments and homogenized in 3–10 volumes of ice cold rigor buffer in a flask chilled with ice. The rigor buffer we use is typically 0.1 M KCl, 2 mM MgCl$_2$, 1 mM EDTA, 0.5 mM dithiothreitol, 10 mM potassium phosphate, pH 7.0 measured at 0°. Using a 10-ml flask with an MSE homogenizer, 30 sec at top speed usually suffices to break up much of the muscle into myofibrils; however, the efficacy of homogenization is best checked by inspection in a light microscope, preferably using phase contrast optics. When a suspension containing mostly single myofibrils has been produced, these are sedimented at about 2000 g for 5 min (a gentle centrifugation producing an easily resuspended pellet). The turbid supernatant (containing many nuclei and other cell debris) is discarded, and the myofibrils are washed by resuspension in 10 volumes of the rigor buffer. Two cycles of sedimentation and resuspension remove most of the nuclei, but several more washes may be required to reduce the levels of contaminating enzymes to insignificance. In the light microscope the myofibrils generally appear straight (this is in contrast to those produced from glycerinated muscle, which are frequently slightly curved), and at high magnification they show clear Z lines and H zones, as well as A and I bands. Twisted and tangled myofibrils, perhaps with indistinct Z lines, indicate a degraded preparation.

Myofibrils from Glycerinated Psoas Muscle.[1] Strips of psoas muscle are tied onto sticks as described in the preceding section, but are then immersed in about 20 ml of a 1 : 1 (v/v) mixture of glycerol and 13.3 mM potassium phosphate, pH 7.0 at 0°. After infiltration of the glycerol overnight at 0°, the solution is changed and the muscle strips are transferred to a freezer at −20°. Here they can be stored for years without significant damage to the contractile assembly. In order to avoid severe contraction of the myofibrils during their preparation, the strips are left at least 5 weeks at −20° before use. Presumably this period is necessary because of the slow rate of breakdown of ATP and phosphocreatine.

To prepare myofibrils from the glycerinated material, the translucent muscle is teased into fine fiber bundles and soaked for 1 hr in rigor buffer (see preceding section for composition). It is then chopped into a homogenizer flask and blended with about 10 volumes of rigor buffer. A more vigorous homogenization is usually required than is necessary for

[1] J. Hanson and H. E. Huxley, *Symp. Soc. Exp. Biol.* **9**, 228 (1955).

fresh muscle. The myofibril suspension is washed as described in the preceding section.

Myofibrils in Bulk from Fresh Muscle. We give here the method of Perry and Corsi.[2] Chilled muscle is minced and immediately blended with 5 volumes of 0.1 M KCl, 5 mM EDTA, 39 mM potassium borate, pH 7.1 at 0°. A household blender, preferably with an antifoaming device, is satisfactory. The myofibrils are sedimented and washed by several cycles of resuspension and centrifugation. The initial blending is done with ice-cold solution to minimize contraction. EDTA, by chelating calcium and magnesium, reduces calcium activation and the MgATPase of the myofibrils. Myofibrils prepared in this way are typically about 2.2 μm in sarcomere length.

Myofibrils in Bulk from Stored Muscle. This method originates from observations of Cooke.[3] Fresh muscle is chilled and minced and immediately stirred into 3 volumes of a 1 : 1 mixture of glycerol and 0.1 M KCl, 1 mM MgCl$_2$, 10 mM potassium phosphate, pH 7.0 at 0°. The mixture is constantly stirred for at least 1 hr (until the mince does not float on standing), and then stored at −20° until myofibrils are needed. The resuspended glycerinated mince is then blended with 3 volumes of rigor buffer (see section on fresh psoas muscle, above) until a myofibril suspension is achieved; a household blender can be used. The myofibrils are sedimented and washed several times as described for fresh psoas muscle.

General Points concerning Preparation of Myofibrils

Residual fiber segments and aggregated debris can be removed by a brief low speed centrifugation of a dilute (e.g., 5 mg/ml) myofibril suspension, discarding the yellowish pelleted debris and retaining the milky supernatant. With longer centrifugation the myofibrils form a white pellet on top of the debris. If a transparent centrifuge bottle is used, the myofibrils can be removed with a spatula without disturbing the debris.

Myofibrils are stable in a wide variety of media. Therefore, after the first sedimentation of the myofibril preparation, the pellet may be resuspended in whatever medium is required. The composition of the rigor buffer is similar to intracellular conditions in terms of ionic strength and pH.

The protein concentration of a myofibril suspension can be estimated by the biuret method or by measuring the absorbance of an aliquot dissolved in warm 1% sodium dodecyl sulfate (SDS). The extinction coefficient at 280 nm is 0.7 ml/mg per centimeter. This value is computed from the weighted actin and myosin contributions and corrected for scattering

---

[2] S. V. Perry and A. Corsi, *Biochem. J.* **68**, 5 (1958).
[3] R. Cooke, *Biochem. Biophys. Res. Commun.* **49**, 1021 (1972).

by subtraction of the absorbance at 310 nm. If the myofibrils are in a buffer containing KCl, dilution with water can prevent the formation of a potassium–SDS precipitate.

Myofibrils tend to aggregate and to settle out of solution on standing, especially above 0°. Care should, therefore, be taken to stir the suspension if several equally concentrated aliquots are required.

A number of methods of skinning muscle fibers "chemically" have been published, and these can also lead to very satisfactory preparations of myofibrils (see, for instance, Magid and Reedy.[4] Methods of preparing myofibrils have also been published by Reedy et al.[5] for insect muscle and by Szent-Györgyi et al.[6] for molluscan muscle.

The sarcomere length of myofibrils can be measured directly in the light microscope with a calibrated graticule. Alternatively, in the case of myofibrils prepared from muscle tied to sticks, the sarcomere length can be measured before homogenization by laser light diffraction.[7] A thin strip of the muscle is irradiated by the unfocused beam, and the regular sarcomere repeat produces diffracted beams whose separation can conveniently be measured at a specimen to screen distance of 20–50 cm. The angle, θ, of the nth order is determined from its tangent, and the sarcomere length, d, is obtained from the relationship $n\lambda = d \sin \theta$. The laser light method, while very useful for indicating the sarcomere length, generally appears to overestimate the value compared with the result of direct measurement of the resulting myofibrils, especially at long sarcomere lengths. This is probably due to preferential breakage and loss of longer sarcomeres during preparation, since the yield from stretched fibers is low and the myofibrils tend to be short.[8]

[4] A. Magid and M. K. Reedy, *Biophys. J.* **30**, 27 (1980).
[5] M. K. Reedy, G. F. Bahr, and D. A. Fischman, *Cold Spring Harbor Symp. Quant. Biol.* **37**, 397 (1972).
[6] A. G. Szent-Györgyi, C. Cohen, and J. Kendrick-Jones *J. Mol. Biol.* **56**, 239 (1971).
[7] E. Rome, *J. Mol. Biol.* **27**, 591 (1967).
[8] S. J. Lovell and W. F. Harrington, *J. Mol. Biol.* **149**, 659 (1981).

[3] Preparation of Native Thin Filaments

By PETER J. KNIGHT and STEPHEN J. LOVELL

Interaction between actin and myosin is central to the production of force in muscle. This interaction is controlled in vertebrate skeletal muscle by the Ca^{2+}-sensitive regulatory protein complex of tropomyosin and troponin located with the actin in the thin filaments.[1] The many investigations of the mechanism of action of this regulatory system have almost invariably employed a synthetic thin filament preparation that is produced by combining separately purified actin, tropomyosin, and troponin. Apart from the labor involved in these several preparations, this approach suffers from the rather incomplete Ca^{2+} sensitivity of the synthetic filaments it produces. It continues to be followed because there is no method available that produces substantial quantities of intact thin filaments directly from muscle. The method given below has been developed by us to produce thin filaments directly from rabbit muscle. Conditions employed throughout the procedure are chosen to reduce the possibility of dissociation of tropomyosin from the filaments. Troponin is degraded by a proteolytic step in the procedure, but replacement by purified troponin consistently produces a filament preparation with at least 95% of full Ca^{2+} sensitivity.

Preparative Procedure

In outline, the method consists of the following steps: (a) preparation of myofibrils; (b) digestion of the myofibrils to cut the head–tail junction of most of the myosin molecules[2] and probably also weaken the Z disk; (c) release of myosin subfragment-1 (S-1) and thin filaments from the fibrils by stirring in the presence of magnesium pyrophosphate; (d) separation of the thin filaments from S-1 by centrifugation. The procedures are performed at 2–5° unless otherwise stated.

Hind leg and back muscles from the rabbit are minced and stirred into 3 volumes of 1 : 1 (v/v) glycerol and stock salt solution (100 mM KCl, 3 mM MgCl$_2$, 0.5 mM DTT, 10 mM imidazole–HCl, pH 7.3 at 5°). After a few hours of gentle stirring, the mince will sink rather than float on standing and can then be used immediately or stored at −20°. Glycerinated mince (250 ml of suspension) is mixed with 3 volumes of stock salt solution and blended until light microscopy shows that the resulting suspension of myofibrils contains very few fiber segments. The fibrils are collected by centrifugation at 10,000 g for 5 min and are resuspended in 500 ml of stock salt solution containing 0.6% Triton X-100 in a household blender at low speed to dissolve membrane vesicles. After extraction for 1 hr, the fibrils are sedimented and washed four times by resuspension in the Triton-containing buffer. The fibrils are washed a few times in stock salt solution to remove the detergent, and then either stored at −20° after addition of an equal volume of glycerol or used immediately in the prep-

[1] A. Weber and J. M. Murray, *Physiol. Rev.* **53**, 612 (1973).
[2] R. Cooke, *Biochem. Biophys. Res. Commun.* **49**, 1021 (1972).

aration of thin filaments. In the latter case, fibrils are washed twice in stock solution lacking $MgCl_2$ and resuspended finally to about 250 ml. The protein concentration of the fibril suspension is measured from the absorbance at 280 nm of a sample diluted with water and then dissolved by 2% sodium dodecyl sulfate (SDS) at 25°. After correction for light scattering by subtraction of the absorbance at 310 nm, the extinction coefficient for a 1% fibril solution at 280 nm is 7 per centimeter.[3]

The protein concentration is adjusted to 20 mg/ml, neutralized EDTA is added to 1 mM, and the suspension is gently warmed to 25°. Digestion with 6 units of α-chymotrypsin per milliliter is initiated; it is terminated after 45 min to 1 hr by addition of an equal volume of cold stock solution containing 2 mM phenylmethylsulfonyl fluoride (PMSF). The fibrils are sedimented and washed five times to free them from peptide material and protease, using 500 ml of stock salt solution containing 2 mM PMSF in each wash, and resuspending finally in 200–250 ml. Neutralized sodium pyrophosphate is then added to 2 mM, and the suspension is stirred gently (with an overhead stirrer rather than a magnetic stir bar) for 80 min. Fibrils are sedimented at 23,000 g for 10 min, and the supernatant is centrifuged at 80,000 g for 3 hr to sediment the thin filaments. Each pellet is covered with 2 ml of stock salt solution containing 2 mM PMSF and 1 mM ATP and allowed to swell overnight. The filaments are dispersed using a glass homogenizer. Aggregates of myosin rod, which usually contaminate the preparation at this point, and S-1 trapped in the pellets may be removed as follows. The swollen pellets are homogenized in 250 ml of 0.20 M KCl, 4 mM $MgCl_2$, 1 mM sodium pyrophosphate, 0.5 mM DTT, 20 mM Tris–HCl, pH 8.3 at 5°, and the pH is adjusted to 8.3 with 1 M Tris base, if necessary, and centrifuged at 80,000 g for 3 hr. The thin filament pellets can be freed of the last traces of myosin rod by repeating this step. Finally the pellets are homogenized in the solution of choice, dialyzed, and clarified by centrifugation at 6000 g for 20 min.

The typical yield of this procedure, estimated by the biuret method, is 50–100 mg of thin filaments, corresponding to 2–4 mg per gram of muscle. On SDS–urea gel electrophoresis the preparation is found to contain actin and tropomyosin in the same ratio as in the intact myofibril and a 20,000-dalton fragment of troponin T. These three species show only slight signs of dissociation when centrifuged at 80,000 g in stock solution at 0.01 mg/ml. The fragment of troponin T is nevertheless quantitatively displaced by whole troponin. The preparation is free from proteolytic activity and shows a markedly lower tendency to gel than do synthetic filaments. There is a small and variable amount of ATPase activity that is activated

[3] K. Sutoh and W. F. Harrington, *Biochemistry* **16**, 2441 (1977).

by EDTA and suppressed by Mg^{2+} and is, therefore, myosin-like. A typical value for the activity would correspond to a 1% contamination by S-1.

Other Methods of Preparation

Kendrick-Jones et al.[4] reported the preparation of intact thin filaments from rabbit muscle by blending myofibrils at a low concentration in a relaxing medium, followed by selective precipitation of thick filaments at pH 6.0. Both the yield and purity of the filaments depend critically on the myofibril concentration at the blending step and on blending time. Similar protocols are also successful for invertebrate cross-striated and catch muscles.[4,5]

Hama et al.[6] described a preparation method for rabbit psoas that includes both initial extraction of myosin at high ionic strength and tryptic proteolysis. The composition of the filament is not described.

A preparation of thin filaments from smooth muscle (pig aorta) has been described.[7] The myosin content of the preparation is reduced by sedimenting the thin filaments at a high salt concentration (0.5 M KCl) in the presence of MgATP. About 11 polypeptides are found to be present in the preparation when it is analyzed by SDS gel electrophoresis; among these, an inactive form of myosin is the most abundant contaminant, at about half the mass of actin.

Acknowledgments

This is contribution No. 1133 from the McCollum-Pratt Institute, The Johns Hopkins University, Baltimore, Maryland. This work was supported in part by Postdoctoral Fellowships from the Muscular Dystrophy Association to P. K. and S. L. and by NIH grant AM04349 to Dr. William F. Harrington.

[4] J. Kendrick-Jones, W. Lehman, and A. G. Szent-Györgyi, *J. Mol. Biol.* **54,** 313 (1970).
[5] W. Lehman and A. G. Szent-Györgyi, *J. Gen. Physiol.* **66,** 1 (1975).
[6] H. Hama, K. Maruyama, and H. Noda, *Biochim. Biophys. Acta* **102,** 249 (1965).
[7] S. B. Marston, R. M. Trevett, and M. Walters, *Biochem. J.* **185,** 355 (1980).

[4] Hybridization and Reconstitution of the Thin Filament

By JOHN M. MURRAY

Preparation of the Troponin– Tropomyosin Complex

Reconstitution of the thin filament is conveniently carried out using pure actin and a complex of troponin and tropomyosin as starting materials. The complex, Ebashi's "native tropomyosin," is prepared from rabbit back and leg muscle by dissociating it from the thin filaments at low ionic strength.[1]

The ground muscle is first extracted in high salt to remove myosin (see this volume [7]). The residue remaining after myosin extraction is washed by resuspending it in 3 ml of 20 mM KCl–0.2 mM NaHCO₃ per gram of muscle, stirring for 10 min at 4°, and then removing as much liquid as possible by filtration through four layers of cheesecloth. The wash is repeated with 20 mM KCl–0.2 mM NaHCO₃, then twice with the same volume of cold distilled water. The washed residue is mixed with an equal volume of cold water and blended in a Waring blender for 5 sec. NaN₃ is added to a final concentration of 1 mM; the mixture is stirred gently at room temperature for 4 hr, then left to stand without stirring overnight in the cold room. In the morning, centrifugation at 16,000 g for 10 min gives a viscous supernatant containing the tropomyosin–troponin complex. The supernatant is lyophilized and stored desiccated at −20°. The pellet from this centrifugation may be used to make a muscle acetone powder from which to prepare actin (see this volume [18]).

The lyophilized material contains, in addition to tropomyosin–troponin, variable amounts of actin, myosin, and other myofibrillar proteins. It is prepared for use by dissolving it at 10 mg/ml in 10 mM Tris pH 7.4, 0.2 mM DTT, then adding KCl to 0.1 M and MgCl₂ to 2 mM. This addition polymerizes some of the contaminating actin, which can be removed along with myosin and insoluble material by centrifuging at 145,000 g for 2 hr. The supernatant contains partially denatured actin that is incapable of polymerization. Somewhat surprisingly, it is possible to renature this actin and polymerize it by adding ATP to 0.1 mM, after which it can be removed by a second spin at 145,000 g.[2] It is important that the ATP not be added until after the first centrifugation, however, as this would dissociate all the actin–myosin rigor bonds and myosin would remain in the supernatant instead of being pelleted with the actin. The final supernatant contains essentially only troponin and tropomyosin at a total concentration of 3–5 mg/ml.

Assembly of the Thin Filament

The thin filament is reassembled by polymerizing actin in the presence of an excess of the regulatory proteins. To purified G-actin (1–5 mg/ml) is added twice its weight of the troponin–tropomyosin complex plus sufficient buffer, KCl, MgCl₂, and ATP to bring the final mixture to 0.1 M KCl, 2 mM MgCl₂, 0.1 mM ATP, pH 7.0. The mixture is stirred briefly, then

[1] S. Ebashi and F. Ebashi, *J. Biochem. (Tokyo)* **55**, 604 (1964).
[2] C. Trueblood, personal communication, 1980.

incubated without stirring at 15° for 45 min. The thin filaments are collected by centrifugation at 145,000 g for 2 hr. The supernatant is decanted, and the pellets and tubes are rinsed with buffer and allowed to drain. In order to resuspend the thin filaments evenly, homogenization will be necessary. However, the quality of the final product is sensitive to the mechanical stresses brought to bear upon it as these pellets are resuspended. For best results, the pellets should be covered with the resuspension medium and allowed to swell and soften for a few hours before gentle homogenization with a loose-fitting glass–Teflon homogenizer.

Thin filaments reconstituted by this method are quite stable. They retain the regulatory proteins when washed repeatedly with buffers of approximately physiological ionic strength and divalent ion content (e.g., $0.1 M$ KCl, 2 mM MgCl$_2$, pH 7.0). Their activation of the ATPase activity of HMM or S-1 is normally at least 90% calcium sensitive (i.e., inhibited by at least 90% in the presence of EGTA).

[5] Preparation of Native Thick Filaments

By JOHN A. TRINICK

In vertebrate skeletal muscle, thick and thin filaments are present in ordered interdigitating arrays. Under suitable circumstances, however, the myofibrillar lattice can be broken down to give a preparation of individual thick and thin filaments. The method described produces undamaged thick filaments. Such thick filaments have been used in a variety of different types of work including electron microscope studies of their structure,[1,2] biochemical of the amounts of myosin and other protein they contain,[3] and ultracentrifuge studies of their sedimentation behavior and the effect of Ca^{2+} on this.[4,5]

Muscle can be broken down into individual thick and thin filaments simply by gentle homogenization, but before this can be done the tissue must be relaxed by detachment of the cross-bridges from the thin filaments; the relaxed state requires MgATP in the absence of calcium. Although vertebrate skeletal muscle normally satisfies these conditions immediately postmortem, homogenization without further treatment does

[1] H. E. Huxley, *J. Mol. Biol.* **7**, 281 (1963).
[2] J. Trinick and A. Elliott, *J. Mol. Biol.* **131**, 133 (1979).
[3] K. Morimoto and W. F. Harrington, *J. Mol. Biol.* **77**, 165 (1973).
[4] K. Morimoto and W. F. Harrington, *J. Mol. Biol.* **83**, 83 (1974).
[5] C. H. Emes and A. J. Rowe, *Biochim. Biophys. Acta* **537**, 125 (1978).

METHODS IN ENZYMOLOGY, VOL. 85

not result in good preparations of separated filaments, possibly because rupture of the sarcoplasmic reticulum causes release of calcium and uncontrolled contraction that is difficult to reverse. The method described by Huxley[1] uses washed rigor myofibrils that, before homogenization, are relaxed by the addition of MgATP and EDTA. Although this procedure has the advantage that it uses as its starting material glycerol-treated muscle that can be stored at subzero temperatures more or less indefinitely, the yields obtained with it are rather low, perhaps because rigor can be difficult to reverse completely in myofibrils. The procedure described below in detail is derived from Huxley's method but does not necessitate glycerination or the production of myofibrils. In addition, the filaments are fresher and the yields tend to be higher.

Preparation of Thick and Thin Filaments from Fresh Muscle[2]

The psoas muscles are dissected from a freshly killed rabbit and placed on ice. Strips of the muscle approximately 2–3 mm in diameter are attached at about their rest length by thin string to wooden sticks or Perspex strips. Each is then placed in a test tube containing approximately 20 ml of a calcium-free Ringer's solution (100 mM NaCl, 2 mM KCl, 2 mM MgCl₂, 1 mM EGTA, 6 mM potassium phosphate, 0.1% glucose, pH 7.0 measured at 0°). (It is convenient to make this and the other buffers described in this chapter in bulk and store them in frozen aliquots until needed.) This solution is meant to mimic, to some extent, the extracellular ionic environment, at the same time depleting the intracellular calcium level. The use of a buffer containing mainly sodium rather than potassium ions avoids potassium contracture during calcium depletion. The strips can be left in ice-cold Ringer's solution for up to about a week, after which there are signs of deterioration, evidenced by the increasing numbers of thin filaments present after homogenization, probably due to Z-line degradation. The minimum time for which the strips need to be left in Ringer's solution is about 12 hr although it can be as little as 1 hr if frog muscle is used. During the incubation in Ringer's solution, the muscle strips become opaque and swell somewhat. After calcium depletion, a muscle strip is cut transversely into small pieces about 3 mm long and placed in a homogenization flask. To the pieces are added about 4 volumes of a relaxing buffer (100 mM KCl, 10 mM MgCl₂, 5 mM ATP, 6 mM potassium phosphate, 1 mM EGTA, 0.1 mM dithiothreitol, pH 7.0 at 0°). For small preparations we use an MSE homogenizer with a 10-ml flask surrounded by ice and operated at a nominal speed of 10,000 rpm for 1 min. After homogenization the separated filaments are clarified, and undisrupted myofibrils are removed by centrifugation at 10,000 g for 15 min, after which the supernatant appears opalescent. At this stage it is convenient to check that thick filaments are present, and mainly unbroken, by examining in the electron microscope a sample of the supernatant negatively stained with 1% uranyl acetate. In order to get a reasonable distribution of filaments on the grid, it is usually necessary to dilute the supernatant 10-fold to a final filament concentration of about 0.1 mg/ml. The fraction of filaments actually liberated by homogenization is small (~5–10%), but few of these are broken. More prolonged or violent homogenization is accompanied by increased fragmentation.

The above method can also be used to give A segments rather than separated thick filaments. The segments, which are bundles of thick filaments held together in register, are valuable becuase they show transverse striations that are the result of structural regularities difficult or impossible to see in single filaments.[6,7] The segments are not easy to produce, and are best obtained by keeping muscle strips in the calcium-depleting Ringer's solution for several days and then homogenizing rather more gently than when making single filaments.

The method of calcium depletion prior to homogenization in relaxing buffer can also be used to make separated thick and thin filaments from insect (bumble bee) flight muscle. In this case the muscle pieces are not physically restrained from contracting by attachment to a stick; otherwise the procedure is identical to that for making vertebrate filaments.

Purification of Native Thick Filaments

The soluble proteins that contaminate the thick and thin filament homogenates can conveniently be removed by reducing the ionic strength of the preparation, since thick filaments aggregate under these conditions. The clarified filament suspension is diluted with 4 volumes of ice-cold distilled water, and the precipitated filaments are collected by centrifugation at 5000 g for 10 min. The pellet is then washed by resuspension in diluted relaxing buffer, after which it is centrifuged again. The pellet is then resuspended in full-strength relaxing buffer. This method is useful also for producing highly concentrated filament suspensions.

Currently there exists no effective method for the purification of large quantities of native thick filaments from thin filaments. Small (milligram) amounts of purified filaments can, however, be isolated on D_2O/H_2O density gradients.[8] These are produced by making up the above KCl/ATP relaxing buffer in both D_2O and H_2O and mixing the solutions to pro-

[6] J. Hanson, E. J. O'Brien, and P. M. Bennett, *J. Mol. Biol.* **58**, 865 (1971).
[7] R. Craig, *J. Mol. Biol.* **109**, 69 (1977).
[8] J. A. Trinick, Ph.D. thesis, University of Leicester, 1973.

duce a linear gradient running from 20 to 100% D_2O. The only difference in preparing the D_2O buffer is that it is titrated to a pH meter reading of 7.4 (pD = pH meter reading + 0.4).[9] Such gradients are stable for several hours, and 1 ml of a 5 mg/ml clarified filament homogenate can be separated into relatively pure thick and thin filaments by centrifugation on a 25-ml gradient in a swing-out rotor running at 20,000 g for 1 hr. After this time, thick filaments are found in a broad band approximately two-thirds of the way down the gradient and the thin filaments are mostly near the surface. This purification can be scaled up and improved to some extent by the use of a zonal rather than a swing-out rotor (20 ml of filament homogenate can be purified on a 600-ml gradient in an MSE BXIV rotor run at 40,000 rpm for 1 hr); however, this is considerably more inconvenient and expensive in D_2O. Morimoto and Harrington[3] have used a purification procedure similar to the one described here but based on sucrose density gradients. They also described a method based on selective adsorption of thin filaments onto DEAE–cellulose. In our hands, neither of these methods has worked as reproducibly as the purification by D_2O gradient.

[9] P. K. Glascoe and F. A. Long, *J. Phys. Chem.* **64**, 188 (1960).

[6] Hybridization and Reconstitution of Thick-Filament Structure

By Jane F. Koretz

The preparation and study of aggregates of myosin and certain of its proteolytic fragments have a venerable history in muscle research, developing in sophistication and information in parallel with improved protein purification and electron microscope procedures. Whereas initial studies of this sort sought to differentiate structurally among crude fractions from a muscle extract,[1] contemporary filament reconstitution work is often used for the exploration of more subtle problems: myosin head mobility,[2] structural differences between myosin molecules of similar molecular weight from different sources,[3] the function of thick-filament associated proteins,[4] etc. The elucidation of these characteristics, and the relative strength of various interactions among myosin aggregates and aggregates

[1] M. A. Jakus and C. E. Hall, *J. Biol. Chem.* **167**, 705 (1947).
[2] R. A. Mendelson and P. Cheung, *Science* **194**, 190 (1976).
[3] T. D. Pollard, *J. Cell Biol.* **67**, 93 (1975).
[4] J. F. Koretz, *Biophys. J.* **27**, 433 (1979).

of its fragments, may lead ultimately to an understanding of natural thick-filament construction and organization.

This chapter is concerned with the preparation and assay of reconstituted myosin filaments, rod aggregates, and light meromyosin (LMM) paracrystals. The bulk of the material is based upon studies of rabbit skeletal muscle myosin and its aggregating fragments, with some procedures for chicken and rat skeletal myosins. Smooth muscle and nonmuscle myosins have also been included, but generally lack the complementary rod and LMM experiments owing to the small quantities of protein obtainable. Unless otherwise noted, the "assay" for each aggregate type will be average length distributions and/or structural descriptions obtained through electron microscopy of negatively stained samples.

Myosin Aggregates

General Considerations

Aggregate preparation basically makes use of the solubility characteristics of myosin in simple salt solutions. As the ionic strength of soluble protein solutions is reduced, an increasing proportion of the myosin molecules dimerize, then form antiparallel nuclei, and finally add on to the nuclei in parallel fashion, resulting in a population of bipolar synthetic myosin filaments. The characteristics of this final population depend strongly on the rate at which ionic strength is reduced, pH, the purity or composition of the myosin solution, and the source of the myosin. Other factors that can influence aggregation under certain special conditions are myosin concentration and final ionic strength for simple salt solutions; other salts, such as calcium salts, will cause aggregation into ordered nonfilamentous aggregates.

Unfortunately, the systematic study of the effects of these factors on rabbit skeletal muscle myosin aggregation was performed in the 1960s and early 1970s, when myosin purification procedures resulted in solutions containing other thick-filament proteins in amounts as great as 15%. Some of these proteins, notably C-protein and possibly F-protein, have been shown to influence both the aggregation process and the final filament population. (Later studies using Sephadex A-50 purification of myosin have been limited to different ranges of conditions and are not directly comparable.) As a result, the length distributions obtained under a given set of environmental conditions with column-purified myosin will differ both in peak location and breadth of distribution from those obtained with less uniform protein solutions. Unless otherwise stated, then, the "assay" results to be expected are only approximate.

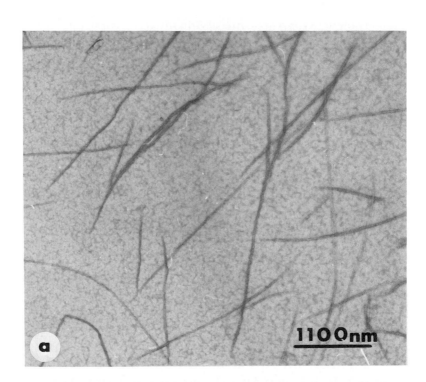

1100nm

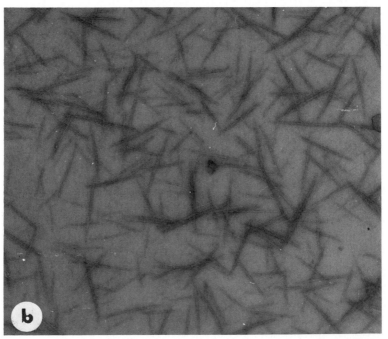

There are two reassuring aspects in synthetic filament preparation that should be mentioned. With careful control of environmental conditions and dialysis tubing preparation, filament population distributions are reproducible both for more than one experiment with a given myosin preparation and from preparation to preparation. We have also found that glycerin storage of myosin at $-20°$ will not affect later myosin aggregation characteristics. This means that a given aggregation procedure can be standardized in the laboratory, and any observed differences due to a change in environmental conditions can be directly attributed to that change.

As will be discussed later, myosins from different sources vary considerably in aggregation properties and structural characteristics. In designing an experiment that makes use of myosin aggregates, one must consider the buffer conditions, protein concentration, and final length distribution in relation to intended function. For example, if a preparation is to be used in electron microscope studies, phosphate buffers should be avoided because of possible crystallization on the microscope grid; thorough washing before staining may eliminate the danger, but could very well eliminate the specimen also. A high final protein concentration could result in overloading of the electron microscope grid, but might be partially compensated by the use of high pH conditions that would result in shorter filaments. For hydrodynamic studies, the use of short filaments and/or crosslinking agents may reduce the possibility of producing shear artifacts.

Basic Formation Procedures

Dialysis

Aggregate preparation by dialysis is the simplest of the various possible procedures, and the easiest to perform reproducibly. In the one-step process generally employed, a small sample of protein in solution is placed in treated dialysis tubing, sealed either with knots or (preferably) dialysis bag clips, and allowed to come to equilibrium with a large volume of low-salt medium over an 18–24-hr period with slow stirring (Fig. 1a). If the sample is intended for electron microscopy, the dialysis medium pro-

FIG. 1. Synthetic thick filaments prepared from column-purified rabbit skeletal myosin. Both pictures are at the same magnification. (a) Filaments prepared by dialysis against $0.10\ M$ KCl, 10 mM imidazole, pH 7.0, using dialysis tubing of 10 mm dry flat diameter and 12,000 M_r exclusion limit. The average filament length is about 3 μm. (b) Filaments prepared by rapid dilution from 0.3 to $0.2\ M$ KCl, 10 mM imidazole, pH 7.0, using 10 mM imidazole as dilutant; $t = 0$ sec. Average filament length is 0.5 μm.

vides a good washing solution, since it will not affect the equilibrium already obtained. It can also be used as a blank for turbidity and protein concentration measurements.

As will be discussed later, the rate of salt removal will help determine the final filament length distribution. For this reason, the selection of a given dialysis tubing can be important, since both the membrane weight exclusion limit and the surface area : volume ratio will affect this process. These parameters, unfortunately, are not usually published, but for small volumes we routinely use tubing with a 12,000 molecular weight (M_r) exclusion limit and 10 mm flat diameter.

One-meter lengths of tubing are boiled twice in double-distilled or distilled demineralized water, then boiled twice in 1 mM EDTA, and finally boiled once again in double-distilled water. These steps humidify the tubing and remove residual glycerin and sulfur compounds left over from manufacture. The prepared tubing can then be stored in the refrigerator or cold room in distilled water until needed. Often a drop of benzoic acid, formaldehyde, or mercaptoethanol is added to inhibit bacterial growth, but we prefer to prepare fresh tubing before every experiment.

There are several advantages in using dialysis bag clips rather than knots. They both provide a secure closure and conserve the amount of dialysis tubing required per sample. When dialyzing a variety of protein mixtures simultaneously, one can easily attach tags to the clips or use a color code with colorfast nylon thread. Further it is a relatively simple procedure to fasten bags using clips within the dialysis container so that they all experience the same gradient environment, an important factor in obtaining reproducible results. Finally, samples can be taken from a single bag at various intervals without necessitating transfer to a new bag or the dialysis of numerous identical samples simultaneously.

Dilution

Preparation of filaments by dilution follows the same principles as the one-step dialysis procedure previously described, except that the resultant length distribution is of much shorter mean length and narrower length distribution. While the effects of time will be discussed in more detail below in the section Effects of Time, it should be noted that there is a direct relationship between filament length and duration of the salt-reduction step; i.e., the more rapidly the solution is diluted, the shorter the filaments and the narrower the distribution. The major problems with this technique are maintenance of a linear gradient of salt reduction and reproducibility of the time interval over which the dilution occurs, but these can be handled with mechanical delivery systems.

The first step is to decide upon the initial and final ionic conditions desired, final protein concentration, and speed of dilution. The initial and final salt concentrations will determine both the nature of the dilution medium and initial protein concentration. For example, dilution of a sample at 0.5 M KCl, 10 mM imidazole to 0.1 M KCl, 10 mM imidazole would involve a 5-fold volume increase if 10 mM imidazole solution is used as a diluter. Then the initial protein concentration must be five times the desired final concentration. For a dilution medium containing 0.05 M KCl, 10 mM imidazole with the same initial and final conditions as above, the volume change would be a factor of 10.

The second step is the determination of the time course of the process. For very fast or very slow dilutions, no special apparatus is required. For example, dilution over a "0" sec time range (actually about 1–2 sec for complete mixing) simply involves dumping in the premeasured diluter while stirring rapidly (see Fig. 1b). Very slow linear dilution can be approximated by a stepwise dilution procedure if each added aliquot is small compared to total volume. The dilution medium can be slowly added at evenly spaced time intervals using a variable-volume automatic pipetter until the total volume has been used.

Since the time intervals of major interest lie between "0" sec and the long period involved in stepwise dilution, other procedures must be employed. Unfortunately, most of those who use dilution for filamentogenesis neither specify their method nor the time over which the process occurs. However, Kaminer et al.[5] described the use of a Sage syringe pump with rapid stirring for a linear dilution time of precisely 2 min, and Pinset-Härström and Truffy[6] used a constant-speed peristaltic pump over 2-, 11.5-, and 23-min time intervals. We have attempted to use a gravity-feed biuret,[7] roughly varying time intervals with the stopcock, but cannot be assured of a linear addition to the protein sample below.

Effects of pH

All other factors being kept constant, pH has a strong effect on filament length distributions, obtained either by dialysis or dilution. The most important study of pH effects on rabbit skeletal myosin was performed by Kaminer and Bell[8] using dilution over a 1–3-min interval. For ionic strengths in the 0.1–0.2 M KCl range, filament length was at a minimum at pH 8.0, rising steadily to a maximum at pH 6.5 and then decreasing somewhat at pH 6.0. At 0.3 M KCl, no significant filament formation is observed above pH 6.0.

[5] B. Kaminer, E. Szonyi, and C. D. Belcher, *J. Mol. Biol.* **100**, 379 (1976).
[6] I. Pinset-Härström and J. Truffy, *J. Mol. Biol.* **134**, 173 (1979).
[7] J. F. Koretz, L. M. Coluccio, and A. M. Bertasso, *Biophys. J.* **37**, 433 (1982).
[8] B. Kaminer and A. L. Bell, *J. Mol. Biol.* **20**, 391 (1966).

Filaments prepared by dialysis exhibit similar pH sensitivity, although their average lengths are much longer. Katsura and Noda[9] reported maximum filament length at pH 7.0, whereas Bertasso and Koretz (unpublished) observed a maximum at pH 6.5 with column-purified myosin in 0.1 M KCl and a distribution curve very similar to that of Kaminer and Bell. The latter results were also obtained with the addition of C-protein, indicating that the sensitivity to pH rests almost wholly in the myosin molecule itself.

There is some indication that filament morphology may be altered by pH. Kaminer and Bell have found small filaments with a smooth central bare zone at high pH and low ionic strength, but filaments are more spindle-shaped under other conditions and rarely exhibit a bare zone. In our own studies, we almost never see a bare zone with filaments prepared either by dilution or by dialysis at pH 7.0,[7] and others working with filament reconstitution have also made this comment. It is probable, then, that observation of a bare zone is a characteristic only of filaments formed at high pH.

Effects of Ionic Strength

As indicated in the preceding section, pH and ionic strength are inextricably linked. Whether filament formation occurs at a given ionic strength will depend strongly on the pH of the solution. The type of filament formed and the average length distribution will also be affected by these factors.

Again the most thorough study of ionic strength effects and pH was performed by Kaminer and Bell,[8] using dilution over a 1–3 min period. For a given pH, the ability of myosin to form "spindle-shaped" filaments was strongly dependent on ionic strength; others have observed similar results under slightly different conditions and/or using dialysis for filamentogenesis. In general, it would appear that the ionic strength at which well-formed filaments appear is about 0.1 M KCl at pH 8.0, rising to about 0.3 M KCl at pH 6.0.

A detailed study of filamentogenesis as a function of ionic strength at pH 7.0 has been performed by Koretz *et al.* [7] using column-purified rabbit skeletal myosin. The ionic strength boundary above which filament elongation does not occur is about 0.21 M KCl, 10 mM imidazole. From 0.2 M KCl to 0.1 M KCl in the presence of 10 mM imidazole, filament length distributions appear to be relatively independent of ionic strength, using dialysis as a formative technique. Below 0.1 M KCl, filaments bundle together into semiordered clumps that have a strong tendency to super-

[9] I. Katsura and H. Noda, *J. Biochem. (Tokyo)* 73, 245 (1973).

precipitate, a well-known observation upon which older myosin purification procedures were based.

By combining these observations, certain generalizations about ionic strength dependence can be made. For a given pH, there is a certain ionic strength (or narrow ionic strength range) above which filamentogenesis is inhibited and below which it occurs. This upper bound is lowest at high pH (8.0–8.5) and highest at low pH (6.0). A lower ionic bound, below which filaments interact to form bundles and may precipitate, is observed at 0.05 M KCl at pH values near 7.0 and may also occur at different ionic strengths for pH values farther from 7.0. Within these bounds, filament length distributions are almost independent of ionic strength when dialysis is the formative technique.

One final point about ionic effects should be noted. Sanger[10] has shown that, for a given set of environmental conditions, monovalent cations will increase filament length in the following order: $K^+ < Na^+ < NH_4^+$. He attributes this effect to the relative strengths with which these ions are bound to myosin, K^+ being the weakest. He also demonstrates that lowering the dielectric constant of the medium (e.g., by the addition of 5% n-propanol) will result in increased filament length.

Effects of Time

As indicated earlier, the speed with which ionic strength is reduced strongly influences the final filament length distribution and may also affect filament structure. This is true not only in comparing dialysis results with those of dilution, but also for dialysis or dilution at varying rates. However, for some of the reasons discussed before (unspecified time courses and/or gradient types during dilution, unspecified dialysis tubing diameter and/or molecular weight exclusion characteristics), a set of relationships between the time course of salt removal and average filament length cannot be provided. One can only state very generally that, for a given set of environmental conditions where filaments form, longer time courses will result in longer average filament lengths. And, along with a shift in the center of the length distribution, there will also be a broadening of the peak with longer times.

Filaments formed by dialysis can be as long as 20 μm or, for smooth-muscle myosin filaments, as short as a fraction of a micrometer. The speed with which salt is removed and the system comes to equilibrium will then depend upon the surface area of the dialysis tubing (and on the ratio of total volume to total surface area) and the molecular weight exclusion limit of the tubing. Thus, salt will be removed faster from a given volume

[10] J. W. Sanger, *Cytobiologie* **4**, 450 (1971).

with narrow-bore tubing of high molecular weight exclusion limit than with wide-bore tubing of the same or lower molecular weight exclusion limit. In our experiments with rabbit skeletal myosin at pH 7.0, using tubing of 10 mm flat dry diameter and M_r 12,000 exclusion limit, we reproducibly achieved filament length distributions of about 3 μm.[7] With other types of tubing, filaments formed under the same environmental conditions would exhibit altered population characteristics.

The rate of salt removal is especially critical in dilution experiments with skeletal muscle myosin. Katsura and Noda[9] found that a dilution time of "0" sec from 0.5 to 0.15 M KCl at pH 6.9 resulted in a sharp length distribution centered around 0.5 μm; if the dilution time over the same salt range is extended to 270 sec, they observed a very broad distribution centered at about 1.5 μm. Similar experiments with column-purified rabbit skeletal myosin at pH 7.0[7] indicate that the most rapid change in average filament length with time occurs in the 0–30-sec time range when ionic strength is decreased from 0.3 to 0.2 M KCl. Between 30 sec and 255 sec, there is little alteration in average length, although some peak broadening occurs. It should also be noted that, with the exception of average length, no morphological differences (such as bare zones) are observed.

Putting dilution and dialysis observations together for skeletal muscle myosin, it appears that the mechanism of filamentogenesis involves two time-dependent steps, a very rapid and a very slow one. Although these steps cannot be unequivocally attributed to intermediates in the aggregation process, and indeed this is not the place for such speculation, it can nevertheless be said that the fast step appears to govern filament length distributions and peak width when salt levels are abruptly reduced. The slower step would then govern distributions and peak widths over longer dilution and dialysis times, such that the average length becomes greater and the peak width broader with increasing time.

It would seem that aggregates of smooth-muscle myosins, and perhaps nonmuscle myosins, are governed by similar, but not identical, rules. That is, dilution and rapid dialysis result in the formation of short filaments, often 0.5 μm or less, while the very slow dilution-dialysis technique of Hinssen et al.[11] results in much longer filaments. In addition, it has been shown that, for some smooth and nonmuscle myosins, the method of filament preparation will affect the morphology observed. The particular details for each myosin type and the dialysis-dilution method will be discussed in more detail in the Assay sections on smooth-muscle and nonmuscle myosins.

[11] H. Hinssen, J. D'Haese, J. V. Small, and A. Sobieszek, J. Ultrastruct. Res. 64, 282 (1978).

Effects of Protein Concentration

The relationship between initial protein concentration and final filament population characteristics has not been extensively studied, even though it is a very important parameter in a number of other aggregating systems. Katsura and Noda[12] noted that their filament population characteristics appear to be independent of myosin concentration when prepared by dilution to 0.15 M KCl at pH 6.75–7.0 down to a protein concentration of approximately 5 μg/ml (1.1 $\times$ $10^{-8} M$). Koretz *et al.*[7] found similar results for intermediate protein concentrations at pH 7.0 with 10 mM imidazole for filaments prepared by dialysis against 0.10, 0.15, and 0.20 M KCl, and a small but linear decrease in average filament length for column-purified myosin at 0.10 M KCl, 10 mM imidazole, pH 7.0, as protein concentration is reduced to 20 μg/ml (4.5 $\times$ $10^{-8} M$) in dialysis experiments.

It is not yet clear whether the results for rabbit skeletal myosin and column-purified rabbit skeletal myosin near physiological pH also hold true at other pH levels and for other skeletal muscle myosins under the same experimental conditions. In addition, almost nothing is known about nonmuscle and smooth-muscle myosin concentration dependence during aggregation. As a result, the best course to take when using a myosin type that has not been characterized in this way is to try to keep final concentrations at or above 0.1 mg/ml.

Assay

Rabbit Skeletal Myosin

Unless otherwise specified, all the work described above in the section on basic formation procedures was performed using rabbit skeletal myosin, which is without doubt the most completely characterized of the various myosins. In general, one may expect short, poorly organized aggregates of about 0.5 μm with a clearly visible bare zone of about 20 nm length at high pH and higher ionic strength.[8] At lower ionic strengths and pH, filaments will present an organized bipolar appearance, with cross-bridge levels and spacing of 14.3 nm clearly visible at high magnification[13,14]; bare zones are rarely if ever seen under these conditions, however, and the length distribution of the population, whether prepared by dialysis or dilution, will be much broader than at higher pH. Intermedi-

[12] I. Katsura and H. Noda, *J. Biochem.* (*Tokyo*) **69**, 219 (1971).
[13] C. Moos, G. Offer, R. Starr, and P. Bennett, *J. Mol. Biol.* **97**, 1 (1975).
[14] J. F. Koretz, *Biophys. J.* **27**, 423 (1979).

ate conditions of pH and ionic strength may result in a mixture of well-formed and less well-organized aggregates, depending upon the conditions and speed of preparation.

Caution must be exercised in evaluating some of the reported results. While the general dependence of myosin aggregation on the various environmental factors holds true, as already discussed, much of the work discussed was performed on myosin preparations that contained other thick-filament proteins (e.g., C-protein and F-protein). Since C-protein[4] and perhaps F-protein[15] affect myosin aggregation, some consideration must be given to the manner in which such contamination might affect experiments.

C-protein (see this volume [14]), a major thick filament component in rabbit skeletal muscle, binds strongly to every portion of the myosin molecule except for the heads.[16] If present in the myosin solution before aggregation, it appears to be incorporated into the filament backbone, either altering or disrupting filament structure, depending upon its concentration. In more recent studies, Koretz et al.[7] have shown that it will also affect length distributions in both dialysis and dilution experiments. When added in a ratio of 1 C-protein molecule per 3.3 myosin molecules (the ratio at which filaments have the most regular appearance), average length distributions from dialysis experiments are 20–25% shorter and the peak width is narrower than with pure myosin. The presence of the same ratio in dilution experiments results in a length distribution that is independent of dilution speed between 0 and 435 sec.

These results differ in almost all particulars from similar studies performed by Miyahara and Noda[17] using a variety of C-protein : myosin ratios. It is possible that the disruption in filament structure observed by Moos et al.[13] at high ratios and by Koretz[4] at low and high ratios may be partially responsible for these differences, but the problem is currently unresolved. What can be stated is that myosin aggregation behavior is affected by C-protein in both dilution and dialysis experiments, and that, because it is incorporated into the filament backbone, it probably binds at an ionic strength above the boundary for filamentogenesis.

F-protein, another thick-filament protein of unknown properties and unknown location in vivo, was investigated by Miyahara et al.[15] F-protein will bind to myosin, and its binding can be inhibited by C-protein, but F-protein will not inhibit C-protein binding to myosin. Further, although F-protein does not appear to affect myosin aggregation behavior, it will

[15] M. Miyahara, K. Kishi, and H. Noda, J. Biochem. (Tokyo) 87, 1341 (1980).
[16] R. Starr and G. Offer, Biochem. J. 171, 813 (1978).
[17] M. Miyahara and H. Noda, J. Biochem. (Tokyo) 87, 1413 (1980).

reduce the effects of C-protein on myosin aggregation. These rather puzzling results may be explainable if C-protein can bind to myosin at a higher ionic strength than can F-protein, but it is advisable to remove F-protein from myosin solutions until further studies can be performed.

Other Skeletal Myosins

Chicken Breast Myosin. Kaminer[18] has characterized the aggregation behavior of chicken breast myosin as a function of pH and ionic strength for a pH range of 6.0–8.0 and ionic strengths of 0.1, 0.2, and 0.3 M KCl using dilution as the formative technique. While average filament length for a given set of conditions might differ slightly from those obtained by Kaminer and Bell[8] for rabbit skeletal myosin, the profiles and general behavior for chicken breast myosin are almost the same. In a later study, at pH 6.5 and 0.1 M KCl,[5] the close similarity is more clearly shown; both average filament length and width of the distribution are almost superimposable for the two myosin types.

Rat Skeletal Myosin. Pinset-Härström and Truffy[6] have characterized the aggregation behavior of rat skeletal myosin as a function of the presence or the absence of ATP, inorganic phosphate, Mg^{2+}, and Ca^{2+} during (primarily) 11.5-min linear dilution experiments from 0.5 M to 0.12 M KCl, at pH 6.8. In the absence of ATP or inorganic phosphate, and with Mg^{2+}, Ca^{2+}, and/or ethylenediaminetetraacetic acid (EDTA) at concentrations below 5 mM, rat skeletal myosin filaments range in length from 5 to 15 μm and in diameter from 30 to 50 nm. Millimolar amounts of ATP or inorganic phosphate in the absence of divalent cations result in irregular aggregates, either long and branched or short (0.2–4 μm) and narrow (10–15 nm). These effects can be neutralized by millimolar amounts of divalent cations; Ca^{2+} produces the 30–50 nm-wide filaments, while Mg^{2+} produces thinner (15–17 nm) ones, but in both cases the lengths range over 5–15 μm. These results hold true for both crude rat skeletal myosin and rat myosin purified on a Sepharose 2B column, which will not remove proteins that are bound to myosin.

Pinset-Härström and Whalen[19] noted that, as rat skeletal myosin ages, it tends to lose light chain LC2 by proteolysis. This results in a gradual loss in the ability of the rat myosin to form long regular bipolar filaments. Removal of LC2 by DTNB treatment results in the same loss of aggregation capacity; addition of excess LC2 does not reverse this loss.

[18] B. Kaminer, *J. Mol. Biol.* **39**, 257 (1969).
[19] I. Pinset-Härström and R. G. Whalen, *J. Mol. Biol.* **134**, 189 (1979).

Smooth Muscle Myosins

General Considerations. Smooth muscle myosins are in many respects similar to skeletal muscle myosins in the effects of environmental conditions on filamentogenesis. Aggregates can be prepared either by dilution or dialysis; the extent of aggregation and final population characteristics will depend on ionic strength, pH, and perhaps protein concentration, and the relative sensitivity of smooth muscle myosin to these parameters is qualitatively similar to that of the skeletal type.

One aspect of smooth muscle myosin aggregation that has been consistently noted is the uniformity of the resultant filament population. Filaments tend to be about 0.3–0.5 μm, whether prepared by dilution or dialysis, and exhibit a narrow length distribution. Although this has been attributed to an intrinsic length determination factor (presumably absent in skeletal myosin), the preparative procedure of Hinssen *et al.*[11] demonstrates rather that smooth myosin aggregation operates in a different time frame. When ionic dilution is stretched over a 24-hr time course, filaments are much longer than previously observed and exhibit important differences in structure. The length results are analogous to the differences in average filament length and length band width for skeletal myosin filaments prepared by dilution or dialysis, while structural differences may be analogous to pH-influenced skeletal myosin structures.

A very recent, and potentially very important, development in smooth muscle myosin studies has been the identification of the link between phosphorylation of the LC2 light chain and increased aggregation capacity. It had been noted earlier that aggregation was increased in crude preparations in the presence of Ca^{2+} and ATP,[20] presumably because of the presence of calmodulin and a specific phosphokinase. This process has been shown[21-23] to be the initial mechanism of actin activation of the myosin ATPases in smooth and some nonmuscle systems, and to affect the rate of actomyosin ATPase in aortic smooth muscle. New work demonstrates an additional effect in increasing myosin aggregation capacity in the nonmuscle calf thymus and porcine platelet myosins and in the smooth muscle chicken gizzard[24] and porcine aortic myosins.[25] The implications of these observations for thick-filament structure and structural lability in fixed and sectioned samples examined in the electron microscope may

[20] C. F. Shoenberg, *Tissue Cell* **1**, 83 (1969).
[21] R. Dabrowska, J. M. F. Sherry, D. K. Aromatorid, and D. J. Hartshorne, *Biochemistry* **17**, 253 (1978).
[22] D. R. Hathaway and R. S. Adelstein, *Proc. Natl. Acad. Sci. U.S.A.* **76**, 1653 (1979).
[23] D. D. Rees and D. W. Frederiksen, *J. Biol. Chem.* **256**, 357 (1981).
[24] J. M. Scholey, K. A. Taylor, and J. Kendrick-Jones, *Nature (London)* **287**, 233 (1980).
[25] J. F. Koretz and D. W. Frederiksen, *Biophys. J.* **33**, 235a (1981).

well be the primary explanation for the variability seen. It is especially noteworthy that the thymus, platelet, and gizzard myosins are in a disassembled state in the presence of 5 mol of ATP per mole of myosin if unphosphorylated, and assembled and stable when phosphorylated.[24] Rabbit and scallop striated muscle myosin, in contrast, form filaments that are stable independent of MgATP concentration up to millimolar quantities.

In the following survey of smooth muscle myosin aggregation behavior, the phosphorylation state of the various myosin species is generally unknown. I will make some assumptions where warranted, based upon protein preparative procedure, dialysis solution composition, and observed filament distributions, but it is safest to assume that the myosin is unphosphorylated unless otherwise indicated.

Chicken Gizzard Myosin. Of the various smooth muscle myosins, chicken gizzard myosin is the one most completely characterized in its phosphorylated and dephosphorylated forms. The latter myosin type has been studied in the electron microscope by Kaminer[18] as a function of both pH and ionic strength when prepared by dilution.

At 0.3 M KCl, decreasing pH results in more and better-formed aggregates. At pH 8.0, relatively few filaments are seen; they are less than 0.2 μm long and 4–7 nm in diameter. As pH is reduced (7.5–6.5), filaments are observed that resemble "broken twigs," about 0.2–0.3 μm by 4–8 nm. At pH 6.0, the "broken twigs" are still seen, along with a second well-organized filament type of 0.5 μm by 10–18 nm.

At 0.2 M KCl, the twiglike aggregates are seen at pH 7–8, whereas at pH 6.0 the filaments appear spindle-shaped and have dimensions of 0.5 μm by 10.5–20 nm. At 6.5 one sees both morphologies. At 0.1 M KCl, a central bare zone of 0.2 μm is seen at all pH values; as pH is decreased filament length increases on either side of this bare zone, and at pH 7.0 and below the shaft appears more regular than at higher pH.

Sobieszek[26] reported a somewhat different filament form for chicken gizzard myosin (and other smooth muscle myosins) of unknown phosphorylation state. When the crude extract is left on ice for 14 hr or dialyzed against a low ionic strength medium at low pH, filaments appear asymmetric, cross-bridges are on only one side at each end of the aggregates, with "bare zones" on the other side and on opposites sides at opposite ends. There is no central bare zone. Filaments of this type have been termed "side-polar" by Craig and Megerman.[27]

Later analysis of gizzard myosin aggregation[11,28] suggests that the

[26] A. Sobieszek, *J. Mol. Biol.* **70,** 741 (1972).
[27] R. Craig and J. Megerman, *J. Cell Biol.* **75,** 990 (1977).
[28] J. V. Small and A. Sobieszek, *Int. Rev. Cytol.* **64,** 241 (1980).

short bipolar filaments with central bare zones are merely the first stage of a more complex growth process. The bare zones themselves are rhomboidal and centrally sited, and apparently rectangular bare zones may possibly be a 90-degree rotation of these rhomboidal structures. Further elongation results in the slippage of the bare zone into oppositely sited bare ends, with myosin heads arranged in a barber-pole type of stripe around the filament axis between the two ends. The longest filaments are side-polar and exhibit 14 nm-spaced cross-bridge levels between the two tips. Small and Sobieszek[28] suggested that such a structure might be obtained simply from the assembly of staggered antiparallel myosin dimers.

Work by Schoenberg and Stewart[29] on chicken gizzard muscle homogenates has explored the time course of aggregation and its dependence on divalent cations and ATP, extending the earlier studies of Schoenberg.[20] Filaments prepared by dilution of the homogenate and subsequent storage on ice went through two phases, a rapid one (5–15 min) in which some filaments are seen and a slower one (1–2 hr) in which most of the filaments form. The presence of millimolar concentrations of magnesium are essential, whether calcium is present or absent; if calcium is also present in at least micromolar quantities, the rapid phase of filament formation is enhanced. At 1 mM concentration, ATP also enhances the rapid phase, but can inhibit it at higher concentrations. The addition of crude gizzard tropomyosin (which contains the light-chain kinase) produces an increase in the rate and extent of filamentogenesis in the presence of 10 mM MgCl$_2$, 2 mM ATP, and 1 μM Ca^{2+}. It is probable that the latter results are due directly to myosin phosphorylation, whereas earlier observations of rate of filament formation might also be dependent on the activation of small amounts of kinase. Otherwise, the earlier observations must be attributed to the high solubility of unphosphorylated myosin in the presence of 1 mM ATP.[24]

Rabbit Uterine Myosin. This myosin type (probably not phosphorylated), which has been studied primarily by Wachsberger and Pepe,[30,31] has been characterized as to ionic strength, divalent cation, and pH sensitivity. (Filaments were formed by dropwise dilution of buffer over 30 min.) In the absence of added magnesium at pH 7.0, filaments are first observed at 0.2 M KCl; they are generally bipolar filaments of variable diameter, tapered ends, and length of 0.3–0.6 μm, which in turn strongly tend to form meshworks due to end-to-end aggregation. Further dilution to 0.1 M KCl reveals a higher proportion of unassociated bipolar filaments and some linear arrays as well as the meshwork.

[29] C. F. Shoenberg and M. Stewart, *J. Muscle Res. Cell Motil.* **1**, 117 (1980).
[30] P. R. Wachsberger and F. A. Pepe, *J. Mol. Biol.* **88**, 385 (1975).
[31] P. R. Wachsberger and F. A. Pepe, *J. Cell Biol.* **85**, 33 (1980).

In the presence of 10 mM MgCl$_2$, filaments are observed at 0.3 M KCl, pH 7.0, primarily in the form of a meshwork. A higher proportion of individual bipolar filaments of the type described above is seen at 0.2 M KCl, as well as linear aggregates and meshwork. At 0.1 M KCl, both the short bipolar aggregates and linear arrays are observed; in addition, long filaments (0.6–1.2 μm) were seen, with projections all along their length or occasionally a bare zone. Lower concentrations of magnesium gave similar results, as did calcium substituted for magnesium. Higher levels of magnesium resulted in side-by-side aggregation of the long filaments at 0.1 M KCl. Filaments prepared by a two-step dilution process to either pH 6.0 or pH 7.0 exhibited the same length distributions (0.44 μm ± 0.07) and the same bipolar structure already described; similar results can be obtained by dialysis.

Guinea Pig Taenia Coli and Vas Deferens. Myosin extracted from these two muscle types demonstrate aggregate structures and ionic strength, divalent cation, etc., sensitivity similar to that previously described for chicken gizzard myosin.[26]

At pH 6.8 their lengths range from 0.3 to 1.5 μm, with diameter of 25–30 nm, and at pH 6.0 maximum length can approach 2 μm. Prolonged exposure to pH 6.0 conditions will result, after several hours, in aggregation of the filaments themselves. Schoenberg and Stewart[29] noted that, in their homogenates, the two types of filaments observed consistently are shorter for taenia coli than for chicken gizzard under the same experimental conditions.

Calf and Porcine Aortic Myosin. Studies of aggregates formed from calf aortic myosin under a variety of conditions have been performed by Craig and Megerman.[27] After dialysis to 0.2 M KCl, pH 8, only small (0.35–0.5 μm) bipolar filaments are observed; lower pH at the same ionic strength results in a mixture of bipolar and side-polar filaments. The latter, observed in filament preparations of other smooth muscle types, have myosin molecules arranged in unipolar fashion on either side of the filament shaft, resulting in opposite polarity on the two sides and a bare region on opposite sides at either end. Only side-polar filaments are observed at 0.3 M KCl, pH 6; these can be as long as 6 μm. Examination of sections of fixed embedded side-polar filaments confirms that the myosin heads are on opposite sides of the shaft.

It is possible that filamentogenesis for calf aortic myosin is similar to that already described for gizzard, i.e., an initial aggregate with central bare region (rhomboidal or rectangular in appearance depending upon angle of observation) that elongates as if by slippage. Unlike gizzard filaments, however, the "barber-pole" effect is not observed for calf aortic filaments. The sides are apparently distinct for long stretches along the aggregate length; spacing of the heads along the filament length is about 15 nm.

Porcine aortic myosin, in contrast, does not appear to form side-polar filaments in either its phosphorylated or dephosphorylated state.[25] At 0.1 M KCl, 10 mM imidazole, 0.1 mM phenylmethylsulfonyl fluoride (PMSF), 69 μM streptomycin sulfate, 0.5 mM dithiothreitol (DTT), pH 7.0 and the presence of at least 1 μM Mg^{2+} (the myosin appears to denature at lower Mg^{2+} concentrations), dephosphorylated myosin forms small bipolar filaments, often without a central bare zone, that taper at either end. Cross-bridges form easily discernible stripes across the backbone, oriented perpendicular to the filament axis, and can be traced to the tips (Fig. 2a).

Under the same formative conditions, phosphorylated porcine aortic myosin forms very long, apparently bipolar filaments that tend to aggregate into oriented bundles (Fig. 2b; compare to rabbit skeletal myosin filaments in Fig. 2c). Individual shafts can be seen, although not traced within the bundle for length determinations; they do not appear to be end-to-end aggregates of the short bipolar filaments, but rather very long myosin filaments. There is also some indication that these filaments are aligned relative to each other in the bundle. Whether these bundles are artifacts of electron microscope grid preparation or not, the difference in filament length between the two myosin types is still further indication of the possible importance of light-chain phosphorylation in structural as well as enzymic functions during smooth muscle contraction.

Nonmuscle Myosins

Nonmuscle myosin filament reconstitution studies have been performed for three distinct cell groups: amoebae (*Acanthamoeba*, *Chaos carolinensis*, and *Amoeba proteus*), slime molds (*Physarum polycephalum* and *Dictyostelium discoideum*), and tissue cells (chick brain, calf thymus, human and bovine platelet, macrophage, guinea pig granulocyte, and tomato). Many of these myosin types require Mg^{2+} for compact and stable filament formation and, like some smooth muscle myosins, can form filaments greater than 1 μm in length under sufficiently long dialysis-dilution conditions. While some appear to form bipolar filaments under the conditions listed, others are capable of forming the side-polar structures already described for some smooth muscle myosin species. Finally, it seems likely that phosphorylation may be an important general mechanism for filament formation and stability under physiological conditions; for calf thymus and porcine platelet myosins,[24] nonphosphorylated myosin will not aggregate into filaments above 10^{-7} to 10^{-6} M ATP, while the phosphorylated species form aggregates stable up to 1 mM ATP.

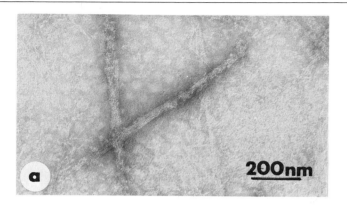

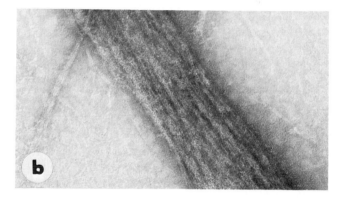

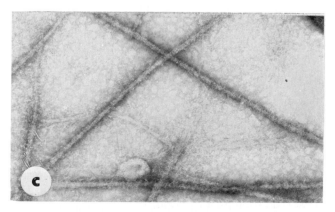

FIG. 2. The effect of phosphorylation on porcine aortic smooth muscle myosin aggregation. Conditions for filamentogenesis are described in the text. All micrographs are at the same magnification. (a) Filaments prepared from dephosphorylated aortic myosin. They are about 0.6–0.7 μm in length, apparently bipolar, and poorly organized. (b) Filaments prepared from phosphorylated aortic myosin. Individual filaments are longer than those ob-

AMOEBAE

Acanthamoeba. Of the three myosin types isolated from this species, only myosin II can form aggregates.[32,33] At pH 6.8 in the presence of 0.1 M KCl, 10 mM imidazole, filaments are 205 ($\pm$15) nm long and exhibit a central bare zone of 97 ($\pm$10) nm length and 6.6 ($\pm$0.5) nm width. Addition of 1 mM EDTA, 1 mM MgCl$_2$, or 1 mM CaCl$_2$ has little or no effect, although filaments tend to be slightly broader if Ca^{2+} is raised to 10 mM. These results are interpreted in terms of a filament model containing 16 myosin molecules: 4 cross-bridge levels on either side of the bare zone, each composed of 2 myosin molecules. The bare zone would then equal the 90-nm myosin tail in length, and filamentogenesis would be regulated by the size of the largest antipolar dimer that could be formed.

Chaos carolinensis and Amoeba proteus. These two carnivorous amoebae contain myosin with almost identical aggregation properties.[34] When dialyzed overnight against a high ionic strength EDTA–EGTA buffer (300 mM KCl, 1 mM EGTA, 1 mM EDTA, 10 mM imidazole, pH 7.0) to remove residual divalent cations, well-formed filaments occur only in the presence of Mg^{2+} (455 nm long, with central bare zone 174 nm long and 16 nm wide). In the presence of Ca^{2+} (or higher concentrations of EDTA and/or EGTA), the filaments are poorly formed and highly branched (328 nm long, with an indistinct bare zone about 110 nm in length and 13 nm wide). Small oligomers of myosin were observed in the background, and there was a tendency for the aggregates to clump. Without predialysis, well-formed filaments were observed independent of the presence or the absence of divalent cations.

The Mg^{2+} requirement for stable filament formation was also indicated in the studies of Hinssen *et al.*[11] for *Amoeba proteus* myosin. Further, with their dilution-dialysis techniques for extending filamentogenesis over at least 24 hr, they observed the relationship between length and formation of side-polar filaments already discussed for gizzard myosin.

SLIME MOLDS

Physarum polycephalum. The studies by Nachmias[35,36] of this myosin indicate also the requirement of Mg^{2+} for aggregation. In crude prepara-

[32] T. D. Pollard, M. E. Porter, and W. F. Stafford, *J. Cell Biol.* **75**, 262a (1977).
[33] T. D. Pollard, W. F. Stafford, and M. E. Porter, *J. Biol. Chem.* **253**, 4798 (1978).
[34] J. S. Condeelis, *J. Cell Sci.* **25**, 387 (1977).
[35] V. T. Nachmias, *J. Cell Biol.* **52**, 648 (1972).
[36] V. T. Nachmias, *J. Cell Biol.* **62**, 54 (1974).

served for nonphosphorylated aortic myosin and do not seem to be end-to-end aggregates of shorter aggregates. They have a strong tendency to aggregate into ordered bundles. (c) Filaments prepared under the same conditions from rabbit skeletal myosin.

tions, no filaments can be seen after dialysis against a low ionic strength medium containing EGTA and no added Mg^{2+}; with Mg^{2+} added to the EGTA (5–10 mM MgCl$_2$, 1 mM EGTA), loose aggregates are observed. In the presence of both Mg^{2+} (5–10 mM) and Ca^{2+} (0.1 mM) thin bipolar filaments about 0.5 μm in length can be seen. This apparent Ca^{2+} dependence is not observed in more highly purified preparations. When dialyzed against 5 mM Mg^{2+}, 50 mM KCl, 5 mM imidazole, pH 7, the *Physarum* myosin will form stable filaments up to 2.5 μm in length whether 1 mM EGTA or 0.1 mM CaCl$_2$ is present. In the presence of EGTA (1 mM) without added Mg^{2+}, no filaments greater than 0.2 μm are visible.

Hinssen and D'Haese[37] have characterized the environmental factors required for *Physarum* filamentogenesis. In addition to Mg^{2+}, pH should be around 7 for filament stability. No filaments can be observed at concentrations greater than 0.15 M KCl or 3 mM ATP; Ca^{2+} up to 0.5 mM will temporarily improve filament formation, but may later affect it.

Using the dilution-dialysis technique for reducing KCl concentration from 0.2 M to 0.05 M over 1–2 days in the presence of 0.5 mM ATP, 2 mM MgCl$_2$, 35 mM histidine, pH 6.9, filament lengths showed a skewed distribution ranging from 0.3 to 2 μm. While the maximum of the distribution was at 0.65 μm, about 70% of the filaments measured were longer. (With the addition of 0.5 mM Ca^{2+} under the same conditions, filaments up to 4 μm in length were observed.) Diameters could be up to 30 nm, especially in filaments larger than 1 μm. Filament structure was of the side-polar type, with cross-bridge banding of 14–15 nm and no central bare zone. In contrast, rapid lowering of ionic strength resulted in short (0.4–0.5 μm) bipolar filaments, primarily a bare zone with tufts of myosin heads at either end (like two mushrooms attached stem-to-stem).

Dictyostelium discoideum. In the presence of 10 mM Mg^{2+} at 0.1 M KCl, *Dictyostelium* myosin will form compact bipolar filaments 0.6–0.8 μm long with a 130–190-nm central bare zone.[38] In the absence of Mg^{2+} under otherwise similar formative conditions, the resultant aggregate is loosely organized and unstructured.

The presence of RNA in the isolated myosin solution appears to affect resultant filament properties as observed in the electron microscope.[39] There is a strong tendency toward both lateral and longitudinal aggregation of the bipolar filaments, and there is also some indication that some of the filaments themselves may be longer. Removal of RNA before filamentogenesis results in a fairly uniform population of short individual bipolar filaments.

[37] H. Hinssen and J. D'Haese, *J. Cell Sci.* **15**, 113 (1974).
[38] M. Clarke and J. A. Spudich, *J. Mol. Biol.* **86**, 209 (1974).
[39] P. R. Stewart and J. A. Spudich, *in* "Motility in Cell Function" (F. A. Pepe, J. W. Sanger, and V. T. Nachmias, eds.), p. 359. Academic Press, New York, 1979.

OTHER NONMUSCLE MYOSINS

Platelet. Niederman and Pollard[40] reported that human platelet myosin forms small bipolar filaments (320 nm in length, 160 nm bare zone length, 10–11 nm bare zone width) when dialyzed against low ionic strength near neutral pH. These results are affected little or not at all by variation of pH between 7 and 8, by ionic strength variation between 0.05 and 0.2 M KCl, by the presence or the absence of 1 mM Mg^{2+} or ATP, or by a variation in myosin concentration between 0.05 and 0.7 mg/ml. (In the presence of 1 mM Ca^{2+} at pH 6.5, however, the filaments tend to be slightly larger.) These results are interpreted in terms of a filament model containing 28 myosin molecules, 14 on either side of a central bare zone whose length is equal to the length of the myosin tail. Heads are arranged at 14–15 nm intervals, with two molecules per level. At very high Mg^{2+} concentrations, these bipolar filaments tend to aggregate into paracrystalline arrays, ordered both end-to-end and laterally in a open meshwork.

Hinssen *et al.*,[11] in their work on bovine platelet myosin, indicated that 10 mM Mg^{2+} is necessary for long stable filament formation. With their dialysis-dilution technique and the presence of Mg^{2+}, they obtained side-polar filaments for those filaments greater than 1 μm in length and bipolar filaments with rhomboidal bare zones for those filaments less than 0.7 μm long. Their theory of assembly is based on the model developed for gizzard myosin.

The function of phosphorylation of the M_r 20,000 light chain of platelet or calf thymus myosin in filamentogenesis under physiological conditions has been shown.[24] Specifically, nonphosphorylated platelet or thymus myosin will not form filaments at low ionic strength in the presence of 1 μM MgATP, while phosphorylated myosin is stable in filamentous form up to millimolar concentrations. In the light of these results, it is probable that the platelet myosin of Niederman and Pollard, which appeared to be insensitive to ATP concentration, was wholly or primarily phosphorylated. It is likely that this process, found in both smooth muscle and nonmuscle systems, is important in structural regulation of actomyosin interactions.

Chick Brain Myosin. When dialyzed against 0.1 M KCl, 10 mM imidazole, pH 7.0, chick brain myosin forms small bipolar filaments (0.3 μm in length, 0.15 μm bare zone).[41] There is some tendency for aggregation of these filaments both laterally and end-to-end; at 25 mM Mg^{2+}, this aggregation results in an ordered lattice, with filaments aligned both laterally and longitudinally in a paracrystalline array. (Similar results are obtained

[40] R. Niederman and T. D. Pollard, *J. Cell Biol.* **67,** 72 (1975).
[41] E. R. Kuczmarski and J. L. Rosenbaum, *J. Cell Biol.* **80,** 341 (1979).

for human platelet myosin.) This type of superaggregation by chick brain filaments was also observed by Burridge and Bray[42] with prolonged dialysis against 0.1 M 2-(N-morpholino)ethanesulfonic acid (MES).

Calf Thymus. (See platelet section above.)

Guinea Pig Polymorphonuclear Leukocytes (Granulocytes). When a solution of this myosin type was diluted to 0.12 M KCl in the presence of 1 mM Ca^{2+} and 10 mM imidazole, pH 7.0, short, thin bipolar filaments were observed (250 nm long, 6–12 nm wide).[43] The presence of divalent cations are apparently essential, since the substitution of EDTA for the Ca^{2+} resulted in inhibition of filamentogenesis.

Macrophage. When formed against 0.1 M KCl, pH 7.0, these filaments were bipolar and about 300 nm in length.[44]

Tomato. At low ionic strength, myosin purified from the common tomato (*Lycopersicon esculentum*) will form bipolar filaments of 0.5–0.6 μm in length with a central bare zone.[45]

Myosin Filament Hybrids (Copolymerization)

A simple method for determining whether myosins from different sources share at least some common determinants along the tail portion of the molecule is hybridization or copolymerization. Mixtures are prepared of the two myosin types in various molar ratios under conditions where both are soluble; filaments are then formed, either by dialysis or dilution, and examined in the electron microscope. This technique can be successful only if the myosins to be compared each form distinctly different filament types. (For example, under the same dialysis conditions, rabbit skeletal myosin will form filaments about 3 μm long, while chicken gizzard myosin filaments are only about 0.5 μm long.) Hybridization will have occurred if filaments formed from the myosin mixtures exhibit, e.g., length characteristics intermediate between the two, such that the distribution is unimodal and peak location is related to the molar proportions of the two proteins. If hybridization has not occurred, a bimodal length distribution is observed; the peak locations for each lobe should be those expected for each myosin type alone.

Successful hybridization studies have been reported by Pollard[3] and Kaminer *et al.*[5] Pollard mixed rabbit skeletal myosin with either human platelet or human uterine myosin in various ratios, and dialyzed against

[42] K. Burridge and D. Bray, *J. Mol. Biol.* **99**, 1 (1975).
[43] T. P. Stossel and T. D. Pollard, *J. Biol. Chem.* **248**, 8288 (1973).
[44] T. P. Stossel and J. H. Hartwig, *in* "Cell Motility," (R. Goldman, T. Pollard, and J. Rosenbaum, eds.), Book B, p. 529. Cold Spring Harbor Laboratory, Cold Spring Harbor, New York, 1976.
[45] M. Vahey and S. P. Scordilis, *Can. J. Bot.* **58**, 797 (1980).

0.1 M KCl, pH 7.0. Kaminer *et al.* mixed chicken gizzard myosin with either rabbit skeletal or chicken breast myosin and rapidly diluted the solutions to 0.1 M KCl, pH 6.5. In all cases, the investigators observed unimodal length distributions whose peaks were intermediate between those expected of either myosin species alone according to the molar proportions of the two myosin types.

Such copolymerization experiments are not always successful, however. While chicken breast and gizzard myosin will form hybrids, mixing experiments by Kuczmarski and Rosenbaum[41] of chick brain myosin and chicken breast myosin resulted in two distinctly different populations. Further, Wachsberger and Pepe[31] have shown that pH may also be an important factor in such studies. They mixed rabbit skeletal and rabbit uterine myosins, and formed filaments by slow dilution at pH 6.0 and 7.0. Although copolymerization occurred at pH 7.0, in agreement with Pollard, it did *not* occur at pH 6.0; rather, a bimodal distribution was observed for each sample containing both myosin types.

Special Procedures

Minifilaments

As discussed earlier for skeletal myosin filamentogenesis, the length distribution obtained by either dilution or dialysis techniques tend to be broad. The narrowest distributions are obtained at high pH and intermediate salt, but the filaments may appear to be poorly formed and possibly disorganized. However, a report by Reisler *et al.*[46] provides the details for formation of short, well-organized myosin filaments of uniform length and high stability at very low ionic strength. These minifilaments are bipolar, with mean length of 303 nm, diameter of about 8 nm, and a 160–180 nm central bare zone; various physical measurements indicate that the filaments are composed of 16–18 molecules and are not accompanied by discernible amounts of larger aggregates or monomer (Fig. 3).

Minifilaments are prepared by a two-step dialysis process. Myosin in 0.5 M KCl, 10 mM phosphate, pH 7.0, at about 5 mg/ml is dialyzed against 5 mM sodium pyrophosphate, pH 8.0, then centrifuged at 27,000 g for 30 min to clarify. The resultant solution is then dialyzed further against 10 mM citrate, 35 mM Tris, pH 8.0, and again clarified at 27,000 g for 15 min. (It should be noted that Reisler's initial myosin solution was not entirely free of C-protein.)

Although the minifilaments are fairly stable in the presence of 5 mM MgATP, they are dissociated somewhat by 5 mM ATP as monitored by

[46] E. Reisler, C. Smith, and G. Seegan, *J. Mol. Biol.* **143**, 129 (1980).

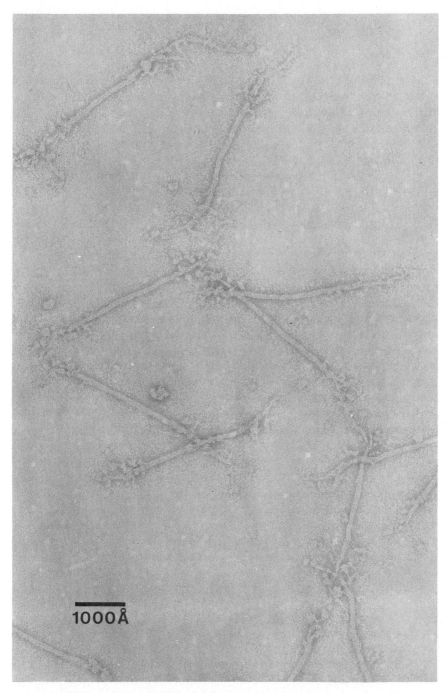

FIG. 3. Minifilaments prepared as described in the text. Electron micrograph by courtesy of Dr. Emil Reisler (E. Reisler and J. Lake, personal communication.)

turbidity. There is no turbidity change upon addition of 30–40 mM KCl at constant pH, but higher salt concentrations result in a reversible turbidity increase. Further studies at slightly higher ionic strength indicate the formation of a homogeneous population of myosin filaments about 0.65 μm long; however, at about 0.2 M KCl, the turbidity indicates filament dissociation. It is possible that minifilaments are an intermediate in the myosin aggregation process. But whether they are or not, their full ATPase activity in the presence of CaATP or MgATP indicates their potential usefulness for future kinetic studies.

Filaments with a 43-nm Axial Period

One of the important similarities between natural thick filaments and reconstituted filaments is the 14 nm axial period of the cross-bridges. As will be discussed later, the tail portion of the myosin molecule can exhibit aggregation patterns of 14 nm or 43 nm or both, but in both natural and reconstituted skeletal myosin thick filaments, the 43-nm period, if present, can be discerned only with optical diffraction. Eaton and Pepe,[47] however, have developed a procedure for the formation of filaments exhibiting only the 43-nm axial period.

Column-purified myosin (2–4 mg/ml in 0.6 M KCl, 0.01 M phosphate, or imidazole, pH 7.0) is diluted rapidly at about 0° by dilution with 0.01 M phosphate or 0.01 M imidazole or 0.1 M KCl, 0.01 M phosphate, 1 mM MgCl$_2$ at pH 7.0 to a final salt concentration of 0.3 M KCl. Alternatively, the myosin solution is dialyzed overnight at 2–4° against 0.3 M KCl containing either 0.01 M imidazole or 0.01 M phosphate with or without 0.6 mM MgCl$_2$ at pH 7.0. The resultant myosin solution is then brought to 0.5 mg/ml with the same buffer, and often left at 2–4° for several days.

The resultant filaments are variable in length, but exhibit a clearly visible axial period of about 43 nm. No bare zone is observed, and it is not possible to determine whether the filaments are bipolar. It should also be noted that these filaments are surrounded in the electron microscope field by a sizable number of very small protofilaments, some of which also exhibit a 43-nm periodicity.

AMP Deaminase Decoration

AMP deaminase (AMP aminohydrolase, EC 3.5.4.6) is a tetrameric enzyme complex found in particularly high concentrations in muscle tissue. While the significance of the reaction it catalyzes (conversion of AMP to IMP and ammonia) to contraction is unclear, its concentration in muscles is directly proportional to actomyosin ATPase activity. It has been

[47] B. L. Eaton and F. A. Pepe, *J. Mol. Biol.* 82, 421 (1974).

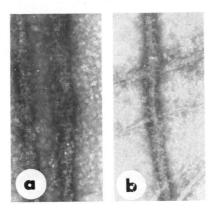

FIG. 4. (a) Synthetic thick filament decorated with AMP deaminase, described in the text. Free tetramers of AMP deaminase can be seen in the background. Spacing between disks is about 14 nm. (b) Undecorated myosin filament at the same magnification for comparison.

shown to bind both to the subfragment-2 portion of the myosin molecule[48] and to the ends of the A band in isolated myofibrils.[49]

A study by Koretz and Frieden[50] has shown that AMP deaminase will bind specifically to synthetic filaments formed from column-purified rabbit skeletal myosin. (Fig. 4a and b) The electron micrographs of the complex demonstrates that it binds every 14.3 nm along the filament axis, forming a sort of collar around the filament. Presumably it is located in the area between cross-bridges on the filament where the backbone is exposed; for steric reasons, it is estimated that no more than three tetramers bind per 14.3-nm interval. Structural analysis using optical diffraction indicates that the underlying filament structure has not been altered by AMP deaminase binding.

Myosin and AMP deaminase are mixed in a molar ratio of one myosin per tetramer or one subfragment-2 region per tetramer at an ionic strength where myosin is soluble (0.5 M KCl, 10 mM imidazole, pH 6.5 or 7.0) and dialyzed against 0.1 M KCl, 10 mM imidazole, pH 6.5 or 7.0. These results suggest that AMP deaminase could easily be used as a specific structural marker in a variety of applications involving myosin filament studies.

Nonfilamentous Myosin Aggregates

In addition to the filamentous aggregates prepared by dilution or dialysis against (generally) KCl, skeletal myosin will form side-by-side

---

[48] B. Ashby and C. Frieden, *J. Biol. Chem.* **252**, 1869 (1977).
[49] B. Ashby, C. Frieden, and R. Bischoff, *J. Cell Biol.* **81**, 361 (1979).
[50] J. F. Koretz and C. Frieden, *Proc. Natl. Acad. Sci. U.S.A.* **77**, 7186 (1980).

antiparallel linear arrays when dialyzed against certain nonphysiological conditions or nonphysiological concentrations of common species. These ribbon-like arrays, which can also be formed from myosin rod (see the next section) or light meromyosin (see final section), have been studied to determine the length of the tail portion of the myosin molecule, the size of the antiparallel overlap, and the substructure of the overlap region.

The procedure of Harrison et al.[51] involves two dialysis steps. Either rabbit or chicken breast myosin is first dialyzed against $1.0\,M$ KCl, $0.05\,M$ Tris-HCl, pH 8.2, and then against 0.075–$0.10\,M$ KSCN, $0.05\,M$ CaCl$_2$, $0.05\,M$-Tris-HCl, pH 8.2. (Lower KSCN concentrations result in the formation of poorly organized filaments.) The resultant antiparallel ribbon-like arrays are fringed on either side by myosin heads and exhibit a tail overlap of about 90 nm.

More precise alignment of the myosin tails is obtained using the technique of King and Young,[52] which involves dialysis against a solution in which the protein remains soluble. Addition of small volumes of distilled water to the dialyzate over a 1.5–4-month period gradually reduces the ionic strength and should be continued to a point just beyond where precipitation occurs. Initial myosin solutions in either $0.5\,M$ barium acetate, calcium acetate, or magnesium acetate worked equally well; end-point concentrations of 0.22–$0.36\,M$ calcium acetate, $0.30\,M$ magnesium acetate, or $0.24\,M$ barium acetate all resulted in identical array formation. Formation was also independent of initial protein concentration in the range of 0.046–2.3 mg/ml, or of prior column purification (Fig. 5).

These aligned segments exhibit a tail overlap region of length 83.2 ± 2.2 nm and overall head-to-head span of approximately 248 nm. The width of the individual arrays are much more variable, 39.3 ± 13.2 nm. King and Young suggested that steric hindrance from the crowded myosin heads serves to limit growth in this direction into long ribbons. Of great interest is the fact that the central overlap region is itself striated, the bands being oriented parallel to the myosin heads. There are 16 of these striations, with a mean spacing of 5.3 nm, and their intensity as a function of distance from a theoretical midline shows bilateral symmetry.

If the procedure used in segment preparation employs lower concentrations of divalent cation (e.g., $1\,M$ ammonium acetate with 0.02–$0.1\,M$ Ca^{2+} or Mg^{2+}), the resultant aggregates more nearly resemble those of Harrison et al. (The overlap region has more ragged boundaries and does not exhibit any striations. At divalent cation concentrations of $0.01\,M$ or less, twiglike filaments are the primary species.[8]) Under conditions where these less ordered segments appear, there is a strong tendency for pairs to associate such that the direction of the tails in one is normal to that of the

[51] R. G. Harrison, S. Lowey, and C. Cohen, J. Mol. Biol. 59, 531 (1971).
[52] M. V. King and M. Young, J. Mol. Biol. 63, 539 (1972).

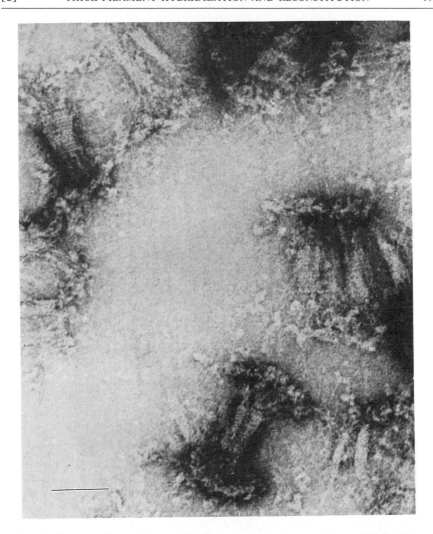

FIG. 5. Segments formed from rabbit skeletal myosin by a technique slightly different from that described in the text. The specimen was prepared by the phosphate-depletion method, where dialysis was performed against two changes of $0.36 M$ calcium acetate so as to dialyze out phosphate while keeping calcium concentration constant. Bar = 100 nm. Electron micrograph by courtesy of Dr. Murray Vernon King.

other. This "crossed segment" has the appearance of a woven square, with myosin heads providing a fringe around the entire perimeter. Addition of KSCN (0.01–0.05 M) to the dialysis medium seems to inhibit crossed-segment formation, as does 0.01 M KI. The latter, however, seems to cause head-to-head association of segments.

Myosin Rod Aggregates

Myosin rod is the largest proteolytic fragment of myosin demonstrating aggregation capability. The conditions governing its assembly and the structural characteristics of its aggregates have not been extensively investigated, but in those cases where studies have been performed, the rod aggregates share some of the structural properties of their original myosin source.

Formation and Assay

Striated Muscle Myosin Rods

Rod aggregates are generally prepared by dialysis as described above for myosin. At pH 7.0, the best rabbit skeletal myosin rod aggregates are observed in the electron microscope at KCl concentrations of 0.10 or 0.15 M[53] (Fig. 6a–c). They form apparently unipolar sheets with irregular edges and a tendency for the sides to curl. The most notable feature of these sheets is a striping pattern perpendicular to the side edges with a spacing of about 14 nm.[13,53] Since the stripes are dark with negative staining, this regular feature indicates a systematic absence of protein. At 0.20 M KCl, only very small aperiodic aggregates are observed, suggesting near solubility under these conditions. At 0.05 M KCl, the aggregates are unipolar, but smaller and less well organized than at 0.10 or 0.15 M. Optical diffraction reveals a very weak 14-nm period, but this repeat is almost impossible to see in the electron microscope.

Addition of C-protein at high molar ratios (one per one or two rod molecules) before dialysis disrupts both sheet formation and the 14-nm periodicity at 0.10 M KCl, 10 mM sodium phosphate, pH 7.0.[13] Instead of curling sheets, one observes irregular filamentous structures. Addition of C-protein after sheet formation appears to have no effect either on aggregate structure or appearance in the microscope. It is not clear in this case whether C-protein is able to bind to preformed rod aggregates.

Ordered bipolar segment formation will occur with chicken breast myosin rod, but not rabbit skeletal rod.[54] Dialysis of chicken rod against 0.05 M KSCN, 0.05 M CaCl$_2$, 0.05 M Tris-HCl, pH 8.2, results in bipolar segments with an overlap region of 130 nm and total width of about 180 nm. Occasional unipolar segments are observed under the same conditions. If the KSCN concentration is raised to 0.075 M, the overlap region is reduced in width to about 90 nm; this is the overlap size observed for

[53] A. M. Bertasso, J. M. Herbst, and J. F. Koretz, *Biophys. J.* **33**, 241a (1981).
[54] C. Cohen, S. Lowey, R. G. Harrison, J. Kendrick-Jones, and A. G. Szent-Györgyi, *J. Mol. Biol.* **47**, 605 (1970).

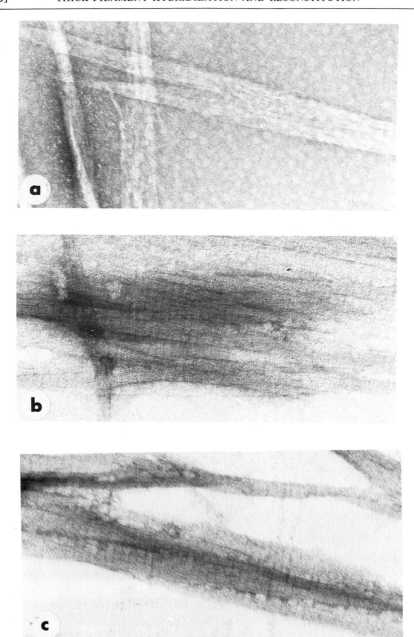

FIG. 6. Aggregates of myosin rod prepared by dialysis of the papain product of column-purified rabbit skeletal myosin at pH 7.0. Note the 14.3 nm striations across the sheets. (a) 0.05 M KCl; (b) 0.10 M KCl; (c) 0.15 M KCl.

chicken breast and rabbit skeletal myosin under similar formative conditions (see Myosin Aggregates section: Assays).

Rods obtained by proteolysis of myosin from the striated adductor muscle of scallop will also form segments when dialyzed against 0.10 M NaCNS, 0.05 M CaCl$_2$, 0.05 M Tris, pH 8.0.[55] However, these segments will themselves aggregate linearly by narrow overlap of the "fringed" regions of adjacent segments. The resultant structure exhibits alternating broad and narrow antiparallel overlaps of the LMM and subfragment-2 portions of the rod, respectively.

Smooth Muscle Myosin Rods

Chicken gizzard myosin rods will form "compound" bipolar segment structures when dialyzed against 0.05 M Ca^{2+}, 0.005–0.10 M KSCN, 0.05 M Tris, pH 8.0.[56] These segments are characterized by three light-staining regions, the central zone of length 43 nm, and the outer two each about 114 nm, for a total length of about 270 nm. The "fringed" borders of the outer two zones are each about 40 nm. Copolymerization with intact gizzard myosin, which will not form segments, results in the observation of myosin heads both at the outer edges of the outer zones and at the outer edges of the fringe. Further digestion of the gizzard rod with trypsin or chymotrypsin will alter the size of the outer zones, but leaves the central 43-nm zone intact. These results can be explained by packing the rods tail-to-tail with a zigzag stagger of 43 nm in the center.

Calf aortic smooth muscle myosin rods will form side-polar filaments when dialyzed against the same dialysis medium used for side-polar aggregation of the myosin (0.3 M KCl, pH 6.0).[27] The basic organization of these aggregates is, in some respects, more clearly visible with the rod than intact myosin. The effect of twist along the filament length is particularly notable.

Human Platelet Myosin Rods

The platelet rod is formed as a by-product of platelet myosin purification and can be separated from myosin by gel filtration.[40] At ionic strengths below 0.3 M KCl at pH 7, platelet rods will aggregate into filamentous structures that tend to be longer and wider than platelet myosin aggregates. In the presence of 1 mM EDTA these rod filaments are thinner and apparently more regular than in the presence of 2 mM MgCl$_2$ or 2 mM CaCl$_2$. If formed in the presence of 1 mM ATP, rod aggregates are smaller than those observed in the absence of ATP.

[55] A. G. Szent-Györgyi, E. M. Szentkiralyi, and J. Kendrick-Jones, *J. Mol. Biol.* **74**, 179 (1973).
[56] J. Kendrick-Jones, A. G. Szent-Györgyi, and C. Cohen, *J. Mol. Biol.* **59**, 527 (1971).

Light Meromyosin Aggregates

Light meromyosin (LMM) is a fragment of myosin tail shorter than myosin rod; it varies in length from 85 to 95 nm depending on the preparative procedure. Like myosin and myosin rod, LMM will form ordered aggregates of various types when dialyzed against or diluted with a low ionic strength medium and will form ordered "segments" under other conditions. However, unlike myosin and rod, different aggregate types can be observed from the same LMM preparation under the same formative conditions on the same electron microscope grid, indicating that differences in striation patterns are not due solely to differences in LMM length.[57]

The most commonly observed LMM aggregates are paracrystals or tactoids, cigar-shaped structures with pointed, sometimes fraying, tips, and staining bands of 14.3 and/or 43 nm spacing perpendicular to the cigar axis (Fig. 7a–c). Also seen are ribbons and sheets, both of which exhibit these striations to a greater or lesser degree. At pH values higher than 7.0, there is an increasing tendency for LMM to aggregate into thin fibers, which in turn form square, open lattices. Finally, under those conditions where myosin and myosin rod form "segment" structures, LMM will also aggregate into segments.

At this time there is no wholly satisfactory model of LMM packing into its various structures. Part of the difficulty in constructing a model lies in the uncertainty of LMM polarity within its aggregates and part in the observation of different structural types in the electron microscope under the same formative conditions. Further, there is some evidence that LMMs of the same molecular weight prepared using different proteolytic enzymes have different aggregation properties, and that certain LMM aggregates exhibit spacing of 3/8, 3/8, and 2/8 of 43 nm rather than the expected 14.3 nm bands.[58,59] Because of these factors, and also because of the variations in proteolytic enzyme type and digestion time in papers on this subject, only the most general guidelines for LMM aggregate preparation can be drawn.

Formation and Assay

Rabbit and Chicken Skeletal LMM Aggregates

The most thorough analysis of the effects of ionic strength, pH, and rate of formation on LMM aggregation were performed by Katsura and Noda[58] using tryptic digests of fresh and old rabbit skeletal myosin. When

[57] Z. A. Podlubnaya, M. B. Kalamkaryova, and V. P. Nankina, *J. Mol. Biol.* **46**, 591 (1969).
[58] I. Katsura and H. Noda, *J. Biochem.* (*Tokyo*) **73**, 257 (1973).
[59] P. M. Bennett, *Proc. Eur. Congr. Electron Microsc., 6th*, Vol. II, p. 517 (1976).

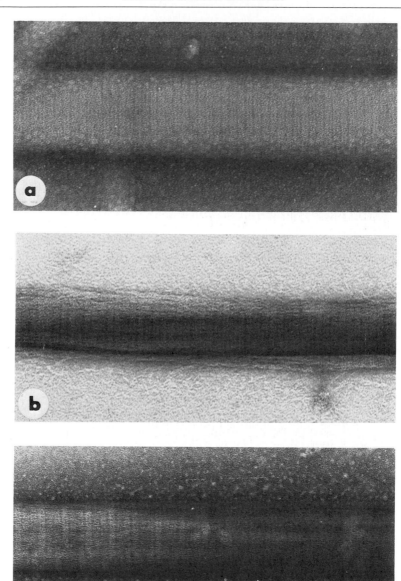

FIG. 7. Aggregates of light meromyosin prepared by dialysis of the trypsin product of column-purified rabbit skeletal myosin at pH 7.0. All micrographs are at the same magnification. (a) 0.5 M KCl, 14.3-nm period; (b) 0.10 M KCl, 14.3-nm period; (c) 0.15 M KCl. This aggregate exhibits a 14.3-nm period on the right and a 43-nm period on the left.

prepared by dialysis, fresh LMM forms tactoids in the pH range of 6.0–7.5 at ionic strengths of 0.10–0.15 M KCl, and in the pH range of 6.0–6.5 at 0.05 M or 0.20 M KCl. Sheet formation is observed at pH 7.0 and 0.15 M KCl, and at gradually lower ionic strengths as pH is raised; nets appear only when the pH is above 8.0 and KCl is less than 0.10 M. Fast dilution, in contrast, eliminates net formation and reduces the occurrence of sheets to 0.05 M KCl at pH 7.5–8.0; conditions for tactoid formation appear to be unaffected. LMM prepared from old myosin will apparently not form sheets. Conditions for tactoid formation, whether by dilution or dialysis, remain unchanged, and nets are observed at higher ionic strengths and lower pH values.

Although LMMs prepared in other ways (e.g., with chymotrypsin) have not been as extensively studied for ionic strength and pH dependence, similar results at neutral pH have been obtained at low ionic strength. That is, the primary structure observed is a tactoid with 14.3-nm and/or 44-nm stripes of varying complexity.[60-62] Even the 95-nm fragment prepared by cyanogen bromide cleavage (LMM-C) forms tactoids at low ionic strength $(3 \times 10^{-5} M$ KHCO$_3)$[63]; these usually exhibit a 14.3-nm period, unlike tryptic LMM, which under the same conditions usually exhibits a 43-nm period, and are almost always accompanied by square nets of 68.5-nm spacing.

An extensive study of the effects of digestion time and proteolytic enzyme alternatives on chicken LMM paracrystal formation has been performed by Chowrashi and Pepe (ionic strength 0.1 M, pH 7.35).[61] All LMM molecules longer than 91 nm form tactoids with a 15-nm repeat; the authors attribute this to superposition of sheets with a 44-nm period, as illustrated in their Fig. 7. Shorter lengths of myosin tail (90 nm or less) result in a variety of periodicities and complex striation patterns that are quite sensitive to small changes in buffer. In addition, nucleic acid contamination, which occurs if neither the original myosin nor the LMM has been purified on an ion exchange column, will cause alterations in the observed striation patterns.

The effect of C-protein on LMM aggregates, whether added before or after formation, has been investigated.[13,61,62] When added to 43-nm period rabbit LMM sheets after sheet formation, C-protein will bind in diffuse stripes separated by about 43 nm; since no other striations are observed in the presence of C-protein, it is presumably binding over the stripes noted in its absence. If added to 14-nm-period chicken paracrystals, it will again demonstrate a 43-nm period. Addition of C-protein before chicken LMM

[60] A. Nakamura, F. Sreter, and J. Gergely, *J. Cell Biol.* **49,** 883 (1971).
[61] P. K. Chowrashi and F. A. Pepe, *J. Cell Biol.* **74,** 136 (1977).
[62] D. Safer and F. A. Pepe, *J. Mol. Biol.* **136,** 343 (1980).
[63] M. V. King and M. Young, *J. Mol. Biol.* **50,** 491 (1970).

paracrystal formation results in less well-formed paracrystals. Where the LMM gave tactoids with a 44-nm repeat, the presence of C-protein (1 C-protein per 5 LMM molecules) also showed this period; where the repeat in the absence of C-protein was 15 nm, the presence of C-protein induced a shift to a 44-nm period, presumably interacting with LMM and altering its interactions.

Rabbit LMM segment formation was studied by King and Young for LMM-C, using either direct dialysis against 0.5 M ammonium acetate or long-term dialysis from 1.0 M to 0.59 M ammonium acetate (as described above for nonfilamentous myosin aggregates).[52] Two types of segments were observed, one with a central overlap of 89 nm and one with an overlap of 83 nm. In addition, if initial conditions of 0.5 M KCl, 0.1 M calcium acetate, 0.01 M Tris, pH 8.18, are reduced to 0.33 M KCl, 0.066 M calcium acetate using the long-term dialysis technique, what is seen is a "twisted festoon," a ribbon 103 nm in width that is twisted into a helix of about 255-nm pitch.

Other Muscle LMM Aggregates

Rabbit Red Skeletal LMM.[60] Tryptic digestion of red skeletal myosin was carried out with more trypsin and for longer periods because of higher resistance to proteolysis. When dialyzed against 0.05 M KCl, 5 mM phosphate, 0.1 mM DTT, pH 6.5, it forms a single paracrystal type with a 43-nm repeat. Unlike all other tactoids studied, the light band consists of five narrow light bands of alternating broader and narrower widths. Chymotryptic digestion results in sheets and small, poorly organized aggregates.

Rabbit Cardiac LMM.[60] When prepared by tryptic digestion as above for red skeletal LMM, this myosin fragment forms tactoids of the same banding type.

Scallop LMM.[55] Preparation of LMM using a tryptic digestion yields a molecule 72.5 nm in length. When dialyzed against 50 mM NaCl, 10 mM phosphate, pH 7.0, it will form loose paracrystals with a 58-nm repeat and 14.5-nm overlap region. Dialysis against 50 mM MgCl$_2$, 50 mM Tris, pH 8.0, yields segments 70–80 nm in width.

Uterine Muscle LMM.[31] Tryptic digestion of this myosin type yields an M_r 85,000 component like that obtained for chicken gizzard myosin. When dialyzed against 0.15 M KCl, 0.01 M imidazole, pH 7.0, uterine LMM forms only small aggregates without apparent period. At pH 6.0 (0.025 M KCl, 0.075 M imidazole), tactoids are observed; they exhibit a 14-nm repeat, but the narrow bands demarcating this repeat are light in the presence of negative stain, rather than dark as in the case of the striated LMMs.

Hybridization Studies

Rabbit Red and White Skeletal LMMs.[60] When mixed together in a 1 : 1 weight ratio and dialyzed against 0.05 M KCl, 5 mM phosphate, 0.2 mM DTT, pH 6.5, the paracrystals formed exhibit striation patterns characteristic of both muscle types.

Rabbit Skeletal, Chicken Skeletal, and Rabbit Uterine LMMs at pH 7.0 and 6.0.[31] At pH 7.0, rabbit skeletal LMM tactoids exhibit a 43-nm period, while chicken skeletal tactoids exhibit a 14-nm period. Mixtures of the two give a single population with 43-nm repeats, indicating hybridization. When uterine LMM is mixed with chicken skeletal LMM, the resultant tactoids have a 43-nm period, again indicating hybridization. Whether rabbit skeletal and chicken uterine LMMs also hybridize at pH 7.0 is unclear, since 43-nm-striped paracrystals are observed in both the presence and the absence of uterine LMM.

At pH 6.0, where both chicken and rabbit skeletal LMMs form paracrystals with an identical 14-nm dark-band appearance, it is not possible to tell whether the two will hybridize. When uterine LMM is mixed with either, two populations are observed, one with the 14-nm dark bands characteristic of the skeletal LMMs and one with the 14-nm light bands characteristic of uterine LMM. Thus, although hybridization seems to occur at pH 7.0, it does not at pH 6.0. (See section on myosin aggregates for similar results obtained for skeletal and uterine myosin assays.)

Acknowledgment

My sincere thanks go to Dr. Frank A. Pepe for making prepublication material available and to Ms. Anne M. Bertasso for her assistance.

[7] Preparation of Myosin and Its Subfragments from Rabbit Skeletal Muscle

By Sarkis S. Margossian and Susan Lowey

Methods for preparing myosin and its subfragments will be described here. The discussion is limited to rabbit skeletal muscle myosin, since this particular myosin has been the most widely used protein in studies of muscle contraction. Myosin consists of two heavy chains, each of ~200,000 daltons, which form an α-helical coiled coil over a length of about 1500 Å and terminate in two globular regions, the subfragment-1 (S1) heads (see Fig. 1).[1]

Myosin (EC 3.6.1.3, adenosinetriphosphatase, ATP-phosphohydrolase), by hydrolyzing ATP, provides the energy necessary for muscle contraction. The essential features of this complex mechanism are a rapid binding of ATP to myosin (binding constant $\sim 10^{11} \ M^{-1}$)[2] and subsequent hydrolysis of ATP into ADP and P_i, followed by a stepwise release of first P_i and then ADP. All of these intermediate states have been identified, and the sequence of events has been elucidated.[3]

At a salt concentration approximating physiological conditions, myosin aggregates to form bipolar thick filaments with a central bare zone; S1 heads protrude from the filament on either side of the bare zone. The insolubility of myosin at $\mu < 0.3$ derives from the rod region, which is insoluble in low salt. The catalytic and actin-binding properties of myosin reside in the water-soluble S1 heads. Both the rod and S1 can be obtained in relatively homogeneous form by proteolytic digestion of myosin. In addition to the above, a number of other subfragments of myosin have been obtained: digestion with trypsin results in light meromyosin (LMM), which like the rod retains the solubility properties of myosin, and in heavy meromyosin (HMM), which consists of the two S1 heads attached to the flexible subfragment-2 (S2) region. Single-headed HMM and single-headed myosin have also been prepared by limited digestion of myosin with papain.[4] The methods described here are those developed in our laboratory over the years; these methods may differ slightly from those reported elsewhere, but the basic approach is essentially the same.

Preparation of Myosin

Solutions

A: 0.30 M KCl, 0.15 M potassium phosphate (pH 6.5), 0.02 M ethylenedinitrilotetraacetic acid (EDTA), 0.005 M MgCl$_2$, 0.001 M ATP

B: 1.0 M KCl, 0.025 M EDTA, 0.06 M potassium phosphate (pH 6.5)

C: 0.6 M KCl, 0.025 M potassium phosphate (pH 6.5), 0.01 M EDTA, 0.001 M dithiothreitol (DTT)

---

[1] S. Lowey, in "Subunits in Biological Systems" (S. N. Timasheff and G. D. Fasman, eds.), Part A, p. 201. Dekker, New York, 1971.

[2] R. S. Goody, W. Hofmann, and H. Mannherz, Eur. J. Biochem. **78**, 317 (1977).

[3] For a review of the mechanism of ATP hydrolysis by myosin, see D. R. Trentham and C. Bagshaw, Q. Rev. Biophys. **9**, 217 (1976) and E. W. Taylor, CRC Crit. Rev. Biochem. **6**, 103 (1979).

[4] S. S. Margossian and S. Lowey, J. Mol. Biol. **74**, 301 (1973).

D: 0.6 M KCl, 0.05 M potassium phosphate (pH 6.5)

E: 0.15 M potassium phosphate (pH 7.5), 0.01 M EDTA

F: 0.04 M sodium pyrophosphate (pH 7.5), 0.001 M DTT

G: 0.02 M sodium pyrophosphate (pH 7.5), 0.001 M DTT

H: 0.04 M KCl, 0.01 M potassium phosphate (pH 6.5), 0.001 M DTT

I: 3 M KCl, 0.01 M potassium phosphate (pH 6.5)

The following method gives a homogeneous preparation by the criteria of gel electrophoresis and ultracentrifugation. A white, male New Zealand rabbit (about 7–8 kg) is stunned and exsanguinated. After quick skinning, the carcass is cooled in ice for about an hour to avoid the risk of extracting largely actomyosin. The leg and back muscles are dissected and ground in a meat grinder that has been rinsed with a solution of 0.02 M EDTA (pH 7.0). About 1 kg of ground tissue is normally obtained from a rabbit that size. All subsequent steps are performed at 4°.

The muscle mince is extracted with 2 liters of solution A for 15 min with constant agitation by a motor-driven overhead stirrer. Care should be taken not to extend the extraction time beyond 15–20 min to minimize actin extraction. Extraction is stopped by diluting the mixture 4-fold with cold water, and the muscle residue is separated by filtration through three layers of gauze. (This residue may be used later for the preparation of either actin or the relaxing factor.) The extract is subsequently precipitated by dilution with cold water to give a final ionic strength of ~0.04. (This is approximately a 10-fold dilution of buffer A.) After settling for at least 3 hr, the supernatant is siphoned off and precipitated protein is collected by centrifugation (7000 rpm, for 15 min in Sorvall GSA or GS3 rotor). The precipitate is dispersed in 220 ml of solution B and dialyzed overnight against 6 liters of solution C.

After dialysis, an equal volume (about 400 ml) of cold, deionized water is slowly added to the protein with constant stirring to precipitate the actomyosin. Care should be taken not to add too much water, since the myosin may also precipitate if the ionic strength is lowered below 0.3. The suspension is stirred for an additional half hour in the cold and then spun at 10,000 rpm (Sorvall GSA or GS3 rotor) for 10 min to remove the precipitated actomyosin. To further clarify the myosin, the supernatant is centrifuged for 1 hr at 19,000 rpm (Beckman 21 rotor), after which the supernatant is diluted to an ionic strength of 0.04 (8-fold dilution), and the myosin is left overnight to settle. About 5–6 liters of a fluffy myosin precipitate is collected by centrifugation (10,000 rpm, for 10 min in Sorvall GSA or GS3 rotor) and resuspended in a minimum volume of solution I. After dialysis overnight against 2 liters of buffer D, the myosin is defatted by centrifugation for 2 hr at 19,000 rpm (Beckman 21 rotor) in cellulose nitrate tubes. The myosin is separated from the lipid layer (and a small pellet of insoluble protein) by puncturing the tubes at the bottom and slowly draining them. Myosin thus obtained can be stored on ice in solution D containing a few crystals of thymol without loss of activity for 3–4 weeks as a 5–6% solution, or it can be stored at −20° in 50% glycerol for over a year.

In order to remove C-protein and traces of contaminating actin, the myosin is further purified by ion-exchange chromatography according to the procedure of Richards et al.[5] as modified by Offer et al.[6] Myosin is diluted to 5–10 mg/ml and dialyzed extensively against solution E with several changes of the dialyzate to ensure adequate equilibration. A 2.5 × 60 cm column of DEAE–Sephadex A-50 (Pharmacia) is equilibrated with the same phosphate–EDTA solution. The conductivity and pH of the myosin and the eluate from the column are checked to ensure adequate equilibration, which is critical for good separation. Myosin (500–600 mg) is applied to the column at a flow rate of about 30 ml/hr. The column is washed with about 350 ml of the starting buffer. Any aggregated myosin and C-protein are eluted in the void volume, and the retained protein is eluted with a linear KCl gradient (0 to 0.5 M KCl in solution E, 1 liter each); 10-ml fractions are collected. Myosin begins to be eluted at about 0.1 M KCl. The myosin-containing fractions are pooled and concentrated either by dilution to $\mu = 0.04$, or by overnight dialysis against solution H. The precipitated myosin is collected by centrifugation (Beckman rotor 30) at 25,000 rpm for 30 min and stored as before. Alternatively, myosin can be purified by chromatography on DEAE-cellulose using pyrophosphate buffers. Myosin, at about 5 mg/ml, is dialyzed against solution F and applied to a 2.5 × 60-cm column of DE-52 (Whatman) equilibrated in solution G. The column is washed with about 350 ml of solution G, then the myosin is eluted with a linear NaCl gradient (0 to 0.5 M in solution G, 1 liter each). The rest of the procedure is as described previously. The myosin thus prepared migrates as a single hypersharp boundary in the Model E analytical ultracentrifuge, and on sodium dodecyl sulfate (SDS)–polyacrylamide gels, it shows essentially the heavy chain and the three light chains: A1, DTNB light chain, and A2, with practically no contamination by other proteins. Because the procedure described here applies to myosin from fast skeletal muscle exclusively, the original, descriptive nomenclature for the light chains will be used, so that the terms alkali 1 (A1), DTNB, 1.c., and alkali 2 (A2) correspond to $LC1_f$, $LC2_f$, and $LC3_f$, respectively.

The extinction coefficients and ATPase activities given in Tables I and II, respectively, were obtained with column-purified myosin. These values do not differ significantly from those of unchromatographed myosin if

[5] E. G. Richards, C. S. Chung, D. B. Menzel, and H. S. Olcott, *Biochemistry* **6**, 538 (1967).
[6] G. Offer, C. Moos, and R. Starr, *J. Mol. Biol.* **74**, 653 (1973).

TABLE I
EXTINCTION COEFFICIENTS OF MYOSIN AND ITS
SUBFRAGMENTS (10^2 CM^2/G)

Subfragment[a]	Extinction coefficient		
	230 nm	236 nm	280 nm
Myosin	—	—	5.3
HMM	52	21	6.0
EDTA · S1	60	25	8.1
Mg · S1	62	26	8.3
S1 · A1	59	24	7.4
S1 · A2	59	24	7.5
Rod	—	—	2.2
LMM	—	—	3.0
S2	—	—	0.7

[a] Protein concentration was determined by the micro-Kjeldahl method.

care is taken in the initial preparative steps to minimize actin contamination.

Preparation of Myosin Subfragments

The large size (M_r 470,000) and relative insolubility of the myosin molecule originally led investigators to look for proteolytic fragments that would retain the properties of the parent molecule and yet be more amenable to physicochemical and enzymic analysis. It was found that both trypsin and chymotrypsin cleave myosin into two distinct subfragments: heavy meromyosin (HMM) and light meromyosin (LMM), where the names reflect the relative molecular sizes. The schematic drawing of the myosin molecule in Fig. 1 illustrates the sites of attack by various proteases to yield the different subfragments.

Heavy Meromyosin

Solutions

A: 0.05% trypsin (Worthington) in 0.001 N HCl
B: 0.10% soybean trypsin inhibitor (Worthington) in 0.3 M KCl, 0.04 M potassium phosphate (pH 7.6)
C: 0.02 M KCl, 0.01 M potassium phosphate (pH 6.5) 0.001 M DTT
D: 0.05% α-chymotrypsin (Worthington) in 0.001 N HCl
E: 0.1 M phenylmethanesulfonyl fluoride (PMSF) in 70% enthanol

TABLE II
Enzymic Activity of Myosin and Its
Subfragments[a]

	ATPase activity (sec^{-1})[b]		
Subfragments	Ca$^{2+,c}$	EDTA[d]	Actin-activated[e]
Myosin[f]	4.2	20.3	18.0
HMM	4.6	20.6	36.3
EDTA · S1	2.4	11.2	17.0
Mg · S1	2.9	9.0	15.0
S1 (A1)	5.0	7.3	14.7
S1 (A2)	4.3	7.5	17.0

[a] Reprinted in a modified form from Margossian and Lowey[29] by the courtesy of the American Chemical Society.

[b] Activities are expressed as moles of ATP hydrolyzed per mole of myosin or subfragment per second.

[c] The reaction mixture contained 0.23 M KCl, 2.5 mM ATP, 2.5 mM CaCl$_2$, 0.05 M Tris · HCl (pH 7.9), and 0.1 mg of protein.

[d] EDTA, 1 mM, instead of CaCl$_2$ and 0.6 M KCl were included in the reaction mixture described in footnote c.

[e] The reaction mixture contained 2.5 mM ATP, 2.5 mM MgCl$_2$, 0.01 M Tris · HCl (pH 7.9), and increasing concentrations of actin. The activities were obtained from Lineweaver–Burk plots of velocity vs actin concentration. All the assays were done at 25° in a final volume of 2.0 ml.

[f] The actin-activated ATPase of myosin was determined in the presence of 25 mM KCl, 2.5 mM ATP, 3.8 mM MgCl$_2$ using a pH stat at pH 7.0. The actin-activated ATPase of myosin ranges from 6 to 18 sec^{-1} (or 3 to 9 sec^{-1} per site). There is a greater variation in the values reported for myosin than for the subfragments, since myosin is a heterogeneous suspension and therefore more influenced by small changes in pH and ionic conditions.

Tryptic Digestion. The preparation of HMM by trypsin is a modification of the procedure described by Lowey and Cohen.[7] To a 2% myosin solution (unless otherwise stated, stock myosin is kept in 0.5 M KCl, 0.05 M potassium phosphate, pH 6.5) at room temperature, trypsin is added to a final concentration of 0.05 mg/ml (1 ml of solution A to 10 ml of myosin).

[7] S. Lowey and C. Cohen, *J. Mol. Biol.* **4**, 293 (1962).

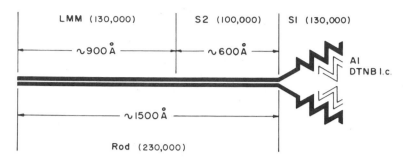

FIG. 1. Schematic representation of the myosin molecule.

Care should be taken to stir the myosin while adding trypsin in order to avoid local precipitation and denaturation of myosin by the acidic solution. After 5 min of stirring, the digestion is stopped by addition of soybean trypsin inhibitor to a final concentration of 0.10 mg/ml (1 ml of solution B to 10 ml of myosin). To separate HMM from LMM and undigested myosin, the digestion mixture is dialyzed against solution C overnight. The dialyzate should be changed at least once (1 liter each change is adequate) to ensure complete precipitation of LMM. HMM is separated by centrifugation at 25,000 rpm for 60 min (Beckman 30 rotor). If a Model E analytical ultracentrifuge is available, it is desirable to check the HMM for the absence of degradation products. Some trypsin inhibitor preparations are less effective at completely inhibiting tryptic activity, in which case doubling the amount of the inhibitor should eliminate the problem. The myosin used in the digestion should be relatively fresh, i.e., 10–15 days after preparation. The HMM prepared from aged myosin may not be homogeneous owing to aggregation of the starting material. The yield of HMM from the digestion conditions described should be 40–50% of the theoretical yield; i.e., if the starting myosin is 20 mg/ml, the yield of HMM should be about 6–7 mg/ml.

Chymotryptic Digestion. Chymotryptic digestion of myosin is performed essentially by the method described for tryptic digestion. α-Chymotrypsin (solution D) is added to a final concentration of 0.05 mg/ml to a 2% myosin solution equilibrated at room temperature.[8] The reaction is stopped after 10 min by the addition of solution E to a final concentration of 0.1–0.3 mM. Care should be taken to stir the digestion mixture while adding PMSF to avoid precipitation of the latter. The mixture is dialyzed overnight against solution C at 4°, and the HMM supernatant is obtained as described for tryptic HMM. Under these conditions the yield is again about 40–50% of the theoretical value.

[8] A. G. Weeds and R. S. Taylor, *Nature (London)* **257**, 54 (1975).

For quantitative kinetic and actin-binding studies, it is advisable to further purify HMM by ion-exchange chromatography. HMM obtained by either tryptic or chymotryptic digestion is extensively dialyzed against 0.05 M Tris-HCl (pH 8.0), 1 mM DTT and applied to a 2.5 × 60 cm column of DEAE-cellulose (Whatman DE-52) equilibrated in the same buffer. After checking the pH and conductivity of the column eluate and that of the HMM preparation, the protein is applied to the column at a flow rate of 40 ml/hr. The column is then washed with 300–500 ml of starting buffer to ensure removal of any aggregated material in the void volume. The retained protein is eluted with a linear KCl gradient (0 to 0.5 M KCl in starting buffer, 500 ml of each) and 5-ml fractions are collected. The fractions from the front two-thirds of the protein peak are pooled and further purified by ammonium sulfate (ultra pure, Schwarz-Mann) fractionation at 0°. Any material that comes out of the solution at 43% saturation is discarded. The protein salting out at 43–57% saturated ammonium sulfate is collected by centrifugation (Sorvall SS34 rotor, 15,000 rpm, 20 min) and dissolved in a minimal amount of 0.01 M potassium phosphate (pH 7.0), 0.10 M KCl, 1 mM DTT, 0.3 mM EGTA and dialyzed exhaustively against this buffer to ensure complete removal of the ammonium sulfate.

Subfragment-1

Solutions

A: 0.2 M ammonium acetate (pH 7.2), 0.002 M EDTA (or 0.002 M MgCl$_2$, in case Mg · S1 is to be prepared)

B: Stock papain (Worthington), diluted to a concentration of 0.5–1.0 mg/ml in 0.005 M cysteine (pH 6.0), 0.002 M EDTA

C: 0.1 M iodoacetic acid (Sigma)

D: 0.05 M Tris-HCl (pH 8.0), 0.001 M DTT, 0.002 M MgCl$_2$ (or 0.002 M EDTA)

E: 0.12 M NaCl, 0.02 M sodium phosphate (pH 7.0), 0.001 M EDTA

F: 0.05 M imidazole (pH 7.0), 0.001 M DTT

G: 0.1 M KCl, 0.01 M imidazole (pH 7.0), 0.001 M DTT, 0.0003 M EGTA

Papain Digestion. The first attempts to obtain S1 by trypsin digestion of HMM[9] resulted in a heterogeneous mixture of proteolytic products. The studies of Kominz *et al.*[10] on the digestion of myosin with papain suggested that this might be a more promising enzyme. In earlier experiments,[11] papain was added to soluble myosin (in 0.6 M KCl) or to

[9] H. Muller and S. V. Perry, *Biochem. J.* **85**, 431 (1962).

[10] D. R. Kominz, E. T. Mitchell, T. Nihei, and C. M. Kay, *Biochemistry* **4**, 2373 (1965).

[11] S. Lowey, H. S. Slayter, A. G. Weeds, and H. Baker, *J. Mol. Biol.* **42**, 1 (1969).

filamentous myosin, and a relatively homogeneous S1 preparation was obtained as judged by sedimentation velocity. With the advent of SDS gel electrophoresis, it was realized that the S1 had multiple cleavages in the light and heavy chain regions. This led to the inclusion of divalent cations in the digestion buffer,[12] as originally described for the preparation of S1 from molluscan myosin.[13]

Myosin at a concentration of 2% is dialyzed overnight against solution A.[12] To ensure complete precipitation, the dialyzate is changed at least once (total volume of dialyzate 2 liters). The myosin suspension is brought to room temperature, and digestion is started by addition of solution B to a final papain concentration of 0.03 mg/ml. The reaction is run for 7 min, during which time a drop in the viscosity of the myosin suspension can clearly be observed. The reaction is stopped by the addition of iodoacetate (solution C) to a final concentration of 0.001 M. The suspension is spun for 90 min at 27,000 rpm (Beckman 30 rotor) to remove insoluble digestion products. The yield of S1 in the supernatant is about 40% of the theoretical value. At higher papain concentrations (0.07–0.10 mg/ml) it is possible to obtain almost 100% of the theoretical yield, but only at the expense of further degradation of the heavy chains. Again, it is advisable to check the S1 in the analytical ultracentrifuge to ensure a monodisperse preparation. The resulting S1 is designated as Mg · S1 if MgCl₂ was included in the reaction mixture, or as EDTA · S1 if EDTA was present. The Mg · S1 is enriched in DTNB light chain, whereas EDTA · S1 is deficient in that light chain.[12]

The S1 can be further purified by ion-exchange chromatography. A solution of S1 (400–500 mg) at 7–10 mg/ml is dialyzed against solution D and applied to a 2.5 × 60 cm column of DE-52 equilibrated in the same buffer. The column is washed with the starting buffer (about 350 ml), and the retained S1 is eluted with a linear KCl gradient (conditions as described for HMM, except that either MgCl₂ or EDTA was present throughout the purification procedure, depending on the type of S1). The fractions from the front two-thirds of the protein peak are pooled and further purified by ammonium sulfate fractionation. Material precipitated at 47% saturation is discarded, and the protein fraction salting out between 47 and 58% saturation is collected (Sorvall SS34 rotor, 15,000 rpm, 20 min) and resuspended in a minimum amount of solution G. Exhaustive dialysis against this buffer is required to ensure removal of traces of ammonium sulfate.

Chymotryptic Digestion. The procedure originally described by Weeds and Taylor[8] is used to obtain chymotryptic S1. In this instance, myosin

[12] S. S. Margossian, S. Lowey, and B. Barshop, *Nature (London)* **258**, 163 (1975).
[13] A. G. Szent-Györgyi, E. M. Szentkiralyi, and J. Kendrick-Jones, *J. Mol. Biol.* **74**, 179 (1973).

filaments are obtained by dialysis of myosin into solution E. Inclusion of EDTA is essential, since replacement of EDTA by MgCl₂ results in the production of HMM, but not S1. The myosin suspension is equilibrated at room temperature, and digestion is achieved by addition of α-chymotrypsin (solution D under Heavy Meromyosin) to a final concentration of 0.05 mg/ml. The reaction proceeds for 10 min, after which it is stopped as in the case of chymotryptic HMM.

Chromatography of chymotryptic S1 is done essentially as described for papain S1, except that S1 is dialyzed into solution F and applied to a 2.5 × 60 cm DE-52 column equilibrated in the same buffer. After washing the preparation with about 350 ml of starting buffer, S1 is eluted with a linear NaCl gradient (0 to 0.12 M in starting buffer, 750 ml each). Chromatography of chymotryptic S1 results in the resolution of the starting material into its isoenzymes with respect to the light chains: S1 (A1) and S1 (A2). The DTNB light chain is completely digested under these conditions. The fractions of each of the two protein peaks are pooled and further purified by ammonium sulfate fractionation as discussed with papain S1. For most purposes, this degree of purification is adequate. For molecular weight measurements,[14] it is desirable to remove trace amounts of high and low molecular weight contaminants from S1 (A1) and S1 (A2) by gel permeation chromatography. This is achieved by applying S1 in a minimal amount of solution G to a 1 × 100 cm column of BioGel A-5M equilibrated in the same buffer.

Preparation of S1 by Other Methods. The methods described above for the preparation of S1 are commonly used procedures. However, one can obtain S1 by a number of other methods that will be briefly mentioned here. One widely used method involves the papain digestion of myofibrils instead of a myosin suspension.[15] The basic principle is the same as with the myosin suspension, in that the exposed cross-bridges are available for proteolytic attack, whereas the LMM–HMM junction is presumably protected in the filament.

Another, more recent, method for preparing S1 is to digest myosin with a calcium-sensitive protease isolated from a genetic strain of myopathic hamsters.[16] The protease, isolated from cardiac tissue, was shown to cleave LC2 specifically in cardiac myosin without attacking the heavy chain.[17] The protease is not as specific with skeletal muscle myosin, where in addition to digesting the DTNB light chain it also causes some

[14] S. S. Margossian, W. F. Stafford, and S. Lowey, *Biochemistry* **20**, 2151 (1981).

[15] R. Cooke, *Biochem. Biophys. Res. Commun.* **49**, 1021 (1972).

[16] A. Bhan, A. Malhotra, V. B. Hatcher, E. S. Sonnenblick, and J. Scheuer, *J. Mol. Cell. Cardiol.* **10**, 769 (1978).

[17] A. Malhotra, S. Huang, and A. Bhan, *Biochemistry* **18**, 461 (1979).

release of S1. This is not a very practical method for S1 preparation because the yields are low and the protease preparation is prohibitively expensive. However, an important advantage of this protease is that it enables one to prepare a myosin that is enzymically active but lacks the DTNB light chain.[18]

Single-Headed HMM and Single-Headed Myosin

Solutions

A: 0.1 M KCl, 0.05 M Tris-HCl (pH 8.0), 0.001 M DTT
B: 0.1 M iodoacetic acid
C: 0.05 M Tris-HCl (pH 8.0), 0.001 M DTT
D: 0.1 M KCl, 0.01 M potassium phosphate (pH 7.0), 0.001 M DTT
E: 0.2 M ammonium acetate (pH 7.2)
F: 0.6 M KCl, 0.05 M potassium phosphate (pH 6.5)
G: 0.15 M potassium phosphate (pH 7.5), 0.01 M EDTA

For experiments in which one wishes to study the effect on actin-binding and enzymic activity of S2 attached to the S1 region, or the effect of a single head attached to the rod in superprecipitation, two additional enzymically active subfragments can be prepared, as described below.[4,19]

Single-Headed HMM. Although either myosin or HMM can be used as the starting material, papain digestion of HMM is the preferred method because of its relative simplicity. A 1% solution of HMM in solution A is digested at room temperature for 10 min with 0.01 mg of papain per milliliter activated in cysteine as described earlier. The reaction is stopped by addition of solution B to a final concentration of 0.001 M. After dialyzing the mixture (700 mg) against solution C, it was applied to a 2.5 × 45 cm column of DE-52 equilibrated in the same buffer. The column is washed with 300 ml of the starting buffer, and the proteins are eluted with a linear KCl gradient (0 to 0.5 M KCl in the starting buffer in a total volume of 1.4 liters). The first protein peak corresponds to S1 and is eluted at 0.05–0.08 M KCl. The second peak elutes between 0.10 and 0.16 M KCl and is single-headed HMM with some contamination by S2. The fractions of the second peak are pooled and further fractionated with ammonium sulfate. The protein salting out between 55 and 60% saturated ammonium sulfate is resuspended in solution D, and ammonium sulfate is removed by dialysis against this buffer. The homogeneity of the preparation can then be established by analytical ultracentrifugation.

Single-Headed Myosin. A 2% myosin solution (95 ml) is dialyzed

[18] S. S. Margossian, A. K. Bhan, and S. Lowey, *Fed. Proc., Fed. Am. Soc. Exp. Biol.* **39**, 2309A (1980).
[19] S. Lowey and S. S. Margossian, *J. Mechanochem. Cell Motil.* **2**, 241 (1974).

against solution E. After equilibration to room temperature, it is digested for 10 min with 0.009 mg of soluble papain per milliliter. The reaction is stopped with iodoacetate (0.001 M), then the soluble supernatant (S1) is removed by centrifugation (25,000 rpm for 60 min; Beckman 30 rotor). The insoluble precipitate is dissolved in solution F (6 mg/ml) and dialyzed against solution G. The mixture (1.0 g) is applied to a 4 × 72 cm column of DEAE-Sephadex A-50 equilibrated in solution G. After washing the column with 1.0 titer of the starting buffer, the retained proteins are eluted with a linear KCl gradient (0 to 0.4 M KCl in the starting buffer in a total volume of 3.0 liters). Fractions of 6 ml are collected at a flow rate of 30 ml/hr. The front of the peak consists mostly of myosin, whereas the back is enriched with rod. The top 10–15 fractions, which consist predominantly of single-headed myosin, are pooled and fractionated with ammonium sulfate. The fraction salting out between 38 and 45% saturated ammonium sulfate is resuspended in solution F. This fraction consists mainly of single-headed myosin contaminated by less than 5% myosin as estimated from ultracentrifuge patterns. This trace amount of myosin has proved to be almost impossible to remove. A variation of this approach[20] uses myofibrils as the starting material for the digestion. The resulting single-headed myosin also contains less than 5% contaminating myosin as estimated from densitometer traces of SDS–polyacrylamide gels. We have not attempted to prepare single-headed myosin by this procedure in our laboratory; however, interested readers should refer to the original report for details of this method.[20]

Light Meromyosin and Rod

Solutions

A: 0.6 M KCl, 0.05 M potassium phosphate (pH 7.0)
B: 0.03 M KCl, 0.01 M potassium phosphate (pH 7.0)
Both LMM and the rod retain the solubility properties of myosin. They are insoluble at physiological salt concentrations, and both precipitate to form ordered aggregates upon precipitation. This property forms the basis of the purification procedures described here.[11,21]

After removal of the soluble HMM (from either tryptic or chymotryptic digestion of myosin) and S1 (from either papain or chymotryptic digests), the insoluble precipitates are dispersed in solution A. This fraction includes undigested myosin and either the LMM or the rod. Three volumes of 95% ethanol previously chilled to 4° are added to the protein solution to denature the residual myosin irreversibly. The mixture is left to stir in the cold for 2–3 hr; the precipitate is then collected by centrifuga-

[20] R. Cooke and E. K. Franks, *J. Mol. Biol.* **120,** 361 (1978).
[21] A. G. Szent-Györgyi, C. Cohen, and D. E. Philpott, *J. Mol. Biol.* **2,** 133 (1960).

tion (10,000 rpm for 30 min; Sorvall GSA rotor), dispersed in solution A, and dialyzed exhaustively against this buffer with frequent changes to ensure complete removal of the alcohol. Under these conditions both LMM and the rod are solubilized, whereas the denatured myosin remains insoluble and is removed by centrifugation. The supernatant is collected and further clarified by high speed centrifugation at 29,000 rpm for 90 min (Beckman 30 rotor). The clarified proteins are purified by one more cycle of precipitation: both LMM and the rod are dialyzed against solution B overnight (with at least one change of buffer, each 1 liter). The precipitated proteins are collected by centrifugation and redissolved in a minimal amount of solution A.

For most purposes this is an adequate degree of purification, but for accurate hydrodynamic experiments, it is advisable to chromatograph these proteins further by gel permeation chromatography on a BioGel A-15M column (1 × 100 cm) equilibrated in solution A. The peak containing the protein fractions is pooled, concentrated by dialysis against low salt, and resuspended in high salt buffer as discussed above.

Subfragment-2

Solutions

A: 0.1 M KCl, 0.02 M potassium phosphate (pH 7.0), 0.002 M EDTA

B: 0.1 M phenylmethanesulfonyl fluoride (PMSF) in 70% ethanol

C: 0.02 M potassium phosphate (pH 6.5), 0.1 mM PMSF

D: 0.1 M KCl, 0.02 M potassium phosphate (pH 7.0) 0.1 mM PMSF

E: 0.5 M NaCl, 0.02 M sodium phosphate (pH 7.0)

F: 0.05% trypsin in 0.001 N HCl

G: 0.1% soybean trypsin inhibitor in 0.3 M KCl, 0.04 M potassium phosphate (pH 7.6)

H: 0.1 M NaCl, 0.02 M sodium phosphate (pH 7.0)

S2 was first made by digesting tryptic HMM with an insoluble derivative of trypsin.[22] The molecular weight reported for this S2 was 60,000. Since then, the widespread use of SDS gels has made it much easier to monitor the extent of proteolysis during the digestion of myosin and its subfragments. Consequently, several groups[23-25] have reported a new preparation of S2 with a molecular weight of 100,000 and have shown that this long S2 is the precursor of the original short S2. Because long S2 is the preferred subfragment, methods to obtain it are described here.

[22] S. Lowey, L. Goldstein, C. Cohen, and S. Mahakian-Luck, *J. Mol. Biol.* **23**, 287 (1967).
[23] K. Sutoh, K. Sutoh, K. Trudy, and W. F. Harrington, *J. Mol. Biol.* **126**, 1 (1978).
[24] S. Highsmith, K. M. Kretzschmar, C. T. O'Konski, and M. F. Morales, *Proc. Natl. Acad. Sci. U.S.A.* **74**, 4986 (1977).
[25] A. G. Weeds and B. Pope, *J. Mol. Biol.* **111**, 129 (1977).

The starting material for obtaining "long" S2 can be chymotryptic HMM, since it shows good preservation of the S2 region.[23,25] HMM (5 mg/ml) in solution A is digested at room temperature with 0.01 mg of α-chymotrypsin per milliliter for 10 min. The reaction is stopped by addition of solution B to a final concentration of 0.3 mM. Three volumes of 95% ethanol are added to precipitate the proteins, and the entire mixture is stirred for 3 hr in the cold. The precipitated protein is collected by centrifugation (15,000 rpm for 30 min; Sorvall SS34 rotor), resuspended in solution C, and dialyzed exhaustively against the same buffer to remove the ethanol. The solution containing S2 is spun for an hour at 27,000 rpm (Beckman 30 rotor) to remove any insoluble material. The crude S2 is further purified by lowering the pH to 4.5. The precipitate is collected by centrifugation (15,000 rpm, for 30 min; Sorvall SS34 rotor), dissolved in solution C against which it is dialyzed overnight, and clarified by centrifugation at 27,000 rpm for an hour (Beckman 30 rotor). The S2 is then purified by chromatography on a Sephacryl S-200 (Pharmacia) column (1.5 × 60 cm) equilibrated in solution D. The fractions containing S2 are pooled, and the protein is concentrated by acid precipitation at pH 4.5. The precipitate is collected by centrifugation and stored in solution D.

An alternative method for obtaining S2 is to digest the rod with a limited amount of trypsin.[23] The rod obtained by digestion of myosin with papain in the presence of MgCl$_2$ is used. Trypsin (solution F) is added to myosin rod (about 0.75 mg/ml) in solution E at room temperature to a final concentration of 0.01 mg/ml. The reaction is stopped after 5 min by addition of solution G, to a final inhibitor concentration of 0.05 mg/ml. To precipitate undigested rod and LMM, the mixture is dialyzed against solution C. The precipitated material is removed by centrifugation for 60 min at 25,000 rpm (Beckman 30 rotor). The supernatant containing S2 is collected, and the volume is reduced by polyethylene glycol (M_r 20,000, Sigma). After overnight dialysis against solution E, S2 is further purified by chromatography on a 1.5 × 60 cm column of Sephacryl S-200. The S2 containing fractions are pooled and concentrated by titration with 1 N HCl to pH 4.5. The precipitate is collected by centrifugation (15,000 rpm for 45 min, Sorvall SS34 rotor) and redissolved in solution H. The S2 thus obtained is homogeneous by the criterion of SDS–polyacrylamide gel electrophoresis.

Properties

This discussion has centered on the preparative procedures for myosin and its subfragments. However, these preparative methods are useful and relevant only if the enzymic and structural features of native myosin are preserved in the subfragments. Below we will describe methods for evaluating the integrity of the fragments.

Enzymic Analysis

The ATPase activity of myosin and its subfragments, in the presence of calcium, EDTA, or actin, provides a good criterion for judging the quality of these protein preparations. In these experiments, care should be taken to specify the conditions under which these assays are performed, since tha salt concentration, pH, and temperature significantly influence the rate of hydrolysis of ATP. Moreover, the demonstration that two sensitive sulfhydryl groups (SH_1 and SH_2) are intimately involved in the myosin ATPase[26,27] activity makes it important to preserve these groups in the reduced state. The Ca^{2+}-ATPase activity is elevated and the EDTA-ATPase activity is depressed when SH_1 is oxidized, whereas oxidation of the second sulfhydryl group, SH_2, results in a complete loss of both Ca^{2+}- and EDTA–ATPase activities. Therefore, Ca^{2+}- and EDTA-ATPase activity measurements give a good indication of the state of oxidation of these sulfhydryl groups. Care should be taken to keep the sulfhydryl groups in a reduced state during the preparative procedures. Inclusion of a reducing agent, such as dithiothreitol (DTT), will meet this requirement. In case of EDTA-ATPase activity measurements, it is necessary to include high concentrations of K^+ in the reaction mixture ($0.6\ M$ KCl is adequate). At concentrations of KCl less than $0.3\ M$, the ATPase activity is significantly reduced. Although the Ca^{2+}- and EDTA-activated ATPase activities are useful *in vitro* indicators of myosin activity, it is only the actin-activated Mg^{2+}-ATPase activity that is of physiological significance. In performing actomyosin ATPase assays, it is desirable to keep the ionic strength below 0.05 to enhance the interaction of actin with myosin subfragments and increase activity. With myosin, however, the assays are usually limited to about 25–50 mM KCl, since at lower ionis strength the solutions become highly viscous and a loss of reproducibility occurs. This problem can be overcome by the use of HMM and S1, the water-soluble subfragments, or by the use of the myosin minifilaments described by Reisler.[28] ATPase data for myosin and its subfragments are summarized in Table II. The ATPase activities are based on the extinction coefficient summarized in Table I. The values reported here are molar activities, and are obtained reproducibly in our laboratory under the conditions defined in Table II.

[26] T. Sekine and W. W. Kielley, *Biochim. Biophys. Acta* **81**, 336 (1964).
[27] M. Yamaguchi and T. Sekine, *J. Biochem.* (*Tokyo*) **59**, 24 (1966).
[28] E. Reisler, *J. Biol. Chem.* **255**, 9541 (1980).

Physical Measurements

When preparing myosin and its subfragments it is important that they should be homogeneous by the criteria of SDS-gel electrophoresis and equilibrium centrifugation. This is especially true of the subfragments, since all of them are products of proteolytic digestion of myosin. When monitored on SDS gels, it is apparent that the chymotryptic subfragments, HMM and S1, are very homogeneous with respect to the heavy chain in which no cleavages can be seen, whereas in tryptic HMM and papain S1 the heavy chains are cleaved.[29] As far as the myosin light chains are concerned, both A1 and A2 appear to be conserved in all the subfragments, and it is only the DTNB light chain (LC2) that seems to be susceptible to proteolytic attack. Thus, in tryptic HMM, the electrophoretic mobility of the DTNB light chain is increased owing to the cleavage of the N-terminal peptide that contains the phosphorylated serine residue.[30] In chymotryptic HMM, the DTNB light chain has the same mobility as in myosin, but when the gels are examined by densitometry it becomes apparent that some of the DTNB light chain has been cleaved and now comigrates with the A2 light chain. In S1 prepared either by chymotrypsin or by papain, in the presence of EDTA, the DTNB light chain is almost completely degraded. However, when S1 is prepared by papain digestion in the presence of $MgCl_2$, the integrity of the DTNB light chain is largely preserved.

These electrophoretic analyses are under denaturing conditions. To establish the homogeneity of myosin and its subfragments under native, nondenaturing conditions, the method of high speed equilibrium (meniscus depletion) centrifugation is used.[31] The results of equilibrium centrifugation reveal a remarkable degree of homogeneity for tryptic HMM, and for both papain and chymotryptic S1s as indicated by the overlap of all molecular weight averages over a wide range of protein concentration. Molecular weights for various subfragments are given in Table III. S1 (A1) and S1 (A2) and EDTA · S1 all have a molecular weight of about 110,000, whereas Mg · S1 has a molecular weight of 130,000; the difference reflects the presence of the DTNB light chain in Mg · S1. The rod and LMM were also homogeneous both on SDS–polyacrylamide gels and by equilibrium centrifugation; the molecular weights for these fragments are 230,000 and 130,000, respectively.[14]

After preparing these proteins, it is necessary to store them without loss of activity and without aggregation. Myosin can be stored in 50%

[29] S. S. Margossian and S. Lowey, *Biochemistry* **17**, 5431 (1978).
[30] D. Stone and S. V. Perry, *Biochem. J.* **131**, 127 (1973).
[31] D. E. Roark and D. A. Yphantis, *Ann N.Y. Acad. Sci.* **164**, 245 (1960).

TABLE III
MOLECULAR WEIGHTS OF MYOSIN SUBFRAGMENTS[a]

Species	SDS[b] gels	Molecular weight[c]
Mg · S1	—	130,000
EDTA · S1	—	110,000
S1 (A1)	90,000 (heavy chain)	112,000
S1 (A2)		106,000
HMM	140,000 (heavy chain)	350,000
LMM	77,000	130,000
Rod	120,000	220,000

[a] Reprinted in a modified form from Margossian, Stafford, and Lowey[14] by courtesy of the American Chemical Society.
[b] SDS, sodium dodecyl sulfate.
[c] The values of the apparent partial specific volume used in these calculations are 0.720 cm³/g for HMM and S1, and 0.701 cm³/g for LMM and rod.

glycerol at $-20°$ for a year or longer. LMM, the rod, and S2 can be freeze-dried directly from $0.6 M$ KCl, $0.01 M$ phosphate (pH 7.0), $0.001 M$ DTT. Similarly, HMM and S1 can also be lyophilized from $0.01 M$ potassium phosphate (pH 7.0), $0.001 M$ DTT, and sucrose at a concentration twice that of the protein by weight. The activity of these subfragments is retained after freeze-drying or after rapid freezing in liquid nitrogen. Nevertheless, it is advisable to chromatograph these subfragments before their immediate use.

In this chapter, we have tried to describe the preparation of myosin and its subfragments in sufficient detail to permit the interested investigator to repeat the preparation without undue difficulty. However, it should be pointed out that no matter how detailed a description is provided, some variability in the product is unavoidable owing, in part, to variations in the proteolytic activity of the commercial enzymes. If the electrophoretic pattern of the subfragment indicates additional bands due to excessive digestion, or if the yield of the subfragment is lower than expected, we suggest reducing or increasing the concentration of protease relative to the amount recommended here. The procedures described above are intended more as a guide than an absolute recipe; in order to obtain subfragments of maximum ATPase activity and structural integrity, the investigator may well have to perform some trial and error experiments before achieving optimal conditions. This is particularly true if a myosin other than rabbit skeletal myosin is used, since the proteolytic digestion will depend on the primary structure of the protein.

[8] Preparation and Fractionation of Myosin Light Chains and Exchange of the Essential Light Chains[1]

By PAUL D. WAGNER

Muscle myosins are made up of two large subunits (molecular weight 200,000) and four small subunits or light chains (molecular weights 15,000 to 30,000). The functions of these light chains and their interactions with the heavy chains are areas of active research in muscle biochemistry. It is useful, when discussing light chains, to divide them into two classes. There is one of each type of light chain associated with each myosin "head" or subfragment-1. Most nonmuscle cells also appear to contain two different types of light chains. The only known exception is the single-headed myosin from *Acanthamoeba castellani*.[1]

One class will be termed the regulatory light chains. The regulatory light chains of vertebrate muscle myosins are phosphorylated by light chain-specific kinases[2] and are, therefore, frequently referred to as the phosphorylatable light chains. While the function of this phosphorylation in striated muscles is not known, in smooth muscle and nonmuscle cells it controls actin interactions[3] and thick-filament formation.[4] Indeed there has been no clear demonstration of a regulatory function for the regulatory light chains of vertebrate striated muscles. The regulatory light chains of molluscan muscle myosins are not phosphorylated but are involved in calcium binding to the thick filament.[5] This calcium binding regulates the actomyosin ATPase in these muscles. The phosphorylatable light chains can replace the scallop regulatory light chain and partially restore calcium sensitivity.[6] The reversible dissociation of the scallop myosin regulatory light chain is a very useful system for examining the function of the regulatory light chains and is described in detail in this volume [9].

Light chains of the second class are thought to be essential, as it has not been possible to prepare active myosin or subfragment-1 that does not contain them.[7] However, as the conditions that dissociate these light

---

[1] H. Maruta, H. Gadasi, J. H. Collins, and E. D. Korn, *J. Biol. Chem.* **253**, 6297 (1978).

[2] W. T. Perrie and S. V. Perry, *Biochem. J.* **119**, 31 (1970).

[3] R. S. Adelstein and E. Eisenberg, *Annu. Rev. Biochem.* **49**, 921 (1980).

[4] J. M. Scholey, K. A. Taylor, and J. Kendrick-Jones, *Nature (London)* **287**, 233 (1980).

[5] A. G. Szent-Györgyi, E. M. Szentkiralyi, and J. Kendrick-Jones, *J. Mol. Biol.* **74**, 179 (1973).

[6] J. Kendrick-Jones, E. M. Szentkiralyi, and A. G. Szent-Györgyi, *J. Mol. Biol.* **104**, 747 (1976).

[7] P. Dreizen and L. C. Gershman, *Biochemistry* **9**, 1688 (1970).

chains also denature the heavy chains, it is difficult to determine the true cause of inactivation.[8] The essential light chains from different vertebrate muscles myosins have similar amino acid sequences[9-13] and can be interchanged without substantial alteration of the myosin ATPase.[14] This exchange procedure is described in detail at the end of this chapter. Little work has been done on the essential light chains of vertebrate cytoplasmic myosins or of any invertebrate myosin. They are clearly different from the regulatory light chains and are generally thought to have the same function as the essential light chains of vertebrate muscle myosins.

The methods given in this chapter have been developed for rabbit and chicken fast-twitch muscle myosins. However, many may be adapted with little or no modification to other myosins. The methods for rabbit and chicken fast muscle myosin will be described in detail and then where applicable expanded to include other myosins.

Fast muscle myosins have three different light chains, LC-1, LC-2 and LC-3 with molecular weights of 20,700, 19,000, and 16,500 respectively.[15] LC-2 is the phosphorylated regulatory light chain. Since it can be selectively dissociated from myosin with 5,5'-dithiobis(2-nitrobenzoic acid) (DTNB) it is frequently referred to as the DTNB light chain.[15] LC-1 and LC-3 are both essential light chains and there is a total of two LC-1 and LC-3 per myosin. The ratio of LC-1 to LC-3 is about 2 : 1 in the adult rabbit and about 1 : 1 in the chicken. LC-1 and LC-3 from the rabbit have been sequenced.[10] The major difference is a 41-amino acid peptide on the N terminus of LC-1. This peptide, which is rich in lysine, allows the separation of LC-1 and LC-3 by ion exchange chromatography. The essential light chains were first dissociated from myosin under alkaline conditions and are frequently referred to as alkali-1 and alkali-2 instead of LC-1 and LC-3.

An $\epsilon_{280}^{1\%}$ of 3.5 can be used to estimate the concentration of mixed light chains and for the DTNB light chain $\epsilon_{280}^{1\%}$ is 6.0.[16] A number of values have been published for the alkali light chains. This author has found that an $\epsilon_{280}^{1\%}$ of 2.2 for both alkali-1 and alkali-2 gives protein concentrations that agree well with those determined by other methods.

---

[8] J. J. Leger and F. Marotte, *FEBS Lett.* **52**, 17 (1975).
[9] A. G. Weeds and B. Pope, *Nature (London)* **234**, 85 (1971).
[10] G. Frank and A. G. Weeds, *Eur. J. Biochem.* **44**, 317 (1974).
[11] A. G. Weeds, *Eur. J. Biochem.* **66**, 157 (1976).
[12] A. G. Weeds, *FEBS Lett.* **59**, 203 (1975).
[13] T. Maita, T. Umegane, Y. Kato, and G. Matsuda, *Eur. J. Biochem.* **107**, 565 (1980).
[14] P. D. Wagner and A. G. Weeds, *J. Mol. Biol.* **109**, 455 (1977).
[15] A. G. Weeds and S. Lowey, *J. Mol. Biol.* **61**, 701 (1971).
[16] J. C. Holt and S. Lowey, *Biochemistry* **14**, 4600 (1975).

MYOSIN LIGHT CHAINS[a]

Source of myosin[b]	Essential light chains		Regulatory light chains	
	Molecular weight	Nomenclature	Molecular weight	Nomenclature
Vertebrate				
Fast-twitch muscle: rabbit, back[10,15,17]; chicken, pectoralis[16,18]	20,700 (25,000)[c] 16,500	LC-1 or alkali-1 LC-3 or alkali-2	19,000	LC-2 or DTNB
Slow twitch: rabbit, soleus[11]; chicken, anterior latissimus dorsi[19]	22,000 (27,000) 22,000 (27,000)	LC-1a LC-1b	19,000	LC-2
Cardiac (ventricle)[d]: beef,[6,9,12,20] chicken[13,19]	21,700 (27,000)	LC-1	18,000	LC-2
Smooth: chicken, gizzard[21–23]	17,000	LC-2	20,000	LC-1
Invertebrate				
Scallop, adductor[6,24]	17,000	SH-LC	17,000	EDTA-LC
Lobster, tail[6,24]	16,000	LC-2	20,000	LC-1

[a] This table is not exhaustive but merely provides a list of the more common sources of myosin light chains. Molecular weights of the light chains are based on their amino acid compositions and, when known, on their sequences.

[b] Superscript numbers refer to text footnotes.

[c] The molecular weight observed on SDS gels is given in parentheses if it differs significantly from the true molecular weight.

[d] Myosins isolated from the ventricle are different from those isolated from the atrium.

General data concerning myosin light chains are given in the table.[6–13,15–24]

Isolation of Mixed Light Chains

There are two general methods for isolating mixed light chains: denaturation in 4 M urea and denaturation in 5 M guanidine-HCl.

[17] J. H. Collins, *Nature (London)* **259**, 699 (1976).

[18] G. Matsuda, Y. Suzuyama, T. Maita, and T. Umegane, *FEBS Lett.* **84**, 53 (1977).

[19] T. Obinata, T. Masaki, and H. Takano, *J. Biochem. (Tokyo)* **86**, 131 (1979).

[20] J. J. Leger and M. Elzinga, *Biochem. Biophys. Res. Commun.* **74**, 1390 (1977).

[21] J. Kendrick-Jones, *Philos. Trans. R. Soc. London Ser. B* **265**, 183 (1973).

[22] R. Jakes, F. Northrop, and J. Kendrick-Jones, *FEBS Lett.* **70**, 229 (1976).

Urea Method. Myosin, 30–40 mg/ml, in 0.6 mM KCl is added to an equal volume of 8 M urea, 50 mM Tris-HCl, pH 8.0, 10 mM EDTA, and 5 mM DTT.[15,16,25] After stirring for 1 hr at room temperature, the heavy chains are precipitated by dilution with 10 volumes of cold water and collected by centrifugation at 15,000 g for 30 min. The light chains remain in the supernatant and are then concentrated on a column of DEAE-cellulose that has been equilibrated in 50 mM Tris-HCl, pH 8.0, and 2 × 10^{-4} M DTT at 4°. A 300-ml column is sufficient to bind the light chains released from 3 g of myosin. The light chains bind to the column, whereas the urea passes through. The light chains are removed in a relatively small volume by a step to 1.0 M KCl. This method of removing the urea avoids dialyzing low concentrations of light chains. A typical yield is about 80 mg of mixed light chain per gram of fast muscle myosin.

Similar dissociation in urea has been used to isolate light chains from vertebrate cardiac,[6,9,19] slow,[24] and smooth muscle myosins.[6,22]

Guanidine-HCl Method. The conditions for denaturation are 15–20 mg of myosin per milliliter, 5 M guanidine-HCl, 5 mM EDTA, 2.5 mM DTT, and 40 mM Tris-HCl, pH 8.0.[2,10,16] It is possible to start with myosin that has been precipitated by dilution or is dissolved in 0.6 M KCl. After stirring for 1 hr at room temperature or overnight at 4°, an equal volume of cold water is added. The heavy chains are then precipitated by slowly adding two volumes of cold ethanol, final ethanol concentration of 66% v/v. After stirring for 30 min at 4°, the heavy chains are collected by centrifugation at 10,000 g for 15 min. The light chains remain in the supernatant. The ethanol is removed by rotary evaporation; the light chain solution is maintained at 20–30°. The solution is then dialyzed overnight against 30–40 volumes of 5 mM potassium phosphate, pH 7.0, at 4°. After this dialysis, residual heavy chains are removed by centrifugation at 10,000 g for 30 min. The light chains are concentrated by freeze drying and finally dialyzed to remove the remaining guanidine-HCl and to prepare them for fractionation. The composition of the dialysis solution depends on which method of fractionation will be used.

Additional light chains can be obtained by reextracting the heavy chains from the ethanol precipitation with guanidine-HCl. The amount of light chain recovered in this second extraction is low, and it is worth doing only if a small amount of myosin is available.

This method has also been used to prepare mixed light chains from vertebrate slow,[11] cardiac,[13,20] and smooth muscle myosins,[23] from some

---

[23] Y. Okamoto and T. Sekine, *J. Biochem.* (*Tokyo*) **87**, 167 (1980).

[24] J. Kendrick-Jones and R. Jakes, *Myocard. Failure* [*Int. Symp.*] *1976*, p. 28 (1977).

[25] J. Gazith, S. Himmelfarb, and W. F. Harrington, *J. Biol. Chem.* **245**, 15 (1970).

invertebrate myosins,[6] and directly from actomyosin and myofibrils.[6] Most proteins, including actin, precipitate in 66% ethanol. Those that do not, such as tropomyosin, are usually removed during the subsequent fractionations.

Denaturation in guanidine-HCl is the more widely used method to prepare mixed light chains from fast muscle myosins. However, the light chains of some other myosins, when prepared by urea denaturation, fractionate better on ion exchange columns than when prepared by denaturation in guanidine-HCl. Light chains from chicken gizzard smooth muscle myosin are an example of this.[6]

DTNB Light Chain Isolation

The DTNB light chains of fast striated muscle myosins can be selectively dissociated from myosin by treatment with 5,5'-dithiobis(2-nitrobenzoic acid).[15,25] Myosin, 15 mg/ml, in $0.5 M$ NaCl is adjusted to pH 8.5 with $1 M$ Tris. DTNB is added to 10 mM from a fresh DTNB stock solution, 100 mM DTNB, 50 mM EDTA, and 200 mM Tris-HCl, pH 8.5. After stirring for 10 min at 4°, 10 volumes of cold water are added and the pH is lowered to 6.5 with acetic acid. The solution is centrifuged at 10,000 g for 30 min. Approximately half of the DTNB light chains remain in the supernatant. The light chain solution is concentrated by freeze drying or by ultrafiltration or on a column of DEAE-cellulose, as described previously for mixed light chains. DTNB is removed by dialysis against 0.5 mM DTT and 25 mM Tris-HCl pH 8.5 at 4°. These light chains may be contaminated by small amounts of alkali light chains, which must be removed if very pure DTNB light chain is required.

While this method selectively dissociates the DTNB light chain from fast muscle myosin, it does not remove the regulatory light chains from slow, cardiac, or smooth muscle myosins. The regulatory light chain from scallop can be selectively dissociated in EDTA.[5,6] (This is described in detail in this volume [9].) The regulatory light chains of some other invertebrate myosins can be preferentially dissociated by a combination of DTNB and EDTA.[6] However, there is no general method to dissociate this class of light chains selectively.

Light-Chain Fractionations

Separation of the DTNB Light Chain from the Alkali Light Chains by Ethanol Fractionation. A major difficulty in fractionating fast skeletal muscle myosin light chains by ion exchange chromatography is contamination of the alkali-1 light chain by the DTNB light chain. To minimize this problem, it is best first to remove most of the DTNB light chain by ethanol fractionation.[26] Rabbit fast muscle myosin mixed light chains (10 mg/ml), prepared either by denaturation in urea or in guanidine-HCl, are dialyzed overnight against 50 mM sodium phosphate pH 7.0, 10 mM β-mercaptoethanol, and 2 mM EDTA and then clarified by centrifugation at 30,000 g for 30 min. Ethanol is slowly added to give 18% v/v. After 3 hr at 4°, the solution is centrifuged at 30,000 g for 15 min. The pellet contains most of the DTNB light chain and almost no alkali light chains. The ethanol concentration of the supernatant is increased to 26%, and after 2 hr the solution is centrifuged at 30,000 g for 15 min. The pellet contains mostly DTNB light chain and a small amount of alkali-1 light chain. There is very little DTNB light chain in the supernatant. This ethanol fractionation does not work as well with light chains from chicken fast muscle myosin as it does with those from rabbit. However, if the chicken light chain solution is left overnight in 30% ethanol, about 70% of the DTNB light chain and only a small amount of the alkali-1 light chain will precipitate. Similar ethanol fractionations have not been reported for the light chains of other myosins. As described previously, it is also possible to remove about half of the DTNB light chain from fast muscle myosin with DTNB prior to denaturation in urea of guanidine-HCl. However, only when the DTNB light chains are removed by ethanol fractionation does this method result in any substantial improvement in the subsequent fraction of the light chains by ion exchange chromatography.

Light Chain Fractionation by Ion Exchange Chromatography. The light chains are dialyzed against 50 mM sodium phosphate pH 6.0 and 10^{-4} M DTT at 4° and then applied to a column of DEAE-cellulose (Whatman DE-52) that has been equilibrated in the same buffer.[15,16] For 200–300 mg of light chains, a 2.5 × 60 cm column is used. Since the alkali light chains have no tryptophan, the column is monitored at 230 nm. The light chains are eluted with a linear gradient of 50 to 400 mM sodium phosphate, pH 6.0 (five column volumes of each). The order of elution is alkali-1, DTNB, and alkali-2. The most common difficulty with this column is that the separation of the alkali-1 light chain from the DTNB light chain is frequently not very good, resulting in poor recovery of these two light chains. If most of the DTNB light chain is first removed by the ethanol fraction described previously, the alkali-1 light chain can be purified in high yields. Occasionally the DTNB light chain separates into two peaks, phosphorylated and unphosphorylated. The essential light chains of cardiac and slow[11] muscle myosins are readily purified on this type of column. These essential light chains elute before the regulatory light chains. However, the regulatory light chains from these two myosins cannot be

[26] W. T. Perrie, L. B. Smillie, and S. V. Perry, *Biochem. J.* **135**, 151 (1973).

purified using this column, as they are contaminated by the essential light chains.

Addition of urea to the ion exchange column buffers sometimes improves the separation of the light chains and facilitates the isolation of the regulatory light chains of cardiac and slow muscle myosin. A column for fast muscle myosin light-chain fractionation is described.[22] Approximately 250 mg of fast muscle myosin light chains in 2 M urea, 25 mM Tris-HCl, pH 7.5, and 10^{-4} M DTT are applied at room temperature to a 2.5 × 60 cm column of Whatman DE-52 cellulose that has been equilibrated in the same buffer. The alkali-1 light chain and the DTNB light chain are eluted sequentially by a linear gradient of 0 to 0.15 NaCl (1 liter of each). The alkali-2 light chain is eluted by a step to 1 M NaCl. This column with only minor modifications has been used to fractionate the light chains from a number of different myosins.[10,22]

Storage

The essential light chains are fairly stable and can be stored either frozen in solution or freeze dried at $-20°$. The regulatory light chains tend to aggregate. This is partly caused by disulfide formation and can be largely prevented by DTT. The DTNB light chain can be stored freeze dried or frozen in solutions containing DTT. However, disulfide formation is only part of the problem; the regulatory light chains of cardiac and slow muscle myosins contain no cysteine residues but aggregate very readily. It is best to store them frozen in solutions containing 2 M urea.

Essential Light-Chain Exchanges

There is currently no method available to recombine purified myosin heavy chains with the essential light chains and to regenerate active myosin. However, two groups[14,27] have developed similar methods to exchange purified essential light chains into myosin or subfragment-1. An excess of purified essential light chain is mixed with myosin or subfragment-1 in the presence of a dissociant. After a suitable period of time, the dissociant is removed by dialysis. During this incubation, some of the added light chain exchanges into the myosin, displacing the originally bound light chain. This technique can be used to make hybrids that contain heavy chains and essential light chains from different myosins or to introduce a chemically modified essential light chain into myosin or subfragment-1.

[27] Y. Okamoto and K. Yagi, *J. Biochem. (Tokyo)* **82**, 17 (1977).

The two dissociants, which have been used with fast muscle myosin, are 4.7 M NH$_4$Cl[14] and 0.6 M KSCN.[27] In addition to promoting light-chain exchange, both sets of conditions can also cause irreversible inactivation of myosin. In this author's experience, inactivation in 0.6 M KSCN is very rapid, making recovery of active myosin difficult. Inactivation by 4.7 M NH$_4$Cl is much slower, and even after 45-min incubations, the myosin retains more than 90% of its normal ATPase activities. Thus, incubation in 4.7 M NH$_4$Cl allows for a reasonable amount of light-chain exchange, approximately 60%, without significant loss of ATPase activity.

Light-Chain Exchanges with Subfragment-1. In recombination experiments with fast muscle myosin subfragment-1, it is advantageous to start with one of the isoenzymes—either S1-A1, subfragment-1 that contains only the alkali-1 light chain, or S1-A2, subfragment-1 that contains only the alkali-2 light chain. These two isoenzymes are easily separated by ion-exchange chromatography.[28]

The procedure for substituting the alkali-1 light chain for the alkali-2 light chain of S1-A2 will be described in detail. The exchange conditions are 20 μM S1-A2, 160–200 μM alkali-1 light chain, 0.1 M imidazole, pH 7.0, 2 mM EDTA, 2 mM DTT and 4.7 M NH$_4$Cl. Either the subfragment-1 or the solid NH$_4$Cl is added last. Since 4.7 M NH$_4$Cl is very close to saturation, some may not dissolve. This has no discernible effect on the exchange reaction. After stirring for 30 min at 4°, the solution is dialyzed overnight against 1 or 2 liters of 50 mM imidazole, pH 7.0, and 10^{-4} M DTT at 4°. The relatively slow removal of the NH$_4$Cl by dialysis seems to be necessary, as rapid lowering of the NH$_4$Cl concentration by dilution inactivates subfragment-1. After dialysis the protein solution is applied to a column of DEAE-cellulose (Whatman DE-52) that has been equilibrated in 50 mM imidazole, pH 7.0, and 10^{-4} M DTT at 4°. For 7.5 ml of exchange solution, a 1.5 × 30 cm column gives reasonable separation. The subfragment-1 is eluted with a linear gradient of 0 to 0.12 M NaCl, 200 ml of each. The order of elution is S1-A1 and then S1-A2. Free light chains remain bound to the column and are removed by a step to 1.0 M NaCl. S1-A1, which is formed by the exchange, makes up about 60% of the total subfragment-1 and can be eluted with almost no contaminating S1-A2. S1-A1 can be concentrated by ultrafiltration and precipitation in 60% saturated (NH$_4$)$_2$SO$_4$. The S1-Al formed by this exchange has ATPase properties identical to those of native S1-A1.[14]

The same exchange and isolation procedure has been used to replace the alkali-1 light chain of S1-A1 with the alkali-2 light chain and to form hybrids between the essential light chains of cardiac and slow muscle

[28] A. G. Weeds and R. S. Taylor, *Nature (London)* **257**, 54 (1975).

myosins and the heavy chain of fast muscle myosin subfragment-1.[14] Since the essential light chains of cardiac and slow muscle myosins have approximately the same charge as the alkali-1 light chain, hybrid subfragment-1, which contain them, elute from the DEAE-cellulose columns close to S1-A1. Thus, to facilitate the isolation of these hybrids, it is best to use S1-A2 rather than S1-A1 in these exchanges. This exchange procedure has also been used to form a hybrid between the cardiac subfragment-1 heavy chain and the alkali-2 light chain.[14] Alkali-1 light chain is not suitable for this exchange, as the hybrid containing it is poorly resolved from cardiac subfragment-1 by ion exchange chromatography.

The alkali light chains contain a single cysteine residue.[10] This allows the introduction of chemical probes at a specific site on the light chain. These modified light chains can be exchanged into subfragment-1 using the procedure outlined above. By suitable choices of subfragment-1 and light chain, it is usually possible to separate the modified subfragment-1 from the unmodified.[29,30] Introduction of chemical probes into subfragment-1, by this exchange procedure, does not appear to alter the ATPase activities of subfragment-1.[14,29,30]

Myosin Exchanges. The exchange conditions are 10 μM rabbit fast muscle myosin (20 μM "heads"), 160–200 μM light chain, 0.1 M imidazole pH 7.0, 0.3 M KCl, 2 mM EDTA, 2 mM DTT, and 4.7 M NH_4Cl.[31] After stirring for 45 min at 4°, the NH_4Cl is removed by dialysis overnight against 0.3 M KCl, $10^{-4} M$ DTT and 10 mM imidazole pH 7.0 at 4°. The myosin is precipitated by dilution with 12 volumes of cold water and centrifugation at 30,000 g for 15 min. The myosin pellet is dissolved in an approximately equal volume of 0.8 M KCl, $2 \times 10^{-4} M$ DTT, and 10 mM imidazole pH 7.0. The myosin is reprecipitated by dilution and centrifuged as before. The second precipitation is necessary to remove residual free light chain. Approximately half of the myosin is lost during these precipitations. There is usually greater than 60% exchange with less than 5% loss decrease in ATPase activity. Exchange with bovine cardiac myosin is performed under the same conditions, except that the time in NH_4Cl is shortened to 30 min, as this myosin inactivates more rapidly in 4.7 M NH_4Cl than does rabbit fast muscle myosin. It is not generally true that fast muscle myosins are more resistant to inactivation by NH_4Cl than are cardiac myosins, as 4.7 M NH_4Cl inactivates chicken fast muscle myosin at the same rate as it does bovine cardiac myosin. This exchange procedure has been used to enrich fast muscle myosin in alkali-1 light chain and in alkali-2 light chain.[31] It has also been used to make a hybrid

[29] D. J. Marsh and S. Lowey, *Biochemistry* **19**, 774 (1980).
[30] D. J. Moss and D. R. Trentham, *Fed. Proc., Fed. Am. Soc. Exp. Biol.* Abstr. 1736 (1980).
[31] B. Pope, P. D. Wagner, and A. G. Weeds, *Eur. J. Biochem.* **117**, 201 (1981).

myosin that contains fast muscle myosin heavy chains and cardiac myosin essential light chains and also a hybrid that contains cardiac myosin heavy chains and fast muscle myosin alkali light chains.[32]

Subfragment-1 has a distinct advantage over myosin in these exchange experiments. While it is usually possible to separate native from hybrid subfragment-1 by ion exchange chromatography, there is no simple method to fractionate myosins. Immunoabsorbance using antibodies specific for one type of essential light chain is the only method presently available.

Acknowledgments

Part of the work described here was supported by USPHS Program Project Grant HL-16683 and NSF Grant PCM-75-22698.

[32] P. D. Wagner, *J. Biol. Chem.* **256**, 2493 (1981).

[9] Preparation of Light Chains from Scallop Myosin

By ANDREW G. SZENT-GYÖRGYI and REGINA NIEBIESKI

Regulatory light chains are regulatory subunits that can be removed reversibly from scallop myosin. The presence of the regulatory light chains in myosin is required for the calcium dependence of the actin-activated MgATPase activity,[1,2] for tension generation of skinned fiber bundles,[3] and for the high affinity specific calcium binding of scallop myosin.[2] Regulatory light chains dissociate from scallop myosin in the absence of divalent cations,[1] and dissociation is complete at elevated temperatures[2] (23° in case of myosin, 35° in case of myofibrils). In the presence of $MgCl_2$, the light chains recombine stoichiometrically with scallop myosin. Scallop myosin freed from their own regulatory light chains readily hybridize with foreign regulatory light chains obtained from a wide variety of myosin species.[4-6] The scallop regulatory light chain has a

[1] A. G. Szent-Györgyi, E. M. Szentkiralyi, and J. Kendrick-Jones, *J. Mol. Biol.* **74**, 179 (1973).
[2] P. D. Chantler and A. G. Szent-Györgyi, *J. Mol. Biol.* **138**, 473 (1980).
[3] R. M. Simmons and A. G. Szent-Györgyi, *Nature (London)* **273**, 62 (1978).
[4] J. Kendrick-Jones, *Nature (London)* **249**, 631 (1974).
[5] J. Kendrick-Jones, E. M. Szentkiralyi, and A. G. Szent-Györgyi, *J. Mol. Biol.* **104**, 747 (1976).
[6] J. Sellers, P. D. Chantler, and A. G. Szent-Györgyi, *J. Mol. Biol.* **144**, 223 (1980).

molecular weight (M_r) of 17,400,[7] contains no cysteine or histidine[5] and has an absorbance, $E_{280\,\text{nm}}^{1\%,1\text{cm}}$ of 1.8.

Preparation of Scallop Regulatory Light Chains

Starting Material. Striated muscles of *Aequipecten irradians* are excised, freed carefully of the adhering pancreatic tissue, rinsed with wash (40 mM NaCl, 5 mM phosphate, pH 7.0) to which 0.1 mM phenylmethylsulfonyl fluoride (PMSF) is added immediately before use. The muscle is placed in an equal volume of ethylene glycol and 40 mM NaCl, 5 mM phosphate, pH 7.0, 1 mM MgCl$_2$, 0.1 mM EDTA, 0.05% sulfadiazine, 3 mM NaN$_3$ to which 0.1 mM PMSF is added. The muscle is stored for 24 hr at 0–4° then at −25° for several months or years.

Preparation of Myosin. A modified method of Focant and Hurieux[8] is used. The muscle is rinsed several times in wash, then homogenized in a Sorvall Omnimixer at full speed for 7 sec in 30–40 volumes of cold 40 mM NaCl, 5 mM phosphate buffer (pH 7.0), 1 mM MgCl$_2$, 0.1 mM dithiothreitol (DTT), 0.1 mM EGTA. The homogenate is centrifuged at about 10,000 g for 5 min. The sedimented residue is resuspended manually in the same solution, washed and centrifuged twice more (DTT, MgCl$_2$, and EGTA may be omitted from the washing solutions).

The washed residue is suspended at a concentration of 5–10 mg of protein per milliliter in the same solution. The solution is made 0.6 M in NaCl and 5 mM in ATP with 4 M NaCl and 0.1 M ATP (pH 7.0). The pH is maintained at 7.0 by the addition of 0.5 M Na$_2$HPO$_4$. After stirring for 1–5 min, the solution is centrifuged at 40,000 g for 15 min.

The supernatant solution containing actomyosin is filtered through gauze. The volume is measured and made 20 mM in MgSO$_4$ or MgCl$_2$ and an additional 5 mM ATP (pH 7.0) is added. pH is maintained at 7.0 with the addition of Na$_2$HPO$_4$.

Neutralized saturated (NH$_4$)$_2$SO$_4$ (enzyme grade, e.g., Schwarz-Mann) at 4° is added slowly with constant stirring until 40% saturation is reached. The pH is maintained at 7.0 by the addition of 1 M Tris base. The solution is allowed to stand for 10 min on ice, then is centrifuged at 15,000 g for 10 min, and the supernatant containing myosin is filtered through gauze. Additional neutralized saturated (NH$_4$)SO$_4$ is added to 65% saturation, and the precipitate is collected by centrifugation. The pellet, containing myosin plus some tropomyosin, is dissolved in 0.6 M NaCl, 5 mM phosphate (pH 7.0), 1 mM MgCl$_2$ and dialyzed overnight against a large volume of 20 mM NaCl, 5 mM phosphate (pH 7.0), 1 mM MgCl$_2$, 0.1 mM EGTA, 0.1 mM DTT.

[7] W. F. Stafford, III and A. G. Szent-Györgyi, *Biochemistry* **17**, 607 (1978).
[8] B. Focant and F. Hurieux, *FEBS Lett.* **65**, 16 (1976).

The next morning the contents of the dialysis sack is diluted if necessary with 2–4 volumes of 5 mM phosphate buffer (pH 6.5) to precipitate the myosin fully. The solution is centrifuged at 15,000 g for 10 min. The pellet containing myosin is washed and centrifuged three times with wash containing no magnesium ions. The pellet is greater than 95–98% pure myosin.

Myosin may be stored at 4° after dialysis against 40% saturated $(NH_4)_2SO_4$ (final saturation) containing 1 mM ATP, 1 mM $MgCl_2$, 0.1 mM EGTA, 10 mM phosphate, pH 7.0.[9] (Check and readjust pH to 7.0 after days 1 and 2.)

Preparation of Regulatory Light Chains

This method is modified according to Chantler and Szent-Györgyi.[2]

The myosin suspension (2–3 mg/ml) in 40 mM NaCl, 5 mM phosphate, pH 7.0 is brought to 23°, and EDTA is added to bring its concentration to 10 mM (0.2 M NaEDTA, pH 7.0). After 5 min of incubation the myosin is centrifuged for 3 min at 15,000 g at 23°, resuspended in 40 mM NaCl, 5 mM P_i (23°) and centrifuged. (The precipitate can be used for the preparation of the essential light chain.)

The combined supernatants are cooled in ice water bath, the regulatory light chain is precipitated by the addition of 2–3% trichloroacetic acid. After 5 min the precipitate is collected by 5 min of centrifugation at 10,000 g and is resuspended in 40 mM NaCl, 5 mM phosphate, pH 7.0. The suspension is neutralized with 1 N NaOH. Regulatory light chain redissolves and is dialyzed against several changes of 40 mM NaCl, 5 mM phosphate, pH 7.0, 3 mM NaN_3. A small amount of precipitate, if any, is removed by 1 hr of centrifugation at 140,000 g. Occasional traces of tropomyosin may be removed by Sephadex G-100 gel chromatography using as solvent 40 mM NaCl, 7.5 mM phosphate, pH 7.5, 3 mM NaN_3.

Purity of the preparation should be checked by 10% urea–acrylamide gel electrophoresis according to Perrie and Perry[10] at pH 8.6, a technique particularly suited for detecting contaminations with essential light chains. For gel buffer 0.132 M Tris-HCl, pH 8.8, and for running buffer 0.125 M Tris-glycine, pH 8.6, is used. Under these conditions the two light chain types separate well, the essential light chains having a greater mobility.[5]

Essential light chain contaminations can be readily removed by preparative urea gel electrophoresis in pH 8.6 Tris-glycine buffer. Samples are run in tubes 1 cm in diameter and 15 cm long at 100–125 V (1 W/tube) for about 9–10 hr; 6–8 mg of light chain may be loaded per tube. The band positions are visualized by Reisner stain on one of the lighter loaded

[9] T. Wallimann and A. G. Szent-Györgyi, *Biochemistry* **20**, 1176 (1981).
[10] W. T. Perrie and S. V. Perry, *Biochem. J.* **119**, 31 (1970).

tubes. Segments about 3 cm in length are cut out and blended briefly in the Tris-glycine running buffer. The suspension is gently shaken for 2 hr, centrifuged, and filtered to remove suspended acrylamide particles. The acrylamide is washed 3 times for 15 min, centrifuged, and filtered. The combined filtrations are lyophilized, dialyzed against 3 mM NaN$_3$, and stored frozen at $-25°$.

Preparation of Scallop Essential Light Chains

To the muscle residue from which regulatory light chains have been removed by EDTA treatment, 5 mM DTT, 2 mM EDTA, 25 mM Tris-HCl, pH 8.0, and 6 M guanidine-HCl are added (enzyme grade, e.g., Schwarz-Mann) and gently stirred overnight at 4°. An equal volume of 40 mM NaCl, 5 mM phosphate, pH 7.0, is added followed by 3 volumes (of the diluted guanidine-HCl extract) of cold 95% ethanol. The solution is well stirred during ethanol addition and brought to room temperature in a water bath. The precipitate is removed by centrifugation, and the supernatant is filtered. Ethanol is removed by flash evaporation. The solution is dialyzed against a large volume of 3 mM NaN$_3$, lyophilized, dissolved in 3 mM NaN$_3$, and stored at $-25°$.

Regulatory light chain contaminations may be removed by preparative gel electrophoresis as described above.

Acknowledgment

This research was supported by grants from the Public Health Service (AM 15963) and the Muscular Dystrophy Association.

[10] Sulfhydryl Modification and Labeling of Myosin

By EMIL REISLER

The myosin molecule contains over 40 thiol residues of which 12 or 13 reside on each of the two myosin heads.[1] Only a few of these are readily available for alkylation under mild conditions. Two sulfhydryl groups, the SH$_1$ and SH$_2$ groups located on the heavy chain of subfragment-1, have been extensively studied since their modification was found to affect dramatically the ATPase activities of myosin.[2] Specific blocking of the highly

[1] S. Lowey, H. S. Slayter, A. G. Weeds, and H. Baker, *J. Mol. Biol.* **42**, 1 (1969).
[2] T. Sekine and W. W. Kielley, *Biochim. Biophys. Acta* **81**, 336 (1964).

reactive SH_1 group inhibits the K(EDTA)-ATPase[2] and the actin activated MgATPase of myosin.[3] In contrast, the CaATPase and MgATPase activities are markedly activated.[2] Subsequent blocking of the less reactive SH_2 group results in complete loss of all the ATPase activities.[4]

The SH_1 and SH_2 thiols of myosin do not appear to be directly involved in ATP binding and hydrolysis.[5,6] Nevertheless, these groups are useful intrinsic probes for monitoring the functional events on the myosin head, and they provide excellent sites for attachment of suitable reporter groups.[7,8] Among the different sulfhydryl reagents used for modifications of the SH_1 and SH_2 groups, the reaction of N-ethylmaleimide (NEM) with myosin from rabbit striated muscle is particularly well documented. Peptide fragments containing the modified thiols have been isolated and sequenced[9,10]; the two critical SH groups are on the p10 CNBr peptide[11] close to the COOH terminus of the heavy chain in subfragment-1.[12] Crosslinking of the SH_1 and SH_2 groups with bifunctional sulfhydryl reagents demonstrates that these thiols, which are 10 residues apart in the primary sequence,[11] are from 2–14 Å apart in the native subfragment-1.[13–16] More recent studies show that thiol crosslinking may result in noncovalent trapping of nucleotide, presumably at the active site.[5,16] These findings generate new interest in the SH_1 and SH_2 groups and their potential uses for structural and functional studies of myosin.

Modification of the SH_1 Groups in Myosin from Vertebrate Striated Muscle

General Procedures

Modification Reaction. The sulfhydryl reagents differ in their reactivity toward the SH_1 groups on myosin. Consequently, for each reagent it is

[3] R. Silverman, E. Eisenberg, and W. W. Kielley, *Nature (London)* **240**, 207 (1964).
[4] M. Yamaguchi and T. Sekine, *J. Biochem. (Tokyo)* **59**, 24 (1966).
[5] J. A. Wells and R. G. Yount, *Proc. Natl. Acad. Sci. U.S.A.* **7**, 4966 (1979).
[6] H. Wiedner, R. Wetzel, and F. Eckstein, *J. Biol. Chem.* **253**, 2763 (1978).
[7] J. Duke, R. Takashi, K. Ue, and M. F. Morales, *Proc. Natl. Acad. Sci. U.S.A.* **73**, 302 (1976).
[8] J. C. Seidel, M. Chopek, and J. Gergely, *Biochemistry* **9**, 3265 (1970).
[9] T. Yamashita, Y. Soma, S. Kobayashi, T. Sekine, K. Titani, and K. Narita, *J. Biochem. (Tokyo)* **55**, 576 (1964).
[10] T. Yamashita, Y. Soma, S. Kobayashi, and T. Sekine, *J. Biochem. (Tokyo)* **75**, 447 (1974).
[11] M. Elzinga and J. H. Collins, *Proc. Natl. Acad. Sci. U.S.A.* **74**, 4281 (1977).
[12] M. Balint, I. Wolf, A. Tarcsafalvi, J. Gergely, and F. A. Sreter, *Arch. Biochem. Biophys.* **190**, 793 (1978).
[13] E. Reisler, M. Burke, S. Himmelfarb, and W. F. Harrington, *Biochemistry* **13**, 3837 (1974).
[14] M. Burke and E. Reisler, *Biochemistry* **16**, 5559 (1977).
[15] J. A. Wells, M. M. Werber, and R. G. Yount, *Biochemistry* **18**, 4800 (1979).
[16] J. A. Wells and R. G. Yount, *Biochemistry* **19**, 1711 (1980).

desirable to establish optimal conditions for specific labeling of the SH_1 groups. This can be done by following the changes in myosin ATPases as a function of increasing incorporation of the reagent. The obvious goal in such experiments is to find reaction conditions under which incorporation of 1 mol of reagent per mole of myosin head results in maximum inhibition (90–95%) of K(EDTA)-ATPase and maximum activation of the Ca-ATPase. These changes in myosin activity are indicative of SH_1 modification.[2] In general, the specific labeling of SH_1 groups is favored at low temperatures, at low ionic strength, and at slightly alkaline pH.[14,17] Thus, when searching for optimal modification conditions the following procedure and reaction conditions can be initially adopted.

Myosin or its active fragments, subfragment-1 (S-1) and heavy meromyosin (HMM), are dialyzed into a buffer (30 mM KCl, 25 mM Tris-HCl, pH 8.0) at 0–5°. Prior to modification the protein concentration is adjusted to 1–10 mg/ml. Modification reactions (at 0–5°) are started by adding a small amount of concentrated sulfhydryl reagent to the protein solution.[18] Several reactions at different molar ratios of reagent to protein, ranging from 1 : 1 to up to as high as 100 : 1, can be conducted almost in parallel. At fixed time points during a 1–2 hr period (or longer if necessary), aliquots are transferred from the reaction mixture to test tubes containing dithiothreitol (DTT) in large excess (50-fold molar) over the reagent.[19] The DTT effectively stops the modification reaction by complexing the unreacted reagent. The different reaction aliquots are then dialyzed at 4° against an appropriate solvent or passed through a Sephadex G-25 column to remove the excess reagent. Subsequently, the amount of covalently attached label (moles of reagent per mole of protein) is quantitated either spectroscopically or, whenever possible, by employing radioactive reagents. Activity assays can normally be carried out directly by diluting the reaction aliquots into CaATP and K(EDTA)-ATPase assay solutions. The results of such measurements are plotted for each modification reaction in the form of residual (relative to control) protein activity as a function of label incorporation. These curves should indicate the optimal labeling conditions defined in the preceding paragraph.

The rate and the specificity of labeling the SH_1 groups on myosin can be significantly increased by including 1 mM MgPP$_i$ in the reaction mixture

---

[17] M. Pfister, M. C. Schaub, J. G. Watterson, M. Knecht, and P. G. Waser, *Biochim. Biophys. Acta* **410**, 193 (1975).

[18] Some sulfhydryl reagents need to be dissolved in organic liquids. In such cases it is desirable to keep the organic liquid content in protein solutions below 5%.

[19] The use of 2-mercaptoethanol instead of DTT is not recommended. On prolonged exposure of myosin to 2-mercaptoethanol, the CaATPase activity is frequently elevated, presumably owing to the presence of impurities in the commercially available reagent [D. Hartshorne and M. F. Morales, *Biochemistry* **4**, 18 (1965)].

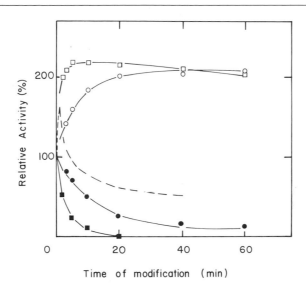

FIG. 1. Labeling of the SH_1 groups on myosin. Typical changes in ATPase activity vs time of modification of myosin (5 mg/ml) in 0.03 M KCl, 0.025 M Tris (pH 8.0). The protein was allowed to react with a twofold molar excess of N-ethylmaleimide per myosin heads either in the presence of 1 mM MgPP$_i$ ($\square$, $\blacksquare$) or in its absence ($\bigcirc$, $\bullet$). CaATPase ($\sqcap$, $\bigcirc$) and K$^+$(EDTA)-ATPase ($\blacksquare$, $\bullet$) activities are given on the ordinate. The dashed curve corresponds to CaATPase activities monitored during modification reaction carried out in the presence of 1 mM MgADP. Adapted from Burke and Reisler.[14]

(Fig. 1).[14,17] However, MgPP$_i$ also increases the labeling of the SH_2 groups, whose modification results in inhibition of the CaATPase activity of myosin (activated in the primary reaction of the SH_1 groups).[4] By closely monitoring the activity changes during the modification reaction and by terminating the reaction at an appropriate time, one can avoid the undesirable labeling of the SH_2 groups.

Determination of the Modification Site on the Basis of ATPase Activity Assays. Labeling of the SH_1 group on the myosin induces remarkable changes in the CaATPase of this protein. The pH and KCl dependence of the CaATPase is altered, and its activation energy is markedly increased.[2] Consequently, the elevation of the CaATPase, which in combination with the inhibition of the K(EDTA)-ATPase indicates labeling of the SH_1 groups, can be detected only under favorable experimental conditions.

A typical assay mixture will contain the following components: 0.5 M KCl, 2–50 mM Tris (pH 7.5), 2 mM ATP, and either 1 mM EDTA for K(EDTA)-ATPase, or 5 mM CaCl$_2$ for CaATPase. The lower Tris concentration is suitable for pH-state assays of activity, whereas the higher concentration is employed in colorimetric phosphate determinations. All measurements are conducted at 25° or at higher temperatures.

Comments. The assignment of the modification site (to the SH_1 or SH_2 groups) on the basis of ATPase properties alone is purely inferential. It is based on the previous identification of the modified peptides in reactions of myosin with NEM and iodoacetamide.[9,20] However, similar inhibition of myosin's K(EDTA)-ATPase and activation of its CaATPase activities is observed upon modification of a single reactive lysine on the myosin head[21] and also as a result of arginine modification.[22] Thus, unless the employed sulfhydryl reagent is known to react exclusively with the thiol groups, it is advisable to verify its specificity toward the SH_1 groups by peptide map analysis.

Reaction with N-Ethylmaleimide- and Maleimide-Based Reagents

Specific and relatively rapid modification of the SH_1 groups on myosin can be achieved by allowing NEM to react with myosin, S-1, or HMM. Purified myosin or myosin fragment is dialyzed against 0.03 M KCl and 0.025 M Tris-HCl (pH 8.0). Stock solutions of NEM are prepared in this solvent, but the addition of few drops of ethanol helps to dissolve the reagent. The concentration of NEM is determined spectroscopically (molar absorptivity is 620 at 305 nm). The modification is carried out at 0–5°, in the presence of 1 mM MgPP$_i$, at a twofold molar excess of reagent over myosin heads, and at about 10 μM myosin concentration. The reaction is terminated after 30 min by adding DTT to a final concentration of 1 mM. The modified protein is then dialyzed against an appropriate solvent to remove the excess reagent and MgPP$_i$.

Reactions carried out at lower protein concentrations or in the absence of MgPP$_i$ require more time for complete modification of the SH_1 groups (Fig. 1). Thus, in all cases, it is important to monitor the modification reaction and its final product by withdrawing aliquots for activity assays.

Comments. Under the above-described modification conditions, little additional NEM is incorporated into myosin after longer reaction times or at greater than twofold excess of reagent per myosin head.[17,20] Similar labeling conditions are also applicable in the reactions of myosin with the maleimide-based fluorescent and spin-label probes. In some cases (the bifunctional maleimide reagents containing an aromatic group being examples), the SH_1 modification can be carried out effectively at close to stoichiometric reagent : protein ratios.

[20] P. A. Kunz, J. T. Walser, J. G. Watterson, and M. C. Schaub, *FEBS Lett.* **83**, 137 (1977).
[21] A. Muhlrad and R. Takashi, *Fed. Proc., Fed. Am. Soc. Exp. Biol.* **39**, 1935 (1980).
[22] D. Mornet, P. Pantel, E. Audemard, and R. Kassab, *Eur. J. Biochem.* **100**, 421 (1979).

With other solvent conditions, the labeling of the SH_1 groups by NEM is less specific. In $0.6\,M$ KCl at $0-5°$ and pH 8.0, an additional "noncritical" thiol group is modified concomitantly with the SH_1 groups.[17,23] Thus, maximum changes in myosin activities are noted after incorporation of 2 mol of NEM per myosin head.[17,23] The presence of $MgPP_i$ in such a reaction system does not improve the specificity of the SH_1 labeling. Modifications with NEM carried out at $25°$ and either low ($0.03\,M$ KCl) or high ($0.6\,M$ KCl) ionic strength are also less specific for the SH_1 groups than comparable reactions carried out in $0.03\,M$ KCl at $0-5°$.

In general, MgADP and MgATP should not be included in the SH_1 modification reaction. The presence of MgADP will increase considerably the reactivity of the SH_2 groups. The presence of MgATP, which protects the SH_1 and SH_2 groups, will lead to the labeling of additional thiol residues on the myosin head.

Reaction with Iodoacetamide- and Iodoacetamide-Based Reagents

Iodoacetamide (IAA). Alkylation of myosin heads with IAA invariably results in the modification of only the SH_1 groups.[20] This highly specific reaction, which can be carried out under different solvent conditions (0 to $25°$, pH 7 to pH 9.0, low and high ionic strength),[20] renders iodoacetamide-based probes and spin labels particularly appropriate for introducing reporter groups into the SH_1 site. The reaction of myosin with two such probes, which have been used to study the protein conformation near the active site,[24] the topography of the myosin head,[25] and the movement of cross-bridges,[26] is described below.

N-(iodoacetylaminoethyl)-5-naphthylamine-1-sulfonic acid (1,5-IAEDANS). 1,5-IAEDANS is commercially available from Aldrich Chemical Co., Molecular Probes Inc. (Plano, Texas), Pierce Chemical Co., Sigma Chemical Co., and other manufacturers. The reagent dissolves easily in ethanol. It has extinction coefficients of 6.8×10^3 and 1.06×10^3 $M^{-1}\,cm^{-1}$ at 337 and 280 nm respectively, which can be used for spectroscopic determination of the amount of reagent incorporated per mole of protein.

The SH_1 groups on myosin are labeled with 1,5-IAEDANS by a modification of the method of Takashi *et al.*[27] Myosin (20 mg/ml) is incubated in the dark with a 40-fold molar excess of reagent over myosin heads. The

[23] M. C. Schaub, J. G. Watterson, and P. G. Waser, *Hoppe-Seyler's Z. Physiol. Chem.* **356**, 325 (1975).

[24] J. C. Seidel and J. Gergely, *Cold Spring Harbor Symp. Quant. Biol.* **37**, 187 (1972).

[25] D. J. Marsh and S. Lowey, *Biochemistry* **19**, 774 (1980).

[26] R. A. Mendelson, M. F. Morales, and J. Botts, *Biochemistry* **12**, 2250 (1973).

[27] R. Takashi, J. Duke, K. Ue, and M. F. Morales, *Arch. Biochem. Biophys.* **175**, 279 (1976).

reaction is carried out at 0° in the presence of 0.6 M KCl (or lower KCl concentrations) and 50 mM N-tris(hydroxymethyl)methyl-2-aminoethanesulfonic acid (TES) buffer (pH 7.0). After 45 min the reaction is stopped by addition of a 50-fold molar excess of DTT over the reagent, and the myosin is precipitated either by ammonium sulfate (50% of saturation) or by 15-fold dilution with 1 mM DTT.

Similar labeling conditions can be also employed to modify myosin with the fluorescent reagents 5-iodoacetamide-fluorescein (IAF)[25] and 5-iodoacetamidosalicylic acid (5-IAS) and with the spin label N-(1-oxyl-2,2,6,6-tetramethyl-4-piperidinyl)iodoacetamide.[16,28] The fluorescent reagents are available from Molecular Probes (Plano, Texas), and the preparation of the spin label has been described by McConnell and Hamilton.[29]

Fluorescent Labeling of Fibers

The procedure for labeling the SH$_1$ groups of myosin in the intact muscle fibers[30] is based on the observation of Duke et al.[7] that the incorporation of 1,5-IAEDANS into myosin is significantly decreased by bound actin whereas the labeling of actin is greatly increased by bound myosin. Consequently, the labeling of myosin in muscle fibers proceeds best under relaxing conditions that minimize the modifications of actin and troponin.

According to Borejdo and Putnam[30] skinned muscle fibers are mounted at resting length in an appropriate chamber and equilibrated with rigor solution (80 mM KCl, 5 mM MgCl$_2$, 2 mM EGTA, and 5 mM sodium phosphate at pH 7.0) for 1 hr at 0°. The fibers are then transferred into a relaxing solution (same composition as above plus 5 mM ATP) for 1 hr at 0° and finally into a relaxing solution containing 0.05–0.1 mg of 1,5-IAEDANS per milliliter for 1 hr at 0°. The incubation is stopped by washing the fibers first with cold relaxing solution containing 1 mM DTT and then with rigor solution. After this procedure Borejdo and Putnam[30] found that the relative labeling of myosin with respect to actin and troponin was 1.0 : 0.10 : 0.06.

In principle, this modification procedure should be also applicable for in situ labeling of myosin with other sulfhydryl reagents.

Reversible Modification of the SH$_1$ Groups

In order to label specifically the SH$_2$ groups on myosin, the more reactive SH$_1$ groups need to be protected during the modification reaction.

[28] S. A. Mulhern and E. Eisenberg, Biochemistry 17, 4419 (1978).
[29] H. M. McConnell and C. H. Hamilton, Proc. Natl. Acad. Sci. U.S.A. 60, 776 (1968).
[30] J. Borejdo and S. Putnam, Biochim. Biophys. Acta 459, 578 (1977).

A reversible protection of the SH_1 groups can be achieved by blocking them in a preliminary reaction with 5,5'-dithiobis(2-nitrobenzoic acid) (DTNB)[31] or with fluorodinitrobenzene (FDNB).[32] Since DTNB may under certain conditions induce the formation of a disulfide bond between the SH_1 and SH_2 thiols,[16] FDNB seems to be a more suitable reagent for the reversible blocking of the SH_1 groups.

Reaction Conditions. Myosin (2–5 mg/ml) in 0.5 M KCl, 50 mM Tris-HCl (pH 8.0) is allowed to react in the presence of 1 mM MgADP with a 4-fold molar excess of FDNB over myosin heads.[33] The reaction is carried out for 20 min at 0° and is terminated by precipitation of the myosin with 20 volumes of cold water.[34] (When soluble fragments of myosin are modified, the excess reagent can be removed on Sephadex G-25 columns.) The precipitated protein is collected by centrifugation, redissolved in 0.5 M KCl, and reprecipitated by addition of water. The dinitrophenylated myosin should have about 80–90% inhibited K(EDTA)-ATPase and about 400% activated CaATPase,[33] and can then be used for specific labeling of its SH_2 groups with NEM or other maleimide reagents. After such modification, the dinitrophenyl group is removed by thiolysis[34]; i.e., by treating the protein for 45–60 min with 5 mM DTT in 0.5 M KCl and 10 mM Tris-HCl (pH 8.5) at room temperature and under nitrogen.[32] The protein is subsequently precipitated with water and dialyzed against an appropriate solvent. Thus, SH_2 alone remains labeled.

Removal of the DNP label from myosin with unmodified SH_2 groups leads to a complete restoration of the original ATPase activities.

Modification of the SH_2 Groups

Myosin (2–10 mg/ml) with reversibly or irreversibly blocked SH_1 groups is allowed to react in the presence of 30 mM KCl, 25 mM Tris-HCl (pH 8.0), and 1 mM MgADP with a 4- to 10-fold molar excess of a sulfhydryl reagent. The reaction is allowed to proceed for 30 min at 0° and is terminated by addition of DTT as described in the preceding sections. The modification of the SH_2 groups is assumed to be complete when the CaATPase activity of the reacted myosin is abolished.

Some sulfhydryl reagents will not react with the SH_2 groups under these or other commonly used conditions. Among such reagents are FDNB,[33] IAA,[20] 1,5-IAEDANS,[27] and the spin label.

[31] J. C. Seidel, *Biochim. Biophys. Acta* **180**, 216 (1969).

[32] E. Reisler, M. Burke, and W. F. Harrington, *Biochemistry* **13**, 2014 (1974).

[33] E. Reisler, M. Burke, and W. F. Harrington, *Biochemistry* **16**, 5187 (1977).

[34] DTT or 2-mercaptoethanol should not be used to stop the reaction since cysteinyl-*S*-dinitrophenyl derivatives are unstable in their presence [S. Shaltiel, *Biochem. Biophys. Res. Commun.* **29**, 178 (1967)].

Modification of the SH₁ and the SH₂ Groups

When unmodified myosin is allowed to react with NEM in the presence of MgADP, the SH_1 and SH_2 groups are sequentially labeled. This is evidenced by the initial activation of the divalent cation-dependent ATPase and its subsequent inhibition (Fig. 1). Reaction conditions for such combined labeling of both the SH_1 and SH_2 groups are the same as in the preceding section except that the molar excess of reagent over protein is larger (10-fold to 16-fold).

Labeling of the Alkali Light Chains

The success of Wagner and Weeds[35] in developing a procedure for dissociating and recombining myosin subunits opened the possibility for specific labeling of the thiol groups located on the alkali light chains A1 and A2. Each light chain contains a single reactive cysteine residue that can be labeled with numerous sulfhydryl reagents. A stoichiometric modification of the A1 light chain with 1,5-IAEDANS has been described[25] and is reported here.

The modification of the A1 light chains with 1,5-IAEDANS is carried out for 2 hr under nitrogen at room temperature and in the dark. The light chains are dissolved in 7 M guanidine hydrochloride, 2 mM DTT, 0.5 M Tris (pH 8.0) and are mixed with an 8-fold molar excess of solid 1,5-IAEDANS. The reaction is stopped by the addition of 2-mercaptoethanol (0.1 M) and the unreacted fluorophore is removed by dialysis and ion exchange chromatography of the light chains on a Whatman DE-52 column. The labeled light chains are then hybridized with the myosin heavy chains.[25] and the extent of labeling is determined spectroscopically by using the absorptivities of 1,5-IAEDANS quoted in the previous section.

Modification of Other Sulfhydryl Groups on Myosin

At present there are no well established procedures for specific modification of SH groups on myosin other than the SH_1 and SH_2 thiols. Preferential labeling of the so-called SH_3 groups, which are probably located in the neck region of myosin, was reported by Schaub *et al.*[36] Modification of a single thiol group located in the LMM region of myosin was observed upon reaction of NEM with actomyosin.[37]

[35] P. D. Wagner and A. G. Weeds, *J. Mol. Biol.* **109**, 455 (1977).

[36] M. C. Schaub, J. G. Watterson, J. T. Walser, and P. G. Waser, *Biochemistry* **17**, 246 (1978).

[37] T. Horigome and T. Yamashita, *J. Biochem.* (*Tokyo*) **82**, 1085 (1977).

Modification of SH Groups on Myosin from Cardiac and Smooth Muscle

Relatively little is known about the properties of the sulfhydryl groups on myosin from cardiac and smooth muscle. The available evidence suggests that direct application to these myosins of the modification procedures developed for myosin from striated muscle does not lead to an equally specific labeling of their sulfhydryls. In myosin from cardiac muscle three thiol groups have to be modified per each myosin molecule (as opposed to two groups in myosin from striated muscle) in order to attain maximum changes in the ATPase activities that are characteristic of the SH_1 labeling.[17,38]

Modification with NEM of myosin from smooth muscle indicates that such reaction elicits different changes in the ATPase activities of this myosin than those normally observed with myosin from striated muscle.[39,40] Obviously, more detailed investigations of myosin from cardiac and smooth muscle are necessary before specific labeling of their thiols can be attempted.

[38] C. Klotz, J. J. Leger, and F. Marotte, *Eur. J. Biochem.* **65**, 607 (1976).
[39] I. Cohen, E. Kaminski, R. Lamed, A. Oplatka, and A. Muhlrad, *Arch. Biochem. Biophys.* **175**, 249 (1976).
[40] J. C. Seidel, *Biochem. Biophys. Res. Commun.* **89**, 958 (1979).

[11] Chemical Modification of Myosin by Active-Site Trapping of Metal-Nucleotides with Thiol Crosslinking Reagents[1]

By JAMES A. WELLS and RALPH G. YOUNT

I. Introduction

It has been discovered that crosslinking of two thiols in myosin in the presence of MgADP leads to stable and stoichiometric trapping of MgADP at the active site.[1a] This has led to the proposal that myosin contains a jawlike active site structure that closes upon binding of

[1] Abbreviations used: pPDM, N,N'-p-phenylenedimaleimide; F₂DPS, 4,4'-difluoro3,3'-dinitrophenyl sulfone; phen, 1,10-phenanthroline; DTNB, 5,5'-dithiobis(2-nitrobenzoic acid); TNB, 5-thio-2-nitrobenzoic acid; F₂DNB, 1,5-difluoro-2,4-dinitrobenzene; SF₁, chymotryptic subfragment-1; NEM, N-ethylmaleimide; εADP, 1,N^6-ethenoadenosine 5-diphosphate; NANDP, N-(4-azido-2-nitroanilino)ethyl diphosphate; AMP-PNP, adenyl-5'-yl-imidodiphosphate; oPDM, N,N'-o-phenylenedimaleimide; NDM, naphthalene-1,5-dimaleimide; DTE, dithioerythritol; sIMP, 6-thioinosine 5'-monophosphate.
[1a] J. A. Wells and R. G. Yount, *Proc. Natl. Acad. Sci. U.S.A.* **76**, 4966 (1979).

MgADP.[1a] Closure of this jaw is proposed to juxtapose two sulfhydryl groups, the so-called SH-1 and SH-2, so that they may be crosslinked and lock shut the active-site cleft.

Active-site trapping can be accomplished with at least eight different thiol crosslinking reagents with rigid crosslink spans that vary from 14 Å to 2 Å.[1-4] Myosin's broad substrate specificity allows this technique to be used to trap a wide range of metal nucleotides.[3-7] Metal nucleotide complexes can be stably trapped and stored under conditions where the rate of dissociation has half-times on the order of several days. This technique can thus be used to probe the microenvironment and structure of the force-generating ATPase site without the need to synthesize ATP analogs that must be covalently attached to the active site.

This chapter focuses first on the general requirements for active-site trapping. Second, the methodology that has been employed to establish thiol crosslinking and to demonstrate concomitant stable trapping of metal nucleotides is discussed. Next, the specific crosslinking reagents and reaction conditions that have been found to be successful or unsuccessful in active-site trapping are described. Finally, some of the applications of this technique are discussed. These include trapping of spectroscopically interesting metals and nucleotides as well as the trapping of a new photoaffinity ATP analog that ensures specific photoincorporation into the active site of myosin.

II. Basic Requirements for Trapping

Thus far the only effective trapping agents are those that can crosslink between thiols. Monofunctional modification of the critical SH groups by N-ethylmaleimide (NEM) does not stably trap MgADP[3] indicating the importance of crosslinking. The most effective thiol crosslinking reagents are hydrophobic and aromatic in nature consistent with the suggestion that the vicinity around SH-1 is hydrophobic.[8]

Myosin's broad substrate specificity is reflected in the broad specificity of Me^{2+} nucleotides that can be trapped. It is possible to trap divalent metal ions such as Mg^{2+}, Ni^{2+}, Co^{2+}, Mn^{2+}, or Ca^{2+} in the presence of

[2] J. A. Wells and R. G. Yount, *Biochemistry* 19, 1711 (1980).
[3] J. A. Wells, M. Sheldon, and R. G. Yount, *J. Biol. Chem.* 255, 1598 (1980).
[4] J. A. Wells, C. Knoeber, M. C. Sheldon, M. M. Werber, and R. G. Yount, *J. Biol. Chem.* 255, 1135 (1980).
[5] R. E. Dalbey, J. A. Wells, and R. G. Yount, *Biophys. J.* 33, 148a (1981).
[6] W. J. Perkins, J. A. Wells, and R. G. Yount, *Fed. Proc., Fed. Am. Soc. Exp. Biol.* 39, 1937 (1980); *Biophys. J.* 33, 149a (1981).
[7] K. L. Nakamaye, J. A. Wells, R. L. Bridenbaugh, and R. G. Yount, *Abstr. Am. Chem. Soc. Meeting (Las Vegas) 1980;* and in preparation.
[8] R. Mendelson, S. Putman, and M. Morales, *J. Supramol. Struct.* 3, 162 (1975).

ADP.[5] Little trapping of metals occurs in the absence of nucleotide.[3] It is possible to trap a variety of di- and triphosphate compounds. For example, all common ribonucleoside di- or triphosphates,[4] nucleotide analogs such as ϵADP[6] and AMP-PNP,[3] a photoaffinity probe NANDP,[7] and pyrophosphate can be trapped.[4] ATP is hydrolyzed and trapped as ADP, so it is likely that this occurs also for all cleavable triphosphates. Trapping of nucleotide requires the presence of a divalent metal.[3] Nucleoside monophosphates such as AMP or IMP[1a,3,4] are not trapped.

There appears to be little or no correlation between the crosslink span of the crosslinking reagent employed and the extent or the stability of MgADP once trapped. The metal ion is slightly less stable than the nucleotide when these are monitored simultaneously.[4] The stability of the trapped species varies from hours to days depending upon the nature of the compound trapped and the salt and temperature conditions. Raising the ionic strength (P. Buzby and R. Dalbey, unpublished results) or the temperature (M. Sheldon, J. Wells, and W. Perkins, unpublished results) leads to less stable trapping. The stability depends upon the sum of the contributions of the divalent metal ($Mg^{2+} > Ni^{2+} > Co^{2+} > Mn^{2+}$) and the nucleotide (AMP-PNP > ADP > ϵADP).[5] So that under conditions (0.1 M KCl, 50 mM Tris, pH 8.0 at 0°) where Mg^{2+}–AMP-PNP is very stable ($t_{1/2} \sim 7$ days) Mn^{2+}-ADP is relatively unstable ($t_{1/2} \sim 4$ hr).

III. Methodology

A. Establishing Intramolecular Crosslinking of SH-1 and SH-2

1. ATPase Inactivations

Modification of SH-1 in myosin is accompanied by a characteristic activation of the divalent metal ion ATPase activity and concomitant inactivation of the K$^+$(EDTA)-ATPase activity[9] (see this volume [10]). In the presence of MgADP, SH-2 becomes more reactive and modification of this thiol after SH-1 alkylation leads to loss of all ATPase activities.[10] These effects may be used to provide suggestive evidence that reaction with the crosslinking reagent leads to modification of these thiols.

1. In the presence of MgADP, reaction with slight excesses of various thiol crosslinking reagents leads to loss of both the Ca^{2+}-ATPase and the $NH_4{}^+/K^+$(EDTA)-ATPase activities.
2. The time course of reaction for the less reactive crosslinking re-

[9] T. Sekine and W. W. Kielley, *Biochim. Biophys. Acta* **81**, 336 (1964).
[10] J. Yamaguchi and T. Sekine, *J. Biochem. (Tokyo)* **59**, 24 (1966).

agents (e.g., DTNB,[2]; F$_2$DNB,[4] shows a transient activation of the Ca^{2+}-ATPase activity followed by its inactivation while the K$^+$(EDTA)-ATPase inactivates monotonically. This suggests reaction first with SH-1 and subsequent crosslinking with SH-2. Even the highly reactive crosslinking reagent pPDM (Wells and Yount, unpublished results) has been shown to exhibit a slight transient activation of the Ca^{2+}-ATPase if the reactions are rapidly quenched with excess 2-mercaptoethanol prior to measuring ATPase activities. Moreover, with the use of thiol compounds to quench crosslinking, it is common to find that the Ca^{2+}-ATPase activity is substantially above the K$^+$(EDTA)-ATPase activity. This suggests that some of the enzyme has SH-1 and SH-2 cross-linked, whereas some has only SH-1 or SH-2 modified.

3. Finally, reactions in the presence of MgADP proceed much more rapidly and to greater extents than do reactions in the absence of MgADP.[4,11,12] Nucleotide-stimulated inactivation suggests that SH-2 is modified.

2. Determination of Thiol Modification

The most convenient method for analysis of total thiols is by reaction with DTNB in 8 M urea,[13] and it is described in detail.[14] The cross-linked enzyme derivative is purified (see Section III,B,1) and diluted to approximately 2 mg/ml. Accurate protein concentrations can be determined by using a dye-binding assay as previously described.[15] Unmodified SF$_1$ is used as a protein standard (M_r = 120,000, footnote 16; $\epsilon_{280}^{1\%}$ = 7.5 cm^{-1}, footnote 17). Samples (10–100 μg) in 0.1 ml total volume are mixed with 5.0 ml of diluted dye reagent and allowed to stand for 5 min; then the absorbance at 595 nm is determined. Chymotryptic SF$_1$ was found to contain approximately 10 thiols. There is only a small difference between modified and unmodified SF$_1$ (i.e., 2 or 3 thiols). Therefore, to be accurate the analysis on crosslinked samples should be performed several times. The method is limited by standard errors in the DTNB analysis ($\pm 2\%$) and protein analysis ($\pm 3\%$) and provides a good estimate of the thiol content to within ± 0.5 SH group.

[11] E. Reisler, M. Burke, S. Himmelfarb, and W. F. Harrington, *Biochemistry* **13**, 3837 (1974).
[12] M. Burke and E. Reisler, *Biochemistry* **16**, 5559 (1977).
[13] G. L. Ellman, *Arch. Biochem. Biophys.* **82**, 1970 (1959).
[14] J. A. Wells, M. M. Werber, and R. G. Yount, *Biochemistry* **18**, 4800 (1979).
[15] J. A. Wells, M. M. Werber, J. I. Legg, and R. G. Yount, *Biochemistry* **18**, 4793 (1979).
[16] A. G. Weeds and R. S. Taylor, *Nature (London)* **257**, 54 (1975).
[17] P. D. Wagner and A. G. Weeds, *J. Mol. Biol.* **109**, 455 (1977).

3. Determination of the Stoichiometry of Crosslink Labeling

The minimal fold excess of crosslinking reagent necessary for reasonably complete and specific inactivation is determined by titrating the enzyme with crosslinking reagent in substoichiometric increments.[2,4,18] The reaction time should be determined from prior time-course studies to determine the time after which no further change in ATPase activity occurs. Then from a plot of the residual ATPase activity versus the fold excess of reagent added, the apparent specificity of labeling and the optimal amount of reagent needed for extensive modification is determined. These studies for skeletal myosin show a linear phase in which inactivation correlates with the excess of crosslinking reagent used. This typically extrapolates to about a onefold excess of dithiol crosslinking reagent being required for loss of all ATPase activities. A second phase of "nonspecific" labeling occurs as the crosslinking reagent concentration is increased. One therefore must decide whether to opt for specific yet incomplete modification using low concentrations of reagent or complete but less specific modification using reagent in excess.

The stoichiometry of the crosslinking reagents may be assessed using either radioactive derivatives, spectral methods, or atomic absorption. The incorporation of crosslinking reagent plotted as a function of the loss of ATPase activities then may be used directly to indicate the extent of specific activity-related labeling. Synthetic procedures are available for making radioactive pPDM and oPDM.[19] This procedure has been modified to simplify preparation of ^{14}C-labeled pPDM and to improve the yield (see Section IV,A,1). It is worth noting that the extent of incorporation of radioactive pPDM closely approximates quantitative reaction indicating the absence of pPDM hydrolysis during reaction with SF$_1$. pPDM has been previously reported to hydrolyze slowly in aqueous solution.[19a]

Determination of the extent of activity-related labeling is especially important for inactivations using cobalt phenanthroline complexes in which greater than stoichiometric amounts of crosslinking reagent are added. The derivatives may be analyzed for cobalt incorporation by atomic absorption as previously described (see Section IV,B,1).

The stoichiometry of labeling for the aromatic disulfide reagents DTNB[2] and thiopurine nucleotide disulfides,[20] may be determined by following the production of free aromatic thiol during ATPase inactivation (see Section IV,B,2).

[18] E. Reisler, J. Mol. Biol. **138**, 93 (1980).
[19] J. S. Nishimura, J. A. Petrich, A. F. Milne, and T. Mitchell, Int. J. Biochem. **9**, 93 (1978).
[19a] P. Knight, Biochem. J. **179**, 191 (1979).
[20] J. A. Wells, C. E. Knoeber, and R. G. Yount, unpublished results, 1981.

4. Establishing Intramolecular Crosslinking

A stoichiometry of crosslinking reagent incorporated to thiols modified to SF_1 concentration of $1:2:1$ provides strong evidence that dithiol crosslinking has occurred. However, the uncertainty of such stoichiometries, especially of thiols modified, is often such that it would be difficult to rule out a small amount of monofunctionally modified enzyme. Monofunctional) modification may be inferred indirectly by the discrepancy between the Ca^{2+}- and EDTA-ATPase activities as discussed in Section III,A,1. For irreversible thiol crosslinking reagents such as pPDM, a sensitive method for detecting the extent of monofunctional reaction is to add an excess of [^{35}S]cysteine to the modified enzyme to block the unreacted half of the reagent. SF_1 is treated with irreversible crosslinking reagents, and the reactions are quenched with a 50- to 100-fold excess of [^{35}S]cysteine (Amersham) over SF_1. After incubation at $0°$ for approximately 30–60 min a 500- to 1000-fold excess of 2-mercaptoethanol over SF_1 is added to reduce any [^{35}S]cysteine–enzyme mixed disulfides that may be formed adventitiously. The enzyme is purified as described in Section III,B,1, and the incorporation of nonexchangeable [^{35}S]cysteine is determined. A non-crosslinked control is included to exclude the possibility of ^{35}S carry-over during the sample work-up. This procedure has been used to determine that over 90% of the enzyme inactivated by a 1 fold excess of pPDM was bifunctionally crosslinked (J. Wells, unpublished results).

A variation of this method may be performed in which the number of DTNB-detectable thiols with and without thiol quenching are compared.[4] SF_1 is allowed to react with crosslinking reagent and divided into two samples. One sample is quenched with 2-mercaptoethanol, and the second is not. Both samples are purified, and DTNB analysis is performed. Free thionitrobenzoate produced and detected in the DTNB reaction can potentially react with the unreacted half of the crosslinking reagent. Thus if a substantial amount of monofunctional modification is present then the 2-mercaptoethanol quenched sample should show a higher apparent thiol content than the nonquenched sample. This difference should be the extent of monofunctional reaction. This method is limited to detecting only relatively large differences between quenched and nonquenched samples for reasons discussed in Section III,A,2. Greater sensitivity can be achieved by using dithioerythritol, which would double the theoretical difference between the thiol quenched and nonquenched samples.

Two methods may be used to demonstrate that the crosslinking is intramolecular: (a) sedimentation velocity; (b) SDS gel electrophoresis. SF_1 crosslinked in the presence of MgADP by pPDM, DTNB,[2] or cobalt phenanthroline complexes show essentially the same s values (5.8 ± 0.1 S in 0.1 M KCl, 50 mM Tris, pH 8.0 at $20°$) as uncross-linked SF_1, confirming the absence of intermolecular crosslinking. These studies were corroborated by SDS gel electrophoretic patterns of pPDM cross-linked SF_1. SF_1 which had incorporated 0.8 equivalent of ^{14}C-labeled pPDM showed that more than 95% of the radioactivity was incorporated into the 95,000-dalton heavy-chain fragment. The position of the heavy chain in gels was indistinguishable from non-crosslinked controls (J. A. Wells and R. G. Yount, unpublished results) confirming the absence of intermolecular crosslinking.

5. Localization of the Crosslinking Reagent

The most definitive method for locating the site(s) of crosslinking is to employ conventional peptide mapping procedures. The 95,000-dalton heavy-chain fragment of chymotryptic skeletal SF_1 gives a very reproducible partial tryptic digestion pattern. Balint and co-workers[21] have demonstrated that chymotryptic SF_1 is split by trypsin into a 50,000-, 25,000-, and 21,000-dalton fragments. The SH-1 and SH-2 thiols were shown to be in the 21,000-dalton carboxyl-terminal fragment of the heavy chain. SF_1 was labeled (0.8 ^{14}C-labeled pPDM per SF_1), and the derivative was subjected to timed tryptic digestion in the presence of 1 mM $CaCl_2$ and 10 mM DTE as described by Balint and co-workers.[21] Digestion samples were quenched between 0 and 60 min by boiling in SDS and then subjected to 5 to 15% gradient polyacrylamide gel electrophoresis.[22] Gels were stained (overnight at 25° in 0.05% Coomassie Blue in 50:50:10 methanol:H_2O:acetic acid) and destained (24 hr in staining solution minus Coomassie Blue). The protein banding pattern obtained was identical to that of non-crosslinked SF_1. This indicated the absence of crosslinks between the NH_2-terminal 25,000-dalton, the middle 50,000-dalton, or the carboxyl-terminal 21,000-dalton tryptic fragments. Gels were dehydrated, infused with PPO, dried under vacuum, and subjected to fluorography to locate the ^{14}C-labeled pPDM according to Laskey and Mills.[23] Essentially all the radioactivity appeared initially in the 95,000-dalton heavy chain of SF_1 and subsequently during tryptic digestion appeared only in the 21,000-dalton fragment. The 21,000-dalton tryptic fragment contains two other cysteines in addition to SH-1 and SH-2.[24] However, a 10,000-dalton CNBr fragment (P_{10}) within the 21,000-dalton tryptic fragment contains only SH-1 and SH-2.[24,24a] It has been observed[25] that a 10,000-dalton CNBr

[21] M. Balint, I. Wolf, A. Tarcsafalvi, J. Gergely, and F. A. Sreter, Arch. Biochem. Biophys. 190, 793 (1978).
[22] U. K. Laemmli, Nature (London) 227, 680 (1970).
[23] R. A. Laskey and A. D. Mills, Eur. J. Biochem. 56, 335 (1975).
[24] M. Gallagher and M. Elzinga, Fed. Proc., Fed. Am. Soc. Exp. Biol. 39, 2168 (1980).
[24a] M. Elzinga and J. H. Collins, Proc. Natl. Acad. Sci. U.S.A. 74, 4281 (1977).
[25] M. Burke and P. Knight, J. Biol. Chem. 255, 8385 (1980).

peptide is labeled by ^{14}C-labeled pPDM (J. A. Wells, E. Huston, and R. G. Yount, unpublished results). Burke and Knight[25] have shown this peptide to be P_{10} by partial squenching.

B. Establishing Trapping of Divalent Metal Nucleotides

1. Enzyme Derivative Purification Procedure

SF_1 that has been allowed to react with crosslinking reagent and quenched appropriately is purified by precipitation with 2.5 volumes of saturated $(NH_4)_2SO_4$ containing 10 mM EDTA, pH 8.0, at 5° and pelleted by centrifugation at 20,000 g for 15 min. The resulting pellet is redissolved in a minimum amount of 0.1 M KCl, 50 mM Tris, pH 8.0, at 0° to give a solution containing ~15 mg/ml of SF_1, which is applied to a Sephadex G-25 column equilibrated in the same buffer. This latter step is conveniently performed by applying no more than 0.8 ml of SF_1 solution to a PD-10 column (Pharmacia) and collecting drops 25 to 55. (There is little variation between these columns, and they can be used after reequilibration several times before discarding.) This work-up procedure recovers 60–70% of the starting protein and can easily be performed in less than an hour. In cases where greater than 200-fold excess of reagents is employed, a second $(NH_4)_2SO_4$ precipitation step should be included to ensure adequate purification. EDTA added in a 2-fold excess over divalent metals with the $(NH_4)_2SO_4$ is an effective means of removing nontrapped Mg-nucleotide. To control for carry-over and adventitious binding of metal nucleotide, a non-crosslinked control should be included. These controls contain typically between 0.02 and 0.1 mol of MgADP per mole of SF_1.

This procedure is superior to dialysis because it can be performed rapidly and normally does not appreciably affect the extent of trapping of Mg^{2+}-nucleotides. However, the high salt $(NH_4)_2SO_4$ precipitation step has been shown to reduce dramatically the extent of Ca^{2+}-nucleotide trapping (see Section III,B,4). For this and other less stable metal nucleotide complexes, $(NH_4)_2SO_4$ precipitations should be avoided and only gel filtration used for purification.

2. Establishing the Trapped Nucleotide Stoichiometry

Three different methods may be used to quantify the stoichiometry of nucleotide trapping. The simplest and most reliable method is to trap radioactive nucleotides. In the absence of the radioactive nucleotide it is often possible to measure the quantity of nucleotide or analog spectrophotometrically. For instance, NANDP, a photoaffinity analog of ADP (see Sections V,A,1 and V,B), in H_2O has an $\epsilon_{474} = 5100$ cm^{-1} M^{-1}. Its spectrum undergoes a slight decrease in molar absorptivity (ϵ_{467} = 4900 cm^{-1} M^{-1}) upon binding to the enzyme. Thus the stoichiometry of NANDP incorporation when crosslinking with weakly chromophoric reagents (e.g., NDM, pPDM, oPDM, DTNB, purine disulfide reagents) could be determined spectrophotometrically in a nondestructive manner.

In cases where a fluorescent nucleotide, e.g., ϵADP, is trapped the purified enzyme may be denatured after trapping by boiling for 2 min or by treating with 0.1 volume of 30% $HClO_4$.[6] The supernatant is removed after centrifugation, adjusted to pH 7 to 8, and the fluorescence intensity measured at appropriate wavelengths. Untrapped standards are treated in a parallel manner for comparison.

3. Establishing the Stoichiometry of the Trapped Metal

Atomic absorption is the most convenient way to analyze for metal trapping. The detection limits for Mg, Ca, Co, Ni, Zn, Fe, Mn, and Cr are all below 0.1 μM for air–acetylene fuel mixtures. A single metal determination can be easily performed in 1.0 ml of SF_1 at 16 μM. At this level of analysis trace amounts of Mg, Co, and Ni in buffers are usually not a problem; Ca^{2+} contamination is sometimes a problem that can be alleviated by passing the buffers through Chelex resins (Bio-Rad) and by performing analyses on higher concentrations of SF_1. A non-crosslinked enzyme control should be run in parallel with crosslinked samples to check for contamination, carry-over, and adventitious metal binding. Matrix effects on metal analyses have not been encountered using KCl-Tris buffers, but they should be checked if different buffers are used. In addition, Na^+ (>5 mM) will depress cobalt atomic absorption measurements. Large viscosity differences between the sample, e.g., myosin solutions >5 mg/ml, and standards can lead to spurious results because of differences in aspiration rates.

Radioactive metals can also be used to increase the sensitivity of detection. Suitable isotopes are available for most of the common divalent metals except Mg^{2+}. In using [57]Co to determine the number of cobalt ions incorporated during crosslinking by cobalt phenanthroline complexes, the following considerations should be made. $[Co(III)(phen)_2CO_3]^+$ and $[^{57}Co(II)(phen)_2(H_2O)_2]^{2+}$ undergo a rapid solution electron exchange reaction that is nearly as rapid as Co(II)phen *in situ* oxidation into SF_1.[15] It is best to premix the [57]$CoCl_2$, 1,10-phenanthroline and $[Co(III)(phen)_2CO_3]^+$ in 0.1 M KCl, 50 mM Tris, pH 8.0, at 25° for approximately 4 hr to allow complete isotopic scrambling to take place. (Note that KCl promotes this exchange process.) After total scrambling, [57]Co(II) specific activity is diminished by a factor equal to its concentration divided by the sum of its concentration and the concentration of $[Co(III)(phen)_2CO_3]^+$. Thin-layer chromatography (TLC) methods can be used to verify that the scrambling is complete.[15]

4. Determining the Stability of the Trapped Me^{2+}-Nucleotide Complex

Two methods have been used for distinguishing between trapped and nontrapped Me^{2+}-nucleotide. Gel filtration of the purified enzyme derivative followed by analysis of the amount of metal and nucleotide copurifying with SF_1 may be used.[1-6] A second method, limited to trapped fluorophores, has been employed to determine the stability of trapped ϵADP.[6] The fluorescence of ϵADP bound or trapped in SF_1 is depolarized to a much lesser degree than is ϵADP free in solution. By continuously monitoring fluorescence depolarization, the dissociation rate of trapped ϵADP can be measured directly.[6]

5. Determining That Trapping Occurs at the Active Site

Several methods may be used to confirm that Mg-nucleotide trapping occurs at the active site, not at any possible second Mg-nucleotide binding site. First, SF_1 turns over MgATP very slowly relative to the rate at which it is crosslinked by pPDM (turnover rate ~ 0.5 min^{-1} in 0.1 M KCl, 50 mM Tris, pH 8.0 at 0°). Labeling experiments demonstrate that when a large excess of $[8\text{-}^{14}C,\gamma\text{-}^{32}P]ATP$ is present, only that which is hydrolyzed to $[8\text{-}^{14}C]ADP$ will be trapped.[1] If trapping had occurred at a nonhydrolytic site, then $[8\text{-}^{14}C,\gamma\text{-}^{32}P]ATP$ should have been trapped, since by the end of inactivation less than 5% of the total ATP was hydrolyzed. Second, the extent of MgADP trapping saturates when the active site has been preequilibrated with increasing concentrations of MgADP before adding crosslinking reagent.[4] This method has also shown that the site at which MgADP stimulates inactivation by cobalt phenanthroline complexes is a tight nucleotide binding site because the rate of inactivation is maximal upon adding essentially stoichiometric amounts of MgADP. Third, it has been shown that the spectral properties of a variety of MgADP analogs are perturbed upon binding to SF_1. The characteristic spectra of $Mg\epsilon ADP$,[6] $MgNANDP$,[7] and $Co(II)ADP$[5] bound to SF_1 were retained after crosslinking. This argues that the site of binding and trapping are the same. The modified spectra of reversibly bound probes are reversed to that of the free probes by adding MgADP, but the spectra of the trapped probes are not reversed because they cannot be displaced. Fourth, the binding of various metal-nucleotides to skeletal SF_1 results in characteristic enhancements in intrinsic tryptophan fluorescence.[26] When MgADP or Mg

[26] M. M. Werber, A. G. Szent-Györgyi, and G. Fasman, *Biochemistry* **11**, 2872 (1972).

pyrophosphate are reversibly bound to SF_1, the fluorescence enhancement changes to that characteristic of MgATP when MgATP is added.[6] However, when MgATP is added to SF_1 containing trapped MgADP or Mg pyrophosphate, no change in tryptophan fluorescence occurs. Finally, equilibrium dialysis experiments have shown that the competitive inhibitor AMP-PNP does not bind to SF_1 that has previously been crosslinked in the presence of MgADP by pPDM, cobalt phenanthroline complexes,[1] or DTNB.[2]

IV. Crosslinking Reagents Used in Active-Site Trapping

A summary of the structures, sources and preparations of reagents employed in active-site trapping is shown in Table I. Table I also describes the conditions used in inactivating SF_1 with these reagents. Finally, Table II summarizes the final ATPase activities, extents of thiol modification and MgADP trapping, as well as the stability of the trapped MgADP in various crosslinked derivatives.

A. Irreversible Crosslinking Reagents

1. Bismaleimides

Modification of SF_1 with pPDM has been the most extensively characterized of all the crosslinking modifications.[2-4,11,14] Commercially available pPDM (Aldrich) is brown and can be purified by sublimation (180°, 3–5 mm Hg for 10–12 hr, 10° cold finger) to yield a light yellow compound shown to be pPDM by infrared and elemental analyses. This procedure slightly improved the potency of pPDM to inactivate SF_1. pPDM is best dissolved in acetonitrile (~ 1 mM) and is stable in solution for weeks when stored at 5°. Acetonitrile is preferred over acetone since it does not quench enzyme tryptophan fluorescence and there is no detectable loss in enzyme activity upon adding up to 3% acetonitrile by volume. The concentration of pPDM may be determined spectrophotometrically using $\epsilon_{313} = 940$ cm^{-1} M^{-1} in acetonitrile (J. A. Wells, unpublished results) or gravimetrically if the pPDM is dried before weighing.

Synthesis of N,N'-p-Phenylenedi[1,4-^{14}C]maleimide. ^{14}C-Labeled pPDM is prepared by coupling p-phenylenediamine with two equivalents of maleic anhydride and subsequent maleimide ring closure by treating with acetic anhydride using a modified procedure of Nishimura et al.[19] Acetic anhydride is redistilled and stored over 3 Å molecular sieves. Tetrahydrofuran (THF) is dried over 3 Å molecular sieves. p-Phenylenediamine (Aldrich) is purified by sublimation at 100°, 1–3 mm Hg using

TABLE I
SUMMARY OF CROSSLINKING REAGENTS USED SUCCESSFULLY IN ACTIVE SITE TRAPPING

Cross-linking reagent (source)	Structure	Cross-link span (Å)	Concentration of stock solution (solvent)	Molar excess over $SF_1{}^a$	Treatment time at 0°	Quench reagent[b]
NDM (molecular probes)		13–14	1 mM (acetonitrile)	1.4	30 min	2-Mercaptoethanol
pPDM (Aldrich)		12–13	1 mM (acetonitrile)	1.3	30 min	2-Mercaptoethanol
oPDM (Aldrich)		4–9	1 mM (acetonitrile)	1.5	30 min	2-Mercaptoethanol
F₂DPS (Pierce)		10	1 mM (methanol)	1.3	30 min	2-Mercaptoethanol

Reagent	Structure		Concentration		Time	Quench
F_2DNB (Pierce)		5	1 mM (methanol)	1.5	22 hr	2-Mercapto-ethanol
Cobalt-phenanthroline complexes	[Co(III)(phen)$_2$CO$_3$] · Cl/Co(II)phen	3–5	50 mM (H$_2$O)	100/10	10 min	EDTA
DTNB[c] (Pierce)		2[d]	1 mM (10 mM phosphate, pH 7.0)	2	15–22 hr	Purification
(sIMP)$_2$[c]		2[d]	1–10 mM (H$_2$O)	2	100 min	Purification

[a] Inactivations were carried out on ~16 μM SF$_1$ in 0.1 M KCl, 50 mM Tris, pH 8.0 at 0° except for (sIMP)$_2$, where Bicine replaced Tris.

[b] One-hundredfold molar excess of 2-mercaptoethanol over SF$_1$ was used. A twofold excess of EDTA over divalent metals was used to quench inactivations by cobalt phenanthroline complexes. Enzyme derivatives were purified as described in Section III,B,1.

[c] Promote the oxidation of SH groups to disulfide bonds.

[d] Sulfur–sulfur distance in a disulfide bond.

TABLE II
SUMMARY OF CHARACTERIZATION OF CROSSLINKED ENZYME DERIVATIVES

Crosslinking reagent	% ATPase		SH/SF$_1$[a]	Trapping ADP/SF$_1$	Stoichiometry Mg/SF$_1$	% Retained after 7 days[b]		Reference
	NH$_4^+$ (EDTA)	Ca^{2+}				ADP	Mg	
NDM	2	7 ± 1	2.0 ± 0.2	—	—	—	—	4
	30	35	—	0.62	0.64	82	58	
pPDM	8 ± 6	9 ± 6	1.8 ± 0.4	—	—	—	—	1a,
	5	—	—	0.69	0.91	85	89	
oPDM	12 ± 6	11 ± 6	2.0 ± 0.4	—	—	—	—	4
	18	—	—	0.73	0.90	75	50	
F$_2$DPS	7 ± 5	11 ± 7	2.0 ± 0.1	—	—	—	—	4
	8	10	—	0.72	0.82	61	31	
F$_2$DNB	4 ± 2	18 ± 3	1.9 ± 0.2	—	—	—	—	4
	8	15	—	0.67	0.80	—	—	
Cobalt–phenathroline	3 ± 1	2 ± 2	2.4 ± 0.7	—	—	—	—	1a, 3,
	4	—	—	0.88	1.03	—	—	
DTNB	3 ± 2	5 ± 3	2.9 ± 0.1	—	—	—	—	2
	—	10	—	0.75	0.73	75	58	
(sIMP)$_2$	9	20	2.7	—	—	—	—	20
	—	10	—	0.70	0.70	—	—	

[a] ΔSH/SF$_1$: Loss in DTNB-detectable thiols after modification as compared to a nontreated control containing 9.5 ± 0.5 SH/SF$_1$. Values reported after modification by DTNB or (sIMP)$_2$ were determined from the release of thionitrobenzoate and 6-thiopurine nucleotide, respectively, by spectroscopic methods. The values reported are averages and standard deviations of two to three separate modifications. Each determination was performed in triplicate.

[b] Percentage of retained MgADP was determined after gel filtration.

[c] Numbers refer to text footnotes.

a Dry-Ice–isopropanol cooled cold finger to yield a flocculent yellow product. Maleic anhydride is purified in a similar manner.

The contents of two 250-μCi vials of [1,4-^{14}C]maleic anhydride (32.6 mCi/mmol; Amersham) are transferred in 1.1-ml total volume into a 3-ml breakseal vial by sequential washings with two 0.5-ml aliquots of THF and one 0.1-ml rinse with THF. p-Phenylenediamine (27 mg, 250 μmol) and 49 mg (500 μmol) of maleic anhydride are added, and the resulting orange mixture is stirred magnetically at 4° for 24 hr. The vial is centrifuged for 5 min at low speed in a clinical centrifuge to pellet the insoluble N,N'-p-phenylenedi[1,4-^{14}C]maleic acid derivative. The supernatant is drawn off, and the precipitate is resuspended in a second 1.0-ml aliquot of THF. The mixture is centrifuged as above, and the supernatant again is discarded. Approximately 30 μCi of the starting [1,4-^{14}C]maleic anhydride is discarded in these steps. The orange precipitate is suspended in 1 ml of THF, and acetic anhydride (0.5 ml) and anhydrous sodium acetate (50 mg) are added. The vial is cooled in a Dry-Ice–isopropanol bath, evacuated to 1–3 mm Hg, and sealed. After warming to room temperature the vial is heated in a 100°-oven for 22 hr. The orange contents will turn yellow and finally brown. The vial is cooled to 0° and opened. H_2O (2 ml) is added, and the mixture is stirred for 10 min at 4° to precipitate the pPDM and hydrolyze unreacted acetic anhydride. The vial is centrifuged as described above, and the supernatant is removed. The brown precipitate is washed with 1 ml of H_2O and collected again by centrifugation. The combined discarded supernatants contained approximately 70 μCi. Acetone (0.5 ml) is added to the brown precipitate, and the slurry is transferred to a small sublimation tube (15 cm $\times$ 2 cm). The vial is rinsed with three more 0.5-ml aliquots of acetone to ensure quantitative transfer. The contents of the sublimation tube are taken to dryness using a stream of argon gas for 45 min at 24°. The brown residue is sublimated at 190°, 1–3 mm Hg onto a 10° cold finger. Three sublimation harvests may be taken over 8 hr. The material on the cold finger is removed by dipping it into a larger tube containing THF. Finally, the THF is removed by drying with a stream of argon gas to yield a pale yellowish powder. The total yield is about 40 mg (60%) and has a specific activity of 3400 cpm/nmol.

The infrared spectrum of pPDM synthesized above is essentially identical to that observed using sublimated commercial pPDM (Aldrich). The ^{14}C-labeled pPDM is equally effective at inactivating SF_1 ATPase and in trapping MgADP. Elemental analyses: analysis, C = 62.75, H = 2.73, N = 10.36; calculated, C = 62.70, H = 2.98, N = 10.45.

Inactivations with pPDM are normally carried out at SF_1 (~17 μM), $MgCl_2$ (0.2–20 mM), ADP (0.1 mM), and pPDM (22 μM) with the reagents added in this order. To measure the Ca^{2+}-ATPase without purifying the enzyme, $MgCl_2$ should not exceed 20 μM in the assay mix compared to 15 mM $CaCl_2$. A 1.3-fold molar excess of pPDM over SF_1 assures essentially complete inactivation. Although inactivations are less complete, equimolar pPDM or less should be employed when more specific labeling is desired. However, such inactivations deviate from pseudo first-order kinetics after 60–70% inactivation. Inactivations are extremely rapid in the first 2 min of treatment ($t_{1/2} \sim$ 0.5 min). The rate of inactivation is greater at pH 8.0 than at pH 7.0,[11] and it appears to be more specific at 0° than 25° (J. A. Wells, unpublished results). MgADP can be routinely trapped with stoichiometries of 0.8 ± 0.1. The nucleotide is remarkably stable at 0° once trapped ($t_{1/2} \sim$ 7 days, Table II), and less than 15% of the MgADP is lost even after 7 hr at 20°. pPDM is thus an excellent choice for an efficient trapping reagent because it reacts rapidly, is relatively non-chromophoric, traps MgADP effectively, and can be easily prepared as the ^{14}C derivative.

Modification of SF_1 with NDM or oPDM is less well characterized, although these reagents are believed to react in a manner identical to that of pPDM. Like pPDM both NDM and oPDM are sparingly soluble in water but dissolve readily in acetonitrile and are stable for weeks. The concentrations were determined gravimetrically. NDM is weakly fluorescent after reaction with thiols. Unfortunately, its emission spectrum overlaps that of protein tryptophans (W. Perkins, unpublished results), which make it undesirable as a fluorescent probe of the crosslinking site. Both oPDM and NDM trapped 0.8 ± 0.1 MgADP per SF_1 upon complete inactivation.

2. Bisnitrofluoroaryl Reagents

Both F_2DPS and F_2DNB are reasonably soluble in methanol, and their concentrations may be determined gravimetrically. Under identical reaction conditions, inactivations with F_2DPS in the presence of MgADP are slower than with bismaleimides ($t_{1/2} \sim 4$ min for F_2DPS vs 0.5 min for bismaleimides). Inactivation ultimately leads to modification of two thiols (Table II) and concomitant trapping of 0.75 ± 0.1 MgADP. For reasons that are not clear the stability of trapped MgADP[4] and MgεADP[6] is less than that for other crosslinking reagents under identical conditions.

F_2DNB is substantially less reactive than F_2DPS.[4,12] The monofunctional reaction occurs rapidly in the presence of MgADP ($t_{1/2} \sim 20$ min) activating the Ca^{2+}-ATPase ($\sim 220\%$ of control). The reaction with a second thiol occurs much more slowly ($t_{1/2} \sim 5$ hr) resulting in Ca^{2+}-ATPase inactivation. Apparently nucleophilic displacement of the first fluorine of F_2DNB reduces the reactivity of the second fluorine. Ultimately two thiol groups are modified and MgADP is trapped (Table II).

Both F_2DPS and F_2DNB are chromophoric and the change in their spectra upon reacting with SF_1 is indicative of thiol modification (W. Perkins, unpublished results). In addition, their spectral properties make them suitable as fluorescence energy-transfer acceptors. F_2DPS in particular has good spectral overlap with the emission spectra of εADP trapped at the active site. Moreover, because the two nitrophenyl rings are attached via a tetrahedral sulfone the dipole movements are not in the same plane. This conveniently restricts the range of values the orientation factor (K^2) can assume that is used in calculating R_0.[27]

[27] L. Stryer, *Annu. Rev. Biochem.* **47**, 819 (1978).

B. Reversible Crosslinking Reagents

1. Cobalt–Phenanthroline Complexes

Inactivation by Co(II)/Co(III)–phenanthroline complexes has been shown to result from incorporation of a single Co(III)phen$_x$ complex into two thiol groups in SF$_1$ [14] (x is as yet uncertain but is probably two). A second site or mixture of sites of cobalt incorporation is also present which lead to loss of an additional noncritical thiol. The mechanism of inactivation has been shown to be *in situ* oxidation of an enzyme-bound Co(II)phen$_x$ complex by [Co(III)phen$_2$CO$_3$]$^+$.[15,28]

[Co(III)(phen)$_2$CO$_3$] Cl is prepared according to Ablov and Palade[29] as previously described.[15] This complex is very soluble (>100 mM) and is stable for days in water; its concentration is determined spectrophotometrically (ϵ_{509} = 133 cm^{-1} M^{-1}). Separate stock solutions (~50 mM) of CoCl$_2$ and 1,10-phenanthroline may be kept for weeks stored at 4°. Inactivations are carried out using SF$_1$ (17 μM) containing 20 mM MgCl$_2$ and 0.1 mM ADP at 0° in KCl-Tris buffer. A 10-fold excess of CoCl$_2$ and 1,10-phenanthroline over SF$_1$ is added, and the solution turns orange owing to formation of an equilibrium mixture of mono-, bis-, and trisphenanthroline Co(II) complexes. A 100-fold excess of [Co(III)(phen)$_2$CO$_3$]$^+$ is added from a 50–100 mM stock solution. Inactivation of Mg^{2+}, Ca^{2+}, and K$^+$(EDTA)-ATPase activities all occur simultaneously in a pseudo first-order fashion for more than three half-times ($t_{1/2}$ in the presence of MgADP ~ 1.5 min).[14] Occasional lags of 5–20 sec occur in the inactivation, and the basis of these lags is not clearly understood.

A 10-fold excess of Co(II)–phenanthroline is needed because the critical site of incorporation in SF$_1$ has a weak affinity for the Co(II)phen$_x$ complex (K_A = 3.7 × 10^3 M^{-1}).[15] The rate of inactivation can easily be slowed and still maintain pseudo first-order kinetics simply by lowering the excess of cobalt phenanthroline complexes added. This makes this system ideal for rate studies of nucleotide stimulation of inactivation. Inactivations are rapidly quenched (but not reversed) by addition of an excess of EDTA over divalent metal ions.[15] Thus, chelating agents in general block inactivation. A 4-fold excess of 1,10-phenanthroline over CoCl$_2$ favors formation of Co(II)(phen)$_3$, and this decreases the rate of inactivation 4-fold. Inactivations go faster as the pH is raised from 7.0 to 9.0 and are dependent on ionic strength.[15] The inactivation is dependent upon the type of Co(II)/Co(III) couple that is used. For example, mixtures

[28] J. A. Wells, M. M. Werber, and R. G. Yount, *J. Biol. Chem.* **255**, 7552 (1980).
[29] A. F. Ablov and D. M. Palade, *Russ. J. Inorg. Chem.* (*Engl. Transl.*) **6**, 306 (1961).

of ethylenediaminediacetic acid (EDDA) Co(II)/Co(III) complexes are ineffective as inhibitors.

The extent of incorporation of cobalt is measured by atomic absorption (see Sections III,A,3 and III,B,2). The cobalt(III)phen$_x$ enzyme complex is reasonably stable at $0°$ ($t_{1/2} \sim 5$ days). However, it is much less stable when heated to $25°$ or when the enzyme is denatured. This latter result has prevented the isolation of Co(III) peptides. The Co(III)phen$_x$ enzyme complex may be easily reduced by brief treatment with a variety of reducing agents in the presence of chelating agents.[15] For instance, Co(III)phen$_x$ SF$_1$ ($\sim 16 \mu M$) in 5–10 mM EDTA, 0.1 M KCl, 50 mM Tris, pH 8.0 at $0°$, was reactivated by addition of a 75-fold molar excess of freshly prepared FeSO$_4$ for 2 min followed by rapid purification using (NH$_4$)$_2$SO$_4$ precipitation and gel chromatography. (Prolonged treatment with Fe(II)EDTA leads to irreversible oxidative damage to SF$_1$.) This gives essentially complete ATPase reactivation ($>90\%$), recovery of modified thiols, and release of trapped MgADP.[1,14,15] Reactivation of ATPase activity is best performed within 1 hr after inactivation by this method. Cobalt(III)-chelated thiols appear to be susceptible to oxidation by dissolved oxygen.[30]

MgADP was trapped quite effectively upon complete inactivation by cobalt–phenanthroline complexes (0.85 ± 0.1 per SF$_1$). It should be noted that a high concentration of MgCl$_2$ (20 mM) is necessary when it is desired to trap only MgADP. This is because free Co(II) can compete with Mg^{2+} for available ADP and it will become trapped.[5] The trapped MgADP is very stable ($t_{1/2} \sim 5$ days) at $0°$, but it rapidly dissociates at $25°$ ($t_{1/2} \sim 2$ hr). The thermal instability results in part from dissociation of the Co(III)(phen)$_x$ enzyme complex.

2. Aromatic Disulfide Reagents

DTNB and the purine nucleotide disulfide analog (sIMP)$_2$ gently oxidize two critical thiols in SF$_1$ to form a disulfide bond and concomitantly trap MgADP at the active site.[2,20] Both of these reagents when added to SF$_1$ in a 2-fold molar excess, in the presence of MgADP, add to what appears to be SH-1. With time, the aryl sulfide portion of the mixed disulfide is displaced by an adjacent thiol (believed to be SH$_2$) to form a new disulfide link. This latter reaction requires the presence of MgADP and takes overnight when DTNB is used but less than 2 hr when (sIMP)$_2$ is used (see Tables I and II). An abbreviated reaction scheme is shown (see p. 111) where AS–SA = DTNB or (sIMP)$_2$. Not shown is a parallel reaction of these reagents to form a mixed disulfide with the single cysteine of the

[30] C. P. Sloan and J. H. Krueger, *Inorg. Chem.* **14**, 1481 (1975).

$$AS-SA \; + \; \underset{SH}{\overset{SH}{\textcircled{E}{<}}} \quad \xrightarrow[1.]{\overset{HSA}{\uparrow}} \quad \underset{SH}{\overset{S-SA}{\textcircled{E}{<}}} \quad \xrightarrow[2.]{\overset{HSA}{\uparrow}} \quad \underset{S}{\overset{S}{\textcircled{E}{|}}}$$

alkali light chains (J. A. Wells, unpublished results). The mixed disulfide can be selectively reduced without reducing the critical cystine disulfide or reactivating enzyme activity. Thus the mixed disulfide appears not to affect the ATPase activity, since reactivation of modified enzyme parallels reduction of the cystine disulfide but not that of the mixed disulfide.[2,20] This mode of crosslinking though slower than that using the cobalt–phenanthroline complexes is an attractive modification in that it is easily reversed by reducing agents. Moreover, there is no possibility of irreversible oxidation of the sulfur atoms involved, as appears to occur over time with reduction of the Co(III)phen complex of SF_1.

Reactions with DTNB. Stock solutions (1 mM DTNB, 10 mM potassium phosphate, pH 7.0 at 5°) are stable for more than 1 week at 5° but should be discarded when they begin to turn yellow. The concentration of DTNB may be estimated gravimetrically and more precisely determined by measuring the change in the absorbance at 412 nm of aliquots treated with excess DTE to produce 2 mol of TNB ($\epsilon_{412} = 13,600$ cm^{-1} M^{-1}). Inactivations are performed by adding a 2-fold excess of DTNB to SF_1 (17 μM) in 0.1 mM ADP, 0.2 mM $MgCl_2$, 0.1 M KCl, 50 mM Tris, pH 8.0, 0°. The Ca^{2+}-ATPase is rapidly activated ($220 \pm 30\%$) ($t_{1/2} \sim 10$ min) and then subsequently inactivated ($t_{1/2} = 2$–3 hr). By approximately 16 hr the Ca^{2+}-ATPase is reduced to 20–30% and the K^+(EDTA)-ATPase to 5–10%. Approximately three TNB groups are produced as measured by the increased absorbance at 412 nm. Because inactivations are slow, they may be stopped simply by purification of the sample (see Section III,B,1). If desired, a rapid quench method may be employed to study the early phase of reaction with DTNB.[2] Two batchwise treatments of reaction mixtures with Dowex 1-X8 (80-fold weight excess over SF_1, 100–200 mesh, Cl$^-$, Bio-Rad) performed rapidly will remove free nucleotide, TNB, and unreacted DTNB quantitatively.

A disulfide, probably between SH-1 and SH-2, is believed to be responsible for inactivation and for trapping MgADP. The trapped MgADP is stable at 0° ($t_{1/2} \sim 5$ days), but it becomes much less stable at 25° ($t_{1/2} \sim 2$ hr), as does the enzyme TNB mixed disulfide. This thermal instability likely results from thiol–disulfide interchange that allows the MgADP to escape.

In several reactivations, treatment of purified DTNB inactivated SF_1 for 4 hr with a 100-fold excess of DTE at 0° resulted in recovery of about

90% of the Ca^{2+}- and NH_4^+(EDTA)-ATPase activities and release of 0.9 ± 0.2 TNB per SF_1. The disulfide bonds can also be cleaved with NaCN. This treatment cyanylates one of the critical thiols, which gives Ca^{2+}-ATPase activation (~230%) and partial recovery of NH_4^+(EDTA)-ATPase activity (35 ± 10%). The rate of ATPase recovery correlates with the cleavage of the protein disulfide, not with the cleavage of the TNB-protein mixed disulfide.[2]

(sIMP)$_2$ Inactivations. This reagent reacts more rapidly than DTNB and therefore should be used whenever it is desired to trap less-stable metal nucleotides via disulfide bond formation. The reaction mechanism is thought to be essentially the same as reaction with DTNB, but disulfide bond formation is about 10 times more rapid. This most likely occurs because the thiopurine moiety of (sIMP)$_2$ is a better leaving group than the thionitrobenzoate moiety of DTNB. The reagent can be easily prepared by oxidation of 6-mercaptopurine riboside 5'-phosphate (sIMP) with NaI_3. For example, 100 mg of sIMP (Sigma) is dissolved in 2 ml of H_2O and adjusted to pH 7. NaI_3 (0.3 M), prepared by mixing 0.3 M NaI and excess I_2, is added dropwise (total volume about 1.0 ml) with mixing until the solution just turns a light yellow. The pH is adjusted back to 7–8 with 0.1 N NaOH, an equal volume of cold ethanol is added, and cold acetone is added to precipitate the sodium salt of (sIMP)$_2$. The precipitate is pelleted by centrifugation in a clinical centrifuge, washed with cold acetone (twice) and cold diethyl ether (twice), and carefully air-dried. The final product is stored desiccated at −20° or lower and is stable for months. Solutions should be made as needed, and amine buffers such as Tris or imidazole should be avoided. Such buffers catalyze the hydrolysis of purine disulfide derivatives.[31] Hindered amine buffers, such as Bicine, are much less nucleophilic and are recommended. The molar concentrations of sIMP and (sIMP)$_2$ are determined spectrophotometrically at pH 8.0. [sIMP] = (5.0 × A_{313} − 0.55 × A_{278}) × 10^{-5}; [sIMP]$_2$ = (3.5 × A_{288} − 1.23 × A_{313}) × 10^{-5}. The production of sIMP is monitored by the increase in absorbance at 313 nm relative to a buffer blank (ϵ_{313} = 21,100). Reactions carried out at 0°, pH 8.0, 50 mM Bicine, 0.1 M KCl, 0.1 mM ADP, 0.2 mM MgCl with a 2-fold excess of (sIMP)$_2$ over SF_1 (17 μM) have half-times of inactivation of about 30 min for the Ca^{2+}-ATPase activity and 8 min for the K^+(EDTA)-ATPase activity. In the presence of 0.1 mM MgADP, 0.80 ± 0.1 mol of MgADP are trapped per mole of SF_1 (Table II).

The time course of cyanide or of DTE reactivation of purine disulfide-inactivated SF_1 is essentially identical to that seen for DTNB-inactivated

[31] R. G. Yount, J. S. Frye, and K. R. O'Keefe, Cold Spring Harbor Symp. Quant. Biol. 37, 113 (1972).

SF_1 when performed under identical conditions. This is expected, since it is believed that both kinds of aromatic disulfide reagents modify the same thiols. As expected, the cleavage of the enzyme 6-thiopurine nucleotide mixed disulfide by DTE or cyanide treatment occurs more rapidly than does the cleavage of the comparable enzyme-TNB mixed disulfide.

C. Crosslinking Reagents That Do Not Trap Effectively

A variety of crosslinking reagents of differing specificities have been investigated that are unsuccessful or only partly successful at trapping MgADP. Dimethyl suberimidate and dimethyl adipimidate, which react specifically with lysine groups, are ineffective at inactivating SF_1 ATPase activity in the presence or the absence of MgADP.[4] Inactivations were attempted in 0.1 M KCl, 50 mM Bicine, pH 8.0, at 0° in the presence of 0.2 mM MgADP using up to a 60-fold excess of crosslinking reagent over a period of 24 hr. As expected, no trapping of MgADP occurs during this time. p-Phenylene diisothiocyanate is relatively ineffective at inactivating SF_1 or trapping MgADP. A 60-fold excess of this reagent over SF_1 in the presence of MgADP is necessary to reduce the NH_4^+(EDTA)-ATPase to 55% after 2 hr. Although partial trapping of MgATP (0.3 ± 0.05) occurs, the large excess of reagent necessary for inactivation makes p-phenylene diisothiocyanate not very useful as a specific active site trapping reagent.

4,4'-Diisothiocyanostilbene-2,2'-disulfonic acid (DIDS), an 18 Å diisothiocyanate reagent, was more effective than p-phenylene diisothiocyanate. A 5-fold excess of DIDS reduced the NH_4^+-EDTA and Ca^{2+}-ATPase to 10 and 20%, respectively. Upon purification 0.8 MgADP per SF_1 was found trapped, and MgADP accelerated the rate of inactivation by DIDS. However, lower-fold excesses of DIDS were not effective for complete inactivation, suggesting other sites of modification. Although isothiocyanates normally react with amino groups, two to three thiols were lost after treatment with this reagent.[4] This is consistent with thiol crosslinking as seen with other reagents.

Arsenite. Adjacent thiols are known to add to arsenite in the presence of 2,3-dimercaptopropanol.[31,32] However, arsenite (7-fold excess over SF_1) with or without 2,3-dimercaptopropanol is ineffective in promoting trapping of MgADP (J. A. Wells, unpublished results). Partial inactivation of the NH_4^+(EDTA)-ATPase and activation of the Ca^{2+}-ATPase occurs, as has been previously reported.[33] However, no further change in ATPase occurred after 30 min, even in the presence of MgATP.

---

[32] A. L. Fluharty and D. R. Sanadi, *J. Biol. Chem.* **236**, 2772 (1961).
[33] B. P. Gaber and A. L. Fluharty, *Q. Rep. Sulfur Chem.* **3**, 317 (1968).

V. Applications

The trapping phenomenon allows a metal nucleotide to be placed in a specific manner at the active site of myosin. An advantage of trapping spectral probes versus simple binding is that the probe reports only from the active site, not from other weaker binding sites. Some of the potential applications of this phenomenon are discussed below.

A. Trapping Spectral Probes

1. Purine Binding Site

εADP bound reversibly to SF_1 or trapped in SF_1 by crosslinking with pPDM undergoes a blue shift and a decrease in intensity of its emission spectrum.[6] The distance between εADP and a chromophoric crosslinking reagent has been determined by fluorescence energy transfer measurements to be 28 ± 5 Å.[6] In this unique situation the donor fluorophore, εADP, is trapped at the active site by a fluorescence acceptor crosslinking reagent, F_2DPS, using a single modification (i.e., trapping) in which site-specific modification at both sites is assured. Thus the two cysteines modified (believed to be SH_1 and SH_2) are too far from the active site to be involved in either the binding or hydrolysis of ATP.

The photoaffinity reagent NANDP has been used as an absorption probe of the active site. Unlike the ε-adenosine moiety, the spectrum of the 4-azido-2-nitrophenyl moiety responds only to changes in the dielectric constant of the medium and is insensitive to pH changes. This probe has been bound reversibly and subsequently trapped in SF_1, and its spectra in both cases are identical. Furthermore, the spectra mimic that of NANDP placed in a medium consisting of a $60:40$ ethanol: H_2O mixture (dielectric constant = 47).

2. Metal Region Probes

Co(II)ADP bound reversibly or trapped in SF_1 shows a characteristic circular dichroism (CD) spectrum[5] suggesting that there are asymmetric binding determinants in the region where the metal binds. The electron paramagnetic resonance (EPR) spectra of Mn(II)AMP-PNP bound reversibly or trapped are identical and show an EPR spectrum indicative of a relatively immobilized Mn(II) similar to that obtained by Bagshaw and Reed.[34] The real advantage of active-site trapping versus simply binding

[34] C. R. Bagshaw and G. H. Reed, *J. Biol. Chem.* **251**, 1975 (1976).

may be in the study of systems that contain intact DTNB light chains. Here, the DTNB light-chain metal-binding site(s) can be probed with spectroscopically interesting metals after the active site has been blocked by trapping spectroscopically silent metal-nucleotides.

B. Photoaffinity Labeling Experiments

NANTP is a substrate for myosin, actomyosin and supports contraction of functionally skinned muscle fibers.[7] Furthermore, the diphosphate analog NANDP binds in a specific manner ($K_D = 1 \mu M$) to a single site on SF_1. However, while photolysis leads to inactivation, photoinactivation has been shown to result from labeling sites other than the active site. [β-^{32}P]NANDP is specifically trapped at the active site of SF_1 by crosslinking with pPDM, DTNB, or cobalt phenanthroline complexes. Once trapped, the half-time of dissociation at 0° is more than 5 days. Subsequent photolysis at 0° allows more than 50% of the trapped NANDP to be covalently incorporated into SF_1. This approach, which may be called "trap and zap," ensures the specific labeling of only active site-related amino acids.

C. Comparative Studies

Many of the trapping techniques applied to skeletal myosin have been applied to chicken gizzard (smooth muscle) myosin and bovine cardiac myosin.[35] pPDM and cobalt phenanthroline complexes have been shown to inactivate both myosins. MgADP stimulates inactivation, although to a lesser degree than seen with skeletal myosin. MgADP is trapped after inactivation but the quantity varies between 0.1 and 0.4 mol per mole of active sites. The suggestion has been made[35] that alternative crosslinking sites exist; one site leads to stable trapping of MgADP, the other does not. However, reaction at either site inactivates the ATPase activity. Incorporation of approximately two equivalents of crosslinking reagent per active site of cardiac myosin correlates with inactivation. It is believed only one equivalent of reagent is responsible for inactivation and the second modification is noncritical. This noncritical modification occurs on the alkali light-chain monofunctionally.[35] Preliminary "trap and zap" experiments with NANDP appear to label the ATPase site on cardiac myosin specifically.

[35] P. R. Buzby, Ph.D. thesis, Washington State University, Pullman, 1981.

Acknowledgments

We wish to thank Mary Sheldon, William Perkins, Ross Dalbey, Cathy Knoeber, Dr. Philip Buzby, and Dr. Kay Nakamaye for making unpublished results available and for many useful discussions. We are also indebted to them and to Dr. Alan G. Weeds and Gary J. Pielak for helpful comments on this manuscript. This research was supported by grants from the Muscular Dystrophy Association, the National Institutes of Health (AM-05195), and the American Heart Association.

[12] Myosin Active-Site Trapping with Vanadate Ion

By CHARLES C. GOODNO

In the ideal world it would be possible to study the actomyosin ATPase by slowing it to the point where each chemical intermediate could be examined at leisure. Although the ideal is unattainable, it is possible to obtain stable complexes of myosin that mimic transient intermediates of ATP hydrolysis. One such model complex, which resembles the central myosin-products intermediate ($M \cdot Pr$), is formed by the binding of ADP and vanadate ion to the myosin active site.[1,1a] This chapter is intended as an introduction to the properties and prospective applications of the complex, as well as a discussion of techniques for its synthesis and analysis.

The stable complex is best understood in the context of the properties of vanadate ion (i.e., orthovanadate, VO_4^{3-}), which is a tetrahedral oxyanion of pentavalent vanadium.[2] By virtue of its geometry, size, and charge, vanadate (abbreviated V_i) is an analog of phosphate ion. The functional aspects of this analogy have been recognized in studies of several phosphatases and phosphotransferases.[3-5] In contrast to its readily reversible interaction with other enzymes, vanadate forms with myosin a complex that appears to be uniquely stable.

Vanadate alone forms a weak reversible complex with myosin, much like phosphate. Yet when ADP is included, a stable, enzymically inactive

---

[1] Abbreviations: V_i, vanadate ion (degree of protonation unspecified); M, single active site of the myosin ATPase; $M \cdot Pr$, myosin-products central complex; $M \cdot ADP \cdot V_i$, reversible ternary complex of myosin; $M^{\dagger} \cdot ADP \cdot V_i$, stable ternary complex (trapped intermediate); P_i, phosphate ion; PP_i, pyrophosphate ion; AMPPNP, adenylyl imidodiphosphate; PAR, 4-(2-pyridylazo)resorcinol; SDS, sodium dodecyl sulfate.

[1a] C. C. Goodno, *Proc. Natl. Acad. Sci. U.S.A.* **76**, 2620 (1979).

[2] M. T. Pope and B. W. Dale, *Q. Rev. Chem. Soc.* **22**, 527 (1968).

[3] E. G. DeMaster and R. A. Mitchell, *Biochemistry* **12**, 3616 (1973).

[4] V. Lopez, T. Stevens, and R. N. Lindquist, *Arch. Biochem. Biophys.* **175**, 31 (1976).

[5] L. Josephson and L. C. Cantley, Jr., *Biochemistry* **16**, 4572 (1977).

complex is formed, which can be isolated free of unbound vanadate and ADP. This complex has a stoichiometry of one ADP and one V_i per myosin active site and a half-life of approximately 3 days at 25°. Although this stability is unaffected by dialysis, chelating agents, and ATP, denaturation of the myosin produces a rapid release of ADP and V_i (<3 min). Thus, the complex appears to be held together by strong secondary forces rather than covalent bonds.

The formation of this inactive complex, symbolized by $M\dagger \cdot ADP \cdot V_i$ (the dagger representing the demise of ATPase activity), is highly specific. If the reaction is carried to various degrees of completion, there is a linear relationship (with a slope of 1.0) between the percentage incorporation of V_i and the percentage of inactivation of the ATPase.[6] This result is consistent with the interpretation that a single vanadate ion inactivates a single myosin ATPase site. The relationship between V_i incorporation and ATPase inactivation remains constant as the concentration of V_i is varied over a range of 4- to 1000-fold excess over myosin active sites. Similar results are obtained when the concentration of ADP is varied. The use of excess ADP and V_i, therefore, does not lead to incorporation at additional sites.

The stable incorporation of V_i is nucleotide-specific in that neither AMP, ATP, AMPPNP, nor PP_i will substitute for ADP. ATP actually protects myosin from modification in the presence of V_i and ADP, causing the rate of V_i incorporation to drop to 3% of its usual value. Since roughly 3% of the steady-state complexes of the myosin ATPase are in the form of $M \cdot ADP$,[7] which can react directly with V_i, this finding suggests that $M \cdot ATP$ and $M \cdot Pr$ complexes are completely protective. These observations, in concert with kinetic studies (see below), lead to the conclusion that V_i and ADP function together as a binary active-site directed reagent for myosin, with the unusual feature that the labeling is apparently not covalent.

Formation of $M\dagger \cdot ADP \cdot V_i$

During the formation of the $M\dagger \cdot ADP \cdot V_i$ complex, ADP and V_i are incorporated into the complex at identical rates, within experimental error. For fixed concentrations of ADP and V_i, the kinetics are pseudo-first order. When the concentrations are varied, the rate of formation of the complex is a hyperbolic function of both ADP and V_i. This hyperbolic dependence is characteristic of active-site-directed reagents, which bind at the enzyme active site prior to the modification reaction. The maximum

[6] C. C. Goodno and E. W. Taylor, in preparation.
[6a] C. C. Goodno and E. W. Taylor, *Proc. Natl. Acad. Sci. U.S.A.* **79**, 21 (1982).
[7] D. R. Trentham, J. F. Eccleston, and C. R. Bagshaw, *Q. Rev. Biophys.* **9**, 217 (1976).

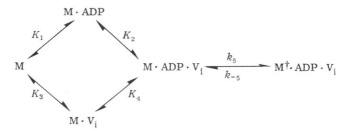

FIG. 1. Mechanism of formation of the stable vanadate complex. Rapid preequilibration leads to a reversible ternary complex (M · ADP · V_i), which slowly isomerizes to the stable complex. The isomerization (step 5) is the rate-limiting step in both forward and reverse directions.

(plateau) rate of formation of the stable complex is approximately 0.01 sec^{-1} (pH 7 or 8.5, ionic strength 0.1, 25°). Half-saturation of the rate occurs with 400 μM V_i and less than 10 μM ADP. These findings are consistent with the formation of a reversible ternary complex of myosin, ADP, and V_i, followed by a slow isomerization step leading to the stable complex (trapped intermediate), as illustrated in Fig. 1.

Although the $M\dagger$ · ADP · V_i complex is exceedingly stable, its formation is slowly and spontaneously reversible, with a first-order rate constant of approximately 2.5×10^{-6} sec^{-1} (pH 8.5, ionic strength 0.1, 25°). The release of V_i and recovery of ATPase activity occur at the same rate, indicating the genuine reversibility of the binding. Thus, the half-life of the complex is approximately 3 days. Combination of the forward and reverse rate constants for the isomerization step gives an equilibrium constant of 4×10^3 in the forward direction. Since the tight binding of V_i is the result of a coupled equilibrium (binding followed by isomerization), the overall binding constant for V_i is given by the product of the two individual binding constants, which is approximately $10^7 M^{-1}$. It is noteworthy that this binding constant is about 100,000-fold higher than the binding constant of myosin for P_i, yet it is similar to the binding constant of P_i in the M · Pr intermediate in ATP hydrolysis.

Interaction of $M\dagger$ · ADP · V_i with Actin

It is well known that actin exerts its activating effect on the myosin ATPase by accelerating the otherwise slow product release step.[8] One would expect, then, that a genuine trapped intermediate of the myosin-products type would interact with actin in such a way as to displace the products. Figure 2 shows that actin indeed displaces V_i from the $M\dagger$ · ADP · V_i complex.[6a] Moreover, increasing concentrations of actin produce higher

[8] R. W. Lymn and E. W. Taylor, *Biochemistry* **10**, 4617 (1971).

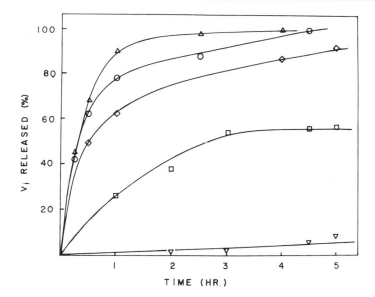

FIG. 2. Kinetics of actin-mediated vanadate displacement from M† · ADP · V_i. The displacement of V_i from the complex by various concentrations of actin was determined colorimetrically. Conditions: 10 μM M† · ADP · V_1, 90 mM NaCl, 20 mM Tris, 5 mM MgCl$_2$, pH 8.5, 25°. Actin concentrations: $\triangledown$, 0; $\square$, 10 μM; $\diamond$, 25 μM; $\bigcirc$, 50 μM; $\triangle$, 75 μM.

rates of V_i displacement, even in the range of actin concentrations that lead to 100% displacement of V_i at equilibrium. Thus, actin not only shifts the equilibrium for the binding of V_i, it also catalyzes the removal of V_i from the myosin active site in much the same fashion as it catalyzes the removal of P_i. The role of actin is corroborated by the observation that ATPase activity is recovered at the same rate as V_i displacement (Fig. 3), with full displacement of V_i leading to complete recovery of ATPase activity.

The binding of actin to M† · ADP · V_i is ionic-strength dependent, as it is with the M · Pr complex. Under the most favorable conditions examined to date (ionic strength 0.03, pH 7.0), the rate of V_i release is a hyperbolic function of actin concentration, with an apparent binding constant of approximately $10^4 M^{-1}$ and a limiting rate constant in the vicinity of 0.5–1.0 sec^{-1}. This rate is approximately 5% of the rate of actin-mediated release of products from the M · Pr complex. Since the rate of spontaneous release of V_i under these conditions is approximately 5×10^{-6} sec^{-1}, actin is able to accelerate the process by a factor of about 100,000. Thus, actin binds to the complex and catalyzes a reaction analogous to product-displacement from the myosin ATPase site.

Conversely, studies[6a] have shown that ADP and V_i together have a

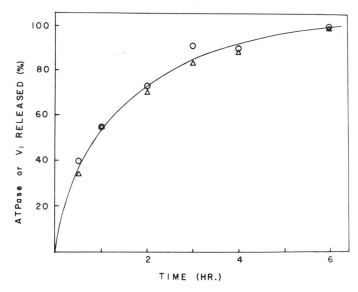

FIG. 3. Correlation of vanadate release and ATPase recovery. 10 μM M$\dagger$ · ADP · V_i was mixed with 17 μM actin (final concentrations). Samples were withdrawn at intervals for measurement of ATPase activity and free V_i. ○, ATPase activity; △, free V_i. Conditions: 90 mM NaCl, 20 mM Tris, 5 mM MgCl$_2$, pH 8.5, 25°.

dissociating effect on the actomyosin complex under conditions where neither species separately has any effect. These results suggest that ADP and V_i may have a relaxing effect similar to that of ATP. Preliminary studies with frog semitendinosus have shown that when V_i is added to a muscle fiber in the presence of endogenously generated ADP (in the approach to rigor), relaxation is obtained under conditions where ADP and V_i have no effect separately.[9] Stiffness is simultaneously reduced to a low level, indicating the almost total dissociation of myosin cross-bridges. These effects are completely reversed by washing out the ADP and V_i.

Methods

Vanadate Analysis

Stock solutions of 100 mM V_i are conveniently prepared by dissolving Na$_3$VO$_4$ or V$_2$O$_5$ in H$_2$O and adjusting the solution to approximately pH 10 with 6 N HCl or 10 N NaOH, respectively. Since V_i exists in a slow, pH-dependent equilibrium with various polymeric vanadate species,[2] this adjustment perturbs the equilibrium. Resulting polymeric species

[9] A. Magid and C. C. Goodno, *Biophys. J.* **37**, 107A (1982).

(orange-yellow) are destroyed by boiling the solution until colorless, followed by rechecking the pH of the cooled solution. Several cycles of adjustment and boiling may be required. Stock solutions can be stored for several months in a stoppered bottle away from reducing agents. Solutions are discarded when they turn noticeably yellow. The final concentration of V_i is checked by measuring the absorbance at 265 nm ($E_{265} = 2925$ M^{-1} cm^{-1}) on an aliquot of stock solution diluted with H_2O.

A solution (2 mM) of the metal-indicator dye 4-(2-pyridylazo)resorcinol (PAR), is prepared by dissolving the dye powder in H_2O, followed by filtration to remove any particulate matter. A stock solution of 1.0 M imidazole buffer, pH 6.0, is also prepared. Vanadate analysis is carried out by the addition of 0.1 ml of 1.0 M imidazole to a 1.0-ml sample of vanadate, followed by addition of 0.1 ml of 2 mM PAR, then mixing. After 5 min of color development at room temperature, the absorbance is measured at 540 nm in a 1-cm semimicro cuvette. The resulting red color is stable for several hours. Calibration curves are prepared over the range of 0–20 μM V_i, the region in which Beer's law is obeyed.

Preparation of Vanadate-Trapped Myosin

For ease in chromatographic separations at low ionic strength, either subfragment-1 or heavy meromyosin is generally used, although whole myosin readily undergoes modification. A typical preparation of M† · ADP · V_i is carried out with a myosin site concentration of 20 μM in 0.09 M NaCl, 5 mM MgCl$_2$, and 20 mM Tris, pH 8.5. (Ionic strength is kept low and pH high to facilitate separation of M† · ADP · V_i by ion-exchange chromatography. Preparation can be carried out at pH 7 with equal ease, although subsequent separation is more complex.) A small aliquot of stock ADP (100 mM) is added to produce a final ADP concentration of approximately 0.2 mM, and the solution is incubated for 5 min in a water bath at 25° to hydrolyze any ATP present in the ADP. Stock V_i (100 mM) is then added to a final concentration of 1 mM, and the modification reaction is allowed to proceed for 5 min.

Separation of M† · ADP · V_i from excess ADP and V_i is readily performed by ion-exchange chromatography. Up to 5 ml of the reaction solution are applied to a 0.5 × 2 cm column of washed Dowex 1-X8 resin (200–400 mesh), which has been fully equilibrated with 0.09 M NaCl, 5 mM MgCl$_2$, and 20 mM Tris, pH 8.5. (Pasteur pipettes make convenient columns.) The reaction mixture is allowed to percolate into the column under gravity flow (approximately 0.2 ml/min), and the resulting effluent is collected. The column is then washed with 2–5 ml of the column buffer, again under gravity flow, and the effluent is collected in the same vessel.

ADP and V_i that are not bound to myosin are quantitatively retained by the column (99.9%), while 95% recovery of the heavy meromyosin or subfragment-1 is obtained in the effluent. The pooled washings, which contain the purified $M\dagger \cdot ADP \cdot V_i$ complex, are then stored at 4°; and the column is regenerated with $1 N$ NaOH. Although Dowex 1 treatment may seem harsh, it produces no observable change in the SDS-gel electrophoresis pattern of the protein and less than a 1% change in the ATPase specific activity.

Analysis of Vanadate-Trapped Myosin

The vanadate content of the isolated complex can be determined directly by treating the complex with 1% sodium dodecyl sulfate to denature the protein, followed by standard colorimetric analysis of the vanadate that is released. However, frequently 5–10% of the bound vanadate is bound in nonspecific (i.e., reversible) fashion, leading to an overestimate of the tightly bound vanadate. For this reason, we prefer a differential method in which vanadate is measured both before and after the addition of SDS. In this procedure, calibration curves are established for V_i in the presence and in the absence of SDS. A 1-ml sample of $M\dagger \cdot ADP \cdot V_i$ is used, and V_i analysis is carried out according to the standard assay procedure, with sample and reagents added directly to the spectrophotometer cuvette. After the absorbance at 540 nm is read, a 63-μl aliquot of 20% SDS is added with mixing, and an additional 5 min is allowed for color development from the released V_i. The absorbance is again read; appropriate blank readings are made with an equivalent amount of protein (without V_i); and the concentration of free V_i before and after denaturation is calculated. Tightly bound V_i (i.e., trapped) is given by the difference between these values. Typical reactions using fresh heavy meromyosin or subfragment-1 give 85–95% modification, as judged by both V_i incorporation and ATPase inactivation. Aging of the protein leads to a decline in V_i incorporation and inactivation, particularly for subfragment-1. Preliminary studies[6] have shown that alkylation of myosin sulfhydryl groups drastically inhibits the incorporation of V_i. Thus, myosin must remain intact to ensure optimal modification.

Applications of Vanadate Labeling

The combination of ADP and V_i produces an affinity label that is completely selective for one site per head of the myosin molecule, and the preponderance of evidence suggests that this site is identical with the ATPase active site. Moreover, currently available data suggest that myosin is the only enzyme that is stably labeled by the combination of ADP and V_i. These properties make ADP/V_i a suitable reagent for active-site titrations of myosin, as well as the presumptive identification of myosin in heterogeneous systems. Using the colorimetric analysis for V_i, one can determine nanomole quantities of myosin sites. With [^{3}H]ADP or other suitable radioactive label in ADP or V_i, the sensitivity can be extended by several orders of magnitude. If desired, the myosin can be recovered unharmed by treatment with actin, which releases the label and restores full ATPase activity.

The $M^+ \cdot ADP \cdot V_i$ complex has a number of possible applications in the investigation of energy transduction by myosin, since it resembles $M \cdot Pr$, the central intermediate in the hydrolysis of MgATP.[7,8] The 20,000-fold longer lifetime of $M^+ \cdot ADP \cdot V_i$ makes it suitable for examination by physicochemical methods that are too slow to be used for the actual intermediate. This long lifetime may also afford a unique opportunity to circumvent the diffusion limitation of muscle fibers by saturating the myosin cross-bridges in a well-defined chemical state (i.e., $M^+ \cdot ADP \cdot V_i$). With blockage of actin by low calcium, this state should have a lifetime in the cross-bridge comparable to its lifetime in isolated myosin. Thus, it may be possible to study the mechanical properties of muscle fibers whose myosin cross-bridges are in a chemically well-defined state, leading to a correlation of chemical and mechanical events involved in muscle contraction.

[13] Assays for Myosin

By Thomas D. Pollard

The first prerequisite in the purification of any protein is reliable simple assays. Fortunately, the distinctive ATPase activities of all the myosins and, in most cases, the unique high molecular weight of the myosin heavy chains and the characteristic shape of the intact myosin molecules make this a simple matter.

ATPase Assays

In most cases myosins have the unique property of having low ATPase activity in Mg^{2+} and high activity in Ca^{2+} and in EDTA, providing that high concentrations of K^+ are present. The high ionic strength also prevents any actin in crude fractions from activating the MgATPase. Most other ATPases are more active in Mg^{2+} than in Ca^{2+} and are inactive in

EDTA. Consequently, one can usually follow myosin purification from the crude extract onward with K(EDTA)-ATPase assay. However, in a few cases a CaATPase assay is useful or essential. For example, *Acanthamoeba* myosin II has a higher CaATPase than K(EDTA)-ATPase activity,[1,2] and *Physarum* myosin has CaATPase activity but no K(EDTA)-ATPase activity.[3]

Determination of Inorganic Phosphate by Pollard and Korn's Modification[4] of the Method of Martin and Doty[5]

Stock Solutions

A: Isobutanol–benzene, 1 : 1. Store at room temperature.

B: Silicotungstic-sulfuric acid. Mix 2 parts of $10 N$ sulfuric acid (made by adding 150 ml of concentrated sulfuric acid to 350 ml of water) with 5 parts of 6% aqueous silicotungstic acid (Fisher No. A-289). Store at room temperature.

C: Ammonium molybdate. Dissolve 10 g of ammonium molybdate (Baker No. 0716) in water to give a final volume of 100 ml. Store at room temperature. Discard when a precipitate forms.

D: Acid alcohol. Mix 1 pint (or 500 ml) of cold 100% ethanol with 15 ml of concentrated sulfuric acid.

E: Stannous chloride. Dissolve 1 g of fresh stannous chloride (Baker No. 3980) in concentrated HCl to give a final volume of 10 ml. Store in a lighttight container in a refrigerator, where it will keep for several months. Also store dry stannous chloride in the refrigerator. If this assay fails to give a linear standard curve, stannous chloride is at fault until proved otherwise.

Working Solutions. These are to be made fresh daily.

F: Stop mixtures. Mix 1 ml of solution A with 0.25 ml of B in small test tubes. Two phases will be present. Alternatively, these two components can be pipetted directly into the assay tubes.

G: Diluted stannous chloride. Dilute 1 part stannous chloride stock solution (E) with 25 parts of $1 N$ sulfuric acid and store in an ice bath on the day of use.

Assay Procedure

1. Pour one tube of stop mixture into each 0.5-ml sample containing 2–300 nmol of inorganic phosphate and immediately vortex for 2

---

[1] H. Maruta and E. D. Korn, *J. Biol. Chem.* **252**, 6501 (1977).

[2] T. D. Pollard, W. F. Stafford, and M. E. Porter, *J. Biol. Chem.* **253**, 4798 (1978).

[3] M. R. Adelman and E. W. Taylor, *Biochemistry* **8**, 4976 (1969).

[4] T. D. Pollard and E. D. Korn, *J. Biol. Chem.* **248**, 4682 (1973).

[5] J. B. Martin and D. M. Doty, *Anal. Chem.* **251**, 965 (1949).

sec. If this is an ATPase assay, the acid will stop the reaction and precipitate the protein.

2. Immediately add 0.1 ml of ammonium molybdate (C) and vortex again for about 10 sec.

3. Let phases separate. The yellow phosphomolybdate complex will be extracted quantitatively into the upper (organic) phase, while organic phosphates such as ATP or ADP will remain in the lower aqueous phase. If the phosphate is radioactive, an aliquot of the organic phase can be used for scintillation or Cerenkov counting. If there is a very large protein precipitate, low speed centrifugation may be necessary to layer it at the interface between the two phases.

4. Using a 1-ml pipette with a propipette, transfer 0.5 ml of the upper phase into a clean test tube and, using the same pipette, add 1.0 ml of acid alcohol (D). This solution is stable.

5. Develop the peacock blue color by reducing the phosphomolybdate complex with 0.05 ml of diluted stannous chloride (G) and vortexing to mix. If the color is navy blue, the stannous chloride is bad.

6. Read the absorbance at 720 nm against a blank without phosphate.

7. Construct a standard curve with 0.1, 0.2 and 0.3 ml of 1 mM phosphate (i.e., 100, 200, and 300 nmol) in 0.5 ml total volume. The extinction coefficient is about 0.007 OD per nanomole.

Myosin ATPase Assay by the Method of Pollard and Korn[4]

Stock Solutions

A: Buffer mixture to give final concentrations of 0.5 M KCl, 10 mM imidazole (pH 7) in the assay (25 ml of 2 M KCl, 1 ml of 1 M imidazole (pH 7), 44 ml of water

B: 100 mM CaCl$_2$

C: 100 mM MgCl$_2$

D: 20 mM EDTA neutralized with KOH

E: 20 mM ATP made by diluting a 100 mM stock solution. This stock is made from disodium-ATP neutralized with KOH immediately after dissolving in water. Store frozen.

Assay Procedure

1. Set up a stop tube for each assay as described under the inorganic phosphate determination.

2. Set up assay tubes by adding 0.35 ml of ATPase mix, 50 μl of 20 mM ATP, and 50 μl of either EDTA, CaCl$_2$, or MgCl$_2$.

3. Start the reaction by adding 50 μl of enzyme to an appropriate tube, shaking gently, and placing the tube in a water bath set at an appro-

priate temperature. Tubes are conveniently started at 30-sec intervals.

4. After a suitable time interval (which will be determined by the activity of the sample), stop the reaction by adding the stop solution and vortexing. Alternatively, the silicotungstic–sulfuric acid and the isobutanol–benzene can be pipetted separately into the assay tube, but this is slower.
5. During the 30 sec before the next tube is to be stopped, add 0.1 ml of 10% ammonium molybdate and vortex for 10 sec.
6. Measure the released inorganic phosphate when all the assays have been stopped and extracted.

Another absolutely unique feature of myosin is that actin filaments stimulate the MgATPase activity providing the ionic strength is low. Nevertheless, this assay is not generally useful for following myosin purification for two reasons. First, the actin-MgATPase activity of myosin is only a small fraction of the total cellular MgATPase activity in nonmuscle cells. Second, many myosins require activating enzymes, such as the *Acanthamoeba* cofactor protein (heavy-chain kinase [4,6]) or the cytoplasmic and smooth muscle myosin light-chain kinases [7,8] for actin-activated ATPase activity. When these activation enzymes are separated from myosin during purification, the myosin loses its actin-MgATPase activity. Thus, although actin-MgATPase of myosin cannot be used for quantitating purification, it is an essential feature of myosin and should be studied as part of myosin characterization.

It is difficult to specify exact conditions for the MgATPase assay because they vary considerably for different myosins. For starters, the following conditions are recommended: 10–30 mM KCl, 2 mM MgCl$_2$, 1 mM ATP, 0.1 mM CaCl$_2$, 10 mM imidazole (pH 7), 0.5 mg of actin per milliliter. The rationale for these conditions is that all actomyosin MgATPases are inhibited by high concentrations of salt and ATP, all require Mg^{2+}, and many are activated directly or indirectly by Ca^{2+}.[see 9]

An important factor to be tested is the dependence of the myosin ATPase activity on actin concentration. If the myosin is sufficiently soluble, the plot of ATPase vs actin concentration is hyperbolic, and a double-reciprocal plot (1/rate vs 1/actin) will be linear. The y intercept is the reciprocal of V_{max}, and the x intercept is $-1/K_{app}$ (the apparent affinity of myosin for actin). Thus, it is possible to test whether any particular parameter affects the activity by altering V_{max}, K_{app} or both.[9]

[6] H. Maruta and E. D. Korn, *J. Biol. Chem.* **252**, 8329 (1977).
[7] R. S. Adelstein and M. A. Conti, *Nature (London)* **256**, 597 (1975).
[8] R. Dabrouska, D. Aromatorio, J. M. F. Sherry, and D. J. Hartshorne, *Biochem. Biophys. Res. Commun.* **78**, 1263 (1977).
[9] C. Moos, *Cold Spring Harbor Symp. Quant. Biol.* **37**, 137 (1973).

Gel Electrophoresis

Myosin polypeptides have been analyzed by polyacrylamide gel electrophoresis using any number of buffer systems, but I have found one simple tube gel formulation and one slab gel system to be particularly useful because both the heavy and light chains are well resolved without using a polyacrylamide gradient. The tube gel method is a variation by Stephens[10] of a system originally used by J. Bryan.

Tube Gel Electrophoresis

Stock Solutions

A: Acrylamide-bis: 30 g of acrylamide (Eastman No. 5521) and 0.8 g of N,N'-bismethylene acrylamide (Eastman No. 8383) to 100 ml with water. Filter and store at 4°.

B: SDS: 10 g of sodium dodecyl sulfate (SDS; Sigma No. L5750) to 100 ml with water. Store at room temperature.

C: Tris-glycine, "250 mM": 30.3 g of tris base (Sigma No. T1503) and 144 g of glycine (Sigma No. G7126) to 1 liter with water. The pH is about 8.6; do not adjust. Store at room temperature.

D: Ammonium persulfate (Bio-Rad): 0.1 g per milliliter of water. Store at $-20°$.

E: Electrode chamber buffer: 100 ml Tris-glycine stock (C) and 10 ml SDS stock (B) per liter. Store at room temperature.

F: 2 × sample buffer: 20 ml of SDS stock (B), 2 ml of Tris–glycine stock (C), 2 ml of 2-mercaptoethanol, 20 ml of glycerol, 10 mg of bromophenol blue to 100 ml with water. Store at room temperature.

G: TEMED: N,N,N',N'-tetramethylethylenediamine (Bio-Rad).

Formulations for Twelve 0.5 × 7 cm Tube Gels

	Acrylamide		
Component (ml)	5%	7.5%	10%
Water	14.25	12.7	11.0
SDS	0.2	0.2	0.2
Tris-glycine	2.0	2.0	2.0
Acrylamide	3.33	5.0	6.67
TEMED	0.01	0.01	0.01
Persulfate	0.20	0.15	0.10

[10] R. E. Stephens, *Anal. Biochem.* **65**, 396 (1975).

Pour the solution into 0.5 (ID) × 10 cm tubes sealed at the bottom with Parafilm. Overlayer with water-saturated butanol during polymerization. The 7.5% gel is particularly suitable for resolution of both myosin heavy chains and light chains.

Samples. Dialyze samples containing 0.2–1.0 mg of protein per milliliter for 2 hr against water to remove salts. Mix samples with an equal volume of 2× sample buffer and immediately heat to 100° in a boiling water bath for 2 min. Apply 10–50-μl samples to the gels and run at 100–150 V. The tracking dye will reach the bottom of the gel in 60–80 min.

Stain. Use the method of Fairbanks *et al.*[11]

Procedure for Slab Gel Electrophoresis

This is a modification by Gibson[12] of the Laemmli method.[13]

Stock Solutions

A: 20% SDS: Dilute 20 g of SDS (BDH No. 44244) to 100 ml with water. Store at room temperature.

B: 3 M Tris-HCl (pH 8.8): Titrate 36.3 g of Tris base to pH 8.8 with HCl and bring to 100 ml with water. Store at 4°.

C: 1 M Tris-HCl (pH 7): Titrate 12.1 g of Tris base to pH 7.0 with HCl and dilute to 100 ml with water. Store at 4°.

D: Resolving gel acrylamide: Mix 28 g of acrylamide (Eastman No. 5521) and 1.09 g of diallyltartardiamide (Eastman No. 11444) with water to 100 ml. Store at 4°.

E: Stacking gel acrylamide: Mix 21.5 g of acrylamide and 3.75 g of diallyltattardiamide with water to 100 ml. Store at 4°.

F: Stacking gel buffer: Mix 19.2 ml 1 M Tris (pH 7) plus 0.8 ml 20% SDS. Store at 4°.

G: Ammonium persulfate: Mix 0.14 g in 15 ml of water. Store at −20°.

H: Electrode chamber buffer: Mix 28.8 g of glycine, 6 g of Tris base, and 2 g of SDS with water to 2 liters. Store at room temperature.

I: 2 × sample buffer: Mix 10 ml of 20% SDS, 10 ml of 2-mercaptoethanol, 5 ml of glycerol, 5 ml of 1 M Tris (pH 7), 5 mg of bromophenol blue, and 20 ml of water. Store at room temperature.

Formulation for a 1 mm × 10 cm × 12 cm Slab Gel with 14% Acrylamide. Mix 1.65 ml of 3 M Tris-HCl (pH 8.8), 6.55 ml of resolving gel acrylamide, 0.065 ml of 20% SDS, and 4.45 ml of water. Then degas in a side-arm

[11] Fairbanks *et al.* (1971).
[12] W. Gibson, *Virology* **62**, 319 (1974).
[13] U. K. Laemmli, *Nature* (*London*) *New Biol.* **227**, 680 (1971).

flask for 2 min using a water pump. Add 0.33 ml of ammonium persulfate and 3 μl of TEMED. Mix gently and pour between plates. Overlayer with 1 ml of 0.1% SDS and allow the acrylamide to polymerize for 1 hr only.

Formulation for the Stacking Gel. Mix 1 ml of 1 M Tris-HCl (pH 7), 40 μl of 20% SDS, 1.6 ml of stacking gel acrylamide, and 5.3 ml of water. Degas for 2 min, and then add 0.4 ml of ammonium persulfate and 2.5 μl of TEMED. Mix gently, drain the 0.1% SDS from the surface of the resolving gel, pour the stacking gel, insert the sample well comb, and polymerize for 60 min only.

Procedure for Electrophoresis. Prepare the samples as described for the tube gels using the 2 × slab gel sample buffer. Apply 10–30-μl samples in the sample wells underneath the electrode buffer and run at 150 V. The tracking dye will reach the bottom of the gel in about 4 hr.

Negative Staining for Electron Microscopy

During the characterization of a new myosin it is usually essential to examine the structure of myosin filaments and of the complex of myosin with actin filaments. Although the preparation of negatively stained specimens is simple, at least in principle, marginal micrographs are occasionally published because some investigators are not familiar with the technical details that improve the chances of making high-quality specimens.

Procedure for Negative Staining

1. Make grids coated with Formvar of colloidin as described in electrom microscopy handbooks and coat with carbon.

2. Just prior to specimen application, make the grid surface hydrophilic by glow discharge in a vacuum evaporator. If the evaporator is not equipped with such a device, an adequate substitute is to use an inexpensive Telsa coil, sold by scientific supply houses as a "gas leak detector." Place the grids on a grounded metal plate inside the evaporator bell jar, evacuate to about 0.1 torr, and then create a discharge inside the bell jar by touching the tip of the active Telsa coil to the external end of one of the metal lead-throughs into the bell jar. If successful, a purple discharge will emanate from the internal tip of the lead-through. Usually 30 sec of discharge will make the grids hydrophilic, although some experimentation will be necessary with an improvised device.

3. Make myosin filaments by dilution or dialysis into a suitable buffer. Start with 50 mM KCl, 2 mM MgCl$_2$, 10 mM imidazole (pH 7), and a myosin concentration of 0.1–1.0 mg/ml. Just before application to the grid, dilute to about 20–80 μg/ml in the buffer.

4. While holding the grid with fine forceps place a small drop of diluted myosin filaments on the grid for about 30 sec. Then withdraw excess sample by touching the edge of the grid to filter paper. If the grid is hydrophilic, a thin film of sample will cover the surface uniformly and will resist removal by the filter paper.

5. The grid can then be stained directly by applying a drop of aqueous 1% uranyl acetate for 15–30 sec. All but a thin film of stain is then removed with filter paper, and the specimen is air-dried.

Alternatively, the background on the grid can be cleaned up and the staining made more uniform by the following steps recommended by Dr. U. Aebi (Johns Hopkins Medical School, Department of Cell Biology and Anatomy). Transfer the grid with the absorbed specimen through 3 drops of buffer on Parafilm for 5 sec each. Merely touch the specimen side of the grid to each drop; do not submerge the grid. Then transfer through 3 drops of water for 5 sec each, withdraw the water by touching the edge of the grid to filter paper. Transfer through 3 drops of 1% uranyl acetate or 0.75% uranyl formate for 5 sec each, and finally withdraw excess stain by touching the edge of the grid to filter paper. It is also helpful to use a glass capillary connected to a vacuum to clean off any droplets of stain from the edge of the grid to aid in the uniform drying of the stain.

It is difficult to obtain uniformly excellent grids, so it is necessary to make multiple specimens and to search each grid for the best staining. If the staining is uniformly poor, first check that the grids are hydrophilic and then dilute the specimen.

Actin filaments can be decorated with myosin either in solution or on the grid itself. The most common cause of failure is the presence of ATP in the actin buffer. Therefore, wait long enough after mixing the proteins to hydrolyze the ATP, or first apply about 50 μg of actin filaments per milliliter to the grid, wash through 3 or 4 drops of buffer without ATP, and then apply the myosin.

[14] Preparation of C-Protein, H-Protein, X-Protein, and Phosphofructokinase

By ROGER STARR and GERALD OFFER

Although myosin is the main component of the thick filaments of vertebrate striated muscle, it is not the only one. In electron micrographs of A bands a set of 11 stripes, 43 nm apart, may be seen in each half of the A band, occupying a region 0.25–0.50 μm from the center.[1] Some, but not all, of the components responsible have now been identified.

Our approach to the isolation of these components is based on the supposition that they might bind to myosin and, therefore, be present in crude preparations of myosin. In sodium dodecyl sulfate (SDS) gels of myosin, prepared by extracting muscle with a high salt concentration and then precipitating at low ionic strength, we found several other bands besides those expected from myosin. The bands of impurities were labeled alphabetically according to their mobility, the main ones being B, C, F, and H.[2] B-protein was later shown to be one of the components of the M line.[3] C-protein was purified, characterized, and shown by antibody labeling of rabbit psoas muscle to be the protein responsible for seven or eight of the stripes.[4–6] In psoas muscle many fibers have C-protein on stripes 4–11, a few have C-protein on stripes 5–11, and a few have no C-protein.[7] F-protein has been identified as the enzyme phosphofructokinase[8]; it is not yet clear whether it is a thick-filament protein. H-protein is responsible for stripe 3 in most fibers in rabbit psoas muscle.

More recently we have shown that another protein, which we have termed X-protein, is also a contaminant of myosin preparations (unpublished results). It was not previously detected because its mobility on SDS gels is only a little less than that of C-protein. Preliminary results suggest that X-protein is not present in all fiber types, but is relatively abundant in slow red and fast red fibers, and absent in fast white fibers.[9] Its relation to "red C-protein" studied by Callaway and Bechtel[10] and Yamamoto and Moos[11] is not yet clear, but they are likely to be the same.

Outline of Preparation

Our method allows us to purify C-protein, F-protein (phosphofructokinase), H-protein, X-protein, and myosin in a single procedure (Fig. 1). Muscle mince is extracted with a high concentration of salt. The thick

[1] R. Craig, *J. Mol. Biol.* **109,** 69 (1977).
[2] R. Starr and G. Offer, *FEBS Lett.* **15,** 40 (1971).
[3] J. Trinick and S. Lowey, *J. Mol. Biol.* **113,** 343 (1977).
[4] G. Offer, *Cold Spring Harbor Symp. Quant. Biol.* **37,** 87 (1972).
[5] G. Offer, C. Moos, and R. Starr, *J. Mol. Biol.* **74,** 653 (1973).
[6] R. Craig and G. Offer, *Proc. R. Soc. Ser. B.* **192,** 451 (1976).
[7] P. Bennett, R. Starr, R. Almond, and G. Offer, unpublished experiments.
[8] J. Trinick and G. Offer, unpublished results.
[9] R. Starr, R. Almond, P. Bennett, and G. Offer, unpublished results.
[10] J. E. Callaway and P. J. Bechtel, *Fed. Proc., Fed. Am. Soc. Exp. Biol.* **39,** 2167 (1980).
[11] K. Yamamoto and C. Moos, *Biophys. J.* **33,** 237a (1981).

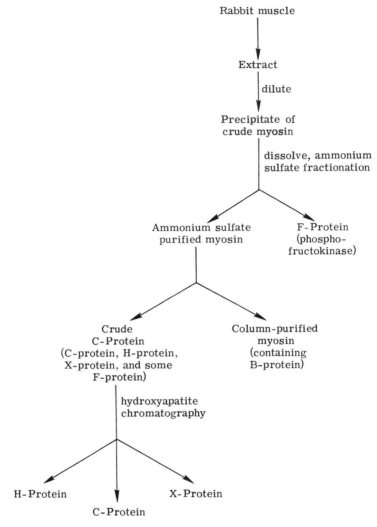

FIG. 1. Preparation of myosin, C-protein, F-protein, H-protein and X-protein.

filament is largely depolymerized under these conditions, and most of its components are extracted in molecular form along with the soluble proteins of the muscle cell. The extract is diluted, whereupon myosin precipitates in the form of aggregates of thick filaments to which the other components bind. Phosphofructokinase and traces of actin may then be removed from the preparation by ammonium sulfate fractionation. Myosin (still contaminated with B-protein) is separated from the other components by chromatography on DEAE–Sephadex. C-protein, H-protein, and X-protein may then be separated on a hydroxyapatite column. The advantage of this form of ion-exchange chromatography is that it is possible to maintain a high ionic strength (provided by KCl) throughout the fractionation and thus to prevent the association of the proteins that occurs at low ionic strength.

Preparation of Crude Myosin

The first part of the procedure is based on that of Perry.[12] A rabbit (New Zealand white and either male or female) is killed by intravenous injection of 3 ml of Expiral (sodium pentobarbitone; Abbot Laboratories Ltd., Queenborough, Kent ME11 SEL) previously warmed to blood temperature. The jugular veins are then cut, and the rabbit is bled and skinned. The back and leg muscles are rapidly excised, placed in six thin-walled polythene bags, and cooled in ice for 30 min. The muscles, after dissection from gross contamination with connective tissue and fat, are then minced in a prechilled electric mincer fitted with a freshly sharpened blade. The minced muscle is weighed and then extracted with 3 volumes (v/w) of Guba–Straub extracting solution (0.3 M KCl, 0.05 M K_2HPO_4, 0.10 M KH_2PO_4), stirring slowly but continuously for 15 min. (Typically 400 g of mince are obtained from a single rabbit; this will eventually yield about 100 mg of C-protein and 10 mg each of H-protein and X-protein.) The resultant slurry is then centrifuged in two polythene 1-liter bottles at 2500 g for 10 min, and the supernatant is carefully decanted. (The extracted mince can be used to make acetone powders for actin preparations if required.) The supernatant is freed from fat particles by filtering under very gentle suction through a cold Büchner funnel 12 cm wide containing a bed of filter paper pulp. This pulp is prepared by blending Whatman No. 1 filter paper (15 sheets, 15 cm in diameter) in about 500 ml of water, pouring it under gentle suction, and finally washing the bed with Guba–Straub solution.

The clarified extract is then put into the bottom of a 25-liter calibrated precipitating jar surrounded by ice. While this solution is being stirred slowly but continuously with a glass paddle driven by a stirring motor, a total of 14 volumes of glass-distilled water at 4° are slowly run in down the side of the precipitating jar over a period of 20 min. After the addition of about 2–3 volumes, a silky sheen develops, and eventually all the myosin precipitates. The contents of the precipitating jar are then allowed to settle for 40 min.

As much as possible of the supernatant is removed by aspiration

---

[12] S. V. Perry, this series, Vol 2, p. 582.

through a wide glass tube. The settled precipitate is transferred to two 1-liter preweighed polythene bottles and centrifuged at 2500 g. The supernatants are discarded, and the same centrifuge bottles are re-used until all the precipitate has been collected. The total mass of the pellet (~200 g) is then obtained by reweighing the bottles. Assuming a pellet density of 1 g/ml, calculated volumes of (a) 2 M KCl, (b) 0.325 M K$_2$HPO$_4$, 0.175 M KH$_2$PO$_4$, (c) 100 mM EDTA, and (d) distilled water are stirred into the pellet to give a solution containing at least 10 mg/ml in 0.5 M KCl, 32.5 mM K$_2$HPO$_4$, 17.5 mM KH$_2$PO$_4$, 1 mM EDTA.[13] The pH of this solution should be 7.0 measured at 5°. (The distilled water should be added after the protein has dissolved). At this point a 1-ml sample is withdrawn and diluted 50-fold with the above buffer, and the concentration of the crude myosin is obtained by spectrophotometry ($E_{280}^{1\,\mathrm{mg/ml}} = 0.533$). Adjustment to a final concentration of 10 mg/ml is then made. The yield of crude myosin at this stage should be about 6 g from about 400 g of muscle mince.

Removal of Actin and Phosphofructokinase

An equal volume of ice-cold 2.80 M ammonium sulfate in 0.5 M KCl, 32.5 mM K$_2$HPO$_4$, 17.5 mM KH$_2$PO$_4$, 1 mM EDTA held in a separating funnel is slowly added with stirring to the crude myosin solution surrounded by ice. The slightly cloudy solution is left for 15 min and then centrifuged at 23,000 g for 25 min. The very small pellet contains most of the actin in the preparation and is discarded; 37.4 g of solid, finely divided ammonium sulfate is then added slowly with stirring to each liter of the supernatant. A heavy precipitate of myosin, C-protein, H-protein, and X-protein results, leaving phosphofructokinase in solution.

If the phosphofructokinase is wanted, it may be obtained by precipitation at an ammonium sulfate concentration between 1.85 and 2.28 M. It is worth noting that this gives a substantially pure preparation of phosphofructokinase within a few hours, and this method is, therefore, much more convenient than previously published procedures.

The precipitate of myosin, C-protein, H-protein, and X-protein is collected by centrifugation at 2500 g for 15 min. In order to free the precipitate of as much entrapped phosphofructokinase as possible, the precipitate is twice washed by thorough dispersion with the maximum convenient volume of 1.73 M ammonium sulfate in 0.5 M KCl, 32.5 mM K$_2$HPO$_4$, 17.5 mM KH$_2$PO$_4$, 1 mM EDTA, and recentrifuge. The washed

---

[13] All solutions containing EDTA mentioned in this paper are prepared by dilution from 100 mM EDTA neutralized to pH 7.0. Otherwise pH values are obtained by composition rather than by adjustment.

pellet is then dissolved in 135 mM K$_2$HPO$_4$, 15.3 mM KH$_2$PO$_4$, 10 mM EDTA. The pH of the solution should be 7.6 at 5°. The total volume (~400 ml) of the solution should be such that the concentration is about 15 mg/ml. The solution, which contains myosin, C-protein, H-protein, X-protein, and some residual phosphofructokinase, is then freed of ammonium sulfate by overnight dialysis on ice against a large volume (10 liters for 400 ml of solution) of the phosphate–EDTA buffer; this stage should be reached on the same day that the rabbit is killed, after about 9 hr.

Next morning the dialysis medium is changed and the dialysis is continued for another 7 hr. It is important for the ion-exchange chromatography that the preparation be nearly free of ammonium sulfate, but it is also important to work as quickly as possible to avoid proteolytic degradation. After dialysis the concentration of the ammonium sulfate-purified myosin preparation is determined by spectrophotometry and adjusted to 10 mg/ml with the phosphate–EDTA buffer.

Separation of Myosin from C-Protein, H-Protein, and X-Protein

The method is based on that of Richards et al.[14] modified by Godfrey and Harrington.[15] The scale of the purification has been increased, however, so that as much as 7 g of protein may be chromatographed.

The column is prepared as follows: 50 g of dry DEAE-Sephadex A-50 (chloride form) (Pharmacia) is allowed to swell in 1 liter of the phosphate–EDTA buffer. The swollen gel is then equilibrated with many changes of this buffer until the supernatant is chloride-free as judged by the silver nitrate test. About 20 liters are required. The DEAE–Sephadex is then in a mixture of the EDTA and phosphate forms. The equilibrated gel is poured to form a 5 × 30 cm column when packed; the column is stabilized for at least 2 hr by downward flow at an operating pressure head of 45–60 cm and kept at a temperature of 2°. Pumped flow is best avoided. Since application of the sample (~500 ml at a concentration of 10 mg/ml kept at 0°) can take about 16 hr, it is convenient to use a valve system, connected to the fraction collector and triggered after an appropriate volume has been collected, so that the flow is switched from sample to buffer.

The protein is eluted by a linear salt gradient applied immediately after the sample. This gradient is made from 1.5 liters of the phosphate–EDTA buffer and 1.5 liters of the same solution containing 0.5 M KCl. The flow rate obtained is 25–30 ml/hr, and 30-ml fractions are collected.

C-protein, H-protein, and X-protein are eluted unretarded in the first

[14] E. G. Richards, C. S. Chung, D. B. Menzel, and H. S. Olcott, *Biochemistry* **6**, 528 (1967).
[15] J. E. Godfrey and W. F. Harrington *Biochemistry* **9**, 894 (1970).

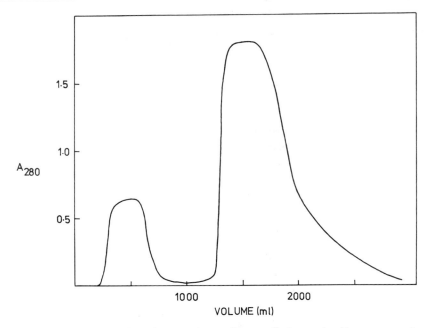

FIG. 2. Chromatography of ammonium sulfate-purified myosin. Four grams of ammonium sulfate-purified myosin were applied to a 5 × 30 cm column of DEAE-Sephadex A-50 and eluted as explained in the text.

peak (Fig. 2) and should immediately be further processed. C-protein is the main component, and collectively we term this crude C-protein. This peak accounts for about 14% of the eluted material absorbing at 280 nm (about 8% by mass).

Myosin is eluted as the large second peak, which trails, particularly at higher loadings. If required, the myosin can be pooled and precipitated by a 14-fold dilution with distilled water after the pH has been reduced to 6.5 with acetic acid. The precipitate may be harvested by centrifugation (2500 *g*) and redissolved by addition of 2 *M* KCl. It is still contaminated with B-protein but is substantially free of adenylate deaminase and adenylate kinase (which are eluted unretarded together with C-protein) and also of nucleoprotein (which remains bound to the column).

Separation of C-Protein, H-Protein, and X-Protein

The mixture of C-protein, H-protein, and X-protein (collectively termed crude C-protein) resulting from the DEAE-Sephadex chromatography is concentrated by precipitation with ammonium sulfate (390 g added to each liter of solution). The resulting precipitate (~150 mg of protein) is dissolved to give ~15 ml of a solution containing $0.3 M$ KCl, 4.8 mM K_2HPO_4, 5.2 mM KH_2PO_4. This has a pH of 7.0 at 5°. The solution is freed of ammonium sulfate by dialysis against two changes of 1 liter of this buffer, and the concentration is adjusted to be ~10 mg/ml. The protein concentration can be determined by spectrophotometry using $E_{280}^{1 \text{ mg/ml}} = 1.09$ for the mixture of proteins.

The hydroxyapatite used should be made by the method of Spencer[16]; most commercial preparations have been unsatisfactory. It is equilibrated with $0.3 M$ KCl, 4.8 mM K_2HPO_4; 5.2 mM KH_2PO_4 by washing with several changes of buffer. It is then poured to form a 2.5 cm × 30 cm column, and buffer is passed to stabilize the packing.

The mixture of C-protein, H-protein, and X-protein is applied at a concentration of 8–12 mg/ml. Subsequently a gradient of phosphate in 0.3 M KCl is applied. This is achieved by using 2 liters of $0.3 M$ KCl, 4.8 mM K_2HPO_4, 5.2 mM KH_2PO_4 in a cylindrical vessel 12.6 cm in diameter containing a stirrer connected at the bottom to a 1-liter cylindrical vessel (diameter 8.7 cm) containing about 900 ml of 0.3 M KCl, 340 mM K_2HPO_4, 160 mM KH_2PO_4 (pH 7.0 measured at 5°). The height of the liquid levels is adjusted so that initially there is no difference in hydrostatic pressures between the two vessels; the height of the liquid column containing the high phosphate concentration needs to be 0.945 times that of the low phosphate. The gradient that results is nearly linear over the region where the proteins of interest are eluted but rises rapidly thereafter; this ensures that all bound protein is eluted prior to re-use of the column. 12-ml fractions of eluate are collected at a flow rate of about 30 ml/hr.

A typical elution profile is shown in Fig. 3. The first (small) peak eluted is discarded. The second peak is H-protein (apparent chain weight 74,000) in a substantially pure form. The large third peak is pure C-protein (chain weight 140,000) again in a high state of purity. Sometimes there is a suggestion of a double peak; this may possibly be due to the existence of phosphorylated and nonphosphorylated forms.[17]

X-protein is eluted as a small broad peak immediately after the C-protein peak. The X-protein (apparent chain weight 152,000) may be shown to run just behind C-protein on SDS gels provided the loads are low and the bands are very sharp. By immunological criteria the X-protein preparation is slightly contaminated by C-protein, although the two proteins are quite distinct species.

[16] M. Spencer, *J. Chromatogr.* **166**, 435 (1978).
[17] S. A. Jeacocke and P. J. England, *FEBS Lett.* **122**, 129 (1980).

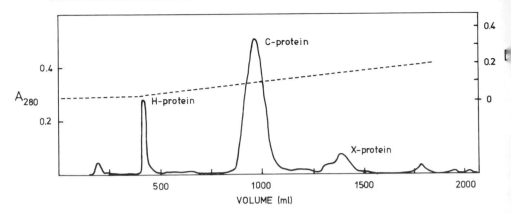

FIG. 3. Chromatography of crude C-protein. Twelve milliliters of crude C-protein (9 mg/ml) were applied to a 2.5 × 30 cm column of hydroxyapatite. Protein was eluted with a phosphate gradient as explained in the text, and 13-ml fractions were collected.

Finally residual phosphofructokinase and myosin are eluted from the column. C-protein, H-protein, and X-protein are all susceptible to proteolytic degradation. As a precaution, all preparative operations including dialyses are conducted as close to 0° as possible, and chromatography is performed on jacketed columns at 2°.

Function of C-Protein, H-Protein, and X-Protein

The function of these proteins has yet to be discovered. The four obvious possibilities are that (*a*) they play some role in the regulation of the interaction between actin and myosin; (*b*) they have some other structural function, for example, in holding the thick filament together or in maintaining the correct distance between neighboring thick filaments; (*c*) they play some role in the process of assembly of the thick filament; (*d*) they have an as yet unidentified enzymic activity.

Since their function is unknown, no assay is currently available although no doubt an immunoassay could be developed. Therefore, their identity is best checked by SDS–polyacrylamide electrophoresis.

[15] Preparation and Properties of M-Line Proteins

By HANS M. EPPENBERGER and EMANUEL STREHLER

Neither the protein composition nor the detailed function of the M-line, the electron-dense structure running transversely through the middle of the H-zone of the sarcomere of cross-striated muscle, is yet fully known. So far only two proteins—the M-type isoprotein of creatine kinase (MM-CK) and the M_r = 165,000 protein now called myomesin—have been identified as true M-line proteins.[1-3] MM-CK has been shown by immunoelectron microscopy to contribute to the electron-dense material forming the visible M-line in chicken skeletal muscle,[3-5] but no MM-CK could be demonstrated in myofibrils without a visible M-line.[6] Myomesin, on the other hand, is present in myofibrils (e.g., from chicken heart) regardless of whether or not a visible M-line is present.[3,7,8] Both M-line proteins have been identified in and purified from species other than chicken.[9,10] The M-line-associated part of creatine kinase represents approximately 5% of the total enzyme present in adult skeletal muscle, whereas myomesin makes up about 0.3% of the muscle proteins extractable under high ionic strength conditions. Both M-line proteins have been characterized biochemically.[1,2,11] Using monospecific antibody the detection of the two antigens is easily performed either by the indirect immunofluorescence technique[5] or by direct visualization of specific IgG decoration in the electron microscope.[3,5-8] Whereas MM-CK is also found in the sarcoplasm, myomesin is detectable exclusively within the M-line

[1] D. C. Turner, T. Wallimann, and H. M. Eppenberger, *Proc. Natl. Acad. Sci. U.S.A.* **70**, 702 (1973).

[2] J. Trinick and S. Lowey, *J. Mol. Biol.* **113**, 343 (1977).

[3] E. E. Strehler, G. Pelloni, C. W. Heizmann, and H. M. Eppenberger, *J. Cell Biol.* **86**, 775 (1980).

[4] T. Wallimann, G. Pelloni, D. C. Turner, and H. M. Eppenberger, *Proc. Natl. Acad. Sci. U.S.A.* **75**, 4296 (1978).

[5] T. Wallimann, D. C. Turner, and H. M. Eppenberger, *J. Cell Biol.* **75**, 297 (1977).

[6] T. Wallimann, H. J. Kuhn, G. Pelloni, D. C. Turner, and H. M. Eppenberger, *J. Cell Biol.* **75**, 318 (1977).

[7] E. E. Strehler, G. Pelloni, C. W. Heizmann, and H. M. Eppenberger, *Exp. Cell Res.* **124**, 39 (1979).

[8] H. M. Eppenberger, J.-C. Perriard, U. B. Rosenberg, and E. E. Strehler, *J. Cell Biol.* **89**, 185 (1981).

[9] R. S. Mani and C. M. Kay, *Biochim. Biophys. Acta* **533**, 248 (1978).

[10] Z. C. Dhanarajan and B. G. Atkinson, *Can. J. Biochem.* **58**, 516 (1980).

[11] K. Morimoto and W. F. Harrington, *J. Biol. Chem.* **247**, 3052 (1972).

METHODS IN ENZYMOLOGY, VOL. 85

region of myofibrils that either have been isolated from muscle tissue or still stay within muscle cells.[3,8]

Here we will describe the purification and some properties of both M-line proteins known to date, but with the notion that more true M-line components may become known in the future.

M_r-42,000 M-Line Protein "MM-Creatine Kinase"

Purification from Chicken Breast Muscle Myofibrils

General Comments. This method is for the purification of the MM-CK specifically bound to the M-line region of chicken skeletal muscle myofibrils. The procedures for the preparation of total creatine kinase from rabbit muscle, from chicken skeletal and heart muscle as well as from chicken brain have been described.[12] The purification of the M-line MM-CK is best achieved by first preparing well-washed isolated myofibrils from which MM-CK can then be extracted. The progress of purification can be followed by assaying the specific enzyme activity as has been described.[12]

Buffers

Washing buffer: 0.1 M KCl, 1 mM EGTA, 5 mM EDTA, 1 mM dithiothreitol (DTT), pH to 7 with NaOH
Low ionic strength buffer: 1 mM DTT, 5 mM Tris-HCl, pH 7.7

Preparation and Washing of Myofibrils. All steps are carried out at 4°. Breast muscle is excised from freshly killed chicken, cleaned from connective tissue, and cut into strips. The strips are immersed in ~20 volumes of washing buffer and then torn into small fiber bundles by means of an injection needle. The fibers are blended three times for 6 sec at position 6 in a Sorvall Omnimixer (Ivan Sorvall Inc.). The resulting suspension is filtered through nylon cloth (NY 250 HD, Scrynel) and then centrifuged for 10 min at 1000 g. The pellet is resuspended in 20 volumes of washing buffer, and the myofibril suspension is washed repeatedly (up to 10 times) by centrifugation (15 min, 1500 g) and resuspension in washing buffer. More than 90% of the total, not specifically bound CK activity is removed from the myofibrillar pellet during these washing cycles (Fig. 1).

Low Ionic Strength Extraction. The last myofibrillar pellet is resuspended in 100 volumes of low ionic strength buffer and extracted overnight with constant stirring. The extracted myofibrils are removed by centrifugation for 10 min at 5000 g.

[12] This series, Vol. 17A [143].

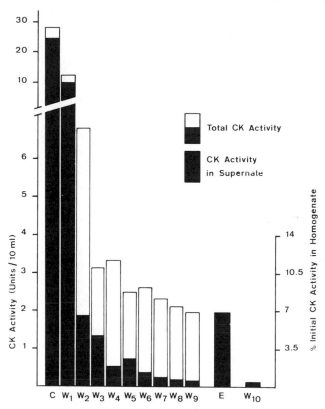

FIG. 1. Effect of washing and extraction on the binding of M-type isoprotein creatine kinase (MM-CK) to the myofibrillar pellet from adult chicken breast muscle. Total CK activity in the suspension (black plus white bars) was measured before, and CK activity released into the supernatant (black bars) was measured after, each centrifugation. C, Crude homogenate (1 : 100 w/v) in washing buffer. W1–W9, Successive washings in washing buffer; little CK activity is released after the first few cycles. E, Extraction in low ionic strength buffer; bound CK is completely released from myofibrils. W10, Final wash in washing buffer; very little CK remains bound. Reproduced from Wallimann et al.[5] with permission of The Rockefeller University Press.

Removal of Contaminating Proteins by Acid Precipitation and Absorption to DEAE–Cellulose. The pH of the supernatant (containing the myofibrillar MM-CK) is lowered to 5 by the addition of 0.2 M acetic acid to precipitate contaminating proteins. The latter are then removed by centrifugation for 15 min at 10,000 g. Swollen DEAE–cellulose (cellulose DE-52, Whatman) that has been washed several times with 0.1 N HCl and 0.1 N NaOH is then added to the supernatant (1 g of DEAE–cellulose/100 ml of supernatant), and the mixture is dialyzed for 12 hr against 1 mM

DTT, 50 mM Tris-HCl, pH 8.6. Finally, the suspension is filtered through a 0.22 μm Millipore filter.

Ammonium Sulfate Precipitation. MM-CK in the clarified solution (to which EDTA has been added to 10 mM) is precipitated by addition of ammonium sulfate to a final concentration of 70%. The resulting precipitate is pelleted for 15 min at 10,000 g, redissolved in 1 mM DTT, 50 mM Tris-HCl, pH 8.6, and dialyzed exhaustively against the same buffer. The preparation is then homogeneous as judged by gel electrophoresis under several conditions.[1,5]

Properties

The specific MM-CK activity increases sharply in the course of preparation (Table I) and approximates the value found for chicken breast muscle MM-CK isolated by a different procedure.[12]

The yield of the structurally bound MM-CK enzyme activity obtained from well washed myofibrils by this low ionic strength extraction procedure is usually between 3 and 7% of the total MM-CK activity in the muscle homogenate (Fig. 1). About 50–100 mg of myofibrillar MM-CK can be obtained from 1 kg of fresh muscle.

By all known criteria, M-line-bound myofibrillar MM-CK and the bulk sarcoplasmic MM-CK are identical in their physicochemical and immunological properties.[1,5,12]

Detection of M-Line-Derived MM-CK by Immunological Techniques

The ability of antibodies to distinguish between the MM and the BB isoenzymes of CK[5] can be used to detect specifically one or the other homodimeric enzyme by several immunological techniques.

Immune Overlay Technique. This method has been successfully applied to the identification of myofibrillar (M-line-bound) MM-CK in extracts of differently incubated chicken myofibrils.[4] To make sure that only the myofibrillar MM-CK (and no sarcoplasmic MM-CK) is assayed, well-washed myofibrils have to be used for the detection of this M-line protein on SDS–polyacrylamide gels. The procedure will be described in detail below in the section on detection of myomesin by the immune overlay technique. Antiserum against MM-CK is used in a 1 : 3 to 1 : 6 dilution in the agarose overlay gel.

Indirect Immunofluorescence Technique. The details of the procedure for the localization of MM-CK in isolated chicken myofibrils have been described.[5] For the localization of MM-CK on cryostat sections or in cultured cells the method given for the labeling in the section on myo-

TABLE I

PURIFICATION OF THE M-LINE-DERIVED CREATINE KINASE
FROM ISOLATED CHICKEN BREAST MUSCLE MYOFIBRILS

Step	Total protein (mg)	Total CK activity (EU)	CK specific activity
1st wash	9620	5500	0.58
2nd wash	1370	7490	5.47
3rd wash	233	1250	5.36
4th wash	47	234	4.98
Low ionic strength extract	318	4440	14.0
Supernatant after acid precipitation	40.2	2500	62.2
Supernatant after DEAE–cellulose absorption	9.2	590	64.1

mesin is employed, taking the appropriate dilutions of antibodies against MM-CK.

Immune Electron Microscopy. The procedure described for the ultrastructural localization of myomesin (see below) can be used by replacing the antibodies against myomesin with antibodies against MM-CK. It has to be noted, however, that incubation of unfixed muscle fibers with the smaller Fab fragments of anti-MM-CK IgG leads to a quantitative extraction of the M-line-bound MM-CK.[4] The Fab fragments of anti-MM-CK IgG therefore cannot be used for the ultrastructural localization of MM-CK in the M-line of untreated (not prefixed) myofibrils.

M_r 165,000 M-Line Protein "Myomesin"

Purification from Chicken Breast Muscle

General Comments. All steps are carried out at 0–4°. As proteolytic degradation can be a serious problem, phenylmethanesulfonyl fluoride (PMSF) and pepstatin A (Sigma) in final concentrations of 0.1 mM and $10^{-6} M$, respectively, must be added to all buffers (washing buffer, extraction buffer, buffers A–C). The following procedure is calculated for 350 g of fresh chicken breast muscle as starting material; 40–50 mg of homogeneous myomesin can be obtained from this much muscle.

Buffers

Washing buffer: $0.1 M$ KCl, 1 mM EGTA, 5 mM EDTA, 1 mM DTT; pH to 7 with NaOH

Extraction buffer: $0.6 M$ KCl, 1 mM EDTA, 1 mM MgCl$_2$, 10 mM Na$_4$P$_2$O$_7$, 0.3 mM DTT, 0.1 M potassium phosphate, pH 6.4

Buffer A: $0.6 M$ KCl, 1 mM EDTA, 0.3 mM DTT, 0.05 M potassium phosphate, pH 6.9

Buffer B: 1 mM EDTA, 0.3 mM DTT, 10 mM potassium phosphate, pH 6.9

Buffer C: 1 mM EDTA, 0.3 mM DTT, 50 mM Tris-HCl, pH 7.9

Preparation of Muscle Mince and Extraction in High Ionic Strength Buffer. Breast muscle from freshly killed chicken is excised on ice, cut into small pieces, and then minced in 700 ml of washing buffer with a meat grinder (Moulinex). The muscle mince is centrifuged for 10 min at 8000 g and resuspended in ~700 ml of washing buffer. The washing procedure is repeated until the OD$_{280}$ remains constant (7–9 washes). The final pellet is then resuspended in 1.2 liters of extraction buffer and extracted for 90 min with constant stirring. The extract is separated from the muscle residue by centrifugation for 30 min at 12,000 g.

Ammonium Sulfate Precipitation and Removal of Actomyosin. The extract (supernatant) is adjusted to pH 7.2 with NaOH. A solution of 91% (NH$_4$)$_2$SO$_4$, 1 mM EDTA, pH 7.2 is slowly added (with stirring) to give a final concentration of 32% ammonium sulfate (91% ammonium sulfate at 4°: 475 g/liter). Stirring is continued for 20 min. The precipitate is pelleted for 15 min at 20,000 g, dissolved in ~400 ml of buffer A, and dialyzed for 24 hr against 10 volumes of buffer A (two changes of buffer). Actomyosin precipitates during a second dialysis step (12 hr against 10 volumes of buffer B, two changes of buffer), while most of the myomesin remains in solution. The actomyosin precipitate is removed from the myomesin-containing solution by centrifugation for 50 min at 40,000 g. Solid ammonium sulfate (Schwarz-Mann, ultrapure) is slowly added (with stirring) to the supernatant to give a concentration of 40% (addition of 225 g of ammonium sulfate per liter of supernatant at 0°). Stirring is continued for 20 min, and the precipitate is then pelleted for 15 min at 8000 g. The pellet is redissolved in ~70 ml of buffer C and dialyzed exhaustively against 2 liters of buffer C (48 hr, three changes of buffer).

DEAE–Cellulose Chromatography. Any precipitates are removed by centrifugation (15 min, 20,000 g), and the supernatant is loaded onto a 2.5 × 35 cm column of DEAE-cellulose (DE-52, Whatman) equilibrated in buffer C. The column is washed with buffer C until the absorbance at 280 nm drops to 0 and then developed at a flow rate of 50 ml/hr with a linear gradient of 0 to 0.3 M NaCl in buffer C. Fractions of ~9 ml are collected.

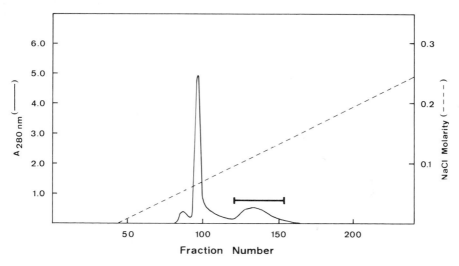

FIG. 2. DEAE–cellulose chromatography of the myomesin preparation after the second ammonium sulfate concentration. Protein, 380 mg dissolved in buffer C, was loaded on a 2.5 × 35 cm cellulose DE-52 column equilibrated in buffer C. Proteins were eluted with a linear gradient of 2 × 1 liter of 0.0 to 0.3 M NaCl in buffer C. The fraction size is 9 ml. Fractions containing myomesin are indicated by the bar.

An elution profile is shown in Fig. 2. Myomesin is eluted at 0.1–0.15 M NaCl together with phosphorylase b.

Separation of Myomesin and Phosphorylase b by Affinity Chromatography on Adenosine 5'-Monophosphate (5'-AMP)-Sepharose 4B. Fractions containing myomesin are pooled, and the proteins are concentrated by vacuum dialysis. About 25 mg of the protein mixture is loaded on a 0.5 × 6 cm column of 5'-AMP-Sepharose 4B (Pharmacia) equilibrated in buffer C. Homogeneous myomesin (as judged from SDS–polyacrylamide gel electrophoresis) is eluted in the breakthrough fraction whereas phosphorylase b is retained by the column owing to its affinity for AMP. The purification protocol is given in Table II.

Properties

Myomesin can be stored in buffer C at −20° for months without considerable degradation; it is soluble both at high and low ionic strength. It is important to note that incorporation of DTT into low ionic strength buffers prevents myomesin aggregation.[9] The molecular weight has been reported to be between 165,000 and 170,000 for all known myomesins. A sedimentation coefficient of 5.1 and an $A_{280 nm}^{1\%}$ value of 12.2 have been determined for the chicken skeletal muscle protein.[2]

TABLE II
PURIFICATION OF MYOMESIN

Step	Volume (ml)	Protein (mg/ml)	Total protein (mg)	Yield (%)
High ionic strength extract	1000	14	14,000	100
After 32% ammonium-sulfate precipitation; proteins dissolved in buffer A	400	30	12,000	85.7
After 40% ammonium-sulfate precipitation; proteins dissolved in buffer C (before DEAE–cellulose chromatography)	75	5	375	2.7
After DEAE–cellulose chromatography (pooled fractions containing mainly myomesin and phosphorylase *b*)	230	0.4	92	0.7
After 5′-AMP-Sepharose 4B chromatography	176	0.25	44	0.3

Detection of M-Line-Derived Myomesin by Immunological Techniques

Detection of Myomesin by the Immune Overlay Technique. Specific antibodies against myomesin[2,3,7] are employed for the detection of myomesin on SDS–polyacrylamide gels[13] by the immune replica technique.[14] Antiserum against myomesin is mixed (at 50–56°) 1 : 1 with a solution of 1.2% (w/v) liquefied agarose (Readyagarose Type BW, Readysystems AG). The solution is then poured over the critical part of the unfixed and unstained SDS–polyacrylamide slab gel. About 100 μl of the antiserum-agarose solution are needed for 1 cm^2 of slab gel. The diffusion of antigens into the solidified immune overlay gel is allowed to proceed for 4–12 hr at 37° in a moist chamber. After careful removal, the overlay gel is washed for about 3 days in PBS (0.15 M NaCl, 0.01 M phosphate buffer, pH 7.2). The antigen–antibody complexes are stained for 3–5 min in 0.05% (w/v) Coomassie Brilliant Blue in 50% (v/v) methanol, 10% (v/v) acetic acid.

This method has been shown to be very effective for identification of myomesin in a complex mixture of proteins extracted from myofibrils under different conditions.[3,7]

[13] U. K. Laemmli, *Nature* (*London*) **227,** 680 (1970).
[14] M. K. Showe, E. Isobe, and L. Onorato, *J. Mol. Biol.* **107,** 55 (1976).

Indirect Immunofluorescence Technique

ON MYOFIBRILS. The details of the procedure have been extensively described for the localization of CK.[5] Rabbit anti-chicken myomesin IgG is used at a concentration of 5–10 μg per milliliter of washing buffer; FITC-labeled goat anti-rabbit IgG (Cappel Laboratories, Inc.) is diluted 1 : 50 to 1 : 100 in the same buffer.

ON CRYOSTAT SECTIONS. Air-dried cryostat sections (collected on coverslips) are rinsed in PBS and then incubated (without prior fixation) for 15–30 min on a drop of an anti-myomesin IgG solution (10–20 μg of anti-myomesin IgG per milliliter of PBS). They are then washed in PBS (15 min) and incubated again for 15–30 min on a drop of a 1 : 50 dilution of FITC-labeled goat anti-rabbit IgG in PBS. After washing in PBS (30 min) the coverslips are mounted on slides with a drop of glycerol and observed under the epifluorescence microscope.

ON CULTURED CELLS. Cells grown on coverslips are rinsed in PBS (5 min), fixed for 30 min in 3% (w/v) paraformaldehyde in PBS, 0.1 mM CaCl$_2$, 1 mM MgCl$_2$, washed in PBS (5 min), and then incubated for 30 min in 0.1 M glycine in PBS (quenching of free aldehyde groups). The cells are permeabilized for 3–5 min in 0.2% Triton X-100 in PBS, washed in PBS (10–15 min), and then incubated with anti-myomesin IgG and FITC-labeled second antibody as described above for the labeling of cryostat sections.

The indirect immunofluorescence method has been used successfully for the localization of myomesin in isolated chicken skeletal[2,3] and heart[3,7] muscle myofibrils, on cryostat sections through chicken cross-striated muscle, as well as for the detection of myomesin in developing cross-striated muscle cells in culture.[8]

Immune Electron Microscopy. The ultrastructural detection of myomesin is most conveniently achieved by the direct visualization of specifically bound antibodies as increased electron density on ultrathin sections. By this technique myomesin has been localized within the M-line region of untreated or differently extracted chicken skeletal[2,3] and heart[7] muscle myofibrils as well as of nascent myofibrils in myogenic cells in culture.[8]

ON MUSCLE FIBER BUNDLES. Fresh or glycerinated slightly stretched muscle fiber bundles are incubated either directly or after prefixation (5 min in 2.5% glutaraldehyde in washing buffer) in a solution containing anti-myomesin IgG or the smaller Fab fragments thereof in washing buffer (1 mg of IgG or Fab per milliliter of washing buffer; incubation for 24 hr at 4°). After antibody incubation the fibers are washed for 1 hr at 4° in washing buffer and then fixed for 30 min at 4° in 2.5% glutaraldehyde in washing buffer. Washing (5 min) is followed by postfixation for 5 min in 1% (w/v) osmium tetroxide in PBS. The washed fibers are then prepared for sectioning by dehydration in ethanol and acetone and embedding in Epon–Araldite.

ON CULTURED CELLS. All steps are carried out at room temperature. Cells grown in small gelatinized culture dishes (diameter 1.5 cm) are prefixed for 30 min with 3% (w/v) paraformaldehyde in PBS, 0.1 mM CaCl$_2$, 1 mM MgCl$_2$, washed with PBS (10 min), and then permeabilized for 30 min in 0.2% Triton X-100 in PBS containing additionally 5 mg of bovine serum albumin (BSA) per milliliter. Thereafter the cells are incubated for 60 min in the same solution containing 50–100 μg/ml of antimyomesin IgG or Fab. Washing (10 min) is followed by fixation in 3% glutaraldehyde in PBS (15 min), washing again (10 min), and postfixation in 1.5% (w/v) osmium tetroxide in PBS (30 min). After extensive washing in PBS (15 min, several changes of PBS) the cells are dehydrated in ethanol and embedded in Epon–Araldite.

Comments

A few additional points should be considered.

Although several protease inhibitors are used during the preparation of myomesin from chicken breast muscle, electrophoresis sometimes reveals a certain heterogeneity resulting in additional protein bands. It is, however, possible to demonstrate that at least the two most prominent faster migrating bands are most likely degradation products of myomesin: *Staphylococcus aureus* V8-protease peptide maps of these proteins are very similar, if not identical to the corresponding peptide maps of the main myomesin band (unpublished observation).

It should be noted that with the immune overlay technique the use of antiserum instead of a purified IgG fraction is recommended in order to get good results because the SDS molecules diffusing out of the polyacrylamide gel have to be absorbed by serum proteins other than the specific IgG molecules. In cases where the use of purified IgG cannot be avoided, serum albumin should be added (to a concentration of approximately 5 mg/ml) to the IgG solution before preparing the immune overlay gel.

The procedures described here have been mainly developed for chicken skeletal muscle, but M-line MM-CK as well as myomesin have been isolated from skeletal muscle of other vertebrate species,[9,10] and there is reason to believe that the list will grow. While the anti-myomesin antibody cross-reacts with an antigen in chicken heart muscle, the procedure described above for the purification of skeletal myomesin fails when used for heart myomesin. The protein from cardiac muscle seems to have an even stronger affinity for myosin than skeletal myomesin. It very probably is lost during the actomyosin precipitation step. Bovine heart extracts also show a myomesin-like protein.[15] No MM-CK, on the other hand, is found in chicken heart muscle.[12] Instead, the BB-type isoenzyme of creatine kinase is found; some of it has been localized by the immunofluorescence technique in the I-band region rather than in the M line of cardiac myofibrils,[6] which do not show a visible electron-dense M line at all.[6,16]

[15] O. S. Herasymowich, R. S. Mani, C. M. Kay, R. D. Bradley, and D. G. Scraba, *J. Mol. Biol.* **136**, 193 (1980).
[16] J. R. Sommer and A. Johnson, *Z. Zellforsch. Mikrosk. Anat.* **98**, 437 (1969).

[16] Preparation and Assay of Paramyosin

By RHEA J. C. LEVINE, MYRA J. ELFVIN, and VITALY SAWYNA

Paramyosin is a 90–100% α-helical, coiled-coil protein[1,1a] composed of two identical polypeptide chains rich in glutamic acid and having a Lys : Arg ratio ≤ 1.[2] The intact molecule contains four sulfhydryl groups (as cysteine residues), which are believed to exist in the reduced form, *in vivo*.[1a] Paramyosin has molecular dimensions of 13 nm (length) by 2 nm (diameter)[1a,3] and a molecular weight of 210,000–230,000.[2–5] The carboxyl-terminal one-third is of a less stable helical structure than the rest of the paramyosin molecule and is more easily cleaved by proteolytic digestion.[6,7]

Limited in distribution to the invertebrates, paramyosin forms the cores of the thick filaments and is covered by a cortical layer of myosin in diverse muscles of every species examined to date, from platyhelminthes through chordates.[2–5,8] The paramyosin content of different muscles varies to an extreme degree, is correlated with the length of the thick filaments, and is also related both to filament diameter and to the maximum active tension that the muscle is able to develop.[9] The smooth "catch"

[1] C. Cohen and K. C. Holmes, *J. Mol. Biol.* **6**, 423 (1963).
[1a] R. W. Cowgill, *Biochemistry* **13**, 2467 (1974).
[2] L. Winkelman, *Comp. Biochem. Physiol. B* **55B**, 391 (1976).
[3] C. Cohen, A. G. Szent-Györgyi, and J. Kendrick-Jones, *J. Mol. Biol.* **56**, 223 (1971).
[4] B. Bullard, B. Luke, and L. Winkelman, *J. Mol. Biol.* **75**, 359 (1973).
[5] M. Elfvin, R. J. C. Levine, and M. M. Dewey, *J. Cell Biol.* **71**, 261 (1976).
[6] J. F. Halsey and W. F. Harrington, *Biochemistry* **12**, 693 (1973).
[7] R. W. Cowgill, *Biochemistry* **14**, 503 (1975).
[8] R. H. Waterson, H. F. Epstein, and S. Brenner, *J. Mol. Biol.* **90**, 285 (1974).
[9] R. J. C. Levine, M. Elfvin, M. M. Dewey, and B. Walcott, *J. Cell Biol.* **71**, 273 (1976).

adductor muscles of lamellibranches and the anterior byssus retractor muscle (ABRM) of *Mytilus* have the highest paramyosin to myosin ratios, and the highly ordered striated muscles of arthropods have the lowest.[9] It has been suggested by some investigators that paramyosin may play a role in molecular interactions associated with either the initiation of, or the release from, the "catch" state, although there has been little agreement on the mechanisms involved.[10,11] However, phosphorylation and dephosphorylation of paramyosin, via a cyclic AMP-dependent protein kinase–phosphatase system, have been implicated as possible events regulating the "catch" type of contraction.[12]

Preparation of Paramyosin

Techniques for isolation of paramyosin from invertebrate muscles take advantage of the protein's resistance to denaturation by organic solvents and its unique solubility properties.[1a] For highest yields, it is useful to use muscles that have higher paramyosin : myosin ratios. Some of these tissues also contain intrinsic proteases or those from bacterial parasites, so appropriate precautions should be taken to avoid proteolysis of paramyosin (see below and this volume [17]).

For most paramyosin isolation procedures, the muscle is carefully dissected away from the rest of the animal, placed on ice, and weighed (except in the case of very small specimens, such as platyhelminthes, where the whole animal is used). This and all subsequent steps should be performed in the cold (0–4°). In a blender or tissue homogenizer, the muscle is homogenized in 3–10 volumes of 0.1 M KCl or NaCl, buffered to between pH 7.0 and 7.5 with either 40 mM Tris or 10 mM phosphate, or in acid solution (see this volume [17]), depending on the subsequent procedure to be followed. When using muscle from the clam *Mercenaria mercenaria*, 10 mM EDTA is kept in all aqueous solutions, to ensure recovery of paramyosin as the intact molecule. Also, as detailed in this volume [17], 0.5 mM DTT is included in all solutions, in order both to maintain the sulfhydryl residues in the reduced state and to prevent proteolysis. The homogenate is pelleted by centrifugation (6000–10,000 rpm), the supernatant is discarded, and the pellet is washed three times by resuspension in large volumes (5–20 X) of the homogenizing solution and recentrifugation. The pellet may be used for the first three extraction procedures outlined below.

[10] W. H. Johnson, J. S. Kahn, and A. G. Szent-Györgyi, *Science* **130**, 160 (1959).
[11] A. G. Szent-Györgyi, C. Cohen, and J. Kendrick-Jones, *J. Mol. Biol.* **56**, 239 (1971).
[12] R. K. Achazi, *Pfluegers Arch.* **379**, 197 (1979).

Acid Extraction of Paramyosin (Technique 1)

The earliest technique used to isolate paramyosin,[13] this method has since been modified by Edwards *et al.* (1977)[14] and W. H. Johnson and co-workers (personal communication) to obtain pure protein from *M. mercenaria* adductors. On the basis of solubility tests at pH 7.0 and migration on SDS–polyacrylamide gels, acid-extracted protein appears to be identical with intact or α-paramyosin obtained by Stafford and Yphantis[15] (see Edwards *et al.*[14]). This procedure is described in detail in this volume [17].

Nondenaturing Extraction of Paramyosin (Technique 2)

Paramyosin was isolated from KCl extracts of fresh muscle by mild, nondenaturing procedures by Bailey.[16] This method was modified by Ruegg[17] and subsequently employed by deVillafranca and Haines[18] and Edwards *et al.*[14] to isolate the protein from horseshoe crab skeletal muscle and *M. mercenaria* adductors, respectively. The yield of pure paramyosin tends to be fairly low using this technique, but the protein remains intact and appears to be identical to acid paramyosin and α-paramyosin. When the protein is to be isolated from clam muscle, it is essential to include a mixture of antibiotics in the media.[14]

Day 1. The washed pellet described above is suspended and allowed to stand for 24 hr in a solution of 0.6 *M* KCl, 0.1 *M* potassium phosphate buffer, pH 7.5 (plus antibiotics) to dissolve myosin and paramyosin.

Day 2. Actomyosin is then preferentially removed in several steps that allow paramyosin to remain soluble. First, the ionic strength of the suspension is slowly lowered to 0.15 and the pH to 6.3 by the addition of ion-free water and 0.1 *N* HCl, while stirring. The precipitate that forms contains actomyosin and some paramyosin, and is separated from the paramyosin remaining in the supernatant by centrifugation. Both the supernatant and the pellet are saved, and the pellet is resuspended by dispersion with a Potter homogenizer, if necessary, in 0.6 *M* KCl buffered to pH 7.5 with phosphate buffer. Actomyosin is again precipitated by slowly lowering the ionic strength to 0.25 by the addition of ion-free water, but maintaining the pH at 7.5, while stirring. The paramyosin-containing

[13] A. J. Hodge, *Proc. Natl. Acad. Sci. U.S.A.* **38**, 850 (1952).
[14] H. H. Edwards, W. H. Johnson, and J. P. Merrick, *Biochemistry* **16**, 2255 (1977).
[15] W. F. Stafford, III and D. A. Yphantis, *Biochem. Biophys. Res. Commun.* **49**, 848 (1972).
[16] K. Bailey, *Pubbl. Stn. Zool. Napoli* **29**, 96 (1956).
[17] J. C. Ruegg, *Proc. R. Soc. London Ser. B* **154**, 209 (1961).
[18] G. W. deVillafranca and V. E. Haines, *Comp. Biochem. Physiol. B* **47B**, 9 (1974).

supernatants are pooled, and the protein is purified by several cycles of recrystallization (by dialysis vs 0.1 M KCl at pH 6.0) and resuspension (in 0.6 M KCl, pH 7.5).

Ethanol Precipitation of Paramyosin (Technique 3)

Ethanol precipitation of muscle tissue was first used by Bailey[16] to preserve material for later extraction of paramyosin. This technique was then modified by Johnson et al.[10] and more recently by Stafford and Yphantis.[15] This procedure has been used to isolate paramyosin from the muscles of wide variety of invertebrates. It is suggested that 10 mM EDTA and 0.5–1 mM DTT be present throughout the preparation in order to minimize proteolytic degradation of paramyosin, which, although reported only for M. mercanaria protein, may be a more prevalent phenomenon. This procedure is relatively quickly done; but the yield obtained may be moderate to low.

Day 1. The pellet obtained from the original homogenization and washes is suspended in 0.6 M KCl, 0.04 M Tris buffer, pH 7.3, rehomogenized if necessary, and extracted for 15 min[19] to 30 min[18] on ice, while stirring, then centrifuged to remove the precipitate. Three volumes of 95% ethanol are slowly added to the supernatant while stirring, to denature the actomyosin preferentially. Both the precipitated paramyosin and denatured, precipitated actomyosin are collected by centrifugation (10,000 rpm, 0°, 30 min), then resuspended in one volume of 0.6 M KCl, 0.04 M Tris buffer, pH 7.3 (dispersed with a Potter homogenizer if necessary) and dialyzed in the cold overnight to 24 hr against 10 volumes of the solvent.

Day 2. The dialyzate is clarified by centrifugation (10,000 rpm, 0°, 10–30 min), and the residue, containing denatured actomyosin, is discarded.

When muscles with high paramyosin contents are extracted in this way, further purification is achieved by additional cycles of recrystallization (by dialysis in 0.6 M KCl, 60 mM phosphate buffer, pH 6.0) and resuspension (in 0.6 M KCl, 60 mM phosphate buffer, pH 7.3). Further purification steps, however, can be added prior to recrystallization, such as those (1–3) given by deVillafranca and Haines.[18]

1. Slow (dropwise) lowering of the pH of the clarified dialyzate to 5.8 by the addition of dilute (0.1 N) HCl, while stirring in the cold, then allowing the solution to stand, for 15 min on ice, prior to collection of the precipitated paramyosin by centrifugation (10,000 rpm, 0°, 15 min). The paramyosin is again suspended in one volume of 0.6 M KCl, 0.04 M Tris

[19] L. M. Riddiford and H. A. Scheraga, Biochemistry 1, 95 (1962).

pH 7.3 (by dispersion in a Potter homogenizer if necessary), then clarified by centrifugation (as before). Cycles of recrystallization can be started at this point (see above).

2. If there is evidence of lipid contamination, as has been found for horseshoe crab preparations,[18] the paramyosin is reprecipitated with an equal volume of 95% ethanol.

3. To concentrate the paramyosin which is, in this case, collected by filtration or centrifugation, the following steps are performed. The paramyosin is resuspended, first in 0.75 volume of 0.6 M KCl, 60 mM phosphate buffer, pH 7.3, 1 mM DTT, dispersed with the Potter homogenizer and allowed to stand on ice for 30 min prior to centrifugation (10,000 rpm, 0°, 15 min). The paramyosin in the supernatant is reprecipitated by again slowly lowering the pH with stirring, in the cold, to 5.8, and is collected by centrifugation (19,000 rpm, 0°, 30 min). The residue is redispersed a second time in 0.5 volume of 0.6 M KCl, 60 mM phosphate buffer, pH 7.3, 1 mM DTT; it is clarified at 10,000 rpm, 0° for 10 min and crystallized by dialysis for 2 hr against 60 mM phosphate buffer, pH 5.1, in the cold. The paramyosin is collected by centrifugation at 19,000 rpm, 0° for 30 min, and is finally suspended in 0.1 volume of 0.6 M KCl, 60 mM phosphate buffer, pH 7.3, 1 mM DTT, dispersed with the Potter homogenizer, and dialyzed against the solvent without DTT overnight.

Day 3. The solution is finally clarified by centrifugation at 30,000 rpm, 0°, for 15 min. The supernatant can be used directly or lyophilized for storage at $-18°$.

Extraction of Paramyosin from Acetone Muscle Powder (Technique 4)

A technique for the isolation of paramyosin from an acetone powder of horseshoe crab skeletal muscle was developed by de Villafranca and Haines[18] and modified by us for use in our laboratory. The acetone powder is prepared as follows:

The muscle is placed on ice, weighed, and homogenized with about one volume of cold 40 mM KCl, 10 mM potassium phosphate buffer, pH 7.3, and centrifuged at 10,000 rpm, 0° for 10 min. The precipitate is collected, suspended in four volumes of cold KCl-buffer (as above), and recentrifuged. This step is repeated three more times. After the final wash the precipitate is suspended by gentle stirring in four volumes of cold acetone and filtered through a Büchner funnel. This step is repeated three more times. The powder remaining after the last filtration step is retained on the filter paper and dried over Dry-Ice, under a hood. The acetone powder should be whitish in color, and is weighed and stored in well-sealed containers at or below $-18°$ until needed.

Preparation of Paramyosin from Acetone Powder

Day 1. Ten grams of acetone powder (see above) are combined with 20 volumes (200 ml) of 0.3 *M* KI, 20 m*M* Tris buffer, 0.1 m*M* DTT, pH 8.3, and allowed to stir in the cold (0–4°) for 1 hr. The mixture is centrifuged at 3500 rpm for 30 min. The supernatant is filtered through two layers of cheesecloth and saved on ice. The precipitate is reextracted in 10 volumes of the same solvent and allowed to stir, in the cold, for 30 min and recentrifuged as before. The residue is discarded, and the supernatant is filtered through two layers of cheesecloth. The supernatants are pooled, and the volume is measured. The pH of the pooled supernatants is slowly lowered to pH 5.3 by the addition of about 50 ml of cold 1 *M* KH_2PO_4 at a rate of 5 ml/min, using a burette, while stirring. A precipitate should form. Then 10 volumes of ice cold ion-free water are added at a rate of 50–75 ml/min, while stirring, to bring the concentration of KI to approximately 0.03 *M*. The mixture is centrifuged at 10,000 rpm, 0° for 20 min; the supernatant is slowly decanted and discarded, and the precipitate is collected, resuspended in about 150 ml of 0.3 *M* KI, 60 m*M* potassium phosphate buffer, pH 8.3, 0.1 m*M* DTT and allowed to stand for 15 min in the cold prior to dispersion with a Potter homogenizer to yield an homogeneous, milky suspension. This is dialyzed overnight in the cold vs 10 volumes of the same solvent.

Day 2. The dialyzate is removed from the dialysis bag and clarified by centrifugation at 19,000 rpm, 0°, for 30 min. The precipitate is discarded, and the volume of the supernatant is measured. The supernatant is brought to 20% saturation with ammonium sulfate, pH 7.3, by the very slow addition (over a 30-min period using a burette) of cold saturated (100%) ammonium sulfate brought to pH 7.3 with KOH just prior to use. The mixture is allowed to stir slowly in the cold for 1 hr, then centrifuged as above. Again the precipitate is discarded and the volume of the supernatant is measured. The latter is brought to 35% saturation with cold saturated ammonium sulfate (pH 7.3, as above) added very slowly (again, over a 30-min period using a burette). The mixture is again allowed to stir slowly in the cold for 1 hr and is centrifuged as before. The supernatant is discarded, and the precipitate, which contains paramyosin, is gently washed off by slowly introducing a small volume of 0.6 *M* KCl, 60 m*M* potassium phosphate buffer, pH 7.4, 0.1 m*M* DTT onto the surface, without agitation, and quickly decanting the liquid. This removes a small amount of the surface layer of the pellet, but ensures removal of all ammonium sulfate in solution. The washed pellet is suspended in 60 ml of the 0.6 *M* buffered KCl solution by dispersion with a Potter homogenizer. The suspension is allowed to stand in the cold for 90 min, or until dissolved, then is clarified by centrifugation at 10,000 rpm, 0° for 15 min. (The small precipitate is discarded.) The pH of the supernatant is lowered (very slowly, over 15 min) to 5.3, using dilute (0.1 N) HCl, and the mixture is allowed to stand on ice for about 15 min, during which time the paramyosin will precipitate out as about 50 ml of an almost solid gel, which is collected by brief centrifugation at 5000 rpm, 0° (the rotor is brought up to speed, then back down again). The supernatant is discarded, and the precipitate is suspended in 40 ml of 0.6 M KCl, 10 mM potassium phosphate buffer, pH 7.3, 0.1 mM DTT by dispersion with a Potter homogenizer, in the cold, and dialyzed overnight in the cold vs 1 liter of the same solution.

Day 3. The paramyosin solution is clarified by centrifugation at 19,000 rpm, 0° for 30 min, the precipitate is discarded, and the paramyosin in the supernatant is purified by a 2-hr dialysis in the cold vs 1.5 liters of 10 mM potassium phosphate buffer, pH 6.0, 0.1 mM DTT (the buffer is changed after 1 hr), followed by centrifugation (10,000 rpm, 0°, 10 min) and resuspension of the precipitate in 25 ml of 0.6 M KCl, 10 mM potassium phosphate buffer, pH 7.3, 0.1 mM DTT. The suspension is redispersed by Potter homogenization and allowed to stand for 20 min in the cold. If the paramyosin is not completely dissolved, then the solution is clarified by brief centrifugation (5 min) at 5000 rpm, 0°. The cycle of recrystallization by dialysis and resolubilization is repeated, but the paramyosin is resuspended in only 10 ml of buffered KCl or NaCl. (Sodium may be substituted for potassium from here on.) It is then dialyzed overnight in the cold against 1 liter of the buffered 0.6 M KCl (or NaCl) solvent.

Day 4. The paramyosin solution is removed from the dialysis bag and clarified by ultracentrifugation (30,000 rpm at 0° for 30 min). The precipitate is discarded, and the paramyosin solution (13–14 ml of 3–4 mg of pure protein per milliliter) is collected. The volume and protein concentration are determined. The pure paramyosin can be kept at 4° for 1–2 days or (preferentially) lyophilized, weighed, and stored immediately at or below −18°. Lyophilized paramyosin is reconstituted by suspension of powder in an appropriate amount of ion-free water. The solution is left for several hours at 4°. If necessary, this can be dialyzed for 2 hr against 0.6 M KCl or NaCl, 60 mM potassium or sodium phosphate buffer, pH 7.3, in the cold.

Assays for Paramyosin

In addition to amino acid analysis, there are three general methods to assay for the presence of paramyosin. These include (*a*) formation and electron microscopic examination of the paracrystalline precipitates of the protein[2–4,18]; (*b*) identification of the polypeptide chains by SDS–polyacrylamide gel electrophoresis; and, alone or in combination with these[2-5]; (c) positive staining of paramyosin chain bands on SDS–polyacrylamide gels and/or paracrystals with labeled antibody specific for paramyosin.

Paramyosin Paracrystals

Formation. Paracrystals, or two-dimensionally regular molecular aggregates of paramyosin, are most commonly formed by precipitation of the protein with the divalent cations barium or magnesium. Because the protein tends to self-aggregate in a more random fashion under a variety of subtle changes in pH and ionic strength, it is often treated with a dispersing agent, such as 1 M urea or 10–100 mM KSCN prior to precipitation.[2,3,18]

Paracrystals Formed in the Absence of Either a Dispersing Agent or Divalent Cations. Paramyosin (1–2 mg/ml) dissolved in 0.6 M KCl, 10 mM potassium phosphate buffer, pH 7.0–7.4, 0.1 mM DTT is dialyzed overnight in the cold vs 10 volumes 0.1 M KCl, 10 mM potassium phosphate buffer, pH 6.0, 0.1 mM DTT.

Paracrystals Formed Using Both a Dispersing Agent and Divalent Cations. Paramyosin (1–2 mg/ml) dissolved in 0.6 M KCl, 10 mM potassium phosphate buffer, pH 7.0–7.4, 0.1 mM DTT is first dialyzed overnight vs either 1 M urea, 50 mM Tris-HCl buffer (pH 8.0) or 10–100 mM (20 mM is a good starting concentration) KSCN, 50 mM Tris-HCl buffer (pH 8.3), then against 50 mM BaCl₂ (or MgCl₂) in 50 mM Tris-HCl buffer pH 8.3.

Examination of Paracrystals. Paracrystals are collected by touching the surface of a carbon-coated electron microscope grid to a drop of the suspension and draining the drop with tangentially applied filter paper. The grids are negatively stained with 1–2% aqueous uranyl acetate, which is likewise immediately drained, and examined and photographed in an electron microscope. Paramyosin paracrystals show a distinct periodic cross-banding at 72.5 ± 2.5 nm, with a variety of subperiodic staining patterns. For a discussion of these, see Cohen *et al.*[3] Frequently, a subperiod of 14.5 nm is also visible, or a Bear–Selby net pattern appears. More than one staining pattern can be present on a single paracrystal; the identifying feature is the 72.5 nm periodicity. (Fig. 1)

SDS-Polyacrylamide Gel Electrophoresis

Either disc or slab gel electrophoresis, using 6 or 10% polyacrylamide, respectively, in the presence of 0.1% SDS can be used to test the purity of the paramyosin preparation and the intactness of its polypeptide chains.

One to 2 ml of paramyosin (1–2 mg/ml) in 0.6 M KCl, 10 mM potassium phosphate buffer, pH 7.0–7.4, is dialyzed for 2 hr in the cold against

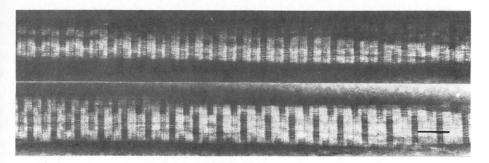

FIG. 1. Electron micrographs of paracrystals formed from *Mercenaria mercenaria* paramyosin. Note the 72.5 nm repeat periodicity. Areas that appear dark reflect regions of increased space between molecules and admit more stain. Several banding patterns, reflecting differences in molecular organization, appear on one paracrystal. Bar = 0.1 μm. Electron micrographs by courtesy of Dr. Carolyn Cohen. Reproduced, with permission, from Plate V of Cohen *et al.*[3]

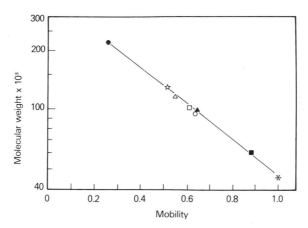

FIG. 2. Graph showing mobility of paramyosin polypeptide chains on sodium dodecyl sulfate–6% polyacrylamide disc gels as compared to the mobility of protein standards. ●, Myosin heavy chains; ☆, β-galactosidase; △, *Limulus* paramyosin (α-paramyosin); □, ▲, *Mercenaria* β- and γ-paramyosins, respectively (extracted in the absence of EDTA); ○, phosphorylase *a*; ■, catalase; *, actin. Reproduced from Fig. 1 of Elfvin *et al.*[5] with copyright permission of The Rockefeller University Press, New York.

20–50 volumes of sample buffer, for either the Weber and Osborn[20] or the Laemmli[21] techniques, then boiled in the presence of 0.1% SDS and 0.1% dithioerythritol and cooled; and 5–20 μl are introduced into the tops of

[20] K. Weber and M. Osborn, *J. Biol. Chem.* **244**, 4406 (1969).
[21] U. K. Laemmli, *Nature (London)* **227**, 680 (1970).

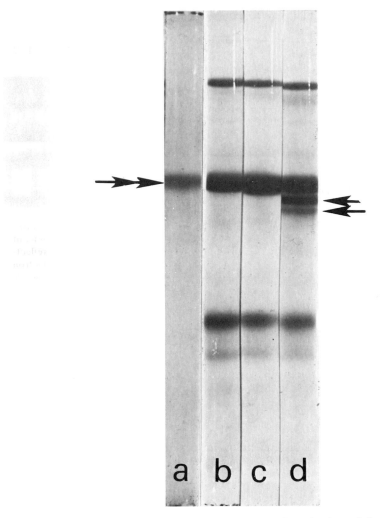

FIG. 3. Sodium dodecyl sulfate–6% polyacrylamide disc gels of muscle homogenates and purified paramyosin. Lane a, purified *Limulus* paramyosin (double arrow); b, *Mercenaria* opaque adductor homogenate; c, same as lane b with added *Limulus* paramyosin; d, same as lane b with added *Mercenaria* β- and γ-paramyosins (single arrows), isolated in the absence of EDTA. Adapted from Figs. 3 and 4 of Elfvin *et al.*,[5] with copyright permission of The Rockefeller University Press, New York.

precast gels in tubes or into sample wells in precast gel slabs. It is important to eliminate any residual KCl from the sample, since potassium forms a precipitate with SDS. (The procedures for casting, loading, running, and staining the gels are found in references cited in footnotes 4, 5, 20, and 21.) The paramyosin should migrate as a single band with chain weight (as

FIG. 4. Immunochemical staining of paramyosin polypeptide chain band on a 50-μm frozen section of a sodium dodecyl sulfate–10% polyacrylamide slab gel of invertebrate myofibrillar homogenates. The section was incubated first with rabbit antibody specific for *Limulus* paramyosin, then with horseradish peroxidase–labeled goat anti-rabbit immunoglobulin G, and was finally stained with diaminobenzidine in the presence of hydrogen peroxide, all according to the technique of Van Ramsdonk *et al.*[22] Arrow points to paramyosin polypeptide chain band.

determined by coelectrophoresis with protein standards) of 105,000–115,000 daltons[2,4,5,15] (Fig. 2). Bands at 100,000 and 95,000 daltons represent degradation products of proteolysis, termed β- and γ-paramyosin, respectively.[15] Contamination with myosin, actin, or tropomyosin will show up as bands at 200,000–210,000 daltons, 43,000 daltons, and 35,000–38,000 daltons, respectively (Fig. 3).

Immunochemical Labeling

Identification of the 105,000–115,000 dalton band as paramyosin can be accomplished by coelectrophoresis with a small quantity of known, pure paramyosin, or, more specifically, by staining of duplicate gels with immunoperoxidase-labeled antiserum (Fig. 4) according to Van Ramsdonk *et al.*[22] This latter technique is extremely specific and takes advantage of the known cross-reactivity among the paramyosins of different invertebrate phyla: antibodies raised against paramyosin from any invertebrate species will bind to paramyosin of any other invertebrate

[22] W. Van Ramsdonk, C. W. Pool, and C. Heyting, *J. Immunol. Methods* **17**, 337 (1977).

species.[5] Similarly, a drop of paracrystals on a glass slide can be stained with anti-paramyosin by the indirect fluorescent antibody technique[5,23] and will appear under dark-field microscopy as brilliant yellow-green needles. Controls for both types of staining include use of anti-paramyosin previously absorbed with paramyosin and use of antibody raised against a different protein.[5,23]

Acknowledgments

We thank H. King and B. Gilfillan for technical assistance and Phyllis Fliegelman for secretarial help. The work described in this chapter was supported by USPHS Grants GM21956 and HL15835 to the Pennsylvania Muscle Institute.

[23] R. J. C. Levine, M. M. Dewey, and G. deVillafranca, *J. Cell Biol.* **55**, 221 (1972).

[17] Preparation Problems Unique to *Mercenaria* Paramyosin

By SONJA KRAUSE and NANCY LETKO MUNSON

As mentioned several times in this volume [16], paramyosin prepared from *Mercenaria mercenaria* is easily degraded during the extraction procedure, so that great care must be taken to obtain the native protein. Although technique 3 from this volume [16] can be used to prepare the native molecule, α-paramyosin,[1] yields are very small. If technique 3 is nevertheless used, it is suggested that not only 10 mM EDTA, but also 0.5 mM dithiothreitol (DTT), be added to all the solutions used during the extraction. The DTT serves not only to keep the sulfhydryl residues in the reduced state as they are presumed to be in the native molecule,[1] but also to prevent degradation. As a matter of fact, evidence exists that the addition of 0.5 mM DTT to the solutions without any EDTA at all prevents the degradation of paramyosin all by itself.[2]

Increased yields of native paramyosin can be obtained by suitable modification of the acid extraction procedure of Hodge,[3] referred to as technique 1 in this volume [16]. In this case, the resulting paramyosin has been called acid-paramyosin, even though it, too, is assumed to be the native molecule. α-Paramyosin and acid-paramyosin have the same mo-

[1] W. F. Stafford and D. A. Yphantis, *Biochem. Biophys. Res. Commun.* **49**, 848 (1972).
[2] B. D. Gaylinn, Ph.D. Dissertation, Rensselaer Polytechnic Institute, Troy, New York, 1980.
[3] A. J. Hodge, *Proc. Natl. Acad. Sci. U.S.A.* **38**, 850 (1952).

METHODS IN ENZYMOLOGY, VOL. 85

lecular weight, using SDS-gel electrophoresis, and similar though not quite identical physical properties.[4] These slight differences in physical properties such as ultraviolet circular dichroism spectra and solubility[4] are probably due to differences in ion adsorption of paramyosin when it is prepared using different methods and solutions.[5] For most purposes, however, α-paramyosin and acid-paramyosin are identical.

The method of extraction detailed below is essentially the method of Hodge[3] as modified by Edwards *et al.*,[6] but with 0.5 mM DTT included in all solutions to ensure that all sulfhydryl groups in the protein are in the reduced state.[7] Note that, in contrast to the procedures used in this volume [16], the muscles (usually the adductor muscles) of the clams are first homogenized in an acid (pH 3.4) solution instead of in an essentially neutral solution.

Solutions (all made up using distilled, deionized water)

Solution A: Aqueous solution (M/6) in citric acid and 0.5 mM in DTT adjusted to pH 3.4 using concentrated KOH.

Solution B: Aqueous solution, (M/6) in citric acid, 0.5 mM in DTT, and 0.1 M in KCl, adjusted to pH 3.4 using concentrated KOH.

Solution C: Aqueous solution, 0.5 mM in DTT, containing 1 ml of glacial acetic acid per 250 ml of solution.

Solution D: Low ionic strength buffer. Aqueous solution, 1 mM in K_2HPO_4, 9 mM in KH_2PO_4, 10 mM in K_2EDTA, and 0.5 mM in DTT, adjusted to pH 6.0 by addition of concentrated KOH or HCl. This solution has contained 3 mM sodium azide in the past, but this appears to have no major effect on the paramyosin preparation.

Solution E: High ionic strength buffer. Aqueous solution, 0.6 M in KCl, 9 mM in K_2HPO_4, 1 mM in KH_2PO_4, 10 mM in K_2EDTA, and 0.5 mM in DTT. The pH should be between 7.0 and 7.5. This solution has contained 3 mM sodium azide in the past, but this appears to have no major effect on the paramyosin preparation.

Preparation of Dialysis Tubing

The dialysis tubing is boiled twice in distilled deionized water; during this procedure, the tubing should rise to the surface of the container. The tubing is then boiled twice in 10 mM K_2EDTA, after which it is again boiled twice in distilled deionized water.

[4] L. B. Cooley, Ph.D. Dissertation, Rensselaer Polytechnic Institute Troy, New York, 1978.

[5] N. Letko Munson, work in our laboratory (1980).

[6] H. H. Edwards, W. H. Johnson, and J. P. Merrick, *Biochemistry* 16, 2255 (1977).

[7] W. F. Stafford, Ph.D. Dissertation, University of Connecticut, Storrs, 1973.

Temperature

All extraction procedures including centrifugation are done at 4°.

Special Notes

1. Seemingly minor changes in the time used for a resuspension of centrifuged pellet or for a dialysis can affect some physical properties of the resulting paramyosin. Addition of sodium azide to solution D and E can have the same effect. Physical properties that appear to be affected included solubility,[5] electro-optic properties,[5] and, quite probably, titration curves.[8] Although evidence exists that paramyosin may be covalently phosphorylated,[9,10] there is also much evidence that the molecule can adsorb considerable phosphate[5,11] and that changes in the number of adsorbed phosphates can change the solubility of paramyosin.[5,11]

2. Centrifugation speeds given are for a Sorvall Superspeed RC 2-B preparative ultracentrifuge. For day 1, the speeds are given for a GSA rotor that holds 250-ml bottles, while for subsequent days, the speeds are given for a SS-34 rotor that holds 50-ml centrifuge tubes.

Day 1

The muscles are dissected away from the rest of the animals and weighed (a dozen clams yield 60–85 g of adductor muscle, both transparent and opaque). The muscles are then placed in a blender with 6 ml of solution A per gram of muscle, alternating between "blend" and "chop" for 1 min; the blend has a milky light pink color. After centrifugation for 20 min at 6000 rpm, the supernatant is discarded and the pellet is blended as before, for 1 min, with 6 ml of solution B per gram of muscle. After stirring for about 3 hr, the blend is centrifuged as before, and the supernatant is again discarded. The pellet is now a light buff color. After blending and chopping the pellet with 4 ml of solution C per gram of muscle for a total of 10 sec, the solution is stirred overnight.

Day 2

The now quite viscous solution is centrifuged at 18,000 rpm for 20 min, and the supernatant is poured into a dialysis tube. The supernatant is dialyzed against 5 times its volume of solution D for 2 hr and then against

[8] L. B. Cooley and S. Krause, *Biophys. J.* **32**, 755 (1980).
[9] R. K. Achazi, *Eur. J. Physiol.* **379**, 197 (1979).
[10] L. Radlick and W. H. Johnson, private communication.
[11] L. B. Cooley, W. H. Johnson, and S. Krause, *J. Biol. Chem.* **254**, 2195 (1979).

the same volume of fresh solution D for another 2 hr. The solution inside the dialysis tubing will have a gel-like texture, and this "gel" is then centrifuged at 12,000 rpm for 20 min, and the supernatant is discarded if possible (a distinct pellet does not always form).

(This is also the return point from day 3 if necessary.) From a total volume of solution E equal to that of *both* batches of solution D used above, about 3 ml per gram of muscle are added to the pellets and non-separated material. The pellets and gel are disrupted manually (a plastic spatula may be used) as much as possible, and the suspension is stirred for 1 hr, after which the pH is adjusted using concentrated KOH to the same pH as solution E. After stirring for another hour, the pH is adjusted again if necessary. The suspension is then dialyzed against the remainder of solution E overnight.

Day 3

The solution in the dialysis tubes, much less gel-like than on day 2 and containing some precipitate, is centrifuged at 18,000 rpm for 20 min, and the supernatant is dialyzed against five times its volume of solution D. The precipitate is saved, at least until the supernatant has dialyzed for about half an hour. If a large amount of fluffy precipitate appears in the dialysis tubes within half an hour, the old precipitate may be discarded. Whether a large amount of fluffy precipitate appears in the dialysis tubes or not, the dialysis is continued and the dialysis tubes are squeezed every half hour to keep the precipitate from adhering to the tubing. After a total of 2 hr, the dialysis solution is replaced by an equal volume of fresh solution D. Two hours later, the suspension is centrifuged at 12,000 rpm for 20 min. If very little precipitate appeared in the dialysis tubes, it is combined with the precipitate saved from the previous centrifugation and the combined precipitates are broken up into a small volume of solution E (see day 2—the return point from day 3), and the procedures involving solution E on day 2 must be repeated, including the overnight dialysis. In this case, an appreciable amount of paramyosin was trapped in the earlier precipitate and a second attempt must be made to extract it.

When a return to the procedures of day 2 becomes unnecessary, the white precipitate is manually dispersed into 70–80% of the volume of solution E that was used to disperse the pellets on day 2 and the suspension is stirred for 2 hr. After adjusting the pH of the suspension to that of solution E using concentrated KOH, the solution is dialyzed overnight against approximately 10 times the volume of solution E in which the precipitate was dispersed.

Day 4

There may be a small amount of flocculent precipitate in the dialysis tubes. After centrifugation at 18,000 rpm for 20 min, the precipitate is discarded and the supernatant is dialyzed for 2 hr against 5 times its volume of solution D; and then for 2 hr more against the same volume of fresh solution D. The resulting milky suspension is centrifuged at 12,000 rpm for 20 min, and the supernatant is discarded. The precipitate is manually dispersed into 10% less solution E than on day 3, and the suspension is stirred for 2 hr. After adjusting the pH of the suspension to that of solution E using concentrated KOH, the suspension is dialyzed overnight against 10 times the volume of solution E used to disperse the precipitate.

Day 5

The material inside the dialysis bag should be a slightly cloudy solution that is centrifuged at 18,000 rpm for 20 min. The precipitate is discarded. The optical density of the clear solution is measured at 280 nm and at 260 nm. If $OD_{280} : OD_{260} < 2.0$, the procedure of day 4 is repeated. In a typical preparation in which the procedure of day 4 does not need to be repeated, 12 large clams yield 84 g of adductor muscle, which yield about 900 mg of paramyosin.

[18] Purification of Muscle Actin

By Joel D. Pardee and James A. Spudich

The opportunity to study the molecular events responsible for muscle contraction and cell motility was made possible in the early 1940s by Banga and Szent-Györgyi[1] and by Straub,[2] who discovered myosin and actin in the extracts of rabbit skeletal muscle. Actin was first isolated when Straub separated the viscous protein from an actomyosin preparation.[2] Subsequent work[3] revealed that actin could be obtained in a nonviscous state (G-actin) by extracting muscle with a low ionic strength buffer, and that addition of salt induced conversion to a viscous form called F-actin (Fig. 1). An improved procedure by Straub and his col-

---

[1] I. Banga and A. Szent-Györgyi, *Studies from the Inst. Med. Chem., Univ. Szeged* **1**, 5 (1941).
[2] F. B. Straub, *Studies from the Inst. Med. Chem., Univ. Szeged* **2**, 3 (1942).
[3] F. B. Straub, *Studies from the Inst. Med. Chem., Univ. Szeged* **3**, 23 (1943).

METHODS IN ENZYMOLOGY, VOL. 85

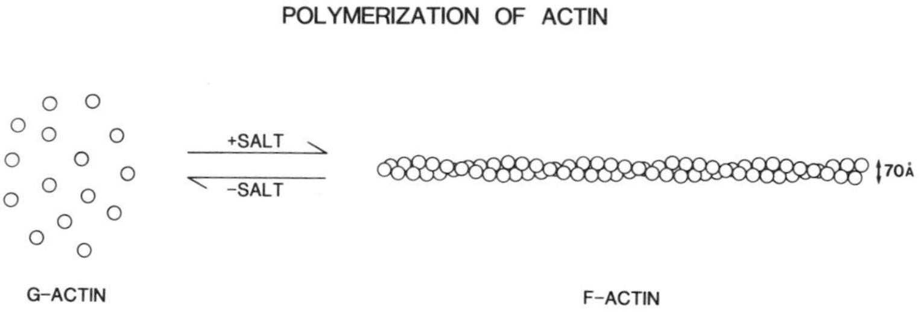

POLYMERIZATION OF ACTIN

FIG. 1. Assembly of muscle actin. G-actin containing one ATP and one tightly bound divalent cation per monomer assembles in the presence of salt into filaments of F-actin 70 Å in diameter.

leagues then incorporated a step that denatured muscle proteins not stabilized in an actomyosin complex; minced muscle was dehydrated with acetone before the actin extraction step.[3,4] After the discovery that bound ATP was important for maintaining the functional integrity of actin,[5-7] the stability of isolated actin was enhanced by inclusion of ATP in the extraction buffers, but the purification protocol of Feuer et al.[4] has remained essentially intact. With the advent of polyacrylamide gel electrophoresis as a highly resolving analytical tool for ascertaining protein purity, it became evident that muscle actin isolated by this classical procedure contained significant amounts of actomyosin-associated muscle proteins, such as tropomyosin[8] and α-actinin.[9] These contaminants promote the gelation of F-actin and greatly affect the physical properties of filaments in solution.[10] Drabikowski and Gergely[11] showed that an enhanced purification could be achieved by extraction of the actin from muscle acetone powder at 0°, and in 1971 Spudich and Watt[12] devised a modification of this method designed to eliminate tropomyosin from muscle actin preparations. Their purification resulted in a single band on sodium dodecyl sulfate (SDS) polyacrylamide gels and has met with widespread use as a general method for obtaining muscle actin.

A subtle problem in establishing methods for actin purification resides in the level of purity acceptable for the investigations at hand. Emerging

[4] G. Feuer, F. Molnar, E. Pettko, and F. B. Straub, Hung. Acta Physiol. 1, 150 (1948).
[5] K. Laki, W. J. Bowen, and A. Clark, J. Gen. Physiol. 33, 437 (1950).
[6] W. F. H. M. Mommaerts, J. Biol. Chem. 188, 559 (1951).
[7] F. B. Straub and G. Feuer, Biochim. Biophys. Acta 4, 455 (1950).
[8] K. Laki, K. Maruyama, and D. R. Kominz, Arch. Biochem. Biophys. 98, 323 (1962).
[9] S. Ebashi and F. Ebashi, J. Biochem. (Tokyo) 58, 1 (1965).
[10] K. Maruyama, M. Kaibara, and E. Fukada, Biochim. Biophys. Acta 371, 20 (1974).
[11] W. Drabikowski and J. Gergely, J. Biol. Chem. 237 3412 (1962).
[12] J. A. Spudich and S. Watt, J. Biol. Chem. 246, 4866 (1971).

experimentation in cell biology, specifically in cytoskeletal biochemistry, requires probing sensitive properties of actin itself and actin associations with other cell proteins. It is therefore the goal of this review to explore some of the pitfalls associated with actin purification and to clarify in some detail the correct use and expected result from each step of the widely used muscle actin purification procedure of Spudich and Watt.[12] Furthermore, additional steps to eliminate trace contaminants are described.

Experimental

Flow diagrams for isolating rabbit skeletal muscle actin are given in Figs. 2 and 3. The detailed procedure for the acetone powder preparation of Feuer *et al.*[4] with minor modifications is presented (Fig. 2) because of its importance in eliminating myosin and proteases from the final product and because the original reference may not be readily available. Adherence to requirements of temperature, buffer conditions, and incubation times are of prime importance for obtaining high-purity actin. Steps at 4° are preferably carried out in a cold room; cold extraction buffers and solvents are prechilled to 4° before use, and buffer pH is determined at 25° before chilling.

Preparation of Acetone Powder (Fig. 2)

1. Preparation of Muscle Mince. The preferred way to kill the rabbit is to let it hang head down by grasping the hind legs with one hand while delivering a sharp blow to the back of the neck, followed by bleeding the animal completely. Immediately after sacrifice, the dorsal lateral skeletal muscles and the hind leg muscles are excised (about 350 g), chilled on ice, washed clear of blood with distilled H_2O, and minced at 4° in a prechilled meat grinder.

2. Extraction with KCl. The mince is quickly extracted with stirring for 10 min in 1 liter of ice-cold 0.1 M KCl, 0.15 M potassium phosphate, pH 6.5. All extracts are filtered by squeezing through four layers of cheesecloth which had been previously boiled for approximately 20 min in distilled H_2O, drained, and brought to 4°.

3. Extraction with $NaHCO_3$. The filtered muscle mince is extracted with stirring for 10 min at 4° in 2 liters of prechilled 0.05 M NaHCO₃ and filtered. Longer extraction times at this stage cause appreciable extraction of actin and are to be avoided.[13]

4. Extraction with EDTA. The filtered residue is extracted with 1 liter of 1 mM EDTA, pH 7.0, by stirring for 10 min at 4°.

5. Extraction with H_2O. The next two extractions are with 2 liters of 4° distilled H_2O for 5 min with stirring.

[13] J. R. Bamburg, personal communication.

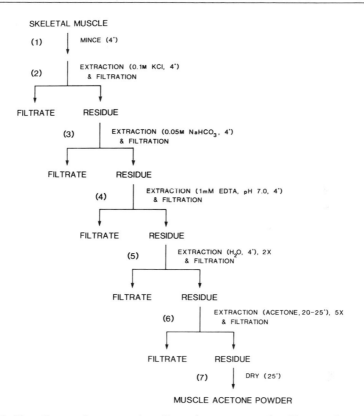

FIG. 2. Flow diagram for preparation of muscle acetone powder. For complete explanation of the protocol, refer to numbered steps in the text under Preparation of Acetone Powder.

6. *Extraction with Acetone.* The final five extractions are with 1 liter of acetone for 10 min each. All acetone extractions are performed at 20–25°. Acetone should be cooled to below 20° or the acetone-mince mixture becomes too warm. Clumps of residue are broken up by stirring during each extraction.

7. *Drying.* The filtered residue is placed in large glass evaporating dishes and air-dried overnight in a hood to obtain dried "acetone powder." The resulting acetone powder is stable for months if stored at −20°.

Isolation of Actin (Fig. 3)

Typical preparations use about 10 g of acetone powder. The minimum yield is approximately 10 mg of actin per gram of acetone powder, but can be as high as 30 mg of actin per gram of acetone powder.

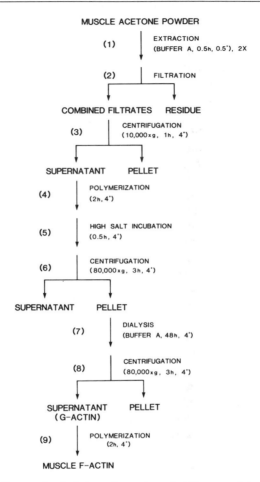

FIG. 3. Flow diagram for isolation of muscle actin. For complete explanation of the protocol, refer to numbered steps in the text under Isolation of Actin.

1. Extraction. The acetone powder is extracted at 0–0.5° for 30 min by stirring with 20 ml of buffer A per gram of acetone powder. The temperature must be kept low during the extraction.[11] Buffer A consists of 2 mM Tris-Cl, 0.2 mM Na$_2$ATP, 0.5 mM 2-mercaptoethanol, 0.2 mM CaCl$_2$, 0.005% azide, final pH = 8.0 at 25°.

Because actin is extracted under depolymerizing conditions, it is desirable to keep the concentrations of Mg^{2+}, K$^+$, and Na$^+$ in the buffer as low as possible. Measurable actin assembly occurs in 2 mM K$^+$ or Na$^+$, and in 0.2 mM Mg^{2+}.[14] Therefore, it is useful to employ reagent powders

[14] J. D. Pardee and J. A. Spudich, (manuscript in preparation).

of Tris-base, K_2ATP or Na_2ATP, and $CaCl_2$. To prepare the buffer, the reagents are dissolved in double-distilled H_2O and titrated to pH 8.0 at 25° with HCl. Add 2-mercaptoethanol after the pH determination, since 2-mercaptoethanol (and dithiothreitol) interferes with accurate pH determination by impairing the sensitivity of the pH electrode. The result is drifting pH readings.

An important caution is that actin is susceptible to proteolysis resulting from even minor bacterial contamination. Therefore, if buffers are prepared from stock solutions rather than from reagent powders, buffer stocks should be stored at 4° with 0.1% sodium azide present to prevent bacterial or mold growth.

2. *Filtration.* The extract is separated from the hydrated acetone powder by squeezing through several layers of sterile cheesecloth; latex gloves are used to avoid contamination. Filtration through a coarse sintered-glass filter under vacuum can also be used, but if sufficient care is not taken, foaming of the protein filtrate will occur, resulting in actin denaturation and reduced yields. If necessary, low-speed centrifugation at $5000-10,000\,g$ for 10–20 min readily removes the bulk of the solids. Reextract the residue by stirring for 10 min in 20 ml of buffer A per gram of acetone powder. Filter and combine extracts.

3. *Centrifugation.* The extract is centrifuged at $10,000-20,000\,g$ for 1 hr at 4°. Decant the supernatant by hand pipetting, leaving the turbid lower layer in the centrifuge tube.

4. *Polymerization.* The KCl concentration of the supernatant is brought to 50 mM, Mg^{2+} to 2 mM, and ATP to 1 mM. Inclusion of 1 mM ATP at this step ensures full polymerization. Allow to assemble for 2 hr at 4°. At this stage of isolation, a visible increase in the solution viscosity should be observed.

5. *High Salt Wash: Tropomyosin Removal.* Solid KCl is slowly added with stirring to a final concentration of 0.6 M, and the solution is stirred gently for 0.5 hr. Some investigators[15] have successfully used 0.8 M KCl at 4° in the wash step, which is useful in the event that the 0.6 M KCl treatment does not eliminate tropomyosin from the actin preparation. However, the actin monomer concentration increases with increasing salt concentration above 0.15 M[16]; thus lower yields of actin may result from washes with higher concentrations of salt.

6. *Sedimentation of Filamentous Actin.* The polymerized actin is centrifuged in 30-ml tubes at $80,000\,g$ (avg) for 3 hr at 4°. To obtain optimal purity it is advisable to remove contaminants trapped in the liquid phase of the F-actin pellet. This can be achieved by homogenizing the total pelleted F-actin into 150 ml of fresh wash buffer (buffer A + 0.6 M KCl, 2

[15] S. MacLean-Fletcher and T. D. Pollard, *Biochem. Biophys. Res. Commun.* **96**, 18 (1980).
[16] M. Kasai, *Biochim. Biophys. Acta* **180**, 399 (1969).

mM MgCl$_2$, 1 mM ATP) and resedimenting the F-actin at 80,000 g (avg) for 3 hr at 4°.[16a] After discarding the supernatant, the intact F-actin pellet is rinsed thoroughly with buffer A.

7. *Depolymerization.* The pellets of F-actin are resuspended by gentle homogenization in 3 ml of cold buffer A per gram of acetone powder originally extracted. Large actin losses can occur owing to incomplete transfer to the homogenizer. A good technique to maximize recovery of F-actin from the centrifuge tube is to allow each pellet to stand on ice in 1 ml of buffer A for 1 hr before transferring to the homogenizer. The softened pellets can then be partially homogenized with a Teflon-coated rod and transferred with a plastic disposable pipette to the homogenizer without significant losses. Dialysis at 4° against 1 liter of prechilled buffer A with one or two changes over a 3-day period gives complete depolymerization of actin, although dialysis times can be shortened considerably if vigorous stirring and dialysis bags having a large surface area are employed. One technique is to divide the homogenate into equal 6-ml aliquots (for 10 grams of acetone powder extracted) and place them into dialysis bags $\frac{1}{4}$ inch in diameter and 12 inches long. The bags are then either mounted on a rapid dialyzer or tied to a magnetic stir bar in a 1-liter graduated cylinder. Rapid rotation of the dialysis bag permits nearly complete exchange of solutes in approximately 6 hr. Three buffer changes at 12-hr intervals results in 90% depolymerization of the F-actin (Fig. 4).

It is not unusual to detect a residual viscosity in the dialyzed actin. This viscosity is due to the presence of nondepolymerized actin that is complexed with myosin or other proteins. The quantity of myosin present can vary from preparation to preparation as a result of differing efficiencies of myosin removal during preparation of acetone powder. Some actin losses are encountered with short dialysis times, but a myosin-free final product is obtained. Extensive dialysis results in eventual dissociation of actin–myosin complexes and appearance of myosin in the final product. Prolonged dialysis can also lead to observable actin proteolysis and should therefore be avoided.

8. *Clarification of G-actin.* The dialyzed actin is centrifuged at 80,000 g (avg) for 3 hr. Shorter centrifugation times can be employed at greater g force; e.g., 150,000 g (avg) is now readily attainable in modern ultracentrifuges with the corresponding clearing time reduced to 1.5 hr. The supernatant fraction is saved. It is convenient to determine protein concentration at this point rather than on the subsequent viscous F-actin final product.

[16a] This additional wash was not included as a step in the original Spudich–Watt report[12] and was not used to purify the actin shown in the figures presented here. This has now been incorporated as a routine step in the procedure.

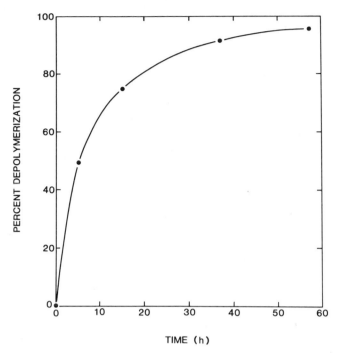

TIME (h)

FIG. 4. Time course of depolymerization. Aliquots of dialyzing actin from step 7 of the isolation procedure (Fig. 3) were centrifuged at 100,000 g (avg) for 2 hr at 4°. Protein was determined in the supernatant (depolymerized actin), and the percentage of depolymerization was calculated. Dialysis was performed in 12 by $\frac{1}{4}$ inch dialysis bags mounted on a rapid dialyzer in 1 liter of buffer A at 4°. The buffer was changed at 5, 15, and 37 hr of dialysis. Total dialysis time was 57 hr.

9. *Polymerization.* G-actin solutions, even in the presence of high concentrations of ATP (1 mM) and stored on ice, begin to lose polymerization activity after 2–3 days. Therefore actin is stored as F-actin. To polymerize, add KCl to 50 mM, MgCl$_2$ to 1 mM, and ATP to 1 mM final concentrations. For storage, also add 0.02% NaN$_3$. The final product can be stored on ice as an F-actin solution or as pellets of F-actin. Pelleting provides additional stability. Actin should not be frozen or lyophilized.

The expected yield for this protocol is 20–30 mg of actin per gram of acetone powder. The resulting actin is generally highly purified (Fig. 5). However, depending upon individual technique, buffer purity, dialysis times employed, and so forth, the actin preparation can contain small amounts of contaminating protein, including proteolysed actin. Consequently, several techniques for further purification of the actin are discussed below in the section Further Purification of Actin.

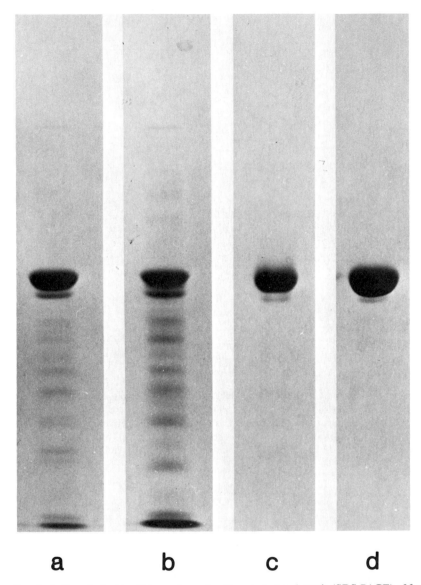

a   b   c   d

FIG. 5. Sodium dodecyl sulfate–polyacrylamide gel electrophoresis (SDS-PAGE) of fractions at various stages in the actin purification. Samples taken during the isolation procedure were mixed 1 : 1 with a solution containing 2% SDS, 1% 2-mercaptoethanol, boiled for 3 min, and applied to a 1.5-mm-thick slab gel containing 12% acrylamide, 0.2% methylene bisacrylamide, and 0.1% SDS in the buffer system of Laemmli.[17] Gels were stained overnight with 0.025% Coomassie Brillant Blue G, 10% acetic acid, and 25% isopropanol and destained in 10% acetic acid by gentle shaking. Each gel lane contains about 6 μg of protein. (a) Muscle extract after 20,000 g clarification (Fig. 3, step 3), (b) supernatant after incubation in 0.6 M KCl and centrifugation at 80,000 g (step 6); (c) sedimented F-actin after incubation in 0.6 M KCl (step 6); (d) final product (step 9).

Analysis of Purity of Final Product

Contaminating proteins are most easily detected by SDS–polyacrylamide gel electrophoresis,[17,18] preferably utilizing slab-type gels of 1.5–2.0 mm thickness. Our experience is that 10–12% acrylamide is an optimal gel concentration for detection of contaminants, since in 8% gels protein species of <25,000 daltons run with the ion front, while 15% gels do not allow discrimination of high molecular weight proteins. The limit of detection for Coomassie Blue-stained protein bands on a 1.5 mm-thick slab gel system is about 0.05 μg per band. Consequently, visualization or densitometry of one 0.15% contaminant requires loading approximately 40 μg of the actin preparation (Fig. 6). Although this constitutes gross overloading of the actin band (actin band staining is linear only to 4 μg per band), minor contaminations of 0.15% can be detected and purity estimated. It is highly desirable to have this degree of sensitivity, since very low levels of other components in actin preparations can significantly alter the properties of actin filaments.[15] For example, a factor that alters the function of actin filaments by specifically binding to filament ends need only represent about 0.2% of the protein in a preparation of filaments about 1 μm long.

Sources of Contamination

Myosin contamination is sometimes observed and can be attributed to incomplete myosin extraction during acetone powder preparation. Proteolysis of actin is indicated by the appearance of a 38,000-dalton protease-resistant core[19] (Fig. 7b). Such proteolysis can arise from either bacterial contamination in extraction buffer or proteolytic activity extracted with actin from some acetone powders (Fig. 7). While proteases are not prevalent in all acetone powder preparations, inspection of purified actin preparations for protease activity is advisable. For those acetone powders that yield proteolytic activity, a slight modification of the actin preparation procedure is recommended. F-actin is sedimented after the 0.6 M KCl treatment (see Fig. 3, step 6), then the F-actin pellet is homogenized thoroughly into 100 ml of 4° buffer A containing 50 mM KCl and 2 mM MgCl$_2$ and immediately recentrifuged at 150,000 g for 1.5 hr at 4°. The soluble protease is fractionated away from F-actin in this step before depolymerization of filaments into protease-susceptible G-actin has been initiated. The washed pellet is then homogenized into buffer A and dialyzed (step 7). An additional precaution when the acetone powder

[17] U. K. Laemmli, *Nature (London)* **227**, 680 (1970).
[18] G. F. C. Ames, *J. Biol. Chem.* **249**, 634 (1974).
[19] G. R. Jacobson and J. P. Rosenbusch, *Proc. Natl. Acad. Sci. U.S.A.* **73**, 2742 (1976).

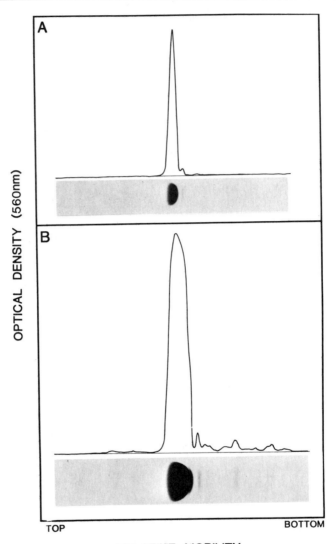

FIG. 6. Detection of actin contamination. Actin prepared as in Fig. 3 was electrophoresed in a 12% sodium dodecyl sulfate–polyacrylamide gel. Stained gels were scanned at 560 nm on an RFT scanning densitometer (Transidyne). (A) Four micrograms of isolated actin. Homogeneity was estimated at 98% by scan area. (B) Forty micrograms of isolated actin. Approximately eleven contaminants containing a total of about 2 μg of protein are detected by densitometry. Homogeneity $\cong$ 95%.

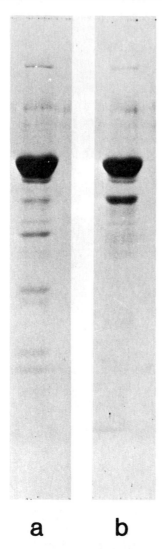

a   b

FIG. 7. Proteolytic activity in acetone powder extracts. (a) Acetone powder extract (10,000 g supernatant; Fig. 3, step 3) taken immediately after extraction. Note the presence of a small amount of 38,000-dalton proteolytic fragment. (b) The same acetone powder extract after 24-hr incubation at 20°. Significant proteolysis of actin has occurred, resulting in a larger amount of the 38,000-dalton fragment. Not all of the acetone powders tested contained an active protease.

preparation contains proteolytic activity is to ensure that the pH of the extraction buffer is 8.0 at 25°; high pH inhibits this protease activity.

Further Purification of Actin

Many current experiments in research on cell motility require actin completely free of trace contaminants such as myosin, tropomyosin, and other factors that are known to alter properties of actin assembly, disassembly, subunit exchange, and ATPase activity. In addition, no specific steps in the purification shown in Fig. 3 are designed to remove efficiently ribonucleotides or polysaccharides, which are not detected by SDS-gel electrophoresis with Coomassie Blue staining. Consequently, we and others have designed additional steps to purify the actin further. The three procedures described below can be considered alternatives, or they can all be used.

Ion Exchange Chromatography. A highly recommended technique for obtaining highly purified muscle actin is ion exchange chromatography. This type of purification offers the considerable advantage of removing both protein and nonprotein contaminants from actin preparations. For further purification of muscle actin, we use the following batch treatment of F-actin with DEAE-cellulose.

1. DE-52 resin (Whatman) is prepared at 25° in 50 mM triethanolamine buffer, pH 7.5, following the instructions provided with the resin. Two milliliters of settled resin are used to further purify 20 mg of actin isolated as shown in Fig. 3.

2. Two milliliters of settled resin are sedimented at 20,000 g for 20 min at 4° and resuspended with 200 ml of DEAE-buffer (10 mM imidazole, pH 8.0 at 4°, 0.1 M KCl, 0.1 mM CaCl$_2$, 1 mM ATP, 0.5 mM 2-mercaptoethanol, and 0.005% NaN$_3$). Equilibration is for 4 hr at 4° with stirring.

3. Equilibrated resin is sedimented and resuspended with 20 ml of 1.0 mg of F-actin per milliliter, which is prepared by diluting F-actin (~5 mg/ml) with cold DEAE-buffer and mixing for 30 min at 4°.

4. The F-actin, DE-52 mixture is stirred in a plastic beaker for 6 hr at 4°. More than 95% of the actin is loaded onto the DE-52 in this step.

5. The resin is sedimented at 20,000 g for 20 min at 4° and the supernatant is discarded.

6. The resin is resuspended by stirring in 20 ml of 10 mM imidazole, pH 6.4 at 4°, 0.3 M KCl, 0.1 mM CaCl$_2$, 1 mM ATP, 0.5 mM 2-mercaptoethanol, and 0.005% azide. Check the pH and titrate to 6.4 if necessary. Actin is eluted for 2 hr at 4° with slow mixing.

7. The resin is sedimented (20,000 g, 20 min, 4°), and the supernatant (20 ml) is immediately dialyzed against 2 liters of buffer A containing 50 mM KCl, 2 mM MgCl$_2$, and 1 mM ATP.

8. The dialyzed F-actin is pelleted at 150,000 g for 1.5 hr at 4°, homogenized into buffer A to a final actin concentration of 6–7 mg/ml, and depolymerized by overnight dialysis.

9. The resulting G-actin is clarified at 150,000 g for 1.5 hr at 4°, assembled with 0.1 M KCl and 1 mM MgCl$_2$, and stored on ice in the presence of 0.02% sodium azide.

The recovery from this procedure is approximately 50% with a final product purity of greater than 99% (Fig. 8).

Depolymerization-Repolymerization. The actin can also be further purified by the following recycling protocol (Fig. 9). We carry out this procedure just prior to using the actin in an experiment.

1. F-actin (stored as a viscous solution at 4–6 mg/ml) is diluted to 0.5 mg/ml and allowed to incubate at 4° for 2 hr in the presence of 0.1 M KCl, 1 mM MgCl$_2$, and freshly added 1 mM ATP. Fresh ATP is always added to stored actin just prior to recycling. Incubated F-actin is sedimented at 150,000 g for 1.5 hr at 4°, the supernatant is decanted, and the tube and pellet are rinsed carefully with buffer A.[20]

2. The resulting F-actin pellet is gently homogenized into cold buffer A[20] (see step 7 in Isolation of Actin) to a final actin concentration of 2–4 mg/ml and dialyzed against 1 liter of buffer through a 10,000-dalton cutoff collodion bag (Schleicher and Schuell) for 6 hr at 4° with rapid stirring. Collodion bags are much more permeable to ATP in low ionic strength buffers than the traditionally used cellulose dialysis tubing.[see 14,21] Consequently, actin depolymerization rates are enhanced, and G-actin denaturation resulting from depletion of ATP within the dialysis bag is minimized. It is important to minimize dialysis time since pure actin depolymerizes quickly whereas actin associated with contaminants such as myosin and gelation factors depolymerizes more slowly; the basis of this purification step resides in these differential depolymerization rates.

3. Dialyzed actin (which still may retain a visible viscosity if actin-associated contaminants are present in the starting material) is centrifuged

---

[20] This recycling protocol also gives good results using modified G-buffers. For example, we have used 2 mM TES, pH 7.2, 25°, 0.5 mM 2-mercaptoethanol, 0.2 mM ATP, 50 μM MgCl$_2$, 0.005% sodium azide in order to reduce Ca^{2+}; J. D. Pardee, P. A. Simpson, L. Stryer, and J. A. Spudich, *J. Cell Biol.*, in press (1982).

[21] A. Martonosi, M. A. Gouvea, and J. Gergely, *J. Biol. Chem.* **235,** 1700 (1960).

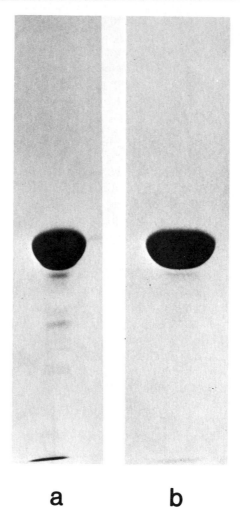

a   b

FIG. 8. DEAE purification of isolated actin. F-actin (20 mg) isolated as in Fig. 3 was treated with DE-52 by the batch method described under Ion Exchange Chromatography. (a) Isolated actin before DEAE treatment. Densitometry of contaminants indicate 1–2 μg of protein contamination per 40 μg of loaded protein, or about 95% actin homogeneity. (b) Isolated actin after DEAE purification. Densitometry of contaminants indicate <0.2 μg of contamination per 40 μg of loaded protein. The molecular weight homogeneity was >99%.

at 150,000 g for 1.5 hr at 4°, the supernatant is decanted, and any pelleted material is discarded.

4. The clarified G-actin is immediately polymerized by addition of ATP, KCl, and MgCl$_2$ to final concentrations of 1 mM, 0.1 M, and 1 mM,

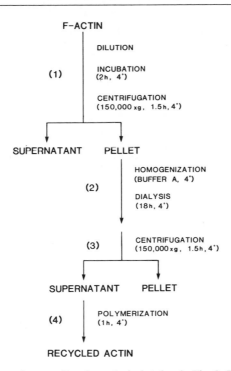

FIG. 9. Flow diagram for recycling the actin isolated as in Fig. 3. For complete explanation of the procedure, refer to the numbered steps in the text under Depolymerization-Repolymerization. Dialysis is performed in 10,000-dalton cutoff collodion bags. Recovery is approximately 70% of starting material.

respectively. Full assembly is complete within 1 hr at 4°. The F-actin solution at this stage should be quite clear and highly viscous. The actin concentration is approximately 2 mg/ml. Purification may be complete at this stage, depending on the purity and type of contaminants originally present. However, a second recycling may be necessary when small amounts of proteolyzed actin and low molecular weight contaminants persist.

The yield from this procedure depends on the level of contamination present in the starting actin material. For highly contaminated starting material (purity = 85–90%), approximately 50% recovery of highly purified actin is obtained. For actin that is 95% pure, recovery is approximately 75%. The value of recycling even apparently clean actin preparations is illustrated in Fig. 10. Although the purity of the actin before recycling is high, the F-actin-associated impurities removed by a single recycling procedure are evident in the depolymerization pellet (Fig. 10b).

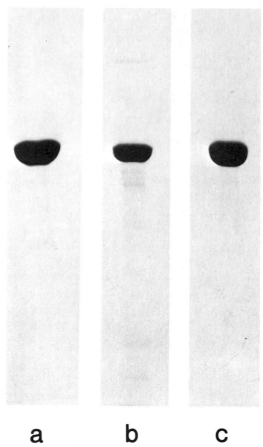

a   b   c

FIG. 10. Removal of F-actin-associated contaminants by recycling. (a) Actin from the isolation procedure shown in Fig. 3. (b) Sedimented material after depolymerization of the F-actin (Fig. 9, step 3). (c) Final product after recycling. Each lane was loaded with 20 μg of protein.

Sephadex Chromatography. Sephadex G-150 chromatography in depolymerization buffer can be used to provide additional purification as well as removal of actin oligomers from G-actin. Polymers are easily separated from the major peak of low molecular weight actin species; if the trailing fractions of the eluted actin peak are pooled, a homogeneous population of monomeric actin can be obtained in addition to increasing the actin purity. The effects of purification by Sephadex G-150 on the properties of muscle actin have been studied by MacLean-Fletcher and Pollard.[15] Polymerization of column-purified monomeric actin results in a significant increase in low-shear viscosity over that observed with actin before column purification. Moreover, some of the fractions from the

Sephadex column show a significant viscosity-reducing activity when added back to the column-purified G-actin before polymerization. These results should be sufficient to warn of dramatic effects on the properties of actin by levels of contaminants heretofore considered negligible.

Concluding Remarks

During the 1970s muscle actin has become a common laboratory reagent; we therefore wished to discuss the details of the purification procedure of Spudich and Watt[12] with special reference to possible problems that may be encountered. The current level of sophistication in cell motility research places new demands on the level of acceptable purity of the actin preparation. Thus several additional steps have been presented for further purification of muscle actin. Particularly useful is a simple batch treatment of F-actin with DEAE-cellulose, which gives actin that is >99% pure.

Actin from many cell types has become widely studied in the 1970s, and these studies have been reviewed.[22-27] Optimal purification of nonmuscle actins generally requires procedures specifically tailored to the cell type used.[28-39] An evaluation of various procedures for the purification of nonmuscle actins can be found in a review by Uyemura and Spudich.[40]

[22] T. D. Pollard and R. R. Weihing, *CRC Crit. Rev. Biochem.* **2**, 1 (1974).
[23] R. Goldman, T. Pollard, J. Rosenbaum, eds., "Cell Motility": *Cold Spring Harbor Conf. Cell Proliferation,* **3**, [Book A-C] (1976).
[24] T. D. Pollard, *in* "CRC Handbook of Biochemistry and Molecular Biology," 3rd ed., Vol. 2, pp. 307–324. Chem. Rubber Co., Cleveland, Ohio, 1977.
[25] R. R. Weihing, *in* "Cell Biology" (P. L. Altman and D. D. Katz, eds.), pp. 341–356. Fed. Am. Soc. Exp. Biol., Bethesda, Maryland, 1976.
[26] M. Clarke and J. A. Spudich, *Annu. Rev. Biochem.* **46**, 797 (1977).
[27] E. D. Korn, *Proc. Natl. Acad. Sci. U.S.A.* **75**, 588 (1978).
[28] Y. Z. Yang and J. F. Perdue, *J. Biol. Chem.* **247**, 4503 (1972).
[29] J. A. Spudich, *J. Biol. Chem.* **249**, 6013 (1974).
[30] J. H. Hartwig and T. P. Stossel, *J. Biol. Chem.* **250**, 5696 (1975).
[31] R. E. Kane, *J. Cell Biol.* **66**, 305 (1975).
[32] T. D. Pollard, K. Fujiwara, R. Niederman, and P. Maupin-Szamier, *in* "Cell Motility" (R. Goldman, T. Pollard, and J. Rosenbaum, eds.): *Cold Spring Harbor Conf. Cell Proliferation* 3 [Book A-C], p. 689 (1976).
[33] D. J. Gordon, E. Eisenberg, and E. D. Korn, *J. Biol. Chem.* **251**, 4778 (1976).
[34] M. P. Sheetz, R. G. Painter, and S. J. Singer, *Biochemistry* **15**, 4486 (1976).
[35] S. Hatano and K. Owaribe, *J. Biochem.* **82**, 201 (1977).
[36] M. R. Adelman, *Biochemistry* **16**, 4862 (1977).
[37] D. G. Uyemura, S. S. Brown, and J. A. Spudich, *J. Biol. Chem.* **253**, 9088 (1978).
[38] J. D. Pardee and J. R. Bamburg, *Biochemistry* **18**, 2245 (1979).
[39] J. P. Weir and D. W. Frederiksen, *Arch. Biochem. Biophys.* **203**, 1 (1980).
[40] D. G. Uyemura and J. A. Spudich, *in* "Biological Regulation and Development" (R. F. Goldberger, ed.), Vol. 2, pp. 317–338. Plenum, New York, 1980.

[19] Methods To Measure Actin Polymerization

By JOHN A. COOPER and THOMAS D. POLLARD

In this chapter we describe several assays of the polymerization of actin. The assays vary widely in terms of their cost, ease of performance, time, sensitivity, and information provided. In some assays the actin solution is undisturbed, and in others the sample is sheared. Some assays use native actin, and others employ actin that has been covalently modified with a label. Some of the key characteristics of the assays are summarized in the table. More details are presented in the section for each assay. The final section discusses how to measure various parameters associated with actin polymerization. We have separated into an accompanying chapter [20] assays of the interactions between actin filaments.

The actin polymerization reaction has been of interest as an example of macromolecular self-assembly and as a step in the formation of the relatively stable thin filaments in skeletal muscle.[1] New work on the polymerization mechanism has been stimulated by studies of actin in nonmuscle cells, which show that actin filaments are important for both cell motility and cytoplasmic structure.[2] Cytoplasmic actin filaments may polymerize and depolymerize more frequently than skeletal muscle thin filaments, since cells show changes in the intracellular distribution of actin during the cell cycle and there is a large pool, about 50% of the total, of unpolymerized actin. In contrast, most of the actin in muscle is polymerized. In an effort to explain these observations, there have been new investigations to examine the conditions and mechanisms of actin polymerization and to search for physiological controls of polymerization in cells. This work has led to exciting discoveries of cellular proteins that act to polymerize or to depolymerize actin, to change the length of actin filaments, and to cap one end of actin filaments.[2] It is important to understand the actin polymerization assays to design and interpret assays for these regulatory proteins.

Mechanism of Actin Polymerization and Properties of Actin Filaments

The polymerization of actin monomers to form 6 nm-wide filaments with double helical substructure is a condensation phenomenon analogous

[1] F. Oosawa and S. Asakura, eds., "Thermodynamics of Protein Polymerization." Academic Press, New York, 1975.
[2] *Cold Spring Harbor Symp. Quant. Biol.* **47** (in press).

METHODS IN ENZYMOLOGY, VOL. 85

METHODS TO MEASURE ACTIN POLYMERIZATION

Method	Signal α[polymer]	Signal:noise ratio	Sensitivity to length	Shear rate	Range	Native actin	Expense	Throughput	Sensitivity to ABPs	Sample size (ml)
1. Capillary viscometry	±	High	Yes	High	Small	Yes	Low	High	Yes	≥0.6
2. ΔOD$_{232}$	Yes	Moderate	No	0	Small	Yes	High	Low	Unknown	≥0.6
3. Flow birefringence	±	High	Yes	Variable	Small	Yes	High	Low	Yes	≥1
4. Fluorescence										
a. NBD–NEM–actin	Yes	Moderate	No	0	Large	No	High	Low	Yes	≥0.5
b. Pyrene–actin	Yes	High	No	0	Large	No	High	Low	Yes	≥0.5
5. Light scattering	Yes	High	No	0	Large	Yes	High	Low	Yes	≥1
6. Electron microscopy	Yes	High	Yes	0	Large	Yes	High	Very low	No	<0.1
7. DNase inhibition	Yes	High	No	0	Large	Yes	Moderate	High	Unknown	<0.1
8. Pelleting	Yes	High	Yes	Moderate	Large	Yes	Moderate	Medium	Yes	0.2
9. Millipore filtration	Unknown	Moderate	Yes	High	Moderate	Yes	Low	High	Probably	<0.1

to the precipitation of a solid from solution.[1] When the concentration of actin monomers is below a certain concentration, called the critical concentration, no polymer forms even under conditions optimal for polymerization. The value of the critical concentration depends on the solution conditions including ionic strength, presence of divalent cations, pH, and temperature. Above the critical concentration actin polymerizes until the monomer concentration falls to the critical concentration, where it remains. At low salt concentrations the critical concentration is high, and actin can be maintained as monomer in high concentrations. The addition of salt such as 20–100 mM KCl or 0.5–5 mM MgCl$_2$ to a solution of monomers induces polymerization. The polymerization is a true self-assembly reaction that requires no factors other than actin and salt.

Actin polymerization from monomers proceeds in at least four reversible steps: activation, nucleation, elongation and annealing.[1] Activation is a conformational change induced by salts which was first detected by resistance to proteolysis.[3] Nucleation is the association of several monomers to form an oligomer from which polymer can grow.[4] These two steps are much slower than elongation and account for the initial lag phase in the time course of pure actin monomer polymerization. During growth of the polymer, monomers bind to the two ends of the filament at different rates,[5] and during or shortly after binding there is a conformational change in the molecule, first detected by an absorbance change.[6] During polymerization the ATP bound to actin monomers is hydrolyzed to ADP and becomes nonexchangeable, but this process lags behind monomer addition.[7]

Actin monomers associate with and dissociate from the ends of actin filaments.[1] Dissociation is a first-order reaction that depends on the concentration of ends. Association is a second-order reaction that depends on the concentration of both monomers and filament ends. The critical concentration is the monomer concentration at which the association rate equals the dissociation rate. Below the critical concentration dissociation exceeds association, and the filaments shorten. Above the critical concentration, the filaments grow until the monomer concentration falls to the critical concentration. The two ends of the filament can be distinguished by the binding of heavy meromyosin to the filament. The rate constants for these reactions differ at the two ends of an actin filament, and elongation is 5 to 10 times faster at the "barbed" end than the "pointed" end.[5] The critical concentration at the two ends may be the same or different.

[3] S. A. Rich and J. E. Estes, *J. Mol. Biol.* **104**, 777 (1976).
[4] M. Kasai, S. Asakura, and F. Oosawa, *Biochim. Biophys. Acta* **57**, 22 (1962).
[5] T. D. Pollard and M. S. Mooseker, *J. Cell Biol.* **88**, 654 (1981).
[6] S. Higashi and F. Oosawa, *J. Mol. Biol.* **12**, 843 (1965).
[7] J. D. Pardee and J. A. Spudich, *J. Cell Biol.* **93**, in press (1982).

Different critical concentrations lead to a steady-state monomer concentration where dissociation exceeds association at one end and association exceeds dissociation at the other end. The result is a flux of actin molecules through the filament at steady state.[8]

The filaments vary in length and are flexible, not rigid, rods. They can break due to thermal or fluid motion in the sample and perhaps join one another end to end, a process called annealing.[9] Breaking is important because many of the assays (see the table) apply a shearing force to the sample, which bends and breaks filaments. The parameters measured by some of the assays, including viscosity and flow birefringence, vary with the shear force because of filament breakage, bending, and orientation by fluid flow.

Polymerization has a number of interesting parameters including the rates of nucleus formation and breakdown, the rates of monomer association and dissociation at polymer ends, the rates of filament breaking and annealing, the critical concentration, and the length distribution of filaments. While some of these parameters, such as breaking and annealing, can be assessed only indirectly, most can be measured accurately using one of the assays. The following section describes assays in detail, concentrating mainly on the measurement of amount of polymer. In the last section we discuss how to use the assays to measure these other parameters.

Assays

Capillary Viscometry

Principle. Actin filaments have a higher viscosity than actin monomers because they are large and asymmetric. In the method described here the viscosity is measured by the time required for a solution to pass through a glass capillary tube. Several other methods to measure viscosity, including rotational devices, are discussed below in the section on flow birefringence and elsewhere in this volume.[10]

Instrument. The capillary viscometer, shown in Fig. 1, is a U-shaped glass tube that includes a section with a narrow diameter through which the sample flows driven by hydrostatic pressure. The length and diameter of the capillary and the hydrostatic pressure head can be varied yielding different shear rates and flow times. Semimicro viscometers size 100 or 150 from the Cannon Instrument Company (State College, Pennsylvania) are commonly used in studies of actin, because they require a small sam-

---

[8] A. Wegner, *J. Mol. Biol.* **109**, 139 (1976).
[9] H. Kondo and S. Ishiwata, *J. Biochem.* (*Tokyo*) **79**, 159 (1976).
[10] T. D. Pollard and J. A. Cooper, this volume [20].

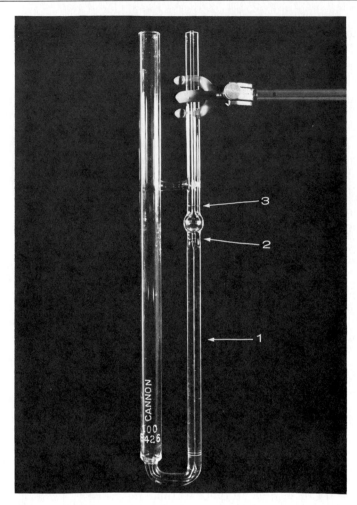

FIG. 1. An Ostwald capillary viscometer. 1, capillary; 2, lower mark; 3, upper mark.

ple (0.6 ml) and have short buffer flow times of about 30 and 60 sec, respectively.

Procedure. Suspend the viscometer in a constant-temperature water bath capable of regulating the temperature within 0.2°, since viscosity and polymerization vary with temperature. We and others often use 25°. Have all solutions, such as actin and buffer, at temperature before use. Introduce a sample of buffer without actin into the viscometer through the wide opening. The volume (0.5–1.0 ml) is critical and must be the same for the buffer and the test sample. Use suction on a rubber tube attached to the small opening of the viscometer to draw up the solution above the upper mark. Release suction and measure the time for the meniscus to move from the upper to the lower mark with a timing device accurate to tenths or hundredths of seconds. Repeat this measurement several times. Remove the buffer with a pipette inserted on the wide side. Introduce into the viscometer a sample with actin in the same buffer and make similar measurements. To follow the time course of actin polymerization, add monomer actin to a polymerizing buffer and make measurements at regular, specified time intervals, such as every minute.

Viscosity is expressed in several ways including relative (η_{rel}), specific (η_{sp}), reduced (η_{red}), and intrinsic [η] viscosity. These terms are defined as follows:

$$\eta_{rel} = t_s/t_b \tag{1}$$

$$\eta_{sp} = \eta_{rel} - 1 \tag{2}$$

$$\eta_{red} = \eta_{sp}/c \tag{3}$$

$$[\eta] = \lim_{c \to 0} \eta_{red} \tag{4}$$

where t_s is the sample flow time, t_b is the buffer time, and c is the actin concentration. Relative and specific viscosity are dimensionless, and the dimensions of reduced and intrinsic viscosity are volume per mass. Relative viscosity, which is kinematic viscosity, can be converted to dynamic viscosity by the following equation: $\eta_s = \eta_{rel}(\rho_s/\rho_b)\eta_b$ where η is the dynamic viscosity in dyne · sec/cm², η_{rel} is relative viscosity above, ρ is density, and s and b denote sample and buffer, respectively. Intrinsic viscosity is the limit of reduced viscosity as actin concentration approaches zero (Eq. 4). This is impossible with actin filaments because of the critical concentration effect. However, an equivalent value can be obtained from the slope of a plot of η_{sp} vs actin concentration above the critical concentration.

Clean the viscometers between each use to prevent irregular wetting and bubbles, which occur to a limited extent even with clean viscometers. Fill each viscometer with nitric acid or chromic–sulfuric acid for several hours, suck 500 ml of distilled water through the viscometer with the vacuum of a water pump, and dry the viscometer in an oven. Additional practical details are available in this series.[11]

Evaluation. Ostwald capillary viscometry is perhaps the most widely used and oldest assay of actin polymerization. It is useful because of its low cost, small sample, ease of performance, reproducibility, and use of native actin. As shown in the classic studies of actin polymerization by Oosawa and his colleagues[12,13] specific viscosity is linearly related to actin

[11] J. E. McKie and J. F. Brandt, this series, Vol. 26, p. 257.
[12] M. Kasai, H. Kawashima, and F. Oosawa, *J. Polym. Sci.* **44**, 51 (1960).
[13] F. Oosawa, S. Asakura, K. Hotta, N. Imai, and T. Ooi, *J. Polym. Sci.* **37**, 323 (1959).

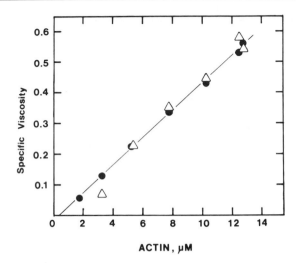

FIG. 2. The dependence of specific viscosity on actin concentration. The critical concentration was 0.5 μM. Conditions: 4 hr, 25°, 20 mM KCl, 10 mM imidazole, pH 7.0, varying concentrations of gel-filtered rabbit skeletal muscle actin in a capillary viscometer type 150. ●, control; △, 2 μM cytochalasin B. From Pollard and Mooseker,[7] with permission.

concentration above the critical concentration (Fig. 2). Results with viscosity were similar to those with light-scattering intensity, sedimentation, flow birefringence intensity, and rigidity. This relation between specific viscosity and actin concentration has been confirmed under many conditions. Specific viscosity for a 1 mg/ml solution of F actin should be about 1.

Capillary viscometry has several drawbacks. The shear rate is high, >1000 sec^{-1}, and cannot be varied as with other techniques. The effects of shear are somewhat paradoxical and are the basis for a major pitfall. High shear breaks actin filaments, deforms them, and orients them with fluid flow, all of which lower viscosity. Flow birefringence extinction angle and rigidity are also decreased with increased shear rates,[12] an observation that supported the classic idea of Oosawa and colleagues that actin polymers are long, semiflexible filaments with multiple crosslinks. Taylor et al.[14] found that even the action of a Pasteur pipette breaks filaments, and we have found that one passage through an Ostwald viscometer lowers viscosity. However, the degree of polymerization and critical concentration presumably do not change.

Paradoxically, the application of shear to actin during polymerization accelerates polymerization, as shown by others[13] and confirmed by our-

[14] D. L. Taylor, J. Reidler, J. A. Spudich, and L. Stryer, *J. Cell Biol.* **89**, 362 (1981).

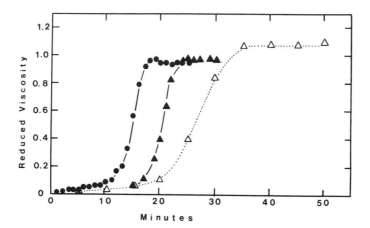

FIG. 3. Effect of frequency of assay on actin polymerization in a capillary viscometer. At time zero, gel-filtered actin in buffer A was mixed with salt to initiate polymerization. ●, Assay every minute; △, assay every 5 min; ▲, assay every 5 min until 15 min and every minute thereafter. Conditions were 12 μM actin, 30 mM KCl, 10 mM imidazole, pH 7.0, 0.2 mM ATP; type 150 Ostwald capillary viscometer; sample size 0.6 ml, 25°.

selves. The data in Fig. 3 show that a sample that is assayed every minute polymerizes faster than one that is assayed every 5 minutes. A likely explanation is that nucleation is the rate-limiting event in polymerization and that shearing creates more nuclei by breaking actin filaments. A solution of actin monomers contains few nuclei. The rate constants for elongation[7] predict that long filaments will grow from these nuclei in a few seconds after adding salt, as shown by the rapid decrease in extinction angle in flow birefringence.[15] As the polymerizing sample flows through the capillary, the shear breaks the long filaments into short pieces, which act as nuclei for elongation. Overall the sample polymerizes more quickly than it would without shear.

Several practical points emerge from the consideration of the effects of shear.

1. The frequency of assaying viscosity must be constant throughout polymerization to examine the time course of polymerization, especially when comparing two samples.
2. To measure the viscosity of a sample that has been sitting undisturbed, one should measure viscosity several times at regular specified time intervals. Viscosity will fall slightly to a stable value.
3. The value of the viscosity depends to a small extent on the force applied to raise the sample up through the capillary tube before

[15] M. Kasai, S. Asakura, and F. Oosawa, *Biochim. Biophys. Acta* **57**, 13 (1962).

allowing it to fall. We have noted discrepancies between persons performing identical experiments that resolved when they made conscious efforts to apply the same force when drawing up the sample.

A second limitation is that viscosity is influenced by the length distribution of filaments, not just the mass of actin polymer. That is, two samples with the same polymer concentration will have different viscosities if the filaments are shorter in one sample. Studies on gelsolin[16] and villin[17] show that these actin regulatory proteins lower viscosity by shortening filaments without affecting the critical concentration. Nunnally et al.[18] found that specific viscosity is related to number average filament length by Staudinger's equation, $\eta = KL^a$, although the exponent, a, was not constant and Staudinger's equation does not theoretically apply to the conditions employed. The important point is that lower viscosity does not imply depolymerization, as often erroneously concluded in the literature.

A third limitation is that actin binding proteins may increase the viscosity of actin filaments if they significantly increase filament dimensions or weight or if they crosslink filaments. Therefore, changes in viscosity in the presence of actin-binding proteins may not reflect the degree of actin polymerization or filament length distribution.

Difference Spectra

Principle. Higashi and Oosawa[6] found that the ultraviolet absorption spectrum of actin filaments is different from that of actin monomers. The difference spectrum was positive from 240 nm to 200 nm, with a relative maximum at about 230 nm. There was also a positive peak at 280 nm.

Procedure. In practice the method is quite easy with only a couple of potential pitfalls. One can use either a single-beam or double-beam spectrophotometer. Ideally one should be able to use a double-beam spectrophotometer with two cells in each beam. The reference cells would contain actin monomer and salt in separate cells, and the sample cells would have actin monomer and salt together in one cell with water in the other. Absorbance would equal zero at time zero. However, it is difficult to pipette actin accurately enough to make this work, because the ΔOD to be measured is small (<4%) compared to the OD of the sample. In practice one can put anything in the reference beam to zero the spectrophotometer on the sample, but a problem exists in extrapolating OD back to time zero when salt is added. A short time is needed to mix

---

[16] H. L. Yin, K. S. Zaner, and T. P. Stossel, *J. Biol. Chem.* **255,** 9494 (1980).
[17] S. Craig and L. Powell, *Cell* **22,** 739 (1980).
[18] M. H. Nunnally, L. D. Powell, and S. W. Craig, *J. Biol. Chem.* **256,** 2083 (1981).

solutions before measurements can begin to be made, and one cannot follow possible OD changes during this time. The time zero value of the absorbance is crucial to calculate ΔOD since ΔOD (time t) = OD (time t) − OD (time zero). One therefore must extrapolate OD back to time zero, assuming a certain shape for the curve.

The measurement of ΔOD_{232} may need to be corrected for light scattering. Higashi et al.[19] found that turbidity in the visible region was proportional to $\lambda^{-3.4}$ and that the absorbance of F-actin at 320 nm could be accounted for solely by turbidity, based on extrapolation from the visible region. Therefore, one should measure the change in absorbance at 320 nm and extrapolate back to 232 nm to determine what proportion of the absorbance change at 232 nm is due to scattering. For actin alone the contribution is usually less than 20% and is ignored; however, actin binding proteins such as myosin may significantly increase scattering and necessitate a correction.

Evaluation. The difference spectra assay is a good one for actin polymerization for several reasons. The ΔOD is proportional to the mass of polymers without effects due to filament length. Higashi and Oosawa[6] found that ΔOD_{234} and degree of flow birefringence increased in parallel in identical samples of polymerizing actin, and Pardee and Spudich[7,20] found similar results comparing ΔOD_{232} and rapid sedimentation. Glenney et al.[21] found that villin, which shortens filaments but has no effect on the amount of polymer, caused no change in ΔOD and a reduction in specific viscosity. The sample is not disturbed, and native actin can be used. Difficulties with the assay include the long time required for unsheared samples to polymerize under some conditions, the difficulty in assaying several samples simultaneously, and limits in actin concentration and sensitivity. The OD of the sample is limited by the spectrophotometer, and therefore only a certain concentration of actin can be used. If one adds other material, in particular protein, which absorbs in the ultraviolet region, the concentration of actin must be even lower. The sensitivity of the assay is also limited by the small signal-to-noise ratio. A 10 μM solution of actin (0.4 mg/ml) has a ΔOD_{232} of about 0.1 on polymerization over a background OD of about 2.5. One order of magnitude below this approaches the limit of resolution of the spectrophotometer. Furthermore, except for villin[21] there are no studies of the effect of actin binding proteins on ΔOD_{232}. These proteins may change the ultraviolet absorption spectrum of actin monomer or polymer or undergo a change in absorption themselves as they interact with actin.

[19] S. Higashi, M. Kasai, F. Oosawa, and A. Wada, *J. Mol. Biol.* **7**, 421 (1963).
[20] J. D. Pardee and J. A. Spudich, *J. Cell Biol.* **87**, 226a (1980).
[21] J. R. Glenney, P. Kaulfus, and K. Weber, *Cell* **24**, 471 (1981).

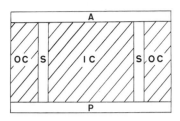

FIG. 4. Cross-sectional schematic diagram through the axis of rotation of a concentric cylinder flow-birefringence apparatus. The light source is below, and the observer is above. A, analyzer; P, polarizer; IC, inner cylinder; OC, outer cylinder; and S, sample.

Flow Birefringence

Principle. The assay of actin polymerization by flow birefringence is based on the orientation of actin filaments in fluid flow. The sample is placed between two concentric cylinders, one of which rotates. Laminar flow in the sample tends to align the filaments in the direction of flow. This alignment is detected by refraction of polarized light.

Theory. Figure 4 is a drawing of the instrument through the axis of rotation. Light is introduced from beneath the apparatus and passes through a polarizer, the sample, and the analyzer, in that order. The observer views the light emerging from the top of the apparatus. A diagram of the apparatus from the top (Fig. 5) shows the orientation of the polarizer and analyzer and how light is transmitted. The polarizer and analyzer are perpendicular to one another and would ordinarily pass no light. The filaments in the sample between them, however, interact with the polarized light to produce light with a component parallel to the analyzer and therefore transmitted. When one cylinder rotates and the filaments align with the flow, there are four positions that do not transmit light. At these positions, the cross of the isocline, the filaments are aligned parallel to the polarizer or analyzer and cannot interact with the polarized light to produce a component that can pass through the analyzer. The extinction angle, χ, is the angle between the polarizer plane and the cross of the isocline where no light passes. The extinction angle, χ, can be related

$$\chi = \frac{\pi}{4} - \frac{\alpha}{12} \left\{ 1 - \frac{\alpha^2}{108} \left(1 + \frac{24 R^2}{35} \right) + \ldots \right\} \tag{5}$$

to the rotational diffusion coefficient, Θ, and the shear rate, G, by Eq. (5), where $\alpha = G/\Theta$, $R = (p^2 - 1)/(p^2 + 1)$ and $p = a/b$, the axial ratio of the molecule.[22] With larger G, χ approaches zero as all the filaments be-

[22] H. A. Scheraga, J. T. Edsall, and J. O. Gadd, *J. Chem. Phys.* **19**, 1101 (1951).

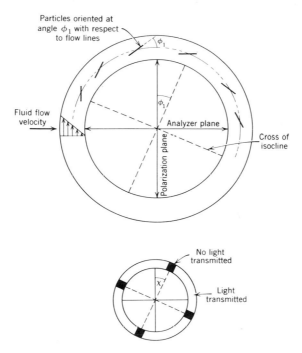

FIG. 5. Experimental manifestation of flow birefringence when all solute particles have the same orientation angle ϕ_1 with respect to the flow lines. The upper diagram shows the cross of isocline, which points to the four locations where the optic axes of solute particles are exactly parallel to the analyzer or polarizer plane. The lower diagram shows the observable result in terms of transmission of light through the annular space between the two cylinders of the apparatus. It is to be noted that the extinction angle χ, which is the angle between the cross of isocline and the polarization planes, is here equal to the angle ϕ_1. From C. Tanford, "Physical Chemistry of Macromolecules" (Wiley, New York, 1963), with permission.

come perfectly oriented. As G approaches zero, χ approaches 45 degrees for rigid particles. Another measurable parameter, the degree of flow birefringence, Δn, is related to the degree of actin polymerization. Δn is the product of an optical constant evaluated independently[23] and an orientation factor, f,[22]

$$f(\alpha, R) = \frac{\alpha R}{15} \left\{ 1 - \frac{\alpha^2}{72} \left(1 + \frac{6R^2}{36} \right) + \ldots \right\} \tag{6}$$

The dependence of Δn on f implies a dependence of Δn on filament length distribution.

[23] A. Peterlin and H. A. Stuart, Z. Phys. **112**, 129 (1939).

Measurements can be made at different shear rates simply by varying the speed of the motor driving the outer cylinder. The shear rate G, in sec^{-1}, can be calculated from the motor speed by Eq. (7).

$$G = \nu 2\pi R_0/(R_0 - R_i) \tag{7}$$

where ν is the motor speed in revolutions per second, R_0 is the outer radius of the sample chamber, and R_i is the inner radius of the sample chamber.[24] This equation holds only for R_0 much greater than $(R_0 - R_i)$.

Procedure. The instrument commonly employed was described by Edsall and co-workers[24] and has been marketed by Rao Instrument Company (Brooklyn, New York). The instrument should be kept at a constant temperature, and all solutions should be at that temperature. Add actin to the salt solution and slowly introduce the sample into the rotor chamber. The size of the chamber varies and may be as small as 2 ml. Turn on the motor, which rotates the outer cylinder. Measure speed of rotation by eye or stroboscopically. Measure the extinction angle simply by observing the isocline and measuring the angle from it to the polarizer plane. To measure the degree of flow birefringence introduce a quarter-wave plate into the light beam and rotate the analyzer until the isocline cross collapses into a single dark line, 45 degrees from the extinction angle. Read directly the angle by which the analyzer was rotated, $\Delta°$, and calculate the degree of flow birefringence,

$$\Delta n = \Delta°\lambda/180d \tag{8}$$

where λ is the wavelength of light employed, and d is the length of the light path through the cell.[24] Multiple measurements can be made on a single sample at a given shear rate to follow the time course of polymerization. Kasai *et al.*[12] employed a homemade rotating cylinder apparatus that allowed measurement of flow birefringence, viscosity, and rigidity on the same sample. The outer cylinder was driven by the motor, and the inner cylinder was suspended on a piano wire, free to rotate. A mirror was fixed to the wire to measure the angle of deflection of the inner cylinder, which was related to viscosity. Rigidity was measured by oscillating the inner cylinder with the outer cylinder held still. These methods are covered in this volume.[10]

Evaluation. Flow birefringence has the advantages of being easy to perform, requiring a small sample, using native actin, and yielding information related to extent of actin polymerization and filament length distribution. The major disadvantages are the cost of the instrument, the shear applied to the sample, and the effects of actin binding proteins.

[24] J. T. Edsall, A. Rich, and M. Goldstein, *Rev. Sci. Instrum.* **23**, 695 (1952).

The degree of flow birefringence depends, as does viscosity, on the degree of polymerization and length distribution. Oosawa *et al.*[13] observed that Δn was zero for actin concentrations below the critical concentration and that Δn increased linearly with increasing actin concentration above the critical concentration. This increase paralleled increases in sedimentation, light scattering, and viscosity and showed that, at a constant shear rate, Δn was proportional to the concentration of actin filaments. However, gelsolin lowers Δn by shortening filaments without affecting the degree of polymerization.[16] Maruyama[25] observed that the application of high shear lowers the Δn of actin filaments measured at low shear, presumably by breaking filaments without altering the amount of polymer. However, Δn of actin filaments increased with increasing rate of shear during measurement, perhaps owing to more efficient orientation of filaments with flow.[12,25]

The extinction angle is inversely related to the rotational diffusion coefficient and therefore the filament length. Kawamura and Maruyama[26] observed that theoretical predictions of filament length based on extinction angle using the equation of Sadron[27] agreed well with lengths measured by electron microscopy. The predictions did not agree so well when actomyosin was used.[28]

Fluorescence

Several studies of the fluorescence of actin and its labeled derivatives have been performed to investigate the process of polymerization and the binding of accessory proteins to actin; however, few investigators have employed fluorescence techniques to quantify polymerization. The intrinsic fluorescence of unlabeled actin, derived from its tyrosine and tryptophan residues, undergoes subtle changes in intensity and spectrum with polymerization,[29] but most studies of actin fluorescence have employed actin labeled with a fluorescent compound. These studies have provided information about the environment of the fluorescent probe during polymerization and upon binding of myosin and other molecules.[30-32] Although the changes accompanying polymerization were noted to be potentially useful for monitoring the state of polymerization, little work was

[25] K. Maruyama, *J. Biochem.* (*Tokyo*) **55**, 277 (1964).
[26] M. Kawamura and K. Maruyama, *J. Biochem.* (*Tokyo*) **67**, 437 (1970).
[27] C. Sadron, *J. Phys. Radium* **9**, 381 (1937).
[28] M. Kawamura and K. Maruyama, *J. Biochem.* (*Tokyo*) **66**, 619 (1969).
[29] S. Lehrer and G. Kerwar, *Biochemistry* **11**, 1211 (1972).
[30] H. C. Cheung, R. Cooke, and L. Smith, *Arch. Biochem. Biophys.* **142**, 333 (1971).
[31] T.-I. Lin, *Arch. Biochem. Biophys.* **185**, 285 (1978).
[32] M. Porter and A. Weber, *FEBS Lett.* **105**, 259 (1979).

done to document the effectiveness of fluorescence in that role. We describe here three efforts to use fluorescence to measure actin polymerization.

NBD–NEM–ACTIN

Principle. Detmers *et al.*[33] found that actin covalently labeled with 7-chloro-4-nitrobenzeno-2-oxa-1,3-diazole (NBD) and *N*-ethylmaleimide (NEM) increases in fluorescence by about twofold upon polymerization.

Procedure. Prepare actin by the method of Spudich and Watt.[34] To a sample of actin (2 mg/ml) in depolymerization buffer A, add NEM (Sigma Chemical Co., St. Louis, Missouri) to a final concentration of 0.3 mM greater than the dithiothreitol from the buffer A, KCl to a final concentration of 0.1 M, and MgCl$_2$ to a final concentration of 1–2 mM. The actin will polymerize rapidly. Hold the solution at room temperature for 15 min, 15° for 15 min, and then 2° for 2 hr before terminating the reaction by adding dithiothreitol to 1 mM. NEM reacts with cysteine-373. Sediment and depolymerize the NEM–actin in the usual manner. Prepare NBD–NEM–actin by mixing 1 volume of polymerized NEM–actin (1 mg/ml in 10 mM Tris, pH 8.0 at 25°, 0.2 mM CaCl$_2$, 0.5 mM ATP, 0.1 M KCl, 2 mM MgCl$_2$) with 1/50 volume of 25 mM NBD-chloride (Pierce Chemical Co., Rockford, Illinois) dissolved in 95% ethanol (spectrograde). Incubate for 5 hr at 15°. Sediment and depolymerize the actin as usual. The labeling ratio is 0.7 NBD per actin, of which 60% is associated with Lys-372.

Measure NBD–NEM–actin polymerization in a fluorescence spectrometer with excitation at 470 nm and emission at 530 nm. To correct for light scattering, subtract the emission signal at 530 nm with excitation at 400 nm from the signal with excitation at 470 nm. The fluorescence will increase 2.2- to 2.3-fold during polymerization. The increase is probably from the label at Lys-372. When measuring the time course of polymerization, make intermittent observations to avoid bleaching of the fluorochrome. Mix unlabeled actin with the fluorescent actin to increase the range of the assay.

Evaluation. This is a rapid assay with a wide range that gives a signal proportional to polymer concentration and is not sensitive to filament length. The sample is not sheared. Fluorescence changes little with binding of myosin subfragment-1 (S-1), DNase I, or tropomyosin-troponin to actin, but other actin-binding proteins have not been examined.

A potential drawback is the double modification of the actin with an NEM group on Cys-373 and NBD groups on Lys-372 and several other

[33] P. Detmers, A. Weber, M. Elzinga, and R. E. Stephens, *J. Biol. Chem.* **256**, 99 (1981).
[34] J. A. Spudich and S. Watt, *J. Biol. Chem.* **246**, 4866 (1971).

residues. However, NBD–NEM–actin copolymerizes with native actin and has a critical concentration similar to that of native actin. There is no experience with this method being employed to measure polymerization.

PYRENE–ACTIN

Principle. Kouyama and Mihashi[35] found that actin covalently labeled with pyrene iodoacetamide increases in fluorescence about 20-fold upon polymerization.

Procedure. Dissolve N-(1-pyrenyl)iodoacetamide (Molecular Probes, Plano, Texas) at an unspecified but presumably high concentration in 33% acetone–67% dioxane. Add the dye in a 1 : 1 molar ratio to a 1 mg/ml solution of actin filaments in 0.1 M KCl, 1 mM MgCl$_2$, 0.1 mM CaCl$_2$, 0.2 mM ATP, 1 mM bicarbonate (pH 7.6), 1 mM NaN$_3$. Incubate at 20° for 20 hr in the dark. One mole of dye binds per mole of actin protomer. The reaction is blocked by pretreatment with NEM, which implies that Cys-373 is the reactive group. To remove unreacted dye, add Whatman CF-11 cellulose to a concentration of 1% (w/w). Remove the cellulose, presumably by low speed centrifugation, and purify the labeled actin filaments by sedimentation and depolymerization. Gel-filter the actin monomer on Sephacryl S-200 before use.[36] The degree of labeling will be 70–95%. The absorption coefficient of labeled actin monomer is 2.2 $\times$ 10^4 M^{-1} cm^{-1} at 344 nm.

The pyrene–actin filaments have an absorption and excitation peak at 365 nm that fluorescent actin monomers do not have, so polymerization can presumably be measured by absorbance at 365 nm. Also, the filaments have fluorescent emission peaks at 386 nm, 397 nm, and 304 nm that are not prominent in the pyrene–actin monomer spectrum. The quantum yield with excitation at 342 nm is 0.083 for pyrene–actin monomer and 0.41 for pyrene–actin filaments, a 5-fold enhancement. However, if emission is observed at 407 nm with excitation at 365 nm, the difference between monomer and polymer is even more pronounced, with a 20- to 25-fold increase in fluorescence upon polymerization.

Evaluation. This is an extremely sensitive, convenient assay. The sample is not sheared. The signal is proportional to polymer concentration, but probably not to filament length. The critical concentration of the labeled actin was 5 μg/ml as measured by fluorescence or viscosity. This is a reasonable value, but the critical concentration of unlabeled actin was not measured. No other evidence about the functional status of the labeled actin was given. By measuring the critical concentration of labeled

[35] T. Kouyama and K. Mihashi, *Eur. J. Biochem.* **114**, 33 (1981).
[36] In our hands actin binds to Sephacryl S-200 in buffer A, eluting after the salt volume. We use Sephadex G-150 to purify actin.

actin and a mixture of labeled actin and unlabeled actin, we have found that labeled actin copolymerizes with unlabeled actin. When heavy meromyosin was added to F-pyrene–actin filaments, the absorption and excitation spectra became identical to those for pyrene–actin monomers. One experiment suggests that heavy meromyosin binds preferentially to unlabeled as opposed to labeled actin protomers in the actin filaments. The effects of other actin-binding proteins were not examined. This method is particularly interesting because of the high sensitivity that it affords, but its general usefulness has yet to be established.

FLUORESCENCE ENERGY TRANSFER

Taylor and co-workers[14] have made the novel application of fluorescence energy transfer to the assay of actin polymerization. In this technique one sample of actin is labeled at Cys-373 with an energy donor (fluorescein), and another sample is labeled at Cys-373 with an energy acceptor (rhodamine or eosin). One then excites with a wavelength that excites the donor but not the acceptor. The excited donor transfers energy to the acceptor, and the acceptor then emits light at its characteristic emission wavelength. The transfer of energy does not involve simple collision or radiation, but probably arises from a vibrational coupling interaction between the excited states of donor and acceptor. Energy transfer is proportional to the reciprocal sixth power of the distance between donor and acceptor and generally occurs only when this distance is less than 50–100 Å. Almost no energy transfer occurs with G-actin, but transfer does occur with F-actin when donor actin and acceptor actin are copolymerized. Polymerization can be monitored by exciting the donor and measuring emission by the acceptor or by measuring quenching of the donor fluorescence. Although such changes have been observed upon polymerization of actin, the correlation of these measurements with the degree of polymerization remains to be established.

Light Scattering

Principle. Actin filaments scatter light more than monomers. Incident light is directed toward a sample, and the intensity of scattered light is measured at an angle away from the incident beam.

Theory. Our theoretical discussion is based on that of Wegner and Engel,[37] who used this technique extensively to study actin polymerization.

---

[37] A. Wegner and J. Engel, *Biophys. Chem.* **3**, 215 (1975).

$$R(\theta) = \frac{i_\theta}{I_0} \cdot \frac{r^2}{(1 + \cos^2\theta)} \tag{9}$$

where $R(\theta)$ is the Rayleigh ratio, r is the distance from the sample to the photomultiplier tube, i_θ is the intensity of scattered unpoliarzed light, I_0 is the intensity of incident light, and θ is the angle between the incident and scattered light. For a heterogeneous solution of rods, Casassa[38] derived the following relationship between light scattering and polymer parameters.

$$R(\theta) = c \cdot K \left[\frac{\pi^2}{2} \frac{(\Sigma M_i w_i/L_i)^2}{\Sigma M_i w_i/L_i^2} + \frac{\lambda' \Sigma M_i w_i/L_i}{4 \sin (\theta/2)} + \cdots \right] \tag{10}$$

c is the total polymer concentration (w/v), M_i is the molecular weight of the ith polymer species, w_i is the weight fraction, L_i is the length, and λ' is the wavelength of light in the medium. This relation holds for polymers with $L_i > \lambda^*$ and diameters much less than λ^* where $\lambda^* = \lambda'/(4 \sin \theta/2)$.

$$K = \frac{2\pi^2 n^2 (dn/dc)^2}{N\lambda^4} \tag{11}$$

where n is the refraction index of the solution, λ is the wavelength of incident light, N is Avogadro's number, and dn/dc is the refractive index increment. dn/dc is calculated for the wavelength employed from the equation of Perlmann and Longsworth,[39]

$$(dn/dc)_\lambda = (dn/dc)_{5780\text{Å}}(0.950 + 2.00 \times 10^6/\lambda^2) \tag{12}$$

where λ is in ångstroms, c is in grams/100 ml, and $(dn/dc)_{5780\text{Å}}$ is 1.86×10^{-3}.

At large values of θ, the first term in the brackets in Eq. 10 is small relative to the second term and is ignored. Solving for c,

$$c = \frac{R(\theta) \, 4 \sin (\theta/2)}{K\lambda'\Sigma(M_i w_i/L_i)} = \frac{R(\theta)}{K \pi \lambda^* \Sigma(M_i w_i/il)} \tag{13}$$

where l is the length of one subunit in the filament. Since $M_i/i = M$, the molecular weight of actin and $\Sigma w_i = 1$,

$$c = \frac{R(\theta)}{K\pi\lambda^* M} \tag{14}$$

M is 42,300 daltons[40] and l is 28 Å. Wegner and Engel[37] showed that the assumptions of length $>\lambda^*$ are valid for polymerizing actin with $\lambda = 546$

[38] E. F. Casassa, J. Chem. Phys. 23, 596 (1955).
[39] G. E. Perlmann and L. G. Longsworth, J. Am. Chem. Soc. 70, 2719 (1948).
[40] J. H. Collins and M. Elzinga, J. Biol. Chem. 250, 5915 (1975).

nm by the linear relation between $1/R(\theta)$ and $\sin(\theta/2)$. Under these conditions light scattering is proportional to polymer mass but insensitive to filament length distribution.[38]

Performance. Centrifuge actin and buffers at 200,000 g for 2 hr to remove dust and polymers. Incubate them at the desired temperature, mix them and introduce a 15-ml sample into the cell. The Cantow scattering photometer and 546 nm are a suitable instrument and wavelength.[37] Measure scattering intensity at 90° with unpolarized light.

Evaluation. Light scattering at 90° is proportional to polymer concentration.[37] Light scattering is insignificant below the critical concentration and insensitive to length distribution if the polymer length is greater than λ^*. This insensitivity is an advantage for measurements of steady-state polymer concentration. However, the fact that scattering may not be proportional to polymer concentration for short polymers is a potential limitation for studies of the early phases of actin polymerization or other short actin filaments. It does use native actin and does not shear the sample. The assay is sensitive to actin-binding proteins, which appreciably increase the cross-sectional mass of the filament. Wegner[41] has used this effect to study the binding of tropomyosin to actin filaments.

Electron Microscopy

Principle. Filament length is measured by direct observation of negatively stained specimens in the electron microscope. By using a morphologically identifiable nucleus, one can measure absolute growth rates at the two ends of the filament. The dependence of these growth rates on monomer concentration yields the rate constants for subunit association and dissociation. Electron microscopy also directly measures filament length distribution.

Procedures

NEGATIVE STAINING. Coat electron microscope grids with a thin plastic film (Formvar or collodion) reinforced with carbon by evaporation.[42] Make the grid surface hydrophilic by glow discharge in a partial vacuum. Gently dilute a sample containing actin filaments to a concentration of ~25 μg/ml in buffer and quickly apply a 5-μl drop to the coated surface of the grid. Alternatively, float the grid on a drop of diluted sample. Adsorb the sample to the grid for a standard time (usually 15–60 sec) and touch the edge of the grid to filter paper, which removes all but a thin film of liquid.

[41] A. Wegner, *J. Mol. Biol.* **131,** 839 (1979).
[42] D. E. Bradley, *in* "Techniques for Electron Microscopy (D. H. Kay, ed.), p. 58. Blackwell, Oxford, 1965.

Before the sample dries, invert the grid onto a drop of 1% uranyl acetate in water for a standard time (usually 15 sec) and then touch it to filter paper to remove all but a thin layer of stain. Allow the remaining stain to air dry. Place the grids in a vacuum desiccator for 30 min before viewing them in the electron microscope.

Where quantitation is not important, the staining can be improved considerably by modifications suggested by Dr. Ueli Aebi (Johns Hopkins Medical School). After removing the bulk of the sample with filter paper, the grid is inverted in succession onto a series of drops on clean Parafilm: 3 buffer drops for 10 sec each, 3 water drops for 10 sec each, and 3 drops of stain for 10 sec each. The grid is touched to filter paper between the buffer and water, between the water and stain, and after the stain. Gentle suction through the constricted tip of a Pasteur pipette is used to remove any remaining stain. Substituting 0.75% uranyl formate, pH 4.2, for uranyl acetate may give superior results. The uranyl formate solution is made by adding solid to water at 100° and stirring in the dark for 30 min at room temperature. The pH is adjusted with NaOH. Even when stored in the dark, this stain will precipitate in 1 or 2 days. These modifications give more uniform staining and a cleaner background.

PREPARATION OF MORPHOLOGICALLY IDENTIFIABLE NUCLEI. The first such nuclei employed were actin filaments decorated with stoichiometric amounts of heavy meromyosin or myosin S-1. Decorate actin filaments with 3 mg of muscle myosin S-1 or 4 mg of heavy meromyosin per milligram of actin in a suitable buffer such as 50 mM KCl, 1 mM MgCl$_2$, 10 mM imidazole, pH 7. The myosin will rapidly hydrolyze the ATP in the actin buffer and then bind to the actin filaments.[43] To shorten the decorated nuclei, pass them through a 25-gauge needle several times. These decorated actin filaments can be used only when the sample of actin monomers to be added to the nuclei contains no free ATP, which will dissociate the myosin heads from the nuclei. Remove free ATP from actin monomers by gel filtration or treatment with Dowex 1 just before the growth experiment. The monomers retain their normal ability to polymerize for several hours when stored at 0–4° in 2 mM Tris, pH 8.0, 0.5 mM dithiothreitol, 0.2 mM CaCl$_2$. Myosin heads will slowly redistribute from the nuclei to undecorated actin filaments formed during the growth experiment, so that the specimens must be prepared for electron microscopy soon after the polymerization.

For experiments with ATP in the buffer, the decorated nuclei can be fixed with 0.1% glutaraldehyde as developed by Dr. M. Runge (Johns Hopkins Medical School). Add an equal volume of 0.2% glutaraldehyde in

[43] D. T. Woodrum, S. Rich, and T. D. Pollard, *J. Cell Biol.* **67**, 231 (1975).

H_2O to the decorated nuclei. Incubate for 7 min at 25°, and quench the glutaraldehyde by adding 0.1 volume of 500 mM ethanolamine in H_2O. Dialyze or gel filter the fixed nuclei to remove the glutaraldehyde and ethanolamine. Shear the nuclei with 20 passes through a 26-gauge needle. These fixed nuclei are stable in ATP for at least 90 min, so the myosin heads are at least partially chemically crosslinked to the actin and/or each other. Most of these fixed decorated filaments are capable of nucleating bidirectional growth of actin filaments. Subfragment-1 does not redistribute rapidly.

Another nucleus is the bundle of actin filaments isolated from the microvilli of the intestinal epithelial cell brush border.[7] The brush borders are isolated as described by Mooseker et al.[44] The microvilli are released by homogenization and isolated by differential centrifugation.[45] The microvillus pellets can be stored for several days at 4° in buffer or for weeks at −20° in buffer with 50% glycerol. The membrane surrounding the microvillar actin filament bundle is solubilized in 1% Triton X-100 in 75 mM KCl, 5 mM $MgSO_4$, 0.1 mM imidazole (pH 7.5) at 0–4° and the filamentous cores are collected by pelleting at 30,000 g for 10 min at 4°. After washing once in buffer without detergent, the filament bundles are suspended in buffer and used within 6 hr.

MEASUREMENT OF ELONGATION RATES. Prepare conventional rabbit skeletal muscle actin and freshly gel-filter it on Sephadex G-150 to remove oligomers that can serve as endogenous nuclei. Mix various concentrations of actin monomers in the range of 50–500 μg/ml (1.2–12 μm) with 50–100 μg of nuclei per milliliter and incubate at 25° for a time just long enough to yield accurately measurable filaments (50–200 sec for low concentrations and 5–20 sec for 500 μg/ml). Stop polymerization by gentle dilution into 25° polymerization buffer to give a monomer concentration close to the predetermined critical concentration for polymerization (~20 to 60 μg/ml depending on the conditions). At this concentration the length of the newly polymerized actin filaments is stable for a time sufficient to prepare a negatively stained grid. Work quickly and yet be gentle when mixing and pipetting in order to minimize filament breakage. The procedure is conveniently performed in a drop on Parafilm. Make samples at three time points and take enough electron micrographs to provide 25–50 filaments for each time point. Measure the length of growth from the ends of the nuclei with a ruler if they are short and straight or with a map reader or digitizer if they are long and curved. Plot average length at each end vs time to establish the growth rate. The identification of the polarity of the filament is obvious when using decorated nuclei. With microvillar cores,

[44] M. S. Mooseker, T. D. Pollard, and K. Fujiwara, J. Cell Biol. 79, 444 (1978).
[45] C. L. Howe, M. S. Mooseker, and T. A. Graves, J. Cell Biol. 85, 916 (1980).

the fast and slow ends have been identified as the barbed and pointed ends, respectively, by decoration with myosin S-1. To determine rate constants, plot the growth rate in monomers per second versus monomer concentration. There are 370 monomers per micrometer of filament.[46] The slope is the association rate constant in M^{-1} sec^{-1}, and the Y intercept is the dissociation rate constant in sec^{-1}. To date this analysis has been carried out only with microvillar actin bundles and fixed decorated filaments as nuclei, but the general approach should be successful with other types of nuclei.

MEASUREMENT OF LENGTH DISTRIBUTIONS. Gently dilute samples of actin filaments to ~20 μg/ml in polymerization buffer and immediately apply to a grid and stain as above. In the electron microscope, search for areas where the filaments are evenly dispersed and well stained. Avoid areas where there are tangles. Take slightly overlapping micrographs at 8000× magnification. Make a montage of ×2.5 enlargements of these micrographs. Measure the length of every distinguishable filament with both ends on the montage as described above. Search carefully with a magnifying glass for short filaments (<0.2 μm), which occur at high frequency but are easy to miss. We have found it impossible to identify these short filaments on micrographs taken at magnifications less than ×8000. Several hundred filaments will give a good picture of the length–frequency distribution from which it is possible to calculate the number average [$\bar{X}_n = (\Sigma N_i L_i)/(\Sigma N_i)$] and weight average [$\bar{X}_w = (\Sigma N_i L_i^2)/(\Sigma N_i L_i)$] lengths where N_i is the number of filaments of a given length, L_i. An alternative approach is to measure the total length of filaments on the montage and divide by half the number of ends on the montage to give the number average length.

Evaluation. Providing that the filaments are not broken during specimen preparation, electron microscopy is the best way to assess the size of actin filaments and the only way to measure the growth rates at the two ends of the filament. It is also possible to calculate the number of filaments in a sample from the length distribution, total actin concentration, and critical concentration.

Electron microscopy has several limitations as a quantitative assay for actin polymerization. It is very slow and expensive. Very short (<0.2 μm) and very long (>10 μm) filaments are apt to be missed or lost in tangles, respectively. Filament breakage during pipetting, mixing, application to the grid and staining is a constant problem that is difficult to control. However, Kawamura and Maruyama[26] found that the length distribution obtained by electron microscopy agreed well with the lengths calculated from the extinction angle of flow birefringence. Pollard and Mooseker[7]

[46] J. Hanson and J. Lowy, *J. Mol. Biol.* **6**, 46 (1963).

also found that the growth of filaments from nuclei was proportional to time, suggesting that the filaments were not being broken, at least when they were relatively short.

DNase I Inhibition

Principle. Bovine pancreatic DNase I binds rapidly and tightly ($K_a = 10^{10} M^{-1}$)[47] to actin monomer or profilactin (a complex of profilin and monomeric actin) in a 1 : 1 complex. The hydrolytic activity of DNase I toward DNA is then inhibited. DNase I binds weakly to actin filaments ($K_a < 10^5 M^{-1}$),[48] and its activity is inhibited to only a small extent.[49] DNase I does cause the depolymerization of actin filaments resulting in a 1 : 1 complex of DNase I and actin, but this process is much slower than the binding to actin monomer. Blikstad *et al.*[49] took advantage of the actin–DNase interaction to devise an assay for actin monomer using inhibition of DNase activity. The assay has been most widely applied to cell extracts to determine the amounts of monomeric and polymeric actin present in cells. These studies showed that, despite ionic conditions favoring polymerization, 50–75% of actin in a cell is monomeric. Cells are extracted with Triton, and actin monomer is measured by the ability of the extract to inhibit DNase I. Treatment of the extract with guanidine depolymerizes actin filaments, and total actin is then assayed. The quantity of actin filaments is the difference between total actin and actin monomer.

Procedures[49]

REAGENTS. Dissolve noncrystalline DNase I from beef pancreas (DN 100; Sigma Chemical Co., St. Louis, Missouri) in 50 mM Tris-HCl pH 7.5, 0.01 mM phenylmethylsulfonyl fluoride (PMSF; Sigma Chemical Co.), and 0.1 M CaCl$_2$ at a concentration of 0.1 mg/ml. Cut the fibers of calf thymus DNA (type 1, Sigma Chemical Co.) into fine pieces with a scissors. Dissolve the DNA at 40 μg mg/ml in 0.1 M Tris-HCl, pH 7.5, 4 mM MgSO$_4$, and 1.8 mM CaCl$_2$ by stirring slowly at room temperature for 24–48 hr. Filter the solution and store at 4°. Absorbance at 260 nm should be 0.5–0.65.

DNASE INHIBITION ASSAY. Mix 10 μl of DNase with a given volume (0.5–30 μl) of sample containing actin. The constituents of the sample or lysate buffer may affect the DNase slightly; therefore, add the same amount of each buffer to all tubes including standards and controls. Rapidly (2–4 sec from mixing of sample with DNase) add 3 ml of DNA

[47] G. Berger and P. May, *Biochim. Biophys. Acta* **139**, 148 (1967).
[48] S. Hitchcock, L. Carlsson, and U. Lindberg, *Cell* **7**, 531 (1976).
[49] I. Blikstad, F. Markey, L. Carlsson *et al.*, *Cell* **15**, 935 (1978).

solution that has been warmed to room temperature. Measure absorbance at 260 nm with time. After an initial lag phase, absorbance increases linearly with time owing to the digestion of the DNA. The slope of the linear portion of the curve gives the DNase rate. Actin inhibits the DNase activity. Calculate the percentage decrease of the slope from a control with no actin. One activity unit is defined as 1% inhibition of DNase. The assay is most accurate between 30 and 70% inhibition, which corresponds to about 0.3–0.7 μg actin. Therefore, by adding 30 μl of lysate the lower limit of the assay is 10 μg actin (0.3 mg/ml); and by adding 0.5 μl, the upper limit is 1.4 mg/ml without dilution. To measure the total actin, mix the sample with an equal volume of 1.5 M guanidine HCl, 1 M sodium acetate, 1 mM CaCl$_2$, 1 mM ATP, and 20 mM Tris-HCl pH 7.5. Incubate at 0°. This will depolymerize up to 1 mg of F-actin per milliliter (2 mg/ml in original sample) in 5 min, but will not inhibit actin binding to DNase or the activity of DNase. The DNase inhibitor activity of actin decreases less than 5% during 2 hr of incubation.

PREPARATION OF CELL LYSATES. Wash cells free of media or serum into phosphate-buffered saline. Suspend cells in a known volume of 5 mM potassium phosphate, pH 7.6, 150 mM NaCl, 2 mM MgCl$_2$, 0.1 mM dithioerythritol, 0.5% Triton X-100, 0.2 mM ATP, and 0.01 mM PMSF. Place lysate on ice.

Evaluation. The DNase assay is widely used because it is an easy and accurate way to measure actin monomer concentration in a solution of pure actin. By using guanidine to depolymerize actin filaments, total actin can also be measured. There are several points of concern about the assay. Although Blikstad *et al.*[49] showed that the assay responds linearly to purified profilactin, actin may exist in other forms or complexes in the cell that do not inhibit DNase I. It is not known what form of actin in cell lysates inhibits DNase, but it seems unlikely to be free monomer, given the actin concentrations and ionic conditions in the extraction solution. The effect of various actin binding proteins is not known. The use of this assay to detect actin "depolymerizing" factors depends on the assumption that the complex of factor with actin inhibits DNase. Obviously, the affinity with which such a complex may bind DNase, and the degree to which it may inhibit DNase activity, is not known. Actin complexed with DNase I in the cell would have no inhibitor activity. Inhibitors of DNase I other than actin may exist in a cell.

Of particular concern is the stability of the monomer and polymer pools after lysis and before assay. If the cell extract is an equilibrium system, there must be a shift from polymer to monomer when it is diluted and cooled to 0°. If a change in ionic conditions accompanies the dilution the critical concentration may increase or decrease, leading to depolymerization or polymerization, respectively. While the critical concentration is likely to be much lower than the concentration of actin, and the amount of polymerization or depolymerization should be small, one should follow the actin pools at times after lysis into several buffers with varying concentrations of salt and divalent cations. Actin-binding proteins and the various cellular compartmentalizations of actin may also influence the stability of the monomer and polymer pools after lysis. Carlsson et al.[50] found depolymerization of actin in lysates with calcium and polymerization in lysates with EGTA even though they claimed that actin filaments and profilactin were stable under these conditions.

Two modifications of the assay that improve its sensitivity have been reported. Snabes et al.[51] used radioactive actin in an immunoprecipitation assay. They prepared ^{125}I-labeled actin at a molar ratio of 0.5 I to 1 actin by the method of Bolton and Hunter. An aliquot of radioactive actin was incubated with a sample of cold actin and DNase I. The DNase I was immunoprecipitated with a rabbit anti-DNase I serum followed by a sheep anti-rabbit γ-globulin serum, and the precipitate was counted. Sensitivity was increased about 1000-fold to 0.24 ng of actin, but the assay requires the special reagents above and takes 2 days. Laub et al.,[52] based on work of LePecq et al.,[53] devised a fluorescent assay for DNA to use in the DNase I assay. Ethidium bromide was added to the DNA. The complex was fluorescent, and as DNase I digested DNA, the fluorescence decreased. Data documenting the increased sensitivity of the assay were not provided.

Ultracentrifugation

Principle. Actin filaments have a much higher sedimentation coefficient than monomers, so that they can be separated by ultracentrifugation. The fraction of the actin in polymer can be measured by analytical ultracentrifugation or, more simply, by pelleting the filaments and determining the monomer concentration in the supernatant.

Procedure. Place a 100-μl sample containing actin filaments over a 50-μl cushion of 30% sucrose in the same buffer in a 5 × 20 mm Airfuge tube; centrifuge for 15 min at 23 psi in a Beckman Airfuge. Remove a 50-μl sample of supernatant from the meniscus and measure its protein concentration. Under these conditions >98% of polymer and <1% of the monomer sediments out of the supernatant sample (Fig. 6).

[50] L. Carlsson et al., Proc. Natl. Acad. Sci. U.S.A. **76**, 6376 (1979).
[51] M. C. Snabes, A. E. Boyd, III, R. L. Pardue, and J. Bryan, J. Biol. Chem. **256**, 6291 (1981).
[52] F. Laub, M. Kaplan, and C. Gitler, FEBS Lett. **124**, 35 (1981).
[53] J.-B. LePecq, P. Yot, and C. Paoletti, C.R. Hebd. Seances Acad. Sci. **259**, 1786 (1964).

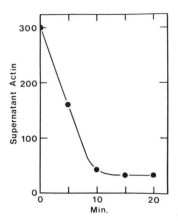

FIG. 6. Centrifugation of actin filaments in a Beckman Airfuge: actin in 50 μl of supernatant (μg/ml) versus time of centrifugation at top speed. Conditions: 300 μg of conventional actin per milliliter, 4 mM MgCl$_2$, 40 mM imidazole, pH 7.0, 2 mM EGTA, 2 mM ATP; sample size, 120 μl, over a 50-μl cushion of 30% sucrose. Data of Peter C. Tseng.

Evaluation. The advantages of this pelleting assay are that it is simple and relatively fast, the sample is small, the actin is native, the range is very large, and, most important, the concentration of monomer and polymer can be determined in the presence of actin binding proteins. Although these proteins will contribute to the protein concentration of the supernatant fraction, the amount of actin can be determined easily by quantitative gel electrophoresis or DNase inhibition. The fraction pelleted is relatively insensitive to the length distribution, although the sedimentation coefficient of very short polymers may be too low to sediment out of the supernatant fraction in 15 min. If the sample contains an appreciable concentration of short polymers, as in the presence of villin, it may be necessary to use another assay such as DNase inhibition which is not sensitive to polymer length.[17]

Millipore Filtration Assay

Grazi and Magri[54] have proposed the use of a Millipore filter to separate monomeric and polymeric actin as the basis for a rapid assay for polymerization. A small sample of 0.2 to 1.5 ml is simply aspirated through a 0.45-μm Millipore filter. One expects small polymers to pass through the filter, since the diameter of the pore is about 150 times the diameter of an actin monomer. Their data, however, indicate that less actin appears in the filtrate than expected by comparison with other meth-

---

[54] E. Grazi and E. Magri, *FEBS Lett.* **123**, 193 (1981).

ods. In particular, they measure a critical concentration of about 100 $\mu g/ml$ by capillary viscometry, but only 25 $\mu g/ml$ passes through the filter. We have found that actin monomer in buffer A does not stick to a 0.2-μm filter unit (Millex-GS, Millipore Corp., Bedford, Massachusetts). The reliability and usefulness of this assay must remain in doubt, although it may be potentially useful as a simple and quick assay.

Approaches to the Measurement of Polymerization Parameters

Nucleation Rate

The absolute rate of nucleation has never been measured because, to our knowledge, there are no direct assays for nuclei. The available indirect assays depend on the ability of nuclei to grow into filaments, the rate of which depends on the rate constants for association and dissociation. Consequently, it is possible to measure relative nucleation rates only under conditions where the elongation rates are the same. The approach is to measure the initial rate of pure actin monomer polymerization. A nonperturbing assay with high time resolution, such as ΔOD_{232}, is better than an assay that shears the sample like viscometry, although even viscometry can be used to obtain crude information about the initial rate from the length of the lag phase or the time to achieve half-maximum viscosity. Gel-filtered actin must be used, since conventional actin contains nuclei. If the initial rate of bulk polymerization of gel-filtered actin is accelerated or slowed by a substance, and in an independent assay it can be established that the elongation rate constants are not changed, then the substance likely affects the nucleation rate. The best way to measure the elongation rate constants is with the electron microscope assay described above. Alternatively, one can assess the effect of the substance on the elongation process by assaying polymerization induced by added nuclei, such as sonicated actin filaments. In addition to potential effects on nucleation rates, substances that affect the nucleation process might alter the number of molecules that form a nucleus. This can be assessed by examining the initial rate of polymerization as a function of actin concentration. Kasai et al.[3] found that the initial rate varied as the third to the fourth power of the actin concentration, evidence that the nucleus contains three to four monomers under the conditions they used.

Elongation Rates—Association and Dissociation

At present the only method suitable for these measurements is the electron microscope assay. Measurements of polymerization in bulk are complicated by the ongoing process of nucleation so that one can never know the number of filament ends that are present at any time. This difficulty may be solved by the use of a buffer in which the nucleation rate is slow and by the addition of a known number of well defined nuclei, perhaps in conjunction with a protein known to cap one end of the filament. Currently, well defined nuclei do not exist.

Breaking and Annealing

These processes are difficult to measure. No quantitative methods exist. Annealing can be detected by the increase of a property related to filament length, such as viscosity or degree of flow birefringence,[25] after disruption of actin filaments, usually by sonication. Whether filaments actually join with each other via their ends during this recovery is unknown. It would seem an improbable event based on the number of filament ends and the rate of diffusion of filaments. Breaking is very difficult to measure. Taylor et al.[14] mixed two populations of actin filaments with different fluorescent labels and followed the rate at which the two populations copolymerized into single filaments by resonance energy transfer, as described in the fluorescence section alone. Sample shear increased the rate of increase of resonance energy transfer, presumably owing to breaking filaments. This assay does not measure breaking directly, but instead it assays the copolymerization of monomers derived from filament ends. Breaking leads to more filament ends and therefore more opportunities for monomers to copolymerize.

Critical Concentration

Several assays can be used to measure critical concentration. One approach is to measure the amount of actin polymer as a function of actin concentration. One plots a parameter such as fluorescence or viscosity versus total actin concentration (Fig. 2). Below the critical concentration the signal is directly proportional to actin concentration with a low slope. At the critical concentration there is an abrupt change in the slope. Above the critical concentration the signal is directly proportional to actin concentration with a high slope. The critical concentration is generally low. Therefore an assay with high sensitivity is important to detect changes in critical concentration.

An alternative approach is to measure the actin monomer concentration in the polymerized sample. The DNase or pelleting assays are suitable for this purpose, although one should correct for the small amount of actin monomer that sediments in the pelleting assay.

The electron microscope assay of elongation rates also yields the critical concentration at *each end*. On a plot of dl/dt, rate of change of length with time, versus actin concentration, extrapolation of the line to $dl/dt = 0$ yields the critical concentration. None of the other assays provide this information about the processes at the two ends directly, although it has been argued[55] that bulk measurement of critical concentration in the presence of 2 μM cytochalasin B, which blocks the barbed end, will give the critical concentration at the pointed end.

Uncertainty exists about the experimental approach to determine the effect of actin-binding proteins on the critical concentration. One way is to hold the concentration of the actin-binding protein constant while varying the actin concentration. This would be preferable for a protein that interacts only with actin monomer. The other approach is to vary the concentrations of both actin and the actin-binding protein, holding their proportions constant.

Length Distribution of Filaments

Electron microscopy is the most direct approach, but several difficulties decrease the reliability of the data. The sample must be handled before it is applied to the grid. Inevitably filaments must be broken during this process. The sample often must be diluted before application so that individual filaments are recognizable. Depolymerization must occur during this time. The basis for filament sticking to grids is not known. If all the filaments were to stick to the grid, which is highly improbable, then the observed length distributions would be correct. If the probability of a filament sticking to the grid is equal for all filament lengths, then the number-average length, but not the weight average, will be correct. If the probability is proportional to weight, the weight-average length, but not the number average, will be correct. Otherwise, neither average will be correct. This bias makes it difficult to be confident about the validity of the observed lengths. The effects of washing and staining techniques performed on filament length and probability of staying attached to grids are also not known.

Other methods, such as flow birefringence, provide only averages for filament lengths and do not reveal the length distribution. Many other methods shear the sample, which alters length distribution. Therefore electron microscopy is the method of choice.

Acknowledgments

We are grateful to Joel Pardee, James Spudich, Lansing Taylor, Patricia Detmers, Anne Marie Weber, Marschall Runge, Peter Tseng, and Ueli Aebi for preprints or discussion of their work. J.A.C. is supported by Medical Scientist Training Program GM 7309. The authors' work is supported by research grants GM-26338 and GM-26132 from the N.I.H.

---

[55] S. L. Brenner and E. D. Korn, *J. Biol. Chem.* **254**, 9982 (1979).

[20] Methods to Characterize Actin Filament Networks

By THOMAS D. POLLARD and JOHN A. COOPER

This chapter covers methods used to characterize actin filament networks. The interest in these networks stems from studies of cell extracts[1-5] and purified proteins[6-11] that show that actin filaments can be crosslinked together to form a gel. Many early studies on intact cells[12] suggested that such gels and the gel–sol transition are important for both cytoplasmic structure and motility.

Actin-filament solutions and gels, like other polymeric materials, have both viscous and elastic properties. When deformed by a stress, as in flow, some of the energy applied to the material is stored by elastic elements, and some of the energy is dissipated by viscous (damping) elements. Such complex materials can usually be modeled as viscous and elastic elements in series and in parallel,[13] though not enough is known about actin gels to do such modeling at the present time. It is known that the viscosity and elasticity of actin filament solutions and gels depend on the shear stress applied to the material.[14]

Gels have properties of both liquids and solids. Like solids they can support their own weight and can resist deformation when the deforming force is less than a critical value, termed the yield point. Above the yield point they can flow like a liquid, although the shear strain, the movement of the liquid, is usually not proportional to shear stress, the force on the liquid, as found in a Newtonian fluid like water or glycerol. This non-Newtonian behavior complicates the characterization of actin-filament solutions and gels.

[1] T. D. Pollard and S. Ito, *J. Cell Biol.* **46**, 267 (1970).
[2] R. E. Kane, *J. Cell Biol.* **66**, 305 (1975).
[3] T. D. Pollard, *J. Cell Biol.* **68**, 579 (1976).
[4] R. R. Weihing, *J. Cell Biol.* **71**, 303 (1976).
[5] J. S. Condeelis and D. L. Taylor, *J. Cell Biol.* **74**, 901 (1977).
[6] T. P. Stossel and J. Hartwig, *J. Cell Biol.* **68**, 602 (1976).
[7] H. Maruta and E. D. Korn, *J. Biol. Chem.* **252**, 399 (1977).
[8] E. A. Brotschi, J. H. Hartwig, and T. P. Stossel, *J. Biol. Chem.* **253**, 8988 (1978).
[9] N. Mimura and A. Asano, *Nature (London)* **282**, 44 (1980).
[10] J. Bryan and R. E. Kane, *J. Mol. Biol.* **125**, 207 (1978).
[11] S. MacLean-Fletcher and T. D. Pollard, *J. Cell Biol.* **85**, 414 (1980).
[12] A. Frey-Wyssling, "Submicroscopic Morphology of Protoplasm and Its Derivatives." Elsevier, Amsterdam, 1948.
[13] J. D. Ferry, "Viscoelastic Properties of Polymers." Wiley, New York, 1961.
[14] K. Maruyama, M. Kaibara, and E. Fukada, *Biochim. Biophys. Acta* **271**, 20 (1974).

METHODS IN ENZYMOLOGY, VOL. 85

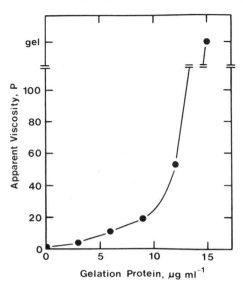

FIG. 1. Gelation of actin filaments by *Acanthamoeba* gelation protein-85. The graph illustrates the dependence of the apparent viscosity (units Poise) measured in the falling-ball viscometer (see Falling-Ball Assay) upon the concentration of the crosslinking protein. Conditions: 0.5 mg of actin per milliliter, 2 mM MgCl$_2$, 1 mM ATP, 1 mM EGTA, 10 mM imidazole (pH 7). Incubation: 10 min at 25°.

Actin-filament solutions are three-dimensional networks of linear actin polymers,[15] which have a high viscosity at low shear rates,[14] presumably due to entanglement or self-associations of the polymers. In the presence of a variety of crosslinking proteins, actin filaments can form gels.[6–11] In every case studied, there is a sharp transition from a viscous liquid to a gel as a function of crosslinker concentration (Fig. 1). This critical gelling concentration of a crosslinker is one of the most characteristic features of a gelling polymer network. According to Flory's[16] network theory of gelation, this critical gelling concentration of crosslinker is the minimal concentration required to connect all the polymers into one continuous network. In theory, the mathematical relationship is given by Eq. (1).

$$P_c = \nu_c/N_0 = 1/\overline{X}_n \qquad (1)$$

where P_c is the number density of crosslinks at the point of incipient gelation, ν_c is the concentration of crosslinker, N_0 is the number of monomers in polymers, and $\overline{X}_n$ is weight-average degree of polymeriza-

[15] F. Oosawa, S. Asakura, K. Hotta, N. Imai, and T. Ooi, *J. Polym. Sci.* **37**, 323 (1959).
[16] P. J. Flory, "Principles of Polymer Chemistry." Cornell Univ. Press, Ithaca, New York, 1953.

tion. Yin *et al.*[17] and Nunnally *et al.*[18] have shown, using different crosslinker proteins and length-shortening proteins mixed with actin, that this equation does hold to a rough extent. The predicted and observed values differ by 3- to 13-fold, but such an error might be expected from uncertainty as to the actual concentration of functional crosslinks in the network. Also, in this theory, it is assumed that all bonds between the polymers are formed by cross-linking units (or molecules), but in the case of actin filaments there is evidence that some of the crosslinks are formed by direct actin–actin bonds.[14,19]

A number of different assays have been used to characterize the physical properties of these actin-filament liquids and gels and to assay the crosslinking and regulatory proteins. The critical gelling concentration can be estimated with very simple assays, while the evaluation of viscous and elastic moduli as a function of shear stress requires more sophisticated equipment. The assays (see the table) fall into three categories: (*a*) test tube inversion, low speed sedimentation, and falling ball viscometry, which are simple enough to use routinely to monitor the purification of gelation factors and regulatory proteins; (*b*) the falling-ball assay and a "gelmeter," which can be used for semiquantitative analysis of gelation; (*c*) several additional assays that require more complex equipment and are available for quantitative analysis of the physical properties of the gels.

Assays

Test Tube Inversion[3]

Principle. This assay tests whether a sample has sufficient rigidity to resist flowing from an inverted tube. A liquid will flow, but a gel will not.

Procedure. The sample is either a cold cellular extract or a mixture of 0.5–1 mg of purified actin per milliliter with a gelation factor. An appropriate buffer is 2 mM MgCl$_2$, 1 mM EGTA, 1 mM ATP, 10 mM imidazole (pH 7). Place 0.2–0.5 ml of sample in a 10 × 75 mm test tube and warm to 25°. In our experience it is preferable to begin with monomeric actin, which is allowed to polymerize without stirring in the presence of the gelation factors. Under these conditions, a homogeneous, optically isotropic gel is formed, whereas a precipitate may form if the sample is stirred. After incubation for 10 min at an appropriate temperature (25°), perform the assay by slowly inverting the tube. A scoring system, such as 0 for a liquid, +1 for a viscous fluid, +2 for a semisolid, and +3 for a gel,

[17] H. L. Yin, K. S. Zaner, and T. P. Stossel, *J. Biol. Chem.* **255**, 9494 (1980).
[18] M. H. Nunnally, L. D. Powell, and S. W. Craig, *J. Biol. Chem.* **256**, 2083 (1981).
[19] S. MacLean-Fletcher and T. D. Pollard, *Cell* **20**, 329 (1980).

TECHNIQUES FOR CHARACTERIZING ACTIN FILAMENT NETWORKS

Technique	Shear rate		Oscillation frequency	Parameters measured[a]	Range	Sample size (ml)	Expense	Commercial instrument	Throughput
	Range	Control							
1. Test tube inversion	~ 0 sec^{-1}	None	0	Gelation		0.5	Low	Not required	High
2. Low speed sedimentation	Low ?	None	0	Sedimentation		1	Moderate	Centrifuge	High
3. Falling ball	$\sim 10^{-1}$ to 10^2 sec^{-1}	Small range depending on angle and viscosity	0	Yield stress Apparent viscosity	30 to 10^4 dyn cm^{-2} 1 to 10^4 cP	0.15	Low	Not required	High
4. Gelmeter	~ 0 sec^{-1}	None	0	Static modulus of rigidity		3	Low	Not required	Low
5. Rotational viscometers	$\sim 10^{-3}$ to 10^3 sec^{-1}	Excellent	0	Yield stress Viscosity	$>5 \times 10^{-3}$ dyn cm^{-2} 1 to $>10^7$ cP	>0.3	Low to high	Available	Low
6. Micromagneto rheometer	$\sim 10^{-3}$ to 10^3 sec^{-1}	Excellent	<0.01–70 Hz	Yield stress G' G''	>10 dyn cm^{-2} ~ 10 to 10^4 dyn cm^{-2} ~ 10 to 10^4 dyn cm^{-2}	0.1	Moderate	Not available	Low
7. Zaner viscoelastometer	$<10^{-3}$ to ~ 1 sec^{-1}	Excellent	0.03–0.45 Hz	Yield stress G' G''	$>10^{-3}$ dyn cm^{-2} $\geq 5 \times 10^{-3}$ dyn cm^{-2} $\geq 5 \times 10^{-2}$ dyn cm^{-2}	0.5	Low	Not available	Low
8. Fukuda viscoelastometer	$\sim 10^{-3}$ to 10^2 sec^{-1}	Excellent	0.01–30 Hz	G' G''	<1 to $>5 \times 10^3$ dyn cm^{-2} <1 to $>5 \times 10^3$ dyn cm^{-2}	0.3	Moderate	Available	Low

[a] G' is the dynamic storage (elastic) modulus; G'' is the dynamic loss (viscous) modulus.

is used to record the results. If the concentration of gelation factor mixed with pure actin is varied, the assay can be used to determine the "critical gelling concentration" of factor.

Evaluation. This assay is inexpensive, simple, and reproducible. The shear rate is zero, the sample size is small, no special equipment is required, and multiple samples can be assayed simultaneously. Because there is usually a sharp transition from viscous liquid to solid gel at the critical gelling concentration of crosslinker, the assay can provide quantitative information about the activity of a test sample. It has even been used to assay the purification of gelation factors (e.g., by Maruta and Korn[7]). On the other hand, the scoring system is subjective, and the assay is not useful for studying the effects of a cross-linker either above or below the critical gelling concentration.

Low Speed Actin Sedimentation [8]

Principle. Actin filaments alone will not sediment when centrifuged at low speed, but actin filaments complexed with crosslinking proteins may sediment. In this assay, sedimentation of actin–crosslinker complexes is measured by measuring the protein concentration at the meniscus before and after low speed centrifugation.

Procedure. Polymerize actin at a concentration of 10 mg/ml in 0.1 M KCl, 2 mM MgCl$_2$, 1 mM ATP, 20 mM imidazole (pH 7.5) for 30 min at 25°. Prepare 1-ml samples at 0° by diluting the actin to 2 mg/ml with samples of gelation factor in the same buffer. Vortex and transfer to 1-ml capacity conical plastic microcentrifuge tubes (Sarstedt, Co., Princeton, New Jersey). Remove 20 μl from each tube for protein determination, and incubate the remainder for various times (usually 1 hr) at 25° room temperature. Then centrifuge the tubes at 10,000 g for 10 min. Carefully aspirate 20 μl from the meniscus of each tube for protein determination. Express actin sedimenting activity as micrograms of protein per 20 μl of sample sedimented per minute of centrifugation. Run control samples without gelation factor for comparison with samples containing gelation factor activity. A similar assay was devised by Mimura and Asano[9] to study a gelation factor from Erhlich ascites tumor cells.

Evaluation. Under the conditions of this assay, the amount of protein sedimented depends on the concentration of macrophage actin-binding protein, smooth muscle filamin, or skeletal muscle myosin added to the actin. The specific activity of these three proteins in this sedimentation assay paralleled their ability to increase the rigidity of actin filament solutions measured with a "gelmeter" (see below).

The main virtues of this assay are that it is simple and inexpensive and many samples can be processed simultaneously. Furthermore, the sedimentation rate depends on crosslinker concentration, so this assay is suitable for identification of active fractions and quantitation of the recovery of activity during purification procedures. It can also be used to characterize isolated actin crosslinking proteins, but sedimentation is not necessarily directly related to gelation. On the one hand, actin will sediment if it is precipitated by a component of the test sample, even if it is not gelled. On the other hand, some gels may be so rigid that they are not sedimented at 10,000 g.

Falling-Ball Assay of Viscosity and Yield Strength[11]

Principle. This assay measures apparent viscosity from the velocity of a small ball moving through the sample. For samples with yield strengths greater than the ball force at 1 g (140 dyn/cm²), it is also possible to measure the yield strength of the sample from the maximum force per unit mass that can be applied to the system without movement of the ball.

Force Balance in the System. When a ball immersed in a medium moves at a constant velocity, the forces acting in the ball and the medium are related as in Eq. (2).

$$F_b = F_y + F_m + F_f \tag{2}$$

F_b, the ball force, is the gravitational or centrifugal force that tends to move the ball down the tube. It equals the sum of the forces that resist the movement of the ball, which are F_y, the static yield force; F_m, the buoyant force; and F_f, the frictional force.

The ball force can be expressed by Eq. (3).

$$F_b = \tfrac{4}{3}\pi R^3 \rho_b \alpha \tag{3}$$

R is the radius of the ball, ρ_b is the density of the ball, and α is the force per unit mass applied to the ball by gravity or centrifugation.

The static yield force, F_y, is that necessary for the ball to deform the sample to move out of the path of the ball.

$$F_y = T_0 \pi R^2 C \tag{4}$$

The static yield strength, T_0, is the maximum force per unit area that the medium will support without displacement, and C is a geometrical constant to correct for the nonhomogeneous distribution of shear stress around the ball. For a sphere, C has been determined empirically to be about 1.75.[20]

The buoyant force arises because the ball displaces its own volume of the medium.

[20] A. M. Johnson, "Physical Processes in Geology." Freeman, San Francisco, California, 1970.

$$F_m = \tfrac{4}{3}\pi R^3 \rho_m \alpha \tag{5}$$

ρ_m is the density of the medium and α is the force per unit mass derived from gravity or centrifugation applied to the medium.

If the ball moves through the medium there will be a frictional force (F_f) resisting its movement.

$$F_f = fu \tag{6}$$

f is the dynamic frictional coefficient and u is the velocity of the ball.

VISCOSITY MEASUREMENTS WITH THE FALLING-BALL DEVICE

Theory. If the medium is not strong enough to support the ball, the ball will accelerate until it reaches a constant velocity. Under these conditions the forces are balanced as given in Eq. (2), and the absolute viscosity of a Newtonian fluid can be determined from the ball velocity using Stokes' equation

$$\eta = f/(6\pi R) \tag{7}$$

If F_y is negligible compared with the other forces in the system, then $F_f = F_b - F_m$ and combining Eqs. (3), (5), and (6)

$$f = [m \cdot \tfrac{4}{3}\pi R^3 \alpha(\rho_b - \rho_m)]/u \tag{8}$$

The calibration constant, m, is required when the ball and the medium are confined to a narrow tube and inclined at an angle. It corrects for drag on the ball resulting from wall effects[21] and friction at the point of contact between the ball and the wall.[22] The value of m is determined empirically and is a function of the radius of the ball, the radius of the tube, and the angle of inclination (see below). If gravity is the only force on the system, the force per unit mass (α) along the axis of a tube inclined at angle θ from the horizontal is $g \sin \theta$. Solving Eq. (7) and Eq. (8) for η

$$\eta = [0.22mR^2 g \sin \theta(\rho_b - \rho_m)]/u \tag{9}$$

Note that viscosity is inversely proportional to ball velocity, u.

The absolute viscosity of a complex non-Newtonian fluid like actin-filament networks varies with the shear rate,[14] so that the observed ball velocity is used to calculate only an "apparent viscosity" from Eq. (9). Note also that the shear dependence of the viscosity of non-Newtonian samples such as actin filaments will amplify differences in the apparent viscosity. A higher shear rate yields a lower apparent viscosity. A sample

[21] J. R. Van Walzer, J. W. Lyons, and K. Y. Kim, "Viscosity and Flow Measurement: A Laboratory Handbook of Rheology." Wiley (Interscience) New York, 1963.
[22] R. H. Geils and R. C. Keezer, *Rev. Sci. Instrum.* **48**, 783 (1977).

with a lower absolute viscosity will have higher ball velocity, which causes a higher shear rate and therefore an even lower apparent viscosity.

Procedure. In a small test tube prepare a sample of 200 μl by mixing in the following order water, 10 × polymerizing buffer (final concentrations 2 mM MgCl$_2$, 1 mM ATP, 1 mM EGTA, 10 mM imidazole, pH 7), unknown sample, and actin monomer to give a final concentration of 0.5 mg/ml. Degas and/or warm to 25° the water, actin, and buffer before use. Immediately after adding actin to initiate the reaction, draw the sample into a capillary tube, 1.3 mm i.d., 12.6 cm long (100-μl micropipette from VWR Scientific, Inc., Univar Corp., San Francisco, California) by capillary action or by using a Clay Adams Pipet Filler (Clay Adams, Division Becton, Dickinson & Co., Parsippany, New Jersey). Some speed is essential in transferring the sample to the capillary before the actin polymerizes. Seal the capillary at one end with Clay Adams Seal Ease and incubate at 25°. Place the capillary on a simple homemade plexiglass stand (Fig. 2) at a fixed angle of inclination in a 25° water bath with transparent walls, such as a 5$\frac{1}{2}$-gallon aquarium equipped with a thermocirculator. Place a stainless steel ball (0.64 mm in diameter, density 7.2 g/cm^2, grade 10, gauge deviation + 0.000064 mm, material 440C (from either the Microball Company, Peterborough, New Hampshire; The New England Miniature Ball Company, Norfolk, Connecticut; or the Winsted Precision Ball Corpora-

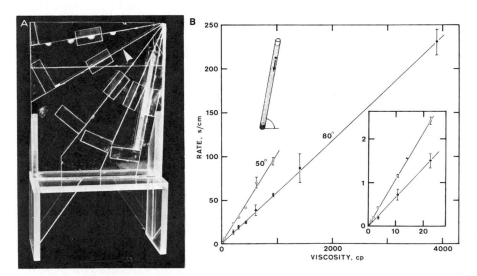

FIG. 2. Miniature falling-ball viscometer. (A) Plastic rack for holding the capillary tubes. The arrow marks a stainless steel ball in a capillary tube. (B) Calibration of the viscometer using glycerol–water mixtures and two different angles of inclination. From MacLean-Fletcher and Pollard.[11]

tion, Winsted, Connecticut 06098) on the meniscus of the sample by hand. A convenient method is to use a pair of fine forceps with magnetized tips. Larger and smaller stainless steel balls can be purchased from the Winsted Precision Ball Corporation. At an appropriate time, push the ball through the meniscus with a thin metal wire to initiate its fall. Do not sharply tap the tube to start the fall. Measure the velocity of the ball by recording the time required for the ball to pass 2-cm intervals beginning about 1 cm below the meniscus. For samples with very high viscosity, a shorter interval, down to 2 mm can be used; and for low-viscosity samples, a longer distance improves accuracy.

Calibrate the viscometer with glycerol–water mixtures that are Newtonian solutions of known viscosity.[23] The ball has a constant velocity throughout the measuring section of the capillary for any given sample over the entire viscosity range tested (1–12,000 cP). At any angle between 10 and 90 degrees, inverse velocity is proportional to viscosity up to the point where the ball will not move (Fig. 2). The measurements are highly reproducible, with a standard deviation from the mean velocity of $<2\%$ over the entire range for an experienced operator. The variation is larger at high viscosity than at low viscosity. Tilting the capillary at different angles (θ) allows one to vary the useful range of the apparatus. Most of our measurements are made at 50 or 80 degrees because our samples have high viscosities. An angle of 10 degrees is more convenient for low-viscosity samples.

The linearity of the relationship between u^{-1} and viscosity demonstrates that the instrumental constant, m, does not vary with the viscosity of the sample, and justifies, for Newtonian fluids, the omission of the F_y term from the viscosity equation.[8] (This result is expected from the magnitude of the forces in the system. At 80°, $F_b - F_m = 0.82$ dyn. For $F_y = 1\%$ of this force, $T_0 = 1.46$ dyn/cm^2. The standards used for calibration have yield strengths even less than this.) These tests establish that the viscometer can be used to measure the absolute viscosity of Newtonian solutions in the range of 1–12,000 cP with an error of $<2\%$.

YIELD-STRENGTH MEASUREMENTS WITH THE FALLING-BALL DEVICE

Theory. When the medium is strong enough to support the ball, there is no acceleration, so $F_f = 0$, and the forces are balanced, $F_b = F_y + F_m$. Under these conditions, it is possible to estimate the static yield strength, T_0, of the medium by determining α_0, the maximum force per unit mass that can be applied to the system without movement of the ball. This is

[23] C. D. Hodgman, "Handbook of Chemistry and Physics." CRC Press, Cleveland, Ohio, 1958.

accomplished by centrifuging a capillary containing a ball immersed in the test medium. This applies to the system a force per unit mass of $\omega^2 R$, where ω is angular velocity in radians per second and R is the radius. Solving for T_0

$$T_0 = \frac{\frac{4}{3}\pi R^3 \alpha_0 (\rho_b - \rho_m)}{1.75\pi R^2} = 0.76 R \alpha_0 (\rho_b - \rho_m) \tag{10}$$

Procedure. If the sample has a yield strength >146 dyn/cm^2, the 0.64 mm ball will fail to move at an angle of 80° under the influence of gravity alone. A measurement of yield strength can be made by increasing the ball force. Push the ball about 0.5 cm below the meniscus, mark its position, and centrifuge in a stepwise fashion for 30 sec at progressively higher speeds. $\alpha_0 = R\omega^2$ for the speed just before the ball moves. For samples with T_0 <146 dyn/cm^2, a rough estimate of yield strength can be made by placing the ball in a horizontal tube and slowly increasing the angle of inclination until the ball moves. In this case $\alpha_0 = g \sin \theta$.

Evaluation. The falling-ball assay for viscosity and yield strength is simple, inexpensive, and highly reproducible. The sample size is small, and the sample is not disturbed before interaction with the ball. The viscosity range is wide, and it is possible to make about one measurement per minute. Kinetic data on a single sample are obtained with ease. By choosing an appropriate angle of inclination, it is possible to keep the maximum shear rate[21] ($D = 1.5 u/R$) for Newtonian fluids below 10 sec^{-1} for samples with viscosities >6 cP. For comparison, typical Ostwald capillary viscometers have shear rates >1000 sec^{-1}. The low shear rate enables one to study weak interactions between actin filaments. The broad range is important because samples vary in apparent viscosity by more than three orders of magnitude. The low shear rate is essential, because the networks are destroyed by shearing.

The ball centrifugation assay is a simple approach for measuring yield strength. Its major limitations at the present time are some uncertainty about the value of the geometrical constant in Eq. (4) and the crude, stepwise application of force by centrifugation. A better method would be to observe the ball continuously as the centrifugal force is gradually increased.

To date this falling-ball apparatus has been used to analyze the interaction of actin filaments with microtubules and microtubule-associated proteins,[24] with immunoglobulins,[25] with *Acanthamoeba* gelation factors,[11,19,26] and with erythrocyte spectrin and band 4.1.[27] In all these

[24] L. M. Griffith and T. D. Pollard, *J. Cell Biol.* **78**, 958 (1978).
[25] M. Fechheimer, J. L. Dias, and J. J. Cebra, *Mol. Immunol.* **16**, 881 (1979).
[26] T. D. Pollard, *J. Biol. Chem.* **256**, 7666 (1981).
[27] V. Fowler and D. L. Taylor, *J. Cell Biol.* **85**, 361 (1980).

studies the added proteins caused the actin filament samples to have a viscosity considerably higher than the viscosity of actin alone, presumably owing to the formation or stabilization of crosslinks between the actin filaments. The apparatus was also used to study the kinetics of *Acanthamoeba* extract gelation.[11] Another type of application has been the detection of very weak interactions between actin filaments and the demonstration that these filament self-associations can be inhibited by cytochalasins,[19] by an unpurified protein factor that contaminates conventional muscle actin preparations,[28] and by an actin filament capping protein which is purified from *Acanthamoeba*.[29] The falling-ball device has also been used to study the consequences of actin-filament shortening by gelsolin[17] or villin[18] on the gelation of actin by high molecular weight actin-binding proteins.

Although the falling-ball method has many advantages, it has three main limitations in studies of gelation. First, only two structural parameters (apparent viscosity and yield strength) can be measured. Second, for a given sample the shear rate can only be varied by a factor of ~12 when using gravitational acceleration and angles of inclination between 10 and 80 degrees. Finally, even at the relatively low shear rates in the falling-ball device, the stress applied to these delicate samples by the falling ball is destructive. For more quantitative evaluation of the physical properties of actin gels, it is necessary to specify the shear rate, to test the dependence of viscosity on shear rate, and to measure elasticity as well.

Gelmeter[8]

Principle. This is a simple hydrostatic device to measure the force required to disrupt a gel and provide a measurement of the modulus of rigidity at the yield point of the gel. The sample is confined in a capillary tube, and the yield point is detected by direct microscopic observation of the capillary as pressure is applied to one end.

Procedure. The gelmeter consists of a capillary tube (i.d. = 1.2 mm, length = 7.5 mm) that is connected at each end to a vertical cylinder (i.d. = 6 mm) (Fig. 3). The contents of the capillary are viewed with a microscope having a 40× objective.

Fill the capillary and the two chambers at 4° with a 3-ml sample of polymerized actin with or without a crosslinking protein. In the case of macrophage actin-binding protein, a gel forms when the device is warmed to 25°. Include a few charcoal particles in the sample to help visualize its microscopic movements.

Determine the yield point by advancing a threaded plunger into one of

---

[28] S. MacLean-Fletcher and T. D. Pollard, *Biochem. Biophys. Res. Commun.* **96**, 18 (1980).
[29] G. Isenberg, U. Aebi, and T. D. Pollard, *Nature (London)* **288**, 455 (1980).

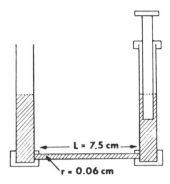

FIG. 3. Gelmeter. Schematic drawing showing the design of the instrument. From Brotschi et al.[8]

the two chambers. This raises the fluid level in that chamber and creates a hydrostatic pressure difference across the capillary. When the force on the sample in the capillary exceeds its yield point, the sample will deform and the charcoal particles will be displaced. The modulus of rigidity (μ) is calculated from the pressure (P), capillary radius (r), capillary length (l), and maximum displacement of particles along the longitudinal axis of the capillary (X) according to Eq. (11).

$$\mu = (Pr^2)/(4lX) \tag{11}$$

Although this procedure is satisfactory for actin filaments alone and actin with filamin, actin-binding protein, or myosin, we suggest two modifications.

1. A more homogeneous gel would form if actin monomers were introduced into the capillary and then allowed to polymerize. The flow of filaments into a capillary may result in their alignment along the axis of the tube and the formation of a nonhomogeneous gel.
2. Inert particles such as polyacrylamide beads might be preferable to charcoal, because charcoal may adsorb some of the constituents of the sample including ATP, which is essential for gelation of some specimens.

Evaluation. This method provides a measure of the static modulus of rigidity of the sample at its yield point. The modulus of rigidity is the elastic (spring) constant of the material at the yield point, the smallest stress that will deform the material. Note that the sample does not flow, so that this definition of the yield point differs from that described above for the falling-ball assay and the rotational viscometers (see next section). The yield point and the modulus of rigidity are important characteristics of a gel, so this is a valuable means of characterizing actin networks. Disadvantages of the method for routine measurements are the large sample size and the low sample throughput. Miniaturization of the apparatus would economize on sample consumed.

Rotational Viscometers

Principle. A cylindrical bob is suspended by a fiber in a concentric cylindrical chamber containing the specimen. The chamber is rotated, applying to the measuring bob a torque that is opposed by a torque generated in the fiber suspending the bob. For a Newtonian fluid the bob will rotate with the chamber and the specimen until the torques are balanced, at which point the bob stops or reverses direction.

Theory.[30] The viscosity (in poise) is calculated from the dimensions of the apparatus, and N is the number of revolutions of the bob.

$$\eta = NC(b^2 - a^2)/2\omega hb^2a^2 \tag{12}$$

where a is the radius and h is the height of the bob, b is the inner radius of the chamber, ω is the angular velocity (rad sec^{-1}) of the chamber, and C is the restoring constant of the suspending fiber.

$$C = (4\pi^2 I)/t_0^2 \tag{13}$$

where I is the moment of inertia of the bob and t_0 is the period of oscillation of the bob in air. The steady-state shear rate for a Newtonian fluid at the radius of the bob is given by

$$D = 2\omega/(1 - a^2/b^2) \tag{14}$$

The shear rate varies across the gap for all fluids. For Newtonian fluids shear rate falls off with the radius squared. For non-Newtonian fluids the situation is much more complex.[21]

For solid-like sample with an appreciable yield strength, the bob will turn with the cylinder until the yield stress, T_B, is reached.

$$T_B = NC/(a^2h) \tag{15}$$

Then the bob will rotate in the opposite direction back to a position where the torques are again balanced and the viscosity can be measured.

A rotational viscometer can also be used to measure the elasticity of a sample.[31] Rather than rotating the chamber or the bob at a constant velocity, the bob is rotated a few degrees and then released. If one assumes that the material is a Voigt element with parallel viscous and elastic elements,

[30] D. W. Kupke and J. W. Beams, *Proc. Natl. Acad. Sci. U.S.A.* **74**, 1993 (1977).
[31] M. Kasai, H. Kawashima, and F. Oosawa, *J. Polym. Sci.* **44**, 51 (1960).

the storage modulus, G' (dynes cm^{-2}) is obtained by simply comparing t, the period of oscillation of the bob in the sample, with t_0, the period of oscillation in air, with the elasticity expressed as

$$G' = C_1(1/t^2 - 1/t_0^2) + C_2\eta^2 \qquad (16)$$

The constants C_1 and C_2 are calculated from the dimensions of the apparatus.

$$C_1 = \pi I(b^2 - a^2)/hb^2a^2 \qquad (17)$$
$$C_2 = \pi^2/C_1 \qquad (18)$$

In the apparatus used by Kasai *et al.*,[31] h was 5.5 cm, a was 0.9 cm, b was 1.1 cm, I was 48.37 g cm^2, so C_1 was 11.25 g · cm^{-1} (F. Oosawa, personal communication, 1981). The oscillation in air, t_0, was a 3.33 sec, and the rate of shear in the sample was less than 1 sec^{-1}.

Instrumentation. A torsion-fiber apparatus can be constructed from simple materials as described in detail by Kupke and Beams[30] (Fig. 4). The torsion fiber in their apparatus was selected to measure very small yield stresses (<0.2 dyn cm^{-2}) and low viscosities. A thicker fiber with a larger restoring constant is needed for gels with higher yield stresses.

Alternatively, commercial instruments are available. However, except for a very expensive low-shear device from Contraves AG (Zurich, SW CH-8052), these commercial instruments are not particularly well suited to the analysis of actin filament solutions and gels because they either have relatively high shear rates or require large samples. The Contraves Low Shear 100 was used by Maruyama[14] for his classic analysis of the shear rate dependence of the viscosity of actin filament solutions. This instrument is available with a control unit that can hold either the torque or the angular velocity constant. Rather than counting the displacement of the bob by direct observation, the rotation of the suspending fiber is recorded electronically with a mirror and photo optical device.

A clever variation of the rotational viscometer employs a cylindrical bob suspended in the sample by a magnetic field.[32] Torque is applied to this bob by rotation of the cylindrical chamber containing the sample. Rotation of the bob is opposed by an external magnetic field, which is varied to keep the bob stationary. Thus the torques are balanced, and the torque applied by rotation of the chamber and sample is measured continuously by the strength of the opposing magnetic field.

Rotational viscometers can also be made or purchased with other types of sample containers. A popular design consists of the sample confined between a plane and a shallow cone. This cone-plate design is widely used in rheology, but has been used little with actin.

[32] J. W. Beams and D. W. Kupke, *Proc. Natl. Acad. Sci. U.S.A.* **74,** 4430 (1977).

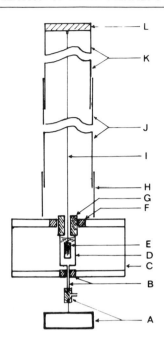

FIG. 4. Rotational viscometer. Schematic drawing of the torsion-fiber apparatus de-scribed by Kupke and Beams[30] A, Reversible, constant-speed motor and shaft to Teflon coupler with side vent for filling; B, hollow shaft to cell and guide bearing, C, thermostated water bath; D, rotatable glass cell; E, immersed cylinder; F, guide bearing for cell top; G, cell extension to fit guide bearing; H, telescoping chimney extension for access; I, thin fiber; J, main chimney; K, telescoping upper chimney for height adjustments; L, fitted disk with center hole for anchoring fiber.

Procedure. Introduce the sample into the chamber and lower the bob into the sample. After a suitable time, rotate the chamber at a constant angular velocity and observe the bob with a 25× microscope. The number of turns of the bob required for the bob to become stationary or to reverse direction is noted and used to calculate the viscosity from Eq. (12) and/or yield stress from Eq. (15). Vary the angular velocity to determine the dependence of viscosity on shear rate. The viscosity of complex, non-Newtonian samples may also vary with time at a constant shear rate. Evaluate the elasticity of the sample by rotating the bob a few degrees, releasing it, and measuring the period of oscillation. With time a Newto-nian fluid will damp the amplitude, but not the period, of the oscillation. A viscoelastic material will also reduce the period of oscillation compared with free oscillation in air. This difference in period of oscillation is used to calculate the elastic modulus from Eq. (16).

Evaluation. Rotational viscometry is one of the best ways to measure the yield stress, the absolute viscosity, and elasticity as a function of shear rate. This approach has been widely used with great success to study actin filaments.[14,31] The shear rate is low and can be controlled. The sample size is small, and the apparatus is relatively simple. The theoretical basis of the measurements is well established. Only one sample can be measured at a time, so it is not particularly useful for routine measurements during protein purification. Although the results are well worth the wait, the microscopic observation of the bob can be tedious. This drawback can be circumvented with a more sophisticated commercial or homemade instrument.

Micromagnetorheometer[33]

Principle. A tiny iron sphere (~50 to 100 μm in diameter) is suspended in 0.1 ml of a viscous material and oscillated by two external electromagnets driven by a sinusoidally varying voltage. From the amplitude and phase of the sphere movement, the dynamic viscous (loss) modulus and elastic (storage) modulus can be measured as a function of the frequency and amplitude of the driving force.

Theory. When the sphere oscillates in a non-Newtonian fluid the energy storing (elastic) elements of the fluid can be described by a storage modulus, G', with units of dynes cm^{-2}.

$$G' = G_0 \cos \delta \tag{19}$$

The shear modulus G_0 is

$$G_0 = T_0/\gamma_0 \tag{20}$$

where T_0 is the amplitude of an oscillating stress, γ_0 is the amplitude of the strain, and δ is the phase angle between the stress and the strain.

The energy dissipating elements in a Newtonian or non-Newtonian fluid are described by a loss modulus, G'', with units of dynes cm^{-2}

$$G'' = G_0 \sin \delta \tag{21}$$

In other words the elastic elements giving rise to strain in the fluid are in phase with the stress, and the viscous elements are 90 degrees out of phase. Another way of saying this is that elastic strain is greatest at the maximum amplitude of the oscillating stress whereas the viscous strain is minimal at the maximum amplitude of the oscillating stress.

[33] R. J. Lutz, M. Litt, and L. W. Chakrin, *in* "Rheology of Biological Systems" (H. L. Gabelnick and M. Litt, eds.), p. 119. Thomas, Springfield, Illinois, 1973.

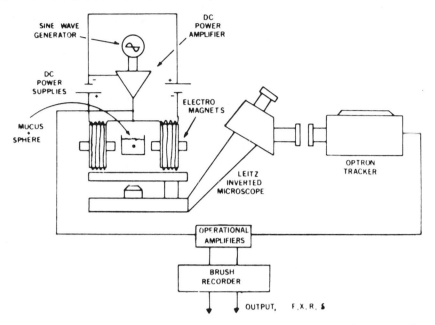

FIG. 5. Micromagnetorheometer. Schematic showing the design of the instrument. From Lutz *et al.*[33]

In the micromagnetorheometer

$$G' = \frac{F_0}{6\pi r X_0} \cos \delta \tag{22}$$

$$G'' = \frac{F_0}{6\pi r X_0} \sin \delta \tag{23}$$

where F_0 is the amplitude of the magnetic force, X_0 is the amplitude of the displacement of the sphere, and r is the radius of the sphere. F_0 is determined empirically using a Newtonian fluid with known viscosity and no elasticity ($G' = 0$).

$$F_0 = 6\pi r \eta (\omega X_0) \tag{24}$$

where ω is the frequency in radians sec^{-1}.

Instrumentation. The original micromagnetorheometer (Fig. 5) used a sine wave generator to drive opposing electromagnets on either side of the specimen, a microscope to image the sphere and a TV tracking device to follow the movement of the sphere. The phase shift was measured electronically using the amplitudes of the F_0 and X_0 as described in detail.[33]

Using suggestions from M. Litt, Tishler *et al.* built an improved instrument that uses a stroboscopic method to determine δ and that has the magnets oriented vertically so that the sphere can be suspended magnetically during the gelation of a sample.[34]

Procedure. Place a sample of about 0.1 ml in the thermostatted chamber and insert a small iron sphere 50–100 μm in diameter. If the yield stress of the sample is not exceeded, the sphere will be stationary. If the sphere does settle slowly, it can be suspended magnetically by adjusting the current through the magnets until time for a measurement. Apply sufficient current to the coils to displace the sphere and measure the amplitude and phase shift of the sphere movements as the driving force is oscillated sinusoidally over an appropriate frequency range. The original instrument had a frequency range of <0.01 to 70 Hz. The instrument can also be used at lower frequencies in a "creep" mode, where a constant force is applied to the sphere and strain (displacement) is followed as a function of time.

Evaluation. This instrument has been used extensively to characterize mucus, a complex, non-Newtonian, viscoelastic substance. The advantages are the small sample size, the broad range and precise control of shear stress, frequency, and shear rate, and the ability to measure both viscosity and elasticity simultaneously. These features suggest that it will be useful for characterizing actin solutions and gels, but only preliminary experiments have been done at this time.

Zaner Viscoelastometer[35]

Principle. This homemade device uses a small planar probe immersed in the sample to measure either static stress–strain information or dynamic viscous and elastic moduli at very low shear rates.

Instrumentation. The viscoelastometer (Fig. 6) consists of a thin sheet of mica (1 × 1.5 × 0.0025 cm) immersed in a small cuvette containing a sample of about 0.5 ml. The mica probe is fastened to the lower end of a 12-cm-long pendulum suspended from its top on a knife edge supported by a rigid metal post. This whole assembly sits on a tilt table driven by an electric micrometer drive. The table is tilted, in increments of 2×10^{-5} rad up to a maximum of $\pm 10^{-2}$ rad, the pendulum tilts in the plane perpendicular to the planar probe, and the probe swings horizontally in a shallow arc through the sample. The tilt of the pendulum in response to tilting of the table is followed by a sensitive metal detector placed near two 6-mm metal disks on the pendulum. The table can either be tilted in steps to record

---

[34] R. Tishler, R. Cone, and T. D. Pollard, unpublished work, 1981.
[35] K. S. Zaner, R. Fotland, and T. P. Stossel, *Rev. Sci. Instrum.* **52,** 85 (1981).

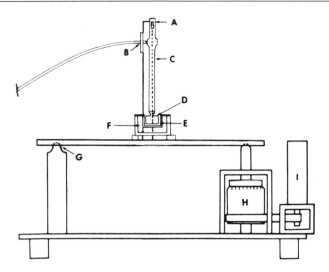

Fig. 6. Zaner viscoelastometer. Schematic of the instrument. A, Knife edge oriented normal to the plane of the diagram; B, metal detector; C, probe; D, mica tip; E, sample chamber; F, water jacket; G, pivot bearings; H, micrometer; I, DC motor. From Zaner et al.[35]

static stress-strain information or sinusoidally at 0.035–0.45 Hz to measure dynamic storage (elastic) and loss (viscous) moduli. In the sinusoidal mode the phase shift between the table and probe tilting is recorded with a counting circuit.

Theory. The equations describing the viscoelastometer are derived in the original paper[35] as follows:

The programmer tilts the table in a sinusoidal manner so that the angle of tilt from the horizontal, θ is

$$\theta(t) = A \sin \omega t \tag{25}$$

where A is the amplitude and ω is the frequency. The motion of the table generates a movement of the probe that is detected by the sensor and is given by

$$\phi(t) = H \sin(\omega t - \delta') \tag{26}$$

where ϕ is the angle between the probe and the metal post, H is the amplitude, and δ' is the phase difference between θ and ϕ.

Assuming that the motion of the mica through the fluid is exerting a pure shearing force, the stress σ at any value of θ and ϕ is given by

$$\sigma = \frac{WR}{2(L_0 - h/2)wh} (\theta - \phi) \tag{27}$$

where W is the weight of the probe, R is the distance from the knife edge to the center of mass of the probe, L_0 is the distance from the knife to the end of the mica tip, w is the width of the mica in contact with the sample, and h is the height of the mica in contact with the sample.

Substituting (25) and (26) into (27), we obtain:

$$\sigma = \frac{WR}{2(L_0 - h/2)wh} (A^2 - 2HA \cos \delta' + H^2)^{1/2} \sin(\omega t - a) \quad (28)$$

where a is

$$a = \tan^{-1} \left(\frac{H \sin \delta'}{A - H \cos \delta'} \right) \quad (29)$$

The shear strain, ϵ, is:

$$\epsilon = \frac{L_0 - h/2}{d} H \sin(\omega t - \delta') \quad (30)$$

where d is the distance from the mica tip to the wall of the cuvette. For viscoelastic materials, the dynamic storage (G') and loss moduli (G'') are

$$G' = \frac{\sigma_0}{\epsilon_0} \sin \delta \quad (31)$$

$$G'' = \frac{\sigma_0}{\epsilon_0} \cos \delta \quad (32)$$

where σ_0 is the amplitude of the stress, ϵ_0 is the amplitude of the strain, and δ is the phase shift between the stress and strain. From Eqs. (28) and (30)

$$\frac{\sigma_0}{\epsilon_0} = \frac{WRd}{2(L_0 - h/2)^2 wh} [(A/H)^2 - 2(A/H) \cos \delta' + 1]^{1/2} \quad (33)$$

and

$$\delta = \delta' - a \quad (34)$$

Substituting Eq. (29)

$$\sin \delta = \frac{\sin \delta'[(A/H) - \cos \delta'] + \cos \delta' \sin \delta'}{[(A/H)^2 - 2(A/H) \cos \delta' + 1]^{1/2}}$$

$$= \frac{A}{H} \frac{\sin \delta'}{[(A/H)^2 - 2(A/H) \cos \delta' + 1]^{1/2}} \quad (35)$$

Defining C by

$$C = \frac{WRd}{2(L_0 - h/2)^2 wh} \quad (36)$$

we obtain

$$G'' = C(A/H \sin \delta')$$ (37)

and, similarly

$$G' = C(A/H \cos \delta' - 1)$$ (38)

Operation. Place a sample of 0.5 ml in the cuvette together with the mica probe. Collect static stress-strain data by tilting the table in steps and recording the steady-state position of the probe arm. Alternatively, rock the table sinusoidally at a range of frequencies under the control of a programmer unit. The probe arm will then swing sinusoidally. The amplitude (H) of the movements is determined by the metal detector, and the phase difference (δ') between the motion of the table and the arm is determined by a counting circuit. Both of these parameters are recorded. The dynamic storage modulus (G') and loss modulus (G'') are calculated from Eqs. (37) and (38).

Evaluation. The viscoelastomer was calibrated with glycerol, and C was ~ 0.75 dyn cm^{-2}. Since a mica probe half as wide as the standard probe gave a value of (A/wH) twice that of the standard probe, it was concluded that the stress applied to the sample is a pure shearing force. This shearing force is less than 0.01 dyn/cm^2, corresponding to a maximum shear rate of 1 sec^{-1} for water and 10^{-3} sec^{-1} for glycerol at 25°. The machine is designed for samples with small dynamic moduli and can easily measure $G' > 0.08$ dyn/cm^2 and $G'' > 0.9$ dyn/cm^2. Modifications of the probe are said to reduce these limits by a factor of 15.

This device was designed specifically for work with actin and it seems well suited for this role due to its sensitivity and small sample volume. Preliminary results with actin and actin-binding proteins[36] seem to bear out this promise.

Fukada Viscoelastometer[37]

Principle. A cylindrical plunger is vibrated up and down in a sample contained in a concentric cylinder (Fig. 7). The shearing force transmitted through the sample causes the outer cylinder to vibrate vertically. The dynamic elasticity and viscosity are determined from the amplitude and phase of the motion of the outer cylinder.

Instrumentation. In the original Fukada instrument (Fig. 7) a stainless steel rod (D) 3 mm in diameter, is immersed 18 mm into a 0.3 ml sample

---

[36] T. P. Stossel, K. Zaner, J. Hartwig, H. Yin, and F. Southwick, *Cold Spring Harbor Symp. Quant. Biol.* **46**, in press (1981).

[37] E. Fukada and M. Date, *Biorheology* **1**, 101 (1963).

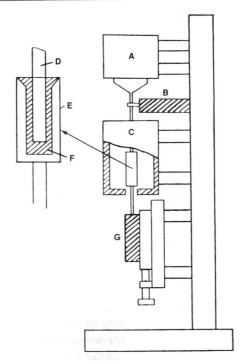

FIG. 7. Fukada viscoelastometer. Schematic of the instrument. A, Electromagnetic vibrator; B, strain gauge for amplitude detection; C, thermostat; D, inner rod; E, outer tube; F, specimen; G, strain gauge for force detection. From Fukada and Date.[37]

contained in the 5 mm in diameter, 20 mm long stainless steel chamber (E). The chamber is tapered at its upper end to minimize surface tension effects. An electromagnetic vibrator (A) oscillates the rod in the range of 0.01–30 Hz and an amplitude of 5–50 μm. The amplitude of these oscillations is detected by a strain gauge (B). At 10 Hz this yields shear rates of 0.5–5 sec^{-1}. The resulting oscillations of the chamber are detected by a second strain gauge (G). Gauge G has a large impedance so that the movement of the chamber is small compared with the rod. The outputs of the two strain gauges are amplified identically and used to drive the X and Y axes of a storage oscilloscope yielding a Lissajous figure revealing the relationship between the stress and strain. The instrument has been available commercially from Sanki Eng. Ltd. (Kyoto, Japan).

Theory. The shearing force, F, can be represented by a complex expression

$$F = F' + iF'' \tag{39}$$

F' and F'' can be read directly from the Lissajous figure on the oscillo-scope. F' is the height on the Y axis between $x = L/2$ and $X = L/2$, where L is the amplitude of the rod motion. F'' is the height on the Y axis between the two points where $X = 0$. The dynamic storage (elastic) modulus is

$$G' = (CF')/L \qquad (40)$$

where C is an instrumental constant. The dynamic loss (viscous) modulus is

$$G' = \omega\eta = (CF'')/L \qquad (41)$$

where ω is the angular velocity and η is the dynamic viscosity. The instru-mental constant, C, is determined experimentally using silicone oils of known viscosity that have no elasticity. $C = 1.2 \times 10^{-2}$ cm^{-1} for the instrument in Fig. 7.

Procedure. As in the case of other viscoelastometers, introduce a sam-ple into the chamber, insert the probe, wait a suitable time, and then measure the relationship between stress and strain as a function of fre-quency of oscillation.

Evaluation. This device allows one to measure both elasticity and vis-cosity simultaneously over a range of frequencies. It may be less versatile than some of the other viscoelastometers described above, but has the virtue that it has been used successfully to study mixtures of actin fila-ments with heavy meromyosin[38] and β-actinin.[39] Similar practical experi-ence is lacking with the Zaner viscoelastometer and the micromag-netorheometer.

Acknowledgment

The authors' work is supported by Research Grants GM-26338 and GM-26132 from the National Institutes of Health. J.A.C. is supported by Medical Scientist Training Program GM 7309. We thank Drs. F. Oosawa, T. P. Stossel, K. Zaner, and S. Craig for their helpful comments on some of the sections.

[38] S. Abe and K. Maruyama, *Biochim. Biophys. Acta* **342**, 160 (1974).
[39] K. Maruyama, S. Abe, and T. Ishii, *J. Biochem.* **77**, 131 (1975).

[21] Preparation and Identification of α- and β-Tropomyosins

By L. B. SMILLIE

Tropomyosins are present in all forms of striated and smooth muscle and probably in many nonmuscle cells as well. Whereas their function in smooth muscle and nonmuscle cells is obscure, their role along with the troponin complex in the actin-linked calcium regulatory systems of skeletal and cardiac muscle is well documented. In all skeletal tissues two major forms of tropomyosin chains (designated α and β) are present. The ratio of α to β is dependent on the muscle source with the α form predominant in fast muscle types, whereas in slower muscles the $\alpha : \beta$ ratio approaches unity.[1,2] Although the two forms are separable on sodium dodecyl sulfate (SDS)–polyacrylamide gels, amino acid sequence data for the α- and β-forms from mixed back and leg muscles of the rabbit have shown that their chain lengths (284 residues) are the same.[3-5] These data have also demonstrated that, although there are a minimum of four different sequence varieties of tropomyosin from this source, the minor variants probably differ from the two major α and β components at only a few amino acid residues.

The physiological significance of the two major α and β components and their minor variants is presently unknown. In cardiac tissue the amount of β component in large and slowly beating hearts (beef, sheep, pig, human) is about 15–20% of the total, whereas in smaller animals (rabbit, guinea pig, dog) it has been reported to be absent.[6] Amino acid sequence analyses have demonstrated that the tropomyosin of rabbit cardiac tissue is pure α component and identical to that of the α component of rabbit skeletal muscle.[7] No indications of sequence heterogeneity were observed. Rabbit cardiac tissue is thus a convenient source for the isolation of homogeneous α,α-tropomyosin without exposure to denaturing conditions.

Essentially all methods described in the literature for the isolation of tropomyosin are modifications of the original procedure described by Bailey.[8] This involves the initial removal of the soluble sarcoplasmic pro-

[1] P. Cummins and S. V. Perry, *Biochem. J.* **133**, 765 (1973).
[2] P. Cummins and S. V. Perry, *Biochem. J.* **141**, 43 (1974).
[3] J. Sodek, R. S. Hodges, and L. B. Smillie, *J. Biol. Chem.* **253**, 1129 (1978).
[4] D. Stone and L. B. Smillie, *J. Biol. Chem.* **253**, 1137 (1978).
[5] A. S. Mak, L. B. Smillie, and G. R. Stewart, *J. Biol. Chem.* **255**, 3647 (1980).
[6] J. Leger, P. Bouvert, K. Schwartz, and B. Swynghedauw, *Pfluegers Arch.* **362**, 271 (1976).
[7] W. G. Lewis and L. B. Smillie, *J. Biol. Chem.* **255**, 6854 (1980).
[8] K. Bailey, *Biochem. J.* **43**, 271 (1948).

METHODS IN ENZYMOLOGY, VOL. 85

teins by extraction of a muscle mince with water or dilute salt solutions and the dehydration of the insoluble residue by washes with alcohol and ether. Acetone is often now substituted for ether. Myosin is denatured by these procedures. The tropomyosin is then extracted from the dried ether or acetone powder with 1 M KCl at neutral pH and purified by repeated cycles of isoelectric precipitation at pH 4.5–4.7 and ammonium sulfate fractionation at pH 7.0–7.8. At this stage the product is sometimes contaminated with low levels of troponin components or their proteolytic degradation products. These may be conveniently removed by chromatography on hydroxyapatite in 1 M KCl with a phosphate gradient at pH 7.0.[9,10]

In many cases it may be expedient to prepare tropomyosin as a byproduct in the preparation of troponin (see this volume [22]). At an early stage in this procedure the muscle extract is adjusted to pH 4.6 to remove the tropomyosin as a precipitate, leaving the bulk of the troponin complex in the supernatant. This precipitate may then be subjected to the repeated cycles of isoelectric precipitation and $(NH_4)_2SO_4$ fractionation as described above.

Since tropomyosin has no easily measurable biological activity, its purity is normally assessed by SDS–polyacrylamide gel electrophoresis in the presence and/or the absence of 6 M urea.[11] In the absence of urea the α chains of rabbit skeletal and cardiac tropomyosin migrate with an apparent M_r of 33,000, close to the true value as established from their known covalent structures. In the presence of 6 M urea the mobility is reduced compared to other proteins with an apparent M_r of 51,000. This unique property of tropomyosin is of great value in its identification in a complex mixture and in establishing the absence of troponin-T contamination in purified preparations. Since troponin T has a similar mobility to tropomyosin in the absence of urea, it can sometimes be difficult to detect. Biological assays for tropomyosin involve measurements of its potentiating effect on the inhibition of the actomyosin ATPase by troponin I and of the calcium sensitivity of a reconstituted actomyosin–tropomyosin–troponin system. Such assays are described elsewhere in this volume [22]. Because of their complexity they are not readily adaptable to routine assays.

As indicated above, purified α,α-tropomyosin can be prepared directly from rabbit cardiac tissue. Alternatively, it can be prepared by chromatography of rabbit skeletal tropomyosin (a mixture of α and β chains) on hydroxyapatite. The protein emerges as two partially resolved

[9] E. Eisenberg and W. W. Kielley, *J. Biol. Chem.* **249**, 4742 (1974).
[10] M. Yamaguchi, M. L. Greaser, and R. G. Cassens, *J. Ultrastruct. Res.* **48**, 33 (1974).
[11] P. M. Sender, *FEBS Lett.* **17**, 106 (1971).

peaks, one of which is α,β-tropomyosin and the other the α,α variety. The preparation of β,β-tropomyosin, however, requires ion-exchange chromatography on CM-cellulose at pH 4.0 in the presence of 8 M urea. If the protein has been adequately reduced at neutral to alkaline pH before application to the column, a good separation of α and β chains can be achieved. Removal of the denaturant by dialysis results in the α,α and β,β species. The recovered α,α-tropomyosin appears to have identical stability and molecular properties to α,α-tropomyosin prepared without denaturation.

Preparation of Acetone Powder from Rabbit Skeletal and Cardiac Muscle

Skeletal muscle is dissected from the backs and hind legs of freshly sacrificed New Zealand white rabbits. Alternatively, frozen skeletal and cardiac muscle can be purchased from Pel-Freeze Biologicals Inc. (Arkansas). In a typical preparation 500 g of fresh tissue from either source is minced in a precooled stainless steel grinder and then stirred in 500 ml of water for 2–3 min. After standing for 20 min the liquid is expressed through two layers of cheesecloth, and the residue is stirred for 2–3 min in 500 ml of 95% ethanol. The liquid is again expressed through cheesecloth. This procedure is repeated three times with 2 liters of 50% ethanol, twice with 2 liters of 95% ethanol, and twice with 2 liters of acetone. All procedures and solutions are at 4°. The grayish white acetone powder (light brown for heart tissue) is then spread thinly over a large sheet of filter paper in a fume hood and allowed to air dry at room temperature. The dried powder can be stored indefinitely at $-20°$. Yield is approximately 60–75 g from skeletal muscle and 55–60 g from cardiac tissue. The procedure can be scaled up or down for larger or smaller amounts and can be applied to muscle tissue from any source.

Purification of Tropomyosin from Acetone Powder

The acetone powder (100 g) is extracted with 1 liter of 1.0 M KCl, 0.5 mM dithiothreitol (pH maintained at 7.0 with 1.0 M NaOH) for 16 hr at room temperature with slow overhead stirring. After expressing the liquid through two layers of cheesecloth, the residue is extracted with 700 ml of the same solution for 2 hr. Subsequent procedures are carried out at 4°. The combined extracts are adjusted to pH 4.6 and stirred for 30 min. After centrifugation at 6000 g for 20 min, the precipitate is dissolved in 1 liter of extraction buffer (pH maintained at 7.0) by stirring for 20 min, and insoluble material is removed by centrifugation at 6000 g for 10 min. The isoelectric precipitation at pH 4.6 and dissolution in extraction buffer at pH 7.0

are repeated twice more except that the final pH 4.6 precipitate is dissolved in 1.5 liter of 0.5 mM dithiothreitol in water (maintaining pH at 7.0). Solid ammonium sulfate is then added to 53% saturation at 0° (31.2 g/100 ml) while maintaining the pH at 7.0. After standing for 30 min, the precipitate is removed by centrifugation at 11,000 g (30 min) and the supernatant is brought to 65% saturation at 0° with solid $(NH_4)_2SO_4$ (7.34 g/100 ml), keeping the pH at 7.0 with 1.0 M NaOH. The precipitate is collected by centrifugation at 11,000 g for 30 min, redissolved in 0.1 liter of 0.5 mM dithiothreitol, and dialyzed extensively against 2 mM 2-mercaptoethanol in water before lyophilization. The yield at this stage is about 1.0–1.2 g of skeletal tropomyosin and 0.9 g of cardiac tropomyosin per 100 g of acetone powder, and the protein can be stored indefinitely in the freeze-dried state.

The product at this stage should be checked for purity by SDS–polyacrylamide electrophoresis at several sample loadings (10–60 μg) in the presence and in the absence of 6 M urea.[11] The skeletal tropomyosin should show a major α component (fastest migrating) and a slower minor β component. The cardiac protein should show only a single α band. If the sample has been incompletely reduced with 2-mercaptoethanol during its preparation for SDS electrophoresis, a higher molecular weight band is often observed corresponding to the disulfide bridged dimer of the tropomyosin chains. If the protein is contaminated with troponin or other impurities at this stage, it may be further purified by additional ammonium sulfate fractionation between 53 and 65% saturation as described above or by chromatography on hydroxyapatite.

Chromatography on Hydroxyapatite

This procedure, first introduced by Eisenberg and Kielley,[9] is useful for removing troponin contamination from tropomyosin. It has also been used by some workers[10] for the preparation of α,α- and α,β-tropomyosins. In our hands, however, the adequacy of the α,α and α,β separation has been variable and the procedure has been used almost exclusively as a final step in the preparation of highly homogeneous mixed chains of tropomyosin.

A column (2.5 × 50 cm) is packed with hydroxyapatite (DNA Grade BioGel HTP, Bio-Rad Laboratories) previously rehydrated with starting buffer as described by the supplier. The column is equilibrated with 2 column volumes (450 ml) of starting buffer: 10 mM sodium phosphate, 1 M KCl, 0.25 mM dithiothreitol, 0.01% sodium azide, pH 7.0. The protein, previously dialyzed against starting buffer, is applied to the column as a solution of 10 mg/ml; a linear gradient (total volume 2.2 liters) to a final

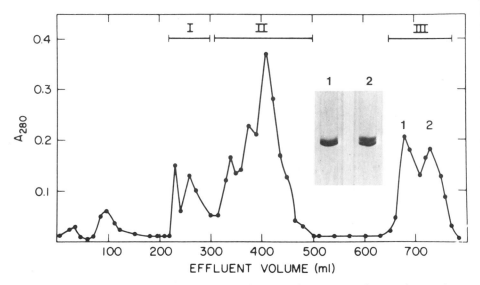

FIG. 1. Hydroxyapatite chromatography of a troponin–tropomyosin complex as described by Eisenberg and Kielley.[9] The column (2.5 × 25 cm) was equilibrated with 1 M KCl, 1 mM PO$_4$ at pH 7.0, 2 mM dithiothreitol, and 200–300 mg of troponin–tropomyosin in 1 M KCl, 2 mM imidazole, 5 mM 2-mercaptoethanol in 20–30 ml applied to the column. The protein was eluted with a linear gradient (total volume 1000 ml) from 1 mM to 200 mM PO$_4$ in 1 M KCl at pH 7.0, 2 mM dithiothreitol. Troponin components are largely eluted in peak II, and tropomyosin is found in peak III, subfractions 1 and 2. The column was operated at 30–35 ml/hr, and 6 samples per hour were collected. Inset shows sodium dodecyl sulfate–polyacrylamide gel electrophoresis of subfractions 1 and 2 of peak III. Subfraction 1 is α,α-tropomyosin; subfraction 2 is α,β-tropomyosin.

concentration of 300 mM sodium phosphate in starting buffer is applied at a flow rate of 30 ml/hr. Tropomyosin is eluted at about 150–200 mM phosphate with the troponin components eluted at 100 mM or lower. Up to 600 mg of protein have been routinely purified in a single column run using this procedure. The column can be regenerated with 1 column volume of starting buffer made 0.4 M with respect to sodium phosphate and then reequilibrated with 2–3 column volumes of starting buffer. In this way a single packed column has been used up to 10 times without apparent loss of resolution or decrease in flow rate.

As previously mentioned, this procedure has been used by some workers for the preparation of α,β and α,α rabbit skeletal tropomyosins. A typical elution profile originally described by Eisenberg and Kielley[9] is shown in Fig. 1. For this purpose lower loadings of protein on the column than those described above are probably required.

Separation of α- and β-Tropomyosins

Although purified rabbit $\alpha\alpha$ tropomyosin can be prepared directly from cardiac tissue or by hydroxyapatite chromatography of the skeletal α and β mixed chains, the purification of purified β-tropomyosin requires ion-exchange chromatography in the presence of 8 M urea. The original procedure described by Cummins and Perry[1] was applied to tropomyosins that had been converted to the carboxymethylated derivatives by alkylation of the sulfhydryl residues with iodoacetic acid essentially as described by Crestfield et al.[12] However, the method is also applicable to nonderivatized rabbit skeletal tropomyosins if precautions are taken to fully reduce disulfide bridges and to prevent their formation during the procedures. If this is not done, the separation of the α and β components will be incomplete. In the following, the procedure used in our laboratory for fractionating the S-β-carboxymethylated α and β chains is described[4] as well as the method for preparing the fully reduced mixed chains that can then be fractionated by the same chromatographic procedure.

The S-β-carboxymethylated protein is prepared by the method of Crestfield et al.[12] The protein (650 mg) is dialyzed overnight against 0.05 M sodium formate buffer, 1 mM dithiothreitol, 8 M urea, pH 4.0, and applied to a column (2.4 × 40 cm) of CM-cellulose, equilibrated with the same buffer. A linear gradient is established with 1.5 liters of starting buffer and 1.5 liters of the same buffer made 0.2 M with respect to NaCl. The column is operated at a flow rate of 30 ml/hr room temperature, and fractions are monitored at 278 nm. A typical elution profile is shown in Fig. 2.

For the preparation of nonderivatized fully reduced tropomyosin the protein (400 mg) is dissolved in 30 ml of starting buffer, the pH is adjusted to 8.0 with 1 M NaOH, and 2-mercaptoethanol is added to a final concentration of 0.36 M. After standing at room temperature for 3 hr the pH is lowered to 4.0 with formic acid and dialyzed overnight against starting buffer. The composition of the latter is the same as above except that EDTA is added to 1 mM and the dithiothreitol concentration is increased to 5 mM. The column is then operated as described above. If reduction of the protein is incomplete, the β peak (Fig. 2) will be contaminated with α chain and an additional peak of mixed chains may appear before the β peak.

Application to Other Muscle Systems

The methods described above are for rabbit skeletal and cardiac muscle. The procedures for the preparation of the acetone powder and of the

---

[12] A. M. Crestfield, S. Moore, and W. H. Stein, J. Biol. Chem. **238**, 622 (1963).

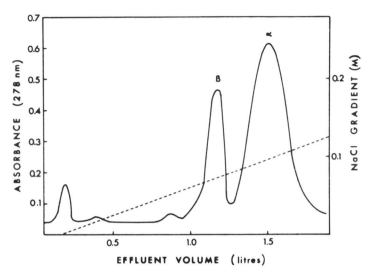

FIG. 2. Fractionation of S-β-carboxymethylated rabbit skeletal α- and β-tropomyosins on CM-cellulose by a procedure adapted from Cummins and Perry.[1] Details are described in the text.

mixed (α plus β) tropomyosin can be applied to any muscle tissue. The final purification on hydroxyapatite also appears to be of general applicability. However, it is unlikely that the latter procedure can be used indiscriminately for the preparation of α,α- and α,β-tropomyosins from a variety of tissue types because the relative ratios of the α and β chains vary significantly. The charge properties of the α and β chains of tropomyosins from various sources can also be expected to be variable. The procedure first described by Cummins and Perry[1] for the fractionation of S-carboxymethylated α- and β-tropomyosin chains does appear, however, to be of more general applicability. It was apparently successfully used by these workers for the separation of derivatized α and β chains from rabbit psoas, chicken skeletal, frog skeletal, and chicken gizzard muscles. In our own laboratory we have been successful in fractionating the derivatized α and β chains of chicken gizzard tropomyosin using a somewhat modified version of the procedure described above. However, we have been unsuccessful in using the procedure for separating the unmodified chains. It is likely that the isolation of purified α and β chains from various muscle types in an unmodified form will require the development of further variations in present procedures or entirely new methods.

When the methods described above are used with muscles from various sources, the recovered tropomyosin appears to be sometimes contaminated with significant amounts of nucleic acid as judged by the ratio of the absorbances at 280 and 260 nm. Such contamination could lead to difficulties in the determination of protein concentrations and in the interpretation of interaction studies of tropomyosin with other proteins by spectral, fluorescence, and other methods. Woods[13,14] has shown that the nucleic acid can be largely removed by ion-exchange chromatography on DEAE-cellulose at pH 7.6, and this procedure is recommended as a final step in the purification of the mixed chains of tropomyosin if nucleic acid contamination is indicated. Although generally unnecessary in the preparation of tropomyosins from rabbit skeletal and cardiac tissues, the isolation of chicken gizzard tropomyosins free of nucleic acid contamination does appear to require such a step (C. Sanders and L. B. Smillie; unpublished experiments). Preliminary evidence indicates that hydroxyapatite chromatography is relatively ineffective for this purpose.

[13] E. F. Woods, *Biochemistry* **8,** 4336 (1969).
[14] E. F. Woods, *J. Biol. Chem.* **242,** 2859 (1967).

[22] Preparation of Troponin and Its Subnits

By James D. Potter

The initial event in the activation (switching on) of muscle contraction is the binding of Ca^{2+} to the regulatory proteins present in the thin actin-containing filaments, which then allows myosin and actin to interact and contraction to occur. During the past two decades much has been learned about the nature of the proteins that confer this Ca^{2+} requirement on the actomyosin system.[1]

Ebashi[1a,2] first showed in the early 1960s that a protein he called "native tropomyosin" was required to confer Ca^{2+} sensitivity to a Ca^{2+}-insensitive actomyosin preparation. It was subsequently shown by Ebashi *et al.*[3] that native tropomyosin consisted of two proteins, tropomyosin and

[1] Abbreviations: Tris, tris(hydroxymethyl)aminomethane; MOPS, morpholinopropanesulfonic acid; ATP, adenosine 5'-triphosphate (disodium salt); DTT, *dl*-dithiothreitol; BME, 2-mercaptoethanol; EDTA, ethylenediaminetetraacetic acid; EGTA, ethyleneglycol bis-(β-aminoethyl ether)-N,N^1-tetraacetic acid; CTn, bovine cardiac whole troponin; STn, rabbit skeletal whole troponin; Tm, tropomyosin; HMM, heavy meromyosin; SDS-PAGE, sodium dodecyl sulfate–polyacrylamide gel electrophoresis; TES, *N*-tris[hydroxymethyl]-methyl-2-aminoethanesulfonic acid.
[1a] S. Ebashi, *Nature* (*London*) **200,** 1010 (1963).
[2] S. Ebashi and F. Ebashi, *J. Biochem.* (*Tokyo*) **55,** 604 (1964).
[3] S. Ebashi, A. Kodama, and F. Ebashi, *J. Biochem.* (*Tokyo*) **64,** 465 (1968).

a new protein named troponin, both of which were required to confer Ca^{2+} sensitivity to a purified actomyosin preparation. These studies also demonstrated that the troponin component was the Ca^{2+} receptor[3] of "native tropomyosin." Subsequent to this, Hartshorne and Mueller[4] found that troponin could be further separated into two components, which were called troponin A and troponin B; both of these, in addition to tropomyosin, were required to confer Ca^{2+} sensitivity to synthetic actomyosin (purified actin and myosin). Schaub and Perry[5] using sulfoethyl (SE)–Sephadex chromatography in the presence of 6 M urea were also able to separate troponin into two components. With the advent of sodium dodecyl sulfate (SDS) gels in the late 1960s and the use of DEAE–Sephadex chromatography in the presence of 6 M urea, Greaser and Gergely[6,7] were able to separate troponin into three subunits that they called troponin T (TnT; the tropomyosin binding subunit) troponin I (TnI; the actomyosin ATPase inhibitory subunit), and troponin C (TnC; the Ca^{2+}-binding subunit), and all three subunits in addition to tropomyosin were required to reconstitute "native tropomyosin" activity.

The procedures we have developed for the purification of troponin and its three subunits over the last several years have evolved from a variety of procedures and will be presented here for both skeletal and cardiac muscle. Although there are significant differences between the purification procedures for the cardiac and skeletal troponins, the same basic procedures are employed. The first step consists of making an ether powder (procedure is identical for both cardiac and skeletal muscle), followed by an high ionic strength extraction and initial fractionation of troponin. This partially purified troponin, which we have called crude troponin, can be further purified to yield pure troponin using Cibacron Blue–Sephacryl chromatography similar to the Cibacron Blue–agarose chromatography procedure introduced by Reisler et al.[8] The crude troponin can also be used to prepare pure troponin subunits by various combinations of chromatography on DEAE–Sephadex and CM–Sephadex in the presence of 6 M urea and 1 mM EDTA.

Preparation of Skeletal and Cardiac Muscle Ether Powder

Triton X-100 Extraction. Diagram 1 illustrates the basic procedure for preparing the ether powder. The procedures for skeletal and cardiac mus-

[4] D. J. Hartshorne and H. Mueller, *Biochim. Biophys. Acta* **175**, 320 (1968).
[5] M. C. Schaub and S. V. Perry, *Biochem. J.* **115**, 993 (1969).
[6] M. L. Greaser and J. Gergely, *J. Biol. Chem.* **246**, 4226 (1971).
[7] M. L. Greaser and J. Gergely, *J. Biol. Chem.* **248**, 2125 (1973).
[8] E. Reisler, J. Liu, M. Mercola, and J. Horowitz, *Biochim. Biophys. Acta* **623**, 243 (1980).

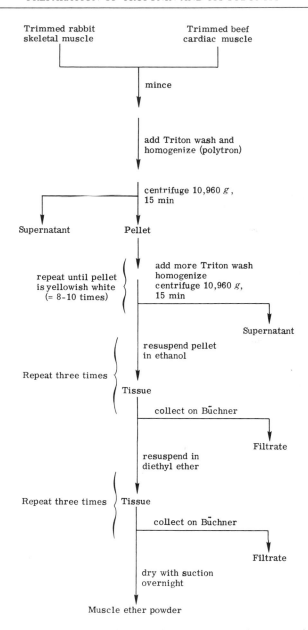

DIAGRAM 1. Rabbit skeletal and beef heart ether powder preparation.

cle are identical. For skeletal troponin, rabbits are sacrificed by a blow to the neck and rapidly bled. The back and hind leg muscles are then trimmed of fat, and the slow skeletal muscles are removed and discarded. For cardiac troponin, beef hearts are obtained from a local slaughterhouse and transported on ice. Only the left ventricle is used, and this is trimmed of fat, valves, and connective tissue. Both types of muscle are then cut into small chunks and ground in a prechilled commercial meat grinder. Five volumes of the wash solution (1% Triton X-100, 50 mM KCl, and 5 mM Tris, pH 8.0) are then added to the ground muscle and further homogenized for 1 min with a Brinkmann Polytron homogenizer set at high speed (5–7). The homogenate is then centrifuged at 10,960 g for 15 min, and the supernatant is discarded. The pellets are resuspended in an equal volume of the wash solution added directly to the centrifuge bottles. These are then rehomogenized with the Polytron in the centrifuge bottles and then recentrifuged. This procedure is repeated 8–10 times, and in the case of cardiac muscle until the residue turns almost white (slight yellow tinge). All these steps are carried out at 4°. It is felt that the purpose of these steps of repeated homogenization and Triton X-100 extraction is to break open lysosomes and other membrane systems allowing soluble proteases to be removed in these low ionic strength washes of the muscle residue. Without these steps, it is difficult to isolate undegraded and stable troponin from either tissue.

Alcohol and Ether Extractions. From this point, the preparation should be continued at room temperature in a fume hood. The pellets are transferred to a 4-liter plastic beaker. Three volumes of prechilled 95% ethanol are added to the pellet, which is then broken up with a gloved hand. The tissue is then collected over a Büchner funnel, diameter 26 cm, containing a Whatman No. 1 filter paper (24 cm), mounted on a 4-liter filter flask to which suction is applied. The filtrate is discarded. The tissue is returned to the beaker and this procedure is repeated three times. Next, the same procedure is conducted as above except that diethyl ether is substituted for the ethanol. After the four washes, the powder is left to dry overnight on the filter paper, with suction. The dry powder is weighed and stored at 4°.

Purification of Skeletal Troponin

The rabbit skeletal ether powder (see Diagram 2) is extracted overnight with a 15:1 (v/w) ratio of 1 M KCl, 25 mM Tris, pH 8.0, 0.1 mM CaCl$_2$, and 0.1 mM DTT. The extract is centrifuged at 10,960 g for 10 min at 4°. This supernatant is set aside, and the pellet is then resuspended by adding 1 M KCl directly to the centrifuge bottles, in a ratio of 7.5:1 (v/w

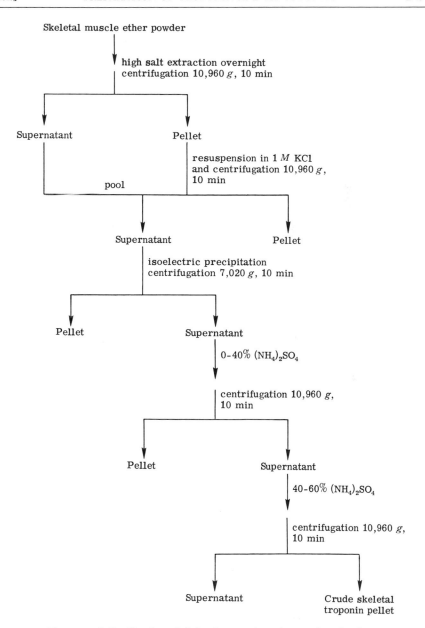

DIAGRAM 2. Purification of skeletal troponin and troponin subunits.

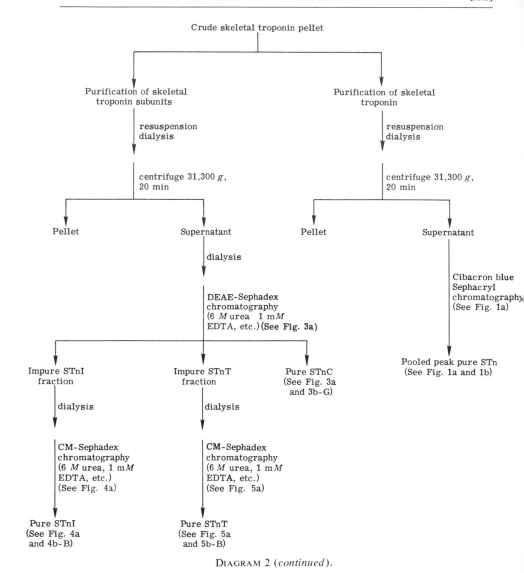

DIAGRAM 2 (*continued*).

of starting tissue). The mixture is then centrifuged at $10,960\,g$ for 10 min at 4°. The pellet is discarded, and the supernatant from this centrifugation is pooled with that of the previous one. The pH of this solution is lowered to 4.6 with 1 N HCl to precipitate tropomyosin. The precipitated tropomyosin is collected by centrifuging at $7020\,g$ for 10 min at 4°. This crude tropomyosin pellet can be stored at $-20°$ to be used later for the purification of tropomyosin. The pH of the supernatant from this centrifu-

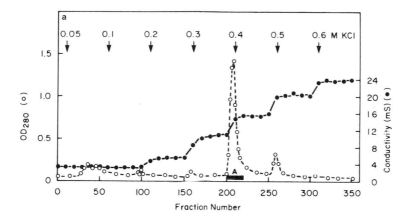

FIG. 1a. Purification of skeletal troponin. The 40–60% $(NH_4)_2SO_4$ pellet from 150 g of rabbit skeletal muscle ether powder was suspended in 10 mM imidazole, pH 7.0, 50 mM KCl, 0.1 mM $CaCl_2$, 0.1 mM DTT, and 0.02% NaN_3 and dialyzed four times against 1 liter of this solution. This was then loaded onto a 5 × 50 cm Cibacron Blue–Sephacryl column equilibrated with the same buffer; 25-ml fractions were collected. The column was eluted with 1 liter of equilibrating buffer until the OD_{280} returned to baseline. The column was then eluted with column buffer containing increasing concentrations of KCl. All volumes were 1 liter of each for 0.1, 0.2, 0.3, 0.4, 0.5, and 0.6 M KCl steps. For re-use the column was first washed with 1.5 liters of equilibrating buffer that contained 6 M urea until the OD_{280} returned to baseline and then with the column buffer until reequilibrated. Pooling tubes 201–220 yielded 612 mg of pure troponin.

gation is adjusted to 8.0 with 1 N KOH, and 230 g of ammonium sulfate per liter (40% saturation) are slowly added with constant stirring, the pH being maintained between 7 and 8. This solution is then centrifuged at 10,960 g for 10 min at 4°; the pellet is discarded. The supernatant is brought to 60% $(NH_4)_2SO_4$ saturation by adding 125 g of ammonium sulfate per liter as described above. The supernatant from this spin is discarded, and the 40–60% $(NH_4)_2SO_4$ pellet (pellet also used to prepare skeletal troponin subunits) is then suspended in a minimal volume of 10 mM imidazole, pH 7.0, 50 mM KCl, 0.1 mM $CaCl_2$, 0.1 mM DTT and 0.02% NaN_3 and then dialyzed and chromatographed as described in the legend to Fig. 1a. The procedure of Reisler et al.[8] has been used with the exception that Cibacron Blue–Sephacryl has been substituted for Cibacron Blue–agarose owing to the better flow rates that can be obtained with the Sephacryl form. Also, several salt steps were employed to elute the Tn from the column. The procedure for linking Cibacron Blue F3G-A to Sephacryl S-200 employed by us is identical to the triazine coupling method of Böhme et al.[9] Pure troponin (A in Fig. 1a, 1b) is eluted from this

[9] H. J. Böhme, G. Kopperschläger, J. Schulz, and E. Hofmann, J. Chromatogr. **69**, 209 (1972).

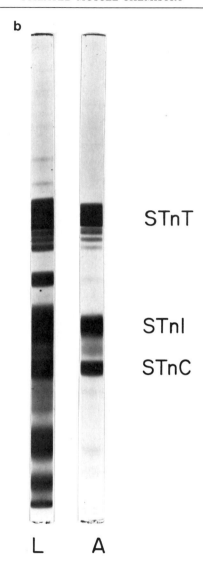

FIG. 1b. Sodium dodecyl sulfate–polyacrylamide gel electrophoresis of Cibacron Blue–Sephacryl purified skeletal troponin. L = crude troponin loaded onto the column; A = fraction A (Fig. 1a; pure troponin).

column at a conductivity between 10.8 and 15.4 mS. The pure troponin can be stored frozen in solution at $-80°$ or lyophilized and stored at $-20°$.

Purification of Cardiac Troponin

Bovine cardiac ether powder (see Diagram 3) is extracted overnight in a 20 : 1 (v/w) ratio of 1 M KCl, 20 mM Tes, pH 7.0, and 15 mM BME which has been homogenized with a Polytron to suspend the powder. The mixture is then centrifuged at 10,960 g for 20 min at 4°, and the pellet is discarded. The supernatant is adjusted to pH 8.0 with 1 N KOH, and

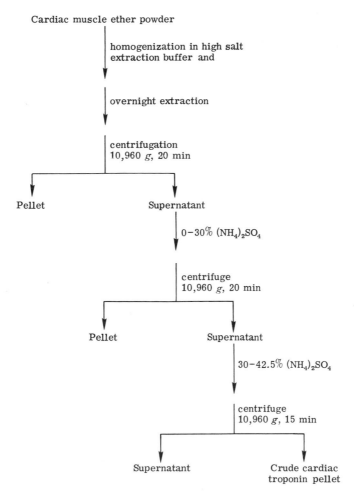

DIAGRAM 3. Purification of cardiac whole troponin and cardiac troponin subunits.

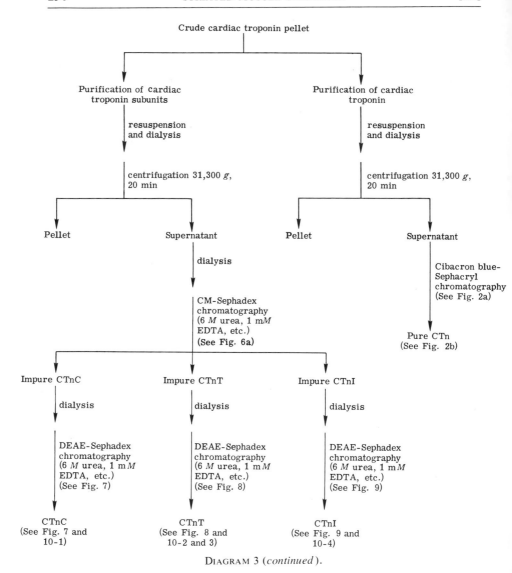

DIAGRAM 3 (*continued*).

brought to 30% ammonium sulfate saturation (167 g/liter added). The pH should be kept between 7 and 8 during the addition of the ammonium sulfate. This solution is then stirred for 1 hr at 4° to ensure that the ammonium sulfate is completely dissolved. It is then centrifuged at 10,960 g for 20 min at 4°, and the pellet is discarded. Next, the supernatant is brought to 42.5% ammonium sulfate saturation (73 g/liter added) and is stirred for 1 hr as above. It is then centrifuged at 10,960 g for 15 min at 4°, and the supernatant is discarded. The pellet is then suspended in a minimal volume of 10 mM imidazole, pH 7.0, 50 mM KCl, 0.1 mM CaCl$_2$, 0.1 mM DTT, and 0.02% NaN$_3$ and dialyzed and chromatographed as described in the legend to Fig. 2a. This pellet is also the starting point for preparing CTn subunits (see section Purification of Cardiac Troponin Subunits). Note that pure CTn is found in fractions F and G in Fig. 2a.

Purification of Skeletal Troponin Subunits

Initial Separation of Skeletal Troponin Subunits and Purification of STnC. The crude troponin pellet (40–60% (NH$_4$)$_2$SO$_4$ pellet) from 130 g of skeletal ether powder (see Diagram 2) is suspended in a minimal volume of, and dialyzed once against 16 liters of, 2 mM Tris, pH 8.0, and 0.1 mM CaCl$_2$. This dialyzed sample is then centrifuged at 31,300 g for 15 min to remove any insoluble material in the solution. The supernatant is dialyzed (with intermittent shaking by hand of the dialysis bag to mix the urea that settles to the bottom of the bag) twice against 2 liters of 6 M urea, 50 mM Tris, pH 8.0, 1 mM EDTA, and 0.1 mM DTT and chromatographed as

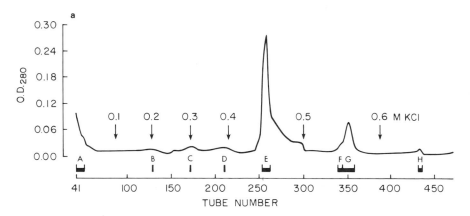

FIG. 2a. Purification of cardiac troponin. The 30–42.5% (NH$_4$)$_2$SO$_4$ pellet from 100 g of beef cardiac muscle ether powder was suspended in 10 mM imidazole, pH 7.0, 50 mM KCl, 0.1 mM CaCl$_2$, 0.1 mM DTT, and 0.02% NaN$_3$ and dialyzed four times against 1 liter of this solution. A total of 619 mg were loaded onto a 5 × 50 cm Cibacron Blue–Sephacryl column equilibrated with same buffer; 25-ml fractions were collected. The column was eluted with 2 liters of buffer until the OD$_{280}$ returned to baseline. The column was then eluted in a stepwise manner with column buffer containing increasing concentrations of KCl. The volumes used were 1 liter for the 0.1, 0.2, and 0.3 M KCl steps and 2 liters for the 0.4, 0.5, and 0.6 M KCl steps. Arrows indicate the changes in salt concentration. Bars indicate pooled fractions. Pure cardiac troponin was found in tubes 340–359 and yielded 70 mg. To re-use the column it was first washed with 2 liters of column buffer that contained 6 M urea and then with column buffer until reequilibrated.

b

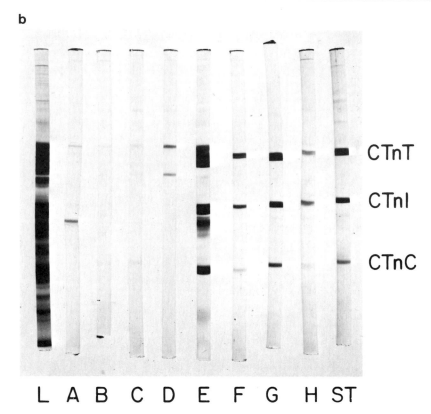

L A B C D E F G H ST

FIG. 2b. Sodium dodecyl sulfate–polyacrylamide gel electrophoresis of Cibacron Blue–Sephacryl chromatography of cardiac troponin. L = crude troponin loaded onto the column; A–H = fractions A–H (Fig. 2a); ST = CTnT, CTnI, and CTnC standards.

described in the legend to Fig. 3a. This chromatography results in the purification of STnC (Fig. 3a, fraction G) and the initial separation of STnI and STnT, which require further purification as described below.

Final Purification of STnI. The impure STnI fraction from the initial separation of the STn subunits (Fig. 3a, fractions A and B) is purified as described in the legend to Fig. 4a. Note that pure STnI is found in fraction B of Fig. 4a.

Final Purification of STnT. The impure STnT fraction from the initial separation of the Tn subunits (Fig. 3a, fraction E) is purified as detailed in the legend to Fig. 5a. Note that pure STnT is found in fraction B of Fig. 5a.

Storage of Subunits. The purified subunits, in their final elution buffers, are adjusted to pH 6.0 and then stored at −80°. To determine the concen-

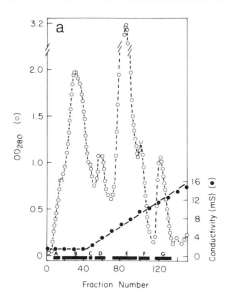

FIG. 3a. Initial separation of skeletal troponin subunits from troponin. The crude troponin prepared from 130 g of skeletal ether powder was dialyzed twice against 2 liters of 6 M urea, 50 mM Tris, pH 8.0, 1 mM EDTA, and 0.1 mM DTT. The protein was loaded onto a 2.5 × 25 cm DEAE-Sephadex A-50 column equilibrated with the same buffer. Fraction size = 6 ml. The column was eluted with a linear gradient produced by 500 ml of column buffer and 500 ml of column buffer plus 0.5 M KCl.

tration of these proteins in urea, small aliquots (~5 ml) of each sample are dialyzed free of urea against 1.0 M KCl, 2 mM MOPS, pH 7.0, and the protein concentration is estimated using the Lowry[10] procedure. The results are corrected for any volume change that occurs during the 1.0 M KCl dialysis.

Purification of Cardiac Troponin Subunits

Initial Separation of Cardiac Troponin Subunits from Troponin. The crude troponin (30–42.5% $(NH_4)_2SO_4$) pellet (prepared as described above for purification of cardiac troponin) is suspended in a minimal volume of 100 mM citrate, pH 6.0, and dialyzed once against 4 liters (see Diagram 3). This protein is then centrifuged at 31,300 g for 20 min at 4° in a Beckman JA-10 rotor to remove any insoluble material. The supernatant is filtered through glass wool and then dialyzed twice against 2 liters of 6 M urea, 50

---

[10] O. H. Lowry, N. J. Rosebrough, A. L. Farr, and R. J. Randall, *J. Biol. Chem.* **193**, 265 (1951).

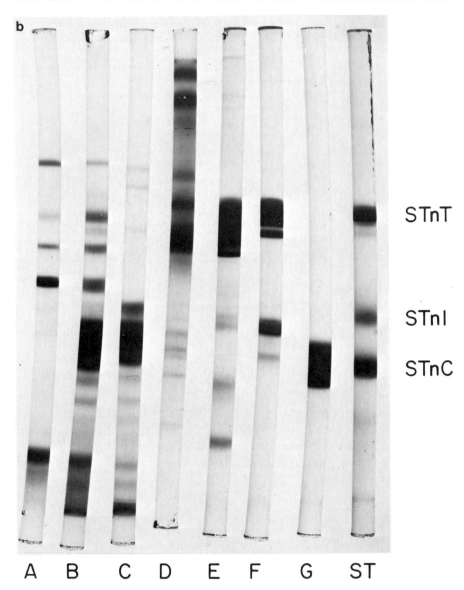

STnT

STnI

STnC

A   B   C   D   E   F   G   ST

FIG. 3b. Sodium dodecyl sulfate–polyacrylamide gel electrophoresis of DEAE-Sephadex chromatography of skeletal troponin subunits. Gels A–G correspond to fractions A–G in Fig. 3a. ST = STnT, STnI, and STnC standards. Note that fraction G contained pure TnC. Fractions A and B were pooled and contained impure STnI, and fraction E contained impure STnT.

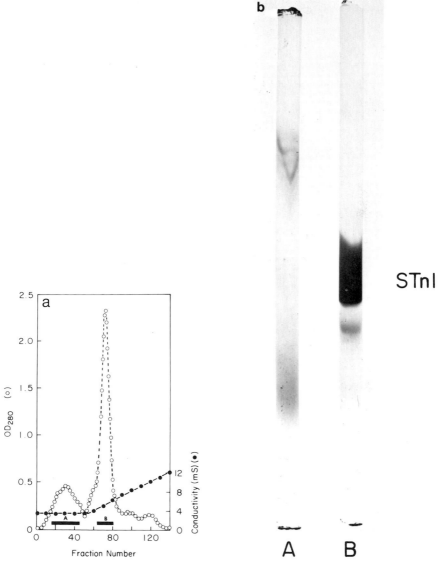

FIG. 4. (a) Final purification of skeletal troponin I. The impure STnI fraction (Fig. 3a, fractions A and B) from the DEAE-Sephadex A-50 column was dialyzed twice against 1 liter of 6 M urea, 50 mM citrate, pH 6.0, 1 mM EDTA, and 0.1 mM DTT. The protein was loaded onto a 2.5 × 25 cm CM-Sephadex C-50-120 column equilibrated with the same buffer. Fraction size = 6 ml. The column was eluted with a linear gradient produced by 500 ml of column buffer and 500 ml of column buffer plus 0.6 M NaCl. (b) Sodium dodecyl sulfate–polyacrylamide gel electrophoresis of CM-Sephadex chromatography of STnI. A and B correspond to fractions A and B in Fig. 4a. Fraction B contained pure STnI.

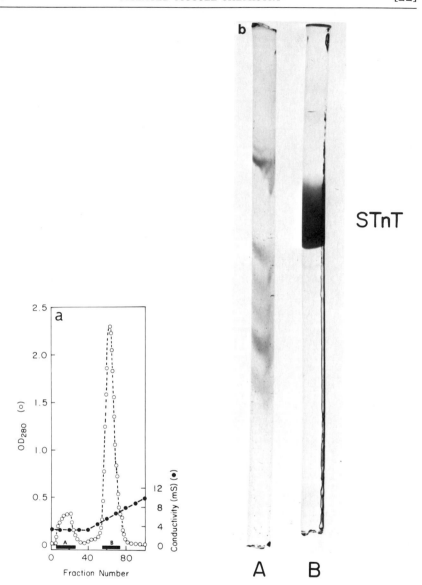

FIG. 5. (a) Final purification of STnT. The impure STnT fraction (Fig. 3a, fraction E) from the DEAE-Sephadex A-50 column was dialyzed twice against 1 liter of 6 M urea, 50 mM citrate (pH 6.0), 1 mM EDTA, and 0.1 mM DTT. The protein was loaded onto a 2.5 × 25 cm CM-Sephadex C-50-120 column equilibrated with the same buffer. Fraction size = 7 ml. The column was eluted with a linear gradient produced by 500 ml of column buffer and 500 ml of column buffer plus 0.6 M NaCl. (b) SDS-PAGE of CM-Sephadex chromatography of STnT. Fractions A and B correspond to fractions A and B in Fig. 5a. Fraction B contained pure STnT.

mM citrate, pH 6.0, 1 mM EDTA, and 0.1 mM DTT and chromato-graphed as described in the legend to Fig. 6a. All the cardiac subunits require further purification after this initial separation.

Final Purification of CTnC. The impure CTnC (Fig. 6a, fraction A) from the initial separation is purified as described in the legend to Fig. 7. Note that pure CTnC is found in fraction B of Fig. 7.

Final Purification of CTnT. The impure CTnT (Fig. 6a, fraction C) from the initial separation is purified as described in the legend to Fig. 8. Note that pure CTnT is found in fractions A and B of Fig. 8.

Final Purification of CTnI. The impure CTnI (Fig. 6a, fraction D) from the initial separation is purified as described in the legend to Fig. 9. Note that pure CTnI is found in fraction A of Fig. 9.

Storage of Subunits. The subunits are stored as described above for skeletal troponin subunits.

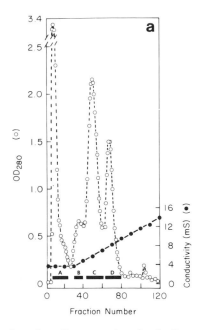

FIG. 6a. Initial separation of cardiac troponin subunits from troponin. Crude troponin was prepared from 80 g of cardiac ether powder as described in the text. The protein was loaded onto a 2.5 × 25 cm CM-Sephadex C-50-120 column equilibrated with 6 M urea, 50 mM citrate, pH 6.0, 1 mM EDTA, and 0.1 mM DTT. Fraction size = 6.4 ml. The column was eluted with a linear gradient produced by 500 ml of column buffer and 500 ml of column buffer plus 0.6 M NaCl.

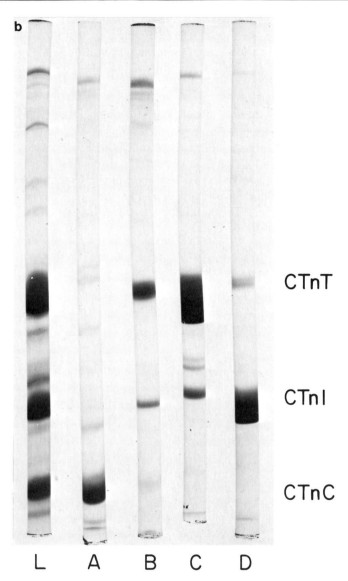

FIG. 6b. Sodium dodecyl sulfate–polyacrylamide gel electrophoresis of CM-Sephadex chromatography of cardiac troponin subunits. Gels A–D correspond to fractions A–D in Fig. 6a. L = crude troponin loaded onto the column. Fraction A contained impure CTnC, fraction C contained impure CTnT, and fraction D contained impure CTnI.

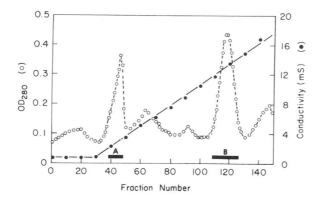

FIG. 7. Final purification of CTnC. The impure CTnC fraction (Fig. 6a, fraction A) from the CM-Sephadex C-50-120 column was dialyzed twice against 1 liter of 6 M urea, 20 mM imidazole, pH 7.0, 1 mM EDTA, and 0.1 mM DTT. The protein was loaded onto a 2.5 × 25 cm DEAE-Sephadex A-50 column equilibrated with the same buffer. Fraction size = 7.5 ml. The column was eluted with a linear gradient produced by 500 ml of column buffer and 500 ml of column buffer plus 0.5 M KCl (see gel Fig. 10).

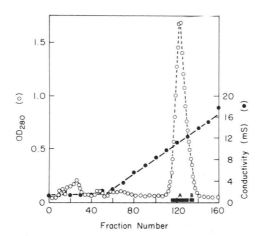

FIG. 8. Final purification of CTnT. The impure CTnT fraction (Fig. 6a, fraction C) from the CM-Sephadex C-50-120 column was dialyzed twice against 1 liter of 6 M urea, 50 mM Tris, pH 8.0, 1 mM EDTA, and 0.1 mM DTT. The protein was loaded onto a 2.5 × 25 DEAE-Sephadex A-50 column equilibrated with the same buffer. Fraction size = 6.0 ml. The column was eluted with a linear gradient produced by 500 ml of column buffer and 500 ml of column buffer plus 0.5 M KCl (see gels Fig. 10).

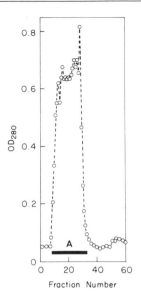

FIG. 9. Final purification of CTnI. The impure CTnI fraction (Fig. 6a, fraction D) from the CM-Sephadex C-50-120 column was dialyzed twice against 1 liter of 6 M urea, 50 mM Tris, pH 8.0, 1 mM EDTA, and 0.1 mM DTT. The protein was loaded onto a 2.5 × 25 cm DEAE-Sephadex A-50 column equilibrated with the same buffer. Fraction size = 6.5 ml. CTnI comes off the DEAE Sephadex A-50 column in the washthrough fraction. No salt gradient was required to elute the protein (see gel, Fig. 10).

Reconstitution of Whole Troponin from Subunits

The procedure for reconstituting skeletal and cardiac troponin complexes from their subunits are the same. To form the various complexes, the subunits that are stored in 6 M urea and 1 mM EDTA (plus other constituents depending on the particular troponin subunit), can be mixed in a 1 : 1 molar ratio directly in these solutions. To facilitate complex formation, the total calcium concentration is then raised to 1.25 mM so that there is an excess of calcium over the chelator. The urea is then removed by dialyzing the complexes against 1.0 M KCl and 10 mM MOPS, pH 7.0. The ionic strength can then be gradually lowered by dialysis using several step changes in KCl concentration (e.g., 0.7, 0.5, 0.3, 0.1) until the desired concentration is reached. Any precipitate formed at the end of the dialysis procedure should be removed by centrifugation. This Tn complex should then be chromatographed on Sephadex G-150 or G-200 to separate the Tn complex from any uncomplexed subunits.

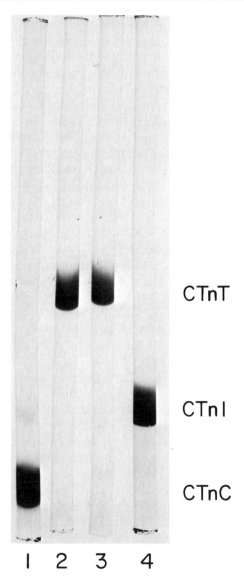

FIG. 10. Sodium dodecyl sulfate–polyacrylamide gel electrophoresis of purified cardiac troponin subunits. 1 = fraction B, Fig. 7; 2 = fraction A, Fig. 8; and 3 = fraction B, Fig. 8; 4 = fraction A, Fig. 9.

Assay of Troponin

Principle. The purpose of the assay is to measure the ability of a Tn sample to confer calcium sensitivity on troponin-depleted actomyosin ATPase. Acto-HMM is reconstituted from purified HMM,[11] F-actin,[12] and Tm.[13] The actin-activated HMM ATPase is measured in the presence and the absence of calcium at increasing concentrations of Tn. As the actin : Tm : Tn molar ratio approaches 7 : 1 : 1, the maximum difference in ATPase activity between plus and minus calcium is achieved. At higher ratios of Tn to actin : Tm, the calcium dependence of the actin-activated ATPase will saturate, whereas at lower Tn ratios the calcium dependence will diminish.

Reagents

Buffer: 10 mM MOPS, 40 mM KCl, 0.01% NaN$_3$, and 0.1 mM DTT adjusted to pH 7.0 with KOH

Stock solutions: (1) 100 mM ATP (disodium salt) adjusted to pH 7.0 with KOH; (2) 50% trichloroacetic acid (TCA), ice cold; (3) 1 M MgCl$_2$; (4) 10 mM CaCl$_2$; (5) 0.2 M EGTA, pH 7.0, with KOH

Protein stocks: 5.0 mg of HMM, 21.4 mg of F-actin, 5.1 mg of Tm, and 3–5 mg of Tn sample, all in the above buffer

Phosphate assay: Reagents for Fiske-SubbaRow[14] determination of inorganic phosphate

PROCEDURE

Reconstituted Acto-HMM. Tn-depleted, Tm-acto-HMM is reconstituted from purified components in a Potter-Elvehjem homogenizer. To a final volume of 15 ml, buffer stock, 0.075 ml MgCl$_2$ stock, the F-actin, Tm, and HMM stocks are added in order, with homogenization after each protein addition on ice. The reconstitution should be performed on the same day as the assay.

Reaction Mixtures. Two-milliliter acto-HMM reaction mixtures are prepared for each data point. Two data points are generated for each Tn addition; additions of Tn are assayed independently in the presence and in the absence of calcium. Each reaction mixture contains 1 ml of Tn-depleted Tm-acto-HMM and either 20 μl of 10 mM CaCl$_2$ (plus calcium) or 20 μl of 0.2 M EGTA (minus calcium) stocks. Aliquots of the Tn sample are added to the plus and minus calcium reaction mixtures, buffer stock is added to the final volume, and the suspension is vortexed to

[11] S. S. Margossian and S. Lowey, this volume [7].
[12] J. D. Pardee and J. A. Spudich, this volume [18].
[13] L. B. Smillie, this volume [21].
[14] C. H. Fiske and Y. SubbaRow, *J. Biol. Chem.* **66**, 375 (1925).

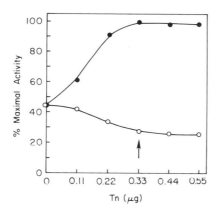

FIG. 11. A typical plot of acto-HMM ATPase, $+Ca^{2+}$ (●) and $-Ca^{2+}$ (○), as a function of STn concentration. Assay conditions are as described in the text. The arrow indicates the Tn addition at which there is an actin : Tm : Tn molar ratio of 7 : 1 : 1. Maximum ATPase activity (100% on the graph) corresponds to 0.25 μmol of P_i liberated per milligram of actomyosin per minute.

ensure mixing. Equilibration of the reaction mixtures is allowed to occur at 25° for 5 min prior to the start of the reaction. Forty microliters of the ATP stock is added to start the reaction, which proceeds for 4 min at 25°. The reaction is stopped by the addition of 0.3 ml of TCA stock. A single blank is made by adding the TCA to a separate reaction mixture before the addition of ATP. Immediately after the TCA addition, the reaction mixtures are chilled on ice.

Phosphate Assay. Precipitated protein is removed from the reaction mixtures after the addition of TCA by centrifugation in a clinical or table-top centrifuge; 1.5 ml of the supernatant is then assayed for inorganic phosphate.

Data Handling. The total net micromoles of phosphate per minute liberated by acto-HMM is determined for individual reaction mixtures after subtracting the phosphate present in the blank from each of the other mixtures. The percentage of activation is calculated for the data points using the highest value for phosphate liberated as 100%. An example of the plotted data is found in Fig. 11.

Acknowledgments

I would like to thank Patricia Walker and Henry Zot for their extensive help in the preparation of this manuscript and Janet Duritsch for the excellent typing of this manuscript.

[23] Purification of Titin and Nebulin

By Kuan Wang

A group of three giant myofibrillar proteins are present in major amounts (12–15%) in a wide range of vertebrate and invertebrate striated muscles.[1–4] Titin, designating a pair of titanic proteins ($M_r \approx 1,000,000$), composes approximately 10% of the myofibrillar mass.[1,2] Nebulin, referring to an ~500,000 M_r protein, represents ~5% of the total myofibrillar proteins.[3,4] These proteins are unusual not only in their giant sizes, but also in their solubility and sarcomeric localization. Myofibrillar proteins are generally associated exclusively with either thin or thick myofilaments or with transverse structures such as Z and M lines. In contrast, titin has been localized by antibody staining at the A-1 junction and, under certain conditions, at the Z line, the H zone, and throughout the A band.[2] Although the ultrastructural organization of titin remains to be elucidated, the intriguing possibility exists that titin may be a component of the putative elastic filament either linking the ends of thick filaments to Z lines or connecting adjacent Z lines of the same myofibril.[5–9] Nebulin is associated with the N_2 line, a nebulous transverse structure that appears to participate in the regulation or maintenance of the changing spatial arrangement of thin filaments from a square lattice at the Z line to a hexagonal array near the A-1 junction.[3,4,10]

Purification Procedures

Two major obstacles face the attempts to purify titin and nebulin. First, they are extremely prone to proteolytic degradation, especially by endogenous calcium-activated proteases. The inclusion of common protease inhibitors in buffers minimizes, but does not completely abolish proteolysis. Since the rate of degradation varies greatly among species

[1] K. Wang and J. McClure, *J. Cell Biol.* **79**, 334a (1978).
[2] K. Wang, J. McClure, and A. Tu, *Proc. Natl. Acad. Sci. U.S.A.* **76**, 3968 (1979).
[3] K. Wang and C. L. Williamson, *Proc. Natl. Acad. Sci. U.S.A.* **77**, 3254 (1980).
[4] K. Wang, *J. Cell Biol.* **91**, 355a (1981).
[5] F. J. Sjöstrand and M. Jagendorf-Elfavin, *J. Ultrastruct. Res.* **17**, 348 (1967).
[6] F. Carlsen, F. Fuchs, and G. G. Knappeis, *J. Cell Biol.* **27**, 35 (1965).
[7] J. Auber and R. Couteaux, *Microscopie* **2**, 203 (1963).
[8] R. H. Locker and N. G. Leet, *J. Ultrastruct. Res.* **52**, 64 (1975).
[9] G. Hoyle, *Am. Zool.* **7**, 435 (1967).
[10] C. Franzini-Armstrong, *Tissue Cell* **2**, 327 (1970).

and muscle types, a careful choice of starting material is crucial for preparing intact proteins. Second, these proteins are not very soluble in aqueous solutions without denaturants. For example, titin tends to associate irreversibly to a gel-like aggregate even in denaturant solutions. Thus, the choice of purification techniques is severely limited.

This chapter describes two procedures for titin and nebulin purification from rabbit back muscle based on the use of by sodium dodecyl sulfate (SDS). The unusual size of these proteins provides the basis for separation by gel filtration. The first method involves the gel filtration of the entirety of myofibrillar proteins solubilized in SDS.[2] The second method makes use of a novel salt fractionation procedure in which the titin–SDS complex is selectively precipitated from SDS-solubilized myofibrils. This procedure, when combined with gel filtration, provides a rapid and efficient method for titin and nebulin purification.[11]

Highly washed rabbit back myofibrils are used as starting material because preparations largely free of endogenous protease can be prepared and stored conveniently. An SDS–polyacrylamide gel electrophoresis system has been developed to allow the resolution and identification of titin and nebulin.[3]

Myofibril Preparation

Myofibrils are prepared from rabbit back muscle by the procedures of Etlinger *et al.*[12] with slight modifications.

Reagents and Buffers[13]

Nembutal (sodium pentobarbitol, 50 mg/ml, Abbott Laboratories)
Low-salt buffer (LSB): 0.1 M KCl, 2 mM MgCl$_2$, 2 mM EGTA, 10 mM Tris-maleate, 0.5 mM dithiothreitol, 0.1 mM phenylmethylsulfonyl fluoride, pH 6.8.
Pyrophosphate relaxing buffer (PRB): LSB supplemented with 2 mM Na$_4$P$_2$O$_7$
Triton X-100 buffer: LSB supplemented with 0.5% (w/v) Triton X-100.
Glycerol (Baker reagent grade).
Procedures. These are carried out at 4° in a cold room. Injected a New Zealand white rabbit with 1 ml of Nembutal per kilogram body weight into the lateral ear vein and then exsanguinate it as soon as it becomes uncon-

---

[11] K. Wang, *J. Biol. Chem.,* submitted.
[12] J. D. Etlinger, R. Zak, and D. A. Fischman, *J. Cell Biol.* **68,** 123 (1976).
[13] pH adjustment of all buffers is made at room temperature. Whenever present, dithiothreitol is added as a solid and phenylmethylsulfonyl fluoride is added in the form of a 100 mM stock solution (in isopropanol), immediately before use.

scious. Remove the back muscle along both sides of the spinal cord with a scalpel. Immediately place the muscle into cold PRB. To trim off the major portion of connective tissue covering one side of the muscle, force a dull knife, at a 45-degree angle to the table, between the sheet of white connective tissue and muscle, keeping some tension on the connective tissue with the other hand. Cut the muscle into small pieces (ca 1 cm³) and weigh. Add fresh PRB (8 ml/g wet muscle) and homogenize the muscle in a multispeed Waring blender (set on ''blend'') for three 15-sec bursts with 1-min intervals. Centrifuge the homogenate in polycarbonate bottles at 3500 rpm for 15 min in a Sorvall GSA rotor (maximum volume per run is 1500 ml). Decant the supernatant and thoroughly resuspend the pellet in a minimum volume of LSB with a plastic kitchen spatula. To remove white connective tissue, pour the suspension through a fine-mesh stainless steel strainer into a large beaker and bring the total volume with LSB up to 8 times the muscle weight. Repeat the washing steps four more times, and then resuspend the pellet in Triton X-100 buffer to solubilize biological membranes. Stir for 10 min and centrifuge at 3500 rpm for 15 min. Wash the pellet with LSB four more times. After the last centrifugation, resuspend the pellet in a minimum volume of LSB until no clumps are visible. Add an equal volume of ice cold glycerol. Swirl or stir gently until an even suspension is obtained. Store at −20°.

Comments

1. Yield: In a typical preparation 20 ml of myofibril suspension (9–12 mg/ml) is obtained from 50 g of wet muscle per kilogram of rabbit. The yield will decrease sharply if care is not exercised in the washing procedure. In particular, excessive blending and vigorous shaking should be avoided; otherwise, many myofibrils adhere to the massive amounts of air bubbles generated and float as a white top layer that is decanted away after centrifugation. This tendency is especially pronounced in the last few washes, when most soluble proteins have already been removed from the preparation. If flotation inadvertently occurs, the myofibrils in the layer can be pelleted by adding Triton X-100 (0.1%, w/v, final concentration) as a wetting agent before centrifugation.

2. Purity: The preparation, consisting of both single and bundled myofibrils, is essentially free of cytoplasmic proteins, connective tissue, and mitochondria. A small number of nuclei are detectable by light microscopy. Membranous vesicles, presumably remnants of the T tubule or sarcoplasmic reticulum, are attached to Z lines.[14]

3. Protease contamination: The present procedure is adopted because it reproducibly yields myofibrils of high purity with low protease contami-

[14] K. Wang and R. Ramirez-Mitchell, *J. Cell Biol.* **83**, 389a (1979).

nation from rabbit back muscle. We routinely carry out all procedures in a cold room to avoid any chance of exposing myofibrils to higher temperature before the bulk of endogenous proteases has been removed. The final preparation, however, still contains sufficient protease activity to degrade part of the titin and nebulin into fragments with slightly faster mobilities within several hours at 37°.[15] Such activities cannot be removed by additional washings. Storage at −20° in 50% glycerol greatly diminishes proteolysis, and only slight degradation is observed in preparations stored under these conditions for up to 2 years.[15]

It should be emphasized that myofibril preparations from other species or muscle types may contain much higher residual protease activity. With certain protease-rich muscles, such as chicken breast muscle, appreciable degradation of titin and nebulin occurs at 4° within a few hours of the first homogenization step.[11,15] In such cases, increasing EGTA to 5 mM in LSB effectively inhibits the bulk of endogenous protease activities. More extensive washings may be necessary to further remove residual proteases.

Dodecyl Sulfate–Polyacrylamide Gel Electrophoresis

The high-porosity gel system described below is a "low bisacrylamide" modification of the continuous buffer system of Fairbanks et al.[16]

Reagents and Buffers

Polyacrylamide gel: 0.6 × 10 cm cylindrical gels of 3.2% acrylamide (acrylamide/bisacrylamide = 50/1) polymerized with 0.1% (v/v) TEMED and 0.06% (w/v) ammonium persulfate
Wash buffer: 5 mM Tris-HCl, pH 8.0
Gel and electrophoresis buffer: 40 mM Tris-acetate, 20 mM sodium acetate, 2 mM EDTA, 0.1% (w/v) SDS, pH 7.4
3× SDS gel sample buffer: 30 mM Tris-HCl, 3 mM EDTA, 3% (w/v) SDS, 120 mM dithiothreitol, 30% (v/v) glycerol, 30 μg of pyronin Y per milliliter, pH 8.0.
Staining solution: 0.1% (w/v) Coomassie Blue R, 25% (v/v) isopropanol, 10% (v/v) acetic acid
Destaining solution: 5% (v/v) isopropanol, 10% (v/v) acetic acid

Procedures. Wash myofibrils twice with 10 volumes of wash buffer at 4° by centrifuging at 5000 rpm for 5 min in a Sorvall SS-34 rotor. Myofibrils swell significantly and can be resuspended easily in a minimal volume of wash buffer. Add an aliquot of myofibrils (5 mg/ml) to 0.5 volume of 3× SDS gel sample buffer preheated to 50° in a plastic microcentrifuge tube

[15] K. Wang and D. Palter, in preparation.
[16] G. Fairbanks, T. L. Steck, and D. F. H. Wallach, *Biochemistry* 10, 2606 (1971).

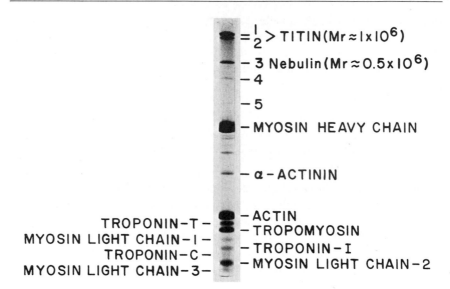

$= \frac{1}{2} >$ TITIN(Mr $\approx$ I x IO^6)

$-$ 3 Nebulin(Mr $\approx$ 0.5 x IO^6)

$-$ 4

$-$ 5

$-$ MYOSIN HEAVY CHAIN

$-$ α - ACTININ

$-$ ACTIN

TROPONIN-T $-$ $\quad$ $-$ TROPOMYOSIN

MYOSIN LIGHT CHAIN-I $-$ $\quad$ $-$ TROPONIN-I

TROPONIN-C $-$ $\quad$ $-$ MYOSIN LIGHT CHAIN-2

MYOSIN LIGHT CHAIN-3 $-$

FIG. 1. Sodium dodecyl sulfate gel pattern of rabbit skeletal myofibrillar protein.

with lid, mix by gentle shaking, close the lid and incubate at 50° for 15–30 min in a water bath. After incubation, centrifuge for 5 min in a Beckman microfuge. Observe whether any gel-like pellet remains unsolubilized.

Apply 30–150 μg of myofibrillar proteins to each gel (load 30–40 μg for better resolution of titin doublets). Electrophorese at 6 mA/gel for 1.25 hr until the tracking dye is 1 cm above the bottom of the gel. Remove the gel and mark the dye front with India ink by piercing the gel with a 22-gauge ink-filled hyperdermic needle. Rinse away any excess ink with water and stain for 10–16 hr with gentle agitation on a platform shaker. Destain in a Bio-Rad diffusion destainer for 10–24 hr.

Comments

1. Gel pattern: A typical gel pattern of rabbit skeletal myofibrillar proteins is shown in Fig. 1. Protein bands with lower mobilities than myosin heavy chain are labeled numerically from the top: band 1 and band 2 are collectively designated as titin; band 3 is designated as nebulin[3,4]; band 4 comigrates with a dimer of myosin heavy chain; band 5 comigrates with filamin. Seven major bands have mobility greater than actin; these correspond to subunits of tropomyosin, troponin, and myosin. This system resolves all major myofibrillar protein subunits from 15,000 to 2,000,000 M_r.

2. Solubilization of titin and nebulin: Titin is unexpectedly difficult to solubilize consistently and quantitatively in SDS. Our best results are

obtained with well-dispersed myofibrils suspended in a very low ionic strength buffer. The addition of SDS to a tightly packed myofibril pellet frequently results in the formation of an insoluble gel-like residue composed mainly of titin. The presence of high salt ($\mu > 0.2$) during solubilization also drastically decreases the amount of titin solubilized. It has been our experience that whenever a gel-like material is present after SDS addition, titin is invariably diminished or completely absent in the solubilized protein solution.

Titin is progressively degraded to a smear on polyacrylamide gels when the SDS–myofibril solution is heated at 100° for more than 2 min. Such degradation is not observed below 75° for up to 30 min. The cause is not yet clear. Proteolysis or nonenzymic hydrolysis of heat-labile peptide bonds is suspected. For this reason we routinely use 50° as a convenient solubilization temperature for highly purified myofibrils. When heating at 100° is desirable (to inhibit proteolysis in muscle homogenate or tissue powder or other protease-rich myofibrillar preparations), the period is limited to 2 min. Nebulin, in contrast, is solubilized quantitatively and consistently in SDS.

3. Handling of gels: The high-porosity gels have little mechanical rigidity and are often frustrating to handle. Avoid any contact of the gel with dry surfaces and never lift the gel from only one end. Transfer of these soft gels is easily accomplished without breakage by using a long, half-cylindrical spatula (Fisher scoopula spatula) to hold the gel along its entire length.

4. Other gel systems: In addition to the cylindrical gel system described above, two slab gel systems have been developed: (a) an agarose–linear polyacrylamide composite gel designed to increase the resolution of these extremely large myofibrillar proteins[17]; (b) a 2 to 12% gradient polyacrylamide gel that provides sharp band of both large and small polypeptides.[18]

Purification

Reagents and Buffers

Wash buffer: 5 mM Tris-HCl, pH 8.0

4× SDS column sample buffer: 200 mM Tris-HCl, 20 mM EDTA, 80 mM dithiothreitol, 20% (w/v) SDS, 4 mM phenylmethylsulfonyl fluoride, pH 8.0

NaCl, 4 M in wash buffer

SDS, 3% (w/v) in wash buffer

---

[17] K. Wang and D. Woodrum, *Anal. Biochem.*, submitted.
[18] L. Somerville and K. Wang, *Biochem. Biophys. Res. Commun.* **102**, 53 (1981).

Gel filtration column buffer: 40 m*M* Tris-HCl, 20 m*M* sodium acetate, 2 m*M* EDTA, 0.1 m*M* dithiothreitol, 0.1% (w/v) SDS, pH 7.4
BioGel A150M, 50–10 mesh (Bio-Rad, Richmond, California)

Procedures. Wash myofibrils (50 mg) twice at 4° with 4 volumes of wash buffer at 5000 rpm for 5 min in a Sorvall SS-34 rotor to remove glycerol and to swell myofibrils. To the pellet, add 1.0 ml of the 4× SDS column sample buffer that has been prewarmed to 50° in a water bath. Stir immediately and transfer to the prewarmed glass vessel of a motor-driven Potter-Elvehjem Teflon–glass tissue grinder. Homogenize the mixture promptly with two or three strokes, incubate at 50° for 15–20 min, and then centrifuge at 28,000 rpm for 1 hr at 20° in a Beckman type 30 rotor. Decant and save the clear supernatant (2.0 ml). The small whitish pellet (a disk a few millimeters in diameter) is discarded. (If a much larger gel-like pellet is obtained, then titin is not solubilized completely.)

METHOD I: GEL FILTRATION OF TOTAL MYOFIBRILLAR PROTEIN. The solubilized protein solution (2 ml) is applied to a 1.6 × 70 cm column of BioGel A-150m equilibriated with gel filtration column buffer. Fractions (2.5 ml each) are collected at a flow rate of 6 ml/hr and monitored by absorbance at 280 nm and by gel electrophoresis (50-μl aliquot of each fraction is mixed and incubated with 25 μl of 3× SDS gel sample buffer as described above). Titin is eluted in the first peak of the elution profile (Fig. 2). An ascending shoulder of the titin peak (gels a and b) consists mainly of nucleic acid ($A_{280}/A_{260} < 0.5$). The descending end of the peak contains smears of degraded titin (gel f). Nebulin is eluted as an overlapping peak near fraction 56 (gel g).

The pooled titin fractions (34 to 45) are contaminated with only a very low level of nucleic acid ($A_{280}/A_{260} \approx 1.3$). The yield varies owing to the degradation and nonquantitative solubilization of titin (see comments below). In general, 3–4 mg of intact titin are obtained from 50 mg of rabbit skeletal myofibrils. The pooled nebulin fractions (51–58) contain 1.5 mg of protein. A larger column (2.5 × 100 cm) can be used to scale up the preparation to 200 mg of myofibrils.

METHOD II: SALT FRACTIONATION OF SDS-MYOFIBRILLAR PROTEINS. In this procedure, titin–SDS complex is specifically and quantitatively precipitated by the addition of NaCl to SDS-solubilized myofibrillar proteins to a final concentration of about 0.64 *M*. The precipitate redissolves easily in SDS column sample buffer and is purified further by gel filtration. The supernatant contains nebulin and other myofibrillar proteins.

Titin Purification. Solubilized myofibrils (200 mg) are prepared as described in Method 1. Dilute the clear supernatant (9 ml) from the 28,000 rpm centrifugation step with 18 ml of wash buffer at room temperature and then add 5.1 ml of 4 *M* NaCl in wash buffer dropwise, with efficient

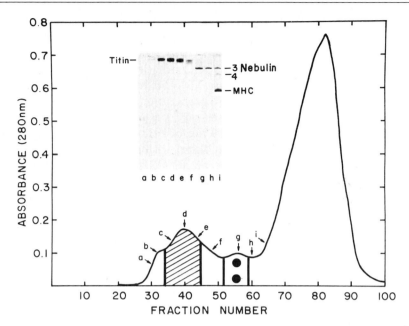

FIG. 2. Purification of titin and nebulin by agarose gel filtration in the presence of sodium dodecyl sulfate (SDS) (Method I). An elution profile of solubilized myofibrillar protein on a BioGel A-150m column is shown. *Inset:* Gel patterns of selected fractions indicated by arrows above the elution profile. Shaded and dotted areas are the pooled titin and the nebulin preparation, respectively. MHC, Myosin heavy chain.

stirring, to bring the final concentration to 0.64 M NaCl. The solution becomes progressively cloudy, and eventually a gelatinous clump appears. Let the mixture stand at room temperature for 30 min and then centrifuge at 14,000 rpm for 20 min in an SS-34 rotor at 20°. Decant and save the supernatant for nebulin purification (see below). Resuspend the pellet with 0.5 ml of 0.64 M NaCl in wash buffer and centrifuge at 5000 rpm for 5 min. Discard the supernatant. To the pellet add 250 μl of 4× SDS column sample buffer prewarmed to 50°. Stir with a glass rod and incubate briefly at 50° for 5 min. Add 2 ml of 3% SDS in wash buffer and incubate further at 50° for 5 min to solubilize pellet completely. Centrifuge the solution at 45,000 rpm for 30 min at 20° in a Beckman type 50 rotor and remove the clear supernatant with a Pasteur pipette. Discard the pellet, if any. Load the clear supernatant (5 ml) to a 2.5 × 90 cm BioGel A-150m gel filtration column equilibrated with gel filtration column buffer. Fractions (10 ml each) are collected at a flow rate of 18 ml/hr, and a 100-μl aliquot of each fraction is analyzed by gel electrophoresis.

Titin is eluted in the first peak (Fig. 3). The first few fractions (31 to 33) of the peak contain a small amount of nucleic acid. The trailing edge (fractions 50 to 60) contains degraded titin appearing on gels as a smear below titin. The second absorbance peak (fractions 75 to 95) is due mainly to dithiothreitol. The titin fractions (35 to 50) are pooled.

The present method yields a titin preparation with protein composition comparable to that obtained by Method I yet is substantially free of nucleic acid contamination. About 15 mg of titin with $A_{280}/A_{260} > 1.5$ are routinely obtained from 200 mg myofibril within 24 hr.

Nebulin Purification. To the supernatant (32 ml) of the salt fractionation step, add 0.5 ml of 4 M NaCl in wash buffer to raise salt concentration to 0.70 M NaCl. After 10 min, centrifuge the solution at 45,000 rpm for 30 min in a Beckman type 50 rotor at 20° to remove any particulate material. Decant and then concentrate the supernatant to ~10 ml by ultrafiltration with an Amicon PM30 filter. Apply the concentrated supernatant to a separate 2.6 × 80 cm BioGel A-150m gel filtration column as described above for titin purification. Nebulin is eluted in the ascending end of the

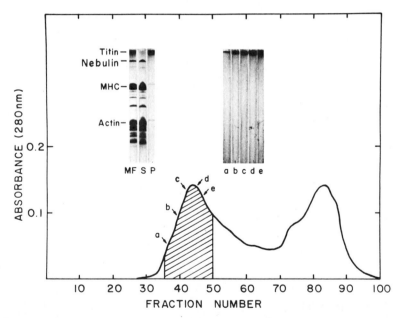

FIG. 3. Purification of titin and nebulin by salt fractionation and agarose gel filtration in the presence of sodium dodecyl sulfate (SDS) (Method II). Myofibrils (MF) are solubilized in SDS and fractionated by the addition of NaCl to 0.64 M. Pellet (P), consisting mainly of titin, is applied to a BioGel A-150m column. The elution profile and gel patterns of selected fractions are shown. The hatched area is the pooled titin preparation. The supernatant (S), containing no titin, is applied to a separate BioGel A-150m column to purify nebulin (elution profile not shown). MHC, Myosin heavy chain.

second peak (cf. Fig. 2), following the first peak, which contains mainly nucleic acids. The nebulin-containing fractions is located by SDS gels and pooled; care is exercised to avoid fractions containing band 4 and myosin heavy chain. About 4–5 mg of nebulin is obtained from 200 mg of myofibrils.

Comments

1. Solubilization of titin: Titin is degraded more extensively and solubilized less efficiently when large amounts of myofibrils are treated with SDS in bulk (compare the sharp titin bands in Fig. 1 with that of the smeared titin in Fig. 3). The degraded smear is, however, completely removed by the gel filtration step.

2. Method I: This method has been applied to the isolation of titin and nebulin from striated myofibrils from various sources.[11] Very similar elution profiles are obtained when myofibrils are solubilized directly in 8 M urea or 6 M guanidinium chloride and eluted on BioGel A-150m gel with the same denaturant.[11]

3. Method II: This method has so far been applied to rabbit skeletal and frog sartorius myofibrils. The novel salt fractionation step separates titin from nebulin and the bulk of other myofibrillar proteins, and, more important, it removes nucleic acid from titin (Fig. 3, inset). The resulting pellet consists, in general, of at least 85% titin. The basis for the specific "salting out" of titin–SDS complex is not clear and perhaps reflects certain unusual molecular structure of titin. The selectivity of the salt fractionation is both protein and salt concentration dependent. High protein and salt concentrations favor the nonquantitative precipitation of the entire range of myofibrillar proteins. The optimal salt concentration at which titin quantitatively precipitates from a 7 mg/ml protein solution varies somewhat from 0.64 to 0.70 M NaCl. If no cloudy precipitate or clump forms after the addition of the desired volume of NaCl, additional NaCl at an increment of 0.01 M should be added until this occurs. Nebulin is purified from the supernatant after the salt concentration is raised further (0.06 M) to precipitate any residual titin and degradation products that would otherwise contaminate nebulin.

4. Titin composition: It should be noted that band 1 and band 2 are partially resolved in the first peak of the gel filtration (Figs. 2 and 3). The pooling of early fractions of the titin peak thus results in a final preparation enriched in band 1. If necessary, the individual fractions can be combined at appropriate ratios to yield a pool with the band 1 to band 2 ratio equal to that found in the original myofibril.

5. Gel filtration column: The 1% agarose (BioGel A-150m) gel filtration beads are fairly compressible, and it is recommended that fine parti-

cles be removed prior to column packing and that a hydrostatic pressure of 30 cm not be exceeded during column packing, sample loading, and elution. Small beads (100–200 mesh), as well as columns with a porous plastic disk or sintered-glass bed support, invariably lead to greatly reduced or halted flow and should therefore be avoided. In our earlier experiments, a 2% agarose (BioGel A-50m) was used with similar, yet slightly inferior, resolution between nucleic acids, titin, and nebulin. The purity and the yield of titin and nebulin are thus less satisfactory.[2]

Concentration and Storage

Purified titin and nebulin in SDS can be concentrated rapidly by ultrafiltration with an Amicon PM30 or PM50 filter. The SDS is also somewhat concentrated by these filters.

Highly purified titin in SDS, with or without added azide, still degrades at a rapid rate at room temperature as evidenced by the appearance of an increasing amount of smear below titin within several days. This degradation is greatly retarded at 4° or below. It should be noted, however, that slow freezing by leaving samples in a freezer frequently, but not always causes the bulk of titin to aggregate upon thawing, into SDS-insoluble gel-like material. Quick freezing in liquid nitrogen and rapid thawing at 50° appear to yield clear titin solution without visible aggregation. In contrast, nebulin is very stable at room temperature, and no irreversible aggregation in SDS has been observed upon repeated freeze-thawing under a variety of conditions.

Molecular Properties

Purified titin consists of two polypeptides of similar size. Band 1 and band 2 have apparent chain weights of 1.4×10^6 and 1.2×10^6, because they comigrate in SDS gels with the heptamer and hexamer of crosslinked myosin heavy chain, respectively.[11] The possibility exists that these two bands are related structurally because limited proteolytic degradation of titin *in situ* converts band 1 into a major fragment that comigrates with band 2.[15] Titin has a strong tendency to aggregate, even in the presence of denaturants, into an elastic gel composed of ultrathin (20–40 Å) filaments.[14]

Nebulin has an apparent molecular weight of $0.5–0.6 \times 10^6$, corresponding in mobility to the trimer of myosin heavy chain. Titin, nebulin, and myosin are chemically and immunologically distinct.

Acknowledgment

This work is supported in part by USPHS Grant AM20270 and CA09182.

Section II

Methods for Smooth Muscle Chemistry

[24] Preparation of Isolated Thick and Thin Filaments

By PETER COOKE[1]

Several detailed inventories of the ultrastructure in the contractile apparatus of various vertebrate smooth muscles have provided an indication of the molecular structures and some functional characteristics of the constituent myofilaments.[2-4] There are three distinct classes of filaments. The most abundant type corresponds to the actin-containing filaments present in striated muscles. They are 6–8 nm in diameter and display a structural organization that is very similar to that of F-actin, and there is evidence to indicate that tropomyosin is a regular and integral component. There is no apparent structural complex associated with the thin filaments in smooth muscle that can be assigned to the troponins (in keeping with the observation that the regulatory events relating to Ca^{2+} are associated with myosin), but the radial shift of tropomyosin occurs during the activation of contractile activity. The pattern of organization of the thin filaments is markedly different from that of the striated muscles: the filaments are grouped into exclusive, small lattice-like bundles with no clear indication of axial registration or regular filament length (Fig. 1).

The thick filaments observed in studies of embedded, thin-sectioned smooth muscle are around 18 nm in diameter and up to 7 μm in length (Fig. 1). They are identified with myosin chiefly because in size, shape, and subunit structure they are in some instances comparable to filamentous aggregates of purified myosin. The thick filaments are parenchymal to the bundles of thin filaments and/or ensheathed by rosettes of thin filaments. In general, their density distribution and pattern of organization in the muscle fibers is variable. These features may indicate that the thick filaments are intrinsically labile and represent the aggregated form of a functionally related dynamic equilibrium.

The structural relationship between the thick and thin filaments and the third class of intermediate, or 10 nm in diameter, filaments is not entirely clear, although on the basis of their insolubility in salt solutions of high ionic strength, detailed ultrastructure, and cytoplasmic organization

---

[1] The author is an Established Investigator of the American Heart Association and supported in part by funds from the Connecticut Heart Association.
[2] C. F. Shoenberg and D. M. Needham, *Biol. Rev. Cambridge Philos. Soc.* **51**, 53 (1976).
[3] J. V. Small and A. Sobieszek, *Int. Rev. Cytol.* **64**, 241 (1980).
[4] A. V. Somlyo, *in* "Handbook of Physiology," Section II; The Cardiovascular System (R. M. Berne, ed.), Vol. 2, pp. 33–67.

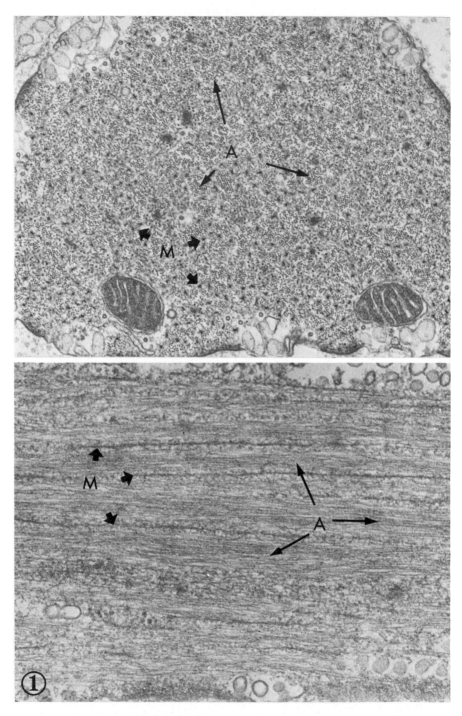

FIG. 1. Electron micrographs showing the pattern of localization of the lattice-like bundles of thin actin-containing filaments (A) and the parenchymal distribution of thick myosin-containing filaments (M) in smooth muscle fibers of the guinea pig taenia coli. ×50,000.

they are treated as a distinctive class of filaments in many cell types. In smooth muscles they are integrated into the system of thin filaments through firm connections at the dense bodies that are homologous to the Z disks of striated muscles, but their contribution to the process of contraction may only be mediated through their influence on the long-range organization within the contractile apparatus.

Sources of Smooth Muscle

Practical sources of fresh smooth muscle in bulk quantities are portions of certain hollow organs from commercially slaughtered birds and mammals: the lateral muscles of avian gizzards, porcine or bovine stomach muscularis, and the uterine myometrium. Small supplies of myofilaments can be obtained from muscular arteries, sperm ducts, and the muscular layers of the intestines (especially, the taenia coli on the cecum of guinea pigs).

General Considerations

Smooth muscle fibers are small, highly asymmetric cells that are integrated into a tissue through a variety of cellular junctions and an *extensive* investment of loose connective tissue. However, the cellular contents are difficult to obtain in high yield without the application of marked mechanical forces to disrupt the tissue and the constituent muscle fibers.

In order to obtain optimal preparations of isolated filaments, there are two stages to be considered: to disintegrate the dissected muscle into a uniform population of particles consisting primarily of muscle fibers, and then to release the myofilaments from a suspension of the muscle fibers. The first stage is accomplished in two steps. The muscle is minced in a worm-drive grinder equipped with a small-bore (1–2 mm) force-plate, then homogenized at low speed in a Waring-style blender in order to produce a slurry consisting of small bundles of muscle fibers that are collected by centrifugation. The myofilaments are obtained by homogenization of a suspension of the minced muscle by a variety of liquid-shear methods, (depending on the volume of the preparation) and isolated by differential centrifugation.

Isolation of Thin Filaments

Five pounds of whole, warm-fresh gizzards obtained by special arrangement from a commercial poultry slaughterhouse will yield around 600 g of completely trimmed lateral muscle. Whole gizzards[5] are placed in

---

[5] In many poultry slaughterhouses the usual practice in dressing the gizzard for retail sale includes transection of one of the lateral muscles.

crushed ice, and the lateral muscles are dissected free of aponeurosis and minced in a standard, worm-drive meat grinder. The mince is weighed then suspended in an equivalent volume of cold buffered solution containing 0.05 M sodium phosphate (pH 7.0) and 0.001 M EDTA and homogenized at low speed in a Waring blender until the mince is transformed to a slurry composed of microscopic fragments of muscle fibers and strands of connective tissue. Upon standing, the muscle fragments settle atop the larger networks of connective tissue, and several hundred milliliters of muscle fragments can be decanted and concentrated by centrifugation at low speed.

The myofilaments are isolated directly from the muscle fragments. Aliquots of muscle are suspended in two volumes of "relaxing solution" (0.1 M KCl, 0.001 M MgCl$_2$, 0.005 M ATP, 0.016 M NaH$_2$PO$_4$, Na$_2$HPO$_4$ at pH 7.3) and homogenized by liquid shear at high rotor speed for 10–30 sec. The homogenate is centrifuged at 500 g to precipitate the cell fragments and to suspend the membranous organelles. Free myofilaments are preferentially localized in the middle, clear solution of the centrifuge tube. Further separation of the myofilament-containing fraction from soluble protein is achieved by removal of the middle layer containing the myofilaments and collection by high speed centrifugation ($>$40,000 g) followed by gentle resuspension.

For small tissue samples ($\sim$0.1 g), the muscle is sliced with a clean razor blade into 1-mm^3 fragments and thoroughly teased apart to microscopic dimensions with fine forceps then homogenized in relaxing solution and isolated as described above.

This method provides a sample of isolated thin filaments that vary widely in length (Fig. 2). If the minced muscle is washed free of sodium phosphate buffer and suspended in 10 volumes of 2 mM Tris buffer (pH 8.5) for 24 hr to dissociate tropomyosin prior to the final homogenization, then the length distribution of the isolated filaments is fairly uniform and short (Fig. 3).

Isolation of Thick Filaments

Owing to the solubility of smooth muscle myosin in salt solutions of low ionic strength, the isolation of a discrete cell-fraction containing thick myofilaments has not been clearly achieved by what might be called routine methods. The thick filaments appear to be very labile either as an intrinsic feature of their functional activity, or their structural integrity is very sensitive to the available methods of isolation. The difficulties encountered in localizing thick filaments *in situ* by electron microscopy suggests that special physicochemical circumstances attend the filamentous form of this protein in smooth muscle. The arrangement of the cross-

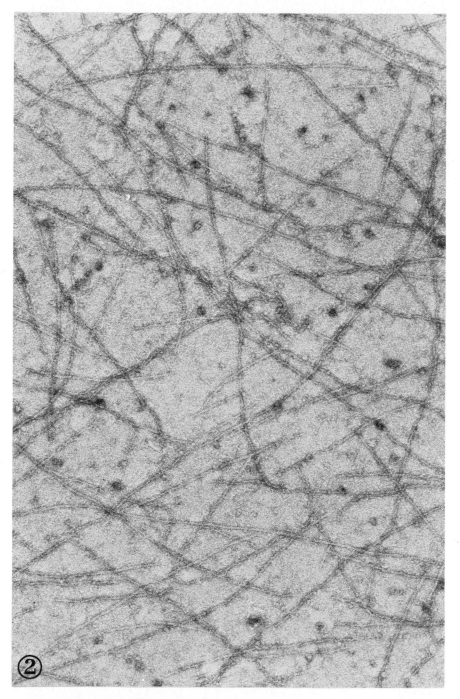

FIG. 2. Preparation of negatively stained thin filaments isolated from chicken gizzard illustrating the range of lengths. ×112,000.

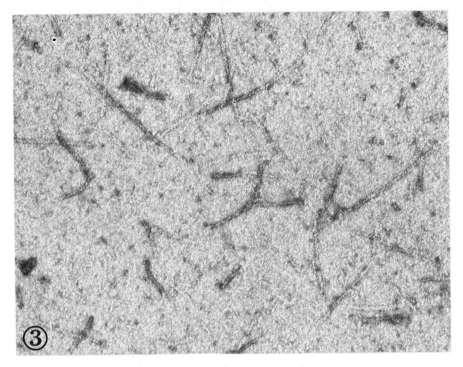

FIG. 3. Preparation of negatively stained segments of thin filaments. ×112,000.

bridge projections observed in sectioned preparations is an obvious guide to evaluate the fidelity of the form of isolated filaments, but the evidence is disparate. Both "bipolar" and "side-polar" molecular arrangements are observed. Methods to isolate both forms are described.

J. V. Small has championed the notion that the thick filaments have cross-bridge projections that have side-polar arrangement.[6] The isolation of this filamentous form starts with a suspension of isolated, intact muscle fibers derived from mammalian sperm ducts, taenia coli, stomach, carotid artery, or avian gizzard.

Dissected muscle strips are immersed in a Ca^{2+}-free Hanks' solution (137 mM NaCl, 5 mM KCl, 1.1 mM Na_2HPO_4, 0.4 mM K_2HPO_4, 4 mM $NaHCO_3$, 5.5 mM glucose, 2 mM $MgCl_2$, 2 mM EGTA, 10 mg of streptomycin per liter, supplemented the 5 mM PIPES buffer). The pH is adjusted to 6.0 for an optimal yield of responsive muscle fibers to chemical induction of contraction and relaxation. The muscle strips are incubated in this solution, fastened to plastic plates, and stored for 2–4 hr at

[6] J. V. Small, *J. Cell Sci.* **24**, 327 (1977).

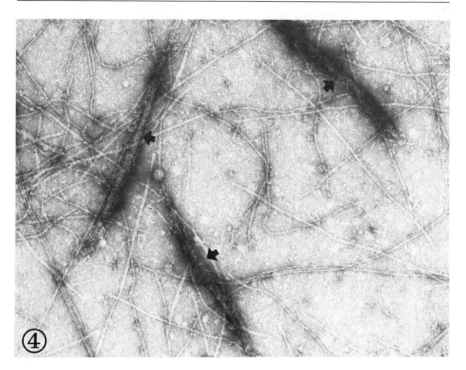

FIG. 4. Preparation of negatively stained myofilaments illustrating isolated thick filaments with a bipolar conformation. ×112,000.

$4°$. Intact, isolated muscle fibers are obtained from the strips by enzymic digestion with collagenase (1 mg/ml) dissolved in a solution containing 137 mM NaCl, 5 mM KCl, 4 mM NaHCO$_3$, 5.5 mM glucose, 2 mM MgCl$_2$, 10 mM PIPES buffer at pH 6.5 with either 0.75 mM ATP or 2 mM xylocaine to maintain the fibers in a relaxed state. The muscle strips are digested at $37°$ for 1 hr and returned to the original modified Hanks' solution (without 2 mM MgCl$_2$), supplemented with 0.5 mM dithiothreitol. After two changes of solution for 1–6 hr at $4°$, the muscle tissue is dispersed by repeated aspiration through the tip of a Pasteur pipette. The isolated fibers are collected from this suspension by low speed centrifugation.

Isolated thick filaments are finally prepared from digested fibers after a brief treatment with 0.2% Triton X-100, washed to remove the detergent, and homogenized to produce cell fragments. The cell fragments are gently dispersed into the constituent filaments by shearing through the tip of a Pasteur pipette to produce a suspension of the thick filaments that range in length from 2 to 8 μm and display a distinctive axial periodicity of 14 nm.

Bipolar thick filaments are prepared from fresh smooth muscle by the methods first developed by R. V. Rice and C. F. Shoenberg. The dimensions of the isolated filaments are dependent on pH, $[Ca^{2+}]$, $[Mg^{2+}]$, and the standing age of the homogenate in a fashion that parallels the results observed for filamentogenesis of purified myosin.[7]

Finely teased or minced smooth muscle in suspended in $0.01-0.1\ M$ KCl, $0.001-0.01\ M$ $MgCl_2$, $0.005\ M$ ATP, $0.016\ M$ $Na_2HPO_4-NaH_2PO_4$ (pH 6.2–7.4). The suspension is briefly homogenized at high speed by liquid shearing and centrifuged at $1000\ g$ for 3 min. The supernatant fluid contains bipolar thick filaments (Fig. 4). The length ranges from 0.2 to 1.0 μm. Long filaments are found at lower pH and ionic strength, and they require 5–10 mM $MgCl_2$. Short filaments predominate at higher pH and ionic strength. Schemes to enrich the fraction containing thick filaments conform to procedures for the isolation of myosin. In general, the isolation of thick filaments has few clearly understood rules. The yield is favored by conditions that prevent formation of the actomyosin complex (1–10 mM $MgCl_2$, 5 mM ATP). This may indicate that the myosin has at least two metastable forms: one dominated by molecular interactions that favor the formation of discrete thick filaments, and the other dominated by interactions involved in the actomyosin complex.

[7] P. R. Wachsberger and F. A. Pepe, *J. Mol. Biol.* **88**, 385 (1974).

[25] Preparation of Individual Smooth Muscle Cells from the Stomach of *Bufo marinus*

By F. S. FAY, R. HOFFMANN, S. LECLAIR, and P. MERRIAM

Studies of the physiological mechanisms underlying the generation and regulation of force in the smooth muscle cell have been performed on strips of tissue from various organs.[1] In interpreting results from such preparations, the assumption has generally been made that the properties of those tissues are a simple reflection of a homogeneous population of cells. This assumption has rarely been tested, let alone verified, and it is quite likely that many of the properties of such preparations at least in part reflect the presence of neuronal elements, the complex mechanical and electrical interconnection of cells, and the presence of a complex

[1] E. Bülbring, A. F. Brading, A. W. Jones, and T. Tomita, "Smooth Muscle." Williams & Wilkins, Baltimore, Maryland, 1970.

extracellular protein matrix representing a significant barrier to diffusion between cells and medium.[1] For this reason, many studies aimed at understanding the cellular mechanisms involved in contraction of smooth muscle are greatly facilitated by using single isolated cells rather than intact tissue. Over the past 8 years, several techniques have been described for the isolation of single smooth muscle cells from various tissue sources.[2-6]

In this chapter, a method is described for the isolation of single smooth muscle cells from the stomach of the toad *Bufo marinus*. This procedure is a modification of a technique originally described by Bagby *et al.*[2] We believe that smooth muscle cells isolated from this animal by our technique are at present the best single smooth muscle cell preparation for many physiological studies. A single animal yields large numbers of cells (6×10^7 cells/animal). The cells appear to be undamaged as a consequence of the isolation procedure based upon a comparison of the structural and functional properties of both isolated cells and intact tissue.[7] The excellent status of the cells is further attested to by the fact that they are relaxed after isolation, yet contract when stimulated by a variety of means.[7-9] The individual cells isolated from the toad are somewhat larger ($6 \times 250 \ \mu$m) than smooth muscle cells from mammalian vasculature and viscera ($3 \times 200 \ \mu$m) thus facilitating studies of various functions at the level of a single cell alone.[10-12] Since smooth muscle cells isolated from the toad share many structural, biochemical, physiological, and pharmacological properties with smooth muscle from mammalian species, results obtained on these cells should be of general significance. Finally, because smooth muscle cells isolated from this animal have been the most widely studied, considerable background data assist an investigator beginning to tackle a new problem. The subsequently described procedure is the one presently utilized in our laboratory and reflects experience of the past 7 years in preparing smooth muscle cells from the stomach of *Bufo marinus*.

[2] R. M. Bagby, A. M. Young, R. S. Dotson, B. A. Fisher, and K. McKinnon, *Nature (London)* **234**, 351 (1971).

[3] F. S. Fay and C. M. Delise, *Proc. Natl. Acad. Sci. U.S.A.* **70**, 641 (1973).

[4] J. V. Small, *J. Cell Sci.* **24**, 327 (1977).

[5] S. Falco, L. Brann and W. Halpern, *Physiologist* **23**(4), 102 (1980).

[6] H. E. Ives, G. S. Schultz, R. E. Galardy, and J. D. Jamieson, *J. Exp. Med.* **148**, 1400 (1978).

[7] F. S. Fay, "Cell Motility." Cold Spring Harbor *Conf. Cell Proliferation* **3** [Book A–C], p. 185 (1976).

[8] J. J. Murray, P. W. Reed, and F. S. Fay, *Proc. Natl. Acad. Sci. U.S.A.* **72**, 4459 (1975).

[9] F. S. Fay and J. J. Singer, *Am. J. Physiol.* **232**, C144 (1977).

[10] F. S. Fay, *Nature (London)* **265**, 533 (1977).

[11] F. S. Fay, H. Schlevin, W. Granger, and S. R. Taylor, *Nature (London)* **280**, 506 (1979).

[12] J. J. Singer and J. V. Walsh, *Am. J. Physiol.* **239**, C162 (1980).

Method of Cell Isolation

Materials

Amphibian physiological saline (APS): 108 mM NaCl, 19.9 mM NaHCO$_3$, 3 mM KCl, 0.975 mM MgSO$_4$, 1.8 mM CaCl$_2$, 0.14 mM NaH$_2$PO$_4$, 0.55 mM Na$_2$HPO$_4$, 11 mM glucose. This solution is made up fresh from refrigerated concentrated stock solutions of all the inorganic constituents, and glucose is added as a dry powder. The CaCl$_2$ must be added slowly and only after the solution has been vigorously bubbled with 95% O$_2$: 5% CO$_2$ for 5–10 min to bring it to pH 7.4. Unless this is done, CaCO$_3$ will precipitate upon addition. The APS must be kept either in a tightly stoppered flask gassed with 95% O$_2$: 5% CO$_2$ or bubbled slowly but continuously with 95% O$_2$: 5% CO$_2$.

Enzymes[13]

Trypsin, type III (Sigma Chemical Company, St. Louis, Missouri)

Collagenase, type I (Sigma). There is considerable variation in the amount of other proteolytic enzymes present within this preparation of collagenase. Thus, prior to purchasing this enzyme we routinely screen several lots of collagenase available at the time in large amounts from Sigma Chemical Company to determine which lot produces from a single population of tissue slices (*a*) the highest percentage of relaxed cells as determined under the microscope; (*b*) the largest contractile response to a standard excitatory stimulus (carbachol, 10^{-5} g/ml) as determined using a Coulter counter[14]; (*c*) the highest total yield of cells as judged by determination of cell density by counting cells in a hemacytometer. The first two criteria are of prime importance, but given two enzyme lots that appear to be equal with respect to these considerations, then cell yield becomes the deciding factor. We generally purchase a sufficient amount of an optimal collagenase preparation to last for at least a year and store it at $-20°$. Batches of collagenase that we found to be optimal for isolation of single smooth muscle cells had the following levels of noncollagenolytic enzymes as assayed by Sigma Chemical Company: clostripain, 1.8 units/mg; trypsin, 0.01 unit/ mg; protease (caseinase), 1.4 unit/mg per minute.

---

[13] *Enzyme activity units:* Trypsin: 1 unit will hydrolyze 1.0 μmol of α-N-benzoyl-L-arginine ethyl ester (BAEE) per minute at pH 7.6 at 25°; collagenase: 1 unit will liberate amino acids from collagen equivalent in ninhydrin color to 1.0 μmol of L-leucine in 5 hr at pH 7.4 at 37° in the presence of Ca^{2+}; clostripain: 1 unit will hydrolyze 1.0 μmol of BAEE per minute at pH 7.6 at 25° in the presence of 2.5 mM dithiothreitol; nonspecific protease: 1 unit will release amino acids from casein equivalent in ninhydrin color to 1.0 μmol of leucine in 5 hr at pH 7.4 and 37°.

[14] J. J. Singer and F. S. Fay, *Am. J. Physiol.* **232**, C138 (1977).

Other chemicals

Bovine serum albumin, fraction V from bovine plasma (Reheis Chemical Company, Phoenix, Arizona)

Soybean trypsin inhibitor (Worthington Biochemical Corporation, Freehold, New Jersey). Stock solution 0.5 mg per milliliter of H_2O is kept frozen until used.

Gentamicin sulfate, 50 mg/ml (Schering Corporation, Kenilworth, New Jersey)

Animals

Toads: Large *Bufo marinus* may be obtained from Mogul Ed (P.O. Box 482; Oshkosh, Wisconsin) or National Reagents (2161 Main Street, Bridgeport, Connecticut). We have found that the quality and quantity of cells obtained from these animals are often poor if they are used immediately after arrival, and thus we generally wait 2 weeks after their arrival before using the animals for isolation of smooth muscle cells. While these toads can survive for prolonged periods without feeding, we find that the yield and responsiveness of the isolated cells are greatly improved if the animals are fed live crickets on a weekly basis. The toads are housed in a room maintained at 80°F in steel bins (34 × 34 × 6 inches) with a layer of wood shavings and water in bowls always available.

Crickets: These may be obtained from Selph's Cricket Ranch (P.O. Box 123, Memphis, Tennessee 38101).

Procedure

Preparation of Tissue Slices. Toads are killed by decapitation using a small animal guillotine (Harvard Apparatus, Millis, Massachusetts). The stomach is removed by transecting it at the pyloric valve and at the stomach-esophageal junction. The mesenteric connections to the stomach are also cut, and the stomach is immediately transferred to a petri dish filled with well oxygenated APS. The stomach is cut open along its greater curvature and rinsed free of its contents by vigorously swirling the stomach in a beaker of APS. The mucosa is then cut away from the stomach wall, and the stomach is bisected along its former lesser curvature. The two strips are cut perpendicular to the former line of curvature into three equal pieces, and the 6 segments of stomach wall are then placed two at a time onto the stage of a Stadie-Riggs Tissue Slicer (A. H. Thomas, Philadelphia, Pennsylvania). Two to three 0.5-mm-thick slices are usually obtained from each segment of stomach wall and are held in well oxygenated APS (95% O_2 : 5% CO_2) until all 0.5 mm-thick slices have been cut.

Incubation Scheme. A summary of the steps involved in the enzymic disaggregation of the tissue slices into single smooth muscle cells is shown in Fig. 1. The procedure begins by draining the stomach slices of APS in

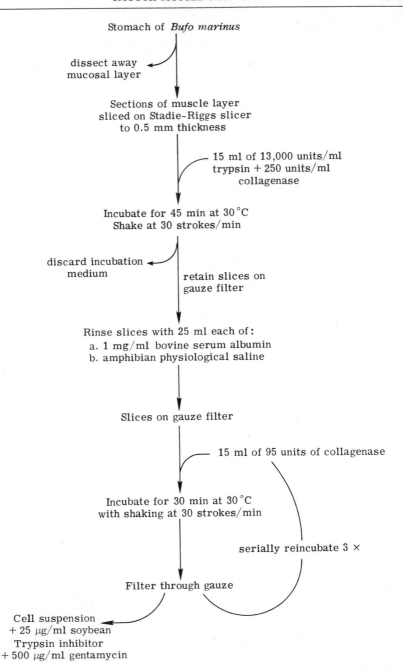

Stomach of *Bufo marinus*

dissect away
mucosal layer

Sections of muscle layer
sliced on Stadie-Riggs slicer
to 0.5 mm thickness

15 ml of 13,000 units/ml
trypsin + 250 units/ml
collagenase

Incubate for 45 min at 30 °C
Shake at 30 strokes/min

discard incubation
medium

retain slices on
gauze filter

Rinse slices with 25 ml each of:
a. 1 mg/ml bovine serum albumin
b. amphibian physiological saline

Slices on gauze filter

15 ml of 95 units of collagenase

Incubate for 30 min at 30 °C
with shaking at 30 strokes/min

serially reincubate 3 ×

Filter through gauze

Cell suspension
+ 25 µg/ml soybean
Trypsin inhibitor
+ 500 µg/ml gentamycin

FIG. 1. Scheme for the isolation of single smooth muscle cells from the stomach of *Bufo marinus*.

which they have been held and placing them into a solution containing trypsin at 13,000 units/ml, and collagenase at 250 units/ml in APS; 15 ml of this collagenase–trypsin mixture are utilized per toad stomach. The suspension of slices in this enzyme mixture are transferred into an Erlenmeyer flask of sufficient size so that the depth of solution is no greater than 1 cm to ensure optimal agitation during the subsequent incubation procedures. [All glassware to which the cells are exposed from this point onward is pretreated with Prosil-28 (PCR Research Chemical Inc., Gainsville, Florida) to prevent loss of cells due to adhesion to the glass.] Then 95% O_2 : 5% CO_2 is flushed into the flask for several seconds, and the flask is stoppered and placed for 45 min in a shaking water bath. The temperature is set at 30.5°, and the shaking rate at 30 strokes/min. At the end of the 45-min incubation period the suspension is poured through cotton gauze spread over the mouth of a beaker. The tissue slices that are retained on the gauze are rinsed first with bovine serum albumin (1 mg/ml) in APS, and then with APS; 25 ml of each of these two rinsing solutions are used per original toad stomach.

The stomach slices are then transferred into a solution of collagenase (95 units/ml) in APS; 15 ml of this solution are used for each original toad stomach. The suspension of slices is transferred to an Erlenmeyer flask of similar size to that used in the previous incubation, and 95% O_2 : 5% CO_2 flushed through the flask; a stopper is inserted and the flask is incubated for 30 min in a shaking water bath adjusted as described above. After this incubation the slices are separated from isolated single cells by passing the suspension through 3–4 layers of cotton gauze stretched over the mouth of a beaker. The solution passing through the gauze contains isolated smooth muscle cells. Soybean trypsin inhibitor at a final concentration of 25 μg/ml is added to stop any further tryptic digestion, and gentamicin sulfate is added to this to a final concentration of 500 μg/ml to halt bacterial overgrowth. The suspensions of smooth muscle cells may be held in a stoppered flask at room temperature for 6–8 hr following enzymic isolation, during which time the cells retain their responsiveness to various agents[9] and maintain stable ion gradients[15] and electrical properties.[12] The cells may also be concentrated or transferred to another suspension medium as will be described subsequently. The material retained on the gauze from the previous incubation is reincubated in an equal volume of fresh collagenase (95 units/ml) in a manner identical to the previous incubation. After this 30-min incubation, the tissue slices are transferred again to fresh collagenase and reincubated as in the previous incubation. The slices are serially incubated twice more for 30 min in a similar manner before being discarded. Soybean trypsin inhibitor and gentamicin sulfate are added to the cells isolated during each incubation immediately after they have been separated from the tissue slices by filtration.

[15] C. R. Scheid and F. S. Fay, *J. Gen. Physiol.* **75**, 163 (1980).

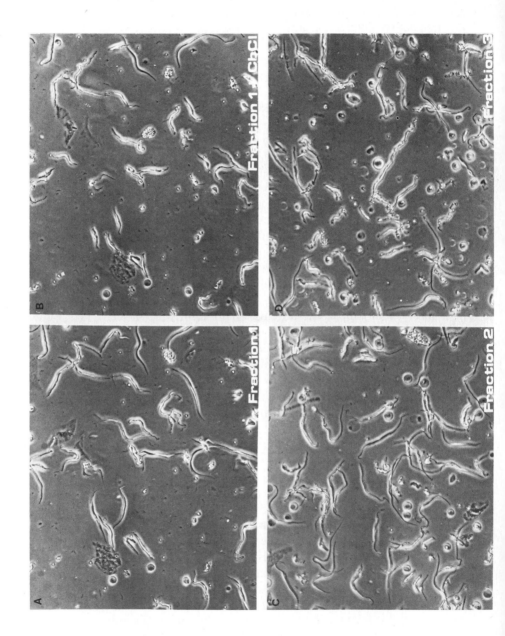

The cells isolated during these four incubations in collagenase may be pooled. We find that the first and second incubations following the initial trypsin/collagenase exposure usually yield a population of smooth muscle cells that contain the highest percentage of relaxed, fully extended cells (see Fig. 2). The density of cells is usually highest, however, in the third and fourth incubations and lowest in the first. Before utilizing any of the cells, we routinely assess their density and physiological state by placing a small volume of cell suspension (~1 ml) on a glass slide and inspecting it with a conventional light microscope equipped with phase contract optics at a magnification of ~100×. After the cells have been allowed to settle for about 10 min, the microscope field should contain principally smooth muscle cells that are spindle shaped, having the appearance shown in Fig. 2. If the cells have been injured during the isolation procedure, the isolated cells contract and appear to be shorter and wider than cells obtained following an optimal isolation. In that case, we do not utilize them for experiments.

As a further test of the physiological integrity of the isolated smooth muscle cells, we routinely test the responsiveness to an excitatory stimulus. This can be done in a quantitative manner by determining the dose-response curve to carbachol utilizing a Coulter counter.[14] A more qualitative, but quite useful assessment of physiological responsiveness, may be obtained by gently adding 10–20 μl of 10^{-2} g of carbachol per milliliter to the slide 1–2 mm away from the field of view and then determining the percentage of cells that contract as the carbachol diffuses into the area being observed. In the experiment shown in Fig. 2, over 70% of the cells contracted after addition of carbachol in this manner, indicating that the cells retained good responsiveness to carbachol after isolation.

Handling of Isolated Cells

The cells may be transferred from the dilute collagenase solution in which they were isolated to another solution by sedimenting them at low speed (10 g) in a centrifuge for 20 min. Higher speeds are to be avoided because the resulting high sedimentation rates appear to traumatize the cells, as judged from the increased percentage of contracted cells that

---

FIG. 2. Phase contrast photomicrographs of smooth muscle cells obtained by enzymic disaggregation of slices of stomach from *Bufo marinus* as described in this chapter. Aliquots (1 ml) of the cell suspension obtained after the first, second, and third successive incubations in collagenase (95 units/ml) were placed on a glass slide, and the cells were allowed to settle for 10 min. Panels A and B show the same field of cells obtained after the first incubation in collagenase: A, before the addition of 20 μl of 10^{-2} g of carbachol per milliliter to the periphery of the field; B, 5 min after adding carbachol. Note that approximately 70% of the cells contracted in response to carbachol. The lower photomicrographs show typical fields of cells obtained after the second (C) and third (D) successive incubation in collagenase. Note the higher cell density obtained after these incubations. ×120.

result. After sedimentation at low speed, the cells may be resuspended in APS or various buffers with modified ionic composition useful for electrophysiological[12] or ion flux studies.[15] For ion flux studies it is often necessary to concentrate the cells for isotope loading prior to initiation of isotope efflux by dilution. The cells may be concentrated in their isolation medium in a very gentle manner by allowing them to settle in a vibration-free area in large Erlenmeyer flasks for 90–120 min. The depth of the cell suspension should not be greater than 1.5 cm. After the settling period, up to 90% of the fluid may be carefully aspirated without significant loss of cells that have settled onto the bottom of the flask. For experiments where high cell densities are not required, the composition of the suspension medium can be substantially altered merely by diluting the cells with a 10–20 times excess of medium with a different composition from APS. When using this approach, consideration must be given to potential precipitates forming due to interaction of ions such as Ca^{2+}, HPO_4, or HCO_3 with the constituents of the diluting medium.

Acknowledgments

The work reported in this paper was supported in part by grants from the NIH (HL 14523) and the Muscular Dystrophy Association.

[26] Preparation of Smooth Muscle Myosin

By Dixie W. Frederiksen and Dianne D. Rees

Myosins have been isolated from a variety of smooth muscles. Almost all of these tissues contain considerable quantities of extracellular collagen and elastin. As a consequence, they are difficult to disrupt. Rather harsh methods are usually necessary for tissue homogenization. Measures to minimize myosin proteolysis and denaturation are, therefore, especially important after the violent disruption of the smooth muscle. All operations are performed in the cold (4–5°); protease inhibitors are included in each buffer; frothing and vigorous stirring are avoided at each step in the procedure. The result is purification of a homogeneous, enzymically active myosin in reasonable yield.

In the development of this procedure[1,2] we have followed suggestions of earlier investigators who have also studied the vascular and other

[1] D. W. Frederiksen, *Biochemistry* **18**, 1651 (1979).
[2] D. D. Rees and D. W. Frederiksen, *J. Biol. Chem.* **256**, 357 (1981).

smooth muscle contractile proteins. The work of Sobieszek[3] of Pollard *et al.*,[4] and of Sparrow and van Bockxmeer[5] has been particularly important.

Details are given here for the purification of myosin from porcine aortic smooth muscle. The same procedures, with occasional modification, can be applied to the preparation of other smooth muscle myosins.

Procedure

Tissue

1. Obtain about twelve 15–20 cm thoracic aorta segments from the abattoir as soon after the slaughter of the hogs as possible (usually 30 min or less) and place them in ~8 liters of ice cold, buffered saline [A: 0.15 M NaCl, 25 mM imidazole (pH 7.0), 0.15 mM phenylmethylsulfonyl fluoride (PMSF)] for transport to the laboratory.

2. Cut the segments in two and mount the whole (7–10 cm) pieces on a plastic rod clamped to a ringstand. Dissect away the adventia by careful slicing, first radially into the outer layer of media and then tangentially to the lumen. The second cut should begin in the outer media exposed with the initial slice and should be rather shallow (3 mm). Peel away the adventia and outer media with the aid of a forceps clamp or hemostat. Invert the aorta and remove the intima in the same way. Since the intima is much thinner than the adventia, this first cut should be no deeper than about 0.5 mm. Slice the cleaned aorta longitudinally and place it in fresh, cold buffer A (Table I). The aortas should look exactly like cooked lasagna noodles; a whitish, fibrous surface on the segment is remnant adventia and indicates a poor job of dissection.

3. Blot and weigh the aorta segments. The following steps are for the preparation of myosin from 200 g of cleaned aortic media. Resuspend the segments in buffer A.

4. Mince the wet segments to fragments about 2 mm on a side (~8 mm^3) by chopping with a household cleaver on a hardwood block. Resuspend the fragments in buffer A.

5. Wash the fragments three times with four volumes of buffer A.

Extraction

1. Decant buffer A from the minced aorta and transfer the mince to a 1-liter plastic bottle. Add 500 ml of wash buffer (buffer B, Table I) to the bottle of mince.

[3] A. Sobieszek, *in* "Biochemistry of Smooth Muscle" (N. L. Stephens, ed.), pp. 413–443. University Park Press, Baltimore, Maryland, 1977.
[4] T. D. Pollard, S. M. Thomas, and R. Niederman, *Anal. Biochem.* **60**, 258 (1974).
[5] M. P. Sparrow and F. M. van Bockxmeer, *J. Biochem.* (*Tokyo*) **72**, 1075 (1972).

TABLE I
BUFFERS AND SOLUTIONS FOR THE PREPARATION OF AORTIC MYOSIN

Buffers[a] and solutions	Composition
A: Isotonic saline (12 liters)	0.15 M NaCl, 25 mM imidazole (pH 7.0), 0.15 mM PMSF
B: Wash buffer (500 ml)	60 mM KCl, 20 mM imidazole (pH 6.8), 1 mM MgSO$_4$, 1 mM DTT, 70 μM streptomycin-SO$_4$, 0.15 mM PMSF
C: Extraction buffer (250 ml)	0.6 M KCl, 20 mM imidazole (pH 7.5), 2 mM EGTA, 1 mM EDTA, 1 mM DTT, 70 μM streptomycin-SO$_4$, 0.15 mM PMSF
D: Assay buffer (2 liters)	35 mM KCl, 50 mM morpholinopropanesulfonic acid (pH 7.0), 4 mM MgSO$_4$, 1 mM DTT, 70 μM streptomycin-SO$_4$, 0.15 mM PMSF
E: Column buffer (5 liters)	0.5 M KCl, 0.1 mM MgSO$_4$, 0.1 mM CaCl$_2$ (or 2.0 mM EGTA), 0.1 mM imidazole (pH 7.0), 1 mM DTT, 70 μM streptomycin-SO$_4$, 0.15 mM PMSF
MgSO$_4$ (25 ml)	1.0 M MgSO$_4$; store at 5°
ATP (10 ml)	50 mM ATP (pH 7.0); store at $-20°$
CaCl$_2$ (25 ml)	10 mM CaCl$_2$; store at 5°
EGTA (25 ml)	0.2 M EGTA (pH 7.0); store at 5°

[a] Make buffers A–E fresh just before use. PMSF, phenylmethylsulfonyl fluoride; DTT, dithiothreitol; EDTA, ethylenedinitrilotetraacetic acid; EGTA, ethyleneglycol bis(β-aminoethyl ether)-N,N'-tetraacetic acid.

2. Homogenize the mince with a Polytron (Brinkmann Instruments, Westbury, New York) equippped with a 20-mm head. Disrupt the tissue with three 30-sec pulses of the Polytron at maximum power. Incubate the mince for 2 min at 0° after each pulse.

3. Centrifuge the homogenate in the Sorvall RC-2B centrifuge, GSA rotor at 9000 rpm (10,000 g) for 15 min at 5° to remove soluble cytoplasmic proteins. Discard the supernatant (Fig. 1A) and transfer the pellets to a 1-liter plastic bottle.

4. Break up the chunks of pellet with a plastic iced-tea spoon. Add 250 ml of extraction buffer (buffer C, Table I) and 10 mM ATP. Polytron once at full power for 10 sec.

5. Centrifuge the homogenate in the Sorvall GSA rotor at 13,000 rpm (22,000 g) for 15 min at 5° to remove solid debris. Discard the pellets. The supernatant is crude actomyosin (Fig. 1B). Decant the fluid into 50-ml polycarbonate centrifuge tubes for the Beckman TI-45 rotor.

6. Centrifuge the crude actomyosin at 35,000 rpm (85,000 g) for 4 hr at 5° in the Beckman TI-45 rotor, L5-50 centrifuge to sediment most of the F-actin and all of the remaining cell fragments. Discard the gelatinous precipitate (Fig. 1C). Decant the actin-depleted actomyosin extract (Fig. 1D) into a 500-ml graduated cylinder.

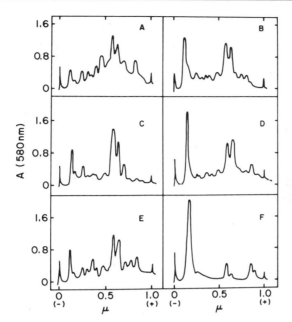

FIG. 1. Polyacrylamide gel electrophoresis in sodium dodecyl sulfate of fractions from preparation of aortic myosin. Protein (15–25 μg) was applied to each gel. (A) Low ionic strength supernatant; (B) high ionic strength-ATP supernatant; (C) high speed ultracentrifugation (UC) precipitate; (D) high speed UC supernatant; (E) supernatant for precipitation with Mg^{2+}; (F) crude myosin. From Frederiksen.[1]

7. Measure the extract volume and add enough 1.0 M $MgSO_4$ to bring the Mg^{2+} concentration at 25 mM. Stir for 5 min. Leave the solution undisturbed at 5° for 9–12 hr. Crude myosin (Fig. 1F) will precipitate.

8. Centrifuge the myosin suspension at 35,000 rpm (85,000 g) in the TI-45 rotor, L5-50 centrifuge for 1 hr at 5° to pellet the protein. Discard the supernatant (Fig. 1E).

9. Suspend the pellets in 10–15 ml of low ionic strength assay buffer (buffer D, Table I). The major contaminants of the myosin at this stage are actin and tropomyosin.

10. Determine the protein concentration[6]; it should be 12–18 mg/ml. Determine the K(EDTA)-ATPase activity[7]; it should be 0.7–1.0 μmol/mg per minute. Determine the MgATPase activity[7] in the presence of added rabbit skeletal muscle actin[8]; it should be 40–80 nmol/mg per minute.

[6] O. H. Lowry, N. J. Rosebrough, A. L. Farr, and R. J. Randall, *J. Biol. Chem.* **193**, 265 (1951).
[7] T. D. Pollard, this volume [13].
[8] J. D. Pardee and J. A. Spudich, this volume [18].

Purification of Phosphorylated Myosin

This stage of the preparation is best carried out as soon as possible after pelleting the crude myosin. The column equilibration should be begun during the 4-hr centrifugation.

1. Equilibrate a 5 × 90 cm of Sepharose 4B column with 2 liters of column buffer (buffer E, Table I, using $CaCl_2$ and 0.5 mM ATP). Use a flow rate of 100–120 ml/hr.

2. Just before application of the sample, allow 200 ml of buffer E mixed with 20 ml of 50 mM ATP (pH 7.0) to drain into the column. The 5.0 mM ATP in this layer will facilitate the separation of actin and myosin. Hint: Start this before the assay in step 10 above.

3. Phosphorylation
   a. Dilute 200 mg of the crude myosin to 56 ml (3.5 mg/ml) with buffer D in a 125-ml Erlenmeyer flask.
   b. Add 0.56 ml of 10 mM $CaCl_2$ and incubate at 25° for 5 min.
   c. Add 1.15 ml of 50 mM ATP (pH 7.0) and incubate at 25° for 5 min.
   d. Place the flask on ice and add 2.16 g of solid KCl. Swirl gently to dissolve the salt.

4. Apply the solution to the Sepharose 4B column; elute at 100–120 ml/hr. Monitor A_{280} of the eluent; collect fractions at 10-min intervals (Fig. 2).

5. Protein in the leading half of the second peak (Fig. 2) is phosphory-lated aortic myosin (Fig. 3) and is ready for concentration and dialysis against an appropriate buffer for experiments.

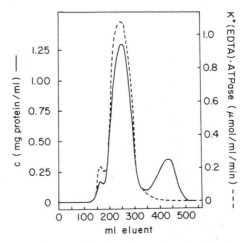

FIG. 2. Agarose chromatography of crude aortic myosin. From Frederiksen.[1]

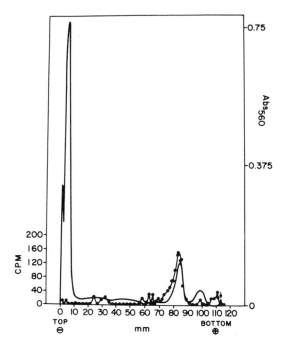

FIG. 3. Polyacrylamide gel electrophoresis of phosphorylated aortic myosin. Crude myosin was phosphorylated with [γ-^{32}P]ATP and used to prepare aortic myosin. The myosin was subjected to electrophoresis in sodium dodecyl sulfate and to subsequent assay for ^{32}P in 2-mm slices of the stained gel. ——, Absorbance of stained gel at 560 nm; ●——●, cpm ^{32}P per 2-mm slice; BPB, bromophenol blue tracking dye position. From Rees and Frederiksen.[2]

TABLE II
ANALYSIS OF AORTIC MYOSIN PREPARATION[1]

Fraction	Protein (mg)	K(EDTA)-ATPase Total (μmol/ min)	Specific (μmol/ mg min)	Purification (fold)
Soluble protein	1890.0	59.1	31.2	1.00
Low-μ extract	1049.0	13.0	12.4	—
High-μ extract (crude actomyosin)	840.0	45.9	54.7	1.75
UC supernatant (actin-depleted actomyosin)	491.0	47.5	96.7	3.10
UC precipitate (crude actin)	311.0	0.5	1.6	—
Mg^{2+}-precipitate (crude myosin)	88.1	35.2	399.0	12.8
Mg^{2+}-supernatant	432.0	14.3	33.1	—
Purified myosin	38.1	30.0	787.0	25.2

Purification of Unphosphorylated Myosin

Prepare aortic myosin without phosphate on the 20,000-dalton light chain from crude myosin exactly as described above for phosphorylated myosin except (*a*) substitute 2.0 mM EGTA for 0.1 mM CaCl$_2$ in buffer E; (*b*) substitute 0.2 M EGTA (pH 7.0) for 10 mM CaCl$_2$ in step 3b.

An analysis of the preparation is given in Table II. The overall yield of myosin is 80–120 mg/200 g of cleaned aorta. The extinction coefficient for the myosins is 0.558 cm^2/mg. The protein is stable for about a week stored at 5° in buffer D.

Acknowledgment

This work was supported in part by NIH Research Grants HL-18516 and NS-15077.

[27] Purification of Smooth Muscle Myosin Light-Chain Kinase

By ROBERT S. ADELSTEIN and CLAUDE B. KLEE

Myosin light-chain kinase catalyzes the transfer of the γ-phosphate of ATP to a serine residue on the light chain of myosin.[1] The enzyme has been isolated from all types of muscles (skeletal,[2] cardiac,[3] and smooth[4,5]) as well as from nonmuscle sources, such as blood platelets[6,7] and brain.[6,8] The enzymes from the various sources resemble each other in one important property: they are completely inactive in the absence of the calcium-binding protein calmodulin. This latter protein, which appears to be present throughout the plant and animal kingdom, modulates the role of Ca^{2+} in regulating the biological activity of a number of proteins.[9] Smooth

[1] R. S. Adelstein and E. Eisenberg, *Annu. Rev. Biochem.* **49**, 921 (1980).
[2] D. K. Blumenthal and J. T. Stull, *Biochemistry* **19**, 5608 (1981).
[3] H. Wolf and F. Hoffman, *Proc. Natl. Acad. Sci. U.S.A.* **77**, 5852 (1980).
[4] R. Dabrowska, D. Aromatorio, J. M. F. Sherry, and D. J. Hartshorne, *Biochem. Biophys. Res. Commun.* **78**, 1263 (1977).
[5] R. S. Adelstein and C. B. Klee, *J. Biol. Chem.* **256**, 7501 (1981).
[6] R. Dabrowska and D. J. Hartshorne, *Biochem. Biophys. Res. Commun.* **85**, 1352 (1978).
[7] D. R. Hathaway and R. S. Adelstein, *Proc. Natl. Acad. Sci. U.S.A.* **76**, 1653, (1979).
[8] D. R. Hathaway, R. S. Adelstein, and C. B. Klee, *J. Biol. Chem.* **256**, 8183, (1981).
[9] C. B. Klee, T. H. Crouch, and P. G. Richman, *Annu. Rev. Biochem.* **49**, 489 (1980).

muscle myosin kinase[10] and human platelet myosin kinase[11] are substrates for cAMP-dependent protein kinase. Phosphorylation of these two enzymes decreases their activity by decreasing their affinity for calmodulin.

Myosin light-chain kinases isolated from a variety of tissues appear to differ in their molecular weights: gizzard smooth muscle = 130,000; bovine brain = 130,000; human platelet = 105,000; bovine cardiac muscle = 92,000; rabbit skeletal muscle = 90,000. Whether the lower molecular weights found for the skeletal muscle, cardiac muscle, and platelet enzymes are due to proteolysis of a larger enzyme has not been determined with certainty. However, proteolysis is a major problem in isolating any of the above species, and numerous precautions to prevent it have been instituted in the procedure outlined below.

We shall describe a method used for isolating a homogeneous species of myosin light-chain kinase in milligram quantities from turkey gizzard smooth muscle. This procedure with slight modification can also be used to isolate myosin kinases from other tissues.

Isolation of Kinase

Materials

All chemicals were reagent grade, unless specified otherwise. Deionized water and dithiothreitol were used throughout the preparation. Proteolytic inhibitors were obtained from the following sources: pepstatin A and leupeptin from Vega Biochemicals; phenylmethylsulfonyl fluoride (PMSF), N-tosylphenylalanine chloromethyl ketone, α-N-benzoyl-L-arginine methyl ester, soybean trypsin inhibitor, and diisopropyl fluorophosphate (the last requires cautious handling) were from Sigma.

Sephacryl S-300, DEAE-Sephacel, Sepharose 4B, AH-Sepharose, and Sepharose 4B CNBr-activated (cyanogen bromide-activated) were obtained from Pharmacia Fine Chemicals. In preparing the calmodulin affinity column, calmodulin was either coupled to Sepharose 4B that was activated with cyanogen bromide (Fluka AG.) as described by Klee and Krinks[12] or Sepharose 4B CNBr-activated (Pharmacia) was used, and the calmodulin was coupled to it using the procedure recommended by the manufacturer.

Ultrapure $(NH_4)_2SO_4$ was obtained from Bethesda Research Laboratories.

[10] M. A. Conti and R. S. Adelstein, *J. Biol. Chem.* **256**, 3178 (1981).
[11] D. R. Hathaway, C. R. Eaton, and R. S. Adelstein, *Nature (London)* **291**, 252 (1981).
[12] C. B. Klee and M. H. Krinks, *Biochemistry* **17**, 120 (1978).

Methods

All procedures were carried out at 4° unless specified otherwise.

Calmodulin was purified from an acetone powder of bovine or porcine brain purchased from Sigma as previously described.[13]

Isolated 20,000-dalton light chain of gizzard smooth muscle myosin is the substrate most commonly used. It is prepared by the method previously described,[5] which is modified from Perrie and Perry.[14] The preparation of isolated myosin light chains results in a mixture of the 20,000-(phosphorylatable) and 16,000-dalton (nonphosphorylatable) light chains, approximately 50% of each. The final step in this procedure is chromatography on DEAE-Sephacel, which removes calmodulin from the light chains. (This step can be omitted in cases where a calmodulin-free assay mixture is not required.)

If the DEAE-Sephacel chromatography is carried out at room temperature in the presence of 6 M urea, the separation between the 20,000-dalton light chain and calmodulin is substantially greater, and it is often possible to separate the phosphorylatable and nonphosphorylatable light chains.

When myosin is to be used as a substrate, it can be prepared as outlined previously[5] or as given in this volume [26].

Preparation of Gizzard Myosin Kinase

An outline of the major steps used in preparing gizzard myosin kinase follows:

Preparation of washed myofibrils
↓
Extraction of myofibrils
↓
Ammonium sulfate fractionation (40–60%)
↓
Sephacryl S-300 gel filtration
↓
DEAE-Sephacel ion exchange chromatography
↓
Calmodulin-Sepharose 4B affinity chromatography

The various buffers used in the isolation procedures are as follows:
Buffer A (wash for preparing myofibrils): 20 mM Tris-HCl (pH 6.8 at 4°), 40 mM NaCl (KCl can be substituted), 1 mM MgCl$_2$, 1 mM

[13] C. B. Klee, *Biochemistry* **16,** 1017 (1977).
[14] W. T. Perrie and S. V. Perry, *Biochem. J.* **119,** 31 (1970).

dithiothreitol, 5 mM [ethylene bis(oxyethylenenitrilo)]tetraacetic acid (EGTA), 75 mg of PMSF per liter, 100 mg of streptomycin sulfate per liter.

Buffer B (kinase extraction): 40 mM Tris-HCl (pH 7.5), 60 mM KCl, 25 mM MgCl$_2$, 1 mM dithiothreitol, 5 mM EGTA, 75 mg of PMSF per liter, 100 mg of streptomycin sulfate per liter, 1 mg of leupeptin and pepstatin A per liter, 10 mg/liter N-tosylphenylalanine chloromethyl ketone, α-N-benzoyl-L-arginine methyl ester, soybean trypsin inhibitor, and 10^{-3} M diisopropyl fluorophosphate.

Buffer C [(NH$_4$)$_2$SO$_4$ pellet solubilization]: 15 mM Tris-HCl (pH 7.5), 0.5 M NaCl, 1 mM ethylenediaminetetraacetic acid (EDTA), 5 mM EGTA, 1 mM dithiothreitol, and the same proteolytic inhibitors as buffer B, except that 10^{-5} M diisopropyl fluorophosphate was used in place of 10^{-3} M.

Buffer D (Sephacryl S-300 column buffer): Same as buffer C, but the only protease inhibitor was 75 mg of PMSF per liter.

Buffer E (DEAE-Sephacel column buffer): 20 mM Tris-HCL (pH 7.8), 1 mM EGTA, 1 mM dithiothreitol, and 75 mg of PMSF per liter. The elution gradient was 20 to 600 mM NaCl, using a 1.2-liter linear gradient for a 2 × 20 cm column.

Buffer F (Calmodulin-Sepharose 4B affinity column): 40 mM Tris-HCl (pH 7.2), 50 mM NaCl, 0.2 mM CaCl$_2$, 3 mM MgCl$_2$, and 1 mM dithiothreitol. Elution was with 2 steps—the same buffer with 0.2 M NaCl in place of 50 mM NaCl and the 0.2 M NaCl buffer with 2 mM EGTA added.

Preparation of Myofibrils. Gizzards are obtained from turkeys killed on the same day, and the gizzards are packed in ice until delivered to the laboratory (within 3 hr). To prepare approximately 50 mg of kinase, it is necessary to start with 500 g of ground gizzard smooth muscle, which can be obtained from 10 average-sized turkey gizzards. In an effort to prevent proteolysis, the preparation is carried through the extraction step, at the least, or through the loading of the Sephacryl S-300 column, at the most, on the day the turkeys are killed. The lobes of smooth muscle are carefully freed from all connective tissue and fat and kept on ice until they are ground in an electric meat grinder producing a finely ground mince of muscle. The mince is suspended in 3–4 volumes of buffer A in a 2-quart Mason jar with 0.05% Triton X-100 added (first wash only) using a Sorvall Omni-Mixer. The mixer is set at $\frac{2}{3}$ full speed and activated three times for 5 sec each. The suspension is thoroughly homogenized with two passes in a Glenco glass–Teflon homogenizer (250 ml), using a Sears $\frac{3}{8}$-inch electric drill to turn the pestle. The mixture is sedimented at 15,000 g for 15 min, and the pellet is resuspended in the same volume of wash solution in the absence of Triton, using the Sorvall Omni-Mixer. The suspension is again sedimented for 15 min at 15,000 g, and the supernatant again is discarded. This step is repeated once again, by which time the myofibrils have lost their red color.

Extraction of Kinase. The pellet of washed myofibrils is resuspended in 3–4 volumes of buffer B using the Omni-Mixer followed by one pass through the Glenco homogenizer and sedimented for 30 min at 15,000 g. The supernatant is filtered through glass wool, and the pellet is discarded. Solid $(NH_4)_2SO_4$ is slowly added to the supernatant, and the protein precipitating between 0 and 40% saturation is sedimented at 15,000 g for 15 min and discarded. The supernatant is then made 60% saturated with $(NH_4)_2SO_4$ and sedimented at 15,000 g for 30 min. This pellet is homogenized by hand in a glass–glass homogenizer in 50 ml of buffer C and is dialyzed either 3 hr (if the Sephacryl column is run on the day of extraction) or overnight against 10 volumes of the same buffer. The protein concentration is about 25 mg/ml, and the total volume is about 60 ml.

Sephacryl S-300 Chromatography. Thirty milliliters (i.e., one-half the volume) of the 40–60% $(NH_4)_2SO_4$ fraction are applied to a 5 × 87 cm column of Sephacryl S-300 (Fig. 1). The column is equilibrated and eluted

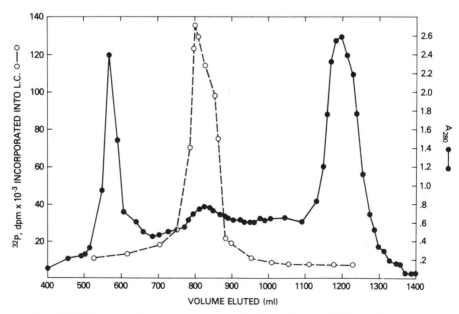

FIG. 1. Gel filtration of smooth muscle myosin kinase. Thirty milliliters (25 mg/ml) of the 40–60% ammonium sulfate fraction are applied to a 5 × 87 cm column of Sephacryl S-300 equilibrated and eluted in a solution of buffer D. The column is eluted at a flow rate of 200 ml/hr. Fractions were assayed for myosin kinase activity (O––O) as outlined in text. L.C., light chain. From Adelstein and Klee.[5]

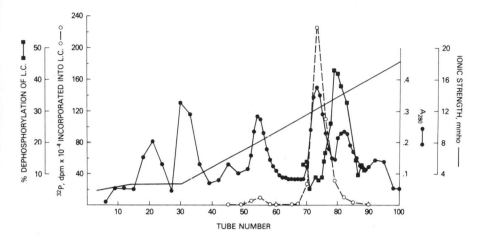

FIG. 2. Ion-exchange chromatography of the pooled Sephacryl S-300 peak of myosin kinase. Myosin kinase (225 ml) was applied to a 2 × 20 cm column of DEAE-Sephacel equilibrated with buffer E. The kinase is eluted at a flow rate of 60 ml/hr with a 1200-ml linear NaCl gradient (20 to 600 mM NaCl) made up in the equilibration buffer. Aliquots of the 15-ml fractions are assayed for myosin kinase (O – – O) and myosin phosphatase (■——■) activity. L.C., light chain. The latter assay is described in this volume [28]. From Adelstein and Klee.[5]

with buffer D at a flow rate of 200 ml/hr. A single peak of myosin kinase elutes with a $K_{av} = 0.31$. This step is repeated with the remaining 30 ml, and both peaks of myosin kinase activity are pooled. The pooled peaks are dialyzed in 4 liters of buffer E. The total volume is 225 ml, and the protein concentration is 1.5 mg/ml.

DEAE-Sephacel Chromatography. A 2 × 20 cm column of DEAE-Sephacel is equilibrated with buffer E, and the Sephacryl S-300 peak is chromatographed at a flow rate at 60 ml/hr. The column is developed with a linear gradient of NaCl in buffer E created by using two identical containers, connected at the bottom, of 600 ml each. The container proximal to the column contains buffer E with 20 mM NaCl and the distal one contains buffer E with 600 mM NaCl. The myosin kinase activity elutes as a single peak of activity at a conductance of 10–13 mmho at 4° (Fig. 2). The pooled peak (150 ml) is dialyzed against 1 liter of 20 mM Tris-HCl (pH 7.8), 50 mM NaCl, 1 mM EGTA, and 1 mM dithiothreitol.

Calmodulin-Affinity Chromatography. The peak of myosin kinase activity eluted from DEAE-Sephacel is applied to a 7-ml column of calmodulin coupled to Sepharose 4B equilibrated with buffer F. Just prior to application, Ca^{2+} and Mg^{2+} are added to the sample to give a final free concentration of 0.5 mM Ca^{2+} and 3 mM Mg^{2+}. The column is washed with three volumes of the equilibrating buffer and then with two or three volumes of

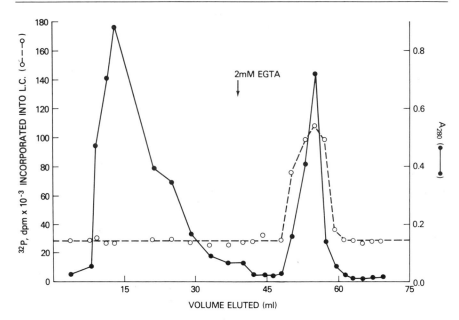

FIG. 3. Calmodulin-affinity chromatography of myosin kinase. Approximately 50 ml of the pooled peak following DEAE-Sephacel chromatography are applied to a 7-ml column of calmodulin–Sepharose 4B equilibrated with buffer F. The column is washed with 3 volumes of equilibrating buffer and then with 2 or 3 volumes of the same buffer made 0.2 M with respect to NaCl. The kinase is eluted by introducing 2 mM EGTA into the buffer containing 0.2 M NaCl. The assay for kinase activity (O--O) is not linear with respect to ^{32}P incorporation. From Adelstein and Klee.[5]

the same buffer made 0.2 M with respect to NaCl. The kinase is eluted from the column with 2 mM EGTA added to buffer F (Fig. 3). The peak of myosin kinase activity is pooled, divided into 0.2-ml aliquots, and kept frozen at $-70°$ until ready for use. Once defrosted, the kinase is stored on ice, since repeated freezing and thawing decreases its activity.

Table I summarizes the steps used in purification, and Table II lists some of the properties of the purified enzyme. Table I was constructed from a representative preparation, starting with 500 g of ground gizzard smooth muscle. Figures 2 and 3 are from preparations where the protein concentrations (but not the volumes) were somewhat less than shown in Table I.

Comments on Purification. The preparation of the myofibrils and the extraction of kinase is based on the procedure first published by Sobieszek and Small.[15] The rest of the procedure is from that published by Adelstein and Klee.[5] The purified kinase has a molecular weight (M_r) of

[15] A. Sobieszek and J. V. Small, *J. Mol. Biol.* **112**, 559 (1977).

TABLE I
PURIFICATION OF TURKEY GIZZARD MYOSIN LIGHT-CHAIN KINASE[a,b]

Fraction	Volume (ml)	Protein concentration [mg/ml (total mg)]	Total activity [μmol/min (%)]	Specific activity (μmol/mg per minute)
Extract supernatant	1500	3 (4500)	450 (100)	0.10
Ammonium sulfate fraction, 40–60%	60	25 (1500)	405 (90)	0.27
Sephacryl S-300	225	1.5 (338)	371 (82)	1.1
DEAE-Sephacel	150	0.7 (105)	231 (51)	2.2
Calmodulin affinity column	20	1.5 (30)	180 (40)	6

[a] From Adelstein and Klee.[5]

[b] Based upon starting with 500 g of ground gizzard smooth muscle.

TABLE II
PHYSICAL AND KINETIC PROPERTIES OF SMOOTH MUSCLE MYOSIN KINASE[a]

Molecular weight	
SDS-polyacrylamide gel electrophoresis	130,000 ± 6000
Sedimentation equilibrium	124,000 ± 2200
Stokes' radius calculated from $s_{20,w}$, M_r, $\bar{v}$	75 Å
$E_{278\,nm}^{1\%}$	11.40 ± 0.2
K_m ATP	50 μM
K_m Myosin light chains (smooth muscle)	5 μM
V_{max}	10–30 μmol/mg of kinase per minute
$K_{0.5}$ Calmodulin[b]	1 nM

[a] From Adelstein and Klee.[5]

[b] Concentration of calmodulin required for 50% activation of a 10^{-10} M solution of myosin kinase.

130,000 as determined by 0.1% SDS–7.5% polyacrylamide gel electrophoresis using the buffer system of Laemmli.[16] It slowly undergoes proteolysis to an M_r 105,000 protein in the absence of proteolytic inhibitors at room temperature.

After DEAE-Sephacel chromatography, the major proteins that can be identified after SDS–polyacrylamide gel electrophoresis appear to be myosin kinase (M_r = 130,000) and tropomyosin (M_r = 43,000 and 38,000). Phosphatases that can dephosphorylate the M_r 20,000 light chain of myosin elute just after the kinase activity (see Fig. 2), and phosphatase activity is removed by the calmodulin–Sepharose 4B affinity column. They cannot be identified after SDS–polyacrylamide gel electrophoresis.

[16] U. K. Laemmli, *Nature* (*London*) 227, 680 (1979).

Occasionally this step must be repeated in order to remove all the phosphatase activity.

The calmodulin columns are quite stable and have been used for over a year. Occasionally, we have passed a solution of 8 M urea, 10 mM EGTA, and 0.8 M NaCl through the column in an effort to remove any residual proteins that might have accumulated.[17] The specificity of the calmodulin column appears to be enhanced by working near the maximum capacity of the column for the binding of myosin kinase. This is important, since applying a sample containing a relatively small amount of kinase to a column of relatively large capacity appears to result in the nonspecific binding of proteins. Occasionally a small peak of myosin kinase, indistinguishable in its properties from that eluted with EGTA, is eluted with 0.2 M NaCl in buffer F, prior to adding EGTA. This may be the result of working near the capacity of the column to bind myosin kinase, since rechromatography of this peak results in all of it binding and subsequently eluting with the EGTA-containing buffer.

Preparation of Myosin Kinase from Human Platelets and Other Nonmuscle Sources

A similar approach to that described for smooth muscle has been used to isolate myosin kinase from nonmuscle cells such as platelets. The main difference is that the amount of contractile proteins is considerably less in these cells and the starting material is a cell extract, not myofibrils. Because we have not succeeded in routinely obtaining a homogeneous preparation with no evidence of proteolysis, our remarks about purification of myosin kinase from nonmuscle cells will be confined to some general principles:

1. Only fresh cells are used and the protease inhibitors are introduced as early as possible.
2. Cells are lysed and extracted under conditions that avoid liberating proteases. Triton X-100 and similar reagents are avoided.
3. The cells are lysed into the extraction buffer, which in addition to the protease inhibitors present in buffer B contains 40 mM Tris-HCl (pH 7.8), 50 mM NaCl, 5 mM EGTA, 10 mM EDTA, and 2 mM dithiothreitol.
4. The crude extract is applied to a column of AH Sepharose 4B and eluted in a single step with 0.5 M NaCl replacing 50 mM NaCl in the extraction buffer.
5. The pooled myosin kinase activity is dialyzed in buffer D and fil-

---

[17] R. Kincaid and M. Vaughan, *Proc. Natl. Acad. Sci. U.S.A.* **76**, 4903 (1979).

tered through a column of Sephacryl S-300 as described above for the smooth muscle enzyme.

6. The myosin kinase is usually calmodulin-free after this step. If this is so, it can be applied directly to the calmodulin–Sepharose 4B affinity column after dialysis against buffer F. If calmodulin is still present in the kinase, the pooled kinase activity is first dialyzed in buffer E and chromatographed on DEAE-Sephacel as described above.

7. The various steps outlined above are scaled down appropriately in relation to the smaller quantities of myosin kinase present in nonmuscle tissues.

Assay of Myosin Light-Chain Kinase Activity

Myosin light-chain kinase is routinely assayed in a volume of 0.1 ml of 20 mM Tris-HCl (pH 7.5), 10 mM MgCl$_2$, 0.2 mM CaCl$_2$ (in excess over EGTA), 0.1 mM [γ-^{32}P]ATP (0.5 Ci/mmol), 0.2 mg of mixed smooth muscle light chains per milliliter (the concentration of 20,000-dalton light chain in the assay mixture is approximately 5 μM, since this constitutes 50% of the mixed light chains) and 10^{-7} M calmodulin.

The assay is initiated by addition of the kinase or a dilution of the kinase in bovine serum albumin. It is terminated by addition of an aliquot of a solution of trichloroacetic acid–sodium pyrophosphate to give a final concentration of 10% trichloroacetic acid and 2% sodium pyrophosphate. The sample is chilled on ice, and transferred to the center of a Millipore filter (type HA, 0.45 μm) which is under vacuum on a Millipore filtration apparatus. The filter is washed five times with 5% trichloroacetic acid–1% sodium pyrophosphate and then counted in a Searle Mark III scintillation spectrometer after immersion in 15 ml of Aquasol (Amersham). Time of the assay varies from 15 sec for 10^{-8} M kinase to 5 min for 10^{-9} M kinase.

Comments on the Assay. As pointed out above, the isolated light chains are usually a mixture of the 20,000- and 16,000-dalton light chains. In order to be certain that the light chains are completely dephosphorylated, as isolated, and that they could be completely phosphorylated by myosin kinase, the extent of phosphorylation is monitored using urea–glycerol polyacrylamide gel electrophoresis.[14] This gel separates the phosphorylated and nonphosphorylated 20,000-dalton light chain as well as the 16,000-dalton light chain of myosin on the basis of charge. Note that the paper by Perrie and Perry[14] describes two gel systems employing urea. The system using 40% glycerol (but not urea) in the gel appears to separate the phosphorylated and nonphosphorylated light chain of smooth muscle myosin, as well as the 16,000-dalton light chain, considerably better than that not employing glycerol.

The rate of incorporation of phosphate into the 20,000-dalton light chain of myosin is linear with time as long as only 20% or less of the total available light chains in the incubation mixture undergo phosphorylation. The concentration of light chains in the routine assay mixture is 5 μM, which is the K_m of myosin kinase for the smooth muscle light chain.

If smooth muscle myosin is used as the substrate for kinase, the conditions for the assay are similar to those described above, but the KCl concentration is raised to 50 mM. When these conditions are used and incorporation into the isolated light chain is compared with that into intact myosin (in this case, using light-chain concentrations of 15–18 μM for both the isolated light chain and the light chains still bound to myosin), the former is found to incorporate the phosphate approximately twice as fast as the latter.[5] On the other hand, if the isolated light chains of cardiac muscle myosin are used as substrate for the smooth muscle enzyme, the rate of incorporation is about one-fifth as fast. When the isolated light chains of skeletal muscle myosin are used as substrate, the rate of incorporation is only one-thirtieth as fast as that with isolated smooth muscle myosin light chains.

[28] Purification of Smooth Muscle Phosphatases

By MARY PATO and ROBERT S. ADELSTEIN

Phosphatases are enzymes that catalyze the hydrolysis of phosphate ester bonds. These enzymes, which are not as well characterized as the kinases, are now beginning to receive deserved attention. Part of the reason they have been neglected is the difficulty in obtaining milligram quantities of intact, homogeneous preparations. In this chapter we describe the purification of two different phosphatases from turkey gizzard smooth muscle. The enzymes described below have been designated as smooth muscle phosphatase I and II (SMP-I and SMP-II). SMP-I and -II have been purified to homogeneity using an affinity column of thiophosphorylated myosin light chains bound to Sepharose 4B.[1] The basis for this technique is the observation that thiophosphorylated proteins are dephosphorylated at a slower rate than phosphorylated proteins.[2,3] These two

---

[1] M. D. Pato and R. S. Adelstein, *J. Biol. Chem.* **255**, 6535 (1980).

[2] J. M. F. Sherry, A. Gorecka, M. D. Aksoy, R. Dabrowska, and D. J. Hartshorne, *Biochemistry* **17**, 4411 (1978).

[3] P. E. Hoar, W. G. L. Kerrick, and P. Cassidy, *Science* **204**, 503 (1979).

enzymes bear important similarities to other phosphoprotein phosphatases purified from different sources.[4]

SMP-I is a trimer composed of three different polypeptide chains (M_r = 60,000, 55,000, and 38,000) in a molar ratio of $1:1:1$.[1] Using sedimentation equilibrium centrifugation under nondenaturing conditions, SMP-I was found to have a molecular weight of 165,000.[5] The activity of SMP-I is partially inhibited by cations.

SMP-II is a single polypeptide chain with a molecular weight of 43,000 determined by sodium dodecyl sulfate (SDS)–polyacrylamide gel electrophoresis. It is inactive in the absence of Mg^{2+}.

Both SMP-I and SMP-II catalyze dephosphorylation of the isolated 20,000-dalton light chain of smooth muscle myosin, but neither enzyme dephosphorylates intact myosin at a significant rate. SMP-I can be dissociated by freezing in 2-mercaptoethanol into a catalytic subunit (M_r = 38,000) and two putative regulatory subunits.[5] The catalytic subunit *is* active in dephosphorylating intact myosin, whereas undissociated SMP-I is active in dephosphorylating myosin kinase that has been phosphorylated by the catalytic subunit of cAMP-dependent protein kinase.[6]

In addition to SMP-I and SMP-II we have also identified a third phosphatase in gizzard muscle, which dephosphorylates intact myosin more rapidly than the isolated myosin light chain. This phosphatase elutes from Sephacryl S-300 just in front of SMP-I (Fig. 1). It differs from SMP-I and SMP-II in that it appears to be substrate (i.e., myosin) specific. To date this enzyme has not been purified to homogeneity from smooth muscle, although Morgan et al.[7] reported a myosin-specific phosphatase purified from skeletal muscle.

Purification of Phosphatases: Materials and Methods

The entire procedure is carried out at 4° unless noted as otherwise. Deionized water is used throughout the preparation.

The following resins were purchased from Pharmacia Fine Chemicals: Sephacryl S-300, DEAE-Sephacel, aminohexyl (AH)-Sepharose 4B, and Sepharose 4B CNBr-activated. Ultrapure $(NH_4)_2SO_4$ was obtained from Bethesda Research Laboratories. Pepstatin A and leupeptin were from Vega Biochemicals. Phenylmethylsulfonyl fluoride (PMSF), L-1-tosylamide-2-phenylethylchloromethyl ketone, soybean trypsin inhibitor, α-N-

[4] E. Y. C. Lee, S. R. Silberman, M. Ganapathi, S. Petrovic, and H. Paris, *Adv. Cyclic Nucleotide Res.* **13**, 95 (1980).
[5] M. D. Pato and R. S. Adelstein, *Biophys. J.* **33**, 278a (1981).
[6] M. A. Conti and R. S. Adelstein, *J. Biol. Chem.* **256**, 3178 (1981).
[7] M. Morgan, S. V. Perry, and J. Ottaway, *Biochem. J.* **157**, 687 (1976).

benzoyl-L-arginine methyl ester, and diisopropylfluorophosphate (DIFP) were from Sigma. All other chemicals used are reagent grade.

The following buffers are used in the purification procedure.

Buffer A: 50 mM Tris-HCl (pH 7.5), 10 mM magnesium acetate, 15 mM dithiothreitol, 75 mg of PMSF per liter, 10 mg of soybean trypsin inhibitor, L-1-tosylamide-2-phenylethylchloromethyl ketone, and α-N-benzoyl-L-arginine methyl ester per liter, 1 mg of leupeptin and pepstatin A per liter, and 10^{-4} M DIFP

Buffer B: 0.5 M NaCl, 20 mM Tris-HCl (pH 7.4), 2 mM ethylenediaminetetraacetic acid (EDTA), 2 mM [ethylenebis(oxyethylenenitrilo)]tetraacetic acid (EGTA), 75 mg of PMSF per liter, 1 mM dithiothreitol

Buffer C: 20 mM KCl, 20 mM Tris-HCl (pH 7.8), 2 mM EDTA, 2 mM EGTA, 1 mM dithiothreitol

Buffer D: Same as C, but with pH adjusted to 7.4

Buffer E: 20 mM KCl, 20 mM Tris-HCl (pH 7.4), 0.5 mM EDTA, 0.5 mM EGTA, 1 mM dithiothreitol

Buffer F: 20 mM KCl, 20 mM Tris-HCl (pH 7.4), 0.5 mM EGTA, 10 mM MgCl$_2$, 1 mM dithiothreitol

Preparation of Thiophosphorylated Myosin Light Chain-Sepharose 4B Affinity Column. The isolated 20,000-dalton light chain of turkey gizzard myosin is coupled to Sepharose 4B CNBr-activated according to the procedure recommended by the manufacturer (Pharmacia Fine Chemicals) and then phosphorylated with ATPγS. The CNBr-activated-Sepharose 4B (10 g), which is swollen in and washed with 1 mM HCl, is added to a solution of myosin light chains (100 mg) in 50 ml of 0.1 M NaHCO$_3$, 0.5 M NaCl. The mixture is shaken gently overnight at 4°. The extent of coupling may be measured by determining the concentration of unbound protein, which is compared to that of the starting material. Complete binding of the ligand to the resin is usually observed. The resin is filtered on a sintered-glass funnel and washed with the coupling buffer. The unreacted activated groups on the Sepharose are blocked by suspending the resin in 50 ml of 1 M ethanolamine (pH 8.0) and gently shaking for about 2 hr at 25°. The resin is washed three times with 100 ml of a 0.1 M sodium acetate buffer (pH 4.0), containing 1 M NaCl followed by 100 ml of a 0.1 M sodium borate buffer (pH 8.0), containing 1 M NaCl and then with 50 mM Tris-HCl (pH 7.4), 0.1 mM CaCl$_2$, 10 mM MgCl$_2$, 1 mM dithiothreitol.

The phosphorylation of the light chain is carried out by suspending all of the above resin in 50 mM Tris-HCL (pH 7.4), 0.1 mM CaCl$_2$, 10 mM MgCl$_2$, 0.5 mM ATPγS, 3×10^{-7} M calmodulin, and 1 mM dithiothreitol and adding turkey gizzard myosin light chain kinase (3×10^{-8} M). The slurry is shaken gently for about 4 hr at 25°. The reaction is terminated by the addition of guanidine hydrochloride to a final concentration of 4 M, after which the resin is filtered and washed with buffer E.

The preparation of smooth muscle myosin,[8] the isolated 20,000-dalton light chain of smooth muscle myosin[9] and smooth muscle myosin kinase[9] has been published. Preparation of myosin kinase is discussed in this volume [27].

The affinity column can be used at least 10 times. After each use, the resin is washed with buffer E containing 1 M KCl to remove proteins bound nonspecifically to the resin. The binding capacity of the resin decreases with use owing to slow dephosphorylation of the myosin light chain. Rephosphorylation of the light chain regenerates its original binding capacity.

Extraction. About 500 g of the muscular portion of fresh turkey gizzards free of fat and connective tissue are ground in an electric meat grinder and suspended in 4 volumes of buffer A. The mixture is homogenized with a Sorvall Omni-Mixer using a 2-quart mason jar and then in a Glenco glass–Teflon homogenizer (250 ml) with a Sears $\frac{3}{8}$-inch electric drill that turns the pestle. The homogenate is sedimented at 10,000 g for 30 min. The supernatant is filtered through glass wool and fractionated with solid $(NH_4)_2SO_4$. Most of the phosphatase activity is precipitated between 30 and 60% saturation and sedimented at 10,000 g for 20 min. The precipitate is suspended in a minimum volume of buffer B to make a thin slurry and is dialyzed against 4 liters of buffer B for at least 3 hr. As this sample dialyzes, the precipitate completely dissolves resulting in a final volume of about 90 ml with a protein concentration of about 60 mg/ml.

Sephacryl S-300 Gel Filtration. The phosphatases are then chromatographed on a Sephacryl S-300 column (5 × 87 cm) equilibrated with buffer B and eluted at a flow rate of 165 ml/hr. This step resolves the smooth muscle phosphatases as shown in Fig. 1. When the column fractions are assayed with the 20,000-dalton myosin light chain in the presence of Mg^{2+}, two peaks of activity are observed (SMP-I and SMP-II). When intact myosin is used for substrate, two peaks of activity are also detected. One peak elutes prior to SMP-I (indicated by an arrow in Fig. 1), and the other peak coelutes with SMP-I. The resolution of these peaks is dependent on the volume of the sample applied to the Sephacryl S-300 column. Application of more than 35 ml of sample to the above column results in a loss of resolution of the various peaks of activity. Therefore three separate Sephacryl S-300 columns were required to chromatograph the entire $(NH_4)_2SO_4$ fraction (90 ml). The different peaks of phosphatase activities are pooled as indicated in Fig. 1 and purified independently. SMP-I and SMP-II have been purified to homogeneity, and thus only their purification will be described.

[8] J. R. Sellers, M. D. Pato, and R. S. Adelstein, *J. Biol. Chem.* **256,** 13137 (1981).
[9] R. S. Adelstein and C. B. Klee, *J. Biol. Chem.* **256,** 7501 (1981).

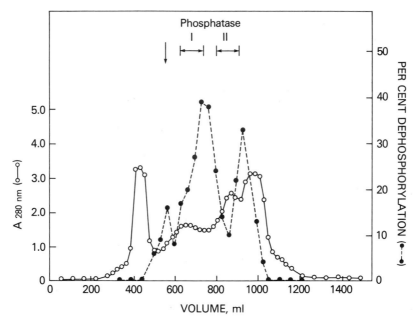

FIG. 1. Gel filtration chromatography of the 30 to 60% $(NH_4)_2SO_4$ fraction of the extract supernatant. A column of Sephacryl S-300 (5 × 87 cm) was equilibrated and eluted with buffer B at 4° at 165 ml/hr using an LKB 2120 Varioperpex II pump. Fractions of 11 ml were collected, of which 5 μl each were used for phosphatase assay using the isolated 20,000-dalton myosin light chain as substrate (●---●). The arrow indicates the position of the peak of intact myosin phosphatase activity. The absorbance at 280 nm was measured (○——○). From Pato and Adelstein.[1]

Smooth Muscle Phosphatase I

DEAE-Sephacel. The pooled fraction from three Sephacryl S-300 columns (about 300 ml) of SMP-I is dialyzed against buffer C overnight. The sample is then chromatographed on a column of DEAE-Sephacel (5 × 9 cm) equilibrated with buffer C at a rate of 125 ml/hr, collecting 10 ml of eluate per tube. After application of the sample, the column is washed with about 2 column volumes of buffer C before starting a linear gradient (total volume = 1200 ml) of 20 to 500 mM KCl. The phosphatase activities toward the intact myosin and the isolated myosin light chain elute as a single broad peak at a conductance of 11–15 mmho. The peak is pooled (about 200 ml) and dialyzed against buffer D overnight.

AH-Sepharose 4B. SMP-I is further purified on a column (1.5 × 5 cm) of AH-Sepharose 4B equilibrated with buffer D. The dialyzed SMP-I is applied to the column and washed with 2 column volumes of buffer D, and then a linear gradient (200 ml) of 20 to 1000 mM KCl is started. The phosphatase activity toward the isolated myosin light chain elutes as a

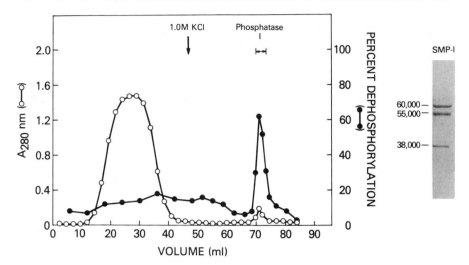

FIG. 2. Affinity chromatography of smooth muscle phosphatase I on thiophosphorylated myosin light chains–Sepharose. The column (1.5 × 11.5 cm) was equilibrated with buffer E at 4° at 35 ml/hr. After loading the sample, the column was washed with 1.5 column volumes of buffer E. Elution of the phosphatase was carried out with buffer E containing 1 M KCl. Fractions of 1.2 ml were collected and assayed for phosphatase activity (●——●). The absorbance at 280 nm was measured (○——○). On the right is a 1% SDS–12.5% polyacrylamide gel of purified SMP-I. From Pato and Adelstein.[1]

single broad peak at a conductance of 20–32 mmho. The front half of the peak is contaminated with activity toward intact myosin, while the back half contains activity only against the isolated myosin light chain (i.e., SMP-I). The latter is pooled and dialyzed overnight against buffer E. Recently we have obtained a complete separation of these two activities using ω-aminooctyl-agarose (Miles Laboratories) (M. D. Pato, S. Shaltiel, and R. A. Adelstein, unpublished observations).

Affinity Column Chromatography. The dialyzed SMP-I sample (about 50 ml) is applied to the affinity column equilibrated with buffer E and then washed with about 2 column volumes of buffer E. The bound phosphatase is eluted with buffer E containing 1 M KCl. Figure 2 shows that a large peak of protein relatively free of phosphatase activity is eluted in the void volume. The phosphatase binds to the resin and is eluted as a sharp peak upon application of 1 M KCl. The SMP-I purified by affinity chromatography is usually homogeneous as judged by SDS–polyacrylamide gel electrophoresis (see Fig. 2). The purified phosphatase fractions are pooled and dialyzed against buffer E overnight.

Concentration and Storage. SMP-I can be concentrated using AH-Sepharose 4B. This resin has very high binding capacity and binds SMP-I tightly. A small column of AH-Sepharose (about 0.5 ml of resin) is constructed with a minimum dead volume. The dialyzed purified SMP-I is passed through the column and washed with buffer E. The SMP-I is eluted with buffer E containing 1 M KCl. About 250-μl aliquots of the eluate are collected in small tubes or 1.5-ml conical Eppendorf plastic tubes. This procedure is best carried out manually. Depending upon the dead volume of the column, the first and second tube are usually free of enzyme activity and the next three tubes contain the highest concentration of phosphatase. The fractions with high phosphatase activity are pooled, divided into small aliquots in Eppendorf tubes, and stored in a $-70°$ freezer. Thawing and refreezing the enzyme causes a partial loss of activity.

Smooth Muscle Phosphatase II

DEAE-Sephacel. The pooled, dialyzed SMP-II fraction from the Sephacryl S-300 column (about 350 ml) is purified on DEAE-Sephacel in the same manner as SMP-I except for the substitution of 10 mM MgCl$_2$ for 2 mM EDTA in the buffer. The phosphatase is eluted as a single peak at a conductance of 8.5–13 mmho.

Affinity Column Chromatography. The phosphatase peak from the DEAE-Sephacel column (about 150 ml) is pooled and dialyzed against buffer F and is directly (i.e., omitting AH-Sepharose chromatography) purified on a thiophosphorylated myosin light chain-Sepharose column.

Prior to introducing SMP-II, the affinity column is washed with buffer F containing 1 M KCl to remove any nonspecifically bound protein and then with buffer F alone. The SMP-II sample is applied to the column, which is then washed with buffer F. SMP-II is eluted from the column with buffer F containing 10 mM EDTA instead of 10 mM MgCl$_2$. SMP-II elutes as a sharp peak (Fig. 3). SMP-II appears to be more stable in the presence of Mg^{2+}, thus the enzyme should not be kept in solutions containing EDTA for prolonged periods. The peak of activity is pooled and dialyzed against buffer F. SMP-II may be concentrated in the same manner as SMP-I or, alternatively, it may be concentrated by dialysis against buffer F containing 50% glycerol. The former is stored in a $-70°$ freezer, and the latter is stored in a $-20°$ freezer.

Assay for Phosphatase Activity

Preparation of Substrates. Turkey gizzard smooth muscle myosin or the isolated 20,000-dalton myosin light chain is phosphorylated with smooth muscle myosin kinase at $25°$ in a reaction mixture containing 50 mM Tris-HCl (pH 7.4), 10 mM MgCl$_2$, 0.1 mM CaCl$_2$, 1×10^{-7} M calmodulin, 0.1 mM [γ-^{32}P]ATP (0.5 Ci/mmol) (for myosin light chains) or 0.2 mM [γ-^{32}P]ATP (for myosin). When the phosphorylation is complete, the reaction mixture is dialyzed against buffer E containing 0.5 M KCl and then buffer E containing 0.1 M KCl (for myosin) or 20 mM KCl (for

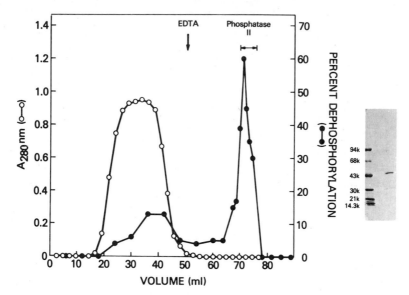

FIG. 3. Affinity chromatography of smooth muscle myosin phosphatase II on a thiophosphorylated myosin light chains–Sepharose column. The procedure was the same as described in the legend to Fig. 2 except that buffer F was used instead of buffer E. The phosphatase was eluted with the same buffer containing 10 mM EDTA instead of magnesium acetate. The fractions (1.2 ml) were assayed for phosphatase activity and the absorbance at 280 nm was measured. On the right is a 1% SDS–12.5% polyacrylamide gel of standard proteins used for determining molecular weight and purified SMP-II; K = ×1000. From Pato and Adelstein.[1]

myosin light chains). The dialysis step removes ATP and thus ensures that the kinase will be inactive during the assay for phosphatase activity.

Phosphatase Assay. The phosphatase activity is monitored by the release of $^{32}P_i$ from the 20,000-dalton myosin light chains. Dephosphorylation is performed at 30° in a reaction mixture (total volume = 50 μl) containing 50 mM Tris-HCl (pH 7.0), and either 1 mM EGTA (for SMP-I) or 10 mM magnesium acetate (for SMP-II). The reaction is initiated by adding 2.5 μl of the enzyme to the reaction mixture incubated at 30°. The reaction is terminated by adding 100 μl of cold 17.5% trichloroacetic acid and then 100 μl of a 6 mg/ml solution of bovine serum albumin.[10] The resulting mixture is kept in ice for at least 5 min before it is sedimented in an Eppendorf 5412 microcentrifuge for 1 min at 15,600 g. An aliquot of the supernatant (200 μl) is pipetted into 10 ml of scintillation fluid (Aqueous Counting Scintillant, Amersham) and counted in a Searle Mark III scintillation spectrometer. The release of $^{32}P_i$ is usually linear with time, provided that no more than 25% of the original counts are released.

[10] J. F. Antoniw, H. G. Nimmo, S. J. Yeaman, and P. Cohen, *Biochem. J.* **162**, 423 (1977).

[29] Preparation of Smooth Muscle α-Actinin

By SUSAN W. CRAIG, CHRISTINE L. LANCASHIRE, and JOHN A. COOPER

Tissue-specific isoforms of α-actinin, an actin-filament crosslinking protein,[1] are found in both muscle and nonmuscle cells. Because of chemical heterogeneity in the isoforms and in the contaminating proteins from different tissues, procedures developed for isolation of α-actinin from one tissue do not necessarily work optimally for another tissue. Consequently, a variety of methods exist for purification of α-actinin from skeletal muscle,[2] cardiac muscle,[3] smooth muscle,[4,5] and nonmuscle cells.[5] These methods vary considerably in the time they require, in their difficulty, and in the purity and yield of the final product.

This paper describes a procedure, used in our previous work,[6,7] that we developed specifically for isolation of α-actinin from chicken gizzard in high yield and purity; it has not been tested for its ability to isolate α-actinin from other sources. Development of this method for smooth muscle α-actinin was greatly facilitated by the prior experience of other investigators in the isolation of cardiac[3] and skeletal[2] muscle α-actinin.

Special Materials

Trimmed, glycerinated, and frozen chicken gizzards (Pel-Freez, Arkansas)

DE-52 preswollen, microgranular anion exchanger (Whatman)

Ultrapure ammonium sulfate (Schwarz-Mann)

Trasylol, a protease inhibitor, (Mobay Chemical Co., New York)

Sepharose 4B (Pharmacia)

Solutions. All buffers are adjusted to appropriate pH at 24° and then are stored and used at 4°.

Extraction medium, 36 liters: 1 mM KHCO$_3$ + 10 U of Trasylol per milliliter; pH as made, 7.4–7.5

---

[1] K. Maruyama and S. Ebashi, *J. Biochem.* (*Tokyo*) **58**, 13 (1965).

[2] A. Suzuki, D. E. Goll, I. Singh, R. E. Allen, R. M. Robson, and M. H. Stromer, *J. Biol. Chem.* **251**, 6860 (1976).

[3] I. Singh, D. E. Goll, R. M. Robson, and M. H. Stromer, *Biochim. Biophys. Acta* **491**, 29 (1977).

[4] B. G. Langer and F. A. Pepe, *J. Biol. Chem.* **255**, 5429 (1980).

[5] J. R. Feramisco and K. Burridge, *J. Biol. Chem.* **255**, 1194 (1980).

[6] S. W. Craig and J. V. Pardo, *J. Cell Biol.* **80**, 203 (1979).

[7] S. W. Craig and C. L. Lancashire, *J. Cell Biol.* **84**, 655 (1980).

METHODS IN ENZYMOLOGY, VOL. 85

DE-52 column buffer, 4 liters: 10 mM Tris-HCl, 1 mM EDTA, 1 mM DTT, 2 mM MgCl$_2$, 10 U of Trasylol per milliliter, pH 7.4 at 24°

Sepharose 4B column buffer, 1 liter: 10 mM Tris-HCl, 0.6 M KCl, 1 mM DTT, 1 mM EDTA, 2 mM MgCl$_2$, 0.02% NaN$_3$, pH 7.4 at 24°

Procedure

Step 1. Preparation of Gizzards. To obtain the purity and yield of α-actinin reported here, it is essential to use chicken gizzards that have been trimmed and stored in 100% glycerol at $-20°$ for at least 1 week prior to use. Fresh muscle does not work nearly as well. If you prepare your own gizzards, cut away all the fat and the thick, horny lining from the muscles before adding the glycerol. Put the muscle into a $-20°$ freezer right after adding the glycerol.

Step 2. Extraction. Rinse the frozen gizzards in distilled water to remove the glycerol. Pass the muscle through a meat grinder and collect 200 g of ground tissue. Take a 20-g portion and suspend it in 40 ml of 1 mM KHCO$_3$ + 10 U of Trasylol per milliliter. Pour the slurry into a length of dialysis tubing (4.5 cm flat width). Do the same for the rest of the muscle. The preceding steps may be done at room temperature; all subsequent steps are done at 4°. Dialyze the 10 bags against 12 liters of the extraction medium. A circular chromatography tank is a convenient dialysis vessel. Change the dialysis medium twice over the next 48 hr. At each change, mix the contents of the bags by inverting them several times. Centrifuge the slurry at 12,000 g for 2 hr. Discard the muscle residue and measure the volume of the supernatant.

Step 3. Ammonium Sulfate Fractionation

a. 0–17%: Place the extract into an ice-water bath on a stirrer. Bring the sample to 17% ammonium sulfate by addition of 9.45 g of ultrapure solid ammonium sulfate per 100 ml of extract. Add the ammonium sulfate slowly, with stirring, and maintain the pH at 7.0. After all the salt has dissolved, let the sample stand on ice for 1 hr. Then centrifuge it at 12,000 g for 15 min. Discard the pellet.

b. 17–30%: Bring the supernatant to 30% saturation by adding 8.15 g of ammonium sulfate per 100 ml of sample. Proceed as described in step 3a. After centrifugation, save the pellet and discard the supernatant.

c. Dissolve the pellet in 10 ml of DE-52 column buffer. Dialyze against 3 × 1-liter changes of DE-52 column buffer for a total of 24 hr. Centrifuge the sample at 48,000 g for 1 hr. Discard the pellet. Analysis of this ammonium sulfate fractionation scheme by sodium dodecyl sulfate (SDS) gel electrophoresis is shown in Fig. 1.

Step 4. Anion Exchange Chromatography. Apply the supernatant from step 3c to a 2 × 16 cm column of DE-52 equilibrated in DE-52 column

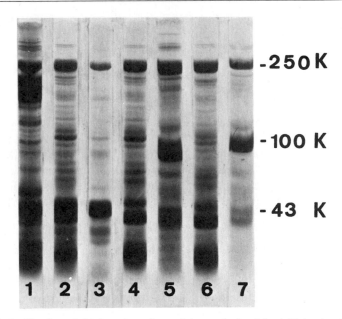

FIG. 1. Purification of chicken gizzard α-actinin: analysis of the initial extract and fractions from the ammonium sulfate fractionation on 7.5% polyacrylamide, sodium dodecyl sulfate (SDS) gels. Gels were loaded with 50 μg of protein and stained with Coomassie Blue. Lanes: 1, Total SDS-soluble proteins from ground gizzard; 2, 1 mM KHCO₃ extract; 3, 0–17% ammonium sulfate pellet; 4, 0–17% ammonium sulfate supernatant; 5, 17–30% ammonium sulfate pellet; 6, 17–30% ammonium sulfate supernatant; 7, proteins in the 17–30% ammonium sulfate pellet that are soluble in the DE-52 column buffer. 250 K = filamin; 100 K = α-actinin; 43 K = actin. K = ×1000.

buffer. You will need to equilibrate about 40 g (as supplied by the manufacturer) of Whatman preswollen DE-52. After unbound material has been washed off with the DE-52 column buffer, elute the column with a linear gradient from 0 to 0.3 M KCl in DE-52 column buffer (total gradient volume of 1200 ml). Set the flow rate at about 40 ml/hr, and collect 4-ml fractions. α-Actinin elutes from Whatman DE-52 between 0.15 and 0.2 M KCl (Fig. 2). Determine which fractions to pool by electrophoresing 50-μg samples from the column on a 7.5% discontinuous SDS slab gel.[8] Use phosphorylase (subunit $M_r = 95,000$) as a marker for α-actinin (subunit $M_r = 100,000$). Pool the major α-actinin-containing fractions as shown in Fig. 2. Precipitate the protein by addition of solid ammonium sulfate to 50% saturation (31.3 g/100 ml). Carry out the precipitation exactly as before. Collect the protein by centrifugation at 48,000 g for 15 min. Dis-

[8] J. V. Maizel, *Methods Virol.* 5, 179 (1971).

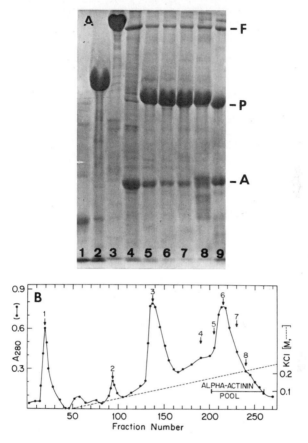

FIG. 2. Purification of chicken gizzard α-actinin: DE-52 chromatography. Panel A: Analysis of DE-52 column profile on 7.5% polyacrylamide-SDS gels. Approximately 50 μg of each sample was loaded onto the gel. Lanes 1–8: Samples from fractions with corresponding numbers as indicated on the column profile in panel B. (Lane 9: Filamin (F), phosphorylase (P), and actin (A) molecular weight markers. Panel B: DE-52 column profile.

solve the pellet in 8 ml of Sepharose 4B column buffer. Remove insoluble material (mostly actin and filamin that do not solubilize after the second ammonium sulfate precipitation) by centrifugation at 48,000 g for 30 min.

Step 5. Gel Filtration. Apply the supernatant from step 4 to a 2.5 × 100 cm column of Sepharose 4B in Sepharose 4B column buffer. Set the flow rate at 20 ml/hr, and collect 4-ml fractions. Residual contaminants of α-actinin elute near the void volume. The major symmetrical peak in the profile (Fig. 3) is 98% homogeneous α-actinin. Between 110 and 130 mg are obtained from 200 g of ground chicken gizzard.

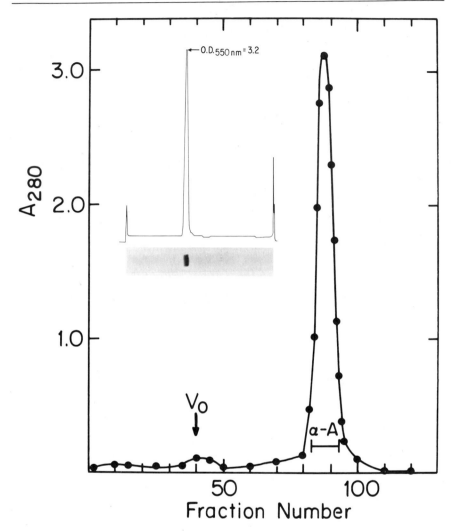

FIG. 3. Purification of chicken gizzard α-actinin: gel filtration on Sepharose 4B. The inset shows a Coomassie Blue-stained SDS polyacrylamide gel and corresponding scan of the α-actinin fractions that were pooled as indicated in the main figure. Ten micrograms of protein were loaded. The inset is from Craig and Pardo.[6]

Step 6. Concentration and Storage of α-Actinin. Precipitate the protein by adding ammonium sulfate to 50% saturation as in step 4. Dissolve the pellet in a buffer of your choice and dialyze extensively to remove ammonium sulfate. α-Actinin is stable and soluble in buffers (pH 7.0–8.0) containing almost no salt (e.g., 1 mM KHCO$_3$) to as much as 0.6 M KCl.

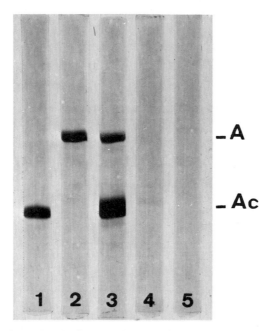

FIG. 4. Crosslinking of actin filaments by smooth muscle α-actinin. Two hundred microliters of monomeric actin (1.5 mg/ml in 2 mM Tris-HCl, 0.2 mM ATP, 0.2 mM CaCl$_2$, 0.5 mM DTT, pH, 8.0) are mixed with 200 μl of α-actinin (0.5 mg/ml in 1 mM KHCO$_3$). KCl is added to 50 mM, and the sample is incubated at room temperature for 15 min and then for 5 min at 0°. Controls are actin alone and α-acintin alone. The mixture of α-actinin and actin filaments (which form with the addition of KCl) form a gel at 0° that can be pelleted by centrifugation at 1000 g for 15 min; α-actinin or actin filaments alone do not pellet under these conditions. Lanes: 1, purified actin (Ac); 2, purified α-actinin (A); 3, 1000 g pellet of α-actinin and actin filament mixture; 4, 1000 g pellet of actin alone; 5, 1000 g pellet of α-actinin alone.

Make the final solution 0.02% in NaN$_3$ and then filter it through a 0.22-μm Millipore filter. Collect the filtrate into sterile tubes and store it at 4°. α-Actinin in 1 mM KHCO$_3$ has been stored in this manner for 1 year with no detectable breakdown or loss in its ability to crosslink and to gel actin filaments (Fig. 4).

Acknowledgment

This work was supported by NIH Grant AI-13700.

[30] Purification and Properties of Avian and Mammalian Filamins

By PETER J. A. DAVIES, YUTAKA SHIZUTA, and IRA PASTAN

Filamin is one of a group of actin binding proteins that appear to play an important role in contractile and cytoskeletal function in smooth muscle cells and nonmuscle cells. The identifying features of filamin are its very high subunit molecular weight ($M_r = 250,000$) and its ability to produce gelation and syneresis of solutions of F-actin. Filamin was first identified in smooth muscle by Wang et al.[1] and has been purified from both avian and mammalian smooth muscle.[2-4] The protein is particularly abundant in smooth muscle (1–5% of total cellular protein) and some nonmuscle cells such as platelets.[4] Lesser amounts of filamin have been detected in skeletal muscle, cultured fibroblasts, and macrophages.[4,5] The immunologic cross-reactivity of filamin and the macrophage actin binding protein[4] discovered earlier by Stossel and his colleagues[6] make it likely that these two proteins are either closely related or identical. Not all cells and tissues appear to contain filamin. Immunochemical methods have not detected filamin in erythrocytes or brain extracts.[4]

The prominence of filamin in contractile cells and the immunochemical localization of filamin in microfilament structures[1] suggests that the protein plays a role in the organization of the cytoskeleton. Filamin binds F-actin[2,7] and, by self-association, crosslinks actin into a three-dimensional lattice or gel.[8] The binding of filamin to actin modifies the interaction of actin with some, but not all, actin binding proteins. Thus, filamin binding to F-actin blocks subsequent tropomyosin binding[9] and actin activation of myosin ATPase[10,11] but has no effect on α-actinin bind-

[1] K. Wang, J. F. Ash, and S. J. Singer, *Proc. Natl. Acad. Sci. U.S.A.* **72**, 4483 (1975).
[2] Y. Shizuta, H. Shizuta, M. Gallo, P. Davies, I. Pastan, and M. S. Lewis, *J. Biol. Chem.* **251**, 6562 (1976).
[3] K. Wang, *J. Biol. Chem.* **251**, 6562 (1976).
[4] D. Wallach, P. Davies, and I. Pastan, *J. Biol. Chem.* **253**, 3328 (1978).
[5] P. J. Bechtel, *J. Biol. Chem.* **254**, 1755 (1979).
[6] T. P. Stossel and J. H. Hartwig, *J. Cell Biol.* **68**, 602 (1976).
[7] K. Wang and S. J. Singer, *Proc. Natl. Acad. Sci. U.S.A.* **74**, 2021 (1977).
[8] P. J. A. Davies, D. Wallach, M. Willingham, I. Pastan, and M. S. Lewis, *Biochemistry* **19**, 1366 (1980).
[9] M. G. Zeece, R. M. Robson, and P. J. Bechtel, *Biochim. Biophys. Acta* **581**, 365 (1979).
[10] P. Davies, P. J. Bechtel, and I. Pastan, *FEBS Lett.* **77**, 228 (1977).
[11] P. J. A. Davies, D. Wallach, M. C. Willingham, I. Pastan, M. Yamaguchi, and R. M. Robson, *J. Biol. Chem.* **253**, 4036 (1978).

ing.[9] What factors may regulate the interaction of actin and filamin is not yet clear, but it has been observed that both avian and mammalian filamins can be phosphorylated by cyclic AMP-dependent protein kinase.[12,13] No evidence has, as yet, been reported to show that phosphorylation of filamin modifies its functional properties.

Ultracentrifugal analysis of chicken gizzard filamin solutions[2,8] and chemical cross-linking studies[3] have shown that the principal filamin species in solution is a dimer (M_w = 500,000 daltons) with variable degrees of higher molecular weight aggregates present. Self-association and aggregation of gizzard filamin can be inhibited by 0.6 M KCl or by conversion of filamin to a monomeric fragment (heavy merofilamin, M_w = 240,000 daltons) that retains the actin-binding properties of native filamin, but does not self-associate.[8,11]

Purification of Filamin

General Considerations

Filamin can be purified from both smooth muscle and nonmuscle cells (see below). However, chicken gizzard is a particularly convenient source, since large quantities of fresh tissue are readily obtainable and inexpensive. Two general problems are encountered during filamin purification: proteolytic degradation, and aggregation of the protein. To reduce proteolysis, we routinely work with fresh tissues, carry out all steps at 0–4°, move as expeditiously as possible through the early steps of purification, and include EDTA in all buffers. Aggregation can be minimized also by working with fresh tissues in the cold, by including reducing agents in the buffers, and by keeping the ionic strength of buffers above 100 mM KCl until all the actin is removed from the preparation.

Purification of filamin is monitored by SDS–polyacrylamide gel electrophoresis on 5% linear polyacrylamide slab gels. Purification samples are adjusted to 1% SDS, 5% glycerol, 10% 2-mercaptoethanol, 20 mM sodium phosphate, pH 6.8, and 0.05% bromophenol blue and electrophoresed in a buffer system containing 0.2 M Tris glycine, pH 8.7, 0.1 mM EDTA, 1% SDS. Filamin is identified by its subunit molecular weight (M_r = 250,000) using as molecular weight standards myosin heavy chain (M_r = 200,000), β and β' subunits of *Escherichia coli* RNA polymerase (M_r = 165,000 and 155,000), phosphorylase A (M_r = 95,000), serum albumin (M_r = 68,000), and ovalbumin (M_r = 43,000). The protein content of

[12] D. Wallach, P. J. A. Davies, and I. Pastan, *J. Biol. Chem.* **253**, 4739 (1978).
[13] P. Davies, Y. Shizuta, K. Olden, M. Gallo, and I. Pastan, *Biochem. Biophys. Res. Commun.* **74**, 300 (1977).

purification fractions is determined by the method of Lowry,[14] and the protein content of column fractions by the OD_{280}. The extinction coefficient of chicken gizzard filamin $E_{280}^{1\%}$ is 7.4.

Purification of Chicken Gizzard Filamin

Tissue Extraction. The starting material is 6–8 medium-sized fresh chicken gizzards. The muscle (55 g) is dissected away from the lining, the fascia, and adherent fat and minced with scissors and a razor blade. The minced muscle is combined with 600 ml of 5 mM EDTA, pH 7.0, containing 5 mM dithiothreitol, freshly made up, and blended for 60 sec using 10-sec bursts in a chilled Waring blender. The pink frothy homogenate is centrifuged at 25,000 g (18,000 rpm) in a Beckman JA 20 rotor for 30 min. The pink supernatant containing mainly myosin, actin, and soluble proteins (1763 mg, 4.3 mg/ml) is discarded, and the sediment containing predominantly filamin, myosin, and actin is resuspended in 600 ml of 50 mM potassium phosphate pH 7.5, 600 mM KCl, 1 mM EDTA, and 1 mM dithiothreitol. The resuspended material is homogenized by a 15-sec burst in the Waring blender and then stirred with a magnetic stirrer for 10 min at 4°. The homogenate is then centrifuged at 100,000 g (29,000 rpm) for 30 min in a Beckman type 30 rotor. The yellowish, viscid supernatant (676 mg, 1.6 mg/ml) containing predominantly filamin and actin with lesser amounts of myosin and other proteins is retained, and the pellets are discarded. Fat is removed from the extract by filtering through a funnel containing glass wool, and then the high salt extract is dialyzed overnight against 20 liters of 50 mM potassium phosphate, pH 7.5, 100 mM KCl, 1 mM EDTA, 5 mM 2-mercaptoethanol. This completes the first day of purification.

Ammonium Sulfate Fractionation. During the overnight dialysis, a gelatinous precipitate containing predominantly myosin and actin with some bound filamin is formed. This precipitate is removed by centrifugation at 100,000 g (29,000 rpm) for 30 min in a Beckman type 30 rotor. The supernatant fraction (610 mg, 410 ml, 1.5 mg/ml) is retained. Solid ammonium sulfate (84.6 g) is slowly added to the supernatant with continuous stirring to give a final concentration of 35% saturation. During addition of ammonium sulfate, the pH is monitored and maintained between pH 7.2 and 7.5 by dropwise addition of 5 N ammonium hydroxide. The flocculate white precipitate is stirred slowly for 30 min and then sedimented by centrifugation at 7800 g (10,000 rpm) in a Beckman JA 20 rotor for 20 min. The supernatant is discarded and the pellets are resuspended in a total of 100 ml of 50 mM potassium phosphate, 100 mM KCl,

---

[14] O. H. Lowry, N. J. Rosebrough, A. L. Farr, and R. J. Randall, *J. Biol. Chem.* **193**, 265 (1951).

1 mM EDTA, and 1 mM dithiothreitol. To facilitate redissolving, the pellets may be treated with a single pass of a motor-driven Teflon pestle in a glass homogenizer. The resuspended pellet is stirred gently for 2 hr and then centrifuged at 25,000 g (18,000 rpm) in a JA 20 rotor for 30 min. Insoluble debris in the pellet is discarded, and the yellowish supernatant is saved. The supernatant is again adjusted to 35% saturation by slow addition of solid ammonium sulfate. Again care is taken to keep the pH at 7.2–7.5. The white flocculate precipitate is stirred for 30 min and then pelleted at 25,000 g (18,000 rpm) for 30 min in a JA 20 rotor. The supernatant is discarded, and the pellet is redissolved in 40 ml of 20 mM potassium phosphate, pH 7.5, 100 mM KCl, 1 mM EDTA, and 1 mM dithiothreitol. At this stage, SDS gels show the preparation (325 mg, 8 mg/ml) to be predominantly filamin with actin being the major impurity.

Ion-Exchange Chromatography. We have found that the next step in filamin purification, ion exchange chromatography, gives us the most problems. Success at this step requires careful monitoring of the pH and conductivity of all solutions. Using our particular conductivity meter, solutions of 1, 10, and 100 mM NaCl at 4° give conductance values of 1.8×10^{-2}, 1.4×10^{-1}, and 1.2 mmho, respectively. At pH 7.5, filamin elutes from DEAE-cellulose between 0.6 and 1.2 mmho (i.e., between 50 mM NaCl and 100 mM NaCl). At the low ionic strengths required to get filamin binding to DEAE, actin and filamin rapidly coprecipitate. To prevent this, we first remove the actin by adsorption to DEAE at higher ionic strength and then lower the ionic strength to permit binding and fractionation of filamin. The fraction from ammonium sulfate precipitation is dialyzed 4 hr at 4° against 1 liter of 20 mM potassium phosphate pH 7.5, 50 mM KCl, 1 mM EDTA, 15 mM 2-mercaptoethanol (conductivity = 1.3 mmho equivalent to 120 mM NaCl). If equilibration with the dialysis buffer is not achieved, then the conductivity of the dialyzed sample is lowered to 1.3 mmho by addition of cold H$_2$O. Any precipitate that forms is removed by centrifugation at 11,400 g (12,000 rpm) for 15 min in a Beckman JA 20 centrifuge. The sample is then pumped rapidly (25 ml/hr) across a 2.6 × 30 cm column of DEAE-cellulose (Whatman DE-52) equilibrated with 20 mM potassium phosphate pH 7.5, 50 mM KCl, 1 mM EDTA, 15 mM 2-mercaptoethanol. Fractions of 5 ml are collected, and the filamin is eluted in the flowthrough fraction. SDS gels at this stage indicate complete removal of actin and a preparation of greater than 80% filamin. This step completes the second day of purification.

The pooled flowthrough (90 ml) is diluted by addition of an equivalent volume of cold H$_2$O to give a conductance of 0.5 mmho (40 mM NaCl equivalent) and the sample is pumped at 25 ml/hr onto a fresh 2.6 × 20 cm column of DEAE-cellulose equilibrated with 10 mM potassium phosphate, 25 mM KCl, 1 mM EDTA, 15 mM 2-mercaptoethanol. Fractions of 5 ml are again collected. The flowthrough fractions contain numerous lower molecular weight impurities. After the OD_{280} has returned to baseline, filamin is eluted by application of a linear gradient using 200 ml of 20 mM potassium phosphate, pH 7.5, 50 mM KCl, 1 mM EDTA, 15 mM 2-mercaptoethanol, and 200 ml of 20 mM potassium phosphate, pH 7.5, 500 mM KCl, 1 mM EDTA, 15 mM 2-mercaptoethanol. Filamin elutes as a single sharp symmetrical peak following the first appearance of the gradient in the column eluate. The purity of the individual fractions is checked by SDS gel electrophoresis, and the purest filamin fractions (8 fractions, 40 ml, OD_{280} 2.56) representing 195 mg of filamin are saved. This sample is then dialyzed overnight against 1 liter of 50 mM potassium phosphate, 100 mM KCl, 1 mM EDTA, 15 mM 2-mercaptoethanol. This completes the third day of the procedure.

Gel Filtration Chromatography. The fourth day involves the last steps, purifications of the filamin by gel filtration on Sepharose 4B. As a matter of convenience, we use a 1.6 × 100 cm column of Sepharose 4B equilibrated with 50 mM potassium phosphate, pH 7.5, 100 mM KCl, 1 mM EDTA, 15 mM 2-mercaptoethanol. The column is run by ascending flow at 10 ml/hr and 2-ml fractions are collected. The maximum sample volume that can be applied to the column is 20 ml, so in the case of the 40 ml of DEAE pool, the sample is divided into two aliquots and run separately. We prefer not to concentrate the filamin prior to gel filtration to reduce the possibility of aggregation of the purified protein. Filamin elutes from the Sepharose 4B column between fractions 40 and 60 (v_e = 100 ml). The profile of the peak shows some tailing. SDS gels indicate that the purity of the filamin is greatest in the leading and peak fractions and that residual impurities are concentrated in the tail. The peaks from the two column runs were pooled, excluding the tail fractions. The final yield was 75 ml of material with an OD_{280} of 1.787. Using the extinction coefficient of $E_{280}^{1\%}$ = 7.4 of chicken gizzard filamin, a typical yield is 183 mg of highly purified filamin at a concentration of 2.4 mg/ml.

Purification of Mammalian Filamins

We have purified mammalian filamin from both smooth muscle and nonmuscle cells. Guinea pig vas deferens can be used as a source of relatively pure smooth muscle, and purification of filamin from this source has been described.[4] The relatively small amount of tissue (8.8 g) obtained from large numbers of animals (50) and the small yield (5 mg) of purified filamin makes this a less than optimal source. As an alternative, we have used bovine aorta as a source of mammalian smooth muscle filamin. The starting material is 25 refrigerated bovine aortas (Pel-Freeze Biologicals).

The intima and adventitia are peeled away from the medial layers of the aorta, and 1 kg of fresh tissue is recovered as a starting material. The media is minced with scissors, and 150-g aliquots are blended for 60 sec with 600 ml of 5 mM EDTA, pH 7.0, containing 5 mM dithiothreitol. The pooled homogenate is centrifuged in aliquots in a Beckman JA-14 rotor at 11,400 g (12,000 rpm) for 30 min, and the supernatant is discarded. The 250-g aliquots of the pellet are homogenized in 600 ml of 50 mM potassium phosphate, pH 7.5, 600 mM KCl, 1 mM EDTA, 15 mM 2-mer-captoethanol for 60 sec in a Waring blender. The filamin is extracted by stirring overnight at 4°, and the high-salt extract is further purified by the same procedures as for chicken gizzard filamin. The final yield of bovine aorta filamin from 1 kg of tissue is 55 mg. The subunit molecular weight and actin binding properties of this filamin appear to be identical to chicken gizzard filamin.

We have also purified filamin from a nonmuscle cell, human blood platelets. Platelet filamin was harder to purify than muscle filamin because of extensive proteolysis during purification. Also, removal of some platelet proteins not present in smooth muscle, particularly platelet band 2,[15] requires modification of the purification strategy. The starting material is 20 units of freshly drawn unrefrigerated human platelets. These are washed and concentrated following the procedure of Majerus.[16] Individual units of platelets are centrifuged in 50-ml polypropylene centrifuge tubes at 100 g for 25 min at 23°. The red cell pellet is discarded, and the platelets are sedimented from the platelet-rich plasma by centrifugation at 2200 g for 15 min at 23°. The plasma is discarded, and the platelet pellets are pooled and gently resuspended in 200 ml of a solution of 103 mM NaCl, 4.3 mm K_2HPO_4, 4.3 mM Na_2HPO_4, 24.4 mM NaH_2PO_4, 5 mM glucose, 5 mM EDTA, pH 6.5. The platelets are resedimented at 2200 g for 15 min at 23° and then resuspended and repelleted once more. The washed platelet pellet is then resuspended in 150 ml of the above solution, chilled on ice, and then disrupted by four 30-sec bursts of sonication using 75 W of power, in a Branson sonifier. Completeness of lysis is checked under the microscope.

To extract platelet filamin, the lysate is adjusted to a final concentration of 50 mM potassium phosphate, pH 7.0, 600 mM KCl, 10 mM ATP, and 5 mM dithiothreitol and a final volume of 250 ml. The extract is stirred at 4° for 2 hr and then centrifuged at 100,000 g for 30 min in a Beckman type 30 rotor. The pellet is discarded, and the supernatant, containing the platelet filamin, is further purified following the scheme for gizzard filamin. The modifications employed are as follows: the two cycles of am-

[15] D. R. Phillips and M. Jakabova, *J. Biol. Chem.* **252**, 5602 (1977).
[16] P. Majerus, this series, Vol. 31, p. 149.

monium sulfate fractionation are to a final concentration of 30%, rather than 35%. DEAE chromatography is carried out at pH 8.0 to improve resolution of filamin from band 2. Gel filtration is carried out on a 1.6 × 63 cm column of Sepharose C1-4B in a buffer containing 50 mM potassium phosphate, pH 7.5, 600 mM KCl, 1 mM EGTA, 1 mM dithiothreitol. The final yield is 5 mg of pure human platelet filamin. The subunit molecular weight and actin binding properties of this nonmuscle filamin appear to be comparable to avian or mammalian smooth muscle filamin.

Section III

Methods for Study of the Motility Apparatus in
Nonmuscle Cells

[31] Purification of Nonmuscle Myosins

By THOMAS D. POLLARD

The procedure for preparation of platelet myosin will be given in detail at the end of this chapter. The general principles of myosin purification, however, will be discussed first in order to provide a better starting point for the investigator who wants to make myosin from a nonmuscle source other than platelet.

Extraction of Myosin from Cells

Extraction of myosin from the cell requires lysis of the cell into a buffer that solubilizes myosin. Three general types of extracting solutions have been used to solubilize myosin: high ionic strength, sucrose, or pyrophosphate buffers (Table I). Both the high ionic strength and pyrophosphate buffers solubilize myosin by dissociating myosin filaments into myosin monomers. To my knowledge, no one has investigated the form of myosin in the sucrose buffers, but it is clear that a number of cytoplasmic myosins (macrophage, *Dictyostelium, Acanthamoeba*) are soluble in sucrose. There are no absolute reasons why one or the other of these general buffer types should be superior for extracting myosin from a given cell and, with the exception of some early work,[1,2] few comparative studies have been made with different extracting solutions. The sucrose buffers seem to have one distinct advantage for nonmuscle cells: They minimize rupture of lysosomes and make it possible to isolate intact myosin from cells, such as macrophages, that are rich in lysosomes.[3] (In concentrated KCl buffers enough lysosomal enzymes are released that most of the macrophage myosin heavy chains are cleaved by proteases.) A further advantage of sucrose extraction is that the extract has a low ionic strength and is, therefore, suitable for ion exchange chromatography.

Besides the high salt, pyrophosphate, or sucrose to solubilize the myosin, the extracting solutions usually contain a pH buffer and a sulfhydryl reducing agent. Some also contain a low concentration of ATP or pyrophosphate to dissociate actin from myosin and a detergent or organic solvent to solubilize membranes.

[1] M. R. Adelman and E. W. Taylor, *Biochemistry* **8**, 4976 (1969).
[2] T. D. Pollard and E. D. Korn, *J. Biol. Chem.* **248**, 4682 (1973).
[3] T. P. Stossel and T. D. Pollard, *J. Biol. Chem.* **248**, 8288 (1973).

METHODS IN ENZYMOLOGY, VOL. 85

TABLE I
MYOSIN EXTRACTION SOLUTIONS

Buffer	Cell types	Reference[a]
High ionic strength		
0.5 M KCl, 0.1 M K$_2$HPO$_4$	Skeletal muscle	Kielley and Harrington[10]
0.6 M KCl, 15 mM Tris (pH 7.5), 10 mM dithiothreitol, 3% Butanol	Platelets	Adelstein et al.[14]
	Vetebrate tissue culture	Adelstein et al.[15]
	Brain	Burridge and Bray[b]
0.9 M KCl, 15 mM sodium pyrophosphate (pH 7.0), 5 mM MgCl$_2$, 3 mM dithiothreitol, 30 mM imidazole (pH 7.0)	Platelets	Pollard et al.[4]
0.6 M Ammonium acetate (pH 6.5), 2 mM magnesium acetate, 2 mM sodium pyrophosphate, 0.5 mM dithiothreitol	Platelets, skeletal muscle	Trayer and Trayer[31]
0.3 M ammonium acetate, 10 mM sodium bicarbonate (pH 7.2), 2 mM CaCl$_2$	Thyroid	Kobayashi et al.[27]
Pyrophosphate		
50 mM sodium pyrophosphate, 10 mM Tris-maleate (pH 6.8)	*Physarum*	Adelman and Taylor[1]
Sucrose		
0.34 M sucrose, 2 mM ATP, 10 mM dithiothreitol, 10 mM Tris-maleate (pH 7.0)	Macrophages, neutrophils	Stossel and Pollard[3]
0.34 M sucrose, 1 mM ATP, 1 mM dithiothreitol, 10 mM imidazole (pH 7.0)	*Acanthamoeba*	Pollard et al.[9]
0.87 M sucrose, 1 mM EDTA, 0.1 mM dithiothreitol, 10 mM Tris (pH 7.5)	*Dictyostelium*	Clarke and Spudich[41]

[a] Superscript numbers refer to text footnotes.
[b] K. Burridge and D. Bray, *J. Mol. Biol.* **99**, 1 (1975).

Cell lysis techniques vary considerably. Striated muscle cells are large enough to be broken with a meat grinder (this volume [7]). Most cells can be broken gently in "Potter" Teflon–glass or "Dounce" glass homogenizers or by nitrogen cavitation in a "Parr" bomb. Platelets are too small to break mechanically, so they are lysed by freeze-thawing or by membrane dissolution with butanol or Triton X-100. When the specimen is very tough, such as smooth muscle (this volume [26]), it may be necessary to resort to more violent means of cell lysis, such as a blade-type homogenizer. However, it is wise to be as gentle as possible and to avoid bubbles during homogenization and all subsequent steps, because myosin is particularly sensitive to surface denaturation at air–water interfaces.

Various investigators have used extraction times from a few minutes to many hours. In those cases where the extent of extraction has been measured as a function of time,[e.g.,4] extraction was complete in less than 30 min, so there is probably no need for prolonged extraction.

Once the myosin is solubilized, the insoluble cell remnants are invariably removed by centrifugation. High speed (100,000 g), long (60 min) centrifugation has the advantage of pelleting microsomes as well as larger membrane fragments and yields an extract of soluble components. Low speed centrifugation (25,000 g) for 15 min is sufficient for muscle and is commonly used for other cells as well. If a pellicle of lipid floats to the top of the centrifuge tube, it can be removed with a Pasteur pipette connected to a vacuum via a trap. Large chunks of fat can be filtered out with glass wool.

A completely different approach is to solubilize as much of the cell as possible except for the molecules of interest. This has been done for years with skeletal muscle where insoluble myofibrils can be separated from the cytosol by cell lysis and extraction under conditions that stabilize the actin and myosin filaments.[e.g.,5] Myofibrils have also been isolated from cardiac[6] and smooth[7] muscles. To purify the myosin, the myofibrils are extracted by one of the methods used to solubilize myosin.

Fractional Precipitation Methods

Actomyosin Precipitation

Precipitation of the specific combination of myosin filaments with actin filaments is one of the original methods for purifying myosin[8] and remains one of the most useful. This precipitation requires low ionic strength for myosin filament formation and the absence of ATP for the combination of myosin with actin filaments. This is usually accomplished by dilution or dialysis. Dilution is faster, but dialysis may give a higher yield in some cases where the myosin and/or actin concentrations are low. The final ionic strength required to precipitate actomyosin is in the range of 50–100 mM depending on the properties of the myosin. The pH is often an important consideration. Usually it must be slightly acidic. For example, platelet actomyosin fails to precipitate above pH 7 but precipitates quan-

[4] T. D. Pollard, S. M. Thomas, and R. Niederman, *Anal. Biochem.* **60**, 258 (1974).
[5] J. D. Potter, *Arch. Biochem. Biophys.* **162**, 436 (1974).
[6] R. J. Solaro, D. C. Pang, and F. N. Briggs, *Biochim. Biophys. Acta* **245**, 259 (1971).
[7] A. Sobieszek and R. D. Bremel, *Eur. J. Biochem.* **55**, 49 (1975).
[8] A. Szent-Györgyi, "The Chemistry of Muscle Contraction," 2nd ed. Academic Press, New York, 1951.

titatively at pH 6.4.[4] ATP in the extract can be removed enzymically with hexokinase and glucose[9] if the myosin ATPase activity is too low to hydrolyze the ATP quickly.

Ammonium Sulfate Precipitation

This has been used successfully for years in muscle myosin purification[10] and has also been widely used for nonmuscle myosin purification. Solid ammonium sulfate can be added directly to the sample, but the evolution of gas bubbles may lead to some myosin denaturation. Consequently, I usually use a saturated solution of ammonium sulfate containing EDTA to bind minor heavy metal contaminants.

Procedure for making saturated ammonium sulfate (from Wayne Kielley)

1. Add 780 g of ultrapure ammonium sulfate (Schwarz-Mann No. 1946) to 1 liter of water and heat until dissolved.
2. Add 0.2 M EDTA (pH 7) to give a concentration of 10 mM.
3. Cool to 4° overnight to crystallize excess ammonium sulfate.
4. Neutralize by adding ammonium hydroxide to give an apparent pH of 8.2. To confirm that the actual pH is 7.0, dilute a small sample 1 : 10 with water and read the pH.

The ammonium sulfate solution can be used to precipitate myosin in two different ways: Usually the desired volume of ammonium sulfate solution is added slowly to the protein solution with gentle stirring. After about 15 min, the precipitate is pelleted at 25,000 g for 15 min. This is satisfactory when the protein concentration is relatively high. When the protein concentration is low, the sample may be dialyzed against the ammonium sulfate. In this case, use the same volume of saturated ammonium sulfate that one would add to the protein solution to give the desired salt concentration. For example, for a final concentration of 1.9 M ammonium sulfate (50% saturation) either add 1 volume of saturated ammonium sulfate solution to 1 volume of protein solution, or dialyze 1 volume of protein vs 1 volume of saturated salt solution. The time required for completion of dialysis will depend on the diameter of the dialysis bag. About 4 hr is sufficient for a bag 2 cm in diameter.

Usually the concentration of ammonium sulfate is raised in steps, and the precipitated protein is pelleted after each step. The exact concentration required to precipitate myosins varies, but most precipitate between 1 M and 2 M. Pilot experiments using steps of 0.2 M are recommended to determine the optimal conditions.

[9] T. D. Pollard, W. F. Stafford, and M. E. Porter, *J. Biol. Chem.* **253**, 4798 (1978).
[10] W. W. Kielley and W. F. Harrington. *Biochim. Biophys. Acta* **41**, 401 (1960).

Chromatography of Myosin

Column chromatography has been used as a final step to remove minor adherent contaminants such as C-protein from muscle myosin preparations[11] and is an absolutely essential step in the purification of all cytoplasmic myosins. The theory of each type of column is well covered in other volumes of this series, so I will emphasize the specific advantages and disadvantages of each for myosin purification.

Gel Filtration Chromatography

Gel filtration has been used very successfully in myosin purification because few other cellular components share the large Stokes' radius of myosin. Oligomeric actin is one of the few cellular proteins that is not separated from myosin on these columns, and it can be eliminated by complete depolymerization of the actin in KI.[4,12,13]

Beaded 4% agarose gels are usually selected for myosin gel filtration. As in all other forms of chromatography, higher resolution is achieved if the gel particles are small and uniformly sized. For this reason the resolution of 200–400 mesh 4% agarose beads from Bio-Rad Laboratories (A-15m) is much better than the larger (100–200 mesh) A-15m beads from Bio-Rad or the widely used 4% agarose beads from Pharmacia (Sepharose 4B). Although the peaks are sharper, there is a penalty for this high resolution; the flow rates are lower than with the smaller beads.

These gel filtration columns are usually equilibrated with a high concentration of KCl plus a pH buffer and a sulfhydryl reducing agent. When KI is used to depolymerize actin in the column sample, one can equilibrate the whole column with KI,[3] but the prolonged exposure to KI usually denatures the myosin. Consequently, many laboratories now use a two-buffer system in which the myosin passes through KI zone into KCl while the much smaller actin monomers move more slowly down the column and remain in the zone of KI applied ahead of the sample.[4]

Procedure (see Fig. 1)

1. A 2.6 × 90 cm column of Bio-Rad A-15m (200–400 mesh) is equilibrated with 600 ml of 0.6 M KCl, 1 mM dithiothreitol, 10 mM imidazole (pH 7). Using 55 cm of hydrostatic pressure, the flow rate at 4° is about 15–20 ml/hr.
2. Before preparing the sample, the KCl buffer is removed from the top of the column and 50–60 ml (~14% of column volume) of 0.6 M KI, 5 mM ATP, 5 mM dithiothreitol, 1 mM MgCl$_2$, 20 mM imidazole (pH 7) are allowed to run into the column bed.

[11] G. Offer, C. Moos, and R. Starr, *J. Mol. Biol.* **74**, 653 (1973).
[12] S. Puszkin and S. Berl, *Biochim. Biophys. Acta* **256**, 695 (1972).
[13] V. T. Nachmias, *J. Cell Biol.* **62**, 54 (1974).

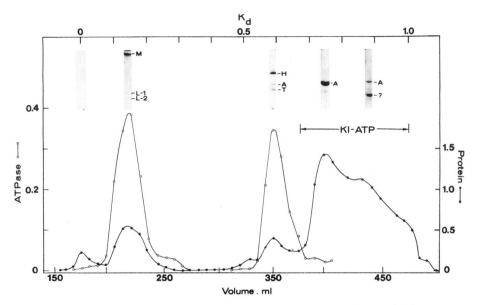

FIG. 1. Gel filtration of platelet actomyosin on a 2.6 × 90 cm column of 4% agarose (Bio-Rad A-15m, 200–400 mesh) according to Pollard *et al.*[4] The zone of KI–ATP buffer is indicated. ○——○, K(EDTA)ATPase activity (μmol/min per milliliter; ●——●, protein concentration (mg/ml). The protein composition of selected fractions is shown by tube gel electrophoresis in sodium dodecyl sulfate. Myosin heavy chain (M); myosin light chains (L-1 and L-2); head fragment of myosin (H); actin (A); tropomyosin (T). From Pollard *et al.*[4]

3. An actomyosin sample containing about 40–120 mg of protein is dissolved by gentle Dounce homogenization in 10–15 ml of 0.6 M KI, 5 mM ATP, 5 mM dithiothreitol, 1 mM MgCl$_2$, 20 mM imidazole (pH 7) and immediately clarified by centrifugation at 100,000 g for 20–30 min.

4. The clarified sample is applied to the column immediately behind the KI zone.

5. KCl buffer is layered over the sample and used to elute the column.

6. Myosin elutes with a partition coefficient of about 0.25 (depending on the particular batch of agarose) and can be detected by absorbance at 280 nm, protein assay, and ATPase assay. Actin elutes with a partition coefficient of about 0.75. Since it is, at least in part, contained in the KI zone, the actin will be denatured unless the KI is removed by dialysis.

7. The column is stored in 0.6 M KCl, 0.05% sodium azide, 2 mM EDTA.

There are also successful procedures for myosin gel filtration without KI. Adelstein *et al.*[14,15] and Ostlund and Pastan[16] have purified myosins from platelets and from cultured cells using agarose columns equilibrated with 0.6 M KCl, 1 mM ATP, 1 mM MgCl$_2$, 5 mM dithiothreitol, 15 mM Tris (pH 7.5). I have had more success separating myosin from actin using the two-buffer system.

Anion Exchange Chromatography

Chromatography on DEAE columns is one of the better ways to remove minor contaminants such as C-protein from muscle myosin[11] and is an important step in the purification of *Acanthamoeba* myosins.[2,9,17] It has also been used to separate muscle myosin subfragment-1 isozymes.[18] My personal experience has been that DEAE-cellulose is easy to work with and gives better resolution of myosins than DEAE-Sephadex or DEAE-agarose.

A limitation of ion exchange chromatography is the insolubility of most myosins at low ionic strength where they bind to the column. This can be overcome by using pyrophosphate[19] or phosphate[11] buffers that dissociate myosin filaments at relatively low ionic strength. For cytoplasmic myosins, sucrose extracting solutions can be used to solubilize the myosin at low ionic strength.

Procedure (see Fig. 2)

1. Wash fresh or previously used DEAE-cellulose according to the manufacturer's instructions. For Whatman DE-52, DE-53, or DE-32, this involves the following steps: (*a*) stir 30 min in 0.5 M NaOH to release tightly bound materials; (*b*) wash with 20 volumes of water on a Büchner funnel; (*c*) stir 30 min under vacuum in 0.5 M HCl to release bound bicarbonate; (*d*) wash with 20 volumes of water on a Büchner funnel; and (*d*) titrate to the desired pH using the free base used in the buffer (e.g., 1 M Tris base or imidazole) and monitoring the pH by filtering small aliquots and returning the filtered DEAE-cellulose to the sample.

2. If the column is to be equilibrated with ATP,[e.g.,20] add it at this point.

3. Pour the column using a high flow rate (achieved by 1–2 m hydro-

[14] R. S. Adelstein, T. D. Pollard, and W. M. Kuehl, *Proc. Natl. Acad. Sci. U.S.A.* **68**, 2703 (1971).

[15] R. S. Adelstein, M. A. Conti, G. Johnson, I. Pastan, and T. D. Pollard, *Proc. Natl. Acad. Sci. U.S.A.* **69**, 3693 (1972).

[16] R. E. Ostlund and I. Pastan, *Biochim. Biophys. Acta* **453**, 37 (1976).

[17] H. Maruta and E. D. Korn, *J. Biol. Chem.* **252**, 6501 (1977).

[18] A. G. Weeds and R. S. Taylor, *Nature (London)* **257**, 54 (1975).

[19] E. G. Richards, C. S. Chung, D. B. Menzel, and H. S. Olcott, *Biochemistry* **6**, 528 (1967).

[20] D. J. Gordon, E. Eisenberg, and E. D. Korn, *J. Biol. Chem.* **251**, 4778 (1976).

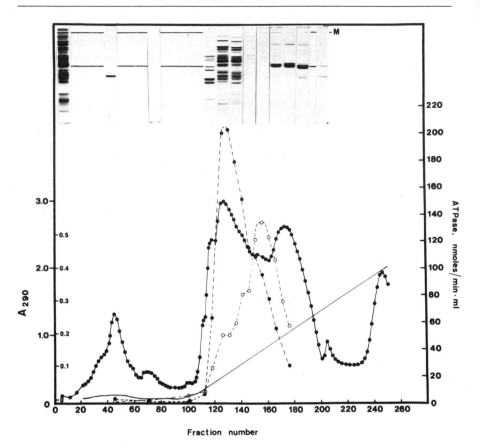

FIG. 2. Ion exchange chromatography of 230 ml of *Acanthamoeba* sucrose extract on an column (4 × 16 cm) of DEAE-cellulose as described by Pollard *et al*.[9] ●——●, A_{290}, ●---●, K(EDTA)-ATPase; ○---○, CaATPase. KCl gradient (——) is given in moles per liter on the scale on the left. Gel electrophoretic analyses of selected fractions are shown at the top, starting with the sample of extract applied to the column on the far left. The photograph of each gel is centered over the fraction analyzed. M marks the mobility of the *Acanthamoeba* myosin II heavy chain, and A marks the mobility of actin. From Pollard *et al*.[9]

static pressure, by pumping, or by N_2 pressure) to pack the bed firmly (see Peterson[21] for details). Once packed, the column is run at lower flow rates on the order of 1 column volume or less per hour. Choose a bed volume appropriate for the sample; for example, 1 ml of DE-52 per 20 mg of *Acanthamoeba* extract protein.[9] DEAE-Sephadex A50 has a capacity of about 5 mg of muscle

[21] E. A. Peterson, *in* "Laboratory Techniques in Biochemistry and Molecular Biology" (T. S. Work and E. Work, eds.), pp. 225–400. North-Holland Publ., Amsterdam, 1970.

myosin per milliliter in 0.15 M phosphate (pH 7.5)[11] or in 0.04 M sodium pyrophosphate (pH 7.5).[19] The column dimensions will depend on the sample size. I generally use a bed with a diameter about one-sixth to one-tenth the length (e.g., 5 cm in diameter, 30 cm long) to achieve high flow rates.

4. Equilibrate the column with 3 or more column volumes of the desired buffer (e.g., 0.04 mM sodium pyrophosphate, pH 7.5, as used by Richards et al.[19] or 1 mM ATP, 0.2 mM CaCl$_2$, 0.5 mM dithiothreitol, 20 mM imidazole (pH 7.5) as used by Gordon et al.[20] Equilibration can be carried out much faster on a Büchner funnel than in a chromatographic column. Check the pH and conductivity of the eluate to confirm that equilibration has been successful.

5. Apply the sample. If it is very large, it may cake up the top of the column and lead to fissuring and channeling. This will destroy the performance of the column. The problem can be avoided by dividing the DEAE-cellulose into two parts. About one-third to one-half of the total is poured into the column and packed as usual. The remainder is mixed with the sample for 15–30 min and then carefully layered and packed on top of the DEAE-cellulose in the column.

6. Wash the column with 1–3 column volumes of starting buffer.

7. Elute the bound myosin by raising the salt concentration of the buffer. Most often this is done with a gradient of KCl or NaCl, but steps may be successful in some cases. I generally use a concave gradient because more proteins elute at low than at high salt concentrations and they will be spread out more evenly among the fractions made by a concave gradient. Such a gradient can be made using three identical chambers: one (a) containing the final salt concentration in the starting buffer and two (b and c) containing the starting buffer. Flow is from (a) to (b) to (c) to the column. Chambers (b) and (c) are stirred. In every case I know, myosin elutes between 0 and 300 mM KCl, so these would be appropriate limits for the gradient.

8. Assay fractions by measuring conductivity, protein concentration, and ATPase activity.

9. The DEAE-cellulose can be re-used many times providing it is washed carefully and bacterial growth is inhibited during storage by azide or toluene.

Hydroxyapatite Chromatography

Hydroxyapatite (an insoluble crystalline form of calcium phosphate) is one of the most useful chromatographic media for myosin; myosin binds to hydroxyapatite even in high concentrations of KCl or NaCl. For exam-

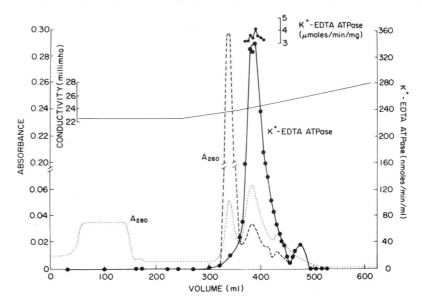

FIG. 3. Hydroxyapatite chromatography of partially purified *Acanthamoeba* myosin I.[2] A 100-ml sample containing about 4 mg of myosin I was applied to a 1.5 × 26 cm column equilibrated with 0.5 M KCl, 1 mM dithiothreitol, 1 mM phosphate, and 5 mM imidazole (pH 7) and eluted at a flow rate of 30 ml/hr with a 600 ml, concave, 1 to 300 mM potassium phosphate gradient in the column buffer. · · · · ·, A_{280}; ----, A_{260}; ●——●, K(EDTA)-ATPase activity. The specific activity of peak fractions is shown at the top. From Pollard and Korn.[22]

ple, *Acanthamoeba* myosin I is purified on hydroxyapatite in 0.5 M KCl.[22,23] Proteins binding to the column are eluted with phosphate buffer. The mechanism by which hydroxyapatite binds proteins is still somewhat mysterious.[24]

In my experience, homemade hydroxyapatite[24] is superior to commercial preparations, but Bio-Rad BioGel HT is satisfactory for many applications.

Procedure (see Fig. 3)

    1. Wash about 1 ml of hydroxyapatite per 2 mg of myosin in 5 volumes of starting buffer (e.g., 0.5 M KCl, 1 mM dithiothreitol, 10 mM imidazole; pH 7.5) by suspension and settling with gravity. Repeat 3 times, removing any fines from the supernatant each time. Handle the hydroxyapatite with great care. Do *not* use magnetic stirring

[22] T. D. Pollard and E. D. Korn, *J. Biol. Chem.* **248**, 4691 (1973).
[23] H. Maruta and E. D. Korn, *J. Biol. Chem.* **252**, 8329 (1977).
[24] G. Bernardi, this series, Vol. 22, p. 325.

bars, nor scrape between two hard objects. The crystals are easily broken and the resulting fines will greatly reduce the flow rate.

2. Pour a 1 : 1 slurry of hydroxyapatite and buffer into a column and equilibrate with 3 column volumes of starting buffer. To achieve an adequate flow rate use a relatively short (e.g., 1.5 × 10 cm) column and 1–2 m of hydrostatic pressure or a peristaltic pump.

3. Apply the sample and wash with 1–2 column volumes of starting buffer.

4. Elute the bound proteins with a phosphate gradient. Potassium phosphate is more soluble than sodium phosphate, so it has advantages in the cold room. As in the case of DEAE, concave gradients are probably the best. In every case examined, myosin elutes between 0 and 200 mM phosphate.

5. Assay fractions for conductivity, protein concentration, and ATPase activity (using [γ-^{32}P]ATP). Simply count an aliquot of the organic phase in the ATPase assay (see this volume [13]). The high concentration of phosphate in the fractions will inhibit the extraction of P_i in the ATPase assay, but the position of the peak can be determined before dialysis or precipitation of myosin to remove the phosphate.

Agarose Adsorption Chromatography

This is not a common chromatographic procedure, but Pollard and Korn[2] found that *Acanthamoeba* myosin I binds to Bio-Rad A-1.5m agarose beads. Since few of the contaminating proteins bound, this proved to be the most important step in the purification procedure. Later, another batch of A-1.5m was found to separate two myosin I isozymes.[25] In experiments with another batch of A-1.5m, I have also been able to separate these two isozymes, but both elute at much higher concentrations of KCl than in the earlier work (Fig. 4). The mechanism is not known, but binding is probably due to traces of carboxyl groups and/or sulfate groups on the agarose. Different batches of A-1.5m can differ considerably in their affinity for myosin I. The manufacturer attributes this to variability in the crude agar received from different suppliers. Consequently, it is impossible to specify exact conditions for chromatography. The original method employed a buffer consisting of 0.2 mM ATP, 1 mM dithiothreitol, 2 mM Tris (pH 7.6). The myosin I bound to the column and was eluted with a gradient of KCl.[2] A later modification by Maruta *et al.*[25] employs a column equilibrated with 42 mM KCl, 1 mM dithiothreitol, 10 mM imidazole (pH 7.5), and the sample is dissolved in

[25] H. Maruta, H. Gadasi, J. H. Collins, and E. D. Korn, *J. Biol. Chem.* **254**, 3624 (1979).

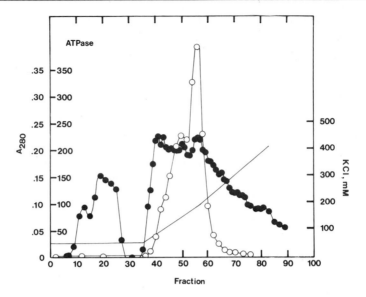

FIG. 4. Agarose adsorption chromatography of partially purified *Acanthamoeba* myosin I by the method of Maruta *et al.*[25] The sample, obtained by DEAE-cellulose chromatography, ammonium sulfate precipitation, and ADP-agarose chromatography, was applied to a 2.5 × 27 cm column of Bio-Rad A-15m (200–400 mesh) equilibrated with 50 mM KCl, 1 mM dithiothreitol, 15 mM imidazole (pH 7.5). The myosin was eluted with a 400 ml, 50 to 500 mM KCl gradient (——). ●——●, A_{280}; ○——○, K(EDTA)-ATPase activity (nmol/min ml^{-1}). The fraction volume was 6 ml (Pollard, unpublished data)

the same buffer with 0.34 M sucrose. The bound proteins are eluted with a gradient of KCl (Fig. 4).

Affinity Chromatography

Affinity chromatography of myosins is still in its infancy, but already a number of ligands immobilized on agarose beads have been shown to bind myosin and myosin fragments (Table II). These affinity columns are useful for (*a*) separation of myosin from other cellular proteins; (*b*) separation of active from inactive myosin; (*c*) separation of myosin isozymes; and (*d*) concentration of dilute myosin solutions. As more experience is gained, I expect these techniques to gain in importance for the purification of myosins.

The immobilized ligands used for myosin include Cibacron Blue, adenine nucleotides, pyrophosphate, actin, and antibodies (Table II). Of these ligands, actin and antibodies are the most specific for myosin, but they have been little used for preparative purposes. In contrast, although

TABLE II

Affinity Chromatography Materials for Myosin

Material	Reference[a]
1. Cibacron Blue	
Dextran Blue–Sepharose	Kobayashi et al.[27]
Cibacron Blue–Sepharose	Toste and Cooke[26]
2. Nucleotides	
Sepharose–adipic acid hydrazide–ATP	Lamed et al.[29]
Sepharose–sebacic acid hydrazide–ATP	Lamed and Oplatka[30]
Sepharose–adipic acid hydrazide–ADP	Wagner[33]
Sepharose–N^6-(6-aminohexyl)–ADP	Trayer et al.[28]
Sepharose–8-(6-aminohexyl)–ADP	Trayer et al.[28]
3. Sepharose–aminohexylpyrophosphate	Trayer et al.[28]
4. Actin	
Sepharose–G-actin	Bottomley and Trayer[34]
Sepharose–F-actin	Bottomley and Trayer[34]
Sepharose–crosslinked F-actin	Winstanley et al.[35]
Sepharose–crosslinked F-actin–tropomyosin	Winstanley et al.[35]
Sepharose–phalloidin stabilized F-actin	Winstanley et al.[32]
Phalloidin F-actin trapped in Sepharose	Grandmont-Leblanc and Gruda[37]
5. Antibodies	
Sepharose–anti-Acanthamoeba myosin I	Pollard, unpublished
Sepharose–anti-Acanthamoeba myosin II	Pollard, unpublished

many enzymes bind to adenine nucleotides, pyrophosphate, and Cibacron Blue, they each have been used successfully for myosin purification. There are many different immobilized myosin ligands available, so I will select one procedure from each class to illustrate the principles of these methods and, where appropriate, will make brief comments about the relative merits of the alternatives.

In the development of these affinity columns several general principles were important. It is necessary to keep these same principles in mind when these columns are used for new applications, because it is by no means certain that the conditions used for the binding and elution of one myosin will be optimal (or even successful) for another myosin. The principles include the following.

First, the ligand must be attached to the insoluble matrix (usually agarose beads) in a way that myosin can bind to the ligand. This may require a hydrocarbon spacer between the agarose and the ligand. Small differences in the length of this spacer may determine whether the myosin will bind.

Second, the conditions for myosin binding and elution may not conform to one's preconceived notions about myosin–ligand interaction.

Therefore, it may be useful to experiment with unusual conditions. The following variables should be tested: the ionic strength; the concentration and type of anion; the concentration and type of divalent cation; the concentration and type of nucleotide (or pyrophosphate); and the temperature. Every one of these variables has been used in the selective binding and elution of myosin in affinity chromatography.

CIBACRON BLUE

The blue dye Cibacron Blue F3GA is a structural analog of the adenine nucleotides and binds to a number of enzymes with binding sites for NADH, ADP, or ATP. The dye binds with high affinity ($K_i \sim 0.2 \ \mu M$) and inhibits myosin ATPase activity.[26] This dye can be attached directly to agarose beads or first conjugated to dextran, which is then attached to cyanogen bromide-activated agarose beads.[27] Both types of immobilized Cibacron Blue have been used successfully for affinity chromatography of myosin. Blue Sepharose can be purchased from Pierce Biochemicals, Bio-Rad, or Pharmacia Fine Chemicals.

Blue Sepharose binds muscle myosin subfragment-1 and heavy meromyosin at low ionic strength [0.5 mM dithiothreitol, 3.2 mM sodium azide, 20 mM N-tris[hydroxymethyl]methyl-2-aminoethanesulfonic acid (TES) (pH 7)] with a capacity of about 3.5 mg/ml.[26] No conditions were found where muscle myosin was soluble and would bind quantitatively to Blue Sepharose. Subfragment-1 and heavy meromyosin are eluted by KCl in the 10–300 mM range by 20 mM TES (pH 9) and by Cibacron Blue at 1 mM, but are not eluted by 1 mM NADH, MgATP, CaATP, Mg pyrophosphate, or EDTA. Actin does not bind, so application of a mixture of subfragment-1 and actin to Blue Sepharose in 1 mM MgATP results in a clean separation of the two proteins. Blue Sepharose also can be used to fractionate isolated muscle myosin light chains. The A-1 light chain does not bind; the A-2 light chain binds very weakly; and the DTNB light chain binds tightly.

Given these properties, Blue Sepharose would appear to be ideal for purification of some cytoplasmic myosins that are more soluble at low ionic strength than muscle myosin. The ability to separate actin and myosin cleanly would appear to be particularly valuable. A limitation will be that other proteins (dehydrogenases and kinases) also bind to Cibacron Blue, but selective elution of some of these enzymes can be achieved with their nucleotide substrates.[26]

The partial purification of thyroid myosin on Blue Dextran–Sepharose by Kobayashi et al.[27] is a good example of how these columns can be used preparatively.

[26] A. P. Toste and R. Cooke, *Anal. Biochem.* **95,** 317 (1979).
[27] R. Kobayashi, R. Goldman, D. Hartshorne, and J. Field, *J. Biol. Chem.* **252,** 8285 (1977).

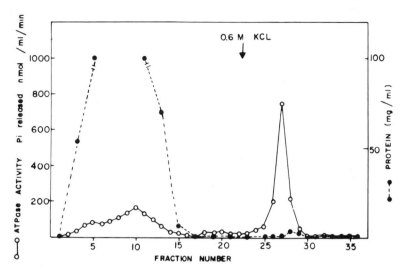

FIG. 5. Blue Dextran-Sepharose affinity chromatography of 20 ml of a crude extract of thyroid gland on a 1 × 12 cm column by the method of Kobayashi *et al.*[27] The bound myosin was eluted with 0.6 M KCl. ●---●, Protein concentration (mg/ml); ○——○, K(EDTA)-ATPase activity. The fraction volume was 3 ml. From Kobayashi *et al.*[27]

Procedure (see Fig. 5)

1. A crude extract is prepared by homogenizing the tissue in 3 volumes of ice cold 0.3 M ammonium acetate, 2 mM CaCl$_2$, 10 mM NaHCO$_3$ (pH 7.2) and centrifuging at 22,000 g for 1 hr.
2. Extract supernatant, 10–20 ml, is applied to a 0.9 × 12 cm column of Blue Dextran-Sepharose equilibrated with the 0.3 M ammonium acetate buffer.
3. After washing the column with 30 ml of buffer, the myosin is eluted with 0.6 M KCl. In the case of thyroid myosin, the specific ATPase activity increases about 50-fold by this procedure. The recovery is about 60%.
4. Final purification is achieved by gel filtration with the discontinuous buffer system of Pollard *et al.*[4]

NUCLEOTIDES

Both immobilized ATP and ADP have been used successfully for myosin affinity chromatography. The most thoroughly tested immobilized nucleotide for myosin chromatography is ADP linked by nitrogen-6 of the adenine via a six-carbon spacer to Sepharose 4B.[28] Another successful

[28] I. P. Trayer, H. R. Trayer, D. A. Small, and R. C. Bottomley, *Biochem. J.* **139**, 609 (1974).

immobilized nucleotide is ADP linked by the ribose via a six-carbon chain to Sepharose by the method of Lamed *et al.*[29] Both are commercially available from PL Biochemicals and have been used for chromatography of muscle myosin and its proteolytic fragments[28,30] and of cytoplasmic myosins from platelets[31] and *Acanthamoeba.*[25]

Procedure[31]

1. A small column (0.8 × 8 cm) of Sepharose-N^6-ADP is equilibrated at 4° C with 0.6 M ammonium acetate, 5 mM EDTA, 2 mM sodium pyrophosphate, 0.25 mM dithiothreitol (pH 7.5).
2. Myosin is extracted from 1 g of minced muscle for 30 min at 4° with 5 volumes of 0.6 M ammonium acetate, 2 mM MgCl$_2$, 2 mM sodium pyrophosphate, 0.5 mM dithiothreitol (pH 6.5) and insoluble material removed by centrifugation at 2000 g for 15 min. After dilution with 5 volumes of column equilibration buffer, the crude extract is applied to the column at a flow rate of 15–20 ml/hr.
3. The column is washed with equilibration buffer until A_{280} reaches zero. The unbound fraction contains virtually all the proteins in the extract except myosin.
4. The bound myosin is eluted by either replacing the ammonium acetate with ammonium chloride or by adding 50 mM ATP to the equilibration buffer. The eluted material is essentially pure myosin and has a higher specific activity than myosin prepared by conventional means. The yield is 10–20 mg/g of tissue.
5. The column is washed with 6 M urea, 2 M KCl. I store my columns of ADP-agarose for months at −20° after first equilibrating them with 50% glycerol and 0.05% sodium azide.

Note that the use of EDTA in the buffer has the theoretical advantage that most other ADP binding proteins are thought to require divalent cations; however, Trayer and Trayer[31] found similar results with muscle extracts with 5 mM EDTA or 5 mM MgCl$_2$.

The same procedure can be used for purification of platelet myosin with the following modifications.[31] Extract frozen–thawed platelets with 10 volumes of extracting solution, centrifuge at 80,000 g for 15 min, and dilute the supernatant with 1 volume of water and 3 volumes of column equilibration buffer containing 0.3 M ammonium acetate. The bound fraction is highly enriched in myosin but contaminated with actin. The actin can be removed by gel filtration with the two-buffer system described above.

[29] R. Lamed, Y. Levin, and A. Oplatka, *Biochim. Biophys. Acta* **305**, 163 (1973).
[30] R. Lamed and A. Oplatka, *Biochemistry* **13**, 3137 (1974).
[31] H. R. Trayer and I. P. Trayer, *FEBS Lett.* **54**, 291 (1975).

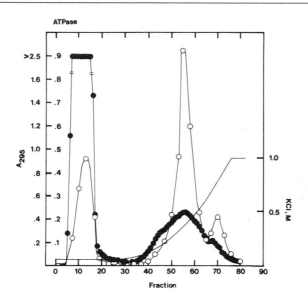

FIG. 6. Sepharose–adipic acid hydrazide–ADP affinity chromatography of partially purified *Acanthamoeba* myosin I by the method of Maruta *et al.*[25] The sample, obtained by DEAE-cellulose chromatography and ammonium sulfate precipitation, was applied to a 1.5 × 15 cm column of Sepharose-ADP (PL Biochemicals No. 5568) equilibrated with 50 mM KCl, 1 mM dithiothreitol, 15 mM imidazole (pH 7.5). Myosin was eluted with a 240 ml, concave, 50 to 1000 mM KCl gradient (——). ●——●, A_{295}; ○——○, K(EDTA)-ATPase activity (mol/min ml⁻¹). The fraction volume was 3 ml. Pollard, unpublished data.

The Sepharose-N^6-ADP column can also be used to separate myosin subfragment-1 isozymes.[32] When the crude subfragment-1 is applied to the column in 5 mM triethanolamine (pH 7.5), denatured material fails to bind and the isozymes with the A-1 and A-2 light chains can be eluted separately with a gradient of 0 to 10 mM ADP.

Two other forms of immobilized ADP have been prepared. One is linked from adenine C-8 to Sepharose via a six-carbon hexane chain,[28] and the other from the ribose to Sepharose by a six-carbon adipic acid dihydrazide.[33] Using the latter column, Wagner was able to separate the muscle myosin subfragment-1 isozymes with a pyrophosphate gradient. Another application of the ribose-linked ADP-Sepharose is in the purification of *Acanthamoeba* myosin I.[25] This myosin is extremely difficult to purify, and the KCl elution from immobilized ADP provides an essential 4- to 10-fold purification at an intermediate stage of purity (Fig. 6).

[32] M. A. Winstanley, D. A. Small, and I. P. Trayer, *Eur. J. Biochem.* **98,** 441 (1979).
[33] P. D. Wagner, *FEBS Lett.* **81,** 81 (1977).

Immobilized ATP was actually the first nucleotide used for affinity chromatography of myosin.[29,30] It was found that muscle myosin, heavy meromyosin, and subfragment-1 all bound to ATP attached from the ribose to Sepharose with a ten-carbon spacer, but that only heavy meromyosin and subfragment-1 bound to ATP attached to Sepharose with a six-carbon spacer. Binding was enhanced by divalent cations, and in their presence the gamma phosphate of ATP was hydrolyzed. Consequently, the ATP columns could only be used one or two times before their properties changed. The bound myosin or its fragments could be eluted with gradients of salt or ATP. The columns successfully separated active from denatured myosin and subfragment-1 from heavy meromyosin. Myosin and single-headed myosin were partially separated by elution with a KCl gradient.

Sepharose–Aminohexylpyrophosphate

Pyrophosphate can be attached to Sepharose at the end of a six-carbon spacer.[28] The resulting material is excellent for affinity chromatography of muscle myosin and its proteolytic fragments (Fig. 7). To my knowledge, it has not been used for cytoplasmic myosin purification, but it is used by Trayer's laboratory (in deference to Sepharose-ADP or -actin) for the

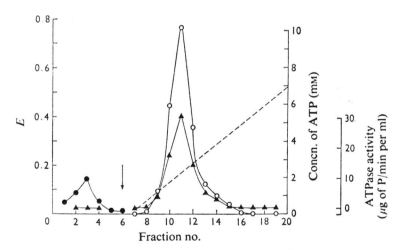

Fig. 7. Sepharose-6-aminohexylpyrophosphate affinity chromatography of crude heavy meromyosin according to Trayer et al.[28] About 4 ml of protein were applied to an 0.8×10 cm column equilibrated with 2 mM MgCl$_2$, 2 mM 2-mercaptoethanol, 5 mM Tris (pH 7.6) and eluted at 20 ml/hr with a linear gradient of 0 to 10 mM ATP (---). Protein concentration by A_{280} (●——●) or turbidity at 360 nm after precipitation with 4% trichloroacetic acid (○——○); CaATPase activity (▲——▲). The fraction volume was 2 ml. From Trayer et al.[28]

single-step purification of subfragment-1 from myosin digests. Owing to its stability and modest expense, it would seem to be an ideal addition to, or substitute for, the nucleotide affinity columns for cytoplasmic myosin purification. It is not commercially available, but the key reactant, aminohexylpyrophosphate can be synthesized relatively easily from 6-aminohexanol and pyrophosphoric acid and then conjugated with cyanogen bromide-activated agarose beads.

ACTIN–AGAROSE

Two strikingly different approaches have been taken for preparing actin–agarose for affinity chromatography. On one hand, actin monomers or polymers can be coupled directly to cyanogen bromide-activated agarose beads. On the other, actin monomers can be equilibrated with agarose beads and then polymerized inside the beads. Providing that the actin filament stabilizing alkaloid phalloidin is present during polymerization, the filaments become permanently trapped inside the pores of the beads, even during prolonged washes with actin-free buffers. Both types of actin–agarose beads can be used for myosin affinity chromatography.

Actin is chemically coupled to agarose beads according to Bottomley and Trayer[34] as follows: Sepharose 4B is activated with 100 mg of cyanogen bromide per gram of wet gel and mixed with 2–4 mg of actin per milliliter in an appropriate buffer. For actin monomers, 5 mM triethanolamine (pH 8.5) is used. For actin filaments, 50 mM KCl and 2.5 mM MgCl$_2$ are included. After reacting overnight at 4°, the beads are washed in a Büchner funnel with 2 M KCl and then with the coupling buffer. The coupling efficiency is about 70%, and 1 to 2 mg of actin are bound per milliliter of beads. Actin filaments have been stabilized by glutaraldehyde crosslinking[35] or by phalloidin,[32] and glutaraldehyde crosslinked actin-tropomyosin filaments[35] have also been coupled to cyanogen bromide-activated agarose beads. Except for the actin filament–agarose conjugate, these various forms of immobilized actin are stable and useful for affinity chromatography.

Procedure (see Fig. 8)

1. Equilibrate a 1.2 × 10 cm column of actin monomer–Sepharose with 5 mM triethanolamine, 2 mM MgCl$_2$ (pH 7.5) at 20° and a flow rate of 20 ml/hr.
2. Apply 10 ml of 1.5 mg/ml of myosin digested with Sepharose–papain and then dialyzed against column buffer.
3. Wash with column buffer until the absorbance of the eluate is zero.

[34] R. C. Bottomley and I. P. Trayer, *Biochem. J.* **149**, 365 (1975).
[35] M. A. Winstanley, H. R. Trayer, and I. P. Trayer, *FEBS Lett.* **77**, 239 (1977).

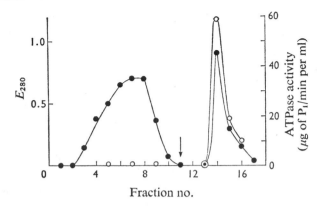

FIG. 8. Sepharose–G-actin affinity chromatography of crude muscle myosin subfragment-1 according to Bottomley and Trayer.[34] About 10 mg of protein in 2 mM MgCl$_2$, 5 mM triethanolamine (pH 7.5) was applied to a 1.2 × 10 cm column equilibrated with the same buffer at 20° with a flow rate of 20 ml/hr. The bound subfragment-1 was eluted with 0.25 M KCl, starting at the arrow. ●——●, A_{280}; ○——○, CaATPase activity. The fraction volume was 2 ml. From Bottomley and Trayer.[34]

4. Elute the bound subfragment-1 with 0.25 M KCl in column buffer. Alternatively, subfragment-1 can be eluted with 2 mM magnesium pyrophosphate, 3 mM MgATP, or 4 mM MgADP. With the nucleotides and pyrophosphate, the elution is "more efficient" if the column is cooled to 4°, and the eluted peaks are sharper when MgCl$_2$ is omitted from the buffer.

5. The column is washed with 1 M KCl and stored in the column buffer. I would include 0.04% sodium azide to retard bacterial growth. The actin bound to the column is stable through repeated use over a period of months.

These actin monomer–Sepharose columns also bind heavy meromyosin, but improved resolution of heavy meromyosin isozymes is possible on crosslinked actin filament–tropomyosin Sepharose. Using a KCl gradient, HMM-A-1 and HMM-A-2 homodimers can be separated from heterodimers.[36] After digestion of muscle myosin with chymotrypsin, subfragment-1 isozymes with either A-1 or A-2 light chains can be separated very cleanly on the actin–tropomyosin columns using the 5 mM triethanolamine, 2 mM MgCl$_2$ buffer described above. Only subfragment-1 with the A-1 light chain binds to the column.[34]

It has not been possible to find a concentration of KCl or ammonium acetate where muscle myosin is soluble and will bind to immobilized actin

[36] H. R. Trayer, M. A. Winstanley, and I. P. Trayer, *FEBS Lett.* **83**, 141 (1977).

columns.[34] For cytoplasmic myosins, it would be worth trying sucrose buffers to solubilize the protein at a low ionic strength, where it might bind to an immobilized actin column.

Actin filaments stabilized with phalloidin can be trapped in agarose beads for use in affinity chromatography by the method of Grantmont-Leblanc and Gruda.[37]

Procedure

1. Wash Sepharose 4B with 0.2 mM ATP, 0.5 mM 2-mercaptoethanol, 0.2 mM $CaCl_2$, 2 mM Tris (pH 8).
2. Mix 3–4 mg of actin monomer in the above buffer with each milliliter of packed Sepharose. (Although not specified in the original article, I would use a solution of actin monomer with a high concentration, 8–10 mg/ml.)
3. Add KCl, $MgCl_2$, and phalloidin to bring the concentrations to 0.1 M KCl, 1 mM $MgCl_2$, and 0.100 mM phalloidin; allow the actin to polymerize for 90 min at room temperature.
4. Wash the Sepharose with enough 0.1 M KCl, 1 mM $MgCl_2$, 0.5 mM 2-mercaptoethanol, 20 mM Tris (pH 8) to bring the A_{280} of the filtrate to zero. The concentration of trapped actin is determined by Pronase digestion and protein determination. The content of trapped actin is 0.1–0.3 mg per milliliter of packed gel and is capable of binding 1.3 to 2.2 mg of muscle myosin per milligram of actin. The myosin binding capacity declines by about 50% in 7 weeks.

The trapped actin filament columns can be used for affinity chromatography of muscle myosin, heavy meromyosin, and subfragment-1. The method for myosin from a crude extract follows.

Procedure for myosin from crude extract

1. Equilibrate a 20-ml trapped-actin column with 0.5 M KCl, 1 mM $MgCl_2$, 0.5 mM 2-mercaptoethanol, 20 mM Tris (pH 7.6) at 4°.
2. Apply a sample consisting of about 20 mg of protein extracted from muscle with a buffer similar to the column buffer.
3. Wash with column buffer until A_{280} is <0.05. Essentially all the extract proteins except myosin fail to bind to the column.
4. Elute the bound myosin with 1 mM sodium pyrophosphate in column buffer. The resulting myosin is very clean and has a high ATPase activity.

A similar procedure can be used for the purification of crude muscle myosin subfragment-1. This remarkable affinity column has an advantage

---

[37] A. Grandmont-Leblanc and J. Gruda, *Can. J. Biochem.* **55**, 949 (1977).

over the other types of immobilized actin. It alone seems to be capable of binding muscle myosin in a high concentration of KCl. Consequently, it should be quite useful for purification of cytoplasmic myosins as well. A disadvantage is its low capacity.

Antibody Columns

In most cases conventional methods can be used to purify myosin, but there are difficult cases, such as *Acanthamoeba* myosin I, where immunological techniques may be useful. For example, antibodies specific for either *Acanthamoeba* myosin I or myosin II can be coupled to agarose beads and used for affinity chromatography (Pollard, unpublished work).

Procedure

1. Homogenize *Acanthamoeba* in 4 volumes of ice cold $0.2 M$ NaCl, 20 mM sodium pyrophosphate (pH 7), 1 mM ATP, 10 μg of benzamidine per milliliter (a protease inhibitor); clarify at 120,000 g for 60 min at 4°.
2. Mix 4 ml of antibody–Sepharose equilibrated with extracting buffer with 2 ml of the extract supernatant overnight at 4°.
3. Pour into a column 1.5 cm in diameter and wash with 10 ml of extracting buffer at about 1 ml/min.
4. Elute with 8 ml of 1 M NaCl.
5. Elute with 8 ml of 2.5 M MgCl$_2$.
6. Elute with 8 ml of 4 M MgCl$_2$.
7. Elute with 8 ml of 7 M urea.
8. Wash the column with extracting buffer.

In the case of the anti-myosin II column, the myosin II eluted in both the 2.5 and 4 M MgCl$_2$ fractions, and none was found in the final urea wash (Fig. 9). With the anti-myosin I column, a polypeptide with the electrophoretic mobility of myosin IB eluted with 2.5 M MgCl$_2$, and another with the mobility of myosin IA eluted with 4 M MgCl$_2$ (Fig. 9). Thus, it would appear that >50-fold purification of the three major myosins from *Acanthamoeba* is now possible with a simple, single-step procedure. The major limitation is the small capacity of these columns, but this will be overcome by using monoclonal antibodies.

Immobilized antibodies to myosin light chains have also been used by S. Lowey and her colleagues (personal communication) for the purification of a muscle myosin subfragment-1 and heavy meromyosin isozymes.

Concentration

There are a variety of mild methods to concentrate myosin solutions. The two that I favor are precipitation with ammonium sulfate and pelleting

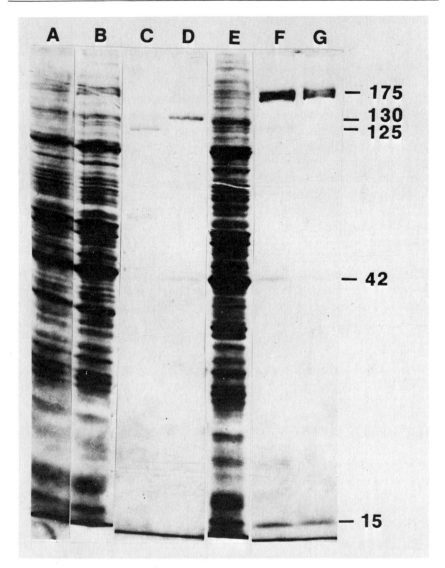

FIG. 9. Anti-myosin affinity chromatography of *Acanthamoeba* myosin I and myosin II as described in the text. The polypeptide composition of various samples is shown by polyacrylamide slab gel electrophoresis (see this volume [13]). Lanes: A, Crude extract; B, polypeptides that did not bind to an anti-myosin I column; C, polypeptides eluted from the anti-myosin I column with 2.5 M MgCl$_2$; D, polypeptides eluted from the anti-myosin I column with 2.5 M MgCl$_2$; D, polypeptides eluted from the anti-myosin I column with 4 M MgCl$_2$; E, polypeptides that did not bind to an anti-myosin II column; F, polypeptides eluted from the anti-myosin II column with 2.5 M MgCl$_2$; G, polypeptides eluted from the anti-myosin II column with 4 M MgCl$_2$. Molecular weights of the major polypeptides are indicated on the right in thousands.

of myosin filaments, because they are mild and the recovery is excellent. Dialysis against 1.5 volumes of saturated ammonium sulfate with EDTA will remove some water from the sample and bring the final concentration of ammonium sulfate to 2.3 M (60% saturation), a concentration sufficient to precipitate all myosins that I know. The precipitate is collected by centrifugation and resuspended gently in an appropriate buffer. Filaments can be formed from most myosins by dialysis against 50 mM KCl, 2 mM MgCl$_2$, 10 mM imidazole (pH 6.5) and then collected by centrifugation. Since some cytoplasmic myosin filaments are small, ultracentrifugation may be necessary.

Another mild, efficient method of concentration is binding to one of the affinity columns described in the preceding section, followed by step elution. This procedure has the additional feature that any denatured myosin will not bind to the column and will be separated from the concentrated, active myosin.

Myosin can also be concentrated by various membrane filtration methods. I have had better success with osmotic pressure than hydrostatic pressure methods. For example, excellent recovery is possible with "dialysis" against dry Sephadex G-150 or polyethylene glycol (Aquacide from Calbiochem). Dry sucrose[17] or 60% sucrose solution[38] can also be recommended for osmotic pressure dialysis. In contrast, I have had more difficulty recovering myosin concentrated by vacuum dialysis in cellulose dialysis sacks or by pressure ultrafiltration on several different Amicon filters. It is possible that the new generation of Amicon filters or similar products from other manufacturers will give better results.

Finally, it is possible to lyophilize muscle myosin subfragment-1 and heavy meromyosin after freezing with an equal quantity of sucrose.[39] Upon rehydration in a smaller volume of water, the protein will be concentrated. I am not aware that any intact myosin has been successfully lyophilized and rehydrated, but my experience in this matter is limited.

Storage

Each myosin has a different stability, but many (including muscle myosin, platelet myosin, and *Acanthamoeba* myosin II) can be stored at concentrations of >1 mg/ml in 0.6 M KCl, 1 mM dithiothreitol, 10 mM imidazole (pH 7) for 5–10 days at 0° with only slight loss of activity. The inclusion of 0.02% sodium azide in the buffer will prevent bacterial growth and does not, as far as I know, alter the myosin.

For longer-term storage, any of the following methods can be used.

[38] J. H. Collins and E. D. Korn, *J. Biol. Chem.* **255**, 8011 (1980).
[39] R. G. Yount and D. E. Koshland, *J. Biol. Chem.* **238**, 1708 (1963).

1. Precipitate with 60% ammonium sulfate, 10 mM EDTA (see II, Fractional Precipitation Methods above) and store the pellets under the ammonium sulfate supernatant at 4° for periods of months.
2. Mix the myosin solution with an equal volume of glycerol and store the solution at −20° for months.[8]
3. Dialyze the myosin against 60% sucrose and store at −20°.

I would not recommend such long-term storage for myosin to be used for the most critical enzymic analysis, but for routine applications, these methods preserve the activity of myosins remarkably well.

If it is possible to lyophilize a particular myosin, this would be the ideal way to preserve the protein.

Human Platelet Myosin

This procedure with modifications given below has been used to purify myosin from many nonmuscle cells.

Procedure[4]

1. Mix washed frozen platelets with 2 ml of extracting solution (0.9 M KCl, 15 mM pyrophosphate (pH 7), 30 mM imidazole (pH 7), 5 mM MgCl$_2$, 3 mM dithiothreitol) per gram of platelets and stir on ice for 30 min. The platelets will lyse during thawing.
2. Centrifuge in a Beckman J-20 or Sorvall SS-35 rotor at 16,000 rpm for 30 min at 4°. Save the supernatant.
3. The pellet can be reextracted with 2 volumes of extracting solution (diluted with 0.5 part water) to increase the yield.
4. Take the supernatant(s), measure the volume and carefully add 3 volumes of ice cold 2 mM MgCl$_2$ with stirring. Then adjust the pH to 6.4 with 0.5 M acetic acid being careful not to overshoot (the pH meter is sluggish in the cold). Stir for 15 min.
5. Pellet the actomyosin precipitate by centrifuging in the J-20 rotor at 4° for 30 min at 16,000 rpm.
6. Discard the supernatant and save the pellet, which is *crude actomyosin*.
7. Gently dissolve the crude actomyosin in a small volume of KI-ATP buffer [0.6 M KI, 5 mM ATP, 5 mM dithiothreitol, 1 mM MgCl$_2$, 20 mM imidazole (pH 7)] using a Dounce homogenizer.
8. Centrifuge down undissolved material at 16,000 rpm for 10 min at 4°.
9. Ammonium sulfate fractionation. Use cold, saturated ammonium sulfate with 0.01 M EDTA (pH 7). Add ammonium sulfate slowly with stirring. After 10 min spin down the precipitate in a J-20 rotor

at 16,000 rpm for 15 min at 4°. Then measure the volume of the supernatant and use it for the next step. Use the following amounts of ammonium sulfate per liter: 0.0 to 1.0 M ammonium sulfate, 360 ml; 1.0 to 1.9 M ammonium sulfate, 478 ml (enriched in myosin); 1.9 to 2.5 M ammonium sulfate, 466 ml (enriched in myosin rod).

10. Dissolve precipitates in a small volume (12 ml for a 450-ml agarose column) of cold KI-ATP buffer.

11. Clarify by centrifugation at 4° for 30 min in a J-20 rotor at 16,000 rpm or in an ultracentrifuge at 100,000 g.

12. Apply either the myosin-enriched or myosin rod-enriched fraction to a 4% agarose column behind a zone of KI-ATP buffer as described above. The myosin will elute from the column with a partition coefficient of about 0.2–0.3.

This method, with modifications, can be used to purify myosin from smooth muscle and many nonmuscle cells. The following are examples of modifications that improve results with specific tissues.

1. For squid brain, extract with 0.6 M KCl rather than 0.9 M KCl to reduce lipid extraction.[40]

2. For macrophages, extract with a 0.34 M sucrose buffer rather than KCl to avoid rupture of lysosomes.[3]

3. For *Dictyostelium*, extract with 30% sucrose buffer and precipitate actomyosin by dialyzing out the sucrose at low ionic strength.[41]

4. For thyroid, extract with 0.3 M ammonium acetate buffer and substitute Blue Dextran–Sepharose chromatography for actomyosin precipitation to remove thyroglobulin.[27]

5. For *Acanthamoeba* myosin II extract with a 0.34 M sucrose·buffer, fractionate on a DEAE-cellulose column (as described in the section on anion exchange chromatography), and precipitate actomyosin II by hexokinase digestion of the ATP in the pooled actin and myosin fractions.[9]

Other modifications will, no doubt, prove to be useful for other cytoplasmic myosins.

Acknowledgments

I am grateful to Dr. D. Frederickson who modified this chapter from an earlier version in "Methods and Perspectives in Cell Biology" (L. Wilson, ed.), Vol. 24. Academic Press, New York, 1982. The author's work is supported by NIH Research Grants GM-26132 and GM-26338.

[40] Y. P. See and J. Metusals, *J. Biol. Chem.* **251**, 7682 (1976).
[41] M. Clarke, and J. A. Spudich, *J. Mol. Biol.* **86**, 209 (1974).

[32] Myosins from *Acanthamoeba castellanii*

By EDWARD D. KORN, JIMMY H. COLLINS, and HIROSHI MARUTA

A minimal definition of a myosin would be an ATPase that, under physiological conditions, is enzymically active only as a complex (actomyosin) with F-actin. By this criterion, *Acanthamoeba castellanii* contains three myosin isoenzymes referred to as myosins IA, IB, and II.

Assay Methods

General Considerations

Most myosins express their ATPase activities under three different ionic conditions: (*a*) K^+ and EDTA (to remove inhibitory divalent cations); (*b*) Ca^{2+} (Mg^{2+} is inhibitory); and (*c*) Mg^{2+} in the presence of F-actin. Although the actin-activated Mg^{2+}-ATPase activity is the physiologically relevant one, it is usually more convenient to use one of the other two assays during a purification procedure. For myosins IA and IB, measuring the K^+(EDTA)-ATPase activity is the method of choice because it is less interfered with by other proteins and ions and because the myosins express maximal activities in this assay. For myosin II, the Ca^{2+}-ATPase assay is preferred because it is much more active under these conditions than when assayed as a K^+(EDTA)-ATPase. In all the assays, the release of P_i from ATP is measured, most conveniently by radioactivity measurements using $[\gamma\text{-}^{32}P]ATP$ as substrate, but the P_i can also be measured colorimetrically.

$K^+(EDTA)$-ATPase Activity

Reagents

KCl, 2 M
Imidazole, 1 M, pH 7.5
EDTA, 0.1 M
ATP, 0.1 M, pH 7.5 (store frozen)
$[\gamma\text{-}^{32}P]ATP$, 2 m$M$, 0.2 mCi/ml (store frozen)
2-Butanol–benzene, 1 : 1
Silicotungstic acid, 4% in 3 N H_2SO_4

METHODS IN ENZYMOLOGY, VOL. 85   ISBN 0-12-181985-X

Ammonium molybdate, 10% in water

Sulfuric acid, 0.73 N in ethanol (for colorimetric procedure only)

Stannous chloride: 1 ml of 10% stannous chloride in 12 N HCl mixed with 25 ml of 1 N H_2SO_4 (for colorimetric procedure only)

Procedure. The assay mixture is prepared in batches of 100 ml by mixing 1 ml of 0.1 M ATP, 25 ml of 2 M KCl, 2 ml of 0.1 M EDTA, and 1 ml of 1 M imidazole with 71 ml of water and readjusting the pH to 7.5 at room temperature. This solution, which is 1 mM ATP, 0.5 M KCl, 2 mM EDTA, and 10 mM imidazole, pH 7.5, is stored frozen in batches of 10 ml and thawed when needed. For the radioactive assay, 0.1–0.2 ml of 2 mM [γ-^{32}P]ATP is also added. Aliquots of 0.5 ml are transferred to small test tubes, one for each assay, to which about 5–25 μl of the myosin preparation are added. Samples are incubated at 25° for 15–60 min. The reaction is stopped by adding 1 ml of the butanol–benzene mixture and 0.25 ml of the silicotungstate solution directly to the assay tube and vortexing vigorously for several seconds. Then 0.3 ml of the ammonium molybdate solution is added, and the vortexing is continued for 15 sec. After phase separation has occurred, 0.5 ml of the upper, organic phase, which contains the phosphomolybdate complex, is added to an appropriate scintillation cocktail, and the radioactivity is measured. For the colorimetric assay, the radioactive ATP is omitted from the assay solution and 0.5 ml of the organic phase containing the phosphomolybdate is transferred to a tube containing 1 ml of 0.73 N H_2SO_4 followed by 0.05 ml of the solution of stannous chloride. The blue color is quantified by its absorbance at 720 nm. The radioactive assay is much easier for routine use.

Ca $^{2+}$-ATPase Activity

Reagents

Imidazole, 1 M, pH 7.5

KCl, 2 M

$CaCl_2$, 1 M

ATP, 0.1 M, pH 7.5 (store frozen)

[γ-^{32}P]ATP, 2 mM, 0.2 mCi/ml (store frozen)

Procedure. The assay mixture is prepared in batches of 100 ml by mixing 2 ml of 1 M imidazole, 2.5 ml of 2 M KCl, 1 ml of 1 M $CaCl_2$, 1 ml of 0.1 M ATP, and, for the radioactive assay, 0.1–0.2 ml of 2 mM [γ-^{32}P]ATP with 93.5 ml of water and adjusting the pH to 7.5 at room temperature. The final concentrations are 20 mM imidazole, 50 mM KCl, 10 mM $CaCl_2$, and 1 mM ATP. The remainder of the assay procedure is identical to that described for the K$^+$(EDTA)-ATPase assay.

Actomyosin Mg²⁺-ATPase Activity

Reagents

Imidazole, 1 M, pH 7.0
MgCl₂, 1 M
ATP, 0.1 M pH 7.0 (store frozen)
[γ-³²P]ATP, 2 mM, 0.2 mCi/ml (store frozen)
F-actin, 10 mg/ml

Procedure. The assay mixture is prepared in batches of 100 ml by mixing 1 ml of 1 M imidazole, 0.5 ml of 1 M MgCl₂, 1 ml of 0.1 M ATP, and, for the radioactive assay, 0.1–0.2 ml of 2 mM [γ-³²P]ATP with 97.5 ml of water and adjusting the pH to 7.0 at room temperature. The final concentrations are 10 mM imidazole, 5 mM MgCl₂, and 1 mM ATP. The assays are carried out exactly as in the other two procedures except that 10–15 μl of F-actin are added for a final concentration of 0.2–0.4 mg/ml.

Purification of Myosins IA and IB[1]

Extraction. Approximately 1 kg of cells is extracted with 2 volumes of buffer containing 75 mM KCl, 12 mM sodium pyrophosphate, 5 mM dithiothreitol, and 30 mM imidazole chloride, pH 7.0. The supernatant obtained after centrifugation at 100,000 g for 3 hr is adjusted to pH 8.0 with Tris and dialyzed overnight against 10 volumes of 7.5 mM sodium pyrophosphate, 0.5 mM dithiothreitol, and 10 mM Tris chloride, pH 8.0. The solution is then applied to a DEAE-cellulose column (5 × 80 cm) previously equilibrated with 10 mM KCl, 1 mM dithiothreitol, and 10 mM Tris chloride, pH 8.0, and eluted with a linear gradient of KCl between 10 and 250 mM in 4.5 liters of the same buffer. Fractions are analyzed for K⁺(EDTA)-ATPase activity, and the peak of activity that is eluted at about 10 mM KCl is pooled. This contains a mixture of myosins IA and IB but is entirely free of myosin II, which elutes later from the column.

Ammonium Sulfate Concentration. All the ATPase activity is concentrated by adding ammonium sulfate to 2 M and collecting the precipitate and dissolving it in 40 ml of a solution containing 15 mM imidazole chloride, pH 7.5, 1 mM EGTA, 1 mM dithiothreitol, and 0.34 M sucrose. The solution is dialyzed overnight against the same buffer and fractionated by addition of ammonium sulfate. The precipitate obtained between 0.9 and 1.5 M ammonium sulfate is dissolved in 20 ml of the buffer just described and dialyzed overnight against the same buffer.

---

[1] H. Maruta, H. Gadasi, J. H. Collins, and E. D. Korn, *J. Biol. Chem.* **254**, 3624 (1979).

ADP–Agarose Chromatography. The dialyzed solution containing myosins IA and IB is applied to an ADP–agarose affinity column (1.6 × 25 cm) equilibrated with 15 mM Tris chloride, pH 7.5, 50 mM KCl, and 1 mM dithiothreitol. The column is washed with 100 ml of the equilibrating buffer, and a mixture of myosins IA and IB is dialyzed overnight against a buffer of 10 mM imidazole chloride, pH 7.5, 1 mM dithiothreitol, and 0.5 M KCl.

Hydroxyapatite Chromatography. The dialyzed solution is applied to a column of hydroxyapatite (2.5 × 25 cm) equilibrated with 10 mM Tris chloride, pH 7.5, 1 mM dithiothreitol, and 0.5 M KCl. The column is washed with the same buffer and then eluted with 200 ml of a gradient of potassium phosphate, pH 7.5 (0 to 0.2 M) at a flow rate of 10 ml/hr. Myosins IA and IB are eluted in a broad peak beginning at about 80 mM potassium phosphate and extending to about 160 mM potassium phosphate. The column can be regenerated for further use by washing it with 0.4 M potassium phosphate in the same buffer. The myosin solution is dialyzed against a buffer containing 15 mM imidazole chloride, pH 7.5, 1 mM dithiothreitol, 42 mM KCl, and 0.34 M sucrose.

BioGel A-1.5m Adsorption. The final purification of myosins IA and IB and their separation from each other is obtained by adsorption chromatography on BioGel A-1.5m. The dialyzed mixture of myosins IA and IB obtained from the ADP-agarose column is adsorbed onto a column of BioGel A-1.5m (200–400 mesh; 1.5 × 85 cm) equilibrated with 10 mM imidazole chloride, pH 7.5, 42 mM KCl, and 1 mM dithiothreitol. The column is then eluted with a linear gradient of KCl between 42 and 300 mM in 200 ml of the same buffer. Myosin IB is eluted at about 60 mM KCl, and myosin IA is eluted at about 120 mM KCl.

Table I shows results of a typical purification of myosins IA and IB.

Properties of *Acanthamoeba* Myosins IA and IB

By the above procedure, myosin IA and myosin IB are each obtained in about 90% purity as judged by SDS–polyacrylamide electrophoresis. Both enzymes have native molecular weights of about 150,000 by sucrose gradient and equilibrium centrifugation.[2] Myosin IA consists of one heavy chain of molecular weight 130,000, one light chain of molecular weight 17,000, and, as isolated, always less than equimolar amounts of a second light chain of molecular weight 14,000. Myosin IB contains one heavy chain of molecular weight 125,000, one light chain of molecular weight 27,000 and less than equimolar amounts of a second light chain of molecular weight 14,000. Myosin IA has a K$^+$(EDTA)-ATPase activity of about 3

---

[2] H. Gadasi and E. D. Korn, unpublished data.

TABLE I
PURIFICATION OF *Acanthamoeba* MYOSINS IA AND IB

Fraction	Myosin present	K⁺ (EDTA) ATPase			
		Protein (mg)	Total (μmol/min)	Yield (%)	Specific activity (μmol/min mg^{-1})
Extract	IA, IB	25,000	400	100	0.016
DEAE-cellulose	IA, IB	4,500	220	55	0.05
Ammonium sulfate	IA, IB	1,000	160	40	0.16
ADP-agarose	IA, IB	279	140	35	0.5
Hydroxyapatite	IA, IB	48	83	21	1.7
BioGel adsorption					
60 mM KCl	IB	7	27	7	3.9
120 mM KCl	IA	15	45	11	3.0

μmol/min mg^{-1} and a CaATPase activity of about 1 μmol/min mg^{-1} while the corresponding activities for myosin IB are about 4 and 1.2 μmol/min mg^{-1}, respectively. Both myosins IA and IB have very low MgATPase activities, about 0.1 μmol/min mg^{-1}, that are not activated by F-actin. However, the heavy chain of both enzymes can be phosphorylated[3,4] by a myosin I heavy chain kinase, that has been only partially purified, to the extent of about 1 mol of P per mole of heavy chain, and the phosphorylated myosins show considerable actin-activated MgATPase activity: 1.7 μmol/min mg^{-1} for myosin IA and 2.4 μmol/min mg^{-1} for myosin IB. By a slight modification of the purification procedure,[1] a portion of myosin IA can be isolated in 70% purity with an additional associated polypeptide of molecular weight 20,000. In this form, all of its ATPase activities, but especially its CaATPase activity, are reduced. It has not yet been possible to form bipolar filaments, typical of myosins from other sources, from these unusually low molecular weight myosins. It has been possible to demonstrate that the ATP catalytic site,[5] the actin-binding site,[4] and the phosphorylation site[3,4] are all on the heavy chains of the myosins I.

Purification of Myosin II[6,7]

Extraction. The cells (150 g) are washed twice with 10 mM triethanolamine hydrochloride, pH 7.5, in 0.1 M NaCl and are

[3] H. Maruta and E. D. Korn, *J. Biol. Chem.* **252**, 8329 (1977).
[4] H. Maruta, H. Gadasi, J. H. Collins, and E. D. Korn, *J. Biol. Chem.* **253**, 6297 (1978).
[5] H. Maruta and E. D. Korn, *J. Biol. Chem.* in press.
[6] J. H. Collins and E. D. Korn, *J. Biol. Chem.* **255**, 8011 (1980).
[7] J. H. Collins and E. D. Korn, *J. Biol. Chem.* submitted.

homogenized in 2 volumes of buffer, pH 7.5, containing 10 mM triethanolamine hydrochloride, 40 mM sodium pyrophosphate, 30% sucrose, 0.4 mM dithiothreitol, and the following inhibitors of proteolysis: 0.5 mM phenylmethylsulfonyl fluoride, pepstatin A (10 μg/ml), and leupeptin (0.8 μg/ml). The extract obtained after centrifuging for 3 hr at 100,000 g is dialyzed for 24 hr against two changes of 15 liters each of 10 mM triethanolamine hydrochloride, 0.1 M KCl, 1 mM EDTA, 1 mM dithiothreitol, pH 7.5, to allow actomyosin to form.

Myosin-Enriched Actomyosin. The actomyosin pellet is collected by centrifugation for 1 hr at 100,000 g and is homogenized in 75 ml of buffer containing 3 mM triethanolamine hydrochloride, 0.1 mM ATP, 0.1 mM CaCl$_2$, 1 mM dithiothreitol, pH 7.5, and dialyzed for 20 hr against two changes of 1 liter each of the same buffer. This procedure allows the excess F-actin to depolymerize to G-actin. The myosin-enriched actomyosin pellet is recovered by centrifugation for 1 hr at 100,000 g and homogenized in 5 ml of buffer containing 10 mM TES, 20 mM sodium pyrophosphate, 5% sucrose, 1 mM dithiothreitol, 0.5 mM ATP, pH 7.5, the suspension is left on ice for 2 hr and then mixed with an equal volume of a solution containing 1.2 M KI, 10 mM MgCl$_2$, 10 mM ATP, 2 mM CaCl$_2$, 1 mM EDTA, 1 mM dithiothreitol, pH 7.5, to dissolve the actomyosin.

BioGel Chromatography. The solution is immediately centrifuged for 30 min at 200,000 g, and the supernatant is immediately applied to a column of BioGel A-15m (200–400 mesh, 2.5 × 90 cm) previously equilibrated with 10 mM TES, 20 mM sodium pyrophosphate, 5% sucrose, 1 mM dithiothreitol, 0.5 mM ATP, pH 7.5. The column is eluted with the same buffer and the Ca^{2+}-ATPase peak that elutes just after the void volume of the column is pooled and adjusted to 40 mM in sodium pyrophosphate.

TABLE II

PURIFICATION OF *Acanthamoeba* MYOSIN II

Fraction	Protein (mg)	Ca^{2+}-ATPase		
		Total (μmol/min)	Yield (%)	Specific activity (μmol/min mg^{-1})
Extract	4400	43.5	100	0.01
Crude actomyosin	658	24.0	69	0.04
Actomyosin enriched in myosin	123	6.5	18	0.05
MgATP Supernatant	48	2.3	6	0.05
BioGel A-15m	9.8	8.4	24	0.86
DEAE	9.4	8.2	23	0.87

DEAE Chromatography. The fraction from the BioGel column is applied to a column of DEAE-cellulose (5 ml in volume) equilibrated with 10 mM TES, 40 mM sodium pyrophosphate, 5% sucrose, 1 mM dithiothreitol, pH 7.5. After washing the column with the same buffer, myosin II is eluted with the same buffer containing 0.2 M KCl. This step removed RNA which remains on the DEAE-cellulose. The myosin II is then dialyzed for 48 hr against two changes of 1 liter of 10 mM imidazole chloride, pH 7.5, 0.1 M KCl, 1 mM dithiothreitol to remove the pyrophosphate and concentrated by dialysis for 12 hr against solid sucrose.

Table II shows results of a typical purification of myosin II.

Properties of *Acanthamoeba* Myosin II

The isolated enzyme is about 95% pure. It has a native molecular weight of about 400,000 and consists of two heavy chains of molecular weight 170,000, two light chains of molecular weight 17,500, and two light chains of molecular weight about 17,000. The enzyme forms bipolar filaments[8] typical of those of most myosins. When isolated by the above procedure, myosin II has the following specific activities: K$^+$(EDTA)-ATPase, 0.2 μmol/min mg^{-1}; CaATPase, 0.8 μmol/min mg^{-1}; MgATPase, less than 0.01 μmol/min mg^{-1}; actin-activated MgATPase, 0.2 μmol/min mg^{-1}. By a totally different procedure,[8] myosin II can be isolated in a form that is indistinguishable in native molecular weight and subunit composition and has the same K$^+$ (EDTA)-ATPase, CaATPase, and MgATPase activities but is not activated by F-actin. The preparation that cannot be actin-activated contains 2 mol of P per mole of heavy chain whereas the actin-activatable form of myosin II contains only 1 mol of P per mole of heavy chain.[6] Both forms of myosin II can be dephosphorylated by potato acid phosphatase[6] with no effect on the K$^+$(EDTA)-ATPase or CaATPase activities, but with a coincident increase in the specific activity of the actin-activated MgATPase activity to 0.8–0.9 μmol/min mg^{-1}.

[8] T. D. Pollard, W. F. Stafford, and M. E. Porter, *J. Biol. Chem.* **253**, 4798 (1978).

[33] Phosphorylation of Nonmuscle Myosin and Stabilization of Thick Filament Structure

By JOHN KENDRICK-JONES, K. A. TAYLOR, and J. M. SCHOLEY

In vitro studies indicate that actomyosin-dependent contractility in vertebrate nonmuscle cells[1-3] and in vertebrate smooth muscles[4,5] may be regulated by the level of phosphorylation of a myosin "regulatory" light chain. The widely accepted view is that, in the absence of calcium, this 20,000 molecular weight (M_r) light chain is not phosphorylated and the myosin is unable to interact with actin. Calcium activates a specific calmodulin-dependent kinase[6-8] that phosphorylates the light chain and initiates myosin interaction with actin. Most studies have concentrated on the role of phosphorylation in the regulation of actin–myosin interaction as measured by the actin-activated myosin MgATPase activity. However, evidence[9-11] suggests that phosphorylation of the 20,000 M_r light chain also seems to control the assembly of vertebrate smooth muscle and nonmuscle myosins into filaments.

The effect of myosin phosphorylation on the stability of vertebrate nonmuscle myosin filaments can be monitored by turbidity measurements, electron microscope observation, and ^{32}P-labeled phosphate incorporation. Similar experiments have been carried out on vertebrate smooth muscle myosin.[9]

Protein Preparation

Myosin was prepared from calf thymus tissue by a procedure based on that of Pollard *et al.*[12] with the modifications suggested by Yerna *et al.*[2]

[1] R. S. Adelstein and M. A. Conti, *Nature (London)* **256**, 597 (1975).
[2] M.-J. Yerna, M. O. Aksoy, D. J. Hartshorne, and R. D. Goldman, *J. Cell Sci.* **31**, 411 (1978).
[3] J. A. Trotter and R. S. Adelstein, *J. Biol. Chem.* **254**, 8781 (1979).
[4] J. V. Small and A. Sobieszek, *Eur. J. Biochem.* **76**, 521 (1977).
[5] J. M. F. Sherry, A. Gorecka, M. O. Aksoy, R. Dabrowska, and D. J. Hartshorne, *Biochemistry* **17**, 4411 (1978).
[6] R. Dabrowska, J. M. F. Sherry, D. Aromatorio, and D. J. Hartshorne, *Biochemistry* **17**, 253 (1978).
[7] M.-J. Yerna, R. Dabrowska, D. J. Hartshorne, and R. D. Goldman, *Proc. Natl. Acad. Sci. U.S.A.* **76**, 184 (1979).
[8] D. R. Hathaway and R. S. Adelstein, *Proc. Natl. Acad. Sci. U.S.A.* **76**, 1653 (1979).
[9] H. Suzuki, H. Onishi, K. Takahashi, and S. Watanabe, *J. Biochem. (Tokyo)* **84**, 1529 (1978).

About 500 g of fresh calf thymus tissue (rapidly cooled in ice-water after removal from the animal and processed within 1 hr) were used for each myosin preparation. The myosin was stored in $0.6 M$ NaCl, 1 mM MgCl$_2$, 25 mM Tris-HCl buffer, pH 7.5, at 5–15 mg ml^{-1} in ice and used within 3 days. This procedure reproducibly gave a nonphosphorylated myosin (at least 80–90% dephosphorylated as indicated by 10% polyacrylamide gel electrophoresis in 8 M urea[13]).

Myosin was prepared from pig or human platelets using essentially the same procedure, but with the following minor modification. The washed platelets (25–100 g of packed pellet) were lysed by rapid freezing in liquid nitrogen and thawed in three volumes of extraction buffer containing 0.8 M KCl, 10 mM dithiothreitol, 5 mM EDTA, 1 mM azide, 2.5 mM ATP, 0.01 mg ml^{-1} lima bean trypsin inhibitor, 25 μg ml^{-1} L-1-tosylamide-2-phenylethylchloromethyl ketone (TPCK), 1 μg ml^{-1} leupeptin, 0.2 mM phenylmethylsulfonyl fluoride (PMSF), and 25 mM Tris-HCl buffer, pH 7.5.

Vertebrate smooth muscle myosin was prepared from chicken gizzards by a number of different procedures to exclude the possibility that myosin filament stability was dependent on the method of myosin preparation. Myosin was thus prepared by (a) a similar procedure to that used for preparing thymus myosin; (b) by the method described by Hartshorne et al.[14]; and (c) by a procedure modified from that used by Focant and Huriaux[15] to prepare myosin from carp and pike skeletal muscle involving ammonium sulfate fractionation of the actomyosin (37–60% fraction) in the presence of $0.6 M$ KCl, 25 mM imidazole (pH 7.0), 20 mM MgCl$_2$, 5 mM ATP, and 1 mM EGTA. All these myosin preparations were further purified by gel filtration chromatography on Sepharose 4B (90 × 5 cm columns) in $0.6 M$ KCl, 1 mM EDTA, 25 mM Tris-HCl (pH 7.5), 0.1 mM PMSF, 0.2 mM DTT, 1 mM azide. No difference in the stability of the filaments prepared from these different smooth muscle myosin preparations was observed. In all cases, filament stability at physiological ionic strength and MgATP concentrations was dependent on the level of myosin phosphorylation.

Vertebrate smooth muscle light-chain kinase was prepared from

[10] C. F. Shoenberg and M. Stewart, J. Muscle Res. Cell Motil. 1, 117 (1980).
[11] J. M. Scholey, K. A. Taylor and J. Kendrick-Jones, Nature (London) 287, 233 (1980).
[12] T. D. Pollard, S. M. Thomas, and R. Niederman, Anal. Biochem. 60, 258 (1976).
[13] W. T. Perrie and S. V. Perry, Biochem. J. 119, 31 (1970).
[14] D. J. Hartshorne, A. Gorecka, and M. O. Aksoy, In "Excitation Contraction Coupling in Smooth Muscle" (R. Casteels, T. Godfraind, and J. C. Rüegg, eds.), pp. 377–384. Elsevier/North-Holland Publ., Amsterdam, 1977.
[15] B. Focant and F. Huriaux, FEBS Lett. 65, 16 (1976).

chicken gizzards by the procedures outlined by Aksoy *et al.*[16] and Dabrowska *et al.*[6] with the following minor modification. After the initial extraction and centrifugation step, the supernatant was adjusted to 0.8 M KCl, 1 mM EGTA, 1 mM MgCl$_2$, 2 mM dithiothreitol, 0.1 mM PMSF, 1 mM sodium azide, and 10 mM Tris-HCl (pH 7.5) and stirred in ice for 30 min. It was then subjected to ammonium sulfate fractionation to yield the 37–55% saturated ammonium sulfate fraction ("crude tropomyosin" fraction). Using this modification, the step involving gel filtration on Sepharose 4B, described in the above procedures, could be avoided.

The light-chain kinase was prepared from calf thymus tissue (~300 g of starting material) by the procedure described by Dabrowska and Hartshorne.[17] To minimize proteolysis, 0.5 mM sodium azide, 0.2 mM PMSF, 20 μg of p-Tosyl-L-arginine methyl ester-HCl (TAME-HCl) per milliliter, and 0.1 mg of lima bean trypsin inhibitor per milliliter were included in the initial extraction medium. Despite these precautions, when the 35–60% saturated ammonium sulfate fraction was chromatographed on Sepharose 4B (90 × 5 cm column) in 0.8 M KCl, 1 mM EGTA, 0.2 mM dithiothreitol, 0.2 mM PMSF, 1 mM azide, and 25 mM Tris-HCl (pH 7.5), two peaks of light-chain kinase activity were detected. The peak eluted at about 100,000 M_r contained the calcium-sensitive kinase, and the peak eluted at a slightly lower molecular weight contained a "kinase" that was fully active in the absence of calcium. Only the fractions containing the Ca^{2+}-dependent kinase were pooled for further purification. The smooth muscle and thymus kinase fractions were further purified on Sepharose 4B–brain calmodulin affinity columns using the procedure outlined by Watterson and Vanaman.[18] Kinase activity was measured using DEAE-cellulose-purified gizzard 20,000 M_r light chains[19] and the assay conditions described by Frearson *et al.*[20]

Calmodulin was initially prepared from calf thymus by the method[2] based on conventional procedures, but later a rapid procedure involving a trichloracetic acid precipitation step as initially proposed by Yagi[21] was used. Calf thymus tissue (~500 g) was thoroughly homogenized in 3 volumes of 0.34 M sucrose, 5 mM EDTA, 2 mM EGTA, 1 mM dithiothreitol, 1 mM azide, 0.1 mM PMSF, 15 mM Tris-HCl (pH 7.5) and stirred for 15 min. The unextracted material was removed by centrifugation at 20,000 g

[16] M. O. Aksoy, D. Williams, E. M. Sharkey, and D. J. Hartshorne, *Biochem. Biophys. Res. Commun.* **69**, 35 (1976).
[17] R. Dabrowska and D. J. Hartshorne, *Biochem. Biophys. Res. Commun.* **85**, 1352 (1978).
[18] D. M. Watterson and T. C. Vanaman, *Biochem. Biophys. Res. Commun.* **73**, 40 (1976).
[19] R. Jakes, F. Northrop, and J. Kendrick-Jones, *FEBS Lett.* **70**, 229 (1976).
[20] N. Frearson, B. W. Focant, and S. V. Perry, *FEBS Lett.* **63**, 27 (1976).
[21] M. Yazawa, M. Sakuma, and K. Yagi, *J. Biochem.* (*Tokyo*) **87**, 1313 (1980).

for 1 hr, and to the supernatant cold 60% (w/v) trichloroacetic acid was added slowly with stirring to a final concentration of 3%. After stirring for 15 min the precipitated protein was collected by centrifugation and dispersed in 25 mM sodium acetate (pH 5.2) by gentle homogenisation, and the pH was adjusted to 5.2 by the addition of sodium hydroxide. With the pH maintained at 5.2, the suspension was stirred for 30 min, then thoroughly homogenized; the undissolved material was removed by centrifugation at 30,000 g for 20 min. To concentrate the supernatant, trichloroacetic acid was added to a final concentration of 3%, the precipitated protein was collected by centrifugation and dispersed in 25 mM Tris-HCl buffer (pH 7.5); the pH was adjusted to 7.5 (volume ~50 ml). After dialysis against 25 mM Tris-HCl (pH 7.5), 1 mM MgCl$_2$, 0.5 mM EGTA, solid urea was added to 4 M and the solution was clarified by centrifugation at 40,000 g for 30 min. The supernatant was chromatographed on a DEAE-cellulose column (40 × 2.54 cm) equilibrated in 2 M urea, 25 mM Tris-HCl buffer (pH 7.5), 1 mM MgCl$_2$, 0.5 mM EGTA, 0.2 mM DTT running buffer. The calmodulin was eluted with a linear gradient (total volume 1400 ml) of 100 mM to 400 mM NaCl in the above running buffer. Calmodulin eluted at ~0.28 M NaCl and was further purified by gel filtration on a Sephadex G-100 (95 × 1.6 cm) column equilibrated and run in 25 mM Tris-HCl (pH 7.5), 1 mM MgCl$_2$, 0.5 mM EGTA and stored frozen in the presence of a 1 mM azide.

Experimental Procedures

The stability of vertebrate nonmuscle myosin filaments was initially monitored by turbidity measurements. Nonphosphorylated thymus myosin (~0.5 mg ml^{-1}) was dialyzed into 0.15 M KCl, 10 mM MgCl$_2$, 1 mM EGTA, 0.1 mM dithiothreitol, 25 mM imidazole (pH 7.0) to form filaments. This myosin filament "solution" was incubated at 20° with 25 μg of calmodulin per milliliter and myosin light-chain kinase purified from either chicken gizzard or calf thymus (100 μg ml^{-1}), and the stability of the filaments was monitored by measuring the turbidity of the solution at λ = 340 nm in a Perkin-Elmer Model 551 double-beam spectrophotometer. A 0.5 mg ml^{-1} solution of thymus myosin filaments gave an absorption at 340 nm of 0.5–0.6. The turbidity remained constant until [γ-^{32}P]ATP (5 μCi/ μmol) was added to a final concentration of 1 mM, whereupon a rapid drop in turbidity was observed ($A_{340} \sim 0.2$) suggesting that the myosin filaments had disassembled. The turbidity remained low until the addition of Ca^{2+} to a concentration of 2 × 10^{-4} M free Ca^{2+}, which led to a steady increase in turbidity to about the initial level, suggesting that myosin filaments were re-forming. Further additions of ATP (1 mM) had no effect on the measured turbidity, indicating that the myosin filaments were now stable. The same results were observed regardless of the source of kinase used, i.e., vertebrate smooth muscle or thymus.

To verify that the turbidity measurements were monitoring the stability of the myosin filaments, aliquots of the myosin solution were taken at precise times for electron microscope observation. It was found to be essential that the turbidity measurements always be used in conjunction with electron microscope observations in order to verify and to "estimate" the degree of myosin filament disassembly and reassembly. The specimens were prepared for the electron microscope by placing 10 μl of the myosin solution on carbon-coated 400-mesh grids, allowing the drop to stand for exactly 30 sec, then washing with 13 drops of buffer and 10 drops of 1% uranyl acetate. In the electron microscope, the initial myosin solution contained numerous myosin filaments (bipolar, average length 3380 $\pm$ 560 Å) (Fig. 1). After addition of ATP, these filaments were missing and only a dense granular background of dissolved protein could be seen. When Ca^{2+} was added, the steady increase in the observed turbidity correlates with the appearance of bipolar filaments (average length 5000 $\pm$ 640 Å) in the electron microscope. These observations suggest that nonphosphorylated thymus myosin filaments, in the absence of Ca^{2+}, are disassembled by MgATP and are induced to reassemble into filaments when the kinase is activated by Ca^{2+}. Similar results have been obtained with myosins isolated from human platelets and vertebrate smooth muscle (Fig. 1),[9,11] whereas myosin filaments prepared from vertebrate and invertebrate striated myosins are stable under these conditions. Further confirmation of the "stability" of these nonmuscle myosin filaments, in the presence and in the absence of ATP, can be obtained by following their sedimentation profiles in the analytical ultracentrifuge under the same conditions, but with myosin concentrations in the range 1.5–2.5 mg ml^{-1}.

To prove that the nonmuscle myosin filaments are induced to assemble when the myosin is phosphorylated by the calcium-activated calmodulin-dependent kinase, the level and specificity of phosphate incorporation into the thymus myosin 20,000 M_r light chain during the course of the turbidity measurements was determined by the following procedures. Incorporation of ^{32}P was measured by a Millipore filtration procedure similar to that described by Jakes et al.[19] Aliquots (100 μl) of the myosin solution were removed at precise time intervals [two aliquots were also removed for polyacrylamide gel electrophoresis in sodium dodecyl sulfate (SDS) and in 8 M urea] and quenched in 2 ml of ice cold 10% trichloroacetic acid, 1 mM ATP, and 50 mM sodium pyrophosphate. Immediately 0.25 ml of bovine serum albumin (2.5 mg ml^{-1}) were added as a carrier, mixed, and heated at 90° for 10 min to ensure complete release of noncovalently bound phos-

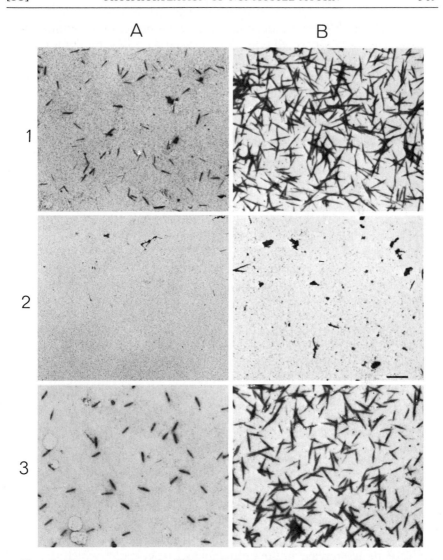

Fig. 1. Electron micrographs illustrating the stability of (A) vertebrate nonmuscle (thymus) and (B) vertebrate smooth muscle (gizzard) myosin filaments. Row 1: Nonphosphorylated myosin filaments prepared by dialysis against $0.15 M$ KCl, 10 mM MgCl$_2$, 1 mM EGTA, 0.1 mM dithiothreitol, 25 mM imidazole (pH 7.0) and incubated at 20° in the presence of calmodulin and kinase. The thymus myosin filaments are ~ 0.35 μm long, whereas those of the vertebrate smooth muscle myosin are ~ 0.75 μm long. Row 2: After the addition of 1 mM ATP, the filaments have completely disappeared and, in their place, a high concentration of dissolved protein can be seen on the grid. Row 3: After addition of $2 \times 10^{-4} M$ free Ca^{2+}, myosin filaments reappear. Bar represents 1 μm.

phate. The samples were cooled in ice, and the protein precipitates were transferred to Millipore filters (type HA, 0.45 μm), the tubes being thoroughly washed to ensure complete transfer of the precipitated protein. The filters were washed under rapid suction with six 4-ml aliquots of the TCA–ATP–pyrophosphate solution and then rapidly with four 5-ml aliquots of cold distilled water. The partially dried filters (the moist filters ensure that the precipitated protein sticks to the filters during transfer to the scintillation vials) were dispersed in scintillation fluid (Aquasol-2, New England Nuclear) and counted in a Beckman (Model LS7000) counter with the appropriate ^{32}P settings (program 10). The ^{32}P incorporation data indicated that, on addition of calcium, reassembly of the myosin filaments was accompanied by a steady increase in ^{32}P incorporation (~0.8 mol of phosphate incorporated per mole of light chain 20,000 M_r). In the absence of kinase or Ca^{2+}, there was no significant ^{32}P incorporation or filament formation.

The specificity of phosphate incorporation into the 20,000 M_r myosin light chain was checked by polyacrylamide gel electrophoresis and subsequent scanning and autoradiography of the dried gels. Aliquots of the myosin solution taken at the same time points as those for the ^{32}P incorporation measurements were quenched in 0.5% SDS sample buffer and in 3% trichloroacetic acid[9] (for 8 M urea PAGE) and analyzed on 10% acrylamide gels using 0.1% SDS–0.1 M Tris-Bicine (pH 8.1) and 8 M urea–0.1 M Tris-glycine (pH 8.6) buffer systems.[13] The relative ratio of the nonphosphorylated to phosphorylated 20,000 M_r light chain content of the myosin was determined by scanning the 8 M urea–acrylamide gels after staining with Coomassie Blue. To check the specificity of ^{32}P incorporation, both types of gels were dried and autoradiographed in cassettes fitted with tungstate intensifying screens (Ilford). There was reasonable agreement between the ^{32}P incorporation and the gel analysis, indicating that only the 20,000 M_r light chain is specifically phosphorylated.

These experiments, carried out under ionic conditions similar to those that may exist in cells, suggest that, in the resting cell, if the myosin is nonphosphorylated it is not assembled into filaments. Stimulation of the cell leads to an increase in the intracellular free calcium, which activates the Ca^{2+}-calmodulin-dependent kinase, resulting in phosphorylation of the light chain leading to myosin filament assembly. However, we will know whether these *in vitro* studies on the regulation of myosin filament assembly have any relevance to the situation inside the nonmuscle cell only when procedures have been developed for observing within the living cell the state of the myosin present at rest compared to periods of contractile activity.

[34] Preparation of Cytoplasmic Actin

By JERRY P. WEIR and DIXIE W. FREDERIKSEN

Assay Procedure

Actin is assayed both by electrophoresis in sodium dodecyl sulfate (SDS) and by its ability to activate the Mg^{2+}-ATPase of rabbit skeletal muscle myosin. Gel electrophoresis in SDS is performed by the method of Weber and Osborn,[1] using both diethylpyrocarbonate crosslinked protein molecular weight markers covering the molecular weight range of 14,300–71,500 and purified skeletal muscle actin[2] as standards. The actin-activation assay is performed as follows. Skeletal muscle myosin is prepared essentially as described by Margossian and Lowey[3] and dialyzed against low ionic strength assay (LISA) buffer (60 mM KCl, 1 mM dithiothreitol (DTT), 1 mM $MgCl_2$, 69 μM streptomycin-SO_4, 40 mM imidazole, pH 7.0). Myosin is diluted with LISA buffer to a final concentration of 1 mg/ml and stored at 4°. Protein solutions to be tested for actin are referred to here as "sample solution." A volume of myosin solution (0.3 ml) is incubated with an equal volume of sample solution diluted with LISA buffer at 25° for 10 min. The relative amounts of sample and LISA buffer will vary depending on the sample's concentration and actin content. ATP (50 mM) is added to a final concentration of 1 mM (12 μl), and incubation at 25° is continued for 200 sec. At 200 sec the reaction is quenched and the release of orthophosphate is determined by one of several established methods (cf. Fiske and SubbaRow[4] or Martin and Doty[5] as modified by Pollard and Korn.[6] Actin activation is taken as the ratio of released orthophosphate by myosin in the presence of a protein sample to the orthophosphate released by myosin alone. Sample solutions that contain ATP and other proteins in addition to actin can be tested in this way and will reveal the presence of functional actin.

Purification Procedure

Six to eight brains from 5- to 10-month-old hogs are obtained from a local slaughterhouse within 30 min of the death of the animals. The brains

[1] K. Weber and M. Osborn, *J. Biol. Chem.* **244,** 4406 (1969).
[2] J. A. Spudich and S. Watt, *J. Biol. Chem.* **246,** 4866 (1971).
[3] S. S. Margossian and S. Lowey.
[4] C. H. Fiske and Y. SubbaRow, *J. Biol. Chem.* **66,** 375 (1925).
[5] J. B. Martin and D. M. Doty, *Anal. Chem.* **21,** 965 (1949).
[6] T. D. Pollard and E. D. Korn, *J. Biol. Chem.* **248,** 4682 (1973).

METHODS IN ENZYMOLOGY, VOL. 85

are placed in ice-cold saline at the slaughterhouse and transported to the laboratory. All subsequent steps are performed at 4–7°. The brains are cleaned of their outer membranes and superficial blood vessels, weighed, and finely chopped. The chopped brains are homogenized in an equal volume of nonpolymerizing (low ionic strength + ATP) extraction buffer (10 mM imidazole, 0.1 mM $CaCl_2$, 0.5 mM ATP, 0.5 mM DTT, 67 μM streptomycin-SO_4, pH 7.5). Streptomycin sulfate is included in all buffers to prevent bacterial growth. The volume of the homogenate is between 1.2 and 1.6 liters.

After extraction for a minimum of 30 min, the homogenate is sedimented by centrifugation at 80,000 g for 90 min (5°). Because of the large volume of homogenate, several centrifuge runs are necessary. The resulting bright red supernatant is pumped onto a 2.6 × 45 cm diethylaminoethyl(DEAE)-cellulose column equilibrated with the extraction buffer containing 0.1 M KCl at a flow rate of approximately 90 ml/hr. The column is eluted first with 200 ml of this buffer, then overnight with 1000 ml of a linear 0.1 to 0.3 M KCl gradient in extraction buffer. Hemoglobin is removed before elution with the gradient. Fractions are collected at 14-min intervals and assayed both by electrophoresis in SDS under reducing conditions and by ability to activate the Mg^{2+}-ATPase of rabbit skeletal muscle myosin as described under Assay Procedure. This assay is performed by addition of an equal aliquot of each fraction to a constant amount of myosin and measurement of phosphate release.

Actin elutes at a salt concentration between 0.15 and 0.21 M. These fractions are pooled and polymerized by addition of $MgCl_2$ to 2 mM and subsequent incubation at room temperature for 2 hr. The polymerized actin is pelleted by centrifugation at 80,000 g for 3 hr (5°) and resuspended in extraction buffer with the help of a small hand-driven homogenizer. Actin is depolymerized by dialysis for 3 days at 4° vs three 1-liter volumes of extraction buffer.

Depolymerized actin is clarified by centrifugation at 80,000 g for 3 hr (5°) and chromatographed on a 2.6 × 90 cm Sephacryl S-200 column equilibrated with extraction buffer. Fractions are collected at 9-min intervals at a flow rate of approximately 45 ml/hr. The major peak eluting from the column is 95% brain actin as judged by electrohporesis in SDS and by the ability to activate the ATPase of muscle myosin. A typical preparative procedure yields 30–40 mg of purified actin from 600 g of cleaned brain.

Properties

Cytoplasmic actins are not as well characterized as actin from skeletal muscle. Purified porcine brain actin comigrates with rabbit skeletal muscle actin in SDS–polyacrylamide with an apparent molecular weight of 48,000. The amino acid composition of brain actin is similar though not identical to that of skeletal muscle actin.[7] Cytoplasmic actins, like their skeletal muscle counterpart, have been shown to contain the amino acid N-methylhistidine.[8] The actins, muscle and cytoplasmic, will polymerize under the appropriate ionic conditions (i.e., in 2 mM $MgCl_2$ or in 0.1 M KCl) to form fibrous actin, F-actin, a 7-nm wide, 35-nm half-pitch, double-stranded helix of actin monomers. The actins will activate the Mg^{2+}-ATPase of myosin.

Actin is a protein found in most, if not all, eukaryotic cells. It is well characterized as part of the highly organized contractile apparatus in mammalian skeletal muscle. Cytoplasmic actin is less well characterized, and its function in nonmuscle cells is not as clearly defined.[9] Although actin from skeletal muscle can be purified in high yield by extraction from acetone powder,[2] this procedure has not been as successful when used with nonmuscle cells, except in the purification of embryonic chick brain actin.[10] We have used a gentle purification procedure similar to that used by Gordon *et al.*[11] for the purification of *Acanthamoeba* actin to isolate actin from porcine brain.[7]

[7] J. P. Weir and D. W. Frederiksen, *Arch. Biochem. Biophys.* **203**, 1 (1980).
[8] M. Clarke and J. A. Spudich, *Annu. Rev. Biochem.* **46**, 797 (1977).
[9] H. G. Mannherz and R. S. Goody, *Annu. Rev. Biochem.* **45**, 427 (1976).
[10] J. D. Pardee and J. R. Bamburg, *Biochemistry* **18**, 2245 (1979).
[11] D. J. Gordon, E. Eisenberg, and E. D. Korn, *J. Biol. Chem.* **251**, 4778 (1976).

[35] Cytoplasmic Tropomyosins

By ISAAC COHEN

After the isolation of tropomyosin from human blood platelets,[1] several other cytoplasmic tropomyosins were purified from a variety of nonmuscle tissues, such as brain, pancreas, fibroblasts,[2] sea urchin eggs,[3] bovine thymus, and SV40 virus-transformed 3T3 tissue culture cells.[4] These all resemble muscle tropomyosin in having a two-chain α-helical coiled-coil structure, in forming paracrystals in the presence of divalent cations, and in the amino acid composition. They differ, however, from

[1] I. Cohen and C. Cohen, *J. Mol. Biol.* **68**, 383 (1972).
[2] R. E. Fine and A. L. Blitz, *J. Mol. Biol.* **95**, 447 (1975).
[3] T. Ishimoda-Takagi, *J. Biochem.* (*Tokyo*) **83**, 1757 (1978).
[4] A. Bretscher and K. Weber, *FEBS Lett.* **85**, 145 (1978).

their muscle counterpart by their size and amino acid sequence. These differences have prompted Fine *et al.*[5] to distinguish two classes of tropomyosins, a muscle type and a nonmuscle type.

Purification

The purification procedure for platelet tropomyosin[1] has been used for the isolation of most nonmuscle tropomyosins. The unusual resistance of tropomyosin and other α-helical fibrous proteins to organic solvents and heat treatment[6] has provided the basis for the isolation procedure of cytoplasmic tropomyosins.

The tissue of interest is first freed from any vascular or contaminating red cell material by either dissection or differential centrifugation. Then it is homogenized in a Waring blender with 20 volumes of absolute ethanol at 0°. After centrifugation the precipitate is resuspended in 20 volumes of anhydrous ethyl ether at 0°. The precipitate collected by centrifugation is then treated once more with ethanol and ether and air-dried. Dithiothreitol (2 mM) should be added in the subsequent steps. The dried residue is extracted overnight at 4° in 1 M KCl adjusted to pH 7.0. The insoluble material is removed by centrifugation, and the extract is placed in a boiling water bath for 10 min. All the following steps are carried out at 2°. After centrifugation at 100,000 g for 1 hr, the supernatant solution is further purified by isoelectric precipitation, adjusting the pH to 4.1 with 0.1 N HCl. The precipitate is collected by centrifugation and dissolved in 0.2 M KCl, 0.03 M sodium phosphate buffer. An ammonium sulfate fractionation is added as a final step.[5] The pellet from a 40% saturation is discarded, and ammonium sulfate is added to 53% saturation. The resulting precipitate is dissolved in 0.6 M KCl, 0.03 M sodium phosphate buffer, pH 7.0, and exhaustively dialyzed against the same solvent. About 30 mg of protein, representing 1.5% at the total protein content, is obtained from 10^{12} human platelets and 0.5 mg per gram of dried brain, corresponding to 0.1% of total brain protein.

Properties

Purity. The purified cytoplasmic tropomyosins prepared thus far generally show a single band under reducing conditions in a continuous sodium dodecyl sulfate (SDS)–polyacrylamide phosphate buffer electrophoresis system.[7] A closely spaced doublet band, possibly representing α and β polypeptide chains, has been obtained with brain and platelet

[5] R. E. Fine, A. L. Blitz, S. E. Hitchcock, and B. Kaminer, *Nature* (*London*), *New Biol.* **245**, 182 (1973).

[6] K. Bailey, *Biochem. J.* **43**, 271 (1948).

[7] K. Weber and M. Osborn, *J. Biol. Chem.* **244**, 4406 (1969).

tropomyosins[4,8] when applied under reduced conditions to discontinuous SDS–polyacrylamide Tris-HCl electrophoresis sytems.[9] The relative electrophoretic mobility of tropomyosin bands was similar in both systems.

Stability. Like their muscle counterpart, cytoplasmic tropomyosins are very stable. They can be boiled at 100°, frozen, or freeze-dried without losing their properties. After dissociation in 8 M urea they can be fully renatured after exhaustive dialysis.[1]

Physical and Chemical Properties. Like muscle tropomyosins, cytoplasmic tropomyosins are fully α-helical two-chain molecules. They have a subunit molecular weight of 30,000, however, and thus are one-seventh smaller than muscle tropomyosins. The molecular weight corresponds well with the 34.5-nm axial period of tropomyosin divalent cation-induced paracrystals. This value, which is close to the length of the molecule, is smaller than the 40-nm axial period of muscle tropomyosin paracrystals.[10]

The amino acid composition of nonmuscle tropomyosins is similar to that of muscle tropomyosins, with over 50% of polar residues and the absence of proline.[1,2,5] The ratio of Lys to Arg is 1.3, similar to that of some invertebrate tropomyosins.[11] The peptide maps show at least 10 peptides with mobilities different from those found in maps of muscle tropomyosin.[2] The NH_2- and COOH-terminal sequences of platelet tropomyosin show some differences from skeletal tropomyosins.[8] Since nonmuscle tropomyosins are six-sevenths the size of muscle tropomyosin, their complete amino acid sequence may reveal six similar regions of 42 residues, each which would span six instead of seven actins.[12] These considerations may account for the weak binding to actin and troponin.[5,13] Indirect immunofluorescence studies of fibroblasts show, however, that most actin and tropomyosin fibers have similar patterns, suggesting a cytoskeletal association of these two molecules.[14] Cytoplasmic tropomyosins can substitute for their muscle counterpart in conferring Ca^{2+} sensitivity to a troponin–actomyosin ATPase system,[5] suggesting the possibility that they may have a calcium regulatory role[15] in addition to their cytoskeletal role.

Acknowledgments

The author is grateful to Dr. Carolyn Cohen (Brandeis University, Waltham, Massachusetts) for careful review of the manuscript.

[8] G. P. Côté, W. G. Lewis, and L. B. Smillie, *FEBS Lett.* **91**, 237 (1978).

[9] U. K. Laemmli, *Nature (London)*, **227**, 680 (1970).

[10] C. Cohen and W. Longley, *Science* **152**, 794 (1966).

[11] G. R. Millward and E. F. Woods, *J. Mol. Biol.* **52**, 585 (1970).

[12] A. D. McLachlan and M. Stewart, *J. Mol. Biol.* **103**, 271 (1976).

[13] G. P. Côté, W. G. Lewis, M. D. Pato, and L. B. Smillie, *FEBS Lett.* **94**, 131 (1978).

[14] E. Lazarides, *J. Cell Biol.* **68**, 202 (1976).

[15] I. Cohen, E. Kaminski, and A. deVries, *FEBS Lett.* **34**, 315 (1973).

[36] Preparation of Tubulin from Brain

By ROBLEY C. WILLIAMS, JR. and JAMES C. LEE

Tubulin is isolated as a heterodimer, composed of two polypeptide chains, each of apparent molecular weight 54,000. We describe two methods for preparing tubulin, each of which yields several hundred milligrams of purified protein. The principal properties of the tubulin prepared by the two methods appear to be identical,[1] and a choice of one route or the other can be made on the basis of available apparatus or of the investigator's familiarity with the manipulations involved. The tubulin that results from either preparation is substantially free of microtubule-associated proteins. It will, under appropriate conditions, polymerize to produce microtubular structures, and it binds colchicine and GTP. Alternative related preparative methods have been described.[2-4] We describe here those with which we are most familiar.

Purification by Cycles of Assembly and Disassembly Followed by Chromatography on Phosphocellulose

This procedure is derived from those originated by Shelanski *et al.*[5] and by Weingarten *et al.*[6] It works well with bovine brain, but can also be used with other mammalian brains.

Materials

PM-4M buffer: 1.5 liters of 100 mM PIPES-NaOH, 2 mM EGTA, 1 mM MgSO$_4$, 4 M glycerol, 2 mM dithioerythritol (DTE), pH 6.9 at 23°

50 mM GTP, 6 ml in water, stored frozen (Sigma type II-S)

PM buffer: 300 ml of 100 mM PIPES-NaOH, 2 mM EGTA, 1 mM MgSO$_4$, 2 mM DTE, pH 6.9 at 23°

[1] J. C. Lee, N. Tweedy, and S. N. Timasheff, *Biochemistry* **17**, 2783 (1978).

[2] C. R. Asnes and L. Wilson, *Anal. Biochem.* **98**, 64 (1979).

[3] G. G. Borisy, J. M. Marcum, J. B. Olmsted, D. B. Murphy, and K. A. Johnson, *Ann. N.Y. Acad. Sci.* **253**, 107 (1975).

[4] D. B. Murphy and R. R. Hiebsch, *Anal. Biochem.* **96**, 225 (1979).

[5] M. L. Shelanski, F. Gaskin, and C. R. Cantor, *Proc. Natl. Acad. Sci. U.S.A.* **70**, 765 (1973).

[6] M. D. Weingarten, A. H. Lockwood, S.-Y. Hwo, and M. W. Kirschner, *Proc. Natl. Acad. Sci. U.S.A.* **72**, 1858 (1975).

PM-8M buffer: 250 ml of 100 mM PIPES-NaOH, 2 mM EGTA, 1 mM MgSO$_4$, 2 mM DTE, 1 mM GTP, 8 M glycerol, pH 6.9 at 23°
Column buffer: 2 liters of 100 mM PIPES NaOH, 2 mM EGTA, 1 mM MgSO$_4$, 2 mM DTE, 0.1 mM GTP, pH 6.9 at 23°

Assembly–Disassembly Cycles

Homogenization. The brains of 4–6 cattle (1.6–2.4 kg) are obtained as soon after slaughter as practicable and transported to the laboratory on ice. (The yield of tubulin decreases rapidly with the time after slaughter.[4] Two hours from slaughter to homogenization is a workable span.) Meninges and superficial blood vessels are removed from the brains at 4°.

The remaining tissue is coarsely minced with scissors and then homogenized, 100 g at a time, with 75 ml of PM-4M buffer in a Sorvall Omni-Mixer, a blade-type homogenizer. Homogenization is for 50 sec at speed 3 and 10 sec at speed 9.

Preparation of Extracts. Two centrifuges are employed simultaneously at early stages of the procedure. The homogenate is poured into 300-ml bottles (approximately 12 are required) and centrifuged 15 min at 4° in two Sorvall type GSA rotors at 8000 rpm (6500 g). The supernatants are decanted, poured into 60-ml tubes (about 12 are required), and centrifuged for 75 min at 4° in two Beckman Type 35 rotors at 35,000 rpm (96,000 g).

Assembly. The supernatants are decanted into a graduated cylinder, the volume (V_1) is noted, GTP is added to 1 mM, and the solution is poured into fresh 60-ml centrifuge tubes (about 10 are required). The filled tubes are balanced for centrifugation and then incubated for 30–45 min in a water bath at 34°. (Polymerization of tubulin can be observed as an increase in viscosity that greatly slows the rate at which small bubbles rise to the top. The solution at this stage is pink. During incubation, the Type 35 rotors are warmed to 30° with tap water, and the centrifuges are warmed. The incubated tubes are centrifuged for 60 min at 27° and 35,000 rpm to pellet microtubules. At room temperature, the supernatants are decanted from the firm and slightly opalescent pellets. (After this point, only a single centrifuge is necessary.)

Disassembly. The pellets are resuspended in a total volume of 0.25 V_1 of ice-cold PM buffer. The pellets are first removed, in pieces, from the bottoms of the tubes by gentle scraping with a rubber-tipped stirring rod. The resulting coarse suspension is poured into an ice-cold Dounce homogenizer (60-ml size) and subjected to about 5 gentle strokes of the pestle (an A-size pestle should be used) to produce a uniform suspension. This suspension is incubated 30 min on ice to depolymerize microtubules, then poured into cold 60-ml centrifuge tubes (about 4 are required) and

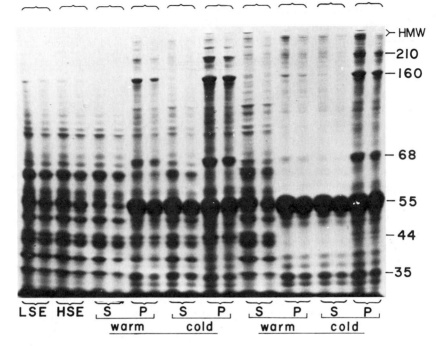

Fig. 1. Microtubule preparation by two cycles of assembly and disassembly. Aliquots of approximately 50 and 33 μg from each fraction (grouped as indicated by the brackets) in the preparative process were loaded on an 8% sodium dodecyl sulfate–polyacrylamide slab gel. Apparent molecular weights, in thousands, are indicated by the numbers. Labels: LSE, low speed extract from first centrifugation; HSE, high speed extract from second centrifugation; supernatants (S) and pellets (P) from warm and cold centrifugations are indicated, from left to right, through two cycles of assembly and disassembly. Reproduced from Runge et al.,[7] with permission of the publishers.

centrifuged in the Type 35 rotor for 60 min at 35,000 rpm and 4°. The supernatants are carefully removed from the pellets, which are sometimes loose, and pooled in a graduated cylinder. The volume (V_2) is noted, and an equal volume of PM-8M buffer is added. The solution can be frozen at −20° and stored for times as long as 72 hr at this point, and it is often convenient to do so.

Second Cycle. A second cycle of polymerization, centrifugation, depolymerization, and centrifugation is carried out by incubating the solution at 34° as described above, centrifuging it, resuspending the resulting pellets in 0.25 V_2 of cold PM buffer, incubating at 0°, and centrifuging again. The resulting cold supernatant contains about 400 mg of protein, of which approximately 85% is tubulin. This material, often referred to as "twice-cycled microtubule protein" is further purified by chromatography. Figure 1 shows the distribution of proteins at each stage of the cycle preparation.[7] It is apparent that the fraction of tubulin (the major protein band at 55,000 apparent molecular weight) increases greatly at each step in the process. The material applied to the phosphocellulose column is represented by the lanes third and fourth from the right, labeled "cold, S."

Phosphocellulose Chromatography. A batch of 20 g of phosphocellulose (Whatman P-11) is precycled in $0.5 N$ NaOH and $0.5 N$ HCl, with careful attention to the precautions specified by the manufacturer to avoid hydrolysis of the phosphate ester bonds. Fine particles are removed. The phosphocellulose is then suspended in 250 ml of column buffer in a beaker and titrated (without the use of a magnetic stirrer) to pH 6.9, washed in two further 250-ml changes of column buffer, and poured to make a 2.0×45 cm column. A further 500 ml of column buffer are passed through the column at $4°$, and both the pH and the conductivity of the eluted buffer are measured to ascertain that they are the same as those of the buffer flowing into the column.

The entire sample is applied to this column at $4°$. Tubulin is rapidly eluted with column buffer at a flow rate of about 15 ml/hr. Fractions of approximately 2 ml are collected, and their protein concentration is assessed by the method of Bradford.[8] Soon after the fractions emerge from the column, a volume of 100 mM MgSO$_4$ sufficient to bring the Mg^{2+} concentration to 1 mM is added to each tube to restore Mg^{2+} that is removed from the buffer by the phosphocellulose.[9] Fractions with concentration greater than 1 mg/ml are pooled. About 200 mg of tubulin are obtained at a concentration near 5 mg/ml. The fraction of tubulin in this final product is greater than 96%, as assayed by SDS-gel electrophoresis. As Fig. 2 shows, the only detectable nontubulin proteins are of molecular weight much less than 50,000.[10]

The phosphocellulose can be readied for re-use by a simple elution of the column with about 500 ml of a buffer containing $3 M$ NaCl, followed by reequilibration in a beaker with column buffer.

Tubulin can be kept for many months in liquid N_2. It is best frozen quickly by allowing small (1–2 mm in diameter) drops of solution to fall directly into liquid N_2. (A convenient arrangement for doing this unattended can be devised with a peristaltic pump.) Before use, such frozen tubulin must be thawed and equilibrated with the buffer in which it is to be used. In our experience, denaturation during the thawing process can be minimized if prolonged contact between frozen and liquid solution is avoided: a single layer of frozen droplets is placed in the bottom of a

[7] M. S. Runge, H. W. Detrich, III, and R. C. Williams, Jr., *Biochemistry* **18**, 1689 (1979).
[8] M. M. Bradford, *Anal. Biochem.* **72**, 248 (1976).
[9] R. C. Williams, Jr. and H. W. Detrich, III, *Biochemistry* **18**, 2499 (1979).
[10] H. W. Detrich, III and R. C. Williams, Jr., *Biochemistry* **17**, 3900 (1978).

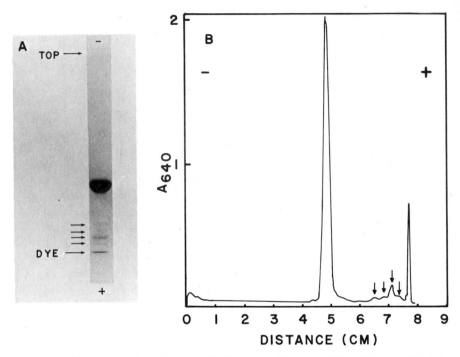

FIG. 2. Electrophoretic analysis of tubulin prepared by assembly and disassembly followed by chromatography. (A) Tubulin (74 μg) was electrophoresed on a 7.5% sodium dodecyl sulfate (SDS)–polyacrylamide slab gel and stained with Coomassie Brilliant Blue R. Electrophoretic migration was from top to bottom. (B) Densitometric scan of a cylindrical SDS–polyacrylamide gel. Tubulin (19 μg) was electrophoresed and stained with Fast Green. Migration was from left to right. The slight deviation from the baseline in the region from 0 to 0.5 cm is produced by refraction of light due to the curvature of the gel surface, and the sharp spike at 7.8 cm corresponds to the dye front. The four impurities of low molecular weight are indicated in both (A) and (B) by the arrows. Reproduced from Detrich and Williams,[10] with permission of the publishers.

beaker, and the beaker is placed in water at 25° until all beads just melt. After thawing, a small amount of turbidity is often present in the solution. It is removed by a brief spin in a clinical centrifuge. The tubulin solution that emerges from the phosphocellulose column is in a poorly defined buffer that contains somewhat unpredictable amounts of glycerol, GDP, and other buffer components.[9] Before it is used, the protein must, therefore, be equilibrated with a known buffer by means of gel filtration or dialysis.

Microtubule-Associated Proteins. Microtubule-associated proteins can be eluted from the phosphocellulose column by the application of column buffer to which has been added 0.75 M NaCl. Their concentration can be assayed by the method of Bradford.[8] A full discussion of these proteins is included in this volume.[11]

Purification by the Modified Weisenberg Procedure

We now describe a modification of the procedure for purifying brain tubulin developed originally by Weisenberg.[12,13] The procedure described is that adopted in the laboratory of J. C. L. and is the one with which he is most familiar.

The isolation procedure was primarily developed for brain tubulin, specifically calf brains, although it has been successfully applied to lamb, dog, and pig brains. All operations are performed in a 4° cold room. Approximately 4 kg of brain tissue are employed as starting material.

Material

Sodium phosphate buffer: 8 liters of $0.24 M$ sucrose in 10 mM sodium phosphate, 0.5 mM MgCl$_2$ buffer at pH 7.0

PMG buffer: 2 liters of 10 mM sodium phosphate, 0.5 mM MgCl$_2$, 0.1 mM GTP buffer at pH 7.0

KCl in PMG: 1 liter each of $0.4 M$ and $0.8 M$ KCl in PMG buffer.

Sucrose in PMG: 250 ml of $1 M$ sucrose in PMG buffer. The sucrose employed in this solution should be ultrapure. The density gradient grade sucrose from Schwarz-Mann has been found to be satisfactory.

DEAE-Sephadex A-50 (10 g) is swollen in 10 mM sodium phosphate buffer at pH 7.0 for approximately 5 hr in a steam bath. The swollen ion-exchanger is equilibrated with PMG buffer by suspending it in 800 ml of the buffer. After the ion-exchanger settles owing to gravity, the supernatant is decanted. The Sephadex is mixed with fresh PMG buffer and allowed to settle again. This procedure is repeated at least three times before the pH of the suspension is adjusted to 7.0.

Sephadex G-25 medium is swollen in deionized water by the same procedure as that for DEAE-Sephadex. The column is packed and washed with at least 500 ml of deionized water before being equilibrated with about 300 ml of PMG buffer.

Purification Procedure

Homogenization. Fresh brains are obtained within 1 hr of slaughter and placed on ice during transportation. The superficial blood vessels and

---

[11] This volume [39].

[12] R. C. Weisenberg, G. G. Borisy, and E. W. Taylor, *Biochemistry* **7**, 4466 (1968).

[13] R. C. Weisenberg and S. N. Timasheff, *Biochemistry* **9**, 4110 (1970).

meninges are removed with forceps, and the tissue is minced with a pair of scissors or a stainless steel spatula with handle. It is then washed three times by suspension in two to three volumes of 0.24 M sucrose in 10 mM sodium phosphate, 0.5 mM MgCl$_2$ buffer at pH 7.0. The wash is decanted by straining through cheesecloth. The mince is resuspended in 1.5 to 2 volumes of the same buffer and homogenized in a domestic Waring blender at the maximum setting for 30 sec. It is important not to denature the protein by excessive blending as might occur with the use of some of the more powerful laboratory blenders.

The cell debris and particulate components are removed by centrifuging the homogenate in a Sorvall GSA rotor at 10,000 rpm (13,000 g) for 30 min. The supernatant is saved, and the pellet is discarded.

Ammonium Sulfate Fractionation. The supernatant is brought to 32% saturation by the addition of 177 g of solid ultrapure ammonium sulfate per liter within an interval of approximately 5 min. An additional period of 10–15 min is allowed for total dissolution of the ammonium sulfate and precipitation of proteins. The precipitate is removed by centrifugation in a Sorvall GSA rotor at 10,000 rpm (13,000 g) for 30 min. The supernatant is saved, and the pellet is discarded.

The supernatant is brought to 43% saturation by dissolving an additional 71 g of (NH$_4$)$_2$SO$_4$ per liter. The precipitate is again removed as described and saved, and the supernatant is discarded.

DEAE-Sephadex Batch Elution. The 43% (NH$_4$)$_2$SO$_4$ precipitate is suspended in 150 ml of PMG at pH 7.0. Resuspension is facilitated by employing a 55-ml Potter–Elvehjem tissue grinder with Teflon pestle. One should avoid using a ground-glass homogenizer and should exercise caution so as not to denature the protein, as would be indicated by foaming.

The solution is mixed with 400 ml of gravity-packed DEAE-Sephadex (A-50) which has been washed thoroughly and equilibrated in PMG buffer. After 10 min for protein adsorption, the slurry is distributed into 50-ml centrifuge tubes (about 16 are required). The Sephadex beads are then pelleted by centrifugation in a Sorvall SS-34 rotor by bringing the rotor briefly up to 5000 rpm. The supernatant is decanted and discarded. Enough 0.4 M KCl in PMG buffer is added to fill each tube, and the Sephadex pellet is resuspended. The tubes are again centrifuged momentarily at 5000 rpm, and the supernatants are decanted and discarded. The washing of Sephadex pellets with 0.4 M KCl is repeated.

Tubulin is then eluted from the DEAE-Sephadex by repeating the washing procedure twice with 0.8 M KCl in PMG buffer. In order to keep the elution volume to a minimum, the amount of 0.8 M KCl employed for each washing procedure should be no more than twice that of the Sephadex pellet.

No extra precaution is necessary to prevent some Sephadex beads from being poured into the 0.8 M KCl eluent. These can be removed by passing the eluent over glass wool. It is essential that at least 90% of the Sephadex beads be removed at this stage lest they clog the G-25 Sephadex column employed at the next stage and decrease the efficiency of desalting.

The eluent should be slightly turbid; the proteins are precipitated by adding 24.8 g of $(NH_4)_2SO_4$ per 100 ml. Again, approximately 10 min are allowed for equilibration after complete dissolution of the salt. The protein precipitate is collected by centrifugation at 14,000 rpm (27,000 g) in a Sorvall SS-34 rotor for 30 min, and the supernatant is discarded.

The protein pellet is redissolved in 15 ml of PMG buffer with the aid of a 55-ml Potter–Elvehjem tissue grinder with Teflon pestle. The precautions mentioned above should again be exercised. The protein solution should be opaque, and the total volume is approximately 50 ml.

Sephadex G-25 Chromatography. The protein solution is applied either to two columns of 2 × 35 cm or to one column of 2.5 × 50 cm of Sephadex G-25. The column is eluted with PMG buffer at a rate of 1.5–2.0 ml/min to remove the ammonium sulfate completely. Fractions of 2 ml are collected. The elution of tubulin can easily be monitored by visual inspection of these fractions. Tubulin solutions are turbid.

MgCl$_2$ Precipitation. To each fraction that is turbid, 1–2 drops of 1.0 M MgCl$_2$ in H$_2$O are added. If the Sephadex column has been successful in removing the $(NH_4)_2SO_4$, the addition of MgCl$_2$ should precipitate tubulin in the fraction. The tubulin solution becomes extremely turbid with a milky appearance immediately after the addition of MgCl$_2$. A slow precipitation usually indicates the presence of $(NH_4)_2SO_4$. Excess MgCl$_2$ should not be added, as it might alter the pH of the solution, leading to low quality of tubulin finally isolated. Fractions that show precipitation with MgCl$_2$ are then pooled, and the precipitate is collected by centrifugation in an SS-34 rotor at 9000 rpm (9000 g) for 5 min. Subjecting the precipitate to excessive centrifugation is to be avoided, since it becomes much more difficult to resuspend the pellet.

The pellet is dissolved in approximately 5 ml of 1 M sucrose in PMG buffer with the aid of a 10-ml Potter–Elvehjem tissue grinder with a grooved Teflon pestle. Gentle handling of the pellet is absolutely essential at this stage. Short and slow strokes of the pestle are recommended so that the tubulin pellet will not form a paste between the bottom of the grinder and the pestle. After resuspending the pellet, the purified tubulin solution is dialyzed overnight against 250 ml of 1 M sucrose in PMG buffer at 4°.

Storage. The protein concentration of the solution is determined spec-

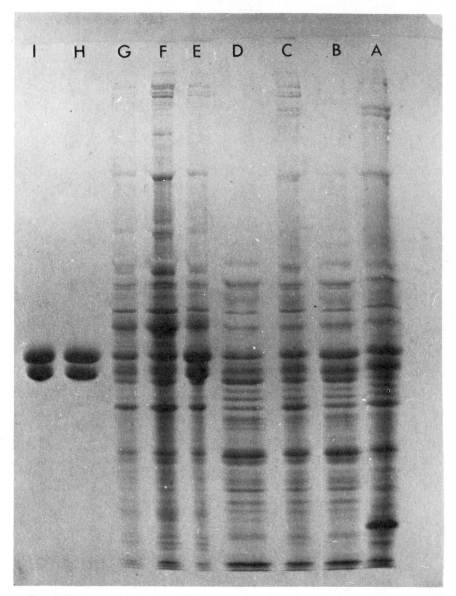

FIG. 3. Sodium dodecyl sulfate (SDS)–urea polyacrylamide gel electrophoresis of tubulin fractions collected during the purification procedure. Electrophoresis was performed according to Laemmli[14] on a slab gel with a linear gradient of 4 to 15% acrylamide and in the presence of 0.1% SDS, 6 M urea. Proteins were stained with 0.1% (w/v) Coomassie Brilliant Blue R. The identity of the sample and amount of protein loaded are as follows: Lane A, brain homogenate, 160 μg; B, supernatant from the 32% $(NH_4)_2SO_4$ fractionation, 110 μg; C, pellet from the 32% $(NH_4)_2SO_4$ fractionation, 100 μg; D, supernatant from the 43% $(NH_4)_2SO_4$ fractionation, 110 μg; E, pellet from the 43% $(NH_4)_2SO_4$ fractionation, 100 μg; F, supernatant from DEAE adsorption, 110 μg; G, 0.4 M KCl wash, 50 μg; H, 0.8 M KCl wash, 20 μg; I, pellet from $MgCl_2$ precipitation, 20 μg. Unpublished data of H. George and J. C. Lee.

trophotometrically in 6 M guanidine hydrochloride at 275 nm with an absorptivity value of 1.03 ml/(mg cm). It is then adjusted to 50–60 mg/ml and rapidly frozen in small aliquots to be stored in liquid nitrogen. Under these conditions the protein can be stored for at least 14 days without observable loss of colchicine binding activity or of its ability to undergo *in vitro* microtubule assembly. Generally about 500 mg of highly purified tubulin can be obtained by this procedure.

Figure 3 presents the results of polyacrylamide gel electrophoresis[14] in the presence of 0.1% SDS–6 M urea, showing the purity of tubulin at various stages of the purification procedure. It is apparent that the present procedure would yield tubulin samples of >95% purity from brain tissues. Experience indicates, however, that brains from mature cows do not yield tubulin of the same purity. Much more contaminating protein remains after elution from DEAE-Sephadex.

Eipper[15] reported that, by substituting a pyrophosphate buffer for the original phosphate buffer, one could obtain a highly purified tubulin, free of carbohydrate and nucleic acid contamination. Our experience shows that tubulin free of contaminants can be obtained when phosphate buffer is employed, although occasionally (approximately 30% of samples tested) less than 0.5% (w/w) of nucleic acids was found to be present. The homogeneous tubulin purified with the use of phosphate buffer can be shown to be as free of contaminant as tubulin prepared with pyrophosphate buffer. At present not enough preparations have been conducted in our laboratory to establish the efficiency of obtaining contaminant-free tubulin by employing the pyrophosphate buffer.

[14] U. K. Laemmli, *Nature* (*London*) **227**, 680 (1970).
[15] B. A. Eipper, *Proc. Natl. Acad. Sci. U.S.A.* **69**, 2283 (1972).

[37] Isolation of Tubulin from Nonneural Sources

By KEVIN W. FARRELL

Historically, the major impetus for isolating tubulin (TU) from non-neural systems did not come until after the successful isolation, by *in vitro* reassembly, of TU from vertebrate brain. Largely for this reason, methodologies developed for the isolation of nonneural TUs are based upon the assembly–disassembly procedures for vertebrate brain TUs.[1,2] Consequently, many of the experimental details I shall consider will be concerned with optimizing TU assembly.

Isolation of Tubulin from Cell Cultures, Nonneural Tissues, and Organs

General Considerations

A major problem encountered when using nonneural cell systems is the lower cell TU content (often $<5\%$) as compared with neural sources (10–40%). This probably accounts for the inapplicability of conventional ion-exchange methods for nonneural TU isolation, which may result in poor TU yields and contamination with other proteins. Given the requirement for a minimum TU concentration for *in vitro* polymerization, the low TU content is probably partly responsible also for the failure of earlier attempts to reassemble these TUs.

To overcome this problem, many workers emphasize the need to use very concentrated protein solutions as starting material. Microtubules (MTs) were successfully reassembled in crude protein solutions of Ehrlich ascites tumor (EAT) cells at 20–40 mg/ml.[3,4] Similarly, protein solutions in excess of 12 mg/ml were suggested for HeLa cell preparations.[5]

Various methods have been used to obtain such high protein concentrations. For EAT cells, 10 g of packed cells were resuspended in 1.0 ml of buffer.[3,4] This contrasts with the more conventional ratio of 1–2 ml of buffer per 1.0 g of tissue for neural sources. Other workers have used ultrafiltration (Amicon UE-20 membrane, $2°$) of soluble protein extracts of renal medulla, prepared at conventional buffer : tissue ratios.[6] For HeLa cells (1.0 g of packed cells per 1.0 ml buffer) gentle sonication (setting No. 2, Heat Systems Sonifier) yielded substantially higher protein concentrations in the soluble extracts than homogenization.[5]

A further problem with cell cultures is the presence of an endogenous activity inhibitory for TU assembly. This activity is not constant for all cell types, and a judicious choice of cell line is indicated. Certain cell types have been ranked in increasing order of inhibitor content: C6

[1] M. L. Shelanski, F. Gaskin, and C. R. Cantor, *Proc. Natl. Acad. Sci. U.S.A.* **70**, 765 (1973).
[2] G. G. Borisy, J. B. Olmsted, J. M. Marcum, and C. Allen, *Fed. Proc., Fed. Am. Soc. Exp. Biol.* **33**, 167 (1974).
[3] K. H. Doenges, B. W. Nagle, A., Uhlmann, and J. Bryan, *Biochemistry* **16**, 3545 (1970).
[4] K. H. Doenges, M. Weissinger, R. Fritzsche, and D. Schroeter, *Biochemistry* **18**, 1698 (1979).
[5] J. C. Bulinski and G. G. Borisy, *Proc. Natl. Acad Sci. U.S.A.* **76**, 293 (1979).
[6] L. D. Barnes, A. G. Engle, and T. P. Dousa, *Biochim. Biophys. Acta* **405**, 422 (1975).

glioma, neuroblastoma line NA, SV40-transformed 3T3 (SV3T3), neuroblastoma line N18, and chinese hamster ovary (CHO) cells.[7] The inclusion of 1–6 M glycerol in the cell extraction buffers ameliorated the inhibitory effect on MT assembly. The concentration of glycerol required was dependent on the cell content of inhibitor.[7] For HeLa cells, the level of inhibitor in cell extracts is a function of the method of cell disruption. Gentle sonication yielded no inhibitory activity, whereas this activity was detected if cells were disrupted using a motor-driven Teflon-on-metal homogenizer.[5]

Initially, all successful attempts to reassemble TU from cell cultures and nonneural tissues employed 4 M glycerol in the cell extraction buffer. In retrospect, it seems that glycerol was important for at least two reasons. First, it may promote MT assembly by decreasing the TU concentration required for polymerization; and second, the presence of glycerol decreases endogenous inhibitory activity of MT assembly.

Subsequent studies have shown that glycerol is not essential for MT assembly in HeLa and EAT cell extracts.[4,5] In view of the above considerations it is not surprising that, for successful MT reassembly under these conditions, use of high protein concentrations and cell disruption methods which minimize the presence of inhibitor, are emphasized.

Tubulin Isolation Buffers

Methodologies for purifying nonneural TUs that are based upon MT assembly–disassembly employ cell extraction–reassembly buffers essentially identical to those developed for vertebrate brain: 0.1 M piperazine-N,N'-bis(2-ethanesulfonic acid (PIPES)/2-(N-morpholino)-ethanesulfonic acid (MES), 0.5–1.0 mM Mg, 1 mM ethyleneglycol-bis-(β-aminoethyl ether)N,N'-tetraacetic acid (EGTA), 1–2 mM GTP, pH 6.4–6.94. Further additions to this basic recipe appear to be important in certain cases. For example, the inclusion of 1 mM dithiothreitol (DTT) in HeLa cell extraction buffers is essential for MT formation.[5] Also, 1 mM EDTA has been included in buffers for pig platelets.[8,9] The inclusion of EDTA reportedly aids TU purification by selectively suppressing copurification of the high concentrations of actin present in platelets.

Tubulin Reassembly

Microtubule assembly can be initiated from 30,000–100,000 g cell supernatants by incubation at 30–37°. The TU may then be purified up to

[7] B. W. Nagle, K. H. Doenges, and J. Bryan, *Cell* **12**, 573 (1977).
[8] A. G. Castle and N. Crawford, *FEBS Lett.* **51**, 195 (1975).
[9] A. G. Castle and N. Crawford, *Biochim. Biophys. Acta* **494**, 76 (1977).

90% or greater purity by cycles of assembly–disassembly as developed for vertebrate brain. To purify the TU away from adventitiously associating protein contaminants, a large number (four or more) of assembly–disassembly cycles may be required. Since TU loss during each cycle is unavoidable, repetitive cycling may be unacceptable in cases where initial cell TU levels are low. In these instances ion-exchange methods may be preferable for the final stages of TU purification after only 1–2 cycles of assembly–disassembly. However, it should be noted that ion-exchange methods may also remove associated proteins apparently required for MT assembly.[5] If TU polymerization is essential, the use of 4 M glycerol may be necessary.

Yields of TU from twice-cycled preparations have been reported as 0.25–0.70 mg per gram of packed cells for C6 glioma and SV3T3,[10,11] and 3–5 mg from 1.5–2.0 × 10^9 HeLa cells.[12]

Tubulin Isolation by Immunosorption

An alternative purification method has been developed for human platelet TU, which has the advantages of speed, high TU yields, and purity.[13] Rabbit anti-TU was raised by injection of two-cycle-purified (to 96%) pig platelet TU. The regimen consisted of an initial intramuscular injection of 400 μg of TU followed by three booster injections each of 400 μg of TU at weekly intervals. The γ-globulin serum fraction was isolated by ammonium sulfate precipitation and preabsorbed with homologous serum proteins and MT-associated proteins (MAPs). Monospecific anti-TU antiserum (10–15 mg) was covalently linked to 1.0 g of N-hydroxysuccinimide-activated BioGel A-15 m (Affi-gel 10) by incubation in 25 ml of 0.1 M phosphate buffer (pH 7.0). From 7 mg of crude platelet extract added to a 12 × 1.5 cm TU affinity column of this resin, 0.2 mg of assembly-competent TU was obtained.

Tubulin solutions have been stored at −70° to −80° without glycerol for several weeks with no apparent loss of assembly competence. Alternatively, the TU may be stored in 4 M glycerol at −20°.

Flagellar and Ciliary Tubulins

Chemical Solubilization of Axonemal Microtubules

I stress at the outset that the procedures listed below for TU isolation, while efficacious for the systems described, may not be applicable to other

[10] G. Wiche and R. D. Cole, *Proc. Natl. Acad. Sci. U.S.A.* **73**, 1227 (1976).
[11] K. Weber, R. Koch, W. Herzog, and J. Vanderkerckhove, *Eur. J. Biochem.* **78**, 27 (1977).
[12] J. A. Weatherbee, R. B. Luftig, and R. R. Weihing, *J. Cell Biol.* **78**, 47 (1978).
[13] Y. Ikeda and M. Steiner, *J. Biol. Chem.* **251**, 6135 (1976).

flagellar systems: significant interspecies variation exists in the solubility properties of axonemal MTs. None of the axonemal MTs are solubilized by exposure to low temperatures.

Central-Pair Tubulin. Tubulin has been isolated from central-pair MTs by dialysis of demembranated whole axonemes at 4° against low ionic strength buffers: 0.1 mM EDTA, 1 mM Tris-thioglycolate, pH 7.5–8.3, solubilizes over 80% of central-pair MTs in *Strongylocentrotus purpuratus* after 7 hr,[14] and virtually all of the central-pair MTs in *Tetrahymena pyriformis* after 2 days.[15] The TU is more than 85% pure in the case of *S. purpuratus* and binds colchicine with a stoichiometry of 0.5–1.0 mol of colchicine per mole of TU dimer. Central-pair MTs may also be solubilized by brief (ca. 1.0–1.5 hr at 4° for *S. purpuratus*[16]) exposure to 0.6 M KCl, 5 mM ethylenediaminetetraacetic acid (EDTA), 1 mM DTT, 1 mM ATP, 10 mM Tris-phosphate pH 8.0. However, this method is less easy to control than the use of low ionic strength buffers, and for *T. pyriformis* axonemes contamination with outer doublet TU can occur.[15]

Outer-Doublet Tubulin. Outer-doublet MTs, freed of central-pair MTs by one of the above procedures, can be solubilized by high ionic strength buffers. For *T. pyriformis* 0.6 M KCl, 10 mM Tris-HCl, pH 8.3, solubilized about 90% of the outer doublet MTs after 18 hr at 4°.[15] More rapid depolymerization of outer doublets was obtained when KI substituted for the KCl. A similar effect was observed for 0.6 M KI and 0.3–0.4 M KSCN on outer doublets from *Echinus* and *Strongylocentrotus*.[17] However, 25–30% of the outer doublets remained insoluble in the form of three-protofilament "ribbons."

Other methods include extraction of acetone powders of *T. pyriformis*[18] and *S. drobachiensis*[19] outer doublets with 10 mM Tris-HCl pH 8.0, 1 mM GTP, and solubilization of 4–6 mg of *S. drobachiensis* outer doublets in 1.0–1.5 ml of 0.5% Sarkosyl, 1 mM GTP, 10 mM Tris-HCl, pH 8.0, and 0.01% 2-mercaptoethanol for 30 min on ice.[20] The latter method solubilized a minimum of 90% of the outer doublets. Also, 1 mM Salygran, 1 mM GTP, 10 mM Tris-HCl, pH 8.0 dissolved 80–90% of *S. drobachiensis* outer doublets after extraction overnight on ice.[19]

Selective solubilization of B-subfiber MTs was achieved in *S. drobachiensis*[21] and *S. purpuratus*[22] by warming outer doublets to 40° for

[14] M. L. Shelanski, and E. W. Taylor, *J. Cell Biol.* **34**, 549 (1967).
[15] I. R. GIbbons, *Arch. Biol.* **76**, 317 (1965).
[16] K. W. Farrell and L. Wilson, *J. Mol. Biol.* **121**, 393 (1978).
[17] R. W. Linck, *J. Cell Sci.* **20**, 405 (1976).
[18] F. L. Renaud, A. J. Rowe, and I. R. Gibbons, *J. Cell Biol.* **36**, (1968).
[19] R. E. Stephens, *Q. Rev. Biophys.* **1**, 377 (1969).
[20] R. E. Stephens, *J. Mol. Biol.* **33**, 517 (1968).
[21] R. E. Stephens, *J. Mol. Biol.* **47**, 353 (1970).
[22] I. Meza, B. Huang, and J. Bryan, *Exp. Cell Res.* **74**, 535 (1972).

1–10 min in 1 mM Tris-HCl, pH 8.0. The residual intact A-tubules were removed by centrifugation (35,000 g, 30 min) and solubilized by warming to 60° for 6 min or by exposure to pH 2.5.

Isolation of Assembly-Competent Outer-Doublet Tubulin

The above procedures give good yields of highly purified TU, but the successful reconstitution of bona fide MTs from such TU solutions is not achieved. To obtain native, assembly-competent TU from outer doublets, one of two methods have been employed.

Thick suspension of outer doublets in a low ionic strength buffer (5 mM MES, 0.5–1.0 mM Mg^{2+}, 1.0 mM EGTA, 1 .0 mM GTP, pH 6.7) can be disrupted by sonication on ice.[16,23,24] Both continuous (40–50 W, 3 min) and intermittent (four 1-min bursts of 200 W) sonication have been used with success. The degree of outer doublet solubilization is dependent upon low ionic strength; the presence of KCl as low as 50 mM noticeably decreased TU yields. Immediately after sonication, KCl to 150 mM should be added to stabilize the solubilized TU.

Pretreatment of outer doublets with trypsin at a weight ratio of 1 : 1500–6400 for 5 min at 25° significantly increased TU yields.[23] However, in this case the maximum degree of outer doublet solubilization (20%) was still below that obtained by other workers (35–54%) without the use of trypsin.[16,24] For this reason, the use of trypsin appears unnecessary and should probably be avoided.

Sonication apparently solubilized both A- and B-subfiber MTs and has been successfully applied to five species of sea urchin sperm flagella and T. pyriformis cilia.

A second, less widely used method, involves exposing outer-doublet suspensions to 16,000 psi in a French pressure cell.[25] Solubilization of the outer doublets apparently results from shearing, rather than hydrostatic pressure, and 17–35% of the outer doublets are solubilized. The TU is native both by colchicine-binding assay (0.75 mol of colchicine per mole of dimer) and assembly competence.

From both these methods, TU solutions of greater than 80% purity can be obtained. Further purification of the TU to at least 95% can be achieved by cycles of assembly–diassembly, essentially as described for vertebrate brain TU, or by ion-exchange and gel filtration chromatography.

Solubilized outer-doublet TU can be stored in liquid N$_2$, either as

[23] R. Kuriyama, J. Biochem. (Tokyo) **80**, 153 (1976).
[24] L. I. Binder and J. L. Rosenbaum, J. Cell Biol. **79**, 500 (1978).
[25] T. A. Pfeffer, C. Asnes, and L. Wilson, Cytobiologie **16**, 367 (1978).

reconstituted MT pellets or as soluble TU solutions, for several weeks with no significant loss of assembly competence.

Other Systems

The assembly–disassembly procedure for TU isolation has been applied to a variety of other nonneural sources. Since the details of these methods are essentially identical to those above, these references are simply listed: Drosophila eggs and embryos,[26] eggs from the sea urchins *Hemicentrotus,* and *Anthocidaris* and the starfish *Asterias,*[27] and cytoplasmic TU from the alga *Polytomella.*[28]

Isolation of TU by a Copolymerization Method

One modification of the assembly–disassembly procedure deserves comment, since it allows the isolation of TU from sources to which conventional techniques cannot be applied. Tubulin from the fungus *Aspergillus nidulans* was labeled *in vivo* by growing the cells for 24 hr at 32° in 200 ml of 6 g of glucose, 0.6 g of $NaNO_3$, 0.2 g of KH_2PO_4, 0.1 g of $MgCl_2$, 0.1 g of KCl, 0.002 g of $FeSO_4$, 0.002 g of $(NH_4)_2SO_4$, 0.004 g of adenine, and 1 mCi of $Na_2[^{35}S]O_4$ (825 mCi/mmol).[29] One gram wet weight of labeled cells was resuspended in 5 ml of 0.1 M PIPES (pH 6.95), 1 mM EGTA, 1 mM GTP, 1 mM PMSF and disrupted by homogenization.

Tubulin was isolated from 100,000 g cell supernatants by adding 20–25 mg of two-cycle-purified porcine brain TU in an equal volume of 0.1 M PIPES (pH 6.95), 1 mM EGTA, 1 mM GTP, 1 mM $MgSO_4$, and 8 M glycerol and copolymerizing the TUs at 37° through cycles of assembly–diassembly. Analysis of the fungal TU was then accomplished with standard biochemical techniques using the ^{35}S as marker. This technique has also been used to isolate *Saccharomyces cerevisiae* TU.[30]

Since these TUs do not form tight complexes with colchicine *in vitro,* the use of radioactively labeled colchicine to identify the TUs during purification is excluded. Furthermore, TU constitutes approximately 1% or less of the soluble cell proteins in these organisms, a fact that probably largely accounts for the absence of reports documenting the *in vitro* reassembly of these TUs by themselves. The use of the copolymerization

[26] L. H. Green, J. W. Brandis, F. R. Turner, and R. A. Raff, *Biochemistry* **14**, 4487 (1975).
[27] R. Kuriyama, *J. Biochem. (Tokyo)* **81**, 1115 (1977).
[28] M. E. Stearns and D. L. Brown, *Proc. Natl. Acad. Sci. U.S.A.* **76**, 5745 (1979).
[29] G. Sheir-Neiss, R. V. Nardi, M. A. Gealt, and N. R. Morris, *Biochem. Biophys. Res. Commun.* **69**, 285 (1976).
[30] R. D. Water, and L. J. Kleinsmith, *Biochem. Biophys. Res. Commun.* **70**, 704 (1976).

method therefore allows analysis of TUs in such cases, where standard methodology is inapplicable.

Vinblastine-Tubulin Paracrystals

Exposure of a variety of cells to low concentrations of the *Vinca* alkaloid vinblastine results in the formation of intracellular paracrystals,[31,32] of which TU is the sole protein component.[33,34] The TU isolated by this method is native by colchicine-binding assay, but no reports of *in vitro* MT reconstitution from this TU have appeared.

The following method is for vinblastine-TU paracrystal isolation from unfertilized *S. purpuratus* eggs,[35] but can probably be adapted to other systems with only minor modifications.

Eggs (40–100 ml) were incubated with 1 to 3 × $10^{-4} M$ vinblastine at 13–15° for 36 hr with gentle stirring. Egg pellets (120 g, 2–3 min) were transferred to 2–3 volumes of 1 M urea, 10 mM Tris-HCl (pH 8.5) at 4° and repelleted after 1–2 min in this buffer. The pellets were then resuspended in 1% Triton X-100 or Nonidet P-40, 10 mM Tris-HCl (pH 7.2), and the cells were lysed by vortexing for 30 sec.

Two points are important at this stage of the isolation for good paracrystal yield.[36] First, the ratio of lysing buffer to egg suspension should not exceed 2 : 1 (v/v). The paracrystals are unstable in lysing buffer; keeping the paracrystal suspension concentrated reduces protein loss during experimental manipulations. This is also true for subsequent steps in the procedure, even when the paracrystals are in stabilizing buffer (see below).

Second, complete solubilization of the egg membranes should be verified by light microscopy. Intact membranes will copurify with the paracrystals and reduce final paracrystal purity.

After lysis, paracrystals were centrifuged to a pellet (12,000 g, 4°), resuspended at 4° in 2–5 ml of a stabilizing buffer (either 100 mM KCl, 1 mM MgCl$_2$, 10 mM Tris-HCl, pH 7.2, or 300 mM KCl, 10 mM MgCl$_2$, 10 mM Tris-HCl, pH 7.5) and passed through a Nitex-44 filter to remove cell debris. The partially purified paracrystals were again pelleted by centrifuging at 1000 g at 4° and resuspended in fresh stabilizing buffer. Further cycles of Nitex filtration and centrifugation may be necessary to remove residual cell debris.

[31] S. S. Schochet, Jr., P. W. Lambert, and K. M. Earle, *J. Neuropathol. Exp. Neurol.* **27**, 645 (1968).
[32] K. G. Bensch, and S. E. Malawista, *J. Cell Biol.* **40**, 95 (1969).
[33] J. Bryan, *J. Mol. Biol.* **66**, 157 (1972).
[34] J. Bryan, *Biochemistry* **11**, 2611 (1972).
[35] L. Wilson, A. N. C. Morse, and J. Bryan, *J. Mol. Biol.* **121**, 255 (1978).
[36] L. Wilson, personal communication.

Paracrystal stability *in vitro* has been examined[37] and found to be optimal in 300 mM KCl, 10 mM MgCl$_2$, and 10 mM Tris-HCl, pH 7.5. At pH levels in excess of 7.5 the paracrystals rapidly dissolve. Similarly, the presence of Mg^{2+} is crucial: addition of 5 mM EDTA to a paracrystal suspension in stabilizing buffer caused rapid (2–4 hr) paracrystal dissolution.

Yields of TU from this procedure are highly variable, but with care can be 1–2 mg of TU per 1.0 ml of gravity-settled eggs. Preliminary data indicate that TU yields can be increased by including 10^{-4} M podophyllotoxin in the cell suspension with vinblastine during paracrystal induction.

The paracrystals can be stored at 4–10° in stabilizing buffer for 3–4 days with some paracrystal loss. Immediate and complete paracrystal solubilization occurs in distilled water or 1 mM Tris-HCl (pH 8.0).

[37] S. Gominak and L. Wilson, unpublished data.

[38] Physical Properties of Purified Calf Brain Tubulin

By GEORGE C. NA and SERGE N. TIMASHEFF

Tubulin, the subunit protein of cellular microtubules, can be purified to near homogeneity by the modified Weisenberg procedure.[1–3] During the past decade, this protein has been examined in great detail from the point of view of both its physicochemical and its biochemical properties, and today it is probably the most carefully characterized type of tubulin. It has been established that such highly purified tubulin retains many of the key biochemical characteristics that it displays *in vivo*. This includes the ability of the protein to self-assemble into microtubules, the response of the assembly reaction to the inhibiting effects of cold temperature, Ca^{2+}, and specific anti-microtubule agents, such as vinblastine and colchicine,[4,5] and the self-aggregation of the protein upon binding vinblastine leading to the formation of large paracrystalline aggregates.[6] Although over the years other ingenious methods have been developed for the preparation of tubu-

[1] R. C. Weisenberg, G. Borisy, and E. Taylor, *Biochemistry*, **7**, 4466 (1968).
[2] R. C. Weisenberg and S. N. Timasheff, *Biochemistry* **9**, 4110 (1970).
[3] J. C. Lee, R. P. Frigon, and S. N. Timasheff, *J. Biol. Chem.* **248**, 7253 (1973).
[4] J. C. Lee and S. N. Timasheff, *Biochemistry* **14**, 5183 (1975).
[5] J. C. Lee and S. N. Timasheff, *Biochemistry* **16**, 1754 (1977).
[6] G. C. Na and S. N. Timasheff, *J. Biol. Chem.*, in press.

lin, the tubulin obtained by the modified Weisenberg procedure, by virtue of its high purity and intact biological activities, remains the protein of choice for precise studies aimed at the examination and analysis of particular biochemical or biophysical properties of the protein under the simplest and best defined conditions. In this chapter, we summarize some of the most characteristic properties of this protein, which can be used as standards in establishing the purity and integrity of tubulin preparations.

While the methods described in this chapter should be applicable to pure tubulin prepared by a variety of procedures, the experience of the authors is limited to highly purified tubulin isolated by the modified Weisenberg procedure; therefore, the values of all physical parameters presented herein should be regarded as valid only for this protein. The methods described in this chapter are those involved in (a) the handling of the protein for experimentation, starting with the purified tubulin stored in liquid nitrogen; (b) the routine characterization of the protein to establish its homogeneity and biochemical integrity; and (c) the determination of its other physical properties.

Preparation of Tubulin for Experimentation

A detailed description of the modified Weisenberg procedure of tubulin isolation and purification is given in this volume[7] and will not be repeated here. Tubulin purified in this manner is usually stored under liquid nitrogen in small aliquot portions in $1\ M$ sucrose-PMG buffer ($0.01\ M$ NaP$_i$, $5 \times 10^{-4}\ M$ MgCl$_2$, $10^{-4}\ M$ GTP, pH 7.0).[8] Stabilized by the high concentration of sucrose and the low temperature, the protein can be stored in this manner for up to 6 months without detectable changes in its properties. Immediately prior to each experiment, the protein is thawed at $4°$, and all subsequent steps are performed at $4°$ if conditions permit.

The transfer of the tubulin from the sucrose stabilizing buffer to the experimental buffer desired is performed by two gel filtrations. To avoid overdilution of the sample, the first gel equilibration is carried out by a batch-dry column technique. A glass column ($1\ cm \times 8\ cm$) with a fritted glass disk bottom is packed with Sephadex G-25 (F) gel and washed with several volumes of the experimental buffer (this column can be cut out from a standard Pyrex $1\ cm \times 30\ cm$ glass column). Vacuum can be applied to the bottom of the column to speed the flow, taking care that the buffer does not run beneath the gel surface. The excess buffer present in the void volume of the gel is then removed by placing the column in a 12-ml conical glass centrifuge tube and spinning at 3000 rpm ($1000\,g$) for 5

[7] R. C. Williams, Jr. and J. C. Lee, this volume [36].
[8] R. P. Frigon and J. C. Lee, *Arch. Biochem. Biophys.* **153**, 587 (1972).

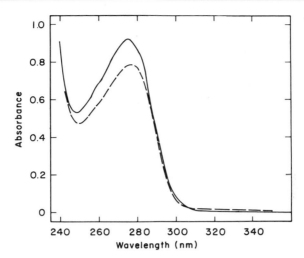

FIG. 1. Ultraviolet spectra of calf brain tubulin prepared by the modified Weisenberg procedure: ---, in PG buffer; ——, in 6 M Gu-HCl.

min in a Beckman J-21 (or similar) rotor fitted with 15-ml tube adapters without covering the rotor with its cap. At this point, a 20–30-mg aliquot of the protein is thawed and loaded onto the column. After 3–5 min of equilibration, the column is again centrifuged in the same rotor at 3000 rpm for 10 min, the protein being collected in a conical centrifuge tube. At this point, the protein sample may display some turbidity, which is probably due to its incomplete resuspension after the Mg^{2+}-induced precipitation in the last step of the tubulin preparation. The solution can be cleared of such large aggregates by centrifugation at 39,200 g for 30 min. The supernatant is then loaded on a second Sephadex G-25 (F) column (1 cm × 12 cm) and eluted with the desired buffer. This second step achieves complete equilibration with the experimental buffer. Control experiments have shown that this procedure allows complete removal of sucrose without significantly diluting the protein sample, results not achieved by using a single larger column.[9]

Concentration Determination

Tubulin concentration can be determined most easily and accurately by measuring its ultraviolet (UV) absorbance. A typical UV spectrum of calf brain tubulin in PG buffer is shown in Fig. 1. This spectrum is charac-

[9] N. Tweedy, Doctoral Dissertation, Brandeis University, 1981.

terized by an absorption maximum at 278 nm and an extinction coefficient of 1.33 ml mg^{-1} cm^{-1}.[3] Protein concentration determination using this extinction coefficient becomes less accurate if the solvent contains magnesium ions, since such solutions of tubulin usually display some turbidity, and the absorption must be corrected for light scattering by the method of Leach and Scheraga.[10] An alternative method that avoids the light-scattering correction involves the measurement of protein absorbance in 6 M Gu-HCl.[11] Since the protein is unfolded in 6 M Gu-HCl and the light scattering of the solution is virtually nil, this method is much more accurate and uses less protein. Ultrapure grade Gu-HCl (Heico, Co.) should be used. However, prior to use, the 6 M Gu-HCl solution must be filtered through a sintered-glass funnel because it always contains some insoluble debris, and it must be checked for no absorption of light at 275 nm. To determine the tubulin concentration, a small volume of the protein (10–50 μl) is added to 1 ml of 6 M Gu-HCl. Since the experimental buffer usually contains 10^{-4} to 10^{-3} M GTP, the proper reference solvent should be the 6 M Gu-HCl solution to which had been added a volume of the experimental buffer equal to the volume of protein solution used. The UV absorption spectrum of tubulin in 6 M Gu-HCl is shown by the solid line of Fig. 1. The absorption maximum of this spectrum is at 275 nm with an extinction coefficient of 1.03 ml mg^{-1} cm^{-1}.[12] This last value has been determined by using the precision dry weight method of Kupke.[13]

On occasion, it may be desired to determine the concentration of tubulin in the presence of ligands, such as vinblastine or colchicine, which themselves absorb in the UV region of interest. If the total concentration of the ligand is unknown, as in cases when the protein is equilibrated with an experimental buffer containing a specific concentration of the ligand that binds strongly to the protein,[14] one must resort to other methods of measurement that are not interfered with by the presence of the ligands. Such methods include the Lowry[15] technique and refractive index determination.[14] When using such methods, one should use tubulin and the above extinction coefficients to establish the standard curve, rather than bovine serum albumin or lysozyme, which are normally employed, but result in incorrect values of tubulin concentration.

[10] S. J. Leach and H. A. Scheraga, *J. Am. Chem. Soc.* **82**, 4790 (1960).

[11] J. C. Lee and S. N. Timasheff, *Biochemistry* **13**, 257 (1974).

[12] G. C. Na and S. N. Timasheff, *J. Mol. Biol.* **151**, 165 (1981).

[13] D. W. Kupke and T. E. Dorrier, this series, Vol. 48, p. 155.

[14] G. C. Na and S. N. Timasheff, *Biochemistry* **19**, 1347 (1980).

[15] O. H. Lowry, N. J. Rosebrough, A. L. Farr, and R. J. Randall, *J. Biol. Chem.* **193**, 265 (1951).

Routine Physical Characterizations of Tubulin

It is a routine practice in our laboratory that each preparation of tubulin be subjected to a number of tests in order to reveal its purity and functional integrity. These include: SDS-gel electrophoresis, ability to assemble into microtubules, and velocity sedimentation in PG buffer, without and with 16 mM MgCl$_2$.

SDS-Gel Electrophoresis. With standard protein loadings, polyacrylamide gel electrophoresis of the Weisenberg tubulin in the presence of 0.1% SDS reveals a single band (Fig. 2, lane A).[3] At very high loadings (200 μg) the gels show some traces of nontubulin proteins.[4] Densimetric tracings of the stained gels should show 98–99% of the protein as tubulin subunits.[9] Addition of urea to the medium leads to resolution of the pattern into two bands corresponding to the α and β subunits of tubulin (Fig. 2, lane B).[3]

Microtubule Assembly in Vitro. Tubulin isolated by the modified Weisenberg procedure has the full capability of self-assembling into microtubules *in vitro*. The microtubule reconstitution experiment is usually performed in an assembly buffer containing 0.01 M NaP$_i$, 1.6 × 10^{-2} M MgCl$_2$, 10^{-4} M GTP, and 3.4 M glycerol, at pH 7.0, although several other buffers have been used equally well.[4,5] The *in vitro* assembly reaction is monitored turbidimetrically at 350 nm. For the assembly experiment, tubulin is equilibrated with the assembly buffer by the two gel filtrations, and the solution is introduced into a jacketed cell of 1 cm light path. The cell is connected to two water baths, set at 10° and 37°, respectively. By controlling the flow of the thermostatted water with two three-way stopcocks, the temperature of the solution can be interchanged swiftly between 10° and 37°.

Figure 3 depicts a typical result of the assembly reaction. The reaction is usually started by raising the cell temperature from 10° to 37°. There is a short lag period without any change in solution turbidity, which is followed by a fast increase of solution turbidity and then a final plateau region. Switching the temperature back to 10° should result in an immediate decline of the solution turbidity to the initial level, indicating complete reversibility of the reaction. At each temperature jump, there is transient perturbation of the absorbance due to schlieren effects in the cell. Figure 3b depicts a plot of the final plateau values of solution turbidity versus the protein concentration. Since the assembly reaction is strongly cooperative, a critical concentration of 0.8 to 1.1 mg/ml is found under these experimental conditions, with no assembly taking place below this concentration level.

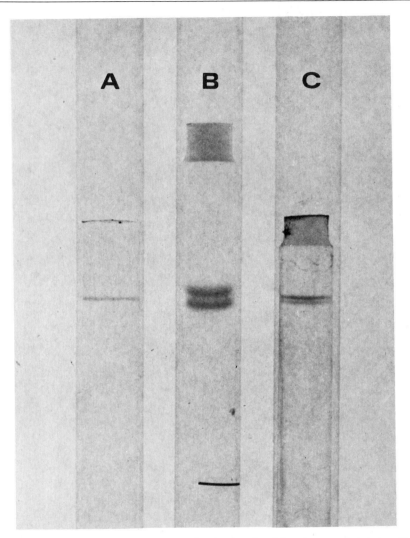

FIG. 2. Polyacrylamide gel electrophoreses of reduced and S-carboxymethylated purified calf brain tubulin; approximately 30 μg of tubulin were applied to each gel; the denaturing media are as follows: Lane A, 0.1% sodium dodecyl sulfate (SDS); B, 0.1% SDS + 8 *M* urea; C, 8 *M* urea. From Lee *et al.*[3]

Microtubule inhibiting or destabilizing agents, such as vinblastine and colchicine, can be added to the assembly mixture either before, during, or after the onset of the assembly reaction, in order to examine their effects on different phases of the process. One should be cautioned, however,

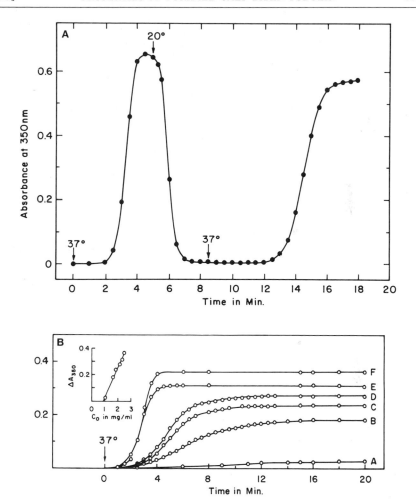

FIG. 3. *In vitro* reconstitution of microtubules from purified calf brain tubulin, as monitored by the solution turbidity at 350 nm. (A) Kinetics and temperature dependence of the assembly–disassembly reaction. (Taken from Lee and Timasheff.[4]) (B) Protein concentration dependence and critical concentration of the assembly reaction (inset); the tubulin concentration varied between 1.0 and 2.5 mg/ml (J. C. Lee and S. N. Timasheff).

that if a stock solution is added to the assembly mixture after its turbidity had plateaued, the simple mechanical shaking of the solution to disperse the ligand usually results in a drop of the solution turbidity. Proper controls (addition of identical volumes of ligand free buffer) should, therefore, be carried out to avoid possible misinterpretation of results.

Velocity Sedimentation in PG Buffer. Tubulin isolated by the modified

Weisenberg procedure sediments in PG buffer (0.01 M NaP$_i$, 10^{-4} M GTP, pH 7.0) as a single, rather symmetrical peak at $s^0_{20,w}$ = 5.8 S (Fig. 4a). This reflects the high homogeneity of the protein and indicates that under these conditions the protein exists predominantly in the form of dimers of 110,000 M_r.[3,16] The hydrodynamic nonideality of the protein gives rise to a negative dependence of the sedimentation coefficient, $s_{20,w}$ on protein concentration, which conforms well to the standard equation[17] $s_{20,w}$ = $s^0_{20,w}$ $(1 - gc)$, where c is the protein concentration (mg/ml) and g is the hydrodynamic nonideality constant of the protein. For purified tubulin g = 0.018 ml/mg[16] so that the sedimentation coefficient at any given protein concentration is equal to $s_{20,w}$ = 5.8 × 10^{-13} (1–0.018 c) sec^{-1}.

Sometimes a small shoulder traveling at approximately 9 S is found at the front edge of the sedimentation boundary (Fig. 4b). This usually happens if the protein preparation procedure had been stretched over too long a period. A typical modified Weisenberg tubulin preparation should take no longer than 10–12 hr from the start of cleaning the brains to the point where the protein is set to dialyze against the 1 M sucrose PMG buffer. There are several pieces of evidence suggesting that the fast-moving shoulder is the result of a slow aggregation of tubulin. A similar broadening of the sedimentation boundary has been observed in purified tubulin 4–5 hr after removal from the sucrose stabilizing solvent, even though the protein had been stored at 0°.[18] Also, the gradual disappearance of the 5.8 S tubulin dimers has been found to be paralleled by a decrease of the colchicine binding activity of the protein, suggesting that these two events are related.[18] Gel electrophoresis and sedimentation equilibrium studies have also revealed the presence of high molecular weight components that can be eliminated by treating the protein with disulfide reducing agents, suggesting that the aggregates are crosslinked tubulin.[3,18] Tubulin in such a state should be discarded.

Velocity Sedimentation in PG–16 mM MgCl$_2$ Buffer. In the presence of high concentrations of magnesium ions, the Weisenberg tubulin can undergo a fast, reversible self-association.[2,16] In velocity sedimentation, the association manifests itself as a single, forward skewed peak at protein concentrations below a threshold value (Fig. 5a–c), and the sedimentation boundary becomes bimodal with a further increase in protein concentration (Fig. 5d and e). Thermodynamic and hydrodynamic analyses of the sedimentation data have shown that, under these conditions, tubulin undergoes a reversible self-association which proceeds by a stepwise

[16] R. P. Frigon and S. N. Timasheff, *Biochemistry* **14**, 4559 (1975).
[17] H. K. Schachman, "Ultracentrifugation in Biochemistry." Academic Press, New York, 1959.
[18] V. Prakash and S. N. Timasheff, *J. Mol. Biol.*, in press.

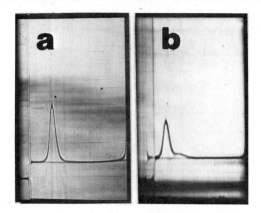

FIG. 4. Sedimentation velocity patterns of purified tubulin in PG buffer. (a) Native tubulin; (b) partially self-aggregated tubulin with 9 S shoulder. The rotor speed was 60,000 rpm.

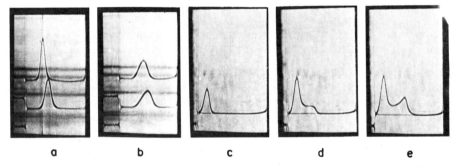

a   b   c   d   e

FIG. 5. Sedimentation velocity patterns of purified tubulin in PG buffer with different concentrations of magnesium chloride. The rotor speed was 60,000 rpm for panels a and b and 48,000 rpm for panels c–e. The protein concentration was about 8 mg/ml for the samples of panels a and b: (a) 41 min after reaching speed; upper, no magnesium; lower, 2.7 mM MgCl$_2$; (b) 38 min after reaching speed; upper, 5.5 mM MgCl$_2$; lower, 8.2 mM MgCl$_2$. Panels c–e: 21 min after reaching speed; 10 mM MgCl$_2$; protein concentration, 4.7, 10.4, and 15.5 mg/ml, respectively. Taken from Frigon and Timasheff.[19]

isodesmic pathway and is then terminated by the formation of an end product in an energetically favorable step.[16,19]

Hydrodynamic and electron microscopic studies have identified the final associated product to be a double-ring structure containing 26 ± 2 tubulin dimers,[19] shown in Fig. 6 and characterized by a sedimentation coefficient, $s_{20,w}^0 = 42$ S. These velocity sedimentation patterns are unique for highly purified tubulin, be it prepared by the Weisenberg procedure or by phosphocellulose chromatography of cycle preparations. These pat-

[19] R. P. Frigon and S. N. Timasheff, *Biochemistry* **14**, 4567 (1975).

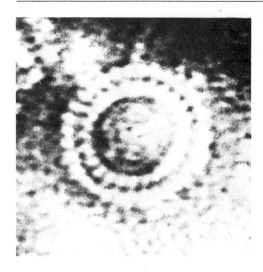

$$O.D. = 47 \pm 3 \, nm$$

$$I.D. = 27 \pm 3 \, nm$$

$$a/b \, (Oblate) = 11.5$$

$$n = 26$$

$$M_{26} = 2.86 \times 10^6$$

$$s = 43 \, S$$

Experimental

$$s_{26} = 42 \, S$$

$$a/b \, (Oblate) = 10.5$$

FIG. 6. Electron micrograph of the magnesium-induced double-ring structure of calf brain tubulin. The image has been rotated about its center of symmetry by 13.8°, followed by a second exposure. Taken from Frigon and Timasheff.[19]

terns are not shared by tubulin made by the cycle procedure. This sedimentation behavior is, therefore, diagnostic of the magnesium-induced rapidly reversible self-association of tubulin into double rings and does not involve the participation of any other macromolecular species. In buffer containing a constituent concentration of $MgCl_2$ of 16 mM, the threshold tubulin concentration value above which bimodality appears is 3–4 mg/ml. In 8 mM $MgCl_2$, this value is 9–10 mg/ml. Failure to generate bimodality above these protein concentrations indicates that the protein has been damaged.

Other Physical Properties

Circular Dichroism (*CD*). Calf brain Weisenberg tubulin gives rise to a characteristic circular dichroism spectrum which is highly sensitive to changes in conformation. The near and far UV spectra are usually obtained from the same protein sample, using fused silica cells with light paths of 0.01 cm (far UV) and 1 cm (near UV). Once the protein is removed from the sucrose-containing stabilizing buffer, its CD spectrum starts to undergo a continuous shift. The magnitude of this shift is not large during the first 2–3 hr and usually falls within the experimental error. However, after longer incubation periods, significant spectral changes do occur. These CD spectral shifts seem to be related to the slow aggregation

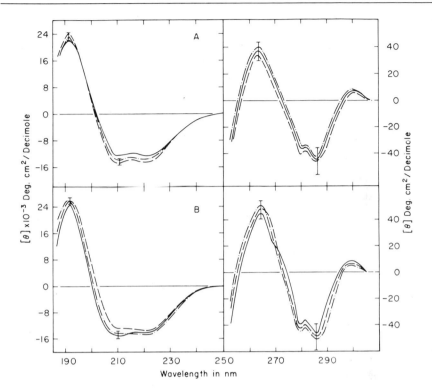

Fig. 7. Circular dichroism (CD) spectra of purified calf brain tubulin in PG buffer. The protein concentration ranges from 1.2 to 1.4 mg/ml. The vertical bars represent the maximum experimental deviations. (A) Taken at different concentrations of cations: $\cdot-\cdot-$, no magnesium; ——, 16 mM Mg^{2+}; ---, 0.1 mM Ca^{2+}. (B) Taken at different temperatures: $-\cdot-\cdot$, at 5°; ——, at 37°; ---, at 37° and with 16 mM Mg^{2+} added.

of the protein and the loss of colchicine binding activity.[18] Therefore, in obtaining the spectra, the operations should be performed rapidly after removal of the sucrose, and the spectral measurements should be repeated to assure reproducibility.

Figure 7 depicts the CD spectra of calf brain tubulin in PG buffer in the presence and in the absence of Mg^{2+} ions and at various temperatures.[20] In the far UV region, the tubulin CD spectrum has two negative bands at 221 nm and 209 nm and a positive band at 191 nm. Simulation of this spectrum by the Greenfield and Fasman procedure[21] gives an apparent secondary structure composition of 26% α-helix, 47% antiparallel pleated sheet

[20] J. C. Lee, D. Corfman, R. P. Frigon and S. N. Timasheff, *Arch. Biochem. Biophys.* **185**, 4 (1978).
[21] N. J. Greenfield and G. D. Fasman, *Biochemistry* **8**, 4108 (1969).

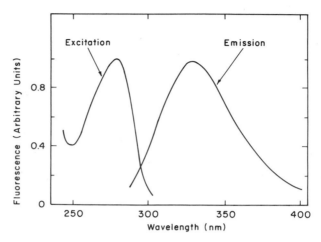

Fɪɢ. 8. Corrected fluorescence spectra of purified calf brain tubulin. The slits for excitation and emission were 2 and 5 nm, respectively. Taken from Andreu and Timasheff.[23]

β-structure, and the rest in a conformation devoid of long-range order. The near UV spectrum consists of two positive bands, a weak one at 300 nm and a stronger one at 265 nm, and two negative bands at 280 and 286 nm.[20] Guanidine hydrochloride denaturation and pH titration studies have led to an assignment of these bands to the aromatic amino acid residues and the single disulfide bond of the protein.[20] This spectrum remains unchanged over the pH range of 6.1 to 7.3, indicating stability of the tubulin over this narrow zone of pH.

Fluorescence Spectra. The intrinsic fluorescence spectrum of tubulin is characterized by excitation and emission maxima at 278 ± 1 and 329 ± 1 nm, respectively (Fig. 8). This intrinsic fluorescence of the tubulin, which is due principally to tryptophan residues,[18] is perturbed by the binding of the anti-microtubule agents vinblastine,[22] vincristine,[22] colchicine,[23], podophyllotoxin,[23] and daunomycin.[24]

Partial Specific Volumes and Preferential Interactions in Mixed Solvents. The partial specific volume of calf brain tubulin has been determined with a precision densimeter in a number of stabilizing and destabilizing solvents, and the magnitude of the preferential interactions of this protein with the solvent components has been calculated. The results are summarized in the table. Tubulin in PG buffer or PSG buffer (0.01 M NaP$_i$, 0.02 M NaCl, 10^{-4} M GTP, pH 7.0) has a partial specific volume of 0.736 ±

[22] J. C. Lee, D. Harrison, and S. N. Timasheff, *J. Biol. Chem.* **250**, 9276 (1975).
[23] H. M. Andreu and S. N. Timasheff, *Biochemistry,* in press.
[24] G. C. Na and S. N. Timasheff, *Arch. Biochem. Biophys.* **182**, 147 (1977).

PARTIAL SPECIFIC VOLUMES AND PREFERENTIAL INTERACTION PARAMETERS OF
TUBULIN IN DIFFERENT MIXED SOLVENTS

Cosolvent	Concentration of cosolvent	Buffer[a]	$\phi_2'^{0b}$ (ml/g)	$\left(\dfrac{\partial g_3}{\partial g_2}\right)_{T,\mu_1,\mu_3}^{c}$ (g/g)	Reference[d]
None	—	PG	0.736	—	11
None	—	PSG	0.735	—	12
Glycerol	1.37 M	PSG	0.742	−0.034	12
Glycerol	2.74 M	PSG	0.748	−0.074	12
Glycerol	4.11 M	PSG	0.754	−0.127	12
Sucrose	0.5 M	PG	0.749	− 0.038	e
Sucrose	1.0 M	PG	0.765	−0.106	e
PEG 1000	10 g/100 ml	PG	0.773	−0.25	25
PEG 4000	10 g/100 ml	PG	0.783	−0.31	25
Gu-HCl	6 M	PG	0.725	0.10	11

[a] PG: 0.01 M NaP$_i$, 10^{-4} M GTP, pH 7.0; PSG: 0.01 M NaP$_i$, 0.02 M NaCl, 10^{-4} M GTP, pH 7.0.

[b] $\phi_2'^0$: Apparent isopotential specific volume of tubulin, extrapolated to zero protein concentration; for method of determination, see J. C. Lee, K. Gekko, and S. N. Timasheff, this series, Vol. 61, p. 26.

[c] $(\partial g_3/\partial g_2)_{T,\mu_1,\mu_3}$: preferential interaction parameter of cosolvent (component 3) with tubulin (component 2); g_i: concentration of component i in grams of i per gram of component 1 (H$_2$O); T: thermodynamic (Kelvin) temperature; μ_i: chemical potential of component i; for methods of determination and detailed meaning, see Lee et al., this series, Vol. 61, p. 26.

[d] Numbers refer to text footnotes.

[e] J. C. Lee, R. P. Frigon, and S. N. Timasheff, Ann. N.Y. Acad. Sci. 253, 284 (1975).

0.001 ml/g.[11,12] The isopotential specific volume of the protein changes upon introduction of the various cosolvents into the buffer. Of the different cosolvents shown in the table, three [sucrose, glycerol, and polyethylene glycol (PEG)] are known to enhance the in vitro microtubule assembly.[4,5,25-27] Furthermore, sucrose and glycerol have been shown to stabilize the native structure of tubulin against denaturation.[8,26] An increase in the concentration of these cosolvents results in increases of the isopotential specific volume of tubulin, namely, in preferential exclusion of the cosolvent from the domain of the protein and the preferential hydration of the latter. Guanidine hydrochloride, on the other hand, denatures the protein and lowers its isopotential specific volume, suggesting prefer-

[25] L. Lee and J. C. Lee, Biochemistry 24, 5518 (1979).
[26] M. L. Shelanski, F. Gaskin, and C. R. Cantor, Proc. Natl. Acad. Sci. U. S. A. 70, 765 (1973).
[27] W. Herzog and K. Weber, Eur. J. Biochem. 91, 249 (1978).

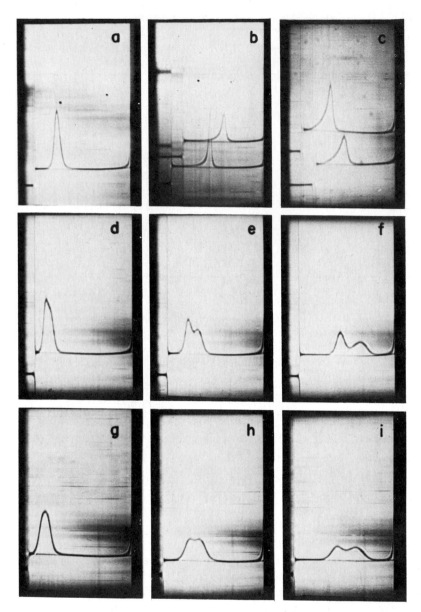

Fig. 9. Sedimentation velocity patterns of purified calf brain tubulin in the presence of various concentrations of vinblastine: (a) tubulin (12 mg/ml) in PG buffer, sedimented at 60,000 rpm for 40 min; (b) tubulin at 6.7 (upper) and 8.3 (lower) mg/ml equilibrated with PG–0.2 mM vinblastine, sedimented at 60,000 rpm for 24 min; (c) tubulin at 8.5 (upper) and 6.8 (lower) mg/ml equilibrated with PG-25 μM vinblastine, sedimented at 60,000 rpm for 16 min; (d–f) tubulin at 10.3 mg/ml equilibrated with PG–10 μM vinblastine, sedimented at 60,000 rpm for 16, 32, 64 min, respectively; (g–i) same as (d–f) except that the speed was 48,000 rpm and the pictures were taken at 32, 64, and 104 min. Taken from Na and Timasheff.[14]

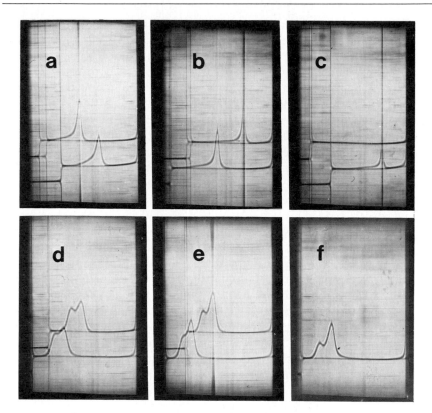

FIG. 10. Effect of magnesium ions on the vinblastine-induced tubulin self-association. (a–c) Velocity sedimentations of tubulin equilibrated with PG 50 μM vinblastine, with MgCl$_2$ added to the following total concentrations: (a) lower, no MgCl$_2$, upper 0.1 mM; (b) lower, 0.5 mM, upper, 1 mM; (c) lower 2.5 mM, upper, 5 mM (note that the last sample pelleted completely to the cell bottom). (d–f) Tubulin equilibrated with PG-10 μM vinblastine, with MgCl$_2$ added to the following total concentrations: (d) lower, no MgCl$_2$, upper, 0.1 mM; (e) lower, 0.5 mM, upper, 1 mM; (f) 2.5 mM. All runs were at 60,000 rpm, and all pictures were taken at 24 min after reaching speed except picture c, which was taken at 16 min.

ential exclusion of water molecules from the protein, or preferential binding of GuHCl to it. From the thermodynamic point of view, these changes in the preferential interaction parameters are consistent with the action of these ligands, either as "thermodynamic boosters" of the *in vitro* microtubule assembly,[7,28] as a structure stabilizing agent, or as denaturants.[5,11]

Vinblastine-Induced Self-Association. Another characteristic property of native pure tubulin is its ability to undergo an isodesmic, indefinite

[28] S. N. Timasheff, *in* "Physical Aspects of Protein Interactions" (N. Catsimpoolas, ed.), pp. 219–273. Elsevier/North-Holland, Amsterdam, 1978.

self-association reaction in PG buffer in the presence of the anti-cancer drug vinblastine.[3,14] Under these conditions, as shown in Fig. 9, at low free ligand concentration, the sedimentation boundary emerges first as a single peak, which resolves gradually into a bimodal pattern. This is characteristic of a Cann–Goad type strong ligand-induced self-association.[29,30] As the concentration of vinblastine is increased, the bimodality becomes less evident, until at high concentrations of free vinblastine ($>10\ \mu M$), the sedimentation boundary retains the form of a single forward-skewed peak throughout the course of the experiment.[14] This is characteristic of an isodesmic, indefinite self-association.[31] Similar behavior is observed when the ligand is vincristine.[32]

The vinblastine-induced tubulin self-association is strongly enhanced by magnesium ions. This is evident from Fig. 10, which shows velocity sedimentation patterns of tubulin in vinblastine–PG buffer and different concentrations of $MgCl_2$.[33] In fact, in the presence of magnesium, tubulin can self-associate into large paracrystalline aggregates[6] that are similar to the vinblastine-tubulin paracrystals observed *in vivo* in cells treated with the drug.[34] This strong enhancement of the aggregation by Mg^{2+} also leads to abnormal looking vinblastine-binding isotherms that can be misinterpreted in terms of two sites with highly different affinities, when in fact, the intrinsic affinities of the two binding sites are identical.[33] Since vinblastine protects the native structure of tubulin,[35] it is recommended, when working with this drug, to add it to the protein at an early time, preferably in the first gel filtration in which the sucrose is removed. This should minimize the slow aggregation of the protein.

Acknowledgments

Publication No. 1399 from the Graduate Department of Biochemistry, Brandeis University, Waltham, Massachusetts 02254. This work was supported by grants from the National Institutes of Health, CA 16707 and GM 14603.

[29] J. R. Cann and W. B. Goad, "Interacting Macromolecules." Academic Press, New York, 1970.
[30] J. R. Cann and W. B. Goad, this series, Vol. 27, p. 296.
[31] R. R. Holloway and D. J. Cox, *Arch. Biochem. Biophys.* **160**, 595 (1978).
[32] V. Prakash and S. N. Timasheff, submitted.
[33] G. C. Na and S. N. Timasheff, *Am. Chem. Soc. Abstr. Biol.* 170 (1980).
[34] K. G. Bensch and S. E. Malawista, *Nature (London)* **218**, 1176 (1968).
[35] L. Wilson, *Biochemistry* **9**, 4999 (1970).

[39] Purification and Assay of Microtubule-Associated Proteins (MAPs)

By ROGER D. SLOBODA and JOEL L. ROSENBAUM

Microtubule proteins have been isolated from a large number of neural and nonneural tissues in the past several years using procedures that are based on the characteristics of warm stability and cold instability of the *in vitro* assembled polymer (this volume [36]). Such *in vitro* assembly–disassembly procedures, described originally by Weisenberg[1] and Borisy and Olmsted,[2] yield semipurified preparations of tubulin, the main structural protein of the microtubule, along with several classes of microtubule-associated proteins (MAPs). The term MAP was originally coined by Sloboda *et al.*[3] to refer to a class of high molecular weight proteins that copurify with tubulin during the *in vitro* assembly–disassembly purification scheme. Quantitation of the relative amounts of these proteins per unit amount of tubulin has shown quite clearly that the MAPs associate stoichiometrically with the tubulin subunit lattice and maintain a constant ratio to tubulin through at least five cycles of assembly and disassembly *in vitro*.[3] More recently, the term MAP has come to refer to any of a number of discrete size classes of polypeptides that associate with and modulate some of the characteristics of brain tubulin.

In this regard, MAPs currently can be divided into three subclasses: (*a*) the high molecular weight MAPs, a group of three or more polypeptides whose subunit molecular weights fall in the range of 260,000 to 350,000, as judged by sodium dodecyl sulfate (SDS)–polyacrylamide gel electrophoresis; (*b*) the Tau polypeptides, a group of four closely spaced electrophoretic bands with apparent molecular weights (M_r) between 58,000 and 65,000; and (*c*) the M_r 210,000 and 125,000 polypeptides isolated from tissue culture cells.

General Considerations

Starting Material. The procedures to be described here begin with microtubules assembled *in vitro* from either brain or tissue culture cells using the methods described in detail by Williams and Lee (this volume [36]). The starting material is homogenized or sonicated in buffers either

[1] R. C. Weisenberg, *Science* **177**, 1104 (1972).
[2] G. G. Borisy and J. B. Olmsted, *Science* **177**, 1196 (1972).
[3] R. D. Sloboda, S. A. Rudolph, J. L. Rosenbaum, and P. Greegard, *Proc. Natl. Acad. Sci. U.S.A.* **72**, 177 (1975).

lacking or containing glycerol, in addition to a sulfonic acid-based buffer [piperazine-N,N'-bis(2-ethanesulfonic acid) (PIPES) or 2-(N-morpholino)-ethanesulfonic acid (MES)], magnesium ion, EGTA, and GTP, a nucleotide required for the stability of the tubulin molecule. The presence of glycerol is helpful, based on our experience, when isolating microtubules from calf or bovine brain; however, tubulin from chick or pig brain will readily assemble in the absence of glycerol. In either case, the solution of microtubule proteins is usually carried through at least two to three complete assembly–disassembly cycles prior to beginning one of the following MAP isolation procedures.

Strategy. The term microtubule proteins has been used in the literature to signify that microtubules are composed of several associated proteins (MAPs), in addition to tubulin. The strategies employed to isolate MAPs from the total microtubule protein preparation are based on several criteria: (*a*) net charge distribution affecting their behavior on ion exchange resins; (*b*) overall molecular size and axial ratio, which affect their elution from molecular sieve columns; and (*c*) varying stability to the effects of elevated temperatures (100°). One or more combinations of these properties have been used as the basis for the isolation procedures described in the following sections.

Column Preparation

Choice of Resin. By far the most widely used resin of choice for column chromatographic separation of MAPs from tubulin is phosphocellulose (PC) (Whatman P-11), used in the method initially described by Weingarten et al. [4] It provides a rapid, essentially complete means for separating MAPs from tubulin provided appropriate care is taken in performing the chromatography. When a solution of microtubule proteins is applied to a suitably equilibrated PC column, the tubulin, a highly acidic protein, does not bind to the column and elutes in the void volume, while the various classes of MAPs can be eluted with a subsequent salt gradient.

Alternatively, an anion exchange resin, such as diethylaminoethyl (DEAE) cellulose (Whatman) or DEAE-Sephadex (Pharmacia) can be used to effect a similar separation. [5] In this case, both the MAPs and tubulin bind to the column and can be separated from one another by a suitable salt gradient. The order of elution, however, is just the opposite from that of the PC column: the MAPs elute at a low salt concentration, whereas the tubulin elutes at high salt concentration, since it binds tenaciously to the DEAE owing to its acidic nature.

[4] M. D. Weingarten, A. H. Lockwood, S. Huo, and M. W. Kirschner, *Proc. Natl. Acad. Sci. U.S.A.* **72**, 1858 (1975).

[5] D. B. Murphy, K. A. Johnson, and G. G. Borisy, *J. Mol. Biol.* **117**, 33 (1977).

Each method has its advantages and disadvantages, and these will be pointed out. However, the authors' experience has led them to conclude that PC is the resin of choice for most applications for the following reasons. First, the tubulin elutes first from the column, uncontaminated by salt since none is used for its elution. As long as care is taken in equilibrating the PC column as described in the following section, the tubulin thus eluted is stable without further additions. Second, the flow characteristics of the PC column are superior to that of the DEAE column, when each is packed, equilibrated, and run as described.

Initially, one of the negative aspects of using PC was that the tubulin was variable with respect to its ability to assemble. Such a behavior, noted by ourselves and others in the literature, was attributed by Cleveland et al.[6] to variations in the quality of commercially available phosphocellulose. However, this instability has since been shown to be due to the sequestration of magnesium ion by the PC. It is known that PC binds magnesium ion[7] and also that magnesium ion is required for microtubule assembly in vitro.[8,9] Williams and Detrich[10] subsequently reported a careful analysis of the effects of PC on the magnesium ion concentration in relation to the stability of tubulin purified by PC chromatography.

Phosphocellulose. One of two things can be done to ensure an adequate supply of free magnesium ion in the PC column fractions. First, as has been suggested previously,[11] one can simply add enough $MgSO_4$ or $MgCl_2$ to the tubulin-containing fractions as they elute from the PC column to adjust their free magnesium ion concentration to 1–2 mM assuming that it is negligible in the effluent. This procedure, however, has the disadvantage that, since one is not sure of the ion concentration either in each fraction or from preparation to preparation, the magnesium ion concentration will always be somewhat variable. The preferred procedure, outlined initially by Williams and Detrich,[10] is to saturate the PC column with magnesium prior to loading the sample. The entire equilibration procedure these authors have suggested is outlined below.

Initially, phosphocellulose is precycled according to the manufacturer's instructions. This includes the following steps. First, the PC is suspended in 0.5 N NaOH or KOH for at least 30 min but no more than 2 hr (1 g of PC per 15 ml of base). Care should be exercised not to stir the solution vigorously (e.g., on a magnetic stirrer); breaking the cellulose backbone will generate increased amounts of "fines," which subsequently

[6] D. W. Cleveland, S.-Y. Hwo, and M. W. Kirschner, J. Mol. Biol. 116, 207 (1977).
[7] T. G. Cooper, "The Tools of Biochemistry," p. 387. Wiley, New York, 1977.
[8] J. B. Olmsted and G. G. Borisy, Biochemistry 14, 2996 (1975).
[9] J. C. Lee and S. N. Timasheff, Biochemistry 16, 1754 (1977).
[10] R. C. Williams, Jr. and H. W. Detrich, III, Biochemistry 18, 2499 (1979).
[11] R. D. Sloboda and J. L. Rosenbaum, Biochemistry 18, 48 (1979).

must be removed. The choice of base (NaOH or KOH) is dictated by (*a*) which base is used initially to titrate the tubulin assembly buffer when isolating the protein from brain; and (*b*) which salt (KCl or NaCl) will be used for eluting bound proteins from the column. For consistency, the same ions should be used in each step.

After the 30–120 min wash in base, the PC is allowed to settle, the supernatant is decanted, and the PC is washed on a Büchner funnel with distilled water until the pH of the effluent is about 8.0. The PC is then suspended in 0.5 *N* HCl for 30 min (1 g/15 ml) and allowed to settle; this step is repeated. The PC is again washed on a Büchner funnel with distilled water until the pH of the effluent approaches 5.0. For storage, the PC is resuspended in column buffer (CB: 50 m*M* PIPES, 1 m*M* MgSO$_4$, 2 m*M* EGTA, 0.1 m*M* GTP, 2 m*M* dithioerythritol, pH 6.9); the slurry is titrated to pH 6.9 with the appropriate base and stored for up to a month at 4°.

To prepare the column, the precycled PC is first saturated with magnesium ion as follows. One volume of the settled PC is suspended in five volumes of 0.1 *M* MgSO$_4$ in H$_2$O, the PC is allowed to settle, and this step is repeated. The PC is then washed with about 20 volumes of CB (4 changes, 5 volumes each). During this treatment, the PC will shrink and become more tightly packed as it settles. This magnesium-saturated, CB-equilibrated, PC slurry is then poured into a column of appropriate dimensions, usually 0.9 cm × 30 cm, and washed with CB until the pH and ionic strength of the effluent are the same as that in the buffer reservoir. More detailed information concerning the pouring, packing, and equilibrating of an ion exchange column can be obtained by consulting Himmelhoch[12] or Reiland.[13] The column is then ready for use.

Phosphocellulose Purification of Tubulin and a Crude MAP Fraction

Two points concerning the loading and eluting of the column are crucial for obtaining a reproducible, adequate purification of high molecular weight MAPs and Tau. First, it is important that the column not be overloaded with protein, even though the tubulin that makes up on the average 80% of the protein flows through the column without binding. A good rule of thumb is that one loads the colunn with no more than 1–3 mg of microtubule protein per milliliter of bed volume. Second, the column flow rate should be adjusted so that it is no more than 0.1–0.5 ml/min, considerably slower than such a column is capable of running. These two precautions are necessary to effect an adequate, reproducible, and complete separation of the tubulin from the MAPs.

[12] S. R. Himmelhoch, this series, Vol. 22, p. 273.
[13] J. Reiland, this series, Vol. 22, p. 287.

Once the peak of tubulin has come off the magnesium-saturated PC column, it is stable for at least 6–10 hr at 4°, or many weeks if drop-frozen in liquid nitrogen. The MAPs can then be removed from the PC column either by a linear 0 to 0.8 M salt (NaCl of KCl) gradient in CB, or by a single wash of CB which contains 0.8 M salt. In the latter case, both the high molecular weight MAPs and the Tau MAPs elute together in the same peak, whereas with the salt gradient, the Tau proteins begin to elute at lower salt concentration (ca. 0.2 M salt) then the high molecular weight MAPs (ca. 0.6 M salt).[11] However, each fraction shows slight cross contamination from some of the components of the other.[6,14] Thus, further modifications of the isolation procedure can be performed as described below if more highly purified Tau vs high molecular weight MAPs are desired.

Further Purification of MAP 2 and Tau

At this writing purification of only one of the high molecular weight MAPs, MAP 2,[3] has been reported.[14-16] To do this, cycled microtubule proteins are first suspended in CB containing 0.75 M salt, and then the solution is rapidly pipetted into tubes previously equilibrated in a boiling water bath. The tubes are kept at 100° for a total of 5 min and then are cooled rapidly by immersing them in an ice water bath. All subsequent steps are carried out at 4°. The heat-denatured protein is first removed by centrifugation at 40,000 g for 30 min, and the supernatant is concentrated to about 10 mg/ml by pressure dialysis with an Amicon cell and filter assembly (PM-30). To enrich further for MAP 2, this solution is then chromatographed on a 1.5 × 50 cm column of BioGel A-1.5m, a molecular sieve resin, equilibrated in CB plus 0.75 M salt. The volume of the concentrated protein applied to the column should be no more than 1% of the bed volume of the column, and each fraction should be approximately 1% of the column volume. Purified MAP 2 will elute first, just after the void volume, and Tau will elute between MAP 2 and the included volume, at an elution volume approximately one-half that of the column volume. Fractions from the column are monitored by absorbance at 280 nm to locate the protein peaks and then by SDS–polyacrylamide gel electrophoresis to identify the polypeptides of interest. The final step in the purification is to concentrate the MAP 2 and Tau peaks to the required protein concentration by pressure dialysis, or precipitation with 50% ammonium sulfate,

[14] W. Herzog and K. Weber, *Eur. J. Biochem.* **92**, 1 (1978).
[15] H. Kim, L. I. Binder, and J. L. Rosenbaum, *J. Cell Biol.* **80**, 266 (1979).
[16] A. Fellous, J. Francon, A.-M. Lennon, and J. Nunez, *Eur. J. Biochem.* **78**, 167 (1977).

and then to desalt the samples into the appropriate buffers by passing them over columns of Sephadex G-25.

Alternatively, one can use a PC chromatography step prior to the molecular sieve chromatography, as has been described by Herzog and Weber.[14] In this procedure, the boiled and concentrated solution of heat-stable proteins, obtained as described above, is first desalted into CB on a column of Sephadex G-25. Next, the resulting sample is loaded onto a PC column using the techniques described above, and the proteins are eluted with a linear salt gradient running from 0 to $0.5\,M$. MAP 2 elutes from the column first, followed by fractions containing both MAP 2 and Tau. The fractions containing MAP 2 and MAP 2 plus Tau are pooled separately, the proteins are concentrated by precipitation with 50% ammonium sulfate, and each sample is redissolved in CB to a concentration of approximately 5 mg/ml. Both samples are then further purified on columns of Sepharose 4B CL (1.6×110 cm) equilibrated in CB. The fractions are again monitored by absorbance at 280 nm and gel electrophoresis, and the fractions containing either homogeneous MAP 2 or Tau are pooled, concentrated, and stored in aliquots at $-70°$ until use.

For an alternative method of purification of MAP 2, the reader is directed to a report by Vallee *et al.*[17] In this paper, procedures are described for the purification of MAP 2 from microtubule protein which has not been subjected to high temperatures. This procedure has the advantage that any heat-labile enzymic activities that may be associated with MAP 2 (i.e., protein kinase activity[3]) are not destroyed by the high temperature used to denature and to precipitate the bulk of the microtubule protein.

In addition, purified MAP 2 can be fragmented by chymotrypsin digestion and separated into what have been termed an assembly promoting fragment and a projecting fragment.[18] Presumably, the assembly-promoting fragment is that portion of the MAP 2 molecule that is associated with the wall of the tubulin subunit lattice; the projecting fragment is that portion of the MAP 2 polypeptide that lies distal to the attachment site, a conclusion based on the observation that the projecting fragment has no assembly-promoting activity (see section on assay methods for MAPs below).

Purification of MAPs from Cells in Culture

Similar procedures to those described above for the purification of MAPs from brain have been used for the purification of MAPs from cells

[17] R. B. Vallee, M. J. DiBartolomeis, and W. E. Theurkauf, *J. Cell Biol.* **90**, 568 (1981).
[18] R. Vallee, *Proc. Natl. Acad. Sci. U.S.A.* **77**, 3206 (1980).

in culture.[19-21] The cells are washed with phosphate-buffered saline, then once with $0.1\,M$ PIPES, pH 6.94 and once with $0.1\,M$ PIPES containing 1 mM dithiothreitol, 1 mM EGTA, 1 mM MgSO$_4$, and 1 mM GTP (PDEMG buffer); a 15–16 ml packed cell volume is combined with an equal volume of PDEMG and sonicated with three, 30-sec pulses at setting 2 (ca. 50 W) with a Heat Systems sonifier. The microtubules are then assembled and disassembled from the tubulin in the extracts according to the procedures described by Borisy et al.[22] (e.g., assembly in the warm, disassembly in the cold) using PDEMG buffer for all steps. To separate the tubulin from the MAPs, NaCl is added to the three times cycled microtubule proteins to a final concentration of $0.1\,M$ NaCl in PDEMG. This is loaded onto a DEAE-Sephadex A-50 column equilibrated with $0.1\,M$ NaCl–PDEMG. The MAP fraction does not bind to the column, and the tubulin elutes at ca. $0.5\,M$ NaCl–PDEMG. The MAPs from HeLa cells purified in this manner typically have molecular weights of ca. 200,000–210,000 and 125,000.[19-21] However, a report suggests that tissue culture cells also contain the high molecular weight MAPs (M_r 300,000–350,000) characteristic of brain.[23] Presumably, the MAPs could also be separated from the tubulin by use of PC, although this has not yet been reported.

Assay for MAPs

Both the high molecular weight MAPs and Tau from brain and the MAPs from tissue culture cells have been assayed by their ability to promote the assembly of tubulin in vitro. Generally, these assays are carried out by determining whether the putative MAP will enhance the initial rate and total amount of assembly of a given concentration of PC or DEAE-purified brain tubulin. The tubulin is usually used at a concentration close to or just below its critical concentration for assembly (ca. 1 mg/ml). The assembly is monitored as described in this volume by Gaskin [41] and by Williams and Lee [36]—e.g., by measuring an increase in turbidity at 350 nm, or increase in viscosity, or by quantitation of microtubule lengths in negatively stained electron microscope preparations.

[19] J. A. Weatherbee, R. B. Luftig, and R. R. Weihing, J. Cell Biol. 78, 47 (1978).
[20] J. A. Weatherbee, R. B. Luftig, and R. R. Weihing, Biochemistry 19, 4116 (1980).
[21] J. C. Bulinski and G. G. Borisy, Proc. Natl. Acad. Sci. U.S.A. 76, 293 (1979).
[22] G. G. Borisy, J. B. Olmsted, J. M. Marcum, and C. Allen, Fed. Proc., Fed. Am. Soc. Exp. Biol. 33, 167 (1974).
[23] R. Pytela and G. Wiche, Proc. Natl. Acad. Sci. U.S.A. 77, 4808 (1980).

Conclusion

Although as yet most MAPs have been isolated and analyzed in this way, there is no reason to suspect that other polypeptides that might associate with tubulin through cycles of assembly and disassembly *in vitro* should also necessarily stimulate assembly. Moreover, there is no reason to believe that all proteins associated with microtubules *in vivo* should coassemble with tubulin *in vitro*. The MAPs that have been studied so far have been defined in part by their ability to coassemble with and stimulate the assembly of tubulin *in vitro;* doubtless there are other associated proteins that do not function in this way.

Acknowledgments

Some of the work summarized here was supported by the following grants: GM25061 (RDS) and GM14642 (JLR) from the National Institutes of Health and CD-45 (JLR) from the American Cancer Society.

[40] Characterization of Tubulin and Microtubule-Associated Protein Interactions with Guanine Nucleotides and Their Nonhydrolyzable Analogs

By DANIEL L. PURICH, BRIAN J. TERRY, ROBERT K. MacNEAL, and TIMOTHY L. KARR

Understanding the linkage between guanine nucleotide interactions with tubulin and the assembly–disassembly properties of microtubules has been a challenging endeavor. These efforts may well provide some new ideas about the self-organizing events in microtubule formation and the regulation of microtubules and related physiological processes. During the last few years, there has been greater recognition of the role that GTP may play in modulating microtubule turnover *in vitro,* and the generality of these conclusions must await widespread characterization of microtubule protein from a variety of microtubule-based organelles. With the ever increasing number of sources for tubulin and microtubule-associated proteins comes the need for some standardized methods and protocols for comparative work. This is especially necessary because certain microtubule protein preparations have characteristic enzymic contaminants, and the levels of these enzymic activities must be measured to avoid faulty conclusions about the interactions of nucleotides in the self-assembly reactions. For this reason, we have summarized here some of the methods for probing tubulin–nucleotide interactions as well as a group of assays for evaluating the associated enzyme activities in a more quantitative fashion.

Preparation of Tubulin–Nucleotide Complexes

The assembly of microtubules induces the hydrolysis of tubulin-bound GTP, but the presence of an ATPase in many brain microtubule preparations obfuscates the kinetics and stoichiometry of GTP hydrolysis relative to tubulin assembly. It is also necessary to devise adequate strategies for making tubulin–nucleotide complexes of known stoichiometry and stability.

GTP Regenerating Systems

Earlier work with bovine brain microtubule protein assembly revealed a poor relationship between the stoichiometry of GTP hydrolysis and tubulin uptake into microtubules. It is now clear that microtubule protein contains a nucleoside 5′-triphosphatase activity[1-4] that is effectively unrelated to the assembly process. As a result of this phosphohydrolase activity, the preparation of microtubule protein by the cycling of protein between warm-sedimentable and cold-soluble forms[5-7] is inefficient in terms of GTP consumption. It is also true that one might choose to work with crude microtubule protein samples lest any regulatory proteins be removed by the recycling protocols. In these cases, an efficient means for regenerating GTP can greatly facilitate such efforts.

MacNeal et al.[4] first proposed the use of the acetate kinase from Escherichia coli as a nearly ideal system for GTP resynthesis from GDP. The reaction is

$$\text{Acetyl-P} + \text{MgGDP}^{1-} \rightleftharpoons \text{acetate} + \text{MgGTP}^{2-}$$

[1] F. Gaskin, S. B. Kramer, C. R. Cantor, C. R. Adelstein, and M. L. Shelanski, FEBS Lett. 40, 281 (1974).
[2] R. G. Burns and T. D. Pollard, FEBS Lett. 40, 274 (1974).
[3] H. D. White, B. A. Coughlin, and D. L. Purich, J. Biol. Chem. 255, 486 (1980).
[4] R. K. MacNeal, B. C. Webb, and D. L. Purich, Biochem. Biophys. Res. Commun. 74, 440 (1977).
[5] R. C. Weisenberg, Science 177, 1104 (1972).
[6] M. L. Shelanski, F. Gaskin, and C. R. Cantor, Proc. Natl. Acad. Sci. U.S.A. 70, 765 (1973).
[7] T. L. Karr, H. D. White, and D. L. Purich, J. Biol. Chem. 254, 6107 (1979).

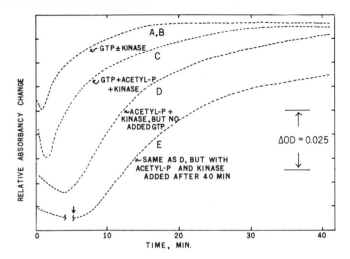

FIG. 1. Acetate kinase regenerating system for microtubule assembly. Recycled microtuble protein (1.2 mg/ml after gel filtration) in MEM buffer was warmed to 37° in the absence and in the presence of 1 mM acetyl-P and 1 IU of acetate kinase per milliliter. Exact conditions for additions are given on the figure.

From early work by Anthony and Spector[8] and in this laboratory,[9] acetate kinase was clearly found to utilize both ADP and GDP as excellent phosphoryl acceptor substrates. It is also clear that the equilibrium constant for this acyl-P dependent reaction is highly favorable ($K_{eq} = 3000$); thus, even when 90% of the acetyl-P is hydrolyzed to acetate, the GTP : GDP ratio will still be greater than 200–300. Because acetate kinase and acetyl-P are of bacterial origin, one gains an added advantage over other potential regenerating reactions (e.g., pyruvate kinase or creatine kinase); one may be assured that there are no contaminating endogenous levels of either substance in even crude animal tissue extracts. Finally, both acetate kinase and acetyl-P are commercially available and relatively inexpensive compared to gram-quantities of GTP required in many tubulin preparations; for large-scale work, one may also readily prepare the acetyl-P by reaction of P_i with either acetic anhydride or isopropylidene acetate.[10] In our work, we have relied on this chemical synthesis for making [32P]-labeled acetyl-P and purine nucleotides.[11]

The usefulness of the acetate kinase system is illustrated in Fig. 1, which also provides additional experimental information in the legend. We

[8] R. S. Anthony and L. B. Spector, *J. Biol. Chem.* **247**, 2120 (1972).

[9] J. A. Todhunter, K. B. Reichel, and D. L. Purich, *Arch. Biochem. Biophys.* **174**, 120 (1976).

[10] E. R. Stadtman, this Series, Vol. 3, p. 228.

[11] R. K. MacNeal and D. L. Purich, *J. Biol. Chem.* **253**, 4683 (1978).

and others[11,12] have found that the final plateau extent of microtubule polymerization depends on the GTP : GDP ratio, which is held exceedingly high in the presence of the acetate kinase. With creatine kinase or pyruvate kinase, one finds that GDP is a poor phosphoryl acceptor, but our kinetic studies indicate that GDP exhibits a relatively low Michaelis constant (around $10^{-5} M$). Thus, one need not add more than 0.1 mM GTP for efficient microtubule assembly in the presence of excess acetyl-P, which may be held at 1–20 mM. To check on the extent of acetyl-P consumption, it is useful to remove an aliquot of reaction mix containing about 0.5 μmol of acetyl-P and incubate this with 0.2 ml of neutral 1 M hydroxylamine (0.1 ml of 2 M KOH + 0.1 ml of 2 M hydroxylamine · HCl, prepared freshly to avoid decomposition) for 20 min, followed by assay with the ferric chloride and hydrochloric acid assay method.[10] A brown complex may be readily measured at 540 nm to estimate the content of acetohydroxamate.

One limitation of the acetate kinase regenerating system is that ADP is also an effective substrate, so the ATP : ADP ratio will also be relatively high. The endogenous ATPase of bovine brain tissue is more effective with ATP, and this can cause loss of long-term stability of the GTP : GDP ratio. Also, the ATP can support the endogenous protein kinase in its phosphorylation of one of the microtubule-associated proteins.[13,14] When these matters are of special concern, the crude extracts may be passed over a Sephadex G-25 (coarse grade) column to remove most of the free nucleotides and other lower molecular weight contaminants. This will not result in the loss of guanine nucleotide from the tubulin exchangeable site because of its high affinity for GTP and GDP.[4,11] Subsequent addition of acetate kinase, acetyl-P, and about 0.1 mM GDP or GTP is then sufficient to hold the GTP : GDP ratio high enough to promote rapid and efficient tubule assembly.

Another practical limitation on the acetate kinase system is its cold sensitivity. The enzyme is apparently reversibly denatured upon cooling, and warming in the presence of nucleotide is most effective in promoting its reactivation prior to use. The most direct procedure for this first involves centrifugation of a small aliquot of the ammonium sulfate suspension of the commercial kinase at 2000 g for 10 min in a small conical tube, then use of a Kimwipe wick to remove the supernatant salt solution, followed by dissolution of the kinase in a 0.1–0.2 ml aliquot of assembly buffer containing $10^{-4} M$ GDP or GTP. After 5 min at 30°, the enzyme is

[12] T. Arai and Y. Kaziro, *J. Biochem. (Tokyo)* **82**, 1063 (1977).
[13] R. D. Sloboda, S. A. Rudolph, J. L. Rosenbaum, and P. Greengard, *Proc. Natl. Acad. Sci. U.S.A.* **72**, 177 (1975).
[14] B. A. Coughlin, H. D. White, and D. L. Purich, *Biochem. Biophys. Res. Commun.* **92**, 89 (1980).

ready for use. We have also found that 5–10% glycerol protects the phosphotransferase from inactivation if it is desirable to preincubate the tubulin with the kinase at 10° to achieve GTP resynthesis without tubule assembly.

As will be discussed later, most microtubule protein contains a slight amount of nucleoside diphosphate kinase, and the addition of ATP will frequently support the assembly reaction as a result of this enzymic activity. This enzyme will utilize ATP, UTP, or CTP as a phosphoryl donor. On the other hand, the acetate kinase also contains an efficient nucleoside diphosphate kinase activity that utilizes only purine nucleotides. It can be added to augment the endogenous kinase only when a purine nucleotide is present.

Labeled Nucleotide Complexes

Tubulin is an α,β-heterodimer with two guanine nucleotide sites: one undergoes exchange readily,[15,16] and the other is effectively nonexchangeable on the time scale used in most experiments with this thermally unstable protein.[15,16] In this section, we describe a reliable means for incorporating labeled nucleotide into the exchangeable nucleotide site. For an account of elegant approaches toward labeling the nonexchangeable site, the reader is referred to a report from Kirschner's laboratory.[17] The procedure described here is that of MacNeal and Purich.[11] The tubulin dimer with 1 mol of exchangeably bound labeled nucleotide is prepared at low, essentially stoichiometric, levels of tubulin and guanine nucleotide to avoid the action of the high K_m ATPase. Acetate kinase and acetyl-P are used to regenerate the GTP whenever GTP is the desired ligand for the protein. A 1–5 mM acetyl-P concentration (including [^{32}P]acetyl-P if labeled GTP is to be used in estimating the hydrolytic reaction) is sufficient for the incubation in a medium containing 2 IU of acetate kinase for each milliliter, and this solution is incubated for 10 min at 30° to phosphorylate any added guanine nucleotide. After cooling to ice-bath temperature, tubulin is added and the reaction mix is maintained at 10° for 2–3 hr to effect quantitative GTP formation, exchange, and tubulin complex formation. A typical reaction mixture is prepared as follows: acetyl-P, 8–10 mM; GDP, 0.1–1.0 mM; and microtubule protein, 2–10 mg/ml (of which 75–80% is typically tubulin and the remainder is composed of the associated proteins). The buffer may be MEM (0.1 M N-morpholinoethanesulfonic acid, 1 mM EGTA, 1 mM $MgSO_4$, pH 6.8) or GEMM

[15] R. C. Weisenberg, G. G. Borisy, and E. W. Taylor, *Biochemistry* **7**, 4466 (1968).
[16] M. Jacobs, H. Smith, and E. E. Taylor, *J. Mol. Biol.* **89**, 455 (1974).
[17] B. M. Spiegelman, S. M. Penningroth, and M. W. Kirschner, *Cell* **12**, 587 (1977).

(0.1 M glutamate, 1 mM EGTA, 20 mM N-morpholinoethanesulfonic acid, 1 mM MgSO$_4$, pH 6.8). Maintenance of the sample at 10° is essential to block assembly of the microtubules, and the long incubation period has been found to achieve the greatest extent of labeled complex formation. If labeled GDP is desired in place of GTP, then the acetate kinase system is unnecessary, and one may follow the above protocol in its absence. After the sample has been treated as outlined above, passage over a Sephadex G-25 (medium) column containing MEM or GEMM buffer is necessary to remove excess nucleotide and acetyl-P. For a 1.2 × 35 cm column, we find a flow rate of 1.0 ml/min should not be exceeded for efficient removal of these components, and such treatment does not remove more than 2–4% of the exchangeable site nucleotide. Tubulin–nucleotide complex elutes with the void volume, and the stoichiometry may be determined by protein assays and liquid scintillation counting in Aquasol. To verify that the labeled nucleotide is fully exchangeable, the bound nucleotide may be chased off of the site by the addition of 1–2 mM GDP. The amount of labeled nucleotide after gel filtration should be less than 1–2%.

Removal of Exchangeable Nucleotide

The tight binding of guanine nucleotides to tubulin was alluded to earlier, and this represents a major experimental hurdle for characterizing more weakly bound nucleotides with tubulin. From our experience, gel filtration and dialysis are practically ineffective for the purpose of freeing tubulin of the exchangeable nucleotide. Two less desirable methods have been perfected for this purpose, and they are described here.

Charcoal Method. The adsorption of nucleotides to charcoal at slightly acidic pH is a useful means to prepare nucleotide-depleted tubulin samples. Penningroth[18] has given a good account of this procedure, which we have modified to account for our own experience. With MEM buffer containing 1 mM 2-mercaptoethanol and adjusted to pH 6.4, tubulin may be first polymerized with 0.05 mM GTP at 37° for 30 min in the presence of the acetate kinase regenerating system. The microtubule fraction is centrifuged in the warm (140,000 g for 1 hr), and the supernatant fluid is completely removed by decantation and wiping the wall of the tube with a Kimwipe. After cold-induced depolymerization, the protein (3–4 mg/ml) is treated with acid-washed charcoal at a rate of 5 mg for each milliliter of protein fluid. After 5 min, the charcoal is removed by centrifugation in a clinical centrifuge or a refrigerated centrifuge at 4°. The process is then

[18] S. M. Penningroth, Roles of ligands in microtubule assembly in vitro. Ph.D. Dissertation, Yale University, New Haven, Connecticut, 1977.

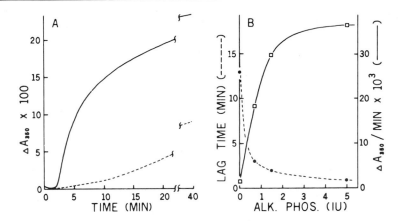

FIG. 2. Polymerization of tubulin at 37° with or without alkaline phosphatase. Tubulin (1.4 mg/ml) was purified by assembly using 0.1 mM acetyl phosphate and 0.1 unit of acetate kinase per milligram of tubulin. After warm centrifugation, the pellet was resuspended in EGTA-free buffer. After cold disassembly the tubulin was centrifuged at 45,000 g for 25 min to remove insoluble aggregates. The supernatant was then gel filtered on a 1.5 × 28 cm Sephadex G-25 column at 2° to remove traces of EGTA. GMP-P(CH₂)P (1 mM), ZnCl₂ (0.03 mM), and alkaline phosphatase (0–5 units) were added at the start of the assay. The volume was 0.9 ml. (A) Assembly as a function of time: upper curve, 5.0 units of alkaline phosphatase; lower curve, no alkaline phosphatase. (B) Lag time and rate of assembly as a function of alkaline phosphatase added: —, rate of assembly; ----, lag time before onset of assembly.

repeated a second time. Analysis of the tubulin-bound nucleotides by thin-layer chromatography suggests that two such treatments reduces the nucleotide content by about 70%, but further treatments are undesirable on the basis of poor yields of the tubulin. The reader should consult the report by Penningroth and Kirschner [19] to learn about the efficiency of this material in assembly induced by several nucleotides and analogs.

Alkaline Phosphatase Method. Purich and MacNeal [20] determined that brief exposure of bovine brain microtubule protein to calf intestinal mucosal alkaline phosphatase can be effective in removing nucleotides held at the exchangeable site. Apparently, the phosphatase converts the GTP and GDP to GMP and guanosine, which are quite feebly bound by tubulin. As shown in Fig. 2A, 5′-guanylyl methylenediphosphate [GMPP(CH₂)P] fails to promote assembly of tubulin · GDP complex even at 1 mM concentrations of the analog. On the other hand, exposure to alkaline phosphatase (3 IU per milligram of tubulin) in the presence of the phosphonate results in microtubule polymerization within 20–30 min.

[19] S. M. Penningroth and M. W. Kirschner, *Biochemistry* **17**, 734 (1978).
[20] D. L. Purich and R. K. MacNeal, *FEBS Lett.* **96**, 83 (1978).

There is a characteristic lag phase (2–3 min) that probably represents the period for desorption of the GDP from the exchangeable site. Similar assembly behavior has also been observed with 5'-guanylyl imidodiphosphate [GMPP(NH)P] and the corresponding analogs of ATP. In all cases, we observe greatest assembly efficiency when the analog is present during the phosphatase treatment, and it would appear that nucleotide- or analog-free tubulin is rather unstable. The plotted data in Fig. 2A suggest that the phosphatase destroys most of the GDP during the lag phase. It might also be noted that small concentrations of Zn^{2+} may be added to stimulate the phosphatase and that the phosphatase may be separated from microtubules by layering on sucrose (50% w/v) followed by centrifugation at 50,000 rpm in a Beckman 50 rotor.

It may be said that the phosphatase treatment is a method for replacing guanine nucleotide by nonhydrolyzable analogs rather than a method for removal of the nucleotide from the tubulin sites. Nonetheless, it is rapid and highly reproducible, and there is no substantial loss of tubulin as observed in the charcoal adsorption methods.

Properties of Tubulin-Nucleotide Complexes

When the exchangeable nucleotide site is occupied by GTP, GDP, or certain related analogs, the assembly–disassembly properties of microtubule protein are altered in characteristic ways. In this section, we describe some of the basic properties of tubulin–nucleotide complexes.

Stoichiometry and Rates of GDP or GTP Release

To determine the rate and extent of guanine nucleotide release from tubulin, one may utilize several coupled enzyme assay procedures. It is essential to restrict the solution conditions to those optimal for tubulin studies, and this has been done in the case of those assays listed below. In any coupled enzyme assay, the auxiliary enzymes and cofactors must be present at levels that do not affect the rate of the primary rate process, here nucleotide release. Thus, if the observed rates of either GDP or GTP are much faster than those presented below, the sufficiency of the auxiliary assay system should be rigorously evaluated.[20]

GDP Assay. Tubulin–GDP complex should first be prepared by incubating recycled, phosphocellulose-treated microtubule protein samples with 1–2 mM GDP at 30° for 30 min. This is followed by gel filtration on a Sephadex G-25 column containing MEM buffer but no added nucleotide. The concentration of tubulin is the estimated by the Lowry method[21]. The

[21] O. H. Lowry, N. J. Rosebrough, A. L. Farr, and R. J. Randall, *J. Biol. Chem.* **193**, 265 (1951).

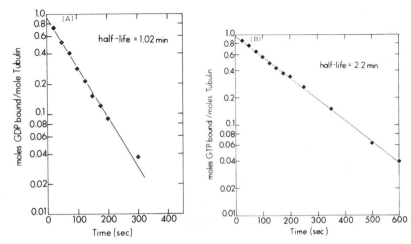

FIG. 3. Assay for the release of GDP and GTP from tubulin. (A) Assay for the release of GDP from tubulin. Phosphocellulose-purified tubulin (1.83 mg/ml; 0.94 ± 0.05 mol of GDP per mole of tubulin) was incubated at 37° with the pyruvate kinase–lactate dehydrogenase assay, and the absorbance at 340 nm was monitored. (B) Assay for the release of GTP from tubulin. Phosphocellulose-purified charged tubulin (1.24 mg/ml; 0.96 ± 0.05 mol of GTP per mole of tubulin) was incubated at 37° with phosphoglycerate kinase–glyceraldehyde-3-phosphate dehydrogenase.

rate assays may then be performed in 1-ml reaction volumes at 37° with the following final concentrations: 5–15 μM tubulin (corrected for the presence of microtuble-associated proteins); 2 mM MgCl$_2$; 0.5–0.6 mM phosphoenolpyruvate; 0.2 mM NADH; 3 IU of rabbit muscle pyruvate kinase; and 15 IU of lactate dehydrogenase in MEM buffer. The assay is initiated by the addition of pyruvate kinase to an assay solution that was previously incubated at 37° for 1 min. The time course of NADH oxidation is followed by observing a decrease in the 340 nm absorbance, and the slope of the semilog replot (Fig. 3A) gives a value for the off-rate constant. The need to know the initial tubulin–GDP complex concentration accurately cannot be overemphasized, because failure to do so may result in a nonlinear plot.[22]

GTP Assay. A modification of the assay conditions first described by Jacobs et al.[16] has been presented by Terry and Purich.[22] The final concentration of a 1 ml of assay sample are: 5–15 μM tubulin-GTP complex; 3 mM magnesium chloride; 6 mM 3-phosphoglycerate; 0.2 mM NADH, 1 IU of phosphoglycerate kinase; and 10 IU of glyceraldehyde-3-phosphate dehydrogenase in MEM buffer. The assays are initiated by the addition of

[22] B. J. Terry and D. L. Purich, J. Biol. Chem. 254, 9469 (1979).

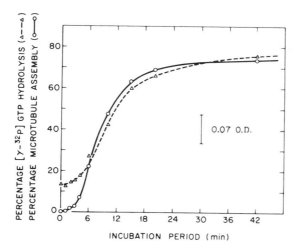

FIG. 4. Assembly of native tubulin–GTP complex and hydrolysis of [γ-³²P]GTP at 37°. Native tubulin–GTP complex was prepared with a final protein concentration of 2.2 mg/ml. Percentage of assembly (O——O) was determined from light-scattering data. △---△, Percentage of hydrolysis.

the kinase to samples preincubated at 37° for 1 min with the otherwise complete assay system. The data shown in Fig. 3B illustrate a typical reaction progress curve in a semilog plot.

When carrying out these assays, the phosphocellulose-treated tubulin has too high a critical concentration to permit assembly. With whole microtubule protein (i.e., tubulin plus associated proteins), assembly will occur and vitiate the experimental findings. We have found that 20 μM podophyllotoxin may be added to block the assembly; colchicine may not be as readily used as a result of its high absorptivity. We should also note that the podophyllotoxin can influence the GTP and GDP off-rates[21]; thus, several determinations of the kinetics at two or three podophyllotoxin concentrations may be necessary to extrapolate out the podophyllotoxin effect.

Assembly and GTP Hydrolysis Kinetics

To evaluate the kinetics of tubulin assembly and GTP hydrolysis, one must first prepare tubulin–[γ-³²P]GTP complex as described earlier. The cold sample is then rapidly warmed and added to a cuvette. As the progress of assembly is recorded in terms of absorbance, one may withdraw protein samples to be quenched and measured for extent of hydrolysis. A typical experimental result is presented in Fig. 4. One may also determine

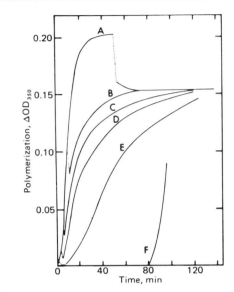

FIG. 5. Effect of added GDP during GTP-supported assembly. Identical samples containing microtubular protein at 1.15 mg/ml and 0.2 mM GTP were assembled at 30°. At the indicated times (dotted lines) the samples were made 2.0 mM GDP and 2.33 mM MgSO$_4$ (curves A–E). Curve F represents the polymerization profile in which the GDP was added to the protein sample at 4° for 2 min, warmed to 30° for 5 min, and recooled to 4° for 25 min, followed by warming to 30°. At 80 min the protein was made 1.5 mM in GTP.

the amount of P_i released during the course of assembly.[11] By the criteria of exchangeability set forth earlier and the observed release of only 1 mol of orthophosphate per tubulin dimer incorporated into the polymerized form, we have concluded that hydrolysis occurs only at the exchangeable nucleotide site.

If GDP is added to tubulin-GTP complex in the cold, then it acts as a potent inhibitor of the microtubule self-assembly events.[11,15] Interestingly, however, we[23] found that, although GDP does not support nucleation, it does promote the elongation phase of the assembly process. If GDP is added at any point after warming tubulin–GTP complex, there is a momentary cessation of assembly followed by a resumption of the elongation reaction. As will be shown below, the critical concentration for microtubule protein in the presence of GDP is about twofold greater than with GTP. Thus, the momentary cessation of assembly reflects the lag resulting from reestablishing a new critical concentration behavior. This behavior is shown in Fig. 5.

[23] T. L. Karr, A. E. Podrasky, and D. L. Purich, *Proc. Natl. Acad. Sci. U.S.A.* **76,** 5475 (1979).

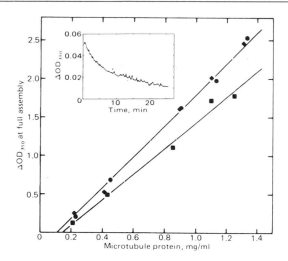

FIG. 6. Stability of microtubules formed in the presence of 0.5 mM GuoPP[NH]P (–◆–), GuoPP[CH$_2$]P (–■–), or GTP (–●–) after dilution into buffer containing 0.5 mM of their respective nucleotide or nucleotide analog at 0.5 mM. *Inset:* Time course for the depolymerization of microtubules formed in the presence of 0.5 mM GuoPP[NH]P after dilution into a rapid-mixing cell.

Critical Concentration Measurements

The condensation equilibrium model of Oosawa and Kasai[24] may be used to define an important experimental parameter commonly known as the critical protein concentration. This parameter establishes the lower bound for coexistence of monomer and polymer; below it, only the monomer exists. When the protein concentration exceeds the critical concentration, all additional protein will polymerize with no increase in the monomer concentration. We may consider the equilibrium between unpolymerized protein X and the polymerized form MT,

$$[X] + [MT_n] \underset{k_-}{\overset{k_+}{\rightleftharpoons}} [MT_{n+1}]$$

in which n and $(n + 1)$ indicate the degree of polymerization with respect to the number of subunits X in each polymeric form. The kinetics of this process must reflect the concentration of polymer and monomer forms, and at equilibrium the equilibrium constant ($K = k_-/k_+$) represents the concentration of monomer at equilibrium with the polymer in this indefinite polymerization process.

A typical critical concentration experiment is described[23] by the results shown in Fig. 6 for GTP, GMPP(NH)P, and GMPP(CH$_2$)P acting to

[24] F. Oosawa and M. Kasai, *J. Mol. Biol.* **4,** 10 (1962).

support microtubule assembly. In these measurements, we first polymerized microtubules in the presence of the stated nucleotide or analog, and then a concentrated sample was diluted to the concentrations specified. After depolymerization to establish the monomer–polymer equilibrium, we estimated the polymer weight concentration on the basis of the absorbancy at 350 nm. In the case of GMPP(NH)P-supported assembly, we found that the subsequent depolymerization phase was subject to slow equilibration (45 min). This observation probably explains earlier nonequilibrium behavior exhibited by tubules assembled by the imidodiphosphate nucleotide analog.[25] For this reason, we advise that the full course of polymerization or depolymerization be determined so that a stable end point is clearly evident.

Monomer–Polymer Exchange Measurements

One important phenomenon associated with microtubule assembly is the entrapment of the exchangeable guanine nucleotide into a nonexchangeable state or form. Apparently, the exchangeable site is topologically located such that assembly results in its blockage by steric factors. This phenomenon is important because it may be exploited to determine the dynamics of monomer–polymer exchange reactions and the kinetics of nucleotide exchange relative to microtubule assembly. The former is particularly valuable in view of the considerable interest in head-to-tail polymerization mechanisms of biopolymer assemblies.[26,27] Under suitable conditions, one may define the rate of labeled release or uptake as described below.

Labeled Monomer Exchange with Polymer. After microtubule assembly has reached a stable plateau value, isotope exchange procedures may be utilized to define the kinetics of labeled monomer uptake by assembled tubules. The basic protocol was first given by Margolis and Wilson,[26] and the present form reflects a number of improvements to optimize the measurements.[27]

Microtubule protein is subjected to gel filtration on a Sephadex G-25 column equilibrated with MEM buffer containing no added guanine nucleotide. The initial concentration of the protein should be sufficient to give a concentration of 2–4 mg/ml after filtration. Polymerization is then induced by warming the sample to 30° in the presence of 18 μM GTP, 18 mM acetyl-P, and 0.1 IU per milliliter of acetate kinase. The stock solution may be divided into six or more samples and incubated for an additional

[25] R. C. Weisenberg and W. J. Deery, *Nature* (*London*) **263**, 792 (1976).
[26] R. L. Margolis and L. Wilson, *Cell* **13**, (1978).
[27] B. J. Terry and D. L. Purich, *J. Biol. Chem.* **255**, 10532 (1980).

30 min beyond the time required for a stable turbidity plateau. At this point, [^{14}C]GTP (about 1.1×10^5 dpm of label commercially available at about 50 mCi/mmol) in reverse order to permit simultaneous centrifugation and to facilitate sampling technique. Thus, the shortest time of incubation, say 30 min, would be the last sample initiated, in this case 30 min prior to centrifugation. To quench these reactions and to analyze for tubulin uptake, 0.3–0.4-ml aliquots of each sample are layered onto a 2.5-ml volume of sucrose (50% w/v in MEM buffer) and centrifuged at 120,000 g for 3 hr at 30°. The end point for the incubation of tracer and assembled tubules may be operationally defined as the time at which the centrifuge rotor has fully accelerated (5 min after starting the centrifugation). After the centrifugation, the supernatant fluid contains all of the unpolymerized tubulin and nucleotide, and this must be carefully removed so as not to contaminate the microtubule pellet with extraneous radiolabel. The wall of each tube should be dried with the aid of a Kimwipe before each sample is redissolved with a small aliquot of buffer. The protein content is estimated by the Lowry method,[22] and radioactivity by standard scintillation counting methods.

Although Margolis and Wilson[26] have presented their estimates of monomer uptake as the percentage of total protein undergoing turnover per hour, it is preferable to define such flux in terms of a rate constant for association (units, M^{-1} sec^{-1}). All such rate constants may be computed from the initial specific radioactivity of the nucleotide and the measured specific radioactivity of the polymer-bound nucleotide at time t. The amount of polymer is equal to the total microtubule protein minus the critical concentration, and we assume that the whole microtubule protein is 80% tubulin by weight. These calculations implicitly assume that dilution by nonradioactive pools is sufficient to treat the kinetics as a unidirectional flux.

Labeled Polymer Exchange with Monomer. Here, the label is uniformly incorporated into the microtubules first, and this is then chased with unlabeled GTP. To achieve adequate exchange of the radioactive nucleotide with tubulin, the microtubule protein is assembled in the warm with 18 μM GTP (containing 2.2×10^6 dpm of labeled GTP), 10 mM acetyl-P, and 0.1 IU of acetate kinase. After attainment of the stable plateau turbidity value, 1 mM GTP is added to reduce the specific radioactivity of the nucleotide that is not present in the microtubules. The microtubule aliquots are then incubated in the reverse order as indicated in the preceding section, and the loss of label is evaluated in the same analytical manner described above.

Both of the above methods rely upon the stability of microtubule in the sucrose cushions used in the centrifugation step. We have found that microtubules do not depolymerize to any appreciable extent under these conditions. Through experience, however, we have observed that it is advantageous to determine the protein concentration of each redissolved pellet rather than to assume that the protein content is identical for each pellet.

Enzyme Activities of Microtubule-Associated Proteins

Microtubule protein may be derived from a number of sources, although the principal sources are still brain and flagella. The relative levels of various enzymic activities become a serious problem, especially when it is necessary to make comparative studies. There are several nucleotide-dependent activities that are frequently found associated with microtubules, and their assays are standardized here. Frequently, the presence of some of these activities can profoundly affect tubulin–nucleotide interactions, and an important goal will be achieved by routine surveys based on such assays.

Adenosinetriphosphatase

The hydrolysis of ATP or GTP may be measured by the following adaptation of the pyruvate kinase–lactate dehydrogenase assay of Pullman *et al.*[28] ADP or GDP production[29] is coupled to the oxidation of NADH, and a recording spectrophotometer (0–0.2 absorbance slide wire) is used to follow the changes in absorbance at 340 nm. Reaction mixtures consist of 50 mM Tris acetate (pH 7.4), 1 mM phosphoenolpyruvate, 0.3 mM NADH, 7 mM $MgCl_2$, 5 mM ATP, 4–5 IU of rabbit muscle pyruvate kinase, and 3.2 IU of lactate dehydrogenase in a 1-ml final volume. Podophyllotoxin (15–20 μM) may be added to prevent microtubule assembly in unfractionated microtubule protein upon warming to 30°. Typically, microtubule protein addition to a final concentration of 0.5–1.0 mg/ml is used to initiate the assay. When necessary, effects of contaminating nucleoside 5'-diphosphates in the assay mix may be minimized by preincubation of the assay mix for 3–5 min prior to microtubule protein addition.

Specific activity of the ATPase is reported in milliunits per milligram of protein, where one unit corresponds to 1 μmol of ATP hydrolyzed per minute. In some cases, the assay system displays some basal activity (usually less than 0.002–0.005 absorbance units per minute) in the absence

[28] M. E. Pullman, H. S. Penefsky, A. Datta, and E. Racker, *J. Biol. Chem.* **235**, 3322 (1960).
[29] H. D. White, B. A. Coughlin, and D. L. Purich, *J. Biol. Chem.* **255**, 486 (1980).

of added ATPase, and this should be subtracted from any reported value. The K_m for the bovine brain microtubule-associated ATPase is about 0.8 mM,[29] and a typical specific activity is about 40 milliunits per milligram with three-cycle purified microtubule protein.

Nucleoside Diphosphate Kinase

The content of nucleoside diphosphate kinase may be quantitatively determined by a coupled assay method.[21] The assays contain the following: 3 mM MgCl$_2$, 6 mM 3-phosphoglycerate, 0.2 mM NADH, 0.2 mM GDP, 0.5 mM UTP, 14 μM podophyllotoxin, 0.4 IU of phosphoglycerate kinase, 7 IU of glyceraldehyde-3-phosphate dehydrogenase, and 1 mg of microtubule protein per milliliter, or about 0.2 mg/ml dialyzed, phosphocellulose-purified proteins in the microtubule-associated protein fraction. The reaction is initiated by UTP addition to the sample after preincubation at 37° for 1 min. The oxidation of NADH should be corrected for any low level of the kinase in the auxiliary enzymes by running a blank in the absence of the microtubule protein.

Specific activity (reported with 1 unit equal to the amount of enzyme activity corresponding to 1 μmol of GDP phosphorylation per minute) is typically 1.7 milliunits per milligram for the unfractionated microtubule protein and 16 milliunits per milligram for the microtubule-associated protein fraction. Unfractionated porcine brain microtubule protein has a similar content.

Protein Kinase

A convenient assay for protein kinase activity of microtubule protein was described by Coughlin et al.[14] Interestingly, the protein kinase reaction requires the participation of a high molecular weight component in microtubule protein, frequently called MAP$_2$ because it is the second slowest migrating protein species on SDS gel electrophoresis.[13,14] Each assay reaction mixture is made up to a final volume of 0.1 ml and contains: 50 mM N-morpholinoethanesulfonic acid (pH 6.2), 10 mM MgCl$_2$, 20 μM podophyllotoxin, 10 μM 3′,5′-cyclicAMP, 4 mM theophylline, and 5–10 μM ATP (containing 4×10^5 to 8×10^5 cpm [γ-^{32}P]ATP). Microtubule protein (1 mg/ml) or phosphocellulose-fractionated microtubule-associated protein (0.2 mg/ml) is used to start the reactions at 30°. For a reaction time course experiment, 0.5-ml samples are prepared as above, and 0.1-ml aliquots are taken at 0, 1, 5, and 10 min for quenching. Here, the aliquots are spotted directly onto 10% trichloroacetic acid (TCA)-

soaked Whatman No. 1 paper disks.[30] The zero point is taken prior to addition of the kinase activity, followed by addition of the protein to the disk. These disks are washed sequentially in cold 10% TCA (three times, each time in a fresh uncontaminated volume of the acid), then three times in 5% TCA, once in ethanol, and then in ethyl ether followed by drying. Radioactivity is measured by Cerenkov counting in 12 ml of water in a plastic scintillation vial.

Tubulin: Tyrosine Ligase.

This enzyme catalyzes the ATP-dependent synthesis of a peptide bond between the amino group of tyrosine and the C-terminal glutamate of the α subunit of tubulin.[31-33] The assay measures the nontranslational incorporation of tyrosine into the protein as catalyzed by an endogenous ligase activity. Because it has been shown that unpolymerized tubulin is the substrate for this enzyme,[34] colchicine is added to the assay both to prevent assembly and to promote the ligase activity. Although the enzyme is most active in the pH 7.4–8.0 range,[33-35] a second peak of activity is found at about pH 6.5.[35]

Filter paper disks (Whatman 3 MM, 2.1 cm in diameter) are presoaked in 10% TCA and dried. Two disks are used for each assay, and they are marked with pencil and pinned to a styrofoam block. The standard reaction mixture contains 25 mM N-morpholinoethanesulfonic acid (pH 6.8 with KOH); 0.15 M KCl, 12.5 mM $MgCl_2$, 2.5 mM ATP, 1 mM dithiothreitol, 0.1 mM colchicine, 10% glycerol, 0.1 mM tyrosine (containing [^{3}H]tyrosine at 1 Ci/mmol), 1 mg of microtubule protein and 0.1 mg of ribonuclease per milliliter, and the ligase in a final volume of 0.05 ml. Obviously, the latter enzyme is not added when assaying endogenous ligase activity. An appropriate blank reaction mixture for a zero time point is achieved by addition of a 0.02-ml aliquot of the assay mix to the TCA disk prior to warming from 4° to 37°, where the reaction is measured after a typical 15-min incubation. Each of the filter paper disks is dried under an air stream and processed further by the method of Mans and Novelli.[36] One unit of the ligase catalyzes the incorporation of 1 nmol of tyrosine per minute at 37°.

[30] L. Rappaport, J. F. Leterrier, A. Virion, and J. Nunez, *Eur. J. Biochem.* **62**, 539 (1976).
[31] H. S. Barra, J. A. Rodriguez, C. A. Arce, and R. Caputto, *J. Neurochem.* **20**, 97 (1973).
[32] H. S. Barra, C. A. Arce, J. A. Rodriguez, and R. Caputto, *Biochem. Biophys. Res. Commun.* **60**, 1384 (1974).
[33] D. Raybin and M. Flavin, *Biochem. Biophys. Res. Commun.* **65**, 1088 (1975).
[34] W. C. Thompson, D. L. Purich, and L. Wilson, *J. Biol. Chem.* (in press).
[35] H. Murofushi, *J. Biochem.* (*Tokyo*) **87**, 979 (1980).
[36] R. J. Mans and G. P. Novelli, *Arch. Biochem. Biophys.* **94**, 48 (1961).

Concluding Remarks

In the foregoing discussion of methods for examining nucleotide interactions with microtubule proteins, we have emphasized the developments in our own laboratory. Actually, there are many assay techniques described in the microtubule literature, but it is impossible to attest to the validity of each. Instead, we have concentrated on methods that we find to be routinely helpful in our research. Wherever practicable, we have acknowledged other methods that may be employed for some of these purposes. Each investigator in a new area frequently encounters rather specialized experimental obstacles, and new assays tend to be developed instead of adapting the established methods. On the other hand, it would be most helpful if we could develop some standardized conditions, and this motivation has been the basis for this chapter.

Acknowledgments

We are grateful to Drs. Hillary D. White and William C. Thompson of this laboratory for detailed assay methods for the ATPase and ligase activities, respectively.

[41] Techniques for the Study of Microtubule Assembly in Vitro

By Felicia Gaskin

Microtubule assembly and disassembly *in vitro* is frequently monitored by turbidity (light scattering) and electron microscopy. Useful information concerning the mechanism of microtubule assembly has also been obtained with a variety of other techniques, i.e., viscosity, sedimentation, flow birefringence, dark-field light microscopy, time-resolved X-ray diffraction, laser light scattering, filtration, calorimetry, and immunocytochemistry. The combination of two macromolecular techniques is often necessary, as each technique has limitations in interpretation.

Turbidity Measurements

Turbidity measurements are useful to follow the assembly and disassembly of microtubules *in vitro*. Microtubules assembled *in vitro* are a good approximation to the long rod limit when light (320–600 nm) is used to measure turbidity. Berne[1] has shown that for very long rodlike particles

[1] B. J. Berne, *J. Mol. Biol.* **89**, 755 (1974).

METHODS IN ENZYMOLOGY, VOL. 85

with a small diameter compared to the wavelength (λ) of the incident light, the turbidity is proportional to λ^{-3} and is a function only of the total weight concentration of scattering particles. Three tests have been used to establish that the turbidity of *in vitro* assembled microtubules is proportional to the weight of assembled tubulin in microtubules.[2]

1. A plot of turbidity (absorbance) as a function of λ shows $A \propto \lambda^{-3.3}$.
2. Sonication to 20% of the initial length results in 80% of initial $A_{350\,nm}$.
3. After polymerization, a curve of A_{350} versus initial concentration (C_0) coincides with a curve of mass of pelleted microtubules versus C_0.

Furthermore, the second test shows that turbidity is reasonably insensitive to microtubule length.

Turbidimetric measurements on microtubule assembly are usually made at 350 nm with a recording spectrophotometer. Since the turbidity measurements must be made at a wavelength where the protein and cofactors do not absorb, the absorption spectra of microtubule protein preparations and any cofactors must be analyzed. Electron microscopy can be used to verify microtubule formation and length distribution. When new experimental conditions are used, tests 1 and 3 described above can also be used to ascertain that the measured turbidity is proportional to the weight of tubulin in microtubules. It has been shown that high concentrations of microtubules, mixtures of tubulin, microtubule protein rings and microtubules, and zinc-induced sheets of tubulin[3] do not scatter light proportionally to the weight of tubulin in microtubules.

Advantages of using turbidity to follow tubulin assembly include the facts that (*a*) relatively small quantities of protein are needed (usually 1 mg/ml); (*b*) the system can be probed without perturbation or interaction; (*c*) samples can be monitored continuously; (*d*) the kinetics of assembly of several samples can be examined under exactly the same conditions with an automatic sample changer; (*e*) rapid dilution experiments can be done with a special cuvette that minimizes shearing.[4]

Electron Microscopy

Microtubule formation is routinely checked by electron microscopy. We put 10 μl of sample on a Formvar film on a copper grid for 1 min and stain by rinsing with 6 drops of 1% uranyl acetate and blotting dry with

[2] F. Gaskin, C. R. Cantor, and M. L. Shelanski, *J. Mol. Biol.* **89**, 737 (1974).
[3] F. Gaskin and Y. Kress, *J. Biol. Chem.* **252**, 6918 (1977).
[4] T. L. Karr and D. L. Purich, *Anal. Biochem.* **104**, 311 (1980).

filter paper. If microtubule lengths are to be estimated, the protein can be fixed first with an equal volume of 2% glutaraldehyde in buffer and maintained at the same temperature for 15 min. Then protein is placed on carbon-over-Formvar glow-discharged grids and treated as described by Sloboda *et al.*[5] with 0.2% cytochrome *c* in 1% amyl alcohol for rinsing and 1% uranyl acetate for staining. At least 100 microtubules are measured in a field on two grids. For very long microtubules, the average length is calculated as the (total length/number of ends) × (2 ends/numbers of microtubules examined).[6] The average length of microtubules fixed in glutaraldehyde is 7% longer.[7]

Kirschner *et al.*[8] have modified the technique of Backus and Williams[9] to study microtubule polymerization quantitatively. The entire contents of a spray drop of solution containing microtubule protein aggregates and bushy stunt virus particles was dried on a grid after it was mixed with uranyl acetate using a dual nebulizer. The number of virus particles and the total length of microtubules or number of rings was calculated in the volume of the drop. The mass of tubulin in the various forms was determined from the concentration of virus particles, the molecular weight of the rings, and the mass per unit length of microtubules. Shearing may be a serious problem with this technique.

Viscosity

Microtubule assembly is frequently studied by viscosity measurements that are easy and straightforward.[10] However, the high shear rate of Ostwald and other capillary viscometers can lead to erroneous interpretations, and viscosity is quite sensitive to microtubule flexibility, nicks, and bends; therefore, viscosity is not necessarily proportional to mass concentration.[11] Much lower shear stresses can be obtained using the Zimm–Crothers low-stress rotating cylinder viscometer.[12] MacLean-Fletcher and Pollard described a falling-ball assay that is simple and inexpensive, requires small samples, and has a 10-fold less shear rate than the Ostwald viscometer.[13] This viscometer is particularly useful with a high concentra-

[5] R. D. Sloboda, W. L. Dentler, and J. L. Rosenbaum, *Biochemistry* **15**, 4497 (1976).
[6] K. A. Johnson and G. G. Borisy, *J. Mol. Biol.* **117**, 1 (1977).
[7] J. S. Gethner and F. Gaskin, *Biophys. J.* **24**, 1101 (1978).
[8] M. W. Kirschner, L. S. Honig, and R. C. Williams, *J. Mol. Biol.* **99**, 263 (1975).
[9] R. C. Backus and R. C. Williams, *J. Appl. Phys.* **21**, 11 (1950).
[10] J. B. Olmsted and G. G. Borisy, *Biochemistry* **12**, 4282 (1973).
[11] T. Suzaki, H. Sakai, S. Endo, I. Kimura, and Y. Shiganaka, *J. Biochem. (Tokyo)* **84**, 75 (1978).
[12] B. H. Zimm and D. M. Crothers, *Proc. Natl. Acad. Sci. U.S.A.* **48**, 905 (1962).
[13] S. D. MacLean-Fletcher and T. D. Pollard, *J. Cell Biol.* **85**, 414 (1980).

tion of microtubules that forms gels. In this method samples are drawn into capillary tubes (1.3 cm × 12.6 cm) and warmed to 37° for a specified time, and the viscosity is measured by the rate at which a 0.64-mm stainless steel ball falls through the sample. The velocity of the ball is inversely proportional to the viscosity of the solution in a wide range. Limitations with the falling-ball method are that (a) measurements include only two structural parameters (apparent viscosity and yield strength); (b) shear rate is not easily varied by a large factor; (c) stress applied to samples can be destructive.

Sedimentation

Tubulin in microtubules and large aggregates can be separated from 6 S tubulin by pelleting the protein in a preparative ultracentrifuge or a benchtop airfuge ultracentrifuge and determining the amount of protein in the resuspended pellet. Centrifugation can also be done through microtubule stabilizing buffers i.e., a glycerol gradient to separate by polymer size and shape[14] or a 50% sucrose cushion to separate 6 S tubulin from microtubules.[15] The time of centrifugation depends on the experimental conditions, i.e., with a 200 μl sample, microtubules pellet in 10 min at 100,000 g in a Beckman airfuge ultracentrifuge.[16]

Analytical ultracentrifugation has been used to study the self-association of tubulin and various intermediates in microtubule assembly. Timasheff and co-workers used the Gilbert theory on the area distribution and the sedimentation profile under the Schlieren diagram to study the stoichiometry and equilibrium constant of tubulin assembly.[17] Sedimentation velocity and electron microscopy have been used to build hydrodynamic models for ring oligomers of tubulin and microtubule protein and to study the polymorphism of ring oligomers. See Scheele and Borisy for a review.[18]

Flow Birefringence

Since microtubules are long rods, they can be oriented by flow under the influence of a shear gradient, and the oriented microtubules are birefringent. Flow birefringence measurements have been successfully used to follow and quantitate microtubule assembly.[19]

[14] F. Gaskin, S. B. Kramer, C. R. Cantor, R. Adelstein, and M. L. Shelanski, *FEBS Lett.* **40**, 281 (1974).
[15] R. Margolis and L. Wilson, *Cell* **13**, 1 (1978).
[16] R. E. Ostlund, Jr., J. T. Leung, and S. V. Hajek, *Anal. Biochem.* **96**, 155 (1979).
[17] S. N. Timasheff, R. P. Frigon, and J. C. Lee, *Fed. Proc.* **35**, 1886 (1976).
[18] R. Scheele and G. Borisy, *in* "Microtubules" (K. Roberts and J. S. Hyams, eds.), pp. 175–254. Academic Press, New York, 1979.
[19] T. Haga, T. Abe, and M. Kurokawa, *FEBS Lett.* **39**, 291 (1974).

Dark-Field Light Microscopy

Despite the fact that microtubules are thinner than the resolution of the light microscope, microtubules can be observed by dark-field light microscopy,[20] and the rate of growth at each end of a microtubule can be determined.[21] However, to visualize the growth of microtubules, it was necessary to overcome limitations imposed by the sensitivity of the film, Brownian motion, and background light scattering at high protein concentrations. It was assumed that the glass surface had little effect on microtubule assembly. The light microscope (i.e., Zeiss Axiomat with a 200-W mercury arc light source) was fitted with infrared filters and a dark-field condenser.[21] An image intensifier was usually employed between the photographic screen and the camera. Microscope slides were prepared with grooves: one groove was used as a well to hold various solutions; the other was covered with a sheet of flat rubber pierced with a needle connected by tubing to a microliter pipetting device, and a closed system was formed by using silicone grease. In a typical experiment, polymerized microtubules were introduced, and some attached to the surface of the slide. Depolymerized microtubule protein was then placed in the sample well of the slide, where it was prewarmed and then rapidly flowed into the area of observation. Photographs were taken at 10–20-sec intervals, and treatment of the large volume of data was facilitated by computer processing.

X-Ray Diffraction

Time-resolved X-ray diffraction using synchrotron radiation has been used to study microtubule assembly.[22] This technique requires a high flux of X-rays and suitable X-ray optics, detector, and data acquisition system. The small-angle scattering is visible after a few seconds and can be used to monitor kinetic and structural transitions.[22] The microtubules and their kinetics of polymerization do not seem to suffer significantly from radiation damage. Initial experiments have examined the structural entities present in the several stages of assembly and disassembly and their dependence on both time and chemical conditions.

Laser Light Scattering

Microtubule protein preparations have been studied by dynamic laser light scattering techniques in order to gain information about diffusion

[20] R. Kuriyama and T. Miki-Noumura, *J. Cell Sci.* **19**, 6707 (1975).

[21] K. Summers and M. K. Kirschner, *J. Cell Biol.* **83**, 205 (1979).

[22] E.-M. Mandelkow, A. Harmsen, and E. Mandelkow, *Nature* (*London*) **287**, 595 (1980).

constants and trace aggregates.[7,23,24] Light from a laser is passed through the solution. The light scattered through the angle is detected with a photomultiplier tube, which generates a series of pulses proportional to the intensity of scattered light. The autocorrelation function computer measures the correlation between the intensity of scattered light at one time with that at a later time. An analysis of the mathematical form of the correlation function yields information about the hydrodynamic motions of the particles, in particular the translational diffusion constant. Some advantages of this method are that (a) the system can be probed without perturbation or interaction; (b) measurements are fast; (c) small amounts of sample are required (0.3 mg/0.5 ml of tubulin); (d) analysis is ultrasensitive to trace components of high molecular weight.

Filtration

A filtration assay using glass fiber filters to trap microtubules, but not rings or tubulin, has been reported.[25] Advantages of this method are that small volumes (40–200 μl) are used and that tightly bound labeled cofactors on microtubules can be measured simultaneously. The filtration assay is a destructive procedure, and the determinations of protein in microtubules and the bound cofactors are done after washing the filters and placing them in 0.5 M NaOH for 1 hr at 50° or overnight at room temperature.

Calorimetry

Calorimetry has been used to measure enthalpy and heat capacity changes for microtubule assembly. Calorimetric data is the summation of the heats of all reactions occurring in the system; with a system as complex as microtubule assembly coupled with GTP hydrolysis, it is not surprising that there is disagreement about the heats and processes that are involved. High-sensitivity differential scanning calorimeters are now being used to interpret the pathways for microtubule assembly.[26,27]

Immunochemistry

Antibodies to tubulin and microtubule-associated proteins have been used at the light and electron microscopy levels to study the organization,

[23] J. S. Gethner, G. W. Flynn, B. J. Berne, and F. Gaskin, *Biochemistry* **16**, 5776 (1977).
[24] J. S. Gethner, G. W. Flynn, B. J. Berne, and F. Gaskin, *Biochemistry* **16**, 5781 (1977).
[25] R. B. Maccioni and N. W. Seeds, *Arch. Biochem. Biophys.* **185**, 262 (1978).
[26] H.-J. Hinz, M. H. Gorbunoff, B. Price, and S. N. Timasheff, *Biochemistry* **18**, 3084 (1979).
[27] S. A. Berkowitz, G. Velicelebi, J. W. H. Sutherland, and J. M. Sturtevant, *Proc. Natl. Acad. Sci. U.S.A.* **77**, 4425 (1980).

regulation, and composition of microtubules in tissues, cell cultures, and cytoskeletons obtained by the treatment of cells with nonionic detergents in microtubule stabilizing buffers. Immunocytochemical methods are described elsewhere in this volume [48].

Acknowledgment

This work was supported by National Institute of Health grant NS 12418.

[42] Microtubule Disassembly: A Quantitative Kinetic Approach for Defining Endwise Linear Depolymerization

By DANIEL L. PURICH, TIMOTHY L. KARR, and DAVID KRISTOFFERSON

Linear protein polymers of actin, tubulin, and tobacco mosaic virus protein are formed as products of efficient entropy-driven self-assembly reactions.[1] The dynamics of the polymerization and depolymerization processes is relevant to understanding the biochemistry of subunit–subunit interactions, the pharmacology of significant chemotherapeutic agents, and the cell biology of self-organizing systems. The assembly is complicated by the nucleation, elongation, and, possibly, chain length redistribution[2]; yet developments in theory and experimental technique now permit a quantitative appraisal of the depolymerization reaction.[3-5] Moreover, the studies on microtubule disassembly involve sufficient generality for application to other linear, indefinite polymerization systems. In this chapter, we present the pertinent theoretical aspects and describe the experimental approaches to characterize the mechanism of depolymerization.

Theory

To analyze endwise depolymerization quantitatively, the following premises must hold: (a) the polymer undergoes stepwise disassembly in a series first-order fashion; (b) the off-rate constant is independent of

[1] M. A. Lauffer, "Entropy-Driven Processes in Biology: Polymerization of Tobacco Mosaic Virus Protein and Similar Reactions." Springer-Verlag, Berlin and New York, 1975.
[2] F. Oosawa and M. Kasai, *J. Mol. Biol.* **4**, 10 (1962).
[3] T. L. Karr, D. Kristofferson, and D. L. Purich, *J. Biol. Chem.* **255**, 8560 (1980).
[4] D. Kristofferson, T. L. Karr, and D. L. Purich, *J. Biol. Chem.* **255**, 8567 (1980).
[5] T. L. Karr, D. Kristofferson, and D. L. Purich, *J. Biol. Chem.* **255**, 11853 (1980).

polymer length over the course of depolymerization; (c) the on-rate constant is zero; (d) turbidity is a measure of the remaining polymer weight concentration and is independent of the polymer length distribution; and (e) the concentrations of the various polymer lengths may be estimated by use of electron microscopy. In this section, we examine the kinetics of the series first-order decay of polymer and its expression in terms of remaining polymer weight concentration.

For linear depolymerization of a polymer of length n, we may assume series first-order kinetics:

$$P_n \overset{k}{\to} P_{n-1} \overset{k}{\to} P_{n-2} \to \to \cdots P, \text{ etc.} \tag{1}$$

Because turbidity is a measure of polymer weight concentration, it is useful to define several useful parameters. If we define x as the number of protomers in a polymer and $f(x)$ as the number concentration of the polymer form with x protomers, the following relations result:

$$N = \sum_{x=m}^{n} f(x) \tag{2}$$

$$P = \sum_{x=m}^{n} xf(x) \tag{3}$$

$$W_0 = M_p \sum_{x=m}^{n} xf(x) \tag{4}$$

where N is the total polymer number concentration, P is the total concentration of protomers in polymeric form, W_0 is the total polymer weight concentration, and M_p is the molecular weight of the protomer. Here, the indices m and n represent the number of protomers in the shortest and longest linear polymers. The notion of a lower bound m is imprecise; however, for the purpose of the following derivations, an exact value for m is unnecessary when $n \gg m$. Berne[6] has considered the light-scattering behavior of rods longer than the wavelength of incident light, and he has shown that the turbidity is directly proportional to the weight concentration of polymer. As the length falls shorter than a minimal value, the light-scattering contribution of such short polymers becomes negligible. Thus, m may be experimentally defined as the minimal polymer length that contributes significantly to the light scattering, and the experiments are carried out such that $n \gg m$.

By finding the expression for the time evolution of the total polymer weight concentration, $W(t)$, we will obtain an expression corresponding to the time evolution of the turbidity change upon depolymerization. We

[6] B. J. Berne, *J. Mol. Biol.* **89**, 755 (1974).

begin with an initial polymer length distribution, and it is assumed that this distribution will progress in a leftward direction along the x, or length, axis as depolymerization occurs. As the distribution progresses toward shorter polymer length, the shape of the distribution will change because the interconversion of $P_i \rightarrow P_{i-1} \rightarrow P_{i-2}$, etc., will not occur in unison. The broadening of the distribution and its impact on the turbidity measurements of depolymerization will depend upon the shape of the initial distribution. To determine the time-dependent changes in the distribution and polymer weight concentration, we shall first assume that k is independent of polymer length. If c_i represents the number concentration of polymer with i subunits, then the following kinetic equation holds for all i *not* equal to n:

$$\frac{dc_i}{dt} = -kc_i + kc_{i+1} = k(c_{i+1} - c_i), \text{ for } i \neq n \tag{5}$$

For i equal to n, we get

$$dc_n/dt = -kc_n \tag{6}$$

or

$$c_n = c_n^0 \exp(-kt) \tag{7}$$

where $c_n = c_n^0$ at t equal to zero; no longer species exists to contribute to the concentration c_n^0.

Equation (5) can now be solved for sequentially decreasing values of i beginning with $i = (n - 1)$, $(n - 2)$, etc. Substituting Eq. (7) into Eq. (5) with $i = (n - 1)$, yields

$$(dc_{n-1})/dt = k(c_n - c_{n-1}) = k(c_n^0 \exp(-kt) - c_{n-1})$$

Rewritten in standard form for linear nonhomogeneous first-order differential equations, this expression becomes

$$(dc_{n-1})/dt + kc_{n-1} = kc_n^0 \exp(-kt)$$

An equation of this form may be solved by finding the integrating factor, which is $\exp(kt)$ in this case. When both sides of the above expression are multiplied by this factor, we find

$$\exp(kt) (dc_{n-1})/dt + k \exp(kt)c_{n-1} = kc_n^0$$

The left-hand side of this equation equals $(d/dt) [c_{n-1} \exp(kt)]$; thus

$$\int d[c_{n-1} \exp(kt)] = kc_n^0 \int dt$$

Carrying out this integration under the condition that $c_{n-1} = c_{n-1}^0$ when $t = 0$, we find

$$c_{n-1} = (kc_n^0 t + c_{n-1}^0) \exp(-kt) \tag{8}$$

Equation (8) can be then substituted back into Eq. (5) with i now equal to $n-2$. The solution is then repeated using the same integrating factor as before. After a few trials, the solution for c_{n-j} becomes apparent:

$$c_{n-j} = \left[\sum_{i=0}^{j} \frac{(kt)^{j-i}}{(j-i)!} c_{n-i}^0 \right] \exp(-kt) \tag{9}$$

Because we can now determine the number of polymers with $(n-j)$ protomers at any time t for any value of j from zero to $(n-m)$, we can write the equation for polymer weight at time t.

$$W(t) = \sum_{j=0}^{n-m} M_p (n-j) c_{n-j}$$

$$= M_p \exp(-kt) \sum_{j=0}^{n-m} \left[(n-j) \sum_{i=0}^{j} \frac{(kt)^{j-i}}{(j-i)!} c_{n-i}^0 \right] \tag{10}$$

Where we have used Eq. (9), and we have factored two quantities from the sum to obtain the right-hand side of Eq. (10). With the initial concentration values for each of the polymeric species, Eq. (10) can be evaluated to yield a depolymerization curve for polymer weight concentration versus time. The sums are easily handled by the computer program available upon request. Alternatively, a standard Runge–Kutta method will yield numerical solutions for this series first-order depolymerization.

Data Analysis

The experimental curves for polymer disassembly and those obtained by computer simulation with the initial length distribution data must be fit to the same scale to obtain the rate constant. One may adjust the initial polymer weight on the computer simulation to equal the initial turbidity reading, and the ratio of these values serves to relate the two parameters. In practice, the computer polymer weight units are arbitrary because the program calculates total length of polymers in the length histogram in units of micrometers. The total length is clearly directly proportional to polymer weight. The time axes are matched by using the ratio of times required for loss of half of the initial turbidity amplitude (t_e) and the initial simulated polymer weight (t_c). The progress curve for turbidity loss defines the real time parameter, and the computer simulation is carried out with the rate constant arbitrarily at 5 reciprocal computer time units. The latter reduces the number of iterations for accurate simulation and thus minimizes expense.

After matching the experimental and theoretical progress curves, the next step is to calculate the microscopic rate constant, k_e, from the computer rate constant, k_c. This requires two conversion factors as shown in Eq. (11).

$$k_e = k_c(t_c/t_e)FL \tag{11}$$

where F is the number of protomers per micrometer of polymer and L is the histogram interval length in micrometers. This formula also reveals that we use the histogram intervals as the reacting species and the frequency values as the concentrations in the computer simulations. This again minimizes expenditure of computer time for calculations, and k_c governs the rate of depolymerization in any interval to the next smaller interval. Thus, k_c is slower than the rate constant for protomer release by the number of dimers per interval. One advantage of this data analysis method is that it requires no knowledge of the protein concentration, the determination of which could introduce further experimental error.

The reader should note that the only truly adjustable parameter in this fitting procedure is the value of the rate constant, which is assumed to be length independent for $n \geqslant m$. The single-parameter fitting procedure lacks the freedom of adjustment available in multiparametric fits, but the agreement between theory and experiment is more satisfying.

Experimental Procedures

For indefinite polymerization processes that result from entropy-driven condensation equilibria, there are three ways to effect depolymerization: (a) dilution to below the critical concentration; (b) reduction of the temperature to destabilize the polymer; and (c) addition of a reagent to reduce the concentration of the form of the protomer in equilibrium with the polymer. The methods for bringing about such changes must be rapid relative to the time course of depolymerization; otherwise, the kinetics of effecting the depolymerization process will obscure the kinetics of polymer loss. In principle, one may estimate the rate constant for depolymerization from the initial rate of depolymerization and the initial polymer number concentration. Alternatively, the entire depolymerization process may be evaluated by use of the theoretical treatment above. The chief advantage of the latter is that one may determine any changes in rate behavior that may appear upon more extensive polymer disassembly. In this section, we describe three methods for rapidly inducing polymer disassembly, and we illustrate each method with microtubule disassembly experiments.

Rapid Dilution Method. Karr and Purich[3,4] described the design and

RAPID DILUTION CELL:

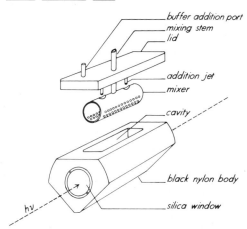

FIG. 1. Diagrammatic representation of the rapid-mixing cuvette.

application of a rapid dilution cuvette for effecting microtubule disassembly. Three goals were satisfied by the design shown in Fig. 1: (*a*) the method is rapid with measurements practical within 5–6 sec of mixing; (*b*) mechanical shearing of polymer is minimal; and (*c*) the amplitude of the absorbance change upon disassembly is optimized by a long pathlength configuration. The unit consists of a long pathlength cuvette with an open top, an immersible mixer fitted to the cuvette cover, a 2-ml glass bulb reservoir for holding concentrated solutions of assembled polymer, and an attached syringe located outside the spectrophotometer sample housing to expel the bulb's contents into the cuvette. The cuvette and all attachments were fabricated from black Delrin to the dimensions noted elsewhere,[7] and these parts were lightly sandblasted to reduce surface reflectance. Quartz end plates were secured by annular, threaded aluminum inserts and sealed with soft polyethylene gaskets made from the red film used to make analytical ultracentrifuge cell gaskets. The perforated mixer is attached by a rod extending to an external handle, and mixing is achieved by several up-down motions. The mixer remains within the cuvette after mixing to prevent splashing into the spectrophotometer, and the mixer's interior is sufficiently large not to obstruct the incident light beam. In our laboratory, the entire cuvette is located in the closed sample compartment of a Cary 118C recording spectrophotometer with the photomultiplier shutter open to defeat the automatic delay mechanism that waits 4 sec

[7] T. L. Karr and D. L. Purich, *Anal. Biochem.* **104**, 311 (1980).

after closure of the sample housing lid before exposing the photomultiplier tube. This may be achieved by means of a screw located in the lid directly above the shutter switch to trip the switch as desired.

In practice, the following protocol works well. Two aliquots of microtubule protein sample are prepared in the cold (0–4°) using dedusted and degassed buffer (by passage first through a 0.45 μm Millipore filter and subsequently boiled for 5 min). One sample is used to follow the course of warm-induced polymerization, and the second is transferred to the glass bulb reservoir and permitted to assemble under otherwise identical conditions. At the same time, 12–18 ml of dilution buffer (maintaining all solution variables as the assembly buffer) may be warmed and added to the rapid dilution cuvette for thermal equilibration. After attainment of the stable assembly plateau, the glass bulb is attached to the side arm of the cuvette (which is connected to the interior by four channels) and the other end of the bulb is fitted to a plastic tube extending to the syringe. The bulb's contents are quickly expelled by air pressurized by the syringe, and the mixer is agitated. A typical dilution-induced disassembly curve for microtubules depleted of associated proteins is shown in Fig. 2A. The experimental points were taken from the spectrophotometer tracing, and the solid line is the theory line estimated on the basis of electron microscopic measurement[8] of the length distribution (see Fig. 2B). The rate constant for tubulin dimer (protomer) release was found to be 105 sec^{-1}.[3]

Rapid Heat-Exchanger Method. The cold sensitivity of microtubule polymers was exploited by Johnson and Borisy[9] to evaluate the kinetics of depolymerization. Karr *et al.*[3] developed a rapid heat exchanger technique to make this approach a quantitative tool for probing depolymerization. The achievement of rapid cooling of aqueous samples requires an appropriate geometry to obviate the relatively poor thermal conductivity of water. The design shown in Fig. 3 demonstrated several important features: (*a*) the high thermal conductivity of the copper block; (*b*) the considerable heat capacity of the 1650 g block and cold-finger insert; (*c*) the redistribution of the aqueous sample into a 0.6 mm layer exposed to two cold surfaces of the exchanger; and (*d*) the gold plating of all surfaces to eliminate contact of the copper block with the protein sample. The introduction of the 2–3 ml of microtubule sample is achieved by means of a syringe, and subsequent expulsion of the sample from the exchanger into the cold cuvette in the spectrophotometer by a second syringe also

[8] D. Kristofferson, T. L. Karr, T. R. Malefyt, and D. L. Purich, *in* "Methods and Perspectives in Cell Biology" (L. Wilson, ed.), Vol. 25, pp. 133–144. Academic Press, New York, 1982.

[9] K. A. Johnson and G. G. Borisy, *J. Mol. Biol.* **117**, 1 (1977).

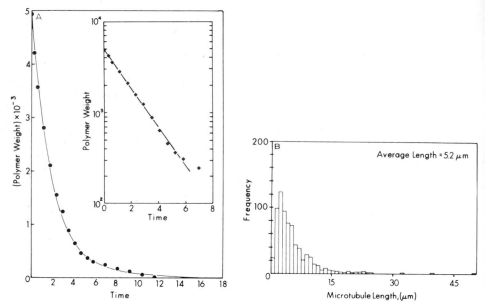

FIG. 2. (A) Plot of theoretical (——) and experimental (●) depolymerization curves for phosphocellulose-purified tubulin. The theoretical curve was calculated from the frequencies in Fig. 2B. A rate constant of 5 reciprocal computer time units was employed in the simulation, and 1 computer time unit equaled 34.5 sec. The experimental curve was obtained by diluting 0.5 ml of 2.4 mg per milliliter of microtubule solution at steady state into approximately 18.2 ml of isothermal (30°) buffer containing 0.5 mM GTP and other buffer components at pH 6.8.   (B) Plot of frequency vs microtubule polymer lengths as determined by electron microscopy.

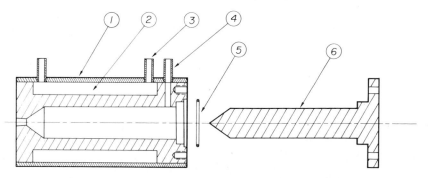

FIG. 3. Sectional assembly diagram of rapid heat exchanger. The overall block length is 10.2 cm with other dimensions drawn to scale. The exchanger block (1) was fabricated to accommodate a syringe inlet with Luer taper. Other items include a water jacket (2), water circulation ports (3), sample exit port (4), O-ring seal (5), and solid copper cold finger insert (6). Total internal volume of assembled exchanger is 3.5 ml.

attached to the three-way nylon Luer valve connected to the cold block. Heat exchange is quite rapid, as indicated by control measurements wherein a 3-ml aliquot at 30° is cooled to 5.8° within 2–3 sec. The pen period of the Cary Model 210 recording spectrophotometer was set at 1 sec to ensure rapid pen response, and a Cary Model 9543 digital temperature readout accessory with thermistor probe was used to sense temperature changes. Finally, a dry nitrogen gas purge of the cuvette housing prevented moisture accumulation on the cuvette faces.

A typical rapid cooling experiment is shown in Fig. 4, and the most significant feature of results is the excellent agreement between theory (solid line) and experiment (data points). These results are clearly at odds with results with a glass-jacketed cuvette that is limited by slow cooling.[9] In the experiment shown in Fig. 4, the process is characterized by a rate constant of 166 sec^{-1}.[3]

Rapid Addition Method. As noted above, the rapid addition of certain effectors may be used to evaluate effector interactions with polymer. For example, one may be interested in the effect on microtubules of calcium ion, colchicine, podophyllotoxin, GDP, etc. The celerity of calcium ion depolymerization of microtubules illustrates the need for rapid mixing; this process is over in about 40–50 sec. To obtain an adequate tracing of

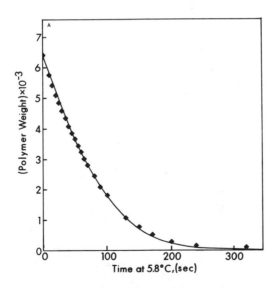

FIG. 4. (A) Cold depolymerization progress curve after temperature jump. Microtubule protein (2.5 mg/ml) was assembled at 30° to stable plateau turbidity, and depolymerization was followed in a Cary 210 spectrophotometer. ◆, Experimental curve; ——, theoretical curve. (B) Microtubule length distribution before cold depolymerization.

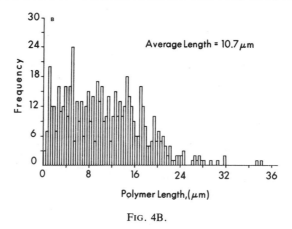

FIG. 4B.

the disassembly time course, we have developed a rapid addition protocol with two components: (*a*) a commercial built-in magnetic stirring mechanism; and (*b*), a spring-loaded plunger in the cuvette housing. The polymer is first assembled by warming in a 1 × 1 cm cuvette in the spectrophotometer to the stable state of assembly. A small aliquot is

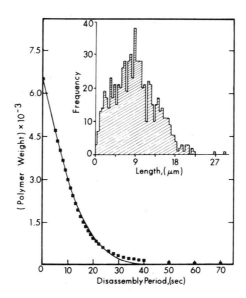

FIG. 5. Plot of microtubule depolymerization vs time after addition of 5 m*M* calcium chloride. The experimental points were obtained from a continuous recording of absorbance at 350 nm, and the zero time value was established by the turbidity value before addition of calcium ion. ■, Experimental curve; ———, theoretical curve. *Inset:* Distribution of microtubule polymer lengths prior to calcium ion addition. Average polymer length was 8.9 μm.

RATE CONSTANTS AND POLYMER NUMBER CONCENTRATION DEPENDENCE FOR DILUTION-,
COLD-, AND CALCIUM ION-INDUCED MICROTUBULE DISASSEMBLY[a]

Condition	Increase in number concentration	Increase in disassembly rate	Rate constant $(\sec^{-1})$	Reference[b]
Rapid dilution	3.12	3.18	113	3
Cold	—	—	166	3
Calcium ion	2.36	2.27	860	5

[a] Values for microtubule protein containing about 75% tubulin and 25% microtubule-associated proteins by weight.
[b] Numbers refer to text footnotes.

removed for electron microscopy, and the plastic or glass rod on the plunger is located with 0.05–0.1 ml concentrated calcium ion solution. After closing the instrument, the chart speed is adjusted to an appropriately fast rate and the plunger is lowered into the cuvette sample. Mixing is efficient even in the relatively viscous fluids containing assembled microtubules, and the tracing can be recorded almost instantaneously. An illustration of the quality of such measurements is shown in Fig. 5, with the solid theory line agreeing well with the experimental points for nearly complete disassembly. A rate constant of 860 $\sec^{-1}$ was obtained in this experiment. Because this value is more rapid than the rate of dilution-induced disassembly, we believe that a direct interaction between calcium ion and the protomers at microtubule ends facilitates the disassembly. If the calcium ion only acted on the monomer to reduce its ability to polymerize, the calcium ion and rapid dilution results should have agreed more closely.

Discussion

The methods outlined in this chapter already have provided valuable kinetic data on microtubule disassembly, and they should provide useful approaches for examining other indefinite polymerization systems. One major limitation of these approaches is that many polymers have a natural polarity with different transitions states (and, therefore different kinetic constants) for the association–dissociation interactions on each polymer end. The present method can only provide an average rate of depolymerization in such cases. Bergen and Borisy[10] have described a semiquantitative method using the polarity of axoneme-induced microtubule assembly to estimate the rates at each end of microtubule polymers. Another limita-

[10] L. G. Bergen and G. G. Borisy, *J. Cell Biol.* **84**, 141 (1980).

tion is that one cannot evaluate the exchange of unbound dimers with polymer resulting from diffusional exchange or head-to-tail polymerization at steady states of assembly. Such exchange reactions require the application of radiotracer methodologies as first recognized by Kasai and Oosawa[11] and analyzed quantitatively by Wegner[12] for F-actin exchange with actin protomers.

The validity of the endwise depolymerization model may also be tested by examining the dependence of the initial disassembly rate on microtubule number concentration. The results of such determinations are shown in the table, and the rate constants are also reported. When the quantitative methods outlined here are used, the correspondence between increased rate of disassembly and increased microtubule number concentration is most satisfying.

Finally, we should note that knowledge of the dynamics of linear depolymerization should be helpful in sorting out some of the rate constants and pathways operating polymer assembly. Presumably, our off-rate constant obtained by rapid dilution and the microtubule critical concentration can be used to evaluate the on-rate constant for the elongation reaction.[3] This on-rate constant appears to be at or near the diffusion limit for macromolecules, suggesting a remarkably high efficiency for microtubule elongation. The challenge now before us is to extend this work and to develop useful approaches for characterizing the nucleation mechanism for initiating the self-organizing events.

[11] M. Kasai and F. Oosawa, *Biochim. Biophys. Acta* **172**, 300 (1969).
[12] A. Wegner, *J. Mol. Biol.* **108**, 139 (1976).

[43] Preparation and Purification of Dynein

By CHRISTOPHER W. BELL, CLARENCE L. FRASER, WINFIELD S. SALE, WEN-JING Y. TANG, and I. R. GIBBONS

On the basis of present knowledge, it appears that the great majority of motile processes in eukaryotes are caused by the action of one of two macromolecular systems: the actomyosin system or the tubulin-dynein system. The systems appear to be generally similar inasmuch as, in both cases, the energy stored in the terminal phosphate of ATP is released by one protein (myosin or dynein) and utilized to perform work on a structural framework constructed of the other protein (actin or tubulin). This is about as far as the comparison can, with any degree of confidence, be taken at present, since although the structure and function of actomyosin are relatively well understood both as a complete system and as separated components, the same cannot be said of the tubulin–dynein system. This is partly because of the more recent identification of the proteins involved in ciliary and flagellar motility, and partly because of the relatively small quantities of material available for study. The latter problem is most acute with dynein, since tubulin comprises a large mass percentage of the proteins of cilia and flagella and is also available in larger quantities from other sources, such as brain, in an at least partly compatible form.

Dynein was first identified in ciliary axonemes of *Tetrahymena* as a high molecular weight protein with MgATPase activity, which could be extracted from the axonemes by exposure to low ionic strength in the presence of EDTA.[1] Selective extraction and recombination followed by electron microscopy indicated that dynein comprises part or all of the arms bridging the gap between the doublet microtubules of the axonemes.[2] These criteria, high molecular weight (sedimentation velocity up to 30 S), possession of ATPase[3] activity, extractability with low ionic strength solutions in the presence of EDTA and relationship to the arms on doublet tubules, became the identifying characteristics of dynein from other sources. The introduction of polyacrylamide gel electrophoresis in the presence of sodium dodecyl sulfate (SDS) led to the recognition that the polypeptide subunits of dynein are very large: estimates of apparent molecular weight have ranged from 300,000 to 500,000.[e.g.,4-6] Further investigation of dynein from a wide variety of sources has revealed multiple isoenzymic forms of the enzyme that are distinguishable by the different electrophoretic mobilities of their high molecular weight subunits. Extraction of axonemes with high concentrations of salt has also been found to release dynein, but whereas both the inner and outer arms of flagellar axonemes from *Chlamydomonas* are extractable in this manner,[7] only the outer arms of, for instance, sea urchin sperm flagella are extracted.[8]

These indications of both inter- and intraspecies differences in dynein are reinforced by reports of different forms of dynein with sedimentation coefficients varying from 10 S to 30 S, depending on the species and the

[1] I. R. Gibbons, *Proc. Natl. Acad. Sci. U.S.A.* **50**, 1002 (1963).
[2] I. R. Gibbons, *Arch. Biol.* **76**, 317 (1965).
[3] Unless stated otherwise, ATPase is to be understood to mean MgATPase.
[4] R. W. Linck, *J. Cell Sci.* **12**, 951 (1973).
[5] R. G. Burns and T. D. Pollard, *FEBS Lett.* **40**, 274 (1974).
[6] G. G. Borisy, J. M. Marcum, J. B. Olmsted, D. A. Murphy, and K. A. Johnson, *Ann. N.Y. Acad. Sci.* **253**, 107 (1975).
[7] G. Piperno and D. J. L. Luck, *J. Biol. Chem.* **254**, 3084 (1979).
[8] I. R. Gibbons and E. Fronk, *J. Cell Biol.* **54**, 365 (1972).

conditions of sedimentation. Furthermore, the increasing resolution of SDS–polyacrylamide gel electrophoresis has shown that the number of distinct bands in the region of the high molecular weight subunits is both large and variable from species to species. At present the largest number of distinct high molecular weight chains identified in a single species is 10 from the flagella of *Chlamydomonas*.[7] At least some of the various high molecular weight chains in flagella of a given species have been shown to derive from multiple isoenzymic forms of dynein.[9,10] Further complexity has arisen in the recognition, in both *Chlamydomonas* and sea urchin sperm flagella, that two or more distinct ATPases may be associated with a single arm.[11,12] It also appears that the dynein–tubulin system is not confined to cilia and flagella. There have been several reports of dynein-like ATPase activity in sea urchin egg cytoplasm,[13–16] and Dentler *et al.*[17] demonstrated the presence of a dynein-like ATPase in the ciliary membrane. These observations suggest that dynein may have a widespread role in cellular motility.

This structural and functional complexity of dynein serves to highlight the differences between this and the actomyosin system, and the difficulties inherent in obtaining a full understanding of flagellar and ciliary motility. Such understanding will depend upon comprehensive enzymic and physicochemical characterization of dynein, a process requiring large quantities of a homogeneous enzyme. Much of the remainder of this chapter describes in detail the preparation, storage, use, and properties of such a dynein ATPase from sea urchin sperm flagella, probably one of the best available sources for providing relatively large quantities of this enzyme. The methods described here relate specifically to our experience in preparing latent activity dynein-1 (LAD-1) from sperm flagella of the Hawaiian sea urchin *Tripneustes gratilla*. However, the general techniques appear to be applicable to sperm flagella of many other animals. The final section presents a brief comparative review of the methods that have been used for the preparation of dynein from various other sources and lists some of the salient properties of these dynein ATPases.

[9] I. R. Gibbons, E. Fronk, B. H. Gibbons, and K. Ogawa, *in* "Cell Motility," *Cold Spring Harbor Conf. Cell Proliferation* **3**, [Book A-C], pp. 915–932 (1976).
[10] K. Ogawa and I. R. Gibbons, *J. Biol. Chem.* **251**, 5793 (1976).
[11] B. Huang, G. Piperno, and D. J. L. Luck, *J. Biol. Chem.* **254**, 3091 (1979).
[12] C. W. Bell, W.-J. Y. Tang, W. S. Sale, and I. R. Gibbons, *J. Biol. Chem.* **257**, 508 (1982).
[13] T. Miki, *Exp. Cell Res.* **29**, 92 (1963).
[14] R. Weisenberg and E. W. Taylor, *Exp. Cell Res.* **53**, 372 (1968).
[15] I. Mabuchi, *Biochim. Biophys. Acta* **297**, 317 (1973).
[16] M. M. Pratt, T. Otter, and E. D. Salmon, *J. Cell Biol.* **86**, 738 (1980).
[17] W. L. Dentler, M. M. Pratt, and R. E. Stephens, *J. Cell Biol.* **84**, 381 (1980).

Preparation of Latent Activity Dynein-1

Collection and Storage of Sea Urchins

Sea urchins (*Tripneustes gratilla*) are collected weekly from sandy reef floor areas at depths of 3–5 m. They are returned to the laboratory in dry (i.e., not water filled) buckets in order to prevent the widespread triggering of gamete shedding that occurs when one or more sea urchins shed into water in close contact with others. The animals are then stored in running sea water at ocean temperature. Ripe *Tripneustes* can be sexed by shaking them to and fro with the gonopore pointing down, which causes small amounts of gametes to appear at the gonopore: eggs appear dark orange in such situations, whereas semen is off-white.

After the gonopores have been carefully rinsed off, males and females may be returned to separate tanks. Complete shedding of semen from male *Tripneustes* can be induced by injecting 0.5 M KCl into the body cavity. A speedier procedure is to remove the Aristotle lantern by cutting around the peristomatous membrane, empty out the body fluid, and fill the body cavity with 0.5 M KCl. In either case, the semen is collected by inverting the animal over a 50-ml beaker filled with sea water containing 0.1 mM EDTA. A single, healthy male *Tripneustes* can yield up to 40 ml of dense semen, although the average is 15–20 ml. If desired, semen may be collected in artificial sea water lacking certain constituents (e.g., Ca^{2+}). It may also be collected dry, without any overlying solution, but we have not found this to be of any particular advantage in the preparation of dynein.

Sea urchins may also be induced to shed gametes by electrical stimulation.[18,19] Unlike injection with KCl, this method does not kill the animal, and it may be returned to the storage tank and with feeding will become ripe again within a few weeks.

Preparation of Sperm Flagellar Axonemes

An important step in the preparation of LAD-1 from the sperm of *Tripneustes* is to isolate flagellar axonemes that are essentially free both of sperm heads and of surrounding membranes that would otherwise prevent direct access to the proteins of the axoneme. The removal of the heads is relatively straightforward and is described in detail later. The removal of membranes from flagellar and ciliary axonemes is a problem

---

[18] E. B. Harvey, "The American *Arbacia* and Other Species of Sea Urchins." Princeton Univ. Press, Princeton, New Jersey, 1956.
[19] K. Osanai, *in* "The Sea Urchin Embryo" (G. Czihak and R. Peter, eds.), pp. 26–40. Springer-Verlag, Berlin and New York, 1975.

that has been approached in two ways. Membranes have been either solubilized by detergents such as digitonin[1] or Triton X-100,[20] or osmotically ruptured and fragmented by exposure of the intact organelle to high concentrations of glycerol or sucrose.[21] Solubilization by detergents has the advantage of disrupting membranes on the molecular level and is generally more likely to effect their complete removal. However, it is often difficult to remove the detergent completely after membrane solubilization, even with extensive washing, since detergent molecules tend to bind to the protein structures of the axoneme.

We formerly used a method of axoneme preparation in which the membranes were solubilized by treatment with 1% w/v Triton X-100,[8,22] but we have since found that LAD-1 prepared by this method sometimes has a partially activated ATPase level, variable from preparation to preparation and similar to, but not as pronounced as, that obtained when LAD-1 is incubated for 10–15 min at room temperature in the presence of 0.05% Triton X-100. We surmise that this elevated ATPase activity is a side effect of Triton in the membrane solubilization step. It has been known for some time[22] that incubation of soluble LAD-1 with Triton X-100 causes profound changes in the physical and chemical properties of the enzyme, not the least of which is a roughly 10-fold increase in ATPase activity over that of the latent form (see Selected Properties of LAD-1 and its Subunits). It is our present opinion that damage to LAD-1 during preparation is manifested by, among other things, increased basal ATPase activity and a consequent decrease in activation ratio (i.e., ratio of ATPase activity after exposure to 0.05% Triton X-100 to basal ATPase activity). By these criteria, solubilization of flagellar membranes by Triton X-100 is not the method of choice (see Table II), and we no longer use it in the preparation of dynein.

The method now used to remove sperm flagella membranes involves osmotic rupture by exposure to 20% w/v sucrose. The detailed procedure for the preparation of axonemes is given below. Compositions of buffers are given in Table I.

About 60 ml of semen are collected and are diluted to 300 ml with EDTA (0.1 mM) sea water. The diluted semen is centrifuged for 5 min at 30 g to remove sand and debris expelled from the urchins along with the semen. The semen supernatant is carefully decanted and centrifuged again at 3000 g for 5–10 min in order to pellet the sperm. The pelleted sperm are resuspended in 300 ml of cold (4° or less) 20% w/v sucrose in glass-distilled water and are homogenized in a Dounce homogenizer with eight strokes of

[20] B. H. Gibbons, E. Fronk, and I. R. Gibbons, *J. Cell Biol.* **47,** 71a (Abstr.) (1970).
[21] E. C. Raff and J. J. Blum, *J. Biol. Chem.* **244,** 366 (1969).
[22] I. R. Gibbons and E. Fronk, *J. Biol. Chem.* **254,** 187 (1979).

TABLE I
COMPOSITION OF BUFFERS USED IN THE PREPARATION AND ASSAY OF CRUDE LAD-1

Isolation buffer	High-salt buffer	ATPase assay buffer
0.1 M NaCl	0.6 M NaCl	0.1 M NaCl
5 mM imidazole-HCl, pH 7[a]	5 mM imidazole-HCl, pH7	0.03 M Tris-HCl, pH 8.1[b]
4 mM MgSO$_4$	4 mM MgSO$_4$	2 mM MgSO$_4$
1 mM CaCl$_2$[c]	1 mM CaCl$_2$	0.1 mM EDTA
1 mM EDTA	1 mM EDTA	1 mM ATP
7 mM 2-mercaptoethanol	7 mM 2-mercaptoethanol	
	1 mM dithiothreitol	

[a] Imidazole is recrystallized from 1 mM EDTA in 80% v/v ethanol.
[b] Tris-base is recrystallized first from 1 mM EDTA and then from 80% v/v methanol.
[c] Ca^{2+} is added because it appears to result in tighter axonemal pellets during preparation. However, preparation in the absence of Ca^{2+} yields crude LAD-1 with essentially identical characteristics.

a tight pestle. This homogenization should be quite vigorous to ensure complete separation of sperm heads and flagellar axonemes, and to effect fragmentation of flagellar membranes. The 20% w/v sucrose is employed as a compromise between the more efficient membrane fragmentation but longer centrifugation times engendered by higher sucrose concentrations and the much less efficient membrane fragmentation and concomitant low yields of dynein obtained with lower sucrose concentrations. This and all subsequent procedures are carried out at 0–4°. The homogenate is then centrifuged at 3000 g for 7 min to pellet most of the sperm heads and membrane fragments. This pellet is generally not very tightly packed and care must be taken to prevent contamination of the supernatant by pellet as the former is decanted. Contamination is usually easy to recognize as the pellet material has a yellow color owing to mitochondrial membrane pigments. The decanted supernatant is centrifuged at about 27,000 g for 15 min. The resulting pellet should display a distinct stratification with a small, compact dark layer at the bottom comprising most of the remaining sperm heads and membrane fragments overlaid by a much thicker off-white layer comprising the axonemes. In resuspending the pellet, the darker bottom layer is discarded. If the pellet displays a large, diffuse yellow layer on the bottom, this indicates significant contamination of the axonemes by membranes and sperm heads, which should be removed by a second low speed centrifugation (1500 g for 5 min) in isolation buffer.

The axonemal pellets are resuspended in isolation buffer (Table I) to a total volume of about 150 ml. Resuspension is achieved with 4 to 5 strokes of the loose-fitting pestle in a Dounce homogenizer. The resuspended axonemes are centrifuged at 12,000 g for 10 min, and the supernatant is

TABLE II
COMPARISON OF YIELDS AND OF PROPERTIES OF CRUDE LAD-1 PREPARED BY TWO DIFFERENT METHODS

Method of preparation[a]	Axonemes[b] (mg/ml semen)	Crude LAD-1[b] (mg/100 mg axonemes)	Latent activity[c]	Triton-activated activity[c]	Activation ratio
Triton X-100 (4)	3.1 ± 0.9	6.3 ± 1.2	0.62 ± 0.10	3.8 ± 0.8	6.1 ± 0.
Sucrose (19)	6.4 ± 1.6	5.7 ± 0.9	0.26 ± 0.04	2.7 ± 0.4	10.4 ± 1.

[a] Main entry refers to method of disrupting sperm flagellar membranes. Otherwise all steps in preparation were identical. Number in parentheses is the number of individual preparations from which data were calculated.
[b] Amount of LAD-1 obtained by single high salt extract; protein measured by method of Lowry.
[c] Specific activity given as micromoles of P_i min^{-1} mg^{-1}.

discarded. The axonemal pellet is gently resuspended by homogenization in axoneme buffer to a volume of about 75 ml; any dark material at the bottom of the pellet is again discarded. At this point the concentration of axonemal protein is determined (we use the method of Lowry, calibrated with bovine serum albumin) and the total yield is calculated. The yield of axonemes from 60 ml of *Tripneustes* semen averages 400 mg when prepared by this method (Table II). Axonemes are generally used immediately after preparation, but they can be stored as a pellet in the cold with a small buffer overlay for at least 24 hr without noticeable deterioration.

Extraction of Crude LAD-1

Crude LAD-1 is extracted by resuspending the pelleted washed axonemes in high-salt buffer at a protein concentration of 3 mg ml^{-1} (Table I). We have found that the lower the concentration at which axonemes are suspended in high-salt buffer, the greater the total yield of crude LAD-1; 3 mg ml^{-1} has been chosen as a compromise between maximizing yield and minimizing final volume and dilution of dynein. Axonemes are extracted for about 15 min at 0°, before being centrifuged at 12,000 g for 15 min. The supernatant is decanted and clarified by centrifugation at 100,000 g for 15 min. The supernatant from this centrifugation constitutes the stock solution of crude LAD-1, and total protein yield is about 20–25 mg (Table II). If a greater quantity of dynein is required, it is most conveniently obtained by extracting the axonemes a second time with high-salt buffer, but this extract is generally contaminated to a greater degree by other heavy chains and by tubulin. Increasing the concentration of NaCl in high-salt buffer does not greatly increase the amount of dynein in each extraction, but there is a fairly sharp decrease in yield if the NaCl concentration is decreased.

The quality of the crude LAD-1 is assessed by measuring the ratio of its ATPase activity after activation by Triton X-100 (i.e., Triton-activated ATPase activity) to that before activation (i.e., latent ATPase activity). ATPase activity is assayed by incubating an aliquot of LAD-1 for 10–40 min (depending on amount added) at room temperature (23°) in ATPase assay buffer (Table I) and determining inorganic phosphate according to the method of Fiske and SubbaRow.[23] Dynein is activated by 10–15 min of preincubation in the presence of 0.05% Triton; activation appears to take place within this time period equally well at either room temperature or 4°. Typical specific activities are: latent—0.25 μmol of P_i min^{-1} mg^{-1}; activated—2.5 μmol of P_i min^{-1} mg^{-1}, for an activation ratio of 10 (Table II). The composition of the LAD-1 may be determined by analysis of its polypeptide content by electrophoresis on sodium dodecyl sulfate–polyacrylamide gels. Typical gel pattern and band designations are shown in Fig. 1. Prominent components are the dynein 1 heavy chains A_α and A_β, and the three intermediate and four light chains. There is also a variable amount of contamination by tubulin, probably deriving mainly from the central tubules of the axoneme, at least one of which tends to disintegrate during extraction with high-salt buffer. Electron microscopic investigation of sectioned axonemes indicates that most of the dynein extracted by high-salt buffer derives from the outer arms.[22] In *Tripneustes,* a single high-salt extraction at 3 mg ml^{-1} generally removes about 40% of the outer arms, and the second high-salt extraction removes about a further 40%. A more complete extraction of outer arms is obtained with high-salt buffer using axonemes of some other species of sea urchin.[24]

Crude LAD-1 may be stored in high-salt buffer at 4° for about a week without a large change in ATPase activity. During this time the activation ratio of the dynein tends to decrease; the half-life for decay of an activation ratio of 10 is usually about a week. In the early stages this decay appears to be caused mainly by an increase in the latent activity of the dynein (i.e., a time-dependent activation), although the Triton-activated activity also decreases gradually. At room temperature, the decay of activation ratio is greatly accelerated, with a half-life in the range of 2–4 hr. Thus, the enzymic activity of LAD-1 is relatively labile, and it should be used as soon as possible. Physically, the enzyme appears to be more robust with little change in sedimentation behavior and polypeptide content after 2 or 3 weeks in storage at 4°.

However, if the dynein is not to be used immediately, it can be stored frozen for longer periods. Solid sucrose is added to concentrated LAD-1 solution (see below) to a final concentration of about 10% w/v and allowed

[23] C. H. Fiske and Y. SubbaRow, *J. Biol. Chem.* **66**, 375 (1925).
[24] B. H. Gibbons and I. R. Gibbons, *J. Cell Sci.* **13**, 337 (1973).

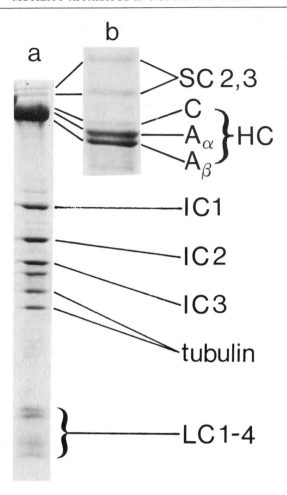

FIG. 1. Polypeptide composition of crude LAD-1 as revealed by polyacrylamide gel electrophoresis in the presence of sodium dodecyl sulfate. Lane a: 35 μg crude LAD-1 electrophoresed on a discontinuous buffer system [U. K. Laemmli, *Nature (London)* **227**, 680 (1970)] 5–15% w/v acrylamide gradient gel. Lane b: 10 μg of crude LAD-1 electrophoresed on a discontinuous buffer system 3–6% w/v acrylamide gradient gel, showing resolution of A_α and A_β heavy chains. SC, sky chain; HC, heavy chain; IC, intermediate chain; LC, light chain.

to dissolve slowly on ice for about 2 hr. The solution is then separated into 2-ml aliquots in Cryotubes (Vanguard International, Neptune, New Jersey) and frozen rapidly in liquid nitrogen for 5 min, followed by long-term storage at −80°. We have recovered about 95% ATPase activity with roughly the original activation ratio after 6 months' storage in this manner.

Attempts to store the LAD-1–sucrose solution by lyophilization have resulted in a significant loss of activity.

Dynein is frequently required at concentrations greater than the 0.25 mg ml^{-1} yielded by high-salt extraction. Concentration can be carried out at 4° in an Amicon ultrafiltration cell using a UM-20 membrane. Concentrations up to 5 mg ml^{-1} can be attained, but at over 2 mg ml^{-1} nonspecific protein aggregation becomes an increasing problem. Although it has not been possible to prevent aggregation completely, it can be minimized by keeping the protein cold and maintaining a sufficient level of reducing agent (dithiothreitol or 2-mercaptoethanol). Aggregated material can be removed by centrifugation at 12,000 g for 5 min. A more delicate method of concentrating small volumes, which seems less prone to causing aggregation, is concentration of the dynein in a dialysis membrane bag against dry Sephadex.

Further Purification of Dynein

Solutions of crude LAD-1 prepared as above are pure enough for use in many basic enzymic investigations and in work involving recombination of dynein arms to extracted axonemes and reactivated sperm.[22,25] However, for many purposes, including detailed enzymological studies and physicochemical characterization (e.g., analytical ultracentrifugation), further purification is required.

Density Gradient Centrifugation

We routinely purify LAD-1 by zonal centrifugation on density gradients of sucrose or glycerol. Density gradients of 5 to 20% w/v sucrose or 8 to 30% w/v glycerol are made up in high-salt buffer. One milliliter of crude LAD-1 solution, often concentrated to about 2–3 mg ml^{-1}, is layered on top of each gradient and centrifuged in an SW41 swinging-bucket Beckman rotor at 35,000 rpm at 4° for 15 hr. After fractionation, the approximate distribution of the protein in the gradient is determined by spotting an aliquot of each fraction onto Whatman 3 MM chromatography paper, drying the paper and staining it for 5–10 min with Coomassie Brilliant Blue (50 ml of distilled water, 50 ml of methanol, 10 ml of glacial acetic acid, 0.05 g of Coomassie Brilliant Blue R-250). After destaining (in 82.5 ml of distilled water, 10 ml of methanol, 7.5 ml of glacial acetic acid), fractions containing protein can be seen as blue spots on the paper. This provides a rapid (ca 30–45 min) method of locating the major protein peaks in the gradient, but to be of most use it should be compared with the

[25] B. H. Gibbons and I. R. Gibbons, *J. Biol. Chem.* **254**, 197 (1979).

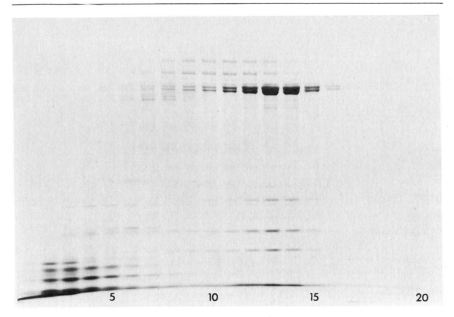

FIG. 2. Sucrose density gradient centrifugation of crude LAD-1. The 20 fractions from a 5 to 20% w/v sucrose density gradient were electrophoresed on a discontinuous buffer system 3–6% w/v acrylamide gradient slab gel, with the fraction from the top of the sucrose density gradient on the left. The 21 S LAD-1 peaks in fraction 13 (see numbers at the bottom of gel), the secondary peak of dynein heavy chains is in fractions 7 and 8, and tubulin peaks around fraction 3. The light chains of 21 S LAD-1 (see Fig. 1) run with the dye front on this gel.

information derived from electrophoresis in SDS–polyacrylamide slab gels. Such a gel of a complete fractionated sucrose gradient is shown in Fig. 2. The major peak, containing the A_α and A_β heavy chains and the intermediate and light chains, is the 21 S form of LAD-1.[22] In the 12–14 S region, there is a faint second peak containing small amounts of the A chains and other dynein heavy chains. The sky chains and C chain tend to span the region between these two peaks. Tubulin and other small polypeptides are found mostly near the top of the gradient. Defining 21 S LAD-1 as comprising the A_α and A_β chains, the intermediate chains, and the light chains (Fig. 2), we find that it accounts for about 50–70% by stain intensity of the initial LAD-1 high-salt extract.[26] The purification of LAD-1 by density gradient centrifugation, although it produces a particle that appears monodisperse by sedimentation velocity and sedimentation equilibrium,[22] does so at the cost of reducing the ATPase activation ratio of this particle relative to that of initial crude LAD-1. The reason for this is uncertain,

[26] C. W. Bell, E. Fronk, and I. R. Gibbons, *J. Supramol. Struct.* **11**, 311 (1979).

although it seems to stem primarily from an increase in the latent ATPase activity of the particle. Recombination of aliquots of all fractions of the sucrose gradient does not reinstate the original latent and Triton-activated ATPase activities, indicating that the change is not caused solely by the separation of some necessary component from the 21 S LAD-1 particle. After it has been fractionated, the 21 S LAD-1 may be stored at 4° (in its sucrose gradient solution) for up to a week without significant further deterioration in activity. Under these conditions, aggregation appears to be insignificant, possibly because of the presence of ~15% w/v sucrose. However, if the sucrose is dialyzed out of purified 21 S LAD-1, and especially if the protein is reconcentrated, aggregation again becomes a problem.

Other Purification Procedures

Gel filtration on cross-linked Sepharose Cl-4B in high-salt buffer may be used to purify 21 S LAD-1. This method has the advantage of being adaptable to larger volumes, but suffers in comparison with sucrose density gradient centrifugation in that the recovery of enzyme off the column is usually less than 50%.

Other purification procedures we have utilized are intended to separate particular polypeptides from the 21 S LAD-1 particle. The method most often used involves dialyzing concentrated LAD-1 (2–3 mg ml^{-1}), originally in high-salt buffer, against a low-salt buffer containing EDTA (Table III). The concentrated, dialyzed LAD-1 is centrifuged at 12,000 g for 5 min to remove aggregated material and is then layered onto 5 to 20% sucrose density gradients made up in low-salt buffer. The gradients are centrifuged in an SW41 swinging-bucket rotor at 35,000 rpm for 15 hr at 4°. Sodium dodecyl sulfate–polyacrylamide gels of samples from fractionated gradients (Fig. 3) show that the A_β and intermediate chains sediment at around 9–10 S and are separated from the A_α chain, which is partially aggregated and sediments at velocities ranging from about 16 to 30 S. The A_β chain and intermediate chain 1 appear to cosediment as a unit, whereas intermediate chains 2 and 3 tend to spread to either side, most often the trailing side of this peak, indicating that in low salt they are not tightly associated with either the A_β or A_α heavy chains.

Another method leading to a similar separation involves chromatography on phosphocellulose. Concentrated LAD-1 is dialyzed against column buffer (Table III), and 1 ml is applied to a 0.5 × 2.0 cm column of phosphocellulose equilibrated in the same buffer. The column is eluted with column buffer, and the proteins in the flowthrough peak are collected. With the best preparations, electrophoresis shows only the A_β chain

TABLE III
COMPOSITION OF BUFFERS AND METHODS USED IN SEPARATION OF LAD-1 POLYPEPTIDES

Low-salt sucrose density gradient centrifugation	Phosphocellulose chromatography	Hydroxyapatite chromatography
Imidazole-HCl, 5 mM, pH 7	Imidazole-HCl, 5 mM, pH 7	Hydroxyapatite column, 0.5 × 5 cm
EDTA, 0.5 mM	NaCl, 10 mM	is equilibrated to 0.01 M sodium
2-Mercaptoethanol, 14 mM	2-Mercaptoethanol 14 mM	phosphate, pH 6.9. LAD-1 (in
LAD-1 is concentrated and	Phosphocellulose column,	original high salt buffer) is applied
dialyzed for 18–24 hr against	0.5 × 2.0 cm, is equilibrated	to column. Column is eluted with
two changes of 100 volumes	in above buffer. LAD-1 is con-	gradient of 0.01–0.5 M
of low-salt buffer, then loaded	centrated and dialyzed against	sodium phosphate, pH 6.9
onto 5 to 20% w/v sucrose	buffer and then applied to	
density gradients prepared in	column	
same buffer		

and intermediate chain 1 in this peak. However, more often, all three intermediate chains emerge with the A$_\beta$ chain. The reason for this variable performance is uncertain, and the procedure requires further development to improve its reproducibility.

Chromatography on hydroxyapatite as introduced by Ogawa[27] for the purification of flagellar ATPase from the sperm of *Pseudocentrotus* can also be used to separate the major LAD-1 polypeptides. LAD-1 in high salt buffer is loaded directly onto a 0.5 × 5 cm hydroxyapatite column equilibrated in 10 mM sodium phosphate, pH 6.9 (hydroxyapatite is prepared according to Bernardi[28]. The protein is eluted with a gradient of sodium phosphate, pH 6.9, from 0.01 M to 0.5 M. Electrophoresis shows that the first polypeptides to be eluted are contaminating tubulin. The next peak includes the A$_\beta$ chain and intermediate chain 1 and following this are peaks containing the sky chains and intermediate chains 2 and 3, the C chain, and the A$_\alpha$ chain, respectively. Although tubulin and, often, the A$_\beta$ chain and intermediate chain 1 are well separated from other peaks, there is considerable overlap among the remaining peaks even when shallow gradients are employed.

Ogawa used chromatography on hydroxyapatite to separate from a low salt extract of *Tripneustes* sperm flagella, previously extracted with high salt, an ATPase other than that now known as LAD-1.[10] This ATPase was named dynein-2 and does not appear to be composed of A-chain polypeptides.

Some of the properties of the separated A$_\alpha$ and A$_\beta$ chains are given in the following section. However, it is of practical import to note here that

[27] K. Ogawa and H. Mohri, *Biochim. Biophys. Acta* **256**, 142 (1972).
[28] G. Bernardi, this series, Vol. 22, p. 325.

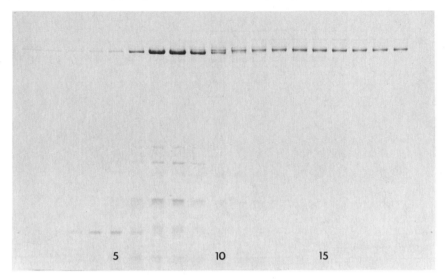

FIG. 3. Separation of the A_α and A_β heavy chains of 21 S LAD-1 by density gradient centrifugation in low-salt buffer. The 19 fractions from a 5 to 20% w/v sucrose density gradient were electrophoresed on a discontinuous buffer system 3–6% w/v acrylamide gradient slab gel with the fraction from the top of the sucrose density gradient on the left. The A_β heavy chain, and the intermediate chains peak in fractions 7 and 8, and the A_α chain spreads from about fraction 10 to fraction 19.

the purified A_β chain appears to maintain its ATPase activity for extended periods and displays essentially no propensity for nonspecific aggregation in conditions ranging from 0.0 to 0.6 M NaCl and pH 6 to 9. On the other hand, the A_α chain displays erratic ATPase activity, apparently because its activity is unstable in the absence of Mg^{2+}.[12] The A_α chain also readily aggregates, particularly at low salt concentration and low pH. The A_α chain may be the major factor in the aggregation of LAD-1.

Selected Properties of LAD-1 and Its Subunits

LAD-1

A summary of the available data on enzymic and physicochemical properties is given in Table IV. The sedimentation velocity and molecular weight yield a frictional ratio (f/f_0) of 1.9 which, as expected from electron microscopic evidence of outer arm structure, suggests a relatively compact particle (cf., myosin, $f/f_0 = 3.5$, myosin subfragment 1, $f/f_0 = 1.4$; calculated from data given by Lowey et al.[29]). It should be noted that

[29] S. Lowey, H. S. Slayter, A. G. Weeds, and H. Baker, J. Mol. Biol. 42, 1 (1969).

TABLE IV
SELECTED PROPERTIES OF LAD-1 AND ITS SUBUNITS

Subunit	Polypeptide Composition[a]	$s^0_{20,w}$ (S)	M_r[b]	f/f_0	K_m[c] (μM)	Specific activity[d]
LAD-1[e]	$A_\alpha \sim 330,000$ $A_\beta \sim 330,000$ IC1 122,000 IC2 90,000 IC3 76,000 LC1-4 24,000–14,000	21.2	1.25×10^6	1.9	Latent, 1.0[f] Activated, ~ 50	0.26 2.7
A_β						
Low ionic strength	$A_\beta \sim 330,000$ IC1 122,000	9.3	—	—	—	—
High ionic strength	$A_\beta \sim 330,000$ IC1 122,000	14.1	—	—	2.5	0.8
A_α	$A_\alpha \sim 330,000$	—	—	—	7.0	0.3

[a] Polypeptide chain designation, followed by apparent molecular weight (M_r).
[b] Sedimentation equilibrium measurement.
[c] For ATP, derived from coupled assay measurements; coupled assay composition is given in Gibbons et al.[32]
[d] Micromoles of P_i min^{-1} mg^{-1}, in 0.1 M NaCl, 30 mM Tris-HCl, pH 8.1, 2 mM MgSO$_4$, 0.1 mM EDTA, 1 mM ATP at 23°.
[e] Crude LAD-1 used for enzymic measurements.
[f] Lineweaver–Burk plots of latent LAD-1 activity sometimes show a second kinetic component at higher ATP concentrations: see text for details.

although we use density gradient centrifugation for routine approximate measurement of sedimentation velocity such measurements are best confirmed on the analytical ultracentrifuge (Beckman Spinco Model E). We have found that analytical centrifugation often displays more heterogeneity than zonal centrifugation on density gradients, either because the presence of sucrose favors disaggregation or because the method is inherently less sensitive to heterogeneity. Densitometric scanning of sodium dodecyl sulfate–polyacrylamide gels stained with Coomassie Brilliant Blue indicates that the A_α and A_β chains are present in equimolar quantities and that the intermediate chains are present in equimolar quantities relative to each other. The stoichiometry between the A chains and the intermediate chains is less certain, but it appears that there is probably 1 mol of intermediate chain 1 per mole of A_β chain. The staining of the light chains is, for reasons that are unclear, too variable to allow meaningful comparisons of this type.

The ATPase activity data in Table IV refer to the conditions of the standard ATPase assay buffer. Variation of NaCl concentration in the assay buffer[22] causes a sigmoid increase in latent specific activity from 0.2 μmol of P_i min^{-1} mg^{-1} at zero salt to a maximum about 3.5 μmol of P_i min^{-1} mg^{-1} at 0.8 M NaCl, above which the activity levels out and then begins to decline. The effect of variation of NaCl concentration on Triton-activated ATPase activity is less dramatic, the activity assayed in 0.5 M NaCl being about twice that assayed in zero NaCl. Substitution of Ca^{2+} for Mg^{2+} in the ATPase assay buffer does not change significantly the dependence of activity on NaCl concentration. However, the ATPase activity at any given NaCl concentration is less in the presence of Ca^+ than in the presence of Mg^{2+}, especially in the case of Triton-activated material, so that at concentrations above about 0.5 M NaCl, the "Triton-activated" Ca^{2+}-ATPase activity is actually less than the "latent" Ca^{2+}-ATPase activity.[22]

Occasionally a second kinetic component will appear at higher ATP concentrations ($>$100 μM) in a Lineweaver–Burk plot for latent LAD-1, giving a biphasic appearance similar to that noted by Takahashi and Tonomura[30] in plots for 30 S *Tetrahymena* dynein. In the case of *Tripneustes*, this biphasic appearance may be caused by the presence of a variable small amount of activated material in the latent preparation.

The specificity of dynein for ATP as substrate is fairly high, and most ATP analogs as well as other nucleoside triphosphates (ITP, GTP, CTP, etc.) are hydrolyzed at less than 15% of the rate for ATP.[30,31] Vanadate, the anionic form of vanadium (V) and a phosphate analog, is a potent uncompetitive inhibitor of dynein ATPase with an apparent inhibition constant (K_i) of 100 nM or less when measured at low salt concentration,[32,33] and it has been used in attempts to probe the steps in the mechanochemical cross-bridge cycle in which dynein produces sliding between flagellar tubules.[34,35]

Separated Subunits

A summary of available data is given in Table IV. The increase in sedimentation velocity of the A_β subunit from 9 S to 14 S between low-salt and high-salt buffers (Table IV) is possibly the result of a self-association of the monomeric subunit to form a dimer. If so, then this, coupled with

[30] M. Takahashi and Y. Tonomura, *J. Biochem.* (*Tokyo*) **84**, 1339 (1978).
[31] I. R. Gibbons, *J. Biol. Chem.* **241**, 5590 (1966).
[32] I. R. Gibbons, M. P. Cosson, J. A. Evans, B. H. Gibbons, B. Houck, K. H. Martinson, W. S. Sale, and W.-J. Y. Tang, *Proc. Natl. Acad. Sci. U.S.A.* **75**, 2220 (1978).
[33] T. Kobayashi, T. Martensen, J. Nath, and M. Flavin, *Biophys. Biochem. Res. Commun.* **81**, 1313 (1978).
[34] W. S. Sale and I. R. Gibbons, *J. Cell Biol.* **82**, 291 (1979).
[35] M. Okuno, *J. Cell Biol.* **85**, 712 (1980).

the equimolarity of A_α and A_β chains in the 21 S particle, would suggest that the latter comprises a heterotetramer containing two each of the A_α and A_β chains. However, the occasional presence of an intermediate 12 S component suggests that the situation may be more complex. The present data are insufficient to warrant a choice between dimerization and/or a conformational change as the basis for the change in sedimentation rate. Regardless of whether the LAD-1 particle is considered to be a dimer or a tetramer of the A heavy chains, there is an apparent discrepancy between the molecular weight of LAD-1 derived from sedimentation equilibrium[22] and the sum of apparent molecular weights (derived from electrophoretic mobilities in the presence of sodium dodecyl sulfate) of the probable polypeptide composition of LAD-1,[26] which suggests that the results of one or both methods are inaccurate.

Preparation and Properties of Dynein ATPases from Other Sources

In this section, the preparation and properties of dynein ATPases from other species will be briefly summarized from a comparative viewpoint. For more detailed treatment, reference should be made to the original work. Some of the data in this section are tabulated in Table V.

Tetrahymena

The cilia of *Tetrahymena* were the source from which the first well defined axonemal ATPase was isolated and described by Gibbons.[1] The procedure developed by Gibbons has subsequently been used with only minor alteration by many workers[e.g.,30,36,37] and has also been applied to other species (see below). Cilia are first detached from the cell bodies by one of a variety of methods that include treatment with ethanol and calcium[1,2] or with glycerol[38] or exposure to the local anesthetic dibucaine in the presence of calcium[39] and are then harvested by differential centrifugation. Prior to solubilization of dynein ATPase, the cilia must be demembranated with detergent, 0.5% w/v digitonin[1] or 1% w/v Triton X-100.[39] Dynein ATPase is extracted from the demembranated cilia by dialysis against 0.1 mM EDTA, 1 mM Tris-thioglycolate, pH 8.3[1,2] or 0.1 mM EDTA, 1 mM dithiothreitol, 1 mM Tris-HCl, pH 8.3,[30] for approximately 18 hr at 0–4°, which solubilizes about 90% of total axonemal ATPase. Raff and Blum[21] used a somewhat different procedure involving extraction of dynein ATPase from glycerinated *Tetrahymena* cilia by incubation with 20

---

[36] J. J. Blum, *Arch. Biochem. Biophys.* **156**, 310 (1973).
[37] I. Mabuchi and T. Shimizu, *J. Biochem.* (*Tokyo*) **76**, 991 (1974).
[38] I. R. Gibbons, *J. Cell Biol.* **26**, 707 (1965).
[39] G. A. Thompson, L. C. Baugh, and L. F. Walker, *J. Cell Biol.* **61**, 253 (1974).

mM ATP in 20 mM imidazole-HCl buffer, pH 8.3, containing 2.5 mM MgSO$_4$. Warner et al.[40] have solubilized dynein ATPase from Triton-demembranated (0.2% w/v Triton X-100) Tetrahymena cilia by extraction in 0.5 M KCl, 2 mM HEPES, 4 mM MgSO$_4$, 0.5 mM EDTA, 1 mM dithiothreitol, and 0.1 mM ATP at pH 7.8. In all cases, differential centrifugation was used to separate the solubilized ATPase from insoluble axonemal residue.

Gibbons[1] found that extraction of Tetrahymena axonemes with 1 mM Tris buffer, 0.1 mM EDTA, pH 8.0, solubilized most of the axonemal ATPase activity and removed both rows of arms from the doublet tubules, suggesting that most of the axonemal ATPase is localized in these arms. Studies with the analytical ultracentrifuge[41] showed that the dynein ATPase exists in two forms sedimenting at 30 S and 14 S, and suggested that the former was a polymer of the latter. However, Mabuchi and Shimizu[37] reported that the 30 S and 14 S forms of dynein have electrophoretically distinct polypeptide subunits suggesting that the two forms represent different isoenzymes rather than being related as monomer and polymer. The 30 S form can rebind to extracted axonemes and restore the appearance of arms, but the 14 S form does not possess this ability.[2] It is not possible to determine in a similar manner the location of the dynein ATPase extracted from Tetrahymena cilia by solutions containing ATP,[21,40,42] since in these cases the cilia disintegrate during extraction. In preparations from most species, such disintegration of ciliary and flagellar axonemes in the presence of ATP is seen only after mild proteolysis,[43] suggesting that autoproteolysis may occur during the isolation of Tetrahymena cilia.

Chlamydomonas

Chlamydomonas possesses two main advantages over Tetrahymena as a source for the investigation of dynein ATPase. First, being an autotrophic organism, its proteins are less likely to be degraded by endogenous proteolytic enzymes during isolation. Second, a wide variety of paralyzed flagellar mutants that lack specific morphological components are available, and some of these have been used to investigate axonemal doublet microtubule arm structure.[11,44,45]

Flagella are generally isolated by pH shock or by the "STEEP-Ca"

[40] F. D. Warner, D. R. Mitchell, and C. R. Perkins, J. Mol. Biol. 114, 367 (1977).
[41] I. R. Gibbons and A. J. Rowe, Science 149, 424 (1965).
[42] F. D. Warner and N. C. Zanetti, J. Cell Biol. 86, 436 (1980).
[43] K. E. Summers and I. R. Gibbons, Proc. Natl. Acad. Sci. U.S.A. 68, 3092 (1971).
[44] R. A. Lewin, J. Gen. Microbiol. 11, 358 (1954).
[45] D. J. L. Luck, G. Piperno, Z. Ramanis, and B. Huang, Proc. Natl. Acad. Sci. U.S.A. 74, 3456 (1977).

TABLE V

SELECTED PROPERTIES OF DYNEIN ATPASE FROM VARIOUS SOURCES

Source	Preparation method[a]	Sedimentation velocity[b] (S)	Molecular weight	Polypeptide composition[c]	Activators	Specific activity[d]	K_m^e (μM)	References[f]
Tetrahymena	Tris-EDTA	14 (A)	600,000	—	Mg^{2+}, Ca^{2+}, Mn^{2+},	3.5	35	2, 31, 41
	20 mM ATP	30 (A)	5.4×10^6	—	Fe^{2+}, Co^{2+}, Ni^{2+}	1.3	11	2, 31, 41
		14 (D)	—	—	Mg^{2+}, Ca^{2+}, EDTA	0.5	—	21
		30 (D)	—	—	Mg^{2+}, Ca^{2+}	2.0	—	21
	Tris-EDTA	14 (D)	—	520,000	—	—	—	37
		30 (D)	—	560,000	—	—	—	37
	0.5 M KCl	—	—	360,000	—	—	—	40
Chlamydomonas	Tris-EDTA	30 (D)	—	—	Mg^{2+}	0.12	1†	30[g]
		12 (D)	—	—	Mg^{2+}, Ca^{2+}	—	—	47
		18 (D)	—	—	Mg^{2+}, Ca^{2+}	—	—	47
	0.5 M NaCl	12 (D)	—	315,000*	Mg^{2+}, Ca^{2+}	6	—	7
		18 (D)	—	~320,000*	Mg^{2+}, Ca^{2+}	5.4	—	7
		13 (D)	—	310,000*	Mg^{2+}, Ca^{2+}	6	—	48
		10–11 (D)	—	315,000*	Mg^{2+}	4	—	48
Gill cilia								
Aequipecten	Tris-EDTA	14 (A)	—	500,000, 450,000	Mg^{2+} Mg^{2+}	—	—	4, 49 4, 49
Unio	0.5 M KCl	—	—	360,000 320,000	—	—	—	40

Source	Method[a]	S value[b]	Molecular weight	MW of heavy chains[c]	Cations	ATPase[d]	ATP[e]	References[f]
Starfish sperm	Tris-EDTA	12 (D)	—	600,000	Mg²⁺, Ca²⁺	0.8	23	50
	0.6 M KCl	20 (D)	—	600,000	Mg²⁺, Ca²⁺	0.2	16	50
Sea urchin sperm								
Pseudocentrotus	Tris-EDTA	10 (A)	350,000	—	Mg²⁺, Ca²⁺	0.38	160	51a
	0.6 M KCl	25 (A)	—	—	—, —	—	—	51a
Hemicentrotus	Tris-EDTA	10,13 (A)	—	—	Mg²⁺, Ca²⁺	0.5	—	52
Pseudocentrotus	—		—	—	Mg²⁺, Mn²⁺ (Ni²⁺, Ca²⁺)	2.5	50	27
Colobocentrotus	0.5 M KCl	13 (D)	—	h	Mg²⁺, Ca²⁺	2.8	50	8, 54
	Tris-EDTA	13 (D)	—	—	Mg²⁺, Ca²⁺	—	—	8, 54
Tripneustes[i]	0.6 M NaCl	21 (A)	1.25 × 10⁶	~330,000*	Mg²⁺, Ca²⁺	0.25	1†	12, 22

[a] Method of preparation of dynein ATPase from isolated axonemes. Tris-EDTA = low-salt dialysis.

[b] Values derived from (A) analytical or (D) density gradient centrifugation.

[c] Apparent molecular weight of high molecular weight chains given; those marked ~ contain more than one heavy chain; those marked * also contain lower molecular weight chains.

[d] Mg²⁺-ATPase, μmol P_i min⁻¹ mg⁻¹: conditions generally 0.1 M salt, 1–4 mM Mg²⁺, 1 mM ATP, pH 8, 20–25°.

[e] For ATP, values marked † display second kinetic component.

[f] Numbers refer to text footnotes.

[g] Also T. Shimizu, I. Kimura, H. Murofushi, and H. Sakai, *FEBS Lett.* **108**, 215 (1979).

[h] Although the apparent molecular weight of *Colobocentrotus* dynein heavy chains was originally estimated to be in the range of 500,000 (Kincaid *et al.*[54]), more recent work (W. S. Sale and W.-J.Y. Tang, unpublished results) has shown that these chains nearly comigrate with the heavy chains of *Tripneustes* dynein 1. Within a given electrophoretic system, dyneins from most sources nearly comigrate, so that the large scatter in apparent molecular weights of the heavy chains appears to be caused by differences between electrophoretic buffer systems and the lack of adequate high molecular weight standards.

[i] Refer to Table IV for a more detailed description.

procedure[46] followed by differential centrifugation. The isolated flagella may be demembranated by exposure to detergents, such as Nonidet P-40 at a final concentration of 0.5% w/v.[11] Dynein ATPase has been extracted from *Chlamydomonas* flagellar axonemes by an essentially unmodified Gibbons low-salt dialysis procedure[47] and by extraction of axonemes with 0.5 M NaCl for 10 min at 4°.[7] The latter method extracts virtually all the dynein ATPase from demembranated axonemes and leaves the doublet microtubules essentially free of both inner and outer arms.[7] One disadvantage engendered by the use of *Chlamydomonas* is that only relatively small quantities of flagellar proteins are available for biochemical studies. However, Luck and his co-workers[7,11] have alleviated this problem considerably by growing the organism in media containing $^{35}SO_4$ of high specific activity in order to produce highly labeled proteins.

Both low-salt dialysis and high-salt extraction release two major forms of dynein that sediment at 18 S and 12 S on sucrose density gradients made up in low ionic strength buffers.[7,47] However, examination of outer and inner arm mutants strongly suggests that both these forms of dynein ATPase are localized in the outer arms.[11] Chromatography on hydroxyapatite coupled with SDS-polyacrylamide gel electrophoresis shows that the two outer arm dynein ATPases contain different high molecular weight polypeptides (apparent molecular weight 300,000–330,000), thus ruling out the possibility that the 18 S particle is a dimer of the 12 S particle. These two dyneins also differ in their content of intermediate and low molecular weight chains. The two dyneins have a similar specific ATPase activity, but it is not clear whether this is latent or activated.[7] The inner arm dynein ATPases contain high molecular weight polypeptides that are different from those in both the 18 S and 12 S dyneins and may form a particle of 10–13 S at low ionic strength.[11,48]

Lamellibranch Gill Cilia

The gill cilia of lamellibranch mollusks have been used in the investigation of dynein ATPase. Linck has isolated gill cilia from *Aequipecten* by three different methods, using twice-concentrated sea water, 10% ethanol–10 mM CaCl or 60% glycerol, with apparently equivalent results.[49] After demembranation of the cilia in 1% w/v Triton X-100, low-salt dialysis of the ciliary axonemes for 48–60 hr extracted half of the ciliary ATPase.[4] The extracted dynein ATPase apparently corresponds to the outer doublet arms. The remaining ATPase, possibly representing the

[46] G. B. Witman, K. Carlson, J. Berliner, and J. L. Rosenbaum, *J. Cell Biol.* **54**, 507 (1972).
[47] T. Watanabe and M. Flavin, *J. Biol. Chem.* **251**, 182 (1976).
[48] G. Piperno and D. J. L. Luck, *Cell* **27**, 331 (1981).
[49] R. W. Linck, *J. Cell Sci.* **12**, 345 (1973).

inner arms, is extractable in active form after brief digestion with trypsin.[49] In contrast, Linck[4] found that low-salt dialysis of sperm flagella of the same species extracted essentially all the dynein ATPase. Warner *et al.*,[40] using the mollusk *Unio,* isolated gill cilia by exposing excised gill tissue to a solution containing 20 mM dibucaine. After demembranation in 0.2% w/v Triton X-100, extraction of the ciliary axonemes in the same high salt buffer which was used by this group in the study of *Tetrahymena* dynein (see above) resulted in the solubilization of most of the outer arm dynein.

Both the low salt-extracted and trypsin-released dynein ATPase isolated from gill cilia by Linck sediment in 10 mM Tris, pH 8, as 14 S particles, and electrophoresis in the presence of SDS reveals two large polypeptides with apparent molecular weights of 450,000 and 500,000.[4] However, Warner *et al.*[40] using crosslinked bovine serum albumin as an electrophoretic standard, estimated the molecular weight of the large polypeptide subunits from *Unio* dynein to lie in the region of 300,000 to 460,000 and found up to six different chains.

Starfish Sperm

Mabuchi and co-workers[50] have studied the flagellar dynein of starfish sperm. Sperm were obtained from the dissected testes of the starfish *Asterias.* The flagella were separated from the sperm heads by homogenization and then demembranated with 1% w/v Triton X-100. Dynein ATPase was extracted by dialysis for 18 hr at 0° against either a low-salt buffer (1 mM Tris-HCl, pH 7.8, 0.1 mM EDTA, 0.5 mM dithiothreitol) or a high-salt buffer (0.6 M KCl, 10 mM Tris-HCl, pH 7.8, 0.1 mM EDTA, 0.5 mM dithiothreitol). Low-salt dialysis solubilized over 90% of the axonemal ATPase activity, whereas slightly less than 50% was solubilized by high-salt dialysis.

These flagellar dyneins were examined in some detail by Mabuchi *et al.*[50] The dynein ATPase extracted in high salt sediments as a single peak at 20 S in high-salt sucrose density gradients, whereas the low-salt dynein ATPase exhibits a single peak at about 12 S when sedimented in low-salt sucrose density gradients. The 20 S form is completely converted to the 12 S form when dialyzed into low salt, and in the converse experiment there is partial conversion of the 12 S form into the 20 S form. Electron microscopic examination showed that extraction of axonemes with low salt appeared to remove both inner and outer arms, whereas extraction with high salt apparently removed only outer arms. On return to moderate ionic strength (20 mM Tris-HCl, pH 7.6, 2 mM MgSO$_4$, 0.5 mM dithio-

[50] I. Mabuchi, T. Shimizu, and Y. Mabuchi, *Arch. Biochem. Biophys.* **176,** 564 (1976).

threitol), the 20 S dynein recombines with extracted flagella and the outer arms reappear on the doublet tubules. Sucrose density gradients run at this ionic strength suggest that the species of dynein that actually recombines is a 24 S particle. Such recombination and restoration of arm structure could not be shown with the 12 S low salt dynein. Electrophoresis of whole axonemes in the presence of SDS shows that five polypeptide chains migrate in the region expected of dynein heavy chains: the 20 S and 24 S dyneins contain only the upper two chains, whereas all five chains are found in the 12 S dynein. Both the 12 S and 20 S dyneins show roughly similar changes in Mg^{2+}- and Ca^{2+}-ATPase activity in various conditions.

Sea Urchin Sperm

The sperm of many species of sea urchin have been used for the investigation of dynein ATPase. In general, the techniques used to collect sperm and to isolate flagellar axonemes are similar to those described above for *Tripneustes*. Variations from these procedures include the use by Ogawa and Mohri[27] of sonication rather than homogenization for separating the heads from the tails of sperm of the sea urchin *Pseudocentrotus depressus* and the use, in earlier work, of glycerol to disrupt flagellar membranes.[27,51] Gibbons et al.[20] introduced the use of Triton X-100 as a demembranating agent for sea urchin sperm flagella, and this has since been used widely. Various methods have been used to solubilize dynein ATPase from sea urchin sperm flagellar axonemes, but all are essentially variations of either the low-salt dialysis method originally described[1] for *Tetrahymena* ciliary axonemes[27,51,52] or the high-salt extraction procedure.[2,8,51,53]

Generally speaking, depending on the conditions under which they are isolated, the sperm flagella dynein ATPases from different species of sea urchin appear to be fairly similar in properties. Extraction of axonemes in low-salt solutions yields dynein ATPases that sediment at between 10 and 13 S (*Pseudocentrotus* and other species,[51a] *Hemicentrotus*,[52] *Colobocentrotus*[8]) and appears to solubilize most of the total axonemal ATPase. On the other hand, extraction of axonemes with high salt appears to solubilize specifically the outer arms (*Colobocentrotus*,[8] *Anthocidaris*,[53] and *Tripneustes*[22]) and solubilizes only about half of the total axonemal ATPase activity. In all cases where they have been thus analyzed, dynein ATPases

[51] C. J. Brokaw and B. Benedict, *Arc... Biochem. Biophys.* **142**, 91 (1971).
[51a] H. Mohri, S. Hasegawa, M. Yamamoto, and S. Murakami, *Sci. Pap. Coll. Gen. Educ. Univ. Tokyo* **19**, 195 (1969).
[52] M. Hayashi and S. Higashi-Fujime, *Biochemistry* **11**, 2977 (1972).
[53] K. Ogawa, T. Mohri, and H. Mohri, *Proc. Natl. Acad. Sci. U.S.A.* **74**, 5006 (1977).

from sea urchin sperm flagella axonemes, regardless of extraction method, have been shown by electrophoresis in the presence of SDS to contain the high molecular weight polypeptides customarily associated with dynein.[10,22,54] Enzymic characterization of dynein ATPases from sea urchin sperm flagella has been extensive: selected data are given in Table V.

Ogawa[55] has prepared a well-defined fragment, fragment A, by digesting with trypsin the dynein of *Hemicentrotus* sperm flagella. This fragment of molecular weight 380,000 retains high ATPase activity but has lost the ability to recombine with low salt-extracted flagellar axonemes, thus suggesting a functional separation of doublet microtubule binding site and ATPase active site in dynein.

Trout Sperm

Ogawa and his colleagues have described the preparation of a dynein ATPase from the sperm of the rainbow trout, *Salmo gairdneri*.[56] Flagellar axonemes were prepared from the sperm in essentially the same way as from sea urchin sperm, and the dynein was extracted in high salt. The dynein appears to be localized in the outer arms. Physicochemical characterization of this dynein in the intact state has been limited so far to SDS–polyacrylamide gel electrophoresis, which shows the presence of high molecular weight chain(s) that approximately comigrate with the high molecular weight polypeptide subunits of sea urchin dynein 1. An immunological similarity between these two dyneins is also indicated.

One advantage of this organism is that a ripe male trout is relatively prolific, holding from 20–50 ml of sperm. However, Ogawa cautions that proteolytic enzyme(s) present in the trout spermatozoa lead to partial fragmentation of the dynein during purification.

Summary

The following characteristics seem to be common to dyneins from all sources so far investigated and serve to distinguish dynein from myosin. First, dyneins are ATPase proteins that can be activated to a similar extent by either Mg^{2+} or Ca^{2+}, and in this respect they differ from myosin, in which the Mg^{2+}-ATPase activity in the absence of actin is up to two orders of magnitude less than the Ca^{2+}-ATPase activity.[57] Furthermore,

---

[54] H. L. Kincaid, B. H. Gibbons, and I. R. Gibbons, *J. Supramol. Struct.* **1**, 461 (1973).
[55] K. Ogawa, *Biochim. Biophys. Acta* **293**, 514 (1973).
[56] K. Ogawa, S. Negishi, and M. Obika, *Arch. Biochem. Biophys.* **203**, 196 (1980).
[57] S. Ebashi and Y. Nonomura, *in* "The Structure and Function of Muscle" (G. H. Bourne, ed.), 2nd ed., Vol. 3, pp. 285–362. Academic Press, New York, 1973.

dynein ATPases display a much greater substrate specificity for ATP than does myosin.[27,31] Second, all dyneins appear to exist in, or to be convertible to active forms that sediment in the range 10–14 S. No form of dynein displaying ATPase activity has been found that sediments at the 6 S velocity of myosin. Third, all dyneins contain very large polypeptide subunits, with apparent molecular weights in the range of 300,000 to 500,000, which are electrophoretically distinct from the heavy chain of myosin. Finally, all ATPases with the above characteristics that have been isolated from cilia and flagella appear to be localized in the doublet microtubule arms.

Another property that is likely to assume greater importance in the future is the question of the functional capability of preparations of dynein as opposed to merely their ATPase activity. Gibbons and Gibbons[25] found that only the latent form of LAD-1 had the ability to recombine functionally and increase the beat frequency of reactivated, dynein-depleted flagella of sea urchin sperm. Although activation of dynein ATPase from other sources by various chemical procedures has been reported,[50,58–60] none has as yet been demonstrated to have functional activity. The chemical and physicochemical factors affecting latency and activation of LAD-1, and their relationship to the action of the functioning dynein arm are as yet unclear. However, it is necessary to postulate that a functional dynein ATPase, in order to perform work, must possess a rest state in which ATP turn over is significantly less than maximal, and it is possible that the latent form of LAD-1 represents the dynein arm in this rest state. Similarly, it is possible that one of the several activated forms of the enzyme that have been reported[22] is analogous to the state of the dynein arm in which the energy of ATP hydrolysis is used to perform work, leading to the sliding of adjacent doublet tubules. The natural mechanisms controlling the cross bridge cycle, which are by-passed or mimicked by the techniques used to activate LAD-1, are yet to be clarified.

Acknowledgments

This work has been supported in part by Grants HD 06565 and HD 10002 from the National Institute of Child Health and Human Development.

[58] J. J. Blum and A. Hayes, *Biochemistry* **13**, 4290 (1974).
[59] T. Shimizu and I. Kimura, *J. Biochem.* (*Tokyo*) **76**, 1001 (1974).
[60] J. J. Blum and M. Hines, *Q. Rev. Biophys.* **12**, 103 (1979).

[44] Preparation of Spectrin

By W. B. GRATZER

Spectrin is the major component of the protein network that covers the cytoplasmic surface of the vertebrate erythrocyte membrane. So far as is known at present, it is unique to the erythrocyte. It is a protein of high molecular weight, comprising two chains associated with one another in the form of a heterodimer unit; their molecular weights are about 230,000 and 250,000. Contrary to early reports, spectrin does not resemble myosin in any important respect. There seems to be no doubt that spectrin exists in the cell as a tetramer, made up of two heterodimers, oriented head-to-head. This elongated element of the cytoskeleton (some 200 nm extended length, as seen in the electron microscope after shadowing[1]) is associated with actin and a protein referred to as 4.1 in the standard numbering system based on relative migration rates in sodium dodecyl sulfate (SDS)–polyacrylamide gels,[2] and possibly one more minor constituent. The primary mode of attachment of the cytoskeleton to the membrane is by way of sites on the spectrin, which associate with high affinity with an integral membrane protein, termed 2.1 or ankyrin.[3] The association becomes weak at low ionic strength, and the spectrin is then dissociated. This is the basis for its preparation from membranes. In dealing with spectrin it should be kept in mind that it is sensitive to proteolytic degradation and that it becomes insoluble in the vicinity of its isoelectric point (about pH 5).[4] Thus, insufficient buffering of a spectrin solution can lead to precipitation.

Two procedures for the preparation of spectrin from human erythrocytes are available: extraction by dialysis in the cold and extraction at elevated temperature (35°). The first takes at least overnight and yields spectrin tetramer; the second takes only minutes and yields dimer. In either case the spectrin makes up some 70–80% of the total extracted protein, the remainder being almost entirely actin and 4.1. A variable part of the spectrin, moreover, depending evidently on the metabolic state of the cells,[5] appears in the form of an oligomeric complex with the other two

[1] D. M. Shotton, B. Burke, and D. Branton, *J. Mol. Biol.* **131**, 303 (1979).
[2] G. Fairbanks, T. L. Steck, and D. F. H. Wallach, *Biochemistry* **10**, 2606 (1971).
[3] V. Bennett, *J. Biol. Chem.* **253**, 2292 (1978).
[4] W. B. Gratzer and G. H. Beaven, *Eur. J. Biochem.* **58**, 403 (1975).
[5] S. E. Lux, K. M. John, and M. J. Karnovsky, *J. Clin. Invest.* **58**, 995 (1976).

components. This cannot be readily dissociated, and is eliminated by gel filtration through a medium such as Sepharose 4B.

The relation between the tetramer and dimer is complicated by metastability at low temperature[6]: concentration-dependent interconversion is slow at low temperature (say 25° and below), so that at room temperature and below the system is essentially frozen. Thus at low ionic strength, at which the equilibrium strongly favors the dimer, tetramer is nevertheless extracted if the temperature is kept sufficiently low throughout. On warming, dissociation proceeds rapidly. Again, at higher ionic strengths, depending on the protein concentration, the equilibrium balance of dimer and tetramer may be established by warming the solution. It is not in practice possible to attain a concentration at which there is total conversion of dimer to tetramer. Thus if tetramer is required, either the equilibrium mixture may be chromatographed in the cold, or better, the initial extraction is performed in the cold, when dissociation to dimer is suppressed.

Method of Preparation

The following are the steps involved in the preparation of spectrin.

1. Cells, best from blood that is fresh or not more than a few days old, are washed with isotonic buffer with careful removal of white cells.
2. The cells are lysed, and the membranes are washed to remove hemoglobin.
3. The spectrin is extracted by dialysis overnight or warming at 35° for a brief period.
4. The membrane vesicles remaining after release of spectrin are removed by ultracentrifugation.
5. If necessary, the spectrin is concentrated by precipitation with ammonium sulfate, dialysis against polyethylene glycol or Ficoll, or vacuum dialysis.
6. For purification the solution is chromatographed on Sepharose 4B or its equivalent.

Cells

The yield of spectrin from stored cells, tends to be poor.[5] Such cells can be incubated with metabolites required for the resynthesis of ATP,[5] but the usual practice is to use fresh, or relatively fresh, blood. Blood banks will frequently have available "short units," in which for some reason, such as the collapse of the doner's vein, the bag was not filled.

[6] E. Ungewickell and W. B. Gratzer, *Eur. J. Biochem.* **88**, 379 (1978).

Such samples (or those from patients with polycythemia or a history of malaria, etc.) are not used for transfusion. From fresh cells a total yield of spectrin approaching 1.5 mg per milliliter of packed cells (say 2 ml of blood) can be expected. A convenient scale of preparation starts with, say, 50 ml of blood.

Wash and lysis buffers, which are respectively 0.15 M sodium chloride, with 20 mM Tris (or 5 mM phosphate) buffer, pH 7.6, and 5 mM Tris, (or again phosphate), pH 7.6, are prepared; about 500 ml and 1 liter, respectively, will be needed. They can be prepared by dilution of stock solutions of 10 times the concentration and are allowed to cool in ice. The blood (50 ml) is distributed between four 40-ml centrifuge tubes (e.g., transparent polycarbonate tubes (DuPont) for use with the Sorvall SS34 rotor). They are made up to volume with the isotonic buffer and mixed by inversion, using Parafilm. The tubes are spun, preferably at this stage in a swing-out rotor (Sorvall HB4) at 5000 rpm for 5 min. The supernatant is removed with a Pasteur pipette attached to a suction pump, and the "buffy coat" layer of white cells is carefully removed. (The volume of blood given allows for generous wastage in the interest of complete removal of the white cells, which are otherwise a damaging source of proteases). The cells are then resuspended by repeated, but not too vigorous, inversion (to avoid lysis) or stirring with a glass rod, and the centrifugation is repeated. Two further washes are desirable; for the last one the cells can be bulked in a single tube. The hemolysis step should then be done without too much delay. (If the cells are left in the packed state overnight, for example, they will tend to clump together during hemolysis.)

Preparation of Ghosts

Two milliliters of packed cells are sampled into each of up to eight 40-ml centrifuge tubes to fill the SS34 rotor. An automatic pipette with the tip shortened to give an enlarged orifice can conveniently be used. The tubes are then filled with the chilled lysis buffer, and each is immediately and thoroughly mixed by inversion. They are centrifuged at 20,000 rpm for 5–10 min, gently removed from the rotor, and placed in an ice-bucket. The supernatant is removed with a Pasteur pipette attached to an aspirator. At this stage a light behind the tube is helpful, since the hemolysate is dark red and the pellet, which is not firmly packed down, may be hard to see and is easily disturbed. Its volume is generally several milliliters. Ice-cold hemolysis buffer is then again added, and the centrifugation is repeated. The supernatant is again removed by aspiration. At this stage (or after the first spin) a small pellet of sticky consistency, much more firmly packed, can be seen below the loose membrane layer at the bottom of the tube. The tube may be rotated to expose this pellet at the top, when it can be cleanly removed by aspiration. The washing procedure with the hemolysis buffer is repeated once more. The ghosts should then be almost white. Sometimes, and especially from old cells or many kinds of abnormal, fragile cells, the hemoglobin is less readily dislodged and the ghosts remain reddish. For optimal extraction of spectrin it has been found advantageous to ensure that the ionic strength is efficiently reduced. This can be achieved by bulking the ghosts from all the tubes in one or two tubes, making up the volume with cold distilled water, and centrifuging once more. Speed at this stage helps to minimize loss of spectrin.

Extraction of Spectrin

Spectrin is extracted from ghosts into a medium of very low ionic strength at a pH above neutrality; a buffer consisting of 0.3 mM phosphate, pH 8.0, is suitable, being just sufficient to avoid adventitious drifts of pH. A further precaution is the incorporation of 0.1 mM EDTA and 0.1 mM phenylmethanesulfonylfluoride to inhibit proteolysis; ghosts are diluted with an equal volume of this buffer. For preparation of *dimers*, the suspension is put into a water bath at 35° and allowed to extract for about 15 min, with periodic swirling. (The limiting factor is probably the rate at which temperature equilibration is reached. At higher temperatures there is a slow onset of irreversible denaturation.) For extraction of *tetramers*, the suspension is dialyzed against the extraction buffer overnight at 4°.[7] The suspension is then centrifuged at 90,000 g for 30 min, when the extracted and vesiculated membranes form a very compact pellet. The supernatant contains some 70–90% of the total spectrin, together with actin, 4.1, and traces of other proteins. Extracts from pink ghosts also yield hemoglobin. To determine the concentrations of crude spectrin by spectrophotometry a correction to the absorbance at 280 nm can be made for hemoglobin, on the basis of the absorbance at 345 nm. Thus the corrected absorbance at 280 nm is $A_{280} - 1.25A_{345}$. The specific absorptivity of spectrin at 280 nm is taken to be $E_{1\,cm}^{1\%} = 10.7$.

Purification of Spectrin Dimer or Tetramer

The membrane extract is applied to column of Sepharose 4B or the equivalent, 2.5 × 90 cm for a total extract on the scale described. The column buffer is conveniently 0.1 M sodium chloride, 0.05 M Tris pH 7.6.

---

[7] It is essential to ensure that the pH remains stable in this barely buffered system. The dialysis tubing can be a source of protons. Equilibration with a sodium bicarbonate solution is the final step of pretreatment (see, e.g., P. McPhie, this series, Vol. 22, p. 23).

Fractions of 5 ml may be collected. The peak coinciding with the void volume is predominantly a stable complex of spectrin, actin, and 4.1, although some F-actin may also be present. This should be well separated from the tetramer, which follows it, or the dimer. Following the pure spectrin dimer or tetramer two more peaks can generally be seen, the second being hemoglobin, the first mainly denatured actin. The spectrin can be screened for purity by gel electrophoresis in the presence of sodium dodecyl sulfate[8] and should be free of contaminants and contain only equal proportions of the two polypeptide chains with no satellite bands reflecting degradation. The concentration is determined spectrophotometrically ($E_{1\,cm}^{1\%} = 10.7$). Spectrin is best stored in the presence of EDTA (0.1 mM), a reducing agent (0.1 mM dithiothreitol), and sodium azide (20 μg/ml). Even with azide alone to prevent bacterial contamination the properties of a solution of purified spectrin are in general unchanged after several days at 0°.

Concentration

Spectrin is stable at concentrations as high as 50 mg/ml with no or little aggregation. It can be concentrated by precipitation with an equal volume of cold saturated ammonium sulfate,[9] although this sometimes generates some irreversibly aggregated material (and is especially unsatisfactory for the crude protein). It is important that precipitation and resuspension be performed very rapidly. Any of the standard concentration procedures may be applied. Vacuum dialysis and dialysis against Ficoll or polyethylene glycol (Calbiochem Aquacide) are satisfactory. Under all circumstances pH control should be maintained, for, as already remarked, spectrin allowed to drift to its isoelectric pH will precipitate.

Phosphorylation

Spectrin is phosphorylated by an endogenous cAMP-independent kinase at four sites, clustered together at the C-terminal end of the smaller subunit.[10] One may take advantage of the turnover of these phosphoryl groups to introduce a natural radioactive label into spectrin for binding and other studies. This can be done in either of two ways: extracted spectrin can be treated with labeled ATP and a crude preparation of the

[8] U. K. Laemmli, *Nature (London)* **227**, 680 (1970).
[9] G. B. Ralston, *Aust. J. Biochem. Sci.* **28**, 259 (1975).
[10] H. W. Harris and S. E. Lux, *J. Biol. Chem.* **255**, 11512 (1980).

red cell kinase (a procedure is given by Pinder *et al.*[11]), or, more conveniently, the label can be introduced into the spectrin in the intact cell.[10,12] The procedure is as follows: cells washed as already described are suspended in a medium containing $0.12\ M$ sodium chloride, 5 mM potassium chloride, 20 mM sodium bicarbonate, 2 mM magnesium chloride, 1 mM calcium chloride, 10 mM glucose, 1 mM adenosine, 0.1 mg of streptomycin and 0.1 mg of penicillin G per milliliter, 0.1 mCi [^{32}P]sodium phosphate, pH 7.5, at a hemotocrit of about 20%. The cells are incubated at 37° in a bacteriological water bath, with gentle agitation, for 24 hr. They are then washed with cold isotonic saline, and extraction and purification of spectrin proceeds as before. The spectrin thus prepared contains almost its full complement of phosphoryl groups, with a labeling level of 200–300 dpm per milligram of spectrin. Dephosphorylation can be accomplished by incubation in the same medium, but without glucose, adenosine, or phosphate for a period of some days, or on isolated spectrin, using a high concentration of bacterial alkaline phosphatase.[11]

Acknowledgment

I am grateful to Jennifer Pinder for a critical reading of the manuscript.

[11] J. C. Pinder, D. Bray, and W. B. Gratzer, *Nature (London)* **270**, 752 (1977).
[12] J. M. Anderson and J. M. Tyler, *J. Biol. Chem.* **255**, 1259 (1980).

[45] Macrophage Actin-Binding Protein

By JOHN H. HARTWIG and THOMAS P. STOSSEL

High molecular weight proteins capable of binding to and crosslinking actin filaments have been isolated from a variety of cell types (see this volume [30]). The available evidence suggests that this family of large proteins creates and maintains the cortical architecture of cells in a rigid lattice composed of short, branching struts of actin filaments.

Formation of a rigid actin filament network depends on a number of factors, which include: (*a*) the concentration of filaments; (*b*) the length of the filaments; (*c*) the functionality of the crosslinker: i.e., how many filaments it crosslinks per molecule; (*d*) the affinity of the crosslinker for the filaments; and (*e*) the orientation of the filaments. The theory of gelation is definitive,[1] permitting networks to be characterized quantitatively

[1] H. L. Yin, K. Zaner, and T. P. Stossel, *J. Biol. Chem.* **255**, 9494 (1980).

in a variety of ways. One of these methods is to determine the number of crosslinking molecules that must be added to the polymers to cause gelation. Since the transition between the sol and gel states is very abrupt, this parameter is easily determined. The gel point can be measured by various assays, including the tipping of test tubes,[2] viscometry,[1] determination of yield pressures with a simple hydrostatic apparatus[3] or with more complex and accurate machinery.[4,5] Such measurements have been done with actin-binding protein crosslinking actin filaments (F-actin). At the gel point the quantity of actin-binding protein molecules bound to actin filaments of known length distribution is very close to that predicted by theory.[1,6,7] The experimental data fit the idea that actin-binding protein is a tetrafunctional crosslinking agent; i.e., each molecule binds two filaments. This conclusion has been verified by direct examination of crosslinked filaments in the electron microscope.[7] The data also indicate that actin-binding protein promotes the formation of an isotropic network, because the theory requires that crosslinking can occur wherever filaments overlap. If crosslinking were to decrease the isotropy of the filaments, for example by promoting the formation of filament bundles, crosslinking would tend to become redundant and involve segments that were already linked into the network, rather than recruiting new filaments into the lattice.

The structure of actin-binding protein is well suited for the type of crosslinking characterized by the gel point studies. Actin-binding protein is a large, asymmetrical, and freely flexible protein that has binding sites for actin located near its free ends.[7] This size and flexibility allows it to span the distance between actin filaments in solution and to find specific binding sites on actin monomers in randomly overlapping filaments.

We describe here the purification of actin-binding protein from rabbit lung macrophages. During isolation, the purification of the protein is monitored by its ability to crosslink actin filaments.

Sedimentation Assay for Actin-Binding Protein Crosslinking Activity

Principle. We have found sedimentation to be a useful assay for the quantitation of the crosslinking. The sedimentation assay measures actin filament aggregation. Filaments polymerized from purified muscle actin

[2] P. J. Flory, "Principles of Polymer Chemistry." Cornell Univ. Press, Ithaca, New York, 1953.
[3] E. A. Brotschi, J. H. Hartwig, and T. P. Stossel, *J. Biol. Chem.* **253**, 8988 (1978).
[4] K. Maruyama, M. Kaibara, and E. Fukada, *Biochim. Biophys. Acta* **371**, 20 (1974).
[5] K. Zaner, R. Fotland, and T. P. Stossel, *Rev. Sci. Instrum.* In press.
[6] J. H. Hartwig and T. P. Stossel, *J. Mol. Biol.* **134**, 539 (1979).
[7] J. H. Hartwig and T. P. Stossel, *J. Mol. Biol.* **145**, 563 (1981).

are extremely long but sufficiently dispersed under physiological conditions of ionic strength and pH that they sediment slowly at low g forces. Therefore, the aggregation of these filaments by the addition of small quantities of a crosslinking agent can bring about the more rapid sedimentation of a relatively large quantity of actin protein at relatively low g forces. The assay accurately reflects the capacity of proteins to crosslink actin filaments into gel networks, and the results concur with direct measurements of the critical gel point for solutions of actin.[3] The aggregation of the filaments is directly proportional to the crosslinker density.[2] Therefore, sedimentation can be measured at concentrations of actin-binding protein both below and above the gel point, and the sedimentation rate is directly proportional to the amount of actin-binding protein in the assay[3] and is, therefore, more suitable for quantitation than gelation assays in which an abrupt sol–gel transition occurs.

Reagents

Actin, prepared from the back and leg muscles of rabbits by the method of Spudich and Watt[8] except that $0.8 M$ KCl is substituted for $0.6 M$ KCl during the first polymerization cycle. G-actin is passed through a sterile $0.45 \mu m$ Millipore filter and stored at $4°$. The G-actin is polymerized at a concentration of 10 mg/ml by the addition of $3 M$ KCl to a final concentration of $0.1 M$ at $25°$ prior to use in sedimentation assays.

Solution: $0.1 M$ KCl, 0.5 mM EGTA, 10 mM imidazole-HCl; pH 7.5

Supplies

Plastic microcentrifuge tubes, 1.5 ml (Sarstedt Co., Princeton, New Jersey; No. 39/10)

Micropipettes, $20 \mu l$

Procedure. Samples to be tested for actin-sedimenting activity are dialyzed into $0.1 M$ KCl solutions at $4°$. Each 0.2-ml assay system contains $40 \mu l$ of F-actin (10 mg/ml), and $160 \mu l$ of the sample or of $0.1 M$ KCl solution (control). If the sample volume is less than $160 \mu l$, the volume difference is made up with the $0.1 M$ KCl solution. The protein solution is placed in a microcentrifuge tube at $4°$ and immediately swirled in a vortex mixer for 1 sec. A 20-μl sample is removed immediately from the solution in the tube, and its protein concentration is determined. The tube is incubated at $25°$ for 30 min and centrifuged at $10,000 g$ for 10 min at $25°$. After centrifugation, $20 \mu l$ of the protein solution are carefully aspirated from the meniscus of the solution in the tube, and the protein concentration is

[8] J. A. Spudich and S. Watt, *J. Biol. Chem.* **246**, 4866 (1971).

determined and compared to the initial protein concentration in the solution prior to centrifugation.

Units of Actin Sedimentation Activity. Actin-sedimentation activity is expressed as micrograms of actin protein sedimented from a 20 μl volume per minute of centrifugation at 10,000 g per milligram of protein in the test sample. To quantitate the rate of actin sedimentation, the amount of protein sedimented at different crosslinker concentrations is determined.

Purification of Macrophage Actin-Binding Protein

Reagents

Solutions
  A: 0.6 M KCl, 0.5 mM ATP, 0.5 mM dithiothreitol, 1 mM EDTA, 10 mM imidazole-HCl, pH 7.5
  B: 1.2 M KI, 5 mM ATP, 5 mM dithiothreitol, 1 mM EGTA, 20 mM imidazole-HCl, pH 7.5
  C: 20 mM KPO$_4$, 0.5 mM dithiothreitol, 0.5 mM EGTA, pH 7.5
  D: Homogenizing buffer: 0.34 M sucrose, 5 mM EGTA, 5 mM ATP, 5 mM dithiothreitol, and 20 mM imidazole-HCl, pH 7.5, containing 0.25 mg of soybean trypsin inhibitor per milliliter
  E: Saturated solution of ammonium sulfate at 4° in 10 mM EDTA, pH 7.0
Columns
  A: 4% agarose, 2.5 × 95 cm (A-15, 200–400 mesh, BioGel, Bio-Rad Laboratories). The bottom 9/10 was equilibrated with solution A. The top 1/10 was equilibrated with solution B.
  B: DEAE-Sepharose (Pharmacia Fine Chemicals), 1.5 × 20 cm equilibrated with solution C

Preparation of Cells. Macrophages are obtained from the lungs of New Zealand white rabbits injected with 1.25 ml of complete Freund's adjuvant (Cappel Laboratories, Cochranville, Pennsylvania) into the marginal ear vein 2–3 weeks prior to being sacrificed.[9] The cells are washed from the lungs by intratracheal lavage with 0.15 M NACl (6 × 50 ml) and collected by centrifugation at 280 g for 10 min at 4°. The pelleted cells are resuspended in 0.15 M NaCl, passed through cheesecloth, and collected a second time by centrifugation at 280 g. The average yield is 3 ml of packed cell per 5 kg rabbit.

All of the following purification steps are done at 4° unless indicated otherwise.

[9] Q. N. Myrvik, E. S. Leake, and B. Fariss, *J. Immunol.* **86**, 128 (1961).

Treatment of Macrophages with Diisofluorophosphate. The treatment of intact macrophages with diisofluorophosphate prior to lysis of the cells increases the amount of actin sedimentation activity in the extracts by 50%, presumably by inhibiting the activity of serine proteases prior to their release into the cytoplasm.[10] Since diisofluorophosphate is a volatile and potent inhibitor of acetylcholinesterase, it is extremely toxic. Utmost care is therefore required in handling this chemical. Precautions include wearing gloves, working in fume hood, and having sodium hydroxide or sodium bicarbonate nearby to inactivate the DFP in case of spillage. A supply of atropine and a hypodermic syringe should be available.

The cells are suspended with 1 volume of 0.15 M NaCl, and this suspension is placed in 50-ml plastic tubes, 20 ml per tube, having caps that can be securely tightened. To each tube 1 ml of 0.1 M diisofluorophosphate in propylene glycol is added in the fume hood. The diisofluorophosphate-treated cells are incubated at ice-bath temperature for 5 min, and the tubes are filled with ice-cold deionized water, tightly capped, and centrifuged at 990 g for 10 min. The fluid expressed by centrifugation is decanted into 5 M NaOH to inactivate residual unreacted diisofluorophosphate. The cell pellets are then resuspended in 0.075 M NaCl in the hood, mixed, and centrifuged again at 990 g. Once again, the supernatant fluid is carefully decanted and inactivated in NaOH.

When chymotrypsin is added to broken cells at this point, its activity is not inhibited, indicating that no unreacted DFP is present.

Homogenization of the Cells. The diisofluorophosphate-treated cell pellets are suspended with 1 volume of homogenization solution, placed in a Dounce glass tissue grinder, and broken, using 50 strokes with a type B pestle. Cell rupture is monitored by phase contrast microscopy. Nuclei, membranes, and organelles are removed from the homogenate by centrifugation at 100,000 g for 1 hr.

Extract Gelation. The supernatant liquid (S1) obtained by centrifugation is removed, made 50 mM with 3 M KCl and 2 mM with 1 M MgCl$_2$, and incubated for 1 hr in a 25° water bath. During this time, the extract solidifies, because under these conditions 50% of the actin in the extract assembles into filaments, which are then crosslinked by molecules of actin-binding protein into a rigid gel lattice. The gel is then centrifuged at 27,000 g for 10 min at 25°, yielding a compressed pellet of gelled material (P2) and a clear supernatant fluid (S2). Actin-binding protein comprises 5.6% of the protein in the compressed gel pellet, a 3.5-fold enrichment over its S1 concentration, where it is 1.6% of the total protein (see the table). However, only 50% of the actin-binding protein present in the S1

[10] P. C. Amrien and T. P. Stossel, *Blood* **56**, 442 (1980).

FRACTIONATION OF ACTIN-BINDING PROTEIN (ABP)
ACTIN-SEDIMENTATION ACTIVITY OF MACROPHAGE EXTRACTS

Fraction	Volume (ml)	Conc. (mg/ml)	Total protein (mg)	% ABP	Actin sedimentation activity			Yield (%)
					Specific[a]	Total[b]	Per mg ABP[c]	
Homogenate	126	26.4	3326	—	—	—	—	—
1	114	7.8	886	1.6	3.9	3648	256	100
2	108.5	5.6	640	0.95	2.7	1727	284	48
2	5.5	26.5	146	5.6	14	2100	256	52
% Agarose	53	0.13	6.9	51	196	1350	383	37
EAE-Sepharose	25	0.12	3.0	95	407	1220	407	30

[a] Specific activity is expressed as micrograms of F-actin protein sedimented in 20 μl/min per milligram of protein.

[b] Total activity = specific activity × total protein concentration.

[c] Total activity per total milligrams of ABP in each fraction. The concentration of ABP in each fraction was determined by densitometry on Coomassie Blue-stained polyacrylamide gradient gels electrophoresed in the presence of sodium dodecyl sulfate × total protein concentration of the fraction.

extract becomes incorporated into the gelled mass. This represents the single greatest yield loss during the purification of actin-binding protein. This loss is most likely due to two factors. First, since half of the actin protein does not sediment at this point, molecules of actin-binding protein are probably attached to these actin monomers or small oligomers and remain in the supernatant fluid after centrifugation. Second, the binding equilibrium under these conditions predicts that a large percentage of the actin-binding protein will remain unbound to F-actin. More actin-binding protein can be removed from the S2 extract with the addition and incubation of skeletal muscle F-actin, followed by centrifugation.

Chromatography on 4% Agarose. The P2 pellet is dissolved in 5 volumes of solution B and clarified by centrifugation at 27,000 g for 10 min. The supernatant liquid is decanted, placed on a magnetic stirrer, and diluted slowly while being stirred with an equal volume of saturated solution of ammonium sulfate. The resulting precipitate is collected by centrifugation at 15,000 g for 20 min and dissolved in 5 ml of solution B. This solution is centrifuged at 27,000 g for 10 min, and the liquid phase is removed and applied to a 4% agarose column. The agarose is prepared by equilibration with solution A; 50 ml of solution B are applied to agarose beads and allowed to run into the gel immediately before use. The sample is eluted with solution A. Figure 1A shows the elution profile of a macrophage P2 pellet on a 4% agarose column. The column fractions are

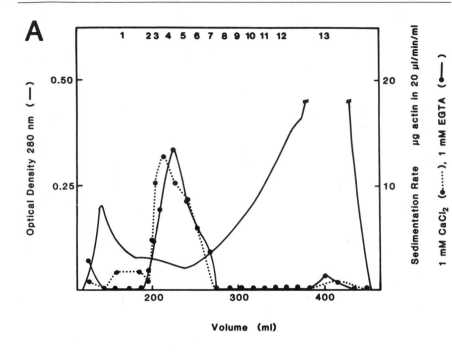

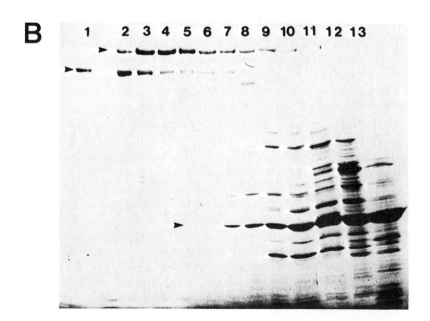

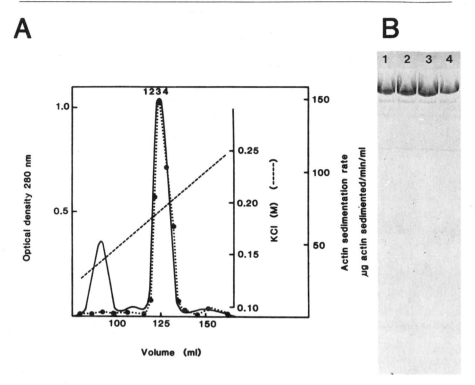

FIG. 2. (A) Ion exchange chromatography of the actin sedimentation activity peak from the 4% agarose column on a DEAE-sepharose colunn. The column was equilibrated as described in the text and eluted with 200 ml of a linear 0 to 0.4 M KCl concentration gradient. Absorbance at 280 nm (——) and actin sedimentation activity (● · · · ●) by the eluted fractions are indicated. The numbers indicate the fractions electrophoresed on polyacrylamide slab gels in (B). (B) Polyacrylamide gradient slab gel (5 to 15%) of the actin sedimentation activity peak from the DEAE-Sepharose column. The numbers correspond to the fractions numbered at the top of (A). The arrowhead shows the migration of the 270,000 subunit of actin-binding protein.

FIG. 1. (A) Gel filtration of the P2 pellet on a 4% agarose column. The column was equilibrated, and the protein was applied and eluted as described in the text. The absorbance at 280 nm (——) and the actin sedimentation activity in the presence (● · · · ●) and in the absence (●——●) of 1 mM CaCl₂ by the eluted fractions is indicated. The numbers at the top of the figure correspond to the fractions analyzed by electrophoresis on polyacrylamide gels in (B). (B) Polyacrylamide gradient slab gel (5 to 15%) of the various fractions eluted from the 4% agarose column in (A). The fractions were denatured and electrophoresed in the presence of sodium dodecyl sulfate and then stained with Coomassie Blue. The numbers correspond to eluted fractions in (A). The arrows, from top to bottom, show the migration of the 270,000 subunit of actin-binding protein, the 200,000 heavy chain subunit of myosin, and the 42,000 actin polypeptide.

monitored for both optical density and actin-sedimentation activity. A single calcium-insensitive peak of actin-sedimenting activity elutes at a K_{av} of 0.31 and as shown in Fig. 1B and contains the actin-binding protein subunit. Fractions with the subunit of actin-binding protein, determined by electrophoresis on polyacrylamide gels in the presence of sodium dodecyl sulfate, or with actin-sedimenting activity are pooled and dialyzed into solution C.

DEAE-Sepharose Chromatography

A small 1.5 × 20 cm column filled with DEAE-Sepharose is equilibrated with solution C. The dialyzed fractions containing actin-binding protein are applied to this column, and the adherent protein is eluted with a linear 0 to 0.4 M KCl concentration gradient. Figure 2A shows the optical density and actin-sedimentation activity profile for this column, and Fig. 2B shows Coomassie Blue-stained polyacrylamide gels of the material that elutes with actin-sedimentation activity. Actin-binding protein elutes at a KCl concentration of 0.18 M and is 95% pure as determined by densitometry of Coomassie Blue-stained polyacrylamide gels.

Storage of Actin-Binding Protein. Actin-binding protein derived from the above columns can be stored in 50% glycerol at $-20°$ without the loss of actin sedimentation activity.

[46] Preparation and Assay of the Intermediate Filament Proteins Desmin and Vimentin

By E. LAZARIDES and B. L. GRANGER

The cytoplasm of many higher eukaryotic cells contains a filamentous system whose individual filaments have an average diameter of 100 Å. Since their diameter is intermediate to those of the 60 Å actin filaments and the 250 Å microtubules in nonmuscle cells and intermediate to those of the 60 Å actin filaments and the 150 Å myosin filaments in muscle cells, these filaments are referred to as intermediate filaments. Recent biochemical and immunofluorescence data have established intermediate filaments as a fibrous system, distinct from other filamentous systems, but composed of chemically heterogeneous subunits (for a review, see Lazarides[1]). Different classes of intermediate filaments are characteristically found in different cell types: keratin filaments in epithelial cells,

[1] E. Lazarides, *Nature (London)* **283**, 249 (1980).

neurofilaments in neurons, glial filaments in glial cells, desmin filaments in muscle cells, and vimentin filaments in cells of mesenchymal origin. It has become apparent, however, that even though these subclasses of intermediate filaments are generally associated with these particular cell types, many cell types have more than one type of filament subunit. For example, epithelial cells contain vimentin as well as keratin filaments,[2,3] glial cells contain both vimentin and glial filaments,[4,5] muscle cells contain vimentin and desmin[6,7] (as well as a newly identified intermediate filament protein referred to as synemin[8]), and many undifferentiated fibroblast-like cells [c.g., chick embryonic fibroblasts and baby hamster kidney (BHK-21) cells] contain desmin in addition to vimentin.[9,10] Distributions of filaments coexisting in the same cell are sometimes distinguishable by immunofluorescence, such as those of keratin and vimentin filaments in epithelial cells;[2,3,11] in other cases, as with desmin, vimentin, and synemin in smooth and skeletal muscle, their localization is coincident.[6-8]

The general approach to identifying and characterizing subunits of intermediate filaments has been to isolate or enrich for them in a given cell or tissue type (as monitored by electron microscopy) and then assay the resulting proteins with electrophoresis to identify the major components. The production of antibodies to a presumptive filament subunit then allows immunofluorescent visualization of its distribution in situ. If a network of cytoplasmic filaments is seen, compatible with the electron microscopic evidence, then that protein is tentatively identified as an intermediate filament subunit. To distinguish the immunofluorescent pattern of intermediate filaments from that of microtubules and microtubule-associated proteins, cells can be observed after treatment with agents that depolymerize microtubules (e.g., Colcemid; see below). To date, this approach has yielded the complicated picture of intermediate filaments described above.

Below we describe in detail techniques for the isolation, biochemical

[2] W. W. Franke, E. Schmid, D. Breitkreutz, M. Loder, P. Boukamp, N. E. Fusenig, M. Osborn, and K. Weber, Differentiation 14, 35 (1979).
[3] W. W. Franke, E. Schmid, K. Weber, and M. Osborn, Exp. Cell Res. 118, 95 (1979).
[4] S.-H. Yen and K. L. Fields, J. Cell Biol. 88, 115 (1981).
[5] A. Paetau, I. Virtanen, S. Stenman, P. Kurki, E. Lindner, A. Vaheri, B. Westermark, D. Dahl, and M. Haltia, Acta Neuropathol. 47, 71 (1979).
[6] B. L. Granger and E. Lazarides, Cell 18, 1053 (1979).
[7] D. L. Gard and E. Lazarides, Cell 19, 263 (1980).
[8] B. L. Granger and E. Lazarides, Cell 22, 727 (1980).
[9] D. L. Gard, P. B. Bell, and E. Lazarides, Proc. Natl. Acad. Sci. U.S.A. 76, 3894 (1979).
[10] G. P. Tuszynski, E. D. Frank, C. H. Damsky, C. A. Buck, and L. Warren, J. Biol. Chem. 254, 6138 (1979).
[11] W. W. Franke, E. Schmid, M. Osborn, and K. Weber, Proc. Natl. Acad. Sci. U.S.A. 75, 5034 (1978).

characterization, and assay of desmin and vimentin from muscle cells. Assays involve identification of these two proteins on two-dimensional gels, production of specific antisera, and immunofluorescence. Testing the specificity of the antisera and detection of subunits in different cell types can be accomplished with the technique of immunoautoradiography.

Presence and General Properties of Desmin and Vimentin in Different Cell Types

Desmin and vimentin coexist in many cell types. A comparison of different cell types shows that the ratio of the two molecules is quite variable. For example, in cultured chicken embryo fibroblasts, 3T3, and CHO cells, vimentin predominates or is the sole filament subunit,[9,10,12] while in BHK-21 cells, desmin and vimentin are present in similar amounts.[9,10,13] Avian erythrocytes contain vimentin but no detectable desmin.[14,15] Smooth, skeletal, and cardiac muscle cells contain predominantly desmin, but also contain lesser amounts of vimentin[6-8]; an exception to this is vascular smooth muscle, which contains predominantly if not exclusively vimentin.[16] No cell type has yet been found that contains desmin but not vimentin with the possible exception of the Purkinje fibers in the heart.[17] However, the extremely low quantities of one or the other subunit often makes detection difficult, so that one must exercise caution in concluding that a particular subunit is absent from a given cell type.

An indispensable tool for the characterization of desmin and vimentin in muscle and nonmuscle cells is sodium dodecyl sulfate–polyacrylamide gel electrophoresis (SDS-PAGE). Although desmin and vimentin can be resolved and tentatively identified by their mobilities in SDS-PAGE, unambiguous identification and resolution of these proteins is possible if SDS-PAGE is combined with isoelectric focusing (IEF) in two-dimensional gels, as described by O'Farrell.[18] Proteolytic fragments of these molecules are characteristic and easily recognized on 2D gels, but virtually unidentifiable on 1D gels because of their comigration with other proteins.

[12] F. Cabral, M. M. Gottesman, S. B. Zimmerman, and P. M. Steinert, *J. Biol. Chem.* **256**, 1428 (1981).
[13] J. M. Starger, W. E. Brown, A. E. Goldman, and R. D. Goldman, *J. Cell Biol.* **78**, 93 (1978).
[14] C. L. F. Woodcock, *J. Cell Biol.* **85**, 881 (1980).
[15] B. L. Granger, E. A. Repasky, and E. Lazarides, *J. Cell Biol.* **92**, 299 (1982).
[16] G. Gabbiani, E. Schmid, S. Winter, C. Chaponnies, C. de Chastonay, J. Vandekerckhove, K. Weber, and W. W. Franke, *Proc. Natl. Acad. Sci. U.S.A.* **78**, 298 (1981).
[17] A. Eriksson and L. E. Thornell, *J. Cell Biol.* **80**, 231 (1979).
[18] P. H. O'Farrell, *J. Biol. Chem.* **250**, 4007 (1975).

Avian desmin usually appears on the 2D IEF–SDS-PAGE as two isoelectric variants (α and β, with isoelectric points of 5.65 and 5.70) and has a molecular weight of 50,000. Mammalian desmin usually appears as a single variant that is of slightly higher molecular weight and more acidic than the avian counterpart ($M_r = 51,000$, p$I = 5.6$).[19,20] Avian vimentin has a predominant variant with a pI of 5.3 ($M_r = 52,000$) and a less abundant variant that is slightly more acidic.[9] Mammalian vimentin is also higher in molecular weight and more acidic than its avian counterpart ($M_r = 54,000$, p$I = 5.2$).[21]

The more acidic variants of avian desmin and vimentin are phosphorylated forms of the most basic variants.[22,23] The presence of phosphatase inhibitors during the preparation of cell extracts greatly enhances the presence of the acidic variants, suggesting that dephosphorylation by intrinsic phosphatases occurs during cell fractionation. α and β desmin copurify through several cycles of filament depolymerization and polymerization, indicating that they have similar solubility properties and are both integral components of intermediate filaments.[24]

In all cultured cell types examined thus far (except mature myotubes—see Fig. 3C), desmin and vimentin filaments are aggregated into cytoplasmic or perinuclear bundles upon prolonged exposure to the microtubule-depolymerizing drug Colcemid. The significance of this aggregation phenomenon has not been determined, but it appears to be a characteristic feature of these filaments. It is useful in that it can be used to corroborate the identity of a given protein as an intermediate filament subunit or associated protein; i.e., any protein that is similarly rearranged by Colcemid is probably associated in some way with desmin or vimentin filaments. Synemin, for example, is a 230,000 dalton protein that colocalizes by immunofluorescence with desmin and vimentin filaments in cultured muscle cells, even if the cells are first treated with Colcemid; it therefore appears to be an intermediate filament-associated protein. That synemin shares most if not all of its solubility properties with desmin and vimentin suggests that the association is a tight one.[8]

Desmin and vimentin have virtually identical solubility properties. Both remain insoluble under physiological conditions, as well as in the presence of high salt and nonionic detergents. Both appear to be soluble under certain conditions of low ionic strength and at pH extremes (<pH 2

[19] J. G. Izant and E. Lazarides, *Proc. Natl. Acad. Sci. U.S.A.* **74**, 1450 (1977).

[20] E. Lazarides and D. R. Balzer, *Cell* **14**, 429 (1978).

[21] Unpublished observations.

[22] C. M. O'Connor, D. R. Balzer and E. Lazarides, *Proc. Natl. Acad. Sci. U.S.A.* **76**, 819 (1979).

[23] C. M. O'Connor, D. L. Gard, and E. Lazarides, *Cell* **23**, 135 (1981).

[24] B. D. Hubbard and E. Lazarides, *J. Cell Biol.* **80**, 166 (1979).

or >pH 10). Desmin and vimentin also copurify through cycles of depolymerization and polymerization.[8,9]

The relative insolubility of these proteins under normal conditions makes enrichment with respect to total protein in the starting material fairly simple: the tissue is extracted with high salt and detergent, for example, and the residue becomes rich with intermediate filaments. Enriched residues are easily assayed on two-dimensional gels for the presence of filament subunits. Complete purification from the enriched residue is more difficult, as solubilization and complete isolation appear to require denaturing solvents.

Since smooth muscle from chicken gizzard is the richest and most convenient source of desmin thus far known, use of this tissue for desmin isolation will be described in detail. The procedure can readily be applied to other tissue sources of desmin and vimentin, such as skeletal and cardiac muscle,[25] as well as other sources of smooth muscle, such as the muscular layers of blood vessels, digestive and urogenital tracts, and mammalian uteri.[26,27] It is also generally applicable in modified form to cultured cells and applies also to vimentin, since its solubility properties are the same as those of desmin.

Purification of Desmin from Smooth Muscle

Purification of desmin from gizzard smooth muscle involves extraction of most other proteins from the tissue, followed by solubilization of desmin from the cytoskeletal residue with acetic acid.[24,26] Neutralization of the acetic acid solution leads to precipitation of the desmin; this precipitate can be redissolved in acetic acid and reprecipitated. The result of this cycling procedure is a preparation consisting primarily of desmin, with minor amounts of actin, vimentin, and synemin.

As with any protein enrichment procedure, care should be taken to minimize proteolysis, denaturation, and modification of the proteins. Processing is carried out at 0–4°, and appropriate protease inhibitors are added to solutions.

Tissue residues are extracted and washed by centrifugation; the volume ratio of supernatant to pellet and the frequency of washes should be maximized (for efficient removal of proteases and other extraneous material) within the limits of convenience.

The following is a typical protocol. Fresh or frozen gizzard muscle (500 g) is dissected free of mucosa and connective tissue. It is

[25] M. Price and E. Lazarides, unpublished observations.
[26] J. V. Small and A. Sobieszek, J. Cell Sci. 23, 243 (1977).
[27] P. Johnson and J. S. Yun, Int. J. Biochem. 11, 143 (1980).

homogenized first in a meat grinder and then in a Waring blender, and suspended in 5–10 volumes of a buffer containing 140 mM KCl, 10 mM EGTA, 10 mM Tris, pH 7.5. The homogenate is pelleted (10,000 g) and resuspended several times in this buffer (the concentration of EGTA can be reduced to 1 mM subsequent to the first or second spin). A large proportion of the desmin can be solubilized at this stage with low salt and EGTA. If this is desired, then the tissue pellet is washed two or three times with 10 mM Tris, 1 mM EGTA (pH 7.5) to remove salts and readily solubilized proteins such as actin and tropomyosin. This is followed by one to three prolonged (0.5–1 day each) extractions with 10 mM Tris, 10 mM EGTA (pH 7.5). Desmin in these latter extracts can be concentrated and enriched further by precipitation with ammonium sulfate at 35% saturation and is useful for bulk preparations of desmin intended for antigen preparation, antibody adsorptions, or peptide analyses.

More complete purification of desmin is achieved from the EGTA-insoluble residue. The insoluble material is freed of most of the remaining actomyosin by extraction with high salt; for example, several washes with a solution containing 0.6 M KI, 10 mM Na$_2$S$_2$O$_3$, 10 mM Tris, 1 mM EGTA (pH 8) leaves little intracellular protein other than desmin. This desmin can be solubilized with urea (see below) or prepared for solubilization with acetic acid: the high salt residue is washed with water to reduce the salt concentration to less than 1 mM, extracted several times with acetone, air dried, and stored at $-20°$. For the final purification of desmin, the acetone powder can be washed with water to remove additional actin and tropomyosin, or it can be suspended directly in 1 M acetic acid. The acetone powder is extracted 2–3 times with acetic acid, and the combined extracts are neutralized by dropwise addition of a 10 M NaOH solution. The precipitate that forms is collected by centrifugation (5000–10,000 g) and redissolved in acetic acid; this cycle is repeated 2–4 times, resulting in a preparation of fairly constant composition. An alternative to precipitation is gelation by dialysis of the acetic acid extract against water or a neutral buffer; this allows polymerization of the desmin into filaments, which can be collected by high speed centrifugation. Electrophoretic analysis of this material reveals that it is composed primarily of desmin, but also contains relatively constant but low amounts of actin, vimentin, and synemin.

Huiatt et al.[28] have reported a procedure for obtaining desmin in pure form, free of contaminating actin. Gizzard muscle that has been extracted with high salt and nonionic detergent is dissolved in a solution containing 6 M urea. This desmin-rich cytoskeletal extract is subjected to chromatog-

[28] T. W. Huiatt, R. M. Robson, N. Arakawa, and M. H. Stromer, J. Biol. Chem. 255, 6981 (1980).

raphy on hydroxyapatite and DEAE-Sepharose in the presence of urea. The final purified desmin polymerizes into 100 Å filaments upon dialysis against a physiological buffer, but remains soluble when dialyzed against 10 mM Tris-acetate, pH 8.5. It is not clear whether these preparations also contain vimentin and synemin in relatively low amounts.

Desmin has also been purified from bovine pulmonary artery by extraction with acetic acid, solubilization with urea, and chromatography on Bio-Rex 70 in urea.[27]

Modification of the above-described method for the purification of desmin from smooth muscle can be used for the rapid enrichment of vimentin in cells grown in tissue culture.[9,13] Cultured cells [e.g., chick embryonic fibroblasts, baby hamster kidney (BHK-21) cells, or mouse 3T3 cells] are washed in phosphate-buffered saline (150 mM NaCl, 10 mM sodium phosphate, pH 7.4), scraped off the plates, and collected in a clinical centrifuge. The cell pellet is then suspended on ice in a buffer containing 0.6 M KI (or KCl), 0.02 M Tris HCl, pH 7.4, 5 mM MgCl$_2$, 1 mM EGTA, 1% Triton X-100. The insoluble material, which contains nuclear remnants and intermediate filaments, is collected in a clinical centrifuge and washed once in the extraction buffer and once with distilled water. These extracts can then be analyzed by 2D IEF–SDS-PAGE for the presence of vimentin or desmin.

One- and Two-Dimensional Polyacrylamide Gel Electrophoresis

One-dimensional electrophoretic analysis of proteins is performed on high-resolution SDS-polyacrylamide slab gels (SDS-PAGE) by a modification of the discontinuous Tris-glycine buffer system.[29] The stacking gel contains 5% acrylamide, 0.13% N,N'-methylene bisacrylamide, 0.125 M Tris-HCl, pH 6.8, and 0.1% SDS. The quantities of acrylamide and bisacrylamide in the separating (lower) gel are provided by a hyperbolic relationship: % acrylamide × % bisacrylamide = 1.3. Gels containing 12.5% acrylamide are used most often because of their high resolution in the molecular-weight range of actin and desmin. A 12.5% analytical gel contains: 12.5% acrylamide, 0.107% bisacrylamide, 0.386 M Tris-HCl, pH 8.7, and 0.1% SDS. Polymerization is catalyzed by the addition of 100 μl of 10% ammonium persulfate and 15 μl of N,N,N',N'-tetramethylethylenediamine per 30 ml of gel solution. The same running buffer is used in both the upper and lower reservoirs: 0.025 M Tris base, 0.112 M glycine, and 0.1% SDS, final pH 8.5. Sample buffer (2X) contains 0.1 M dithiothreitol (or 1% 2-mercaptoethanol), 0.1 M Tris-HCl, pH 6.8, 10% glycerol, 2% SDS,

[29] U. K. Laemmli, *Nature (London)* **227**, 680 (1970).

and bromophenol blue. After electrophoresis, gels are stained at least 3 hr at room temperature in 47.5% ethanol, 10% acetic acid, and 0.1% Coomassie Brilliant Blue R-250, and then destained in 12% ethanol, 5% acetic acid. A piece of polyurethane foam or a Kimwipe can be put in with the gel to serve as a dye sink. Gels are photographed over a light box with Polaroid PN-55 film, using an orange-colored filter to enhance contrast.

Two-dimensional electrophoresis is carried out according to the system of O'Farrell.[18] The first dimension (isoelectric focusing) is prepared and prerun as described,[18] but with the following modifications: the gels (2.5 × 120 mm) contain 0.2% (w/v) Ampholines, pH range 3.5–10; 1% Ampholines, pH range 4–6; and 2% Ampholines, pH range 5–7 (each supplied as 40% w/v solution). Ampholines are not added to overlay solutions or to lysis buffers. Samples are dissolved in 8–10 M urea, 1–2% NP-40, and 1% 2-mercaptoethanol, overlayed directly with 0.02 M NaOH, and run at 450 V for 16 hr and then at 800 V for 1 hr. Most nonmembrane proteins focus into sharper bands if NP-40 is eliminated from both the sample and the gel; this holds true even if the sample has been solubilized in SDS. To avoid proteolysis of desmin and vimentin during urea denaturation, samples are suspended in 1 volume of 1% SDS and immediately placed in a boiling water bath for no more than 1 min. They are then dissolved at room temperature in 10 volumes of 9 M urea and 1% 2-mercaptoethanol (and 1–2% NP-40, if desired) prior to isoelectric focusing.

Antibody Preparation

The primary concern in obtaining an antiserum specific for a particular peptide is purity of the antigen. Antigens differ in their immunogenicity, so that use of a poorly immunogenic antigen contaminated with a small amount of highly immunogenic antigen can lead to erroneous conclusions. This was often the case before the advent of techniques adequate for assessing antigen purity and antiserum specificity. For this reason, SDS-PAGE is routinely used as a final purification step for most antigens. Ideally, prior to SDS-PAGE, the protein of interest should be purified as much as possible by other means (for example, differential extraction or precipitation, ion exchange chromatography), as whole cell extracts most likely contain many proteins with mobilities on SDS gels that are very similar to that of the desired protein. Also, most if not all proteins streak slightly in SDS-PAGE, and contamination from this phenomenon should be minimized.

For production of antibodies against desmin, we have used desmin that is present in the 10 mM Tris–10 mM EGTA extract of gizzard as well as desmin that has been partially purified by cycles of precipitation and solubilization in acetic acid, both as described above. Synemin is a by-product of the latter procedure. A rich source of vimentin is embryonic chicken skeletal muscle, from which high salt-detergent cytoskeletons can be made and run directly on SDS gels. Somewhat cleaner preparations of vimentin (and less likely to be contaminated with desmin and other muscle proteins) can be obtained from cytoskeletons of cultured fibroblasts or chicken erythrocytes. Triton-KCl cytoskeletons of embryonic chicken muscle for the purification of vimentin are prepared as follows: leg and breast muscles from 14- to 16-day-old embryos are dissected out and homogenized in a Dounce homogenizer in $0.1 \ M$ KCl, 50 mM Tris HCl, pH 8.0, 5 mM EGTA, 1 mM O-phenanthroline, 0.5 mM phenylmethylsulfonyl fluoride (PMSF). The muscle homogenate is washed (by pelleting at 10,000 g and resuspension) twice in this buffer, filtered through cheesecloth, and extracted three times in $0.6 \ M$ KCl, 2 mM $Na_4P_2O_7$, 0.5% Triton X-100, 20 mM Tris-HCl, pH 7.5, 1 mM EGTA, 10 mM 2-mercaptoethanol, 0.5 mM O-phenanthroline, 0.1 mM PMSF. The final insoluble residue is washed with water, dissolved in hot solubilization buffer (2% SDS, 50 mM Tris-HCl, pH 6.8, 0.5% 2-mercaptoethanol 10% glycerol, 0.004% bromophenol blue) for 3 min and loaded onto preparative slab gels. Desmin purified by depolymerization and polymerization or from the EGTA-soluble lyophilized material is similarly dissolved in solubilization buffer prior to electrophoresis.

After electrophoresis on preparative SDS gels, protein bands are visualized by staining for 10–15 min in 0.25% Coomassie blue, 47.5% ethanol, 10% acetic acid and destaining in 12% ethanol, 5% acetic acid. There are various other ways to visualize bands or determine their positions, but this is probably the most precise way; exact localization is very important if there are other protein bands in close proximity. The band of interest is carefully excised with a razor blade, equilibrated in a physiological salt solution, and homogenized in a motor-driven Potter–Elvehjem Teflon–glass homogenizer. This last step is made easier and safer by using a laboratory jack with a large rubber stopper on top to apply pressure to the bottom of the glass vessel. Primary injections are usually given with Freund's complete adjuvant, whereas boosts contain no adjuvant. Alternatively, adjuvant can be omitted entirely. For the primary injection with adjuvant, the homogenate is coagulated with 0.1 volume of 10% $AlCl_3$, and then emulsified with 0.5–1 volume of adjuvant by repeated passage between two linked syringes. It is injected subcutaneously at several sites on the back of a female New Zealand white rabbit. Booster injections of just gel homogenate are administered at intervals of 2–6 weeks thereafter. Small amounts of material can also be injected intradermally, intramuscularly, or into the footpads, if desired. The smallest particles in the gel homogenate (those sedimentable only by relatively high speed centrifugation) can be injected without adjuvant into an ear vein. Injections of 0.1–1 mg of protein in a volume of 1–10 ml are most common. Blood is collected from the marginal ear vein 1–2 weeks after booster injections. The blood is allowed to clot 2–4 hr at room temperature and contract for 12–24 hr at 4°, and the serum is clarified by centrifugation. γ-Globulins are partially purified by precipitation at 0° with ammonium sulfate at 50% saturation, dialyzed against Tris- or phosphate-buffered saline containing 10 mM NaN$_3$ and 1 mM ε-aminocaproic acid (a plasmin inhibitor), and stored at −70°. Antiserum specificities and titers are usually tested by double immunodiffusion analysis and indirect immunofluorescence; anti-desmin activity is routinely assayed for by the labeling of Z lines in isolated chicken myofibrils, while antivimentin is tested for by the staining of intermediate filaments in cultures of chicken fibroblasts. Since double immunodiffusion is not a sensitive test of specificity, the antisera are assayed by immunoautoradiography for cross-reaction with any protein other than the antigen of interest.

Immunoautoradiography

Antiserum specificity is assayed by immunoautoradiography, an extremely sensitive technique for determining which antigens in a given cell extract react with a given antiserum. Immunoautoradiography is a solid-phase assay in which proteins are separated by high-resolution SDS-PAGE and then immobilized in the gel; antiserum is applied to the gel, and those proteins that specifically bind antibody are detected by autoradiography using a radiolabeled tag.[30] The details of a modification of this technique are presented below.

This technique is applicable to many different gel systems, but we routinely use one-dimensional discontinuous slab gels and two-dimensional IEF–SDS gels (see above). Two-dimensional gels have a much greater resolving capacity and are therefore preferred. In addition, they allow easier identification of proteolytic fragments and undissociated multimers of the proteins under study. Isoelectric focusing is most useful for acidic proteins, but basic proteins can be visualized on nonequilibrium pH gradient electrophoresis (NEPHGE) gels.[31] Alternatively, isoelectric focusing gels can be run on the second-dimension slab gel in conjunction with an aliquot of the original sample at the side, in order to display all of the proteins in a given system on the same slab. Whereas only one sample

[30] K. Burridge, this series, Vol. 50, p. 54.
[31] P. Z. O'Farrell, H. M. Goodman, and P. H. O'Farrell, *Cell* **12**, 1133 (1977).

can be run on a two-dimensional gel, many different samples can be run on a single one-dimensional gel; this simplifies the task of following a scarce antigen through a purification procedure (by simultaneously assaying for the antigen in aliquots of different fractions), and also allows rapid screening of different cell types for immunoreactive forms of the antigen.

The size of the gel depends on the degree of resolution desired and the amount of reagent available (larger gels require more antiserum, etc.). Gel thickness should be minimized to facilitate diffusion of reagents into and out of the gel, but must be sufficient to allow easy handling and to prevent fragmentation during any of the manipulations. We routinely use gels that are 1.6 mm thick and consist of 12.5% acrylamide and 0.107% N, N'-methylene bisacrylamide (MBA). We have not determined the extent of penetration of antibodies into these gels, but have found that gels of 15% acrylamide, 0.5% MBA do not work, suggesting that gels are not being strictly surface-labeled. Polypeptides are fixed in the gel by precipitation with ethanol and acetic acid; their resistance to elution and loss during processing in physiological salt solutions is probably a function of the size and solubility properties of the polypeptides, and perhaps also of the acrylamide and bis concentrations. We have noted the disappearance of very few proteins during processing, most notably a myosin light chain and tropomyosin. Gel porosity also affects the size range of proteins that are resolved in the separating gel. If proteins smaller than 20,000 daltons, for example, run with the dye front, and if the dye front is labeled with antibody, then a different technique must be used to display these proteins for immunoautoradiography (see below).

The protein sample that is run on the gel should be from the system under study. The technique is sensitive enough generally to allow whole cells or tissues to be run on the gel, without having to enrich for the antigen of interest. Polypeptide components that are not present in sufficient quantity to be seen by Coomassie Blue staining can be readily detected with immunoautoradiography. This has the added advantage of allowing one to avoid artifactual proteolysis of the antigen during processing, which is very difficult if not impossible to eliminate completely in any sort of nondenaturing isolation or enrichment procedure. Also, if the technique is being used to demonstrate specificity of an antiserum, then a positive result is more convincing if the antigen is a minor component of the sample. Intrinsic proteases can be inhibited in a variety of ways. Nondenaturing methods (i.e., the addition of certain chemical compounds, such as PMSF and EGTA, to the solutions used) are generally not completely effective. Since all the proteins will be denatured prior to electrophoresis, a main concern in preparing a whole cell or tissue sample is to inactivate proteases without physically degrading or modifying other proteins. Many proteases are active in 8 M urea or in 1% SDS at room temperature, so these denaturing conditions are not stringent enough; boiling in SDS, however seems to be effective. Cultured cells can be washed free of medium, treated with ethanol to precipitate protein and extract lipid, scraped from the plates and pelleted in ethanol, and then boiled directly in SDS (or stored in a freezer or lyophilized until ready for use). Whole tissue can be pulverized to a fine powder in liquid nitrogen, thawed in ethanol, pelleted, and treated as above. SDS does not seem to affect the isoelectric focusing of most proteins; we have had good luck with this approach even in the absence of NP-40 in the IEF gel and sample (see above).

The following protocol is used routinely in this laboratory. It is not necessarily the best procedure, since many parameters of the technique have not been investigated for efficacy or optimization, but it is very reproducible and has generally given us unambiguous results.

Polyacrylamide gels are run as described above and then fixed for 4–16 hr in 50% ethanol, 10% acetic acid. All incubations and washes are carried out at room temperature on a rocking platform in 8 × 8-inch Pyrex baking dishes tightly covered with plastic food wrap or in polystyrene boxes with a fine nylon mesh underneath the gel to prevent sticking; solutions are removed exclusively by aspiration so that the gels always remain flat and never have a chance to tear. After fixation, gels are washed with several changes of distilled water to remove the ethanol and acetic acid, and then equilibrated in buffer I [140 mM NaCl, 10 mM Tris-HCl (pH 7.5), 5 mM NaN$_3$, 0.1 mM EGTA, 0.1% gelatin (Difco; dissolved by heating the solution to 60°)]. Precipitates of gelatin occasionally form, but these do not affect the final result. From fixation to equilibration in buffer I takes 1–2 days, depending on the frequency with which the solutions are changed. Gels are then incubated for 1 day with antiserum diluted in buffer I. An antiserum that gives a strong precipitin line in double immunodiffusion can be diluted at least a thousandfold; low-titer sera may require less dilution. Whole sera, ammonium sulfate-fractionated globulins, and DEAE-purified IgG have all been used successfully. Normally, we use serum fractionated with ammonium sulfate at 50% saturation. For a standard-sized two-dimensional gel, 100 μl of antiserum are diluted in 100 ml of buffer I for the incubation; after 1 day, the antiserum is removed and the gel is washed for 3 days in several changes (150–200 ml each) of buffer I. Radioiodinated protein A (see below) is applied for 1 day, and then removed by 2 days of washing in buffer I and 1 day in buffer I without gelatin. Gels are stained in 0.1% Coomassie Brilliant Blue R-250, 47.5% ethanol, 10% acetic acid, destained in 12% ethanol, 5% acetic acid, and dried onto filter paper. Autoradiograms are made by exposure to Kodak

X-Omat R XR5 film with a DuPont Cronex Lightning-Plus intensifying screen at $-70°$ for a few hours to a few days, or, for increased resolution, at room temperature without an intensifying screen for up to a month.

Staphylococcus aureus protein A (Pharmacia) is iodinated by the chloramine-T method,[32] using an excess of tyrosine to quench the reaction. All solutions are made up in 0.5 M potassium phosphate, pH 7.5, and used at room temperature. One millicurie of $Na^{125}I$ (high specific activity, carrier-free, in NaOH solution) is mixed with 100 μl of 0.5 M potassium phosphate (pH 7.5); 20 μl of protein A (5 mg-ml) and 20 μl of chloramine-T (2.5 mg/ml) are then added. After 2 min, 150 μl of tyrosine (0.4 mg/ml) are added. The mixture is passed centrifugally[33] through a 3-ml bed of Sephadex G-25 equilibrated (prespun several times) in the KPO_4 buffer, and the void fraction is used for labeling; it is usable for at least a month when diluted 10-fold in buffer I and stored at 4°. The centrifugal desalting step can be performed by packing a disposable 3-ml syringe barrel with Sephadex, hanging it by its flanges inside a conical centrifuge tube, and spinning it for about a minute at a medium speed of a tabletop clinical centrifuge. One-third to two-thirds of the ^{125}I ends up in the void fraction, and this is enough for 20–30 two-dimensional gels, each in 100 ml of buffer I.

Several cautionary aspects of the technique of immunoautoradiography are worth mentioning. Protein A does not bind to all classes of antibody in a given species, and there is also wide variability between species.[34] It is conceivable that this could result in the failure to detect a given antibody specificity by immunoautoradiography. Similar considerations apply if a secondary antibody is used instead of protein A. It seems unlikely that this would be significant for most antisera, but it would certainly be an important consideration if monoclonal antibodies were used.

Faint, nonspecific labeling of certain polypeptides is occasionally observed. Most notable is myosin heavy chain, which sometimes labels slightly with protein A. Stacking gels and perimeters of resolving gels

[32] F. C. Greenwood, W. M. Hunter, and J. S. Glover, *Biochem. J.* **89**, 114 (1963).
[33] M. W. Neal and J. R. Florini, *Anal. Biochem.* **55**, 328 (1973).
[34] G. Kronvall, U. S. Seal, J. Finstad, and R. C. Williams, *J. Immunol.* **104**, 140 (1970).

FIG. 1. Two-dimensional immunoautoradiography. (A) An extract of chick embryo skeletal muscle was run on a two-dimensional gel [isoelectric focusing (IEF) from right (basic) to left (acidic) and sodium dodecyl sulfate–polyacrylamide gel electrophoresis (labeled SDS) from top to bottom] and subsequently processed with anti-vimentin and radioiodinated protein A. The stained, dried gel is in (A) and the corresponding autoradiogram is shown in (B). Some of the major cytoskeletal proteins are labeled. Arrows denote some of the proteolytic fragments of vimentin. From Granger and Lazarides.[6]

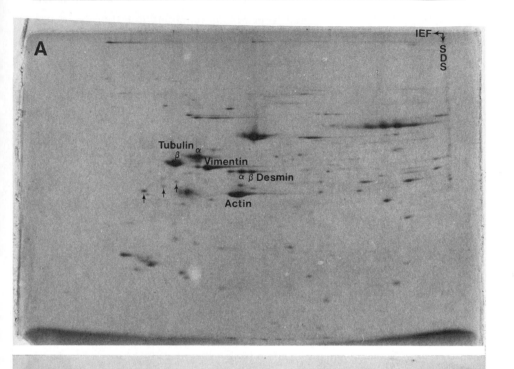

retain more radioactivity than interiors or resolving gels; we have not determined the reason for this, but it is useful in that it serves as a clear reference for aligning the dried gel with the autoradiogram. An example of two-dimensional immunoautoradiography using anti-vimentin is shown in Fig. 1.

Antigen for immunization is usually prepared in this laboratory by excision of a band from a polyacrylamide gel that has been briefly stained with Coomassie Blue. The gel slice is homogenized and equilibrated in a physiological buffer prior to injection. The animal is thus immunized with Coomassie Blue and polyacrylamide in addition to the purified protein. We have found that antibodies to Coomassie Blue may be present in most if not all antisera. These findings are relevant to immunoautoradiography, in that they can be a source of interference, but have no effect in most other applications. Gels can be stained with Coomassie Blue prior to application of antisera and subsequent processing for immunoautoradiography, but lengthy exposures can result in autoradiograms that show all of the proteins in the gel labeled very weakly. This could be due to nonspecific binding immunoglobulins or protein A to the Coomassie Blue-stained proteins, or, as suggested above, it could be a result of the dye acting as a weakly immunogenic hapten. Antibodies to polyacrylamide result in a uniformly labeled gel and must therefore be removed. This can be accomplished by preincubation of the antiserum with a blank polyacrylamide gel homogenate that has been equilibrated in buffer I, removal of the gel particles by centrifugation, and application of the supernatant to the gel to be labeled.

The problem of anti-polyacrylamide can also be overcome by transfer of the proteins from the gel to diazotized paper[35] or nitrocellulose paper[36] for immunoautoradiography. This approach is much less time-consuming than labeling in gels, due to reduced incubation and washing periods, but suffers from a slight loss in resolution and the difficulty of visualizing the exact positions of the transferred proteins. Different proteins may also be transferred with different efficiencies. Transfer to paper would, however, be useful for examining small polypeptides that are not resolved in polyacrylamide gels that are porous enough for immunoautoradiography. This would allow dissection of proteins with proteases and mapping of antigenic peptides, for example, to study the structural relationships of cross-reacting proteins.

The ultimate test of antiserum specificity is the ability of the purified antigen to block all antibody activity. This bypasses the argument that there are antibodies in the serum with specificities for antigenic determi-

[35] J. Renart, J. Reiser, and G. R. Stark, *Proc. Natl. Acad. Sci. U.S.A.* **76**, 3116 (1979).
[36] H. Towbin, T. Staehelin, and J. Gordon, *Proc. Natl. Acad. Sci. U.S.A.* **76**, 4350 (1979).

nants that are lost upon denaturation in the polyacrylamide gels. Blocking can be performed by preincubation of antisera with protein that has been purified by preparative SDS gel electrophoresis; protein bands are cut out, eluted in 0.2% SDS, dialyzed against distilled water, and lyophilized prior to adsorption. Elution can be performed electrophoretically or by diffusion. Appropriate controls, such as adsorptions with unrelated proteins, must also be performed. Complete blocking by this technique suggests that there are not antibodies against determinants that are lost upon denaturation but are detectable in an assay such as immunofluorescence.

Immunofluorescence

A number of cytoskeletal proteins are assayed by indirect immunofluorescence. Since desmin, vimentin, and synemin exhibit indistinguishable cytoplasmic distributions, their exact distributions can be compared by double immunofluorescence. In general, all of our antisera are elicited in rabbits and indirect immunofluorescence is performed using commercially available fluorescein-conjugated IgG fraction of goat anti-rabbit IgG as a labeled secondary antibody.

For direct immunofluorescence, antibodies are usually conjugated with rhodamine [using rhodamine B isothiocyanate (RBITC) or tetramethyl rhodamine isothiocyanate (TMRITC)].[37,38] Immunoglobulin G is partially purified by precipitation with an equal volume of saturated ammonium sulfate, dialyzed against 40 mM NaCl, 10 mM sodium phosphate, pH 7.5, and passed through a column of DEAE-cellulose equilibrated in the same solution. The effluent (nonretained protein peak) is collected and concentrated by ultrafiltration to 7–10 mg of protein per milliliter. It is then incubated with approximately 40 μg of rhodamine-ITC per milligram of protein in 0.15 M sodium carbonate buffer, pH 9.5, overnight at 4° with constant stirring. Unbound dye is removed by passage through a gel filtration column (e.g., Sephadex G-25). Conjugated IgG is then further fractionated on DEAE-cellulose; the conjugates eluting with 40 mM NaCl, 10 mM sodium phosphate, pH 7.5, are used for immunofluorescence (molar dye : protein ratio of approximately 2).

Antibodies to chicken cytoskeletal proteins are routinely assayed by indirect immunofluorescence on primary or secondary cultures of embryonic chicken skeletal muscle.[39] These cultures contain both skeletal muscle cells (myoblasts that fuse to form myotubes) and nonmuscle (fibroblast-like) cells, so that muscle-specific antisera are easily recog-

[37] J. J. Cebra and G. Goldstein, *J. Immunol.* **95**, 230 (1965).
[38] P. Brandtzaeg, *Scand. J. Immunol.* **2**, 273 (1973).
[39] Konigsberg, I. R., this series, Vol. 58, p. 511.

nized. Cells are grown on collagen-coated glass coverslips and can be processed in a variety of ways for immunofluorescence. Cells can be fixed with formaldehyde or with an organic solvent such as ethanol, methanol, or acetone; if fixed with formaldehyde, the membrane can be permeabilized with detergent or organic solvent. Typically, cells are washed very briefly in phosphate-buffered saline (PBS; 150 mM NaCl, 5 mM KCl, 10 mM NaPO$_4$, pH 7.4) and fixed by immersion of the coverslip in PBS containing 2–4% formaldehyde at 37° for 10 min. Alternatively, coverslips are immersed in PBS containing 0.1–0.5% formaldehyde, 0.5% Triton X-100 for 5 min at 37°, and subsequently in PBS containing formaldehyde as above. After fixation, coverslips are washed in PBS or Tris-buffered saline containing 0.5% Triton X-100, and all subsequent washes and incubations are made in this same buffer. If the antigen is an insoluble cytoskeletal component, cells can be extracted prior to fixation with 0.5% Triton X-100 and 0.6 M KCl or 0.6 M KI or with 0.5% Triton X-100 alone (with the inclusion of 5 mM MgCl$_2$ to prevent cell detachment).

Coverslips are incubated with antisera for 30–60 min at 37°. Antisera are usually partially purified by ammonium sulfate fractionation, and occasionally by DEAE-cellulose chromatography, prior to immunofluorescence. Antisera are diluted appropriately in PBS depending on their titer; primary antisera are usually diluted 10- to 50-fold and secondary antisera 100- to 150-fold. Coverslips are washed at least 10 min after each antibody incubation. Double immunofluorescence is performed using a modification of the indirect-direct staining method.[40] Fixed cells are incubated with primary antiserum followed by fluorescein-conjugated secondary antiserum, as described above for indirect immunofluorescence. Unreacted antigen-combining sites of the secondary antibodies are then blocked by incubation with normal rabbit serum (ideally the preimmune serum of the following, final antibody). Cells are then washed and incubated for 30–60 min at 37° with rhodamine-conjugated antisera (see Fig. 2). Cells are finally washed in PBS for 10–30 min and mounted on a drop of Elvanol (or Gelvatol) or 90% glycerol. Cells are observed with a Leitz microscope equipped with phase and epifluorescence optics and filter modules H or K for fluorescein and N2 for rhodamine. Cells are photographed on Kodak Tri-X panchromatic film at ASA 1600 and developed in Diafine (Acufine, Inc.). Color images are recorded on Kodak Ektachrome 200 color slide film and developed commercially.

A related assay for intermediate filament-associated proteins takes advantage of the fact that intermediate filaments in many cells characteristically aggregate into thick, sinuous cytoplasmic bundles when the cells are incubated with Colcemid (10^{-6} to $10^{-5} M$ for 12–24 hr) (see Fig. 3B and C).

---

[40] R. O. Hynes and A. Destree, *Cell* 15, 875 (1978).

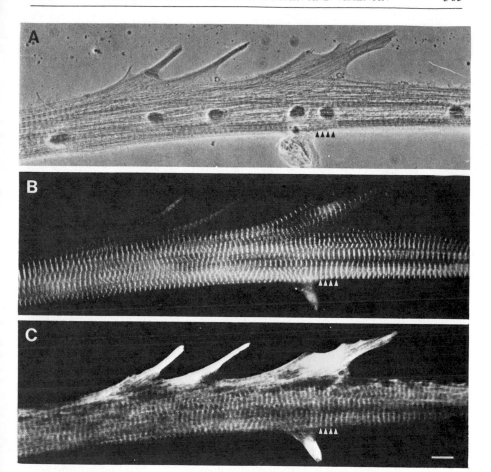

Fig. 2. Double immunofluorescence. (A) Phase contrast micrograph of an embryonic chicken myotube after 7 days in culture. (B) Same myotube labeled with anti-α-actinin followed by a fluorescein-conjugated secondary antibody. α-Actinin is used as a marker for myofibril Z lines. (C) Same myotube labeled with rhodamine-conjugated anti-desmin. Desmin is undergoing its transition from cytoplasmic filaments to a Z-line-associated form at this time. Note asynchrony of transition in different parts of the same myotube. Arrowheads denote Z lines. Bar = 10 μm. From Gard and Lazarides.[7]

If an antigen becomes rearranged in this manner under these conditions, then this is further evidence that it somehow is associated with or is a component of intermediate filaments.

Antisera are also routinely assayed on unfixed fragments of skeletal muscle, namely, myofibrils and Z-disk sheets. Myofibrils are easily recognized arrays of sarcomeres with a distinct and identifiable topology; major proteins of myofibrils have been correlated with particular sections

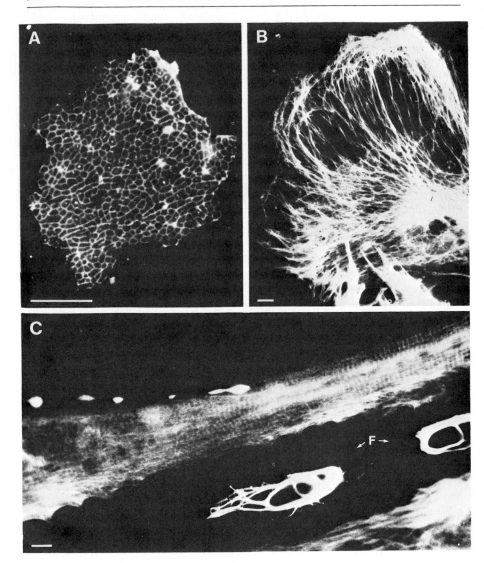

FIG. 3. Assays for desmin and vimentin by indirect immunofluorescence. (A) Z-disk sheet labeled with anti-desmin. Desmin is present at the periphery of each of the more than 500 Z disks in this sheet. (B) Portion of a cultured chick embryo fibroblast stained with anti-vimentin. (C) A week-old myogenic culture treated with Colcemid and stained with anti-vimentin. Intermediate filaments in fibroblasts (F) aggregate in response to Colcemid, while Z-line-associated forms of desmin and vimentin (in myotubes) resist aggregation. Bars = 10 μm. (A) From Lazarides[1]; (C) from Gard and Lazarides.[7]

of these structures. Z-disk sheets are transverse sections of muscle fibers, representing individual planes of laterally registered Z disks.[41] Myofibrils and Z-disk sheets have together allowed the elucidation by immunofluorescence of central and peripheral domains of the Z disks. The intermediate filament proteins, desmin, vimentin, and synemin, are characteristically present at the periphery of each Z disk, whereas α-actinin is found within each disk. Antisera specific for filamin have also shown that filamin is a component of the peripheral domain.[42] Figure 3A shows a Z-disk sheet labeled with antidesmin.

Myofibrils and Z-disk sheets are prepared for immunofluorescence as follows: Thin strips of skeletal muscle are tied to supports (e.g., wooden applicator sticks) to prevent contraction, and stored at $-15°$ in a physiological salt solution containing 50% (v/v) glycerol and 1–5 mM EGTA. After a few weeks, myofibrils can be prepared by vigorous homogenization of the glycerinated tissue in this same solution in a spinning-blade homogenizer; progress of the homogenization is monitored frequently by phase-contrast microscopy in order to minimize the number of unfragmented fibers and maximize the length of the resulting myofibrils. Myofibrils can be purified somewhat by differential centrifugation in the glycerol solution to eliminate connective tissue, unhomogenized chunks of muscle, vesicles, and soluble proteins. Myofibrils are also stored at $-15°$ in the glycerol solution.

Z-disk sheets are prepared by extraction of small strips of glycerinated muscle with buffered (pH 7.5–8.0) 0.6 M KI containing 0.1 mM EGTA and 10 mM sodium thiosulfate to react with free iodine. This is usually performed for a week at 4° or a day at room temperature. The extracted tissue is homogenized in this solution in a small spinning-blade homogenizer (e.g., a Virtis "45"), and progress is monitored by phase-contrast microscopy in an effort to maximize the number and size of the Z-disk sheets.

For immunofluorescence, a drop of myofibrils suspended in the glycerol solution, or a drop of Z-disk sheet homogenate in the KI solution, is spread on a glass coverslip. After rinsing of nonadherent material and equilibrating in the immunofluorescence buffer (the material will remain on the coverslips even in the presence of Triton X-100), the coverslips are treated as described above for fixed cells.

Summary

Desmin and vimentin share similar molecular properties and both appear to function as subunits of intermediate filaments. Their quantity and

[41] B. L. Granger and E. Lazarides, *Cell* **15**, 1253 (1978).
[42] R. H. Gomer and E. Lazarides, *Cell* **23**, 524 (1981).

ratio varies in different cell types, some cell types containing predominantly desmin, others predominantly or exclusively vimentin. Desmin can be purified conveniently from gizzard smooth muscle by cycles of depolymerization and polymerization. Cytologically, desmin and vimentin can be identified in immunofluorescence with antibodies specific for each of them. Antibody specificity is assessed by immunoautoradiography and adsorption controls coupled with immunofluorescence. Desmin and vimentin react characteristically with a filamentous system in the cytoplasm of myogenic and nonmuscle cells grown in tissue culture that is distinct from actin filaments and microtubules. This filamentous network and the respective immunofluorescence of desmin or vimentin is induced to aggregate in cells exposed to Colcemid. Desmin and vimentin are also components of myofibril Z disks and in particular the periphery of the Z disk.

Section IV

Special Techniques for the Study of the
Contractile Protein Complex and the Cytoskeleton

[47] Optical Diffraction and Filtering of Electron Micrographs of Contractile Proteins

By PAULINE M. BENNETT

In 1964 Klug and Berger[1] showed that the electron micrograph of a regularly repeating biological object could act as an optical diffraction grating. Since then the technique of optical diffraction and its subsequent developments, optical filtering and three-dimensional reconstruction, have found wide use in the analysis of biological structures, particularly the arrays formed by the cytoskeletal and contractile fibrous proteins, such as microtubules[2] and muscle filaments.[3]

Optical diffraction can be applied at a number of levels, from the analysis of structures not previously investigated, to the routine assessment of a large number of micrographs of a known structure to select areas for computer analysis. Between these extremes is the use of the pattern as a diagnostic "fingerprint" for the presence of a certain protein. For example, F-actin has a very characteristic diffraction pattern in the presence or the absence of other components.[3-5] Finally, the effect of new components on otherwise well defined structures can be determined so that actin-binding proteins, microtubule-associated proteins (MAPs), and dynein could be located on their parent structures.

There are now a number of review articles as well as original papers describing how to set up the apparatus and to interpret the diffraction patterns from micrographs of biological structures.[6-9] One clear and useful review is to be found in this series.[10] Here I describe briefly the information to be gained from optical diffraction and give some practical hints

[1] A. Klug and J. Berger, *J. Mol. Biol.* **10**, 565 (1964).

[2] L. A. Amos and A. Klug, *J. Cell Sci.* **14**, 523 (1974).

[3] E. J. O'Brien, P. M. Bennett, and J. Hanson, *Philos. Trans. R. Soc. London Ser. B* **261**, 201 (1971).

[4] D. J. De Rosier, E. Mandelkow, A. Silliman, L. Tilney, and R. Kane, *J. Mol. Biol.* **113**, 679 (1977).

[5] L. G. Tilney, D. J. De Rosier, and M. J. Mulroy, *J. Cell Biol.* **86**, 244 (1980).

[6] C. A. Taylor and H. Lipson, "Optical Transforms." Bell, London, 1964.

[7] D. J. De Rosier and A. Klug, *J. Mol. Biol.* **65**, 469 (1972).

[8] R. W. Horne and R. Markham, *in* "Practical Methods in Electron Microscopy" (A. M. Glauert, ed), Vol. I, part II, p. 327. North-Holland Publ., Amsterdam, 1973.

[9] U. Aebi, P. Smith, J. Dubochet, C. Henry, and E. Kellenberger, *J. Supramol. Struct.* **1**, 498 (1973).

[10] H. P. Erickson, W. A. Voter, and K. Leonard, this series, Vol. 49, p. 39.

METHODS IN ENZYMOLOGY, VOL. 85

on how to use a diffractometer once it has been decided that the technique may give valuable information.

Information Gained from Optical Diffraction

The optical or Fraunhoffer diffraction pattern is produced when the micrograph is illuminated by a parallel beam of light, which is subsequently focused. The information in the pattern lies in the position, intensity, and relative phase of the regularly disposed diffraction spots. Analysis of the data reveals the periodic repeats in the structure, the arrangement of the subunits with respect to one another, and the shape of the subunits. Frequently the results are visualized most clearly in the so-called filtered image of the original micrograph. Here the randomly disposed "noise" spots in the diffraction pattern arising from random variations in the stain distribution and specimen damage are blocked while the unimpeded reflections are allowed to pass through the diffraction plane to be recombined and imaged by another lens.[7,10] A valuable development of the filtering principle, applied to helical and other structures where two sides are superimposed in the micrograph, is one in which reflections from either back or front of the structure only are allowed to contribute to the final image.

Use of a Diffractometer

It is important that the diffractometer is properly adjusted so that the diffraction pattern of any object is centrosymmetric.[6] That is, when inverted through the origin the pattern appears the same. Departures from centrosymmetry can be confusing when analyzing diffraction patterns, particularly those with a large number of closely spaced weak reflections. For the best performance, all optical surfaces must be clean and free from dust, the system must be accurately aligned with the optic axes of all components coincident with the axis of the instrument, and, finally, it must be accurately focussed so that the recording film is exactly in the diffraction plane.

The subject for diffraction can be the original micrograph, particularly if a faithful electron dose–optical density recording is necessary, but for routine analysis a positive transparency on film is adequate. Its use prevents damage to the original micrograph and allows the contrast to be enhanced in weakly diffracting subjects, such as low density micrographs. However, care should be taken to keep to the region of linear response to intensity in the copy film since artifacts can be introduced by too high contrast. Inhomogeneities in the subject, such as variations in the thickness of emulsion or film base, can lead to a loss of centrosymmetry in the diffraction pattern. This can be overcome by covering the subject in immersion oil between optical flats.

The most appropriate magnification of the micrograph for the diffractometer must be determined. Values of $\times 30,000-60,000$ are usual. The aperture of the diffracting lens restricts the size of the areas to be diffracted and the recording device limits the size of the diffraction pattern. However, the pattern should be of a reasonable size if filter masks are to be made and manipulated. It is valuable to know the calibration constant of the diffractometer, defined as the value Dd in (units)^2. Where d is a repeating spacing in the micrograph and D the distance from the center of the pattern of the corresponding diffraction spot. Then, knowing the instrumental limit on D, the minimum value of d can be determined and hence the minimum magnification of the micrograph.

The area of interest in the micrograph must be masked off in some way to prevent the pattern becoming swamped by superfluous information from the rest of the micrograph. This can be done directly on the micrograph with masking tape or by using variously shaped holes cut in pieces of card. Some diffractometers have, built in, a set of adjustable slits. Mask shapes generally used are either circular or, more useful for filamentous objects, rectangular. This shape gives rise to strong spikes of intensity in a direction perpendicular to the sides of the rectangle, which often obscure important features in the pattern. A useful trick is to use a parallelogram as a mask shape. By adjusting the precise shape, the spikes of intensity can be maneuvered away from the areas of the pattern of particular interest.

Before optical filtering is commenced, it is essential to interpret the diffraction pattern correctly, since artifacts can be induced in the recombined image if the wrong reflections are selected. The correct interpretation is more likely if a large number of patterns from different examples of the same structure are first analyzed.

A number of methods of making filtering masks are available.[10] The simplest require stiff card or opaque plastic sheet that, when cut, gives clean edges. The mask is made on the same scale as the diffraction pattern and holes cut to allow through the selected reflections including the central beam. For other than the simplest masks use can be made of a pantograph punch,[6] an optical comparator,[10] or photoetching techniques.[10]

The size of the holes cut is important.[10,11] In diffraction patterns from helical structures the reflections are elongated along layer lines and the whole reflection should be allowed to pass through the mask in order to

[11] R. D. B. Fraser and G. R. Milward, *J. Ultrastruct. Res.* **31**, 203 (1970).

preserve the shape of the structure. Crystalline arrays give reflections that are usually more compact, and it is possible to use either very small holes to select beams of peak intensity and phase or larger holes to allow transmission of the whole reflection. With larger holes local variations in the image are apparent, whereas with smaller holes a more averaged image is obtained. However when using small holes the mask has to be very accurately made and positioned. Furthermore, if the reflections are in any way broken up by, for example, noise, then erroneous images can be produced. In this case bigger holes should be used.

If the subject of diffraction is helical or otherwise two sided and the filtered image of either back or front is required then, since it arises from both sides, the undiffracted beam at the origin of the diffraction pattern can be reduced in intensity by half using wire gauze of 50% transmittance. A similar trick can be used to enhance the contrast in a weakly diffracting object, taking care to avoid attenuating the intensity of the origin too much. The mesh of the gauze should be small enough so that multiple images, due to diffraction by the gauze, do not overlap the principal recombined image (400–1000 gauge mesh is usually suitable).

[48] Fluorescent Localization of Contractile Proteins in Tissue Culture Cells

By KUAN WANG, JAMES R. FERAMISCO, and JOHN F. ASH

In contrast to the stable, regular, and paracrystalline contractile apparatus of striated muscle, the motile systems in nonmuscle cells are transitory and highly variable. The study of the spatial and temporal distribution of contractile proteins in these cells thus represents a great challenge.

Two sensitive and versatile localization techniques based on the use of fluorescent probes at the light microscopic resolution are described here.

1. Immunofluorescence microscopy. In this highly specific technique, fluorescently labeled specific antibody is used to visualize and to localize antigen molecules by fluorescence microscopy. It has been applied successfully to the study of organization of muscle proteins in myofibrils and in tissue culture muscle cells (see footnote 4 for a review article). Its application to the investigation of intracellular contractile and structural proteins in cultured nonmuscle cells was begun in 1974 with a study on the distribution of actin in nonmuscle cells (see footnote 5). A wealth of useful

METHODS IN ENZYMOLOGY, VOL. 85

structural information has been accumulated since then by the use of this technique. A shortcoming of the technique as applied to nonmuscle cells is that it yields only a static picture of the highly dynamic motile events, owing to the fact that cells are fixed and permeabilized to allow the antibody to reach intracellular targets.

To circumvent this problem, a second localization technique has been developed (see footnote 3).

2. Molecular cytochemistry. In this technique, fluorescently labeled contractile proteins are microinjected into living cells. The distribution and movement of these injected tracer molecules inside the living cells are monitored continuously by fluorescence microscopy.

These two techniques complement each other and, together, have provided powerful structural information about the molecular and structural basis of motile events in nonmuscle cells.

This chapter is divided into three parts. The first part details the immunofluorescence techniques; the second part describes the microinjection technique; and the last part discusses briefly the limitations and cautions required in the interpretation of fluorescent labeling patterns. The procedures described here are those with which we are personally familiar and may not be the choice for all applications. This chapter should be used as a flexible guide. We have included background and rationale for most procedures so that appropriate modifications can be made as individual need arises.

A vast literature exists on these subjects.[1-13] Several excellent monographs and chapters on immunofluorescence techniques are available.[6-13]

[1] U. Groschel-Stewart, *Int. Rev. Cytol.* **65**, 193 (1980).
[2] B. R. Brinkley, S. H. Fistel, J. M. Marcum, and R. L. Pardue, *Int. Rev. Cytol.* **63**, 59 (1980).
[3] D. L. Taylor and Y. L. Wang, *Nature (London)* **284**, 405 (1980).
[4] F. A. Pepe, *Int. Rev. Cytol.* **24**, 193 (1968).
[5] E. Lazarides and K. Weber, *Proc. Natl. Acad. Sci. U.S.A.* **71**, 2268 (1974).
[6] M. Goldman, "Fluorescent Antibody Methods." Academic Press, New York, 1968.
[7] R. C. Nairn, "Fluorescent Protein Tracing." Churchill-Livingstone, Edinburgh and London, 1976.
[8] J. H. Peters and A. H. Coon, *Methods Immunol. Immunochem.* **5**, 424 (1976).
[9] G. O. Johnson, E. J. Holborow, and J. Dorling, *in* "Handbook of Experimental Immunology" (D. M. Weir, ed.), 3rd ed., Chapter 15. Blackwell, Oxford, 1978.
[10] A. Kawamura, Jr. "Fluorescent Antibody Techniques and Their Application." Univ. Park Press, Baltimore, Maryland, 1969.
[11] W. Hijmans and M. Schaeffer, eds., *Ann. N. Y. Acad. Sci.* **254** (1975).
[12] R. D. Goldman, T. D. Pollard, and J. Rosenbaum, eds. "Cell Motility," *Cold Spring Harbor Conf. Cell Proliferation* **3** [Book A-C] (1976).
[13] A. A. A. Thaer and M. Sernetz, "Fluorescence Techniques in Cell Biology." Dekker, New York, 1971.

Reviews that specifically address the application to nonmuscle contractile proteins have appeared.[1-3]

Fluorescent Staining of Fixed Cells

Preparation of Cells for Labeling[13a]

Prior to labeling, it is generally necessary to prepare cells by first fixing them to preserve structure and then to permeabilize them to allow labeling reagents to reach intracellular targets. A wide range of procedures have been used, with a varying degree of preservation of antigenic sites and cellular morphology. Details of three popular fixation–permeabilization procedures are described here.

Tissue Culture Cells

It is most convenient to grow cells on coverslips that can then be processed and mounted on slides for observation. Some cells (e.g., embryonic chicken fibroblasts) will adhere to and grow on most commercially available coverslips without special treatment; others adhere poorly to glass. Treatments that improve adhesion and growth include cleaning with chromic acid solution or coating the surface with a positively charged polymer or with collagen.[14] Cells, such as lymphocytes or lymphomas, that will grow only in suspension will be discussed later.

Uncoated coverslips are sterilized by autoclaving, by flaming ethanol-dipped coverslips, or by ultraviolet irradiation. Sterile coverslips can be coated with polycations by incubating coverslips in culture dishes for 15–60 min at room temperature with an autoclaved 0.01 mg/ml aqueous solution of poly(L-ornithine) (Sigma, type 11b) or polylysine (Polysciences). After removing the solution, tissue culture medium and cells may be added immediately or the dishes may be stored dry for future use.

A discussion of tissue culture techniques is beyond the scope of this chapter; the reader is referred to other sources.[14-16] In general, cells should be grown for at least 24 hr on the coverslips before fixation to allow recovery from trypsinization (unless the reorganization of cytoskeleton during spreading is being investigated). Cell density can influence individual cell morphology, and high-resolution fluorescence microscopy is dif-

---

[13a] By J. F. Ash.

[14] W. B. Jakoby and I. H. Pastan, eds., this series, Vol. 58.

[15] P. F. Kruse, Jr. and M. K. Patterson, eds., "Tissue Culture: Methods and Applications" Academic Press, New York, 1973.

[16] J. Paul, "Cell and Tissue Culture," 5th ed. Academic Press, New York, and Churchill-Livingstone, Edinburgh and London, 1973.

ficult with crowded fields because of stray light contributed by adjacent cells. Dead cells release intracellular components that may adhere to cell surfaces. Cell debris absorb fluorescent conjugates nonspecifically and thus fluoresce brightly. For these reasons, standard conditions for culture initiation and maintenance should be established.

Fixation and Permeabilization

Fixation of cells should meet the criteria that (*a*) cellular structures must be preserved in an unaltered state; (*b*) antigenic determinants or ligand binding sites of cellular macromolecules must remain active; (*c*) all areas of the cell must be freely accessible to the antibodies or ligands used to localize specific targets. These three criteria are potentially antagonistic: good fixatives, while preserving fine structures, tend to alter or destroy antibody or ligand binding sites as well as to decrease accessibility of probes. Mild fixatives, while less effective in structural preservation, are gentler to binding sites and yield a porous structure that allows easier access of probes. For these reasons, therefore, the effect of several different fixation treatments on the detailed labeling pattern should be compared (cf. Fig. 1 and the section Fluorescent Staining of Living Cells).

Cold Acetone Treatment. Acetone is placed in a glass container and chilled in an explosion-proof freezer to −20°. A coverslip with attached cells is rinsed twice with phosphate-buffered saline (PBS; 0.15 M NaCl, 0.01 M sodium phosphate, pH 7.4) at 37°, placed in cold acetone, and returned to freezer for 30 min. The container is then removed and allowed to warm above 0°, and the coverslips are transferred to dishes containing 5 ml of PBS and rinsed several times to remove acetone. The coverslip is now ready for labeling. Cold acetone fixes and permeabilizes cells simultaneously because it precipitates proteins and extracts membrane lipids. This treatment generally produces the greatest distortion of cell structure, especially fine surface processes. However, it is convenient and not as harsh as chemical fixatives. Other water-soluble organic solvents or mixtures thereof, such as methanol and ethanol, have also been used, generally at low temperatures (−20°).[17–19] Variants of the procedures described above include shorter treatment time (1 min)[20] or air-drying the cells following acetone treatment.[21]

Formaldehyde–Triton X-100 Treatment. A 3.7% (w/v) formaldehyde fixatives solution is prepared by adding 3.7 g of solid paraformaldehyde

[17] J. R. Feramisco and S. H. Blose, *J. Cell Biol.* **86**, 608 (1981).
[18] J. M. Sanger and J. W. Sanger, *J. Cell. Biol.* **86**, 568 (1981).
[19] M. Osborn and K. Weber, *Cell* **12**, 561 (1976).
[20] E. Lazarides, *J. Cell. Biol.* **68**, 202 (1976).
[21] I. M. Herman and T. D. Pollard, *J. Cell Biol.* **88**, 346 (1981).

(Polysciences) to distilled water and heated to 50–60°. A few drops of 1 N NaOH are added, and the solution is stirred until all the paraformaldehyde depolymerizes. The solution is then cooled to room temperature, 10 ml of 10 times concentrated PBS are added, the pH is adjusted to 7.4, and the volume is brought to 100 ml with distilled water. Formaldehyde fixation is optimal with freshly prepared solutions, but for routine work the solution may be stored frozen at −20° and thawed before use.

Prior to fixation, the cells are quickly rinsed twice with PBS at 37° to remove cell debris. The formaldehyde fixative at 37° is added to cover the cells (2 ml in a 35-mm dish), and the cells are allowed to come to room temperature during a 20-min incubation. After fixation the cells are rinsed with 2-ml additions of the following schedule: PBS, 5 min; PBS containing 0.1 M glycine at pH 7.4, 5 min (to quench remaining formaldehyde); PBS containing 0.1% (w/v) Triton X-100 (Sigma) (to render cells permeable), 5 min; PBS, 5 min. Formaldehyde fixation[19] is adequate for most structures observed at the light microscope level while retaining most cellular antigenic determinants. If membrane structure is of interest, cells should be observed during fixation with phase-contrast microscopy to detect the extent of membrane blebing induced by this fixative on some cells.

Formaldehyde-fixed cells can also be permeabilized by cold acetone, cold methanol, or cold 95% ethanol at −20° for a few minutes.[17–19] Formaldehyde–ethanol treatment may preserve many fine ultrastructural details.[19]

Glutaraldehyde–Triton X-100 Treatment. (The authors thank Drs. J. V. Small and J. DeMey for assistance in preparing the following procedures.) Coverslips with attached cells are washed twice with calcium-free Hanks' balanced salt solution (Gibco) and then twice with a cytoskeleton buffer (Hanks' balanced salt plus 2 mM $MgCl_2$, 2 mM EGTA, 5 mM PIPES, pH 6.1), both steps at 25°. The cells are then treated for 1 min at room temperature with a solution containing glutaraldehyde and Triton X-100 [cytoskeleton buffer plus 0.25% (w/v) glutaraldehyde, 0.5% (w/v) Triton X-100, pH 6.1]. They are then rinsed in cytoskeleton buffer and processed further at room temperature according to the following schedule: 1% glutaraldehyde in cytoskeleton buffer, 10–15 min; 0.5% Triton X-100 in cytoskeleton buffer, 10 min; 0.5 mg of $NaBH_4$ per milliliter freshly dissolved in cytoskeleton buffer, three times 5–10 min each (to reduce residual aldehydes)[22]; cytoskeleton buffer, twice, 5 min each time.

A unique feature of this procedure[23,24] is that cells are fixed first with low concentrations of glutaraldehyde in the presence of the permeabilizing

[22] K. Weber, P. Rathke, and M. Osborn, *Proc. Natl. Acad. Sci. U.S.A.* **75** 1820 (1978).
[23] J. V. Small and J. E. Celis, *Cytobiologie* **16** 308 (1978).
[24] J. DeMay, M. Joniau, M. De Brabander, W. Moens, and G. Guens, *Proc. Natl. Acad. Sci. U.S.A.* **75** 1339 (1978).

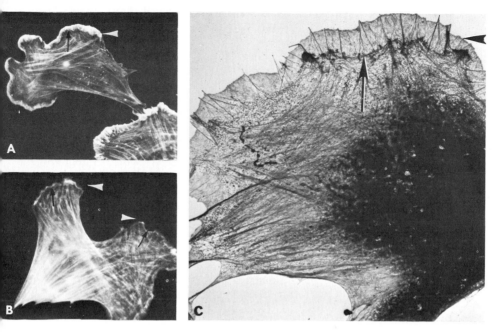

FIG. 1. Effect of fixation conditions on fluorescent staining patterns. The distribution of filamin in chick heart fibroblasts was studied using the indirect immunofluorescent staining method in cells fixed by either of the following two procedures: (A) glutaraldehyde–Triton X-100 mixture method (see text); (B) glutaraldehyde fixation followed by Triton X-100 permeabilization.[23] Note that in (A) the entire ruffle is stained brightly by filamin antibody. In contrast, in (B) only the base portion of the ruffle is stained. The fine structure of cells fixed by the glutaraldehyde-Triton X-100 mixture method is shown in (C) after uranyl acetate staining. Arrows indicate the base of the ruffle; arrowheads indicate the margin of the ruffle. Reproduced by courtesy of J. V. Small, unpublished results, 1981.

agent Triton X-100, followed by additional fixation and permeabilization. Although involved, this procedure appears to preserve gross morphology as well as fine ultrastructure, especially the microtubule network[24] and the delicate network of microfilaments at the leading edge of fibroblasts.[23] This procedure, therefore, is promising for correlating fluorescent images and ultrastructural features.

An analogous procedure[25] uses glutaraldehyde fixation (without Triton X-100), followed by Triton X-100 permeabilization and $NaBH_4$ reduction. A comparison of filamin staining patterns of cells fixed by these two procedures is shown in Fig. 1. The dramatic difference in the stain intensity in the ruffle region demonstrates the need to compare carefully several fixation protocols in pilot experiments.

Suspension Cells. For cells grown in suspension, it is convenient to fix cells in solution, remove the cells from fixative by pelleting in a clinical centrifuge, and then allow the cells to attach to polyornithine- or polylysine-coated coverslips for subsequent processing.

Fluorescent Conjugates[24a]

The preparation and characterization of high quality fluorescent conjugates of antibody of required specificity and affinity is perhaps the most time-consuming and crucial step in fluorescent localization work. The antibody preparation should be monospecific so that only the antigen under study is recognized, and it should have high affinity binding so that nonspecific (nonimmunological) binding to cellular structure is negligible. The conjugation procedures should be efficient, reproducible, and gentle to preserve the specificity and affinity of the antibody. A detailed discussion is beyond the scope of this chapter. The reader is referred to a comprehensive review[1] and to treatises of immunological methods for details.[6-9,25] Only general outlines and comments about procedures are described below.

Preparation of Monospecific Antibody

POLYCLONAL VS MONOCLONAL ANTIBODY

Conventional polyclonal antibodies produced in animals generally consist of a heterogeneous mixture of antibodies with different affinities to the various antigenic determinants present in the antigen molecule. In contrast, a monoclonal antibody produced by hybridomas is a homogeneous preparation with uniform affinity to only one of the many antigenic determinants. The defined and restricted specificity of monoclonal antibodies is highly desirable in antibody labeling techniques.[26] The staining pattern by monoclonal antibodies reflects, however, the distribution of a *single determinant,* not of the entire molecule. Furthermore, such labeling is much more susceptible to conformational and structural changes resulting from experimental manipulations. In contrast, polyclonal antibodies, being a mixture to all determinants, offer a better chance in revealing the distribution of the *molecule* because it is unlikely that all determinants would be damaged equally during the staining process. In this connection, a defined mixture of appropriate monoclonal antibodies may achieve the same goal. The procedures for preparing and characterizing polyclonal antibodies are discussed briefly below; the preparation of monoclonal antibodies can be found in treatises on hybridomas.[27]

[24a] By K. Wang.
[25] H. Van Vunakis and J. J. Langone, eds., this series, Vol. 70.
[26] J. J. Lin, *in* "Cell and Muscle Motility" (R. M. Dowben and J. W. Shay, Eds.), Vol. 2. Plenum, New York, 1982.
[27] R. H. Kennett, T. J. McKearn, and K. B. Bechtol, "Monoclonal Antibodies." Plenum, New York, 1980.

Antigen. Only highly purified proteins should be used for immunization. Even "pure" protein preparations may contain up to a few percent of contaminants that, if strongly antigenic, may elicit unwanted antibodies. For this reason, additional purification with preparative SDS gel electrophoresis is highly desirable. Antibodies elicited to SDS-denatured antigens are reported to be directed to determinants of both native and denatured molecular structure.[28]

Animals. Rabbits, goats, and guinea pigs are popular animals for immunization. Prior to immunization, preimmune sera from these animals should be screened by indirect-staining of cells (see below) to ensure the absence of autoantibodies to cellular structures. A high proportion of rabbits and goats that we and others have screened contain autoantibodies to certain cytoskeletal elements.[29,30]

Antibody Characterization. The specificity and titer of antisera should be established by as many procedures as is feasible. The following methods are particularly useful: (*a*) double diffusion and immunoelectrophoresis in the presence or the absence of denaturants or detergents[31,32]; (*b*) solid-phase immunoassay with microtiter plates[33]; (*c*) antibody staining of gel[34] or of a replica of gel electrophoretic protein patterns (western blot technique)[35,36]; (*d*) immunoprecipitation of antigen from SDS-solubilized cellular proteins in the presence of Triton X-100.[37,38] The last two methods are powerful and informative because both techniques simultaneously establish both the presence of specific antibody and the chain weight of the antigen and any cross-reacting material. All of the above methods circumvent the difficulties encountered in applying conventional immunological techniques to insoluble or aggregation-prone contractile proteins.

Antibody Purification. Either IgG fractions that have been purified by salt fractionation and DEAE-cellulose ion exchange chromatography[6-9] or affinity-purified specific IgG routinely should be used for conjugation or for indirect staining (see below). If serum quantity is limited, IgG fraction

---

[28] K. Weber and M. Osborn, *in* "The Proteins" (H. Neurath and R. L. Hill, eds.), 3rd ed., Vol. 2, pp. 1–45. Academic Press, New York, 1976.
[29] W. G. Gordon and K. Burridge, *Cell* **12** 671 (1977).
[30] K. Wang, unpublished observations.
[31] K. Wang, J. McClure and A. Tu, *Proc. Natl. Acad. Sci. U.S.A.* **76** 3698 (1979).
[32] R. E. Kessler, *Anal. Biochem.* **116** 129 (1981).
[33] T. T. Tsu and L. A. Herzenberg *in* "Selected Methods in Cellular Immunology" B. B. Mischell and S. M. Shiigi, Ed. Chapter 18 (1980). W. H. Freeman, San Francisco.
[34] K. Burridge, this series, Vol. 50, p. 54.
[35] H. Towbin, T. Staehelin, and J. Gordon, *Proc. Natl. Acad. Sci. U.S.A.* **76**, 4350 (1979).
[36] J. Renart, J. Reiser, and G. R. Stark, *Proc. Natl. Acad. Sci. U.S.A.* **76** 3116 (1979).
[37] S. H. Blose and D. I. Meltzer, *Exp. Cell Res.* **135**, 229 (1981).
[38] G. Krohne and W. W. Franke, *Proc. Natl. Acad. Sci. U.S.A.* **77** 1034 (1980).

can be purified in high yield by the Affi-Gel blue method.[39] Crude IgG prepared by salt fractionation is contaminated with serum proteases or complement, which may lead to degradation of cellular structure or complement-mediated cell lysis. When serum is to be used directly, complement should be heat-inactivated (60° for 30 min).[40]

Pure monospecific antibody can be purified from IgG fractions (or directly from antiserum) by affinity chromatography using solid immunoabsorbent.[41] Affinity purification becomes necessary when antibody titer is low or when the antiserum is multispecific; i.e., it contains irrelevant antibodies that stain cellular structures. The procedure is also used to prepare "antigen absorbed" antibody (the flowthrough material of the immunoabsorbent), which serves as an excellent control for nonspecific staining (see Tables IV and V). It is essential that only highly purified antigen be attached to the solid support in preparing the immunoabsorbent because the immunological specificity and purity of the purified antibody are directly determined by the purity of the antigen. Antibody recovery varies widely, depending on the affinity of the antibody–antigen complexes and the strength of the denaturants applied to dissociate them. Some commonly used eluents are 0.1 M glycine-HCl (pH 2.5),[41] 4 M sodium thiocyanate,[41] 2.5 M sodium iodide in 50 mM Tris-HCl (pH 9.0),[40] 4 M guanidine-HCl,[42] 4 M MgCl$_2$, 20 mM Tris-HCL (pH 6.8).[38] A portion of the recovered antibody is invariably inactivated by the eluents. In most cases, the purified antibodies are enriched in low-affinity antibodies because the high-affinity ones are firmly attached to the immunoabsorbent and cannot be dissociated by the eluents. This offsets, to some extent, the large gain in antibody titer. For this reason, if the contaminating antigens of a multispecific antiserum are known, the irrelevant antibodies should be absorbed away by immunoabsorbent attached with these antigens in order to yield a monospecific antiserum without lowering its affinity.

Immunoglobulins and Derivatives. In addition to IgG, other antibody reagents are also useful in special circumstances. The divalent fragment of IgG, F(ab')$_2$, which lacks Fc region,[43] is being increasingly used because the Fc region of IgG tends to bind nonspecifically to cellular structures. Its use is essential in labeling cells with endogenous Fc receptors (e.g., lymphocytes) to avoid staining due to the binding of IgG via the Fc region

[39] Bio-Rad Laboratories, Richmond, California.
[40] P. J. Lachmann and M. J. Hobart, in "Handbook of Experimental Immunology" (D. M. Weir, ed.), 3rd ed., Chapter 5A. Blackwell, Oxford, 1978.
[41] S. Fuchs and M. Sela, in "Handbook of Experimental Immunology" (D. M. Weir, ed.), 3rd ed., Chapter 10. Blackwell, Oxford, 1978.
[42] S. Lowey, personal communication.
[43] D. R. Stanworth and M. W. Turner, in "Handbook of Experimental Immunology" (D. M. Weir, ed.), 3rd ed., Chapter 6. Blackwell, Oxford, 1978.

(which has no antigen-binding site) to cellular receptors. The smaller, monovalent Fab fragment of IgG[43] has not been used extensively, despite the fact that its small size may allow it to probe deeper into dense structures such as stress fibers. The importance of probe size in antibody labeling is nicely illustrated by the finding [44] that Fab is capable of labeling all myosin heads in skeletal myofibrils. The steric limitation for probes such as IgA and IgM is expected to be significant. A careful comparison of labeling patterns by probes of similar binding specificity, yet of different sizes and valences, may lead to useful geometric information about structure surrounding the antigen.

Other Binding Proteins. In addition to antibodies, two nonimmunological binding proteins are widely used in fluorescent localization studies: protein A and avidin. Protein A binds to the Fc regions of many IgG subclasses of a variety of species.[44a] Its fluorescent conjugates are a popular and convenient secondary reagent in indirect staining. Avidin is an egg white protein that binds tightly to biotin and any protein to which biotinyl groups are covalently attached. This constitutes the avidin–biotin complex technique.[45,46]

Fluorophores

Although a wide range of fluorophores are available, fluorescein and rhodamine derivatives are used almost exclusively because they absorb strongly and fluoresce intensely in the visible spectrum. Table I lists some popular fluorophores.

A major drawback of fluorescein is that its fluorescence fades rapidly owing to photochemical decomposition. However, additives such as n-propyl gallate[47] or p-phenylenediamine[48] in the mounting medium may effectively retard the photobleaching of fluorescein fluorescence. Rhodamines are much more resistant to photobleaching.

Preparation and Characterization of Conjugates

Conjugates used in fluorescent localization are of two types, classified according to the staining procedures: (*a*) in the direct staining method, the specific antibody or specific binding proteins are labeled by fluorophores; (*b*) in the indirect staining method (the "sandwich" technique) where two

[44] R. Craig and G. Offer, *J. Mol. Biol.* **102**, 325 (1976).
[44a] H. Spiegelberg, *Adv. Immunol.* **19**, 259 (1974).
[45] H. Heitzman and F. M. Richards, *Proc. Natl. Acad. Sci. U.S.A.* **71**, 3537 (1974).
[46] E. A. Bayer and M. Wilchek, *Trends Biochem. Sci.* (Pers. ed.), **3**, N257 (1978).
[47] J. Sadet, personal communications, 1981.
[48] G. D. Johnson and G. M. De C. Nogueira Araujo, *J. Immunol. Methods* **43**, 349 (1981).

TABLE I

COMMONLY USED FLUOROPHORES

Name (abbreviation)	Structure[a]	Suggested sources[b,c]	Comments
1. Fluorescein isothiocyanate (FITC)	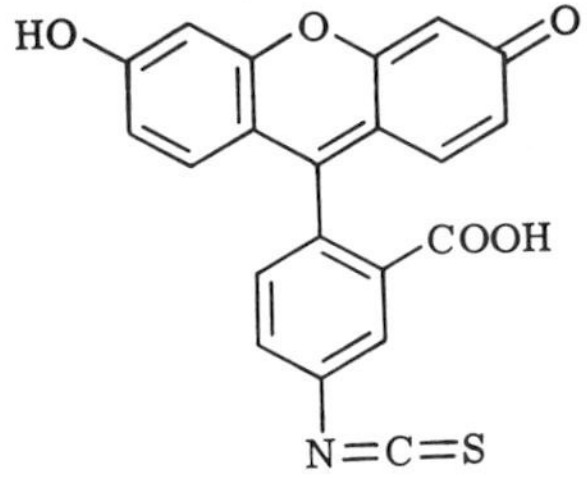 (fluorescein structure with COOH and N=C=S groups)	BBL, MP, S	The fluorescence of fluorescein derivatives Nos. 1 to 4: a. Have high quantum yield b. Increases about twofold when pH is increased from 6 to 8 c. Fades rapidly upon illumination
2. Dichlorotriazinyl-aminofluorescein (DTAF)	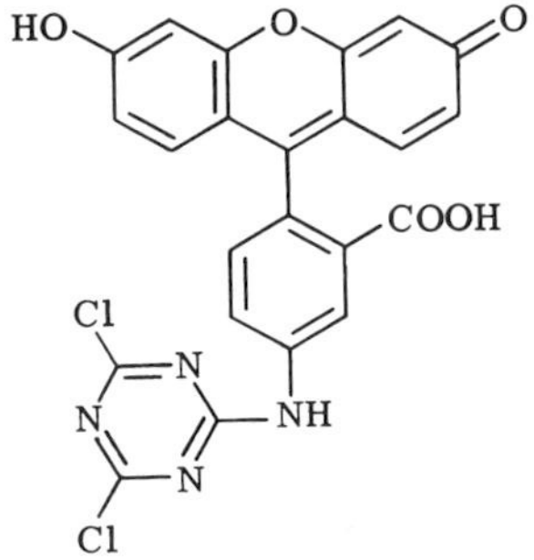 (fluorescein structure with COOH and NH-dichlorotriazine group)	RO, MP	
3. Iodoacetamide fluoroscein (IAF)	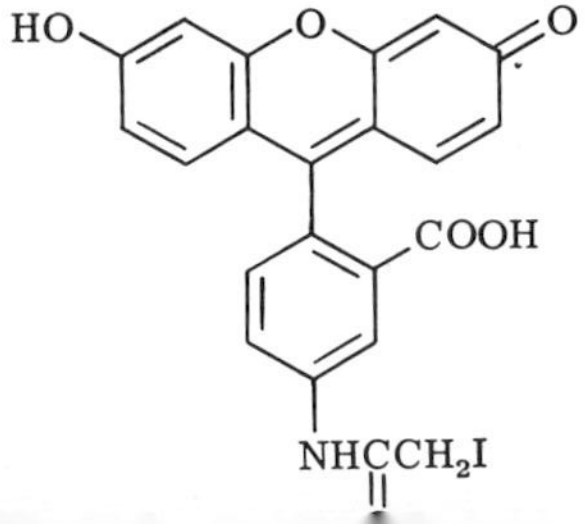 (fluorescein structure with COOH and NHCCH$_2$I group)	MP	Derivatives 3 and 4 (SH specific) are useful for specific labeling of SH-containing proteins for microinjection. (IgG has no free -SH)

4. Fluorescein maleimide

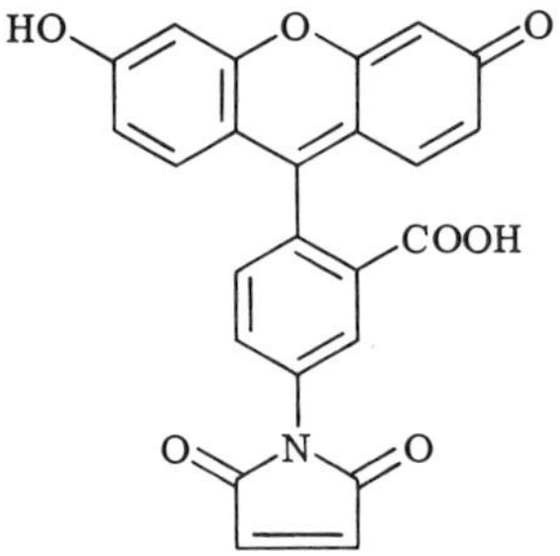

MP

5. Tetramethyl-rhodamine isothiocyanate (TRITC)

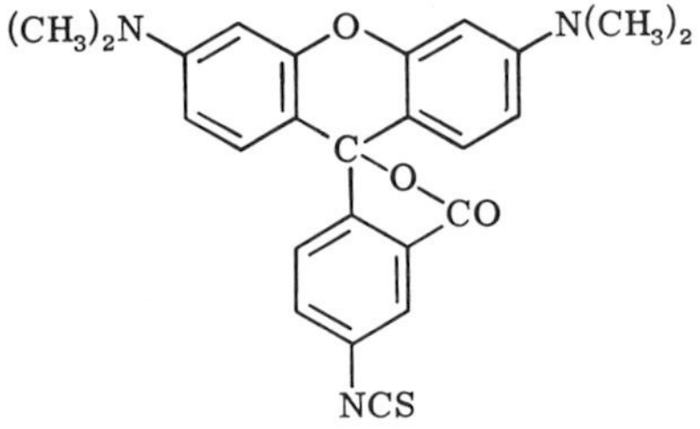

BBL, C

The fluorescence of rhodamine derivatives Nos. 5–8:
a. Have medium quantum yield
b. Does not vary between pH 6 and 8
c. Fades slowly

6. Lissamine rhodamine B sulfonyl chloride (RB200SC)

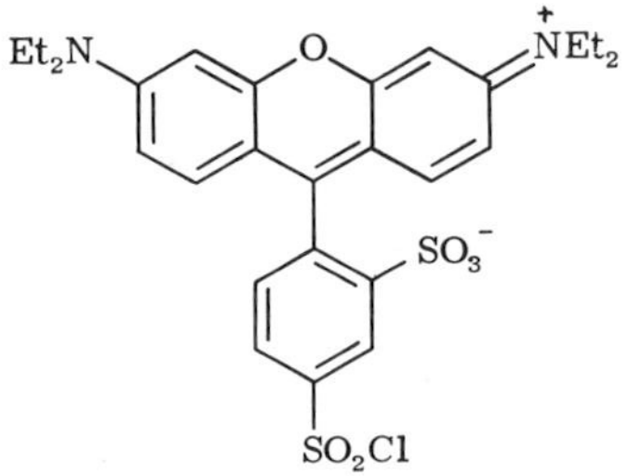

E, PS, MP

(continued)

TABLE I (*continued*)

Name (abbreviation)	Structure[a]	Suggested sources[b,c]	Comments
7. XRITC	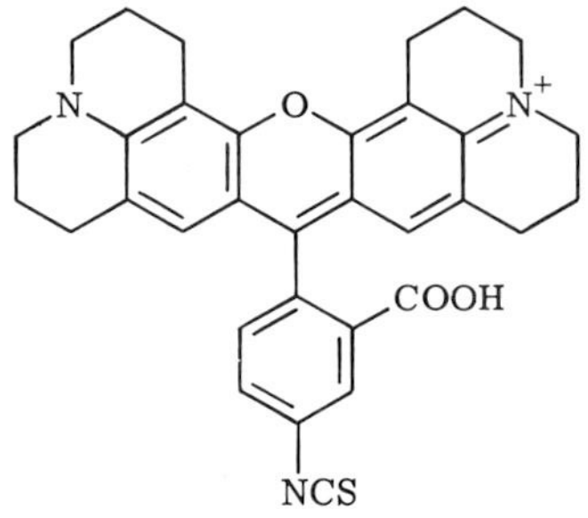	RO	Derivatives 7 and 8 are red-emitting dyes, ideal for dual labeling with fluoroscein
8. Texas Red (TxR)	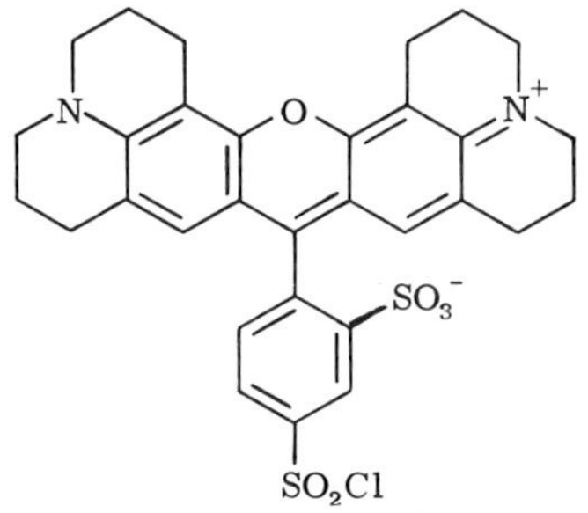	MP	

[a] Only one of the isomers is illustrated here.

[b] The quality of commercial fluorophores varies widely. The suggested sources list only those with which we have had first-hand experience and found satisfactory. Possible batchwise variation from the same source has not been tested: BBL, Baltimore Research Laboratory, Baltimore, Maryland; C, Cappel Laboratories, Cochranville, Pennsylvania; E, Eastman Kodak, Rochester, New York; MP, Molecular Probes, Plano, Texas; PS, Polysciences, Warrington, Pennsylvania; RO, Research Organics, Cleveland, Ohio; S, Sigma, St. Louis, Missouri.

[c] These reactive fluorophores should be stored in an evacuated desiccator at 4° in the dark. A shelf life of at least a year can be expected

reagents are used, the first (or "primary") reagent remains unlabeled whereas the second (or "secondary") reagent, which binds specifically to the primary reagent, is conjugated with fluorophores. In general, the secondary reagent is an antibody directed against the IgG of the species that provides the primary antibody. Thus, fluorescein-conjugated *anti-rabbit* IgG raised in goats is used to direct unconjugated primary antibody raised in *rabbits*. Similarly, fluorescent protein A and avidin are examples of secondary reagents.

The conjugation of proteins with fluorophores involves the addition of appropriate amounts of the fluorophore to a concentrated protein solution at an alkaline pH (8.5–9.5). The reaction conditions are adjusted so that optimal substitution by the dye molecules is obtained. At the end of reaction, excess dye and hydrolysis products are removed by gel filtration. The conjugates are fractionated further by anion exchange chromatography to remove under- and overconjugated proteins. Free dye and overconjugated proteins tend to bind to specimens nonspecifically, whereas underconjugated proteins decrease fluorescence by competing with optimally labeled proteins for binding sites.

The procedures for preparing optimally labeled IgG with FITC, TRITC, and RB200SC are now well established[7–9,49] and will not be presented here.

The conjugation procedures for Texas Red-IgG and fluorescent avidin are described in detail below as examples of conjugation procedures.

PREPARATION OF TEXAS RED CONJUGATED IgG[50]

Place goat IgG solution (10 mg/ml in 0.175 M sodium phosphate, pH 7.0) in a test tube over a magnetic stirrer at room temperature and raise the pH to 9.0 with 1 N NaOH. Open one sealed ampoule of Texas Red (1 mg per vial, Molecular Probes Inc., Plano, Texas) and dissolve the contents in 100 μl of anhydrous dimethylformamide (stored over molecular sieves). Add the fluorophore (4 μl/mg of IgG each time) in two aliquots over a 1-hr period with constant stirring. Readjust the pH to 9.0 if necessary. At the end of reaction, apply the mixture to a long Sephadex G-50 column (approximately 20 times the volume of the reaction mixture is needed for good resolution) equilibrated with 0.0175 M sodium phosphate, pH 6.3. The blue conjugate elutes in the void volume. The protein recovery is ~90%.

Under these conditions (reagent : protein weight ratio = 0.08 : 1) the conjugate has an average $A_{595}/A_{280} = 0.8$ (Table II).

[49] P. Brandtzeg, *Scand. J. Immunol.* **2**, 273 (1973).
[50] K. Wang and C. L. Williamson, unpublished procedure, 1981.

TABLE II

PHYSICAL PROPERTIES OF COMMONLY USED FLUORESCENT IgG CONJUGATES

Conjugate	Absorption maxima (nm)	Emission maxima (nm)	Absorbance ratios of conjugate suitable for staining	Conjugate protein concentration (mg/ml)
FITC-IgG[a]	495	515	$\dfrac{A_{495}}{A_{280}} = 0.5-1.5$	$\dfrac{A_{280} - 0.35\,A_{495}}{1.4}$
TRITC-IgG[a]	550, 515 (minor)	570	$\dfrac{A_{515}}{A_{280}} = 0.2-1.0$	$\dfrac{A_{280} - 0.56\,A_{515}}{1.4}$
RB200SC-IgG[a]	575	595	$\dfrac{A_{575}}{A_{280}} = 1.2-2.0$	$\dfrac{A_{280} - 0.32\,A_{575}}{1.4}$
TxR-IgG[b]	595	615	$\dfrac{A_{595}}{A_{280}} = 0.5-1.5$	—

[a] Data obtained from Brandtzeg.[49]
[b] TxR-IgG data: K. Wang, unpublished results (1981).

A detailed description of the reaction conditions and physical properties of Texas Red conjugated IgGs, lectins, and avidin is given by Titus et al.[51]

PREPARATION OF FLUORESCENT AVIDIN[52]

The protocol for the conjugation of avidin with fluorescein and rhodamine derivatives differs slightly from that of fluorescent IgG preparations owing to the high reactivity of avidin. Variability in both the degree of labeling and yield with different batches of avidin and different fluorophores have been experienced; it is suggested that pilot reactions at pH values from 7 to 9 be analyzed before allowing a large quantity of avidin to react with a particular dye. Useful fluorescein avidin conjugates contain 2–3 mol of dye per mole of avidin.[53]

Procedures. To 1 ml of avidin solution (Sigma, 10 mg/ml in 0.1 M NaCl), add 0.35 ml of the appropriate buffer (1 M sodium bicarbonate adjusted with saturated Na_2CO_3 for reactions from pH 8 to 9; 1 M potassium phosphate adjusted with 5 N NaOH for reactions from pH 7 to 8) and place on ice. Add the fluorophore (a total of 5 mol of dye per mole of avidin; $M_r = 67,000$, $E_{282}^{1\%} = 15.4^{54}$) in three aliquots over a 30-min period, mixing the solution after each addition or with continuous stirring. Add 70 μl of glycine solution (1 M glycine, pH 8.0) to terminate the

[51] J. A. Titus, R. Haugland, S. O. Sharrow, and D. M. Segal, in preparation.
[52] D. Louvard, J. F. Ash, and M. Heggeness, unpublished procedure.
[53] H. Heggeness and J. F. Ash, *J. Cell. Biol.* **73**, 783 (1977).
[54] N. M. Green, *Adv. Protein Chem.* **29** 85 (1975).

reaction, and then dialyze the solution against two 1-liter changes of 0.01 M Tris-HCl, pH 9.0 buffer overnight in the cold. After dialysis, centrifuge the solution at 10,000 g for 10 min to pellet any precipitated protein. If most of the avidin is precipitated at this stage, then conjugation procedures should be repeated with the pH of the reaction mixture reduced. Apply the supernatant to a small 5-ml QAE-Sephadex column equilibrated with 0.01 M Tris-HCl, pH 9.0. Wash the column with several volumes of the same Tris-HCl buffer and then elute the fluorescent avidin with steps of 0.1 M and 0.25 M NaCl in the Tris-HCl buffer. Dialyze the eluted fractions separately against PBS, and centrifuge to remove any precipitated protein.

Comments

1. Reaction conditions: Conjugation procedures frequently require modifications when IgGs from different species, or fluorophores of different batches, are used. Lyophilized proteins from commercial sources may contain chemicals with reactive amino groups or azide preservatives that interfere with the labeling reaction. These protein samples should be dialyzed prior to conjugation. The rapid mixing of amorphous fluorophore powder with protein solution is important for reproducible results, but this is sometimes tricky because the powder traps air. Rapid mixing can be achieved by sprinkling the powder on top of the protein solution in a centrifuge tube, quickly centrifuging the solution for a few seconds in a clinical centrifuge, followed by gentle vortexing or stirring. The reaction rate is highly pH dependent, so proper pH monitoring is important (pH electrode is preferred because pH indicator paper gives erroneous readings in the presence of high concentrations of protein).

2. Stability and storage of conjugates: Freshly prepared protein conjugates, especially those of rhodamines, tend to aggregate upon storage, presumably owing to slow denaturation or aggregation of highly conjugated population. Removal of such aggregates by centrifugation immediately prior to staining is important to avoid nonspecific staining. Conjugates can be stored for at least several months in the dark at 4° with 0.01% merthiolate or 0.02% NaN_3 as preservatives or at $-20°$ to 80° for at least 4 years. It should be noted that slow release of dye molecules upon storage has been observed. Periodic checking by SDS–gel electrophoresis (without tracking dyes) or repurification is advisable.

3. Physical characterization of conjugates: Table II lists some useful physical properties of several fluorescent IgG conjugates. The average degree of labeling, easily measured from the absorbance ratios of fluorophore to protein, is perhaps the most useful parameter to evaluate conjugates for staining. The most useful range of absorbance ratios of various conjugates (which have been fractionated on DEAE-cellulose to

narrow the distribution of degree of labeling) for direct and indirect stainings is included in Table II. It should be noted, however, that overconjugated IgG (higher absorbance ratio) may be usable at low conjugate concentration. Similarly, underconjugated IgG (lower absorbance ratio) may yield strong fluorescence at high protein concentrations.

4. Affinity-purified conjugates and antigen-absorbed conjugates: When conjugates of affinity-purified antibody are needed in the direct staining method, they should be prepared by affinity purification from a *conjugated IgG fraction*. In this manner, the unbound conjugate (after repeated passage through fresh immunoabsorbents) can then be used as the antigen-absorbed control in staining. This procedure yields specific and control conjugates with the same, or closely similar, degree of dye substitution. This uniformity is difficult to achieve by labeling separately affinity-purified antibody and absorbed antibody.

5. Fluorescent conjugates of IgG are commercially available from a number of sources. The authors have found the FITC- and TRITC-conjugated IgG from Miles Laboratories (Elkhart, Indiana) and Cappel Laboratories (Cochranville, Pennsylvania) (the latter also supplies conjugated F(ab')$_2$) to be generally satisfactory at the working dilutions recommended by the manufacturers. Extensive lines of avidin–biotinyl complex reagents are offered by Vector Laboratories (Burlingame, California) and by E-Y Laboratories (San Mateo, California). FITC protein A is available from Sigma (St. Louis, Missouri) and Pharmacia (Piscataway, New Jersey). TRITC protein A is available from E-Y Laboratories. The authors have no first-hand experience with these latter products.

Staining Procedures[54a]

Direct vs Indirect Staining

Direct staining, using directly conjugated specific antibody to an antigen, is simple to perform and easy to control.

Indirect staining, using a conjugated antibody directed against the (unconjugated) primary reagent, has two main advantages. First, it allows the use of only one conjugate for staining all primary antisera raised in the same species and avoids the task of conjugating and characterizing each antiserum. Second, it allows an appreciable gain in sensitivity resulting from the increase in the number of conjugates bound per antigen.

Direct Staining

The coverslip with the specimen is removed from the petri dish with a tweezer, quickly drained of excess buffer by touching the edge of a piece

[54a] By K. Wang.

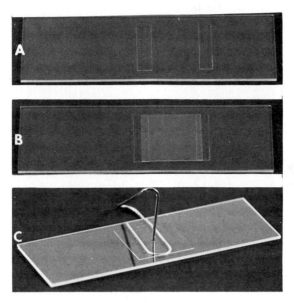

FIG. 2. A staining slide without (A) and with (B) a coverslip: The slide is constructed by gluing two parallel glass strips (cut out along the edge of No. 1 coverslip with a diamond pencil) onto a 1 × 3 inch slide. A minimal (~100 μl) volume of conjugates can be used to stain the entire surface area of the inverted coverslip without drying when the slide is incubated in a moisture chamber. (C) A paper clip press is used in applying gentle pressure to coverslip when viscous mounting media are used. A 1¼-inch paper clip is bent on one side as illustrated. The tension of the lifted arm can be easily adjusted by the degree of bending.

of filter paper, and inverted (with the cell side down) onto a drop (50–100 μl) of fluorescent conjugate in PBS on a staining slide (Fig. 2A and B). The slide is incubated in a moisture chamber at room temperature for 20–60 min. At the end of the incubation, the coverslip is removed from the slide and immersed in 5 ml of PBS in a petri dish. The coverslip is washed three times, 5 min each time, with 5 ml of PBS. (Avoid adding buffers directly to the coverslips or cells will be dislodged.) The coverslip is lifted occasionally to wash the underside. After excess PBS is drained off with a filter paper, the coverslip is mounted on a slide (with the cell side down) on a drop of mounting medium, e.g., Aquamount. A paper clip press (Fig. 2C) is applied, and excess mountant is removed with a piece of filter paper. Sliding motions of the coverslip should be avoided to protect cellular morphology. Sample slides are air-dried with the aid of a blow drier. The upper coverslip surface is then wiped with a cotton swab to remove salt residues or cells that have grown on the underside of the coverslip in the culture dish. The slide is ready for observation and can be stored in a

TABLE III

DIRECT STAINING METHOD: SAMPLE PREPARATIONS AND TROUBLE-SHOOTING[a]

Sample No.	Cell preparation treated with[b]	Desired results	Symptoms	Possible causes
1	Buffer	No fluorescence	Strong fluorescence	a. Autofluorescence b. Inefficient filters c. Fluorescence induced by fixatives
2	Conjugated specific IgG	Bright fluorescence	Dim or no fluorescence	a. Low-titer antibody b. Weakly cross-reacting antibody c. Dilute or/and underconjugated antibody d. Harsh fixation
3	Conjugated normal IgG[c]	No fluorescence	Strong fluorescence	a. Autoantibodies to cellular material b. Fc receptors in cells c. Concentrated or/and overlabeled conjugates d. Drying during staining e. Cell debris
4	Conjugated specific IgG, preabsorbed with antigen	No fluorescence	Strong fluorescence	a. Multispecific antibodies b. Autoantibodies to cellular material c. Fc receptors in cells d. Concentrated or/and overlabeled conjugates e. Drying during staining f. Cell debris
5	Unlabeled specific IgG, followed by (or mixed with) conjugated specific IgG	No (or diminishing) fluorescence	Strong fluorescence	a. Exchange of unlabeled and conjugated antibodies b. Incomplete blocking by the unlabeled antibody c. Concentrated or/and overlabeled conjugates d. Drying during staining e. Cell debris

[a] For single labeling and for each antigen in dual labeling.

[b] All antibody and conjugate solutions are used at the *same* concentration, and all conjugates have a comparable extent of labeling.

[c] Several different batches of normal sera, including preimmune serum, should be used.

light-tight box at room temperature. A series of slides, including appropriate controls, are processed in parallel (Table III).

COMMENTS

1. It is important that the coverslip does not become dry at any stage of the staining procedure, because concentrated or dried reagents and conjugates stain cellular structures nonspecifically. A moisture chamber can be easily prepared by placing wet filter paper in a petri dish or by placing wet paper towels in a transparent plastic box (e.g., Cole Parmer No. 6818-30) with wooden applicator sticks as the support. For large numbers of slides, a glass thin-layer chromatography storage box (Brinkmann, New York) is useful. To prevent bleaching during incubation, the chambers should be protected from sunlight or strong fluorescent light.

2. Occasionally the wrong side of the coverslip may inadvertently be stained or mounted, which results in a slide preparation with either unstained cells or no cells at all. To help keep track of the cell side, the coverslip may be prescratched on one side near a corner, or frequently, the cell side can be easily identified by its mottled appearance under reflection.

3. The optimal concentration of conjugates for direct labeling is determined empirically by staining the cells with a series of concentrations of specific conjugates as well as control conjugates. The lowest concentration at which specific fluorescence is acceptable and nonspecific fluorescence is absent or tolerable is then chosen for further work. Generally, 10–100 μg of affinity-purified antibody conjugate per milliliter and 0.1 to several milligrams of conjugated IgG fraction per milliliter are used.

4. The choice of mounting medium requires some care, as it may affect the fluorescence intensity, the stability of the antibody–antigen complexes, as well as the transmitted light images. A popular one appears to be phosphate-buffered glycerol prepared by mixing 9 volumes of glycerol and 1 volume of PBS, pH 7.2–7.4. Because fluorescein emission is stronger at higher pH (>8.0), glycerol buffered at pH 8.0–8.5 may be desirable in certain cases. The use of semipermanent mounting media that solidify has obvious advantages. Aquamount (pH 8.0, Lemer Laboratory, New Haven, Connecticut) and buffered Gelvatol[55] [16% (w/v) Gelvatol 20–30 (Monsanto, St. Louis, Missouri); 33% (v/v) glycerol, 0.14 M NaCl, 0.01 M potassium phosphate, 0.1% NaN_3, final pH 7.2] are both nonfluorescent, poly(vinyl alcohol)-based mountants. Reagents that retard the photobleaching of fluorescein, e.g., n-propyl gallate[47] or p-phenylenediamine,[48] are generally added, if needed, to the mountants.

[55] J. R. Feramisco and S. H. Blose, *J. Cell Biol.* **86**, 608 (1980).

Indirect Staining

This is the most widely used method of staining because of its sensitivity and convenience. The coverslip is first incubated with unlabeled specific IgG or antiserum, washed thoroughly with PBS and then stained with conjugated antibody directed against IgG of the species in which the primary antibody is produced. The cells are washed and mounted. The incubation and coverslip washing is carried out in exactly the same manner as described for the direct method. Appropriate control samples (Table IV) are processed in parallel.

COMMENT

Optimal concentrations of both the primary antibody and the conjugates are determined by titrations. In general, a concentration range of 0.1 to 0.5 mg of conjugated antiglobulin second reagent (cf. Table II) is satisfactory for most applications. The optimal concentration of the primary antibody varies widely depending on its titer and affinity.

Single Labeling vs Dual-Labeling

The dual-labeling technique is used when the relative distribution of two antigens or proteins is to be compared in the same cell. These techniques are particularly important in fluorescent localization studies of contractile proteins in nonmuscle cells because the contractile machineries in these cells are irregular and constantly changing. Comparisons of widely varying labeling patterns of cells of different shape are difficult and frequently misleading. The application of dual-labeling techniques has made significant contributions in our understanding of transmembrane interactions,[56-58] organelle-cytoskeleton attachment,[59] interactions of microtubules and intermediate filaments,[60] and differential distribution of contractile proteins in membrane ruffles.[61,62]

Dual-Labeling Techniques

To reveal simultaneously the distribution of two antigens in the same cell, cells are stained with two sets of specific reagents labeled with con-

[56] I. I. Singer and P. R. Paradiso, *Cell* 24 481 (1981).
[57] J. F. Ash, D. Louvard, and S. J. Singer, *Proc. Natl. Acad. Sci. U.S.A.* 74 5584 (1977).
[58] B. Geiger and S. J. Singer, *Cell* 16 213 (1979).
[59] M. H. Heggeness and S. J. Singer, *Proc. Natl. Acad. Sci. U.S.A.* 76 2185 (1978).
[60] B. Geiger and S. J. Singer, *Proc. Natl. Acad. Sci. U.S.A.* 77 4769 (1980).
[61] M. H. Heggeness, K. Wang, and S. J. Singer, *Proc. Natl. Acad. Sci. U.S.A.* 74 3883 (1977).
[62] I. M. Herman, N. J. Criscona, and T. D. Pollard, *J. Cell Biol.* 90 84 (1981).

TABLE IV

INDIRECT STAINING METHOD: SAMPLE PREPARATIONS AND TROUBLESHOOTING[a]

Sample No.	Cell preparation treated with[b]		Desired result	Symptoms	Possible causes
	Step 1	Step 2			
1	Buffer	Buffer	No fluorescence	Strong fluorescence	a. Autofluorescence b. Inefficient filters c. Fluorescence induced by fixatives
2	Specific IgG	Conjugated anti-IgG	Bright fluorescence	Dim or no fluorescence	a. Low-titer antibody b. Weakly cross-reacting antibody c. Dilute or/and underconjugated antibody d. Harsh fixation
3	Normal IgG[c]	Conjugated anti-IgG	No fluorescence	Strong fluorescence	a. Autoantibodies to cellular material b. Fc receptors in cells c. Concentrated or/and overlabeled conjugates d. Drying during staining e. Incomplete washing of primary antibody
4	Specific IgG preabsorbed with antigen	Conjugated anti-IgG	No fluorescence	Strong fluorescence	a. Multispecific antibodies b. Autoantibodies to cellular material c. Fc receptors in cells d. Concentrated or/and overlabeled conjugates e. Drying during staining f. Incomplete washing of primary antibody
5	Buffer	Conjugated anti-IgG	No fluorescence	Strong fluorescence	a. Autoantibodies to cellular material b. Fc receptors in cells c. Concentrated or/and overlabeled conjugates d. Drying during staining e. Cell debris

[a] For single labeling and for each antigen in dual labeling.
[b] All primary antibody solutions are used at the same IgG concentration. All conjugate solutions are used at the same concentration.
[c] Several different batches of normal sera, including preimmune serum, should be used.

TABLE V
EXAMPLES OF DUAL-LABELING TECHNIQUES

Method	Staining method	Staining reagents	Protein stained
1[a]	Direct	FITC rabbit anti-myosin	Myosin
	Direct	Rh rabbit anti-tubulin	Tubulin
2[b]	Indirect	Rabbit anti-tubulin, FITC goat anti-rabbit IgG	Tubulin
	Indirect	Guinea pig anti-desmin, Rh goat anti-guinea pig IgG	Desmin
3[c]	Indirect	Rabbit anti-actin, FITC goat anti-rabbit IgG, normal rabbit IgG	Actin
	Direct	Rh rabbit anti-fibronectin	Fibronectin
4[d]	Indirect (nonimmunological)	Biotinyl heavy meromyosin,[e] FITC-avidin	Actin
	Indirect	Rabbit anti-filamin, Rh goat anti-rabbit IgG	Filamin

[a] K. Fujiwara and T. D. Pollard, *J. Cell Biol.* **70** 181a.
[b] B. Geiger and S. J. Singer, *Proc. Natl. Acad. Sci. U.S.A.* **77**, 4769 (1980).
[c] Hynes and Destree.[63]
[d] M. H. Heggeness *et al.*[61]
[e] Prepared from rabbit skeletal myosin, which does not cross-react with nonmuscle myosin.

trasting fluorophores. Each of the two antigens can be stained by either the direct or the indirect method. A major consideration in dual labeling is to ensure that the two antigens are *independently* labeled without immunological, steric, and optical interferences from each other.

So far, fluorescein and rhodamine pairs have been used almost exclusively as the contrasting fluorophores for dual labeling of cytoskeletal elements (see the section Fluorescent Conjugates). Table V lists several examples of dual-labeling techniques. The choice of methods depends not only on the sensitivity desired, but also on the quantity and the host species of antisera.

Direct–Direct Stainings. The use of two direct conjugates is straightforward and is preferred if a sufficient quantity of antisera is available for conjugation and characterization, and if the lower sensitivity of the direct method is acceptable.

Indirect–Indirect Stainings. Indirect methods may be required to increase sensitivity. When both antigens are stained by the indirect methods, four antigen–antibody reactions are involved; the selection of reagents can be fairly confusing and careful planning is called for here to avoid cross-labeling of antigens.

One popular method (Table V, Method 2) involves raising primary antibodies in two different animal species, such as rabbit and guinea pig, whose IgGs do not cross-react immunologically. Antiglobulin antibodies to each of the two first species obtained from a third animal species (e.g., goat) are conjugated and used as the secondary reagents. To ensure that no undesirable cross-labeling would occur, the two conjugates can be affinity purified, and furthermore, be absorbed with heterologous immunoadsorbent. For example, fluorescein-conjugated goat anti-rabbit IgG is affinity purified on a rabbit IgG-containing absorbent and then passed through a guinea pig IgG-containing absorbent. Rhodamine-conjugated goat anti-guinea pig IgG is treated in an analogous fashion.

Direct–Indirect Stainings. When both primary antibodies are raised in the same or immunologically cross-reacting species (e.g., rat and mouse), and only one is sufficient for direct labeling, the following method[63] is applicable (Table V, Method 3): One antigen is stained first by the indirect method *prior to* the application of the direct conjugates to stain the second antigen. Normal (nonimmune) globulin or serum of the species that provides the primary antisera is applied to saturate all available IgG binding sites on the conjugated antiglobulins. Then the second conjugated antibody is applied. As can be seen, the correct sequence—indirect staining, then blocking, then direct staining—is crucial for successful dual staining by this method; reversing the sequence would cause both antigens to be labeled with the conjugated antiglobulin. One technical drawback of this procedure is that antibody–antigen complex may dissociate and interchange, thus resulting in the labeling of the second antigen by the conjugated antiglobulin. In practice, this possible cross-labeling does not seem to occur at a significant rate. Postlabeling fixation may circumvent this difficulty. In any case, specimens labeled by this method should be examined at various times after labeling to assess this possibility.

Nonimmunological Methods. Nonimmunological staining methods, such as the avidin–biotin complex technique,[50,51] fluorescent derivatives of phallacidin[63a] or S_1[63b] for actin localization, can be combined easily with an immunological method for dual labeling without introducing complications of cross-labeling.

Staining Procedures

Staining of the two antigens can be carried out either sequentially or simultaneously with a mixture of reagents at each step, according to the procedures described for single labeling. Two comments are in order.

1. A series of titrations varying the *relative concentrations* of the two sets of reagents are frequently needed in dual labelings. This is due to the

[63] R. O. Hynes and A. T. Destree, *Cell* 15 875 (1978).
[63a] L. S. Barak, R. R. Yocum, E. A. Nothnagel, and W. W. Webb, *Proc. Natl. Acad. Sci. U.S.A.* 77, 980 (1980).
[63b] J. A. Schloss, A. Milsted, and R. D. Goldman, *J. Cell Biol.* 74, 794 (1977).

fact that the filter systems used to distinguish specific fluorescences are not perfect, and strong emissions from one fluorophore may cross over and be detected in the filters designed for the other fluorophore (see section Fluorescence Microscopy and Fig. 7).

2. The *sequence* of labeling should be carefully evaluated because the binding of one set of labeling reagents may impede the binding of the second set of reagents to their target. An example of such a phenomenon in dual labeling is illustrated in Fig. 3. In this experiment,[64] the prior labeling of actin with high concentrations of reagents completely abolished the subsequent labeling of myosin to the "stress fibers," allowing only the more diffuse form of myosin staining to be observed. In contrast, staining with a reversed order or staining simultaneously with mixed reagents led to the labeling of stress fiber by both actin and myosin. This blocking phenomenon immediately suggests that the two antigens are proximate within a distance approximated by the dimensions of the reagents. The molecular "ruler" aspect of the dual labeling deserves further exploration because it provides structural information as to the molecular dimensions, far beyond the limit of resolution of the light microscope.

Staining Controls

Controls are essential for interpreting staining patterns because of the presence of "unwanted" fluorescence, i.e., fluorescence not resulting from the particular antigen–antibody interaction under study. Many sources contribute to such "noise": autofluorescence of the cell; nonspecific (nonimmunological) binding of conjugates to cellular structure; specific (immunological) binding of contaminating antibodies of a multispecific antiserum to cellular antigens; specific binding of conjugates to nonspecifically adsorbed primary antibody (in indirect staining) and binding of conjugates to Fc receptors of certain cells. A variety of diagnostic controls are thus needed to assess the extent of unwanted fluorescence.

Some useful controls for direct staining, indirect staining as well as for dual labeling are listed in Tables III, IV, and VI, respectively. It is recommended that *all* controls should be performed in pilot experiments whenever new batches of conjugates, reagents, and different cells are utilized, and *some* should be carried out routinely.

It is not essential, and indeed it is rare, that all controls are completely unstained when observed with a modern epifluorescence microscope, contrary to the impression conveyed by many black control micrographs

[64] K. Wang, unpublished observations, 1977.

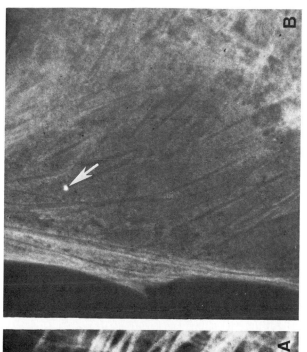

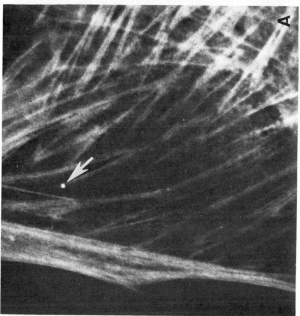

Fig. 3. Effect of labeling sequences on staining patterns of dual-labeled cells. The distribution of actin and myosin in the same human skin fibroblast cell was studied by the avidin–biotinyl heavy meromyosin and indirect anti-myosin procedures (Method 4 in Table V). In the micrographs shown, actin was stained first (A) followed by myosin staining (B). Note that stress fibers are brightly stained in (A) yet are completely dark in (B). Cell debris (arrowhead), brightly stained, serves as a convenient position marker for comparing patterns. In other experiments where myosin was stained first or when actin and myosin were stained simultaneously by mixed reagents, stress fibers were stained brightly by both actin and myosin reagents (K. Wang, unpublished results, 1977).

TABLE VI

OPTICAL AND STERIC CONTROLS FOR DUAL LABELING

Sample No.	Cell preparation treated with[a]	Desired results	Symptoms	Possible causes[b]
1[c]	FITC aX	Bright fluorescence in FITC optics; no fluorescence in Rh optics	Strong fluorescence in both optics	a. Optical crossover due to inefficient Rh filters b. Conjugate too concentrated for the Rh filters
2a	1. FITC aX 2. Rh aY	Identical FITC or Rh images	Different FITC or Rh images	Steric interference due to the close proximity of X and Y
2b	1. Rh aY 2. FITC aX			
2c	A mixture of FITC aX and Rh aY			
3[c]	1. aX 2. FITC algGx	Bright fluorescence in FITC optics; no fluorescence in Rh optics	Strong fluorescence in both optics	a. Optical crossover due to inefficient Rh filters b. Conjugate too concentrated for the Rh filters
4[c]	1. aY 2. algGy (unlabeled) 3. buffer 4. FITC algGx	No fluorescence in FITC optics	Strong fluorescence in FITC optics	Binding of FITC algGx to either aY or/and to algGy

5a	1. aY 2. Rh algGy 3. aX 4. FITC algGx	Identical FITC or Rh images
5b	1. aX 2. FITC algGx 3. aY 4. Rh algGy	Different FITC and Rh images
5c	1. mixture of aX and aY 2. mixture of FITC-a IgGx and Rh algGy	Steric interference due to the close proximity of X and Y

[a] Two antigens, X and Y, are stained in the same cell preparation with the following sets of specific reagents at concentrations that yield specific labeling (as established by tests listed in Tables III and IV): Samples 1 and 2, direct–direct staining: FITC aX, FITC-conjugated IgG directed against X; Rh aY, rhodamine-conjugated IgG directed against Y. Samples 3–5, indirect–indirect staining: aX, unconjugated IgG directed against X; aY, unconjugated IgG directed against Y; FITC algGx, FITC-conjugated IgG directed against the IgG of the species in which aX is raised; Rh algGy, rhodamine-conjugated IgG directed against the IgG of the species in which aY is raised.

[b] Consult Tables III and IV for additional causes related to reagents and cell preparation.

[c] Test Rh-conjugates and FITC filters in a similar fashion.

seen in the literature. Many controls are too dim to be recorded, and low-level staining is occasionally eliminated during reproduction of micrographs. Since it is the intensity *difference* between the sample and the controls that determines the specific staining of the antigen under study, the presence of dim fluorescence in controls is acceptable if specific fluorescence is bright. The manipulation of micrographs is not unjustified so long as both sample and controls are recorded and processed *identically*. However, when unwanted fluorescence is strong and approaching that of specific staining, appropriate measures should be taken to reduce such fluorescence. Tables III, IV, and VI, which list symptoms and possible causes, may be consulted for this purpose.

We have discussed, throughout the chapter, various means to prepare high quality staining reagents that are essential to reduce unwanted fluorescence. Certain manipulations of cell preparations may also be useful at times: varying the fixation or fixation conditions; raising pH of the staining buffer; pretreating the permeabilized cells with inert proteins, such as bovine serum albumin or normal serum (of a species that does not cross-react with *any* of the reagents).

Fluorescence Microscopy[64a]

A fluorescence microscope is used for observing and recording fluorescent images of labeled specimens. The basic instrument consists of a suitable light source, an illumination system, objective lens, filter combinations tailored to detect each fluorophore, and image-recording devices for documentation. Rapid progress has been made in the design of fluorescence optical components. The development of incident light excitation (epifluorescence), the availability of highly selective interference filters, and the introduction of image intensification devices and electronic image analysis systems are among important advances that have greatly improved the sensitivity of detection and the ease of operation and documentation. Most modern fluorescence microscopes can be so sensitive that for many fluorescent labeling and localization studies the undesirable nonspecific emission from some carefully prepared control specimen can be dishearteningly vivid.

A detailed discussion is certainly beyond the scope of this chapter. However, a general understanding of the basic design and operating principles of fluorescence optics is crucial to experimental design, to troubleshooting, and to interpreting fluorescence images; We therefore discuss below the essential features and the operation of each of the major fluorescence optical components.

[64a] By K. Wang.

Fluorescence Microscope

Commercial Source. Most epifluorescence devices are of modular design and can be purchased as a package of accessories for many microscopes from major manufacturers, such as Zeiss, Leitz, Nikon, Olympus, Reichert, and American Optical. The cost of these accessories varies (from $5,000 to $10,000), as do the quality and the ease of operation.

Light Source. The direct current-operated high-pressure mercury arc lamps (HBO100, HBO50) and xenon arc lamps (XBO75) are preferred light sources for a wide range of applications. Unlike the alternating current-operated lamps (such as HBO200) whose intensity fluctuates frequently, these lamps have stable, extremely high intensities and are ideally suited for epifluorescence and quantitative fluorimetry. Mercury bulbs emit a series of high-energy line spectra in the UV and visible ranges (up to about 600 nm) which is superimposed on a continuous light emission spectrum in a similar range. With these sources, fluorescein is selectively excited near 450–490 nm by the continuous mercury emission, whereas rhodamine is selectively excited by either the 546 or 578 nm mercury lines. Xenon lamps emit a continuous spectrum covering the entire UV and visible range, which provides higher light intensity for fluorescein excitation and lower intensity for rhodamine excitation than do mercury bulbs.

Epifluorescence Illuminator and Filter Combination. A schematic diagram of an epifluorescence illuminator is presented in Fig. 4. An exciter filter selects and passes only the light frequency needed for stimulating fluorescence emission of the fluorophore under study. An interference mirror (chromatic beam splitter) is positioned at a 45° angle to the illuminating beam to reflect downward all the incident light below a specified wavelength through the objective lens to excite the specimen. The fluorescent emission, collected by the same objective lens, passes through the beam splitter without reflection because it has higher wavelengths than the specified wavelength of the beam splitter. A third filter (barrier filter), located between the beam splitter and eyepieces, removes any residual excitation light and allows a specified portion of the fluorescence to pass through for observation. The vertical illumination system allows a wide range of transmitted light optics, such as phase contrast, Nomarski differential interference contrast, or interference reflection to be used either simultaneously or alternatively on the same field without exchanging optical components. Good morphological definition and precise fluorescence localization are thereby easily accomplished. More important, since the objective lens serves also as a condenser, a high numerical aperture objective concentrates the excitation light onto a very small

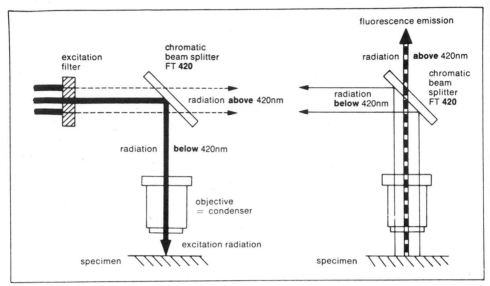

Excitation with chromatic beam splitter   Fluorescence with chromatic beam splitter

FIG. 4. A schematic diagram of the disposition and operating principles of exciter filter and chromatic beam splitter in a vertical illuminator. See text for details. By courtesy of Carl Zeiss, Inc.

sample area as well as collects a large fraction of the emitted fluorescence. An appreciable gain in fluorescence intensity is achieved as a result.

The selection of filters is crucial for successful fluorescence microscopy. Factors such as the absorption and emission spectra of the fluorophores, the spectral property of the light source, and the number of distinct fluorophores to be distinguished in the same field are among the important considerations. The widespread use of mercury or xenon light sources and the application of fluorescein and rhodamine conjugates in labeling have greatly simplified the selection of filters. Many filter combinations, each tailored for a specific application, are offered by microscope manufacturers. Table VII lists filter sets that we found satisfactory in single- or dual-labeling studies of contractile proteins in muscle and nonmuscle cells. For single-labeling studies, a wide latitude in selection is allowed. The selective narrow-band excitation near the absorption maxima of the fluorophore is preferred (sets 1 and 2 for fluorescein; set 3 for rhodamines). In situations where brighter fluorescence is required or in cases where autofluorescence is needed to define cell morphology, wideband excitation is useful (sets 4 and 6).

For dual-labeling studies, filter selection is far more restricted because

TABLE VII
USEFUL FILTER COMBINATIONS FOR EPIFLUORESCENCE

Set	Fluorophore	Exciter filter[a]	Chromatic beam splitter[a]	Barrier filter[a]
1[b]	Fluorescein	BP450-490	FT510	LP520
2[b]	Fluorescein[c]	BP450-490	FT510	LP520 + KP560 or BP520-560
3[b]	Rhodamine[c]	BP546/10	FT580	LP590
4[b]	Rhodamine[d]	BP515-560	FT580	LP590
5[b]	Texas Red-XRITC[e]	BP577/10	FT600	LP600
6	Fluorescein and rhodamine[f]	BG12	FT510	LP520

[a] BP: band-pass filter. The numbers designate the wavelength region (in nm) between which light is transmitted. FT: chromatic beam splitter. The number designates the wavelength region (in nm) below which light is reflected. LP: long-pass filter. The number designates the wavelength region (in nm) above which light is transmitted. KP: short-pass filter. The number designates the wavelength region (in nm) below which light is transmitted.

[b] Heat suppression filters (BG38 and KG1), provided with the illuminators, should be used.

[c] Required for double-labeling of fluorescein with rhodamine.

[d] Wide-band green excitation filters for brighter rhodamine fluorescence.

[e] BP577/10 and LP 600: available from Corion Corp. Holliston, Massachusetts; FT600 available by custom order from Zeiss.

[f] For simultaneous observation of fluorescein and rhodamine fluorescence.

of the necessity to distinguish two fluorophores in the same specimen by selectively exciting and/or specifically detecting each fluorescence emission. A judicious choice of the pair of contrasting fluorophores and the associated filters are thus of crucial importance. As a rule, the absorption and emission spectra of the pair should be as widely separated as possible. Fluorescein and rhodamine pairs appear to be satisfactory for most applications. The more recently developed red-emitting dyes such as Texas Red and XRITC seem to be most promising for use with fluorescein.

It should be noted that for dual labeling with rhodamine (set 2, Table VII), an additional filter (KP560) is present in the fluorescein filters to reduce red detection (compare with set 1 for single labeling). This extra filter is necessary because rhodamine dyes do emit when excited by the 450–490 nm blue light used to stimulate fluorescein. In the absence of a KP560 filter, any emission of rhodamine above 560 nm will be detected by the fluorescein optics, and this constitutes a false signal. The optical spillover is appreciable when TRITC, which absorbs significantly in the blue region, is paired with fluorescein, whereas RB200SC, Texas Red, or XRITC have progressively lower spillover. The reverse spillover of fluorescein to rhodamine optics is negligible with the narrow-band excitation at or above 546 nm (set 3) because fluorescein does not absorb in this spectral range. However, small yet detectable spillover does occur when wide-band green excitation at 515–560 nm is used in the rhodamine filters (set 4) because fluorescein is excited to fluoresce to well beyond 590 nm.

For simultaneous observation of fluorescein and rhodamine, set 6 is used. Both fluorophores are excited by the blue light. Fluorescein appears green, rhodamine appears orange, and the dual-stained region appears greenish yellow with the LP520 barrier filter. The rhodamine can also be selectively observed with a LP590 filter. The simultaneous observation and recording of both emissions without any instrumental adjustment is advantageous. Color film is required for documentation.

The above discussion serves to emphasize the important fact that barrier filters allow any light within the specified spectral range to be detected, irrespective of the sources. The discrimination of fluorophores thus relies completely on selecting a spectral range of the fluorescence that is attributable to only one of the fluorophore pair. The importance of fluorophore and filter selections and the necessity to assess the extent of optical spillover are obvious (see Table VI).

Each filter set is installed in one of the several modules in the vertical illuminator housing. These filter modules can be quickly interchanged to place the desired filter set for the fluorophore in the light path. For double-labeling work, it is recommended that an illuminator housing that can accommodate at least three modules (such as Zeiss IIIRS, Leitz Ploempak) be selected so an empty module is available for transmitted light images.

Most of these filters are interference filters consisting of glass substrates with deposited layers of metal salts that are easily damaged beyond repair by scratches or adverse environmental conditions such as moisture and heat. In particular, the chromatic beam splitter has a very fragile gel layer surface that should not be touched.

Objectives and Eyepieces. A wide range of phase-contrast and brightfield objectives specifically designed for fluorescence microscopy are available. The choice of lens depends on the fluorescence intensity of the sample, total final magnification, and the type of simultaneous transmitted light optics. Since the brightness of perceived images increases with the second power of the numerical aperture and decreases with the second power of magnification, brighter images are obtained with a low-power objective lens with high numerical aperture. For fluorescent localization of contractile proteins in tissue culture cells, high-resolution images are required. For this purpose, the Zeiss oil immersion Planapochromat 63X/NA1.4 lens yields sharp, crisp, and intense fluorescent images at high magnification and appears to be a popular choice (for the well-to-do). It should be noted that this and perhaps other highly corrected lenses contain elements that completely absorb the light below 371 nm and is, therefore, useless as an epifluorescence condenser for excitation below this wavelength. An additional medium power (e.g., 25–40×) dry lens with high numerical aperture is useful for rapid scanning of slides before detailed examination at high power. Objectives with a built-in iris diaphragm are convenient and desirable for photomicrography because the diaphragm reduces the glare caused by stray light.

The eyepieces are selected based on the following guide: for a given total image magnification, the use of a low-power eyepiece with a high-power, high numerical aperture objectives optimize both the excitation and perception of the fluorescence images. The use of a high-power (12.5×) eyepiece may be necessary for critical focusing of fine morphology.

Special Fluorescence Techniques and Optical Components

1. Stereo fluorescence microscopy[65]: Stereo pairs of fluorescence imges can be obtained by the use of a half-covered diaphragm in the back focal plane of the objective lens. This technique allows the three-dimensional organization of the filamentous cytoskeletal network in cells to be analyzed easily, thus alleviating the problem of optical sectioning resulting from shallow depth of focus in most high-resolution light microscopy.

2. Total internal reflection fluorescence[66]: This technique allows the exclusive excitation of fluorescence from regions of contact between cultured cells and the substrate. A fused quartz cube is used to direct and reflect the incident laser illumination. This technique is best suited for examination of the relationship of focal contacts and contractile proteins in a living cell microinjected with fluorescently labeled proteins (see Fluorescent Staining of Living Cells and Fig. 7).

Image Recording. Because fluorescent images are generally weak and tend to fade (some very quickly), rapid and sensitive recording devices are required. While photographic films are still preferred, especially for high resolution work, electronic imaging devices are being increasingly used for recording dynamic changes of fluorescent images of living cells (see Fluorescent Staining of Living Cells) and for intensifying and recording very weak fluorescent images.

[65] M. Osborn, T. Born, H. J. Koitzsch, and K. Weber, *Cell* **14** 477 (1978).
[66] D. Axelord, *J. Cell Biol.* **89** 141 (1981).

Photomicrography

Cameras. Fully automatic 35 mm cameras (such as Leitz orthomate W or the built-in camera of Zeiss PM3) equipped with a photomultiplier for light sensing and aperture and exposure speed control, and a spot meter for optimizing the exposure of a small selected area of the field, are ideal for photographing fluorescent images. The much less costly manual camera systems also work well, although their use requires more practice and patience. A simple camera body, a vibration-free suspension shutter, and a phototube without fixed lens, prisms (which decrease fluorescence) or photochanger (which requires an additional step of focusing, thus causing undue bleaching) are adequate.

The use of 35 mm SRL camera bodies requires care because shutter movement frequently causes vibrations that blur the images, especially when the camera is top-mounted on an upright microscope stand. Such cameras (with viewing port covered) are satisfactory when they are mounted to a sturdy inverted microscope stand [such as Zeiss inverted IM35 (Fig. 5)]. The light-sensing elements (photoiodide) in many cameras (with notable exceptions such as Olympus OM-2) are generally inadequate for measuring low light fluorescence, but are of great value for photographing other types of optical images. Cameras with manual mode of operation that can override the automatic mode are essential in order to photograph images of all samples under identical conditions.

Films. Black and white films are satisfactory for most applications. Compared to color films, they have larger exposure latitude and can tolerate some under- and overexposure. The choice of film speed depends on fluorescent intensities. Fast film such as Kodak Tri-X (ASA 400), which can be push-processed to a higher speed (e.g., by Diafine to ASA 1600) is most popular. Fine grain, low speed film, such as Kodak Pan X (ASA 32) can be used for strong fluorescence and high resolution work. Certain Polaroid films have much higher speeds (ASA up to 13,000); however, the larger formats (4 × 5 inches) reduce the intensities at the film plane and thus somewhat compromise the gain in speed.

Occasionally, color micrographs are required in experiments where the characteristic color of fluorescence is essential for interpretation, e.g., in the simultaneous observation of dual labels. Color photography is technically more involved because the reproduction of the fluorescent spectrum depends on the light source, filter combinations, the type of film, and the exposure time. Color slide film (transparency) is preferred over color print film, because the former is more versatile. Daylight film has better red sensitivity than tungsten film; therefore, it is suitable for photographing both fluorescein and rhodamine fluorescence. Tungsten film is adequate for only fluorescein or blue green emission.

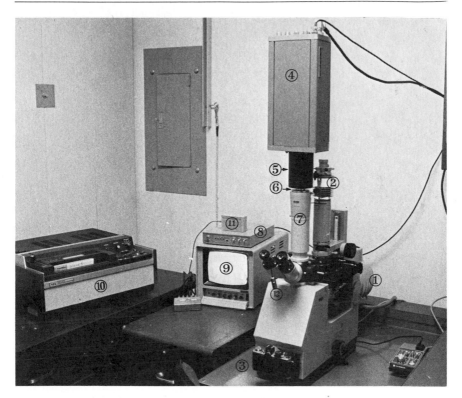

FIG. 5. A video intensification microscope. A Zeiss IM35 inverted microscope is equipped with an HBO100 mercury source (1) for epifluorescence, a tungsten light source (2) for transmitted light optics, and an Olympus OM-2 35 mm camera with motor drive and timer (3) for still photography. A Venus DV-2 TV camera (4) with a built-in two-stage image intensifier (5) is top-mounted to the microscope with a standard C-mount (6) or, as shown here, through an additional phototube (7) that accommodates inside an eyepiece (from 6× to 16×) to allow discrete changes of final magnifications. (Alternatively, the magnification on the monitor can be varied continuously by raising or lowering the TV camera, which is mounted on a separate stand and connected to the microscope with a lighttight expandable bellows.)

The image output of the TV camera is fed through a time-date generator (8) (Vicon V240 TW, which displays date-time to 0.1 sec), displayed on a 9-inch TV monitor (9) (Conrac SNA9-C, 700 line resolution), and recorded on a $\frac{3}{4}$-inch time-lapse video recorder (10) (NEC VC7505, 320 lines resolution, with stop action). A microphone (11) and a speaker (12) are used for audiodubbing.

The Venus DV-2 camera has fiber optics coupling and a dynamic range from 5×10^{-6} fc to 1×10^{-1} fc face-plate illumination. Sensitivity can be controlled either automatically or manually. (The manual control is an important feature for fluorescence microscopy. This is lacking in certain cameras, such as RCA 1040H). The camera has an extended red-sensitive detector and thus is suitable for detecting both fluorescein and rhodamine emissions.

Photography and Microscope Observations. For each type of film optimal exposure times are usually determined empirically by taking a series of exposures, each on a representative new field of equivalent brightness to avoid error due to photobleaching. For Kodak Tri-X film, correct exposure times in the range of 2 sec to 2 min are not uncommon.

Even for fully automatic cameras, trial exposures are necessary to calibrate the light meter for a particular fluorophore, filter set, and film combination. This is because the response curve of the photomultiplier in the camera varies over the visible spectrum, and the light meter is calibrated for only the most sensitive spectral range. Both Zeiss PM3 and Leitz orthomate W are equipped with photomultipliers with much reduced sensitivity in the red region (above 570 nm). If the meter is set at the designated film speed when photographing rhodamine fluorescence, the detector would underestimate the light intensity reaching the film and thus cause an overexposure of the film. To compensate for this insensitivity in the red, the ASA dial should be set to a *higher* value for rhodamine fluorescence. Such adjustment is usually not necessary for fluorescein emission.

In addition to light-meter calibration, several points of practical importance are worth noting.

1. In photographing control samples, the exposure time (and subsequent film processing and printing) should be identical to that of the labeled samples to allow a meaningful comparison of the relative intensities. One should photograph the labeled sample first, override the automatic mode with the manual mode, and then expose the control for the same time interval. If left in the automatic mode, the camera would prolong the exposure until sufficient light is accumulated to yield a bright, poorly contrasted picture of the weakly fluorescent controls.

2. To avoid photobleaching prior to photography, it is wise to use dim transmitted light to localize and to focus on an area of interest before examining it with fluorescence mode. In dual-labeled specimens, one should photograph the fast-bleaching fluorescein first, followed by fade-resistant rhodamine, and then transmitted light images.

3. Slight, yet significant, changes in focal planes or lateral shift of images may occur when filter modules are exchanged, especially when high-power, shallow depth of field objectives are used. Refocusing may be necessary. The thickness of beam splitters of various filter sets may require careful matching to eliminate the problem of lateral-shift images.

4. In observing individual fluorescence of dual-labeled samples, one should pay attention to the fact that, even with preparations in which the overall fluorescence intensities have been carefully matched by adjusting reagent concentrations, significant optical crossover resulting from

strongly fluorescent or light-scattering cellular structures may still occur (see Fig. 3 for an example). This can be easily detected by a shift in the characteristic color observed with each filter set. The crossover information is preserved by color films but is lost completely in black and white recording. Notes should be made for such occurrence to aid subsequent interpretation of micrographs.

5. The light source intensity decreases gradually and then drastically near the end of its useful life (about 200 hr for HBO100). The exposure time thus needs to be adjusted accordingly.

Comparison of Micrographs. A comparison of image details is frequently required in dual-labeling studies. Visual observation is useful only when two images are distinctly different. Further photographic processing may be needed to compare micrographs showing signs of overlapping images. This can be accomplished by printing a composite by overlaying a reversed high-contrast negative of one image onto a slightly underexposed positive print of the second image.[60] Electronic image processing is feasible for special purposes but not yet practical for routine applications.

Video Intensification Microscopy

The use of image-intensifying devices in fluorescence microscopy has many advantages[67,68]: Very dim or nonvisible fluorescent images can be detected and recorded; the excitation light can be greatly reduced to minimize radiation damage to living cells; the photobleaching is greatly reduced; and the interface of intensifier to a time-lapsed video recorder allows slow or fast dynamic events to be recorded continuously.

Video systems have also been applied to other optical methods, such as phase contrast,[69] differential interference contrast,[70] and polarized optics,[70,71] A more detailed discussion of the application of video systems to microscopy can be found in recent publications.[67-71]

Video Equipment. An image intensifier can be used in combination with either a 35 mm camera or a television camera. In the first case, the intensified images that appeared on the fluorescent screen of the device installed in the light path is photographed directly by the camera. However, most of the battery-operated low-cost (<$5000) intensifiers designed for night surveillance have low light yield (50- to 500-fold intensification), and image distortion is severe for several models that one of the authors (K.W.) has tested. In the second case, an image intensifier is built in as

[67] M. C. Willingham and I. Pastan, *Cell* **13** 501 (1978).
[68] D. S. Forman and D. E. Turriff, *Histochemistry* **71** 203 (1981).
[69] R. D. Allen, N. S. Allen, and J. L. Travis, *Cell Motil.* **1** 291 (1981).
[70] R. D. Allen, J. L. Travis, N. S. Allen, and H. Yilmaz, *Cell Motil.* **1** 275 (1981).
[71] S. Inoue, *J. Cell Biol.* **89** 346 (1981).

part of a very low light television camera. The intensified image is displayed on a monitor and can be recorded with a time-lapse video recorder for playback and image analysis. Such a system (termed video intensification microscopy[67]) has proved to be powerful in studying dynamic movements of fluorescent probes in living cells (see section on microinjection below) as well as in conventional microscopy of fixed preparation.[68] A typical setup is illustrated in Fig. 5.

Operation. The system is fairly easy to use. The sensitivity of the camera can be controlled by either manually adjusting the gain or by using the automatic light control that responds to the average light level of the entire field. The contrast and brightness of the images on the monitor can be easily varied to reveal salient image features, a task that usually required hours of darkroom work to bring out similar features from a photographic film.

For a fixed-cell preparation, the entire observation session can be recorded continuously with real-time or time-lapsed mode at very low illumination to avoid bleaching. For a living cell preparation time-lapsed mode is used to record the slow redistribution of fluorescent probes. Prolonged illumination even at low light levels causes bleaching of fluorescence. Light source shutters which open only intermittently at preprogrammed intervals alleviate this problem.[71]

The extreme sensitivity of the video system imposes much greater demands on the proper alignment of the illumination system as well as on the preparation of clean sample slides (dirt, flare, and reflection all become conspicuous). More important, perhaps, is the realization that control samples are also intensified. In this connection, it should be emphasized that, for a meaningful comparison between control and labeled samples, the instrumental settings of the entire system should be identical. In particular, the TV camera should be on manual control (not on automatic) and be on the same gain level. (A manual control is absent in some cameras, such as RCA 1040H.)

Resolution and Documentation. A major shortcoming of the present medium-cost video system is the inferior image quality compared to direct photographic image. For the present setup, the loss of image details due to the limited resolution of the video recorder is the major source of discontent. Micrographs of better image quality can be obtained by directly photographing real-time images on the high-resolution video monitor. An Olympus OM-2 camera, equipped with a Vivitar f/3.5 55 mm macrolens, on a tripod is satisfactory if the shutter speed is at least $\frac{1}{8}$ sec to avoid interference by incomplete scanning. Photographs of playback images from the videotape are obtained in a similar fashion. Videotapes can be edited for long-term storage.

Purchase of Video Equipment. Because video equipment varies widely in design and specifications, the selection of proper components can be frustrating and time consuming. In addition to specifications of each individual component, the compatability of components is also important to consider. Even compounds of identical specifications from different manufacturers may be sufficiently different so the components are not interchangeable (e.g., our NEC tapes cannot be edited by a Panasonic tapeediting machine designed for handling ¾-inch tapes). It is, therefore, prudent to field test the *entire system* on the microscope with a series of experiments *before* purchasing.

In view of the rapid advances in digital image-processing technology, the purchase of a TV camera that is compatible with such techniques is highly desirable.

Fluorescent Staining of Living Cells[71a]

Microinjection of Tissue Culture Cells

There are several techniques available to allow the introduction of molecules and other such elements into living cells, including microneedle injection[72-74] and fusion of loaded red blood cell ghosts[75] or liposomes.[76] This discussion will be limited to the microneedle injection technique. The technique of microneedle microinjection has been used since about 1960.[72] It has been used for the transplantation of organelles,[73,74] nucleic acids,[77-79] virus proteins,[80] antibodies,[81-83] and fluorescently labeled cytoskeleton proteins.[55,84-89] It is in the latter two cases that the first look

[71a] By J. R. Feramisco.
[72] R. Chambers and E. L. Chambers, "Explorations into the Nature of the Living Cell." Harvard Univ. Press, Cambridge, Massachusetts, 1961.
[73] A. Graessmann, *Exp. Cell Res.* **60** 373 (1970).
[74] E. G. Diacumakos, *Methods Cell Biol.* **7** 288 (1978).
[75] R. A. Schlegel and M. C. Rech.
[76] G. Gregoriadis, *N. Engl. J. Med.* **295**, 704 (1976).
[77] M. Graessmann and A. Graessmann, *Proc. Natl. Acad. Sci. U.S.A.* **73** 366 (1976).
[78] D. W. Stacey, V. G. Allfrey, and H. Hanafusa, *Proc. Natl. Acad. Sci. U.S.A.* **74**, 1614 (1977).
[79] M. A. Cappecci, *Cell* **22**, 479 (1980).
[80] R. Tjian, G. Fey, and A. Graessmann, *Proc. Natl. Acad. Sci. U.S.A.* **75**, 1279 (1978).
[81] I. Mabuchi and M. Okuno, *J. Cell Biol.* **74**, 251 (1977).
[82] J. J.-C. Lin and J. R. Feramisco, *Cell* **24**, 185 (1981).
[83] M. W. Klymkowski, *Nature (London)* **291**, 249 (1981).
[84] D. L. Taylor and Y. L. Wang, *Proc. Natl. Acad. Sci. U.S.A.* **75**, 857 (1978).

into the dynamic distributions and functional roles of the contractile apparatus of living cells was made.

The introduction of fluorescently labeled actin into giant protozoans,[84] and of labeled α-actinin,[55,85] actin,[86] vinculin,[87] tropomyosin,[89] and tubulin[88] into mammalian tissue culture cells, has demonstrated the general versatility of this basic approach to observing the contractile apparatus of living cells by showing the apparent incorporation of these conjugates into the cytoskeleton of the living cells. The introduction of antibodies against myosin into sea urchin eggs,[81] antibodies against actin into *Xenopus* oocytes,[90] and monoclonal antibodies against intermediate filaments into mammalian tissue culture cells[82,83] has been used specifically to disrupt cytokinesis, chromosome condensation, and the normal distribution of the intermediate filaments, respectively. This approach of the microinjection of specific antibodies directed against components of the contractile system(s) is going to be a powerful complementary approach to biochemical and localization approaches in determining the function of this system(s).

A wide variety of cell types has been used including gerbil fibroma cells,[55,85,88] 3T3 cells,[87] human lung fibroblasts,[86] and PtK cells.[88]

Microneedle Injection of Proteins

The general techniques of microneedle injection have been described in detail elsewhere,[74,91] and that of Graessmann will be briefly outlined here. The general procedure utilizes a glass capillary needle filled with the substance to be injected into the cell, a micromanipulator to place the needle into the cell, an air-pressure device such as a syringe to transfer the substance from the needle into the cell, and a phase-contrast microscope to allow visualization of the injection process (Fig. 6).

Tissue Culture Cells. Cells for injection can be grown on glass coverslips in petri dishes or on petri dishes themselves. The glass coverslips should be of the size (No. 1.5, 24 mm in diameter) that can fit into the Dvorak-Statler[92] chamber (Nicholson Precision Instruments, Bethesda, Maryland)

[85] J. R. Feramisco, *Proc. Natl. Acad. Sci. U.S.A.* **76**, 3967 (1979).

[86] T. E. Kreis, K. H. Winterhalter, and W. Birchmeier, *Proc. Natl. Acad. Sci. U.S.A.* **76**, 3814 (1979).

[87] K. Burridge and J. R. Feramisco, *Cell* **19**, 587 (1980).

[88] C. H. Keith, J. R. Feramisco, and M. L. Shelanski, *J. Cell Biol.* **88**, 234 (1981).

[89] J. Wehland and K. Weber, *Exp. Cell Res.* **127**, 397 (1980).

[90] D. Rungger, E. Rungger-Brandle, C. Chaponnier, and G. Gabbiani, *Nature (London)* **282** 320 (1979).

[91] A. Graessmann, M. Graessmann, and C. Mueller, this series, Vol. 65, p. 816.

[92] J. A. Dvorak and W. K. Statler, *Exp. Cell Res.* **68**, 144 (1971).

FIG. 6. Microinjection devices: 1, microcapillary needle and holder; 2, Leitz micro-manipulator; 3, 50-ml syringe; 4, Leitz Diavert inverted microscope with 32× objective lens; 5, camera; 6, Leitz heavy base; 7, Dvorak-Statler chamber.

to allow the injected cells to be observed in a controlled environment. Temperature is controlled by an air curtain, and the chemical environment is controlled by pumping medium through the chamber.[67] Analysis of the injected cells can be followed in several ways including phase-contrast microscopy and fluorescence microscopy. No special condensers are required for the Zeiss PM3 when these chambers are used if the stage adaptor for the chamber is not used.

Preparation of Conjugates

The procedures for preparing fluorescent conjugates can vary over a wide range of conditions, but they generally will be determined by the properties of the protein and the reagent. Many readily available fluorescent reagents, such as some of those listed in Table I, are capable of transferring the fluorophore moiety to the protein through a variety of reaction mechanisms utilizing a variety of amino acid targets. Typical reaction conditions employed for the fluorescent labeling of α-actinin with tetramethylrhodamine isothiocyanate (Cappel Laboratories) are outlined below.[85]

Purified α-actinin (1–5 mg/ml in 50 mM sodium phosphate, pH 8–9) is mixed with tetramethylrhodamine isothiocyanate (1 mg/ml, in 0.5 M sodium bicarbonate) to the level of 20 μg of dye per milligram of α-actinin. The reaction mixture is incubated at 4° for 6 hr; it is then passed through a 1 × 15 cm column of Sephadex G-50 equilibrated in 20 mM Tris-OAc, 20 mM NaCl, 15 mM 2-mercaptoethanol, 0.1 mM EDTA, pH 7.6. This procedure yields a functional (as judged by the ability of the conjugate to bind to F-actin) fluorescent conjugate of α-actinin[55] labeled with 2–4 mol of dye per mole of α-actinin.

Proteins, such as actin or tubulin, that undergo self-assembly reactions and interact with a variety of proteins can be labeled either as individual subunits, as polymers, or as complexes with other binding proteins. Actin[86] and tubulin[88] have been labeled with tetramethylrhodamine isothiocyanate and dichlorotriazinylaminofluorescein as polymers. The latter reagent is most useful for the labeling of the relatively unstable tubulin, since this reagent reacts very quickly with the tubulin.[93] The labeled polymers are then purified by cycles of depolymerization–polymerization, thus ensuring that the labeled molecules are capable of self-assembly.

Control Conjugates. Injection of control conjugates should be done to determine the specificity of the incorporation of the fluorescently labeled protein. Controls that have been carried out thus far include the injection of heat-inactivated fluorescently labeled proteins[85,87,88] or fluorescently labeled bovine serum albumin or ovalbumin.[84]

Preparation of Microcapillaries. The glass capillaries are generated by a two-stage pulling process of glass tubes (1.5 mm o.d., 1.2 mm i.d.) (Kimble No. 48465). Twenty-centimeter lengths of the glass tubes are first cleaned in Aquaregia (1 part concentrated HNO_3 · 3 parts concentrated HCl) overnight, washed exhaustively in deionized water, and dried at 60° in a drying oven. The first pull is accomplished by heating the middle of the dried tube over a small flame of a bunsen burner with rotation until the tube can be gently extended to a length of about 25 cm. This "pre-pull" will cause a decrease in the diameter of the tube near the middle to ~0.5 mm. The pre-pulled tube is then clamped into a needle puller (available from, for example, E. Leitz or D. Kopf Instruments) with the narrow area centered within the heating filament. By adjusting the filament temperature and the forces exerted on the glass tube, needles with tip diameter

[93] D. Blakeslee and M. G. Baines, *J. Immunol. Methods* **13**, 305 (1976).

(outer diameter) of 0.2–1.0 μm can be generated. The needles can be treated with siliconizing agents.[91] The siliconization is useful for very large needles (>1.0 μm in diameter), and for very small needles (<0.2 μm in diameter), but is not necessary for medium-sized needles (0.2–1.0 μm in diameter). The needles are hooked up to the 50-ml syringe and then serially rinsed (inside as well, as accomplished by suction and pressure with the syringe) in H_2O, ethanol, tetrahydrofuran, 0.5% dichlorodimethylsilane (in tetrahydrofuran, v/v), tetrahydrofuran, and ethanol. The needles, plain or siliconized, can be stored by embedding them (tips up) in a styrofoam base, which is then covered with a beaker, for months.

Injection of Cells

Buffers. The protein to be injected into the cells must be in an injection buffer that will not have deleterious effects on the cells. A variety of buffers have been used with success.[77,85,86,88] For example, one injection buffer is composed of 0.048 M K_2HPO_4, 0.014 M NaH_2PO_4, 0.0045 M KH_2PO_4, pH 7.2. Another buffer, used for antibody injection, is composed of 0.01 M NaH_2PO_4, 0.07 M KCl, pH 7.2.[82]

Concentration and Volume of Proteins. Protein concentrations for the injections should be adjusted according to the purpose of the experiment: relatively low ones for the introduction of "tracer" amounts (i.e., 10% of the endogenous amount) of protein and high ones for introducing stoichiometric or overwhelming amounts. The former case generally applies to the introduction of fluorescently labeled structural proteins, whereas the latter case generally applies to the injection of antibodies directed against components of the cells. Since the microinjection procedures usually introduce 10^{-14} to 10^{-13} liters of sample into each cell[91,94] the protein concentrations are in the range of 0.1–5.0 mg/ml for "tracer" injections of proteins and in the range of 5–50 mg/ml for injections of large amounts of the proteins, depending upon the properties of each protein and the goal of each experiment. For example, α-actinin is estimated to be about 0.25% w/w of the total protein in a fibroblast (as estimated by two-dimensional gel electrophoresis), and a fibroblast is assumed to contain 0.75 ng of protein. Thus, there would be $\sim 6 \times 10^6$ molecules of α-actinin per cell. The injection of 5×10^{-14} liters of a 5 mg/ml solution of α-actinin would add $\sim 0.6 \times 10^6$ molecules of α-actinin to each cell, an increase in the number of α-actinin molecules of 10%.

The average volume injected into cells can be estimated by injecting a radioiodinated protein of known specific activity into the cells, rinsing the cells with PBS, and scraping the cells into a vial for counting.

[94] J. A. Schloss, A. Milsted, and R. D. Goldman, *J. Cell Biol.* 74, 794 (1977).

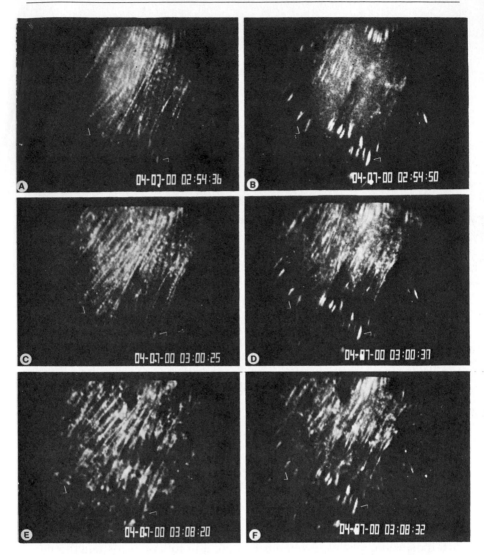

FIG. 7. Microinjection of fluorescently labeled proteins into living cells. The micrographs shown here are selected to illustrate an instrumental artifact which, if undetected, can lead to erroneous interpretation. A mixture of α-actinin (labeled with TRITC) and vinculin (labeled with FITC) was injected into a respreading fibroblast. Two hours after injection, cytochalasin B was added to the culture medium and the distribution of TRITC–α-actinin (A, C, E) and FITC–vinculin (B, D, F) was followed over a period of 15 min by a RCA 1040H image intensification camera. The date-and-time (to 1 sec) is displayed on the micrographs. Note that in (B), vinculin staining reveals the characteristic localization of vinculin on attachment plaques (arrowheads). Note also that cytoplasmic "stress-fiber" staining increases from (D) to (F) and, moreover, these stainings are similar to that of the α-actinin staining in (C) and

Prior to injection, the protein solution is centrifuged for 5 min at 12,000 g in a Beckman microfuge and is filtered through a 0.2 μm filter (Millipore Corporation). The solution (~5 nl) is loaded into the needle by capillary action or by suction with the syringe. Immediately, the cells are removed from the culture incubator and injected with the solution. With practice ~100 cells or more can be injected in 10 min. Using only a 50-ml syringe to pressure inject the cells, one can control the approximate volume injected into each cell by a visual analysis since the cells swell slightly during the injection process and the cytoplasm appears to lose contrast momentarily. Alternatively, a micrometer-calibrated pressure device can be used to control the volume injected.[74] The cells are then returned to the incubator for the appropriate time and processed for immunofluorescence, time-lapse photography, etc. The injection process itself usually does not damage the cells and more than 90% of the injected cells usually survive.

Figure 6 illustrates a microinjection setup.

Analysis of Injected Cells

Direct vs Indirect Fluorescence

There are several choices available for the analysis of the distribution of an injected protein by fluorescence microscopy.

1. When the protein or antibody is injected as the fluorescent conjugate, the distribution of the fluorescence can be visualized directly in the living cell. This protocol is ideally suited for time-lapse analysis of the dynamic aspects of the protein distribution in living cells undergoing changes such as respreading, migration (Fig. 7). Alternatively, the injected cells can be fixed and then stained for another antigen with a contrasting fluorophore (see above, section Staining Procedures). In this manner, the distribution of the injected protein as well as an endogenous protein can be observed simultaneously in the fixed cells.

2. When the injected protein is unlabeled, its distribution can be localized in fixed cells by specific antibody using staining procedures

---

(E), respectively. The apparent "codistribution" of vinculin and α-actinin in stress-fiber, as suggested by these micrographs, is completely artifactual. The artifact arose because (a) FITC-vinculin fluorescence faded rapidly during the course of the experiment; (b) the RCA camera, which is equipped with an automatic light level control only, automatically increased the gain to intensify the diminishing emission from FITC–vinculin; and (c) the small amount of crossover of the bright and fade-resistant rhodamine emission into FITC optics are therefore increasingly exaggerated by the camera until, in (F), it became the predominant feature in FITC optics. J. R. Feramisco, unpublished results, 1981.

described above (Staining Procedures section). This approach does not require the faithful conjugation of the protein, a task that may be difficult for certain proteins.

Microscopy

For time-lapsed analysis of the distribution of conjugated proteins in the living cell, it is necessary to utilize an image intensification system as described above in the section Fluorescence Microscopy to avoid photobleaching of fluorophores and possible light-induced damage to the cell.

Figure 7 shows typical results of such an analysis. Also included in Fig. 7 is an illustration of the technical problem that one may encounter in dual-labeling studies. The low level crossover of rhodamine emission to fluorescein optics is greatly intensified by a television camera (RCA 1040H) by the automatic light control. (This camera has no manual control.) Because the authentic fluorescein emission is now weak and bleached, the "fluorescein" and rhodamine patterns now appear to be identical on the monitor screen.

Interpretation of Fluorescence Patterns

The reliability and effort involved in the interpretation of fluorescence patterns of either fixed cells or living cells depend on whether cellular or molecular structural information is sought. The presence and gross cellular distribution of a protein can be easily established by its specific antibody staining patterns in fixed or microinjected cells. The staining patterns of a large number of cells are easily analyzed. These characteristic patterns are extremely valuable in correlating morphology, motility, and differentiation states of tissue culture cells.

In contrast, the interpretation of labeling patterns in terms of molecular organization of the structure of the contractile apparatus and of the cytoplasm requires careful consideration of technical details. Frequently, results from a variety of complementary and correlative biochemical and ultrastructural techniques are required to substantiate the tentative and speculative conclusions about molecular interactions based on fluorescent labeling studies.

We have throughout this chapter discussed the symptoms and possible remedies of faulty experimental practices that may lead to uninterpretable or misleading results. We discuss below the intrinsic limitations of the fluorescent labeling techniques, which in turn set the limits of interpretation.

Fixation and Permeabilization

The possible deleterious effect of fixation procedures on the antigen and on the cellular structure is difficult to assess and nearly impossible to predict (cf. Fig. 1). A promising method for a systematic study of fixation appears to be the monitoring of the distribution of microinjected fluorescently labeled proteins before, during, and after fixation, and comparing with the antibody labeling patterns of the same protein in the same fixed cell. So far, only one fixation procedure (formaldehyde fixation and acetone permeabilization) has been investigated and found to be satisfactory based on indistinguishable staining patterns of living and fixed cells.[17] Other procedures and other proteins remained to be evaluated by this technique.

Target Accessibility

The accessibility of macromolecular probes to their specific intracellular targets is determined mainly on the probe size relative to the openness or porosity of the cellular structure surrounding the target. Inaccessible intact targets thus evade detection. The same considerations apply to microinjected fluorescent proteins, because cellular compartmentalization or slow turnover of an existing structure may impede the incorporation of labeled proteins into functional structure.

Fluorescence Microscopy

Resolution. The resolution of fluorescence microscopy is limited to about 0.2 μm, at least an order of magnitude larger than the dimensions of most proteins. Thus, the costaining (codistribution) of two proteins in one locus indicates that the two proteins are closer than 0.2 μm. It does not necessarily follow that the two proteins are physically associated at molecular dimensions. Similarly, the continuous staining of cellular structure, such as stress fibers, may result from superimposition of images of nondiscontinuous substructure not resolvable by light microscopy. The elegant demonstration that a single microtubule can be observed as a 0.2-μm fiber by anti-tubulin staining in fluorescence microscopy[95] does not guarantee that all 0.2-μm tubulin staining represents a single microtubule.

Depth of Focus. The depth of focus of the fluorescence microscope at high resolution is frequently smaller than the thinnest dimension of a tissue culture cell. In general, optical sections at several focal planes are

[95] M. Osborn, R. E. Webster, and K. Weber, *J. Cell Biol.* 77, R27 (1978).

obtained. The fluorescence images from each focal plane are thus contaminated by out-of-focus images from other planes. This is important to consider in interpreting "diffuse" staining patterns especially from a thick portion of the cell.

Fluorescence Intensity. Fluorescence intensity of many fluorophores is sensitive to environmental conditions such as pH, ionic compositions, polarity, viscosity, and proximity to other proteins. Many of these factors are easily controlled in fluorescent staining of fixed cells. However, such factors are not easily manipulated in microinjected living cells. This sensitivity of fluorescence to environmental factors renders it difficult to correlate brightness with the amount or concentration of labeled protein in a given cellular loci. For example, it is conceivable that, in extreme cases, labeled protein may not fluoresce at all owing to environmental quenching.

Contrast. It should be noted that unique structural features are detectable by fluorescence staining only if their staining intensity is higher than that of the adjacent areas. For example, stress fiber is visualized by actin-labeling because the intensity of actin-staining of stress fibers is higher than that of the surrounding cytoplasm. A diffuse staining would result if they are equal.[62]

Acknowledgments

K. W. would like to thank Dr. L. J. Reed for encouragement and support, Ms. C. L. Williamson for able and cheerful assistance in fluorescence microscopy, and Ms. G. Donnell for typing the manuscript. The work in K. W.'s laboratory is supported in part by grants from the National Institutes of Health (AM20270 and CA09182) and the American Heart Association Texas Affiliate Inc.

[49] Preparation of Contractile Proteins for Photon Correlation Spectroscopic and Classical Light-Scattering Studies

By CHARLES MONTAGUE and FRANCIS D. CARLSON

The development of photon correlation spectroscopy (PCS) has made possible the precise measurements of translational and rotational diffusion coefficients of proteins, nucleic acids, and other particles of biological origin and the subject has been extensively reviewed.[1-10] It is also possible

[1] B. J. Berne and R. Pecora, "Dynamic Light Scattering." Wiley, New York, 1976.
[2] B. Chu, "Laser Light Scattering," Academic Press, New York, 1974.

to conduct classical light-scattering measurements of the molecular weight, radius of gyration, and second virial coefficient with a photon correlation spectrometer. Because such spectrometers employ lasers, they provide a monochromatic, constant-intensity light source and a highly collimated beam that is ideally suited for light scattering on extremely dilute macromolecular solutions.

Sample Preparation

The reliability of any light-scattering or PCS measurement is critically dependent upon the absence of large, unwanted, contaminants, henceforth referred to as "dust," regardless of origin. Dust may be introduced into a sample as an accidental contaminant of solvents or glassware, or as a result of accidental irreversible aggregation of the sample itself. In either case its effects on any kind of light-scattering measurement can be devastating. We present here brief descriptions of the various procedures that have been used successfully to remove dust, regardless of origin, from solutions of contractile proteins and organelles. Also presented are some techniques for collecting and editing data that reduce the errors due to dust. These techniques are applicable to solutions of macromolecules in general and when properly used will reduce systematic errors to 5% or less.

As a rule dust diffuses much more slowly than the particle of interest. The intensity autocorrelation function, $g^{(2)}(\tau)$, for laser light quasi-elastically scattered from a solution of diffusing particles containing dust is as follows[11]:

$$g^{(2)}(\tau) \equiv \frac{\langle I(0) \cdot I(\tau) \rangle}{\langle I \rangle^2} = 1 + \frac{I_s^2}{\langle I \rangle^2} |g^{(1)}(\tau)|^2 + \frac{2I_s \cdot I_D}{\langle I \rangle^2} |g^{(1)}(\tau)| + \frac{I_D^2 \cdot X}{\langle I \rangle^2} \quad (1)$$

[3] F. D. Carlson, *Annu. Rev. Biophys. Bioeng.* **4**, 243 (1975).

[4] H. Z. Cummins and E. R. Pike, "Photon Correlation and Light-Beating Spectroscopy." Plenum, New York, 1974.

[5] H. Z. Cummins and E. R. Pike, eds., "Photon Correlation Spectroscopy and Velocimetry." Plenum, New York, 1977.

[6] V. Digiorgio, M. Corti, and M. Giglio, eds., "Light Scattering in Fluids and Macromolecular Solutions." Plenum, New York, 1980.

[7] V. A. Bloomfield, *Annu. Rev. Phys. Chem.* **28**, 233 (1977).

[8] J. M. Schurr, *CRC Crit. Rev. Biochem.* **4**, 371 (1977).

[9] V. A. Bloomfield and T. K. Lim, this series, Vol. 48, p. 415.

[10] B. Chu, *Phys. Scr.* **19**, 458 (1978).

[11] H. Z. Cummins and P. N. Pusey, *in* "Photon Correlation Spectroscopy and Velocimetry" (H. Z. Cummins and E. R. Pike, eds.), p. 164. Plenum, New York, 1977.

where I_s is the intensity of the light scattered by the sample of interest, I_D is the intensity of the light scattered by the dust, $I = I_s + I_D$, and $g^{(1)}(\tau)$ is the normalized field autocorrelation function associated with the scatterer of interest. X is defined by

$$1 + X \equiv \langle |h(\text{o})|^2 \cdot |h(\tau)|^2 \rangle / \langle |h|^2 \rangle^2 \cong \langle |h|^4 \rangle / \langle |h|^2 \rangle^2$$

where $h(\tau)$ is the complex amplitude of the fluctuating component of the light scattered by the dust. When the dust diffuses much more slowly than the particle of interest, there are four cases to be considered.

1. No dust: $I_D = 0$ and $g^{(2)}(\tau) = 1 + |g^{(1)}(\tau)|^2$. There is no error due to dust.

2. "Stationary dust." This can arise from scattering by the walls of the cuvette or from a very rigid gel phase in the sample itself. In this case, in general, there is no easy way of determining the relative intensities of the light scattered by the sample and the light scattered by the dust alone, and an accurate measurement of $g^{(2)}(\tau)$ may not be obtainable.

3. If there is enough dust present in the scattering volume at all times, $h(\tau)$ is a Gaussian variable, $X = 1$, and we obtain

$$g^{(2)}(\tau) = 1 + \left[\frac{I_s}{\langle I \rangle} g^{(1)}(\tau) + \frac{I_D}{\langle I \rangle} \right]^2 \qquad (2)$$

While this case is unlikely, it is worth noting because of its possible usefulness in studying the monomer component in the presence of polymeric forms.

4. "More-than-Gaussian dust." When one or two large dust particles pass through the scattering volume infrequently and the fraction of the measurement time that the dust is in the measuring volume is small, then $X \gg 1$ and if $X I_d \gg I_s$, Eq. (1) becomes

$$g^{(2)}(\tau) - 1 \cong \frac{I_s^2}{\langle I \rangle^2} |g^{(1)}(\tau)|^2 + \frac{I_D^2}{\langle I \rangle^2} X \qquad (3)$$

Since the second term on the right is a constant, $g^{(1)}(\tau)$ is easily obtained. However as Cummins and Pusey[11] noted, since $I_d \gg I_s$ the presence of the dust particle is readily detected and data collection can be interrupted until it leaves the scattering volume. Alon and Hochberg[12] have described an automatic switch for this purpose. A scheme for reducing the error due to dust by editing data taken on a series of short experiments is described below.

Clearly it is best to remove as much of the dust as possible so that the conditions specified in case 1 above are obtained and the effects of dust

[12] Y. Alon and A. Hochberg, *Rev. Sci. Instrum.* **46**, 388 (1975).

are effectively eliminated. The procedures described below are directed toward reducing dust in the sample to a minimum and to collecting and editing data in ways that reduce the errors due to dust.

Preparation of "Dust-Free" Solutions

To keep dust contamination at a minimum, a few general procedures must be employed. For aqueous solutions, distilled water should be deionized and filtered with the equivalent of a Millipore Milli-Q reagent grade water system equipped with a Twin-90 filter (Millipore Corporation, Bedford, Massachusetts). This system will remove dust particles greater than 0.22 μm in diameter. Superclean water can be obtained with Millipore filter transfer system consisting of a stainless steel filter holder, a 0.05 μm filter, and a variable flow pump as supplied by the manufacturer. Care must be taken to keep the filtered water from becoming contaminated with dust from the collection container or room air. Superclean water obtained from this system is dust-free in that for long periods the light scattered from the water is due only to the water itself.

It often happens that the purified sample of interest contains aggregates of denatured material, dust particles from contaminated glassware, or other large contaminants such as breakdown products from columns used in the purification procedure. The presence of such dust in a preparation can be easily detected by eye with the aid of a low power He-Ne laser (1 mW and 1 mm or less beam diameter). To do this one must view the light scattered at small angles in the forward direction *making absolutely certain that the main transmitted laser beam cannot enter the eye*. This is easily done by using an opaque beam stop to intercept the direct beam as it leaves the cuvette and thus cannot possibly enter the eye of an observer located beyond the beam stop. As dust particles pass through the laser beam in the cuvette they scatter light most strongly in the forward direction and thus give rise to very bright flashes of light that persist as long as they remain in the laser beam. A completely dust-free solution will exhibit no such bright scatterers against the uniformly bright light scattered by the macromolecules of interest. Small amounts of dust will produce infrequent bright flashes.

The dust-removal procedures employed prior to filling the cuvette are critical and sample-dependent. G-actin can be freed of dust by filtration through a 0.08 μm filter with appropriate prefilter. Native thin filaments and F-actin, particles about 1 μm long, can be freed of dust by centrifugation in stoppered tubes at 35,000 g for 2 hr in an angle-head rotor. Myosin should be centrifuged at 65,000 g for 2 hr before it is transferred to the cuvette. Final transfer to the light-scattering cuvette is done in a dust-free box. This is achieved by flowing moisture-free compressed air through a

0.2 μm filter into a standard laboratory glove box, maintained under 1 inch of water pressure.

Precautions must be taken in the final transfer of protein solutions to the light-scattering cuvette. The sample should be introduced along the sides of the cuvette to avoid drop formation and possible surface denaturation. With myosin or other proteins that bind to glass surfaces, the cuvette should be prefilled with a clean solution, emptied, and refilled with the protein solution. The acetone-washed glass wets very easily and if filled without prewetting, a thin film of the myosin solution will spread along the surface of the dry glass resulting in denaturation, aggregation, or enhanced binding of the myosin to the glass. For highly dilute solutions, binding to the glass will decrease the myosin concentration in solution. To avoid this, the cuvette should be filled with the myosin solution, allowed to stand for several minutes to permit binding to occur, then emptied and refilled. Binding can be monitored by measuring the intensity of the scattered light at a fixed angle. It should remain unchanged if no significant binding occurs or should decrease to a constant value as binding saturates. Binding to the glass and the dust-elimination procedures such as filtration and centrifugation can and often do reduce the concentration of the sample. For concentration studies, the final concentration of the sample must be determined.

Preparation of "Dust-Free" Cuvettes

After taking care that the sample is as dust-free as possible, it must be transferred to a ''clean'' cuvette, i.e., one that has no contaminants adhering to the walls, nor dust that might come free in the sample solution.

To remove surface contaminants, the cuvettes are heated in a solution containing a surface-active agent (NRS-250 from Norell Chemical Co. Inc., Landerville, New Jersey) for 30 min, avoiding boiling. The cuvettes are next washed extensively with distilled water, heated in distilled water to remove any remaining surface-active agent, rinsed with distilled water again, and finally rinsed with n-propanol before being placed inverted in an oven at 50° to dry. Heating and soaking can also be done in a heated sonicator followed by sonication for 2 min. Longer sonication can damage the cuvettes. Sonication and/or longer heating may be necessary for more resistant stains.

To remove dust that remains inside these washed cuvettes, a still, similar to that described by Thurmond,[13] is used with some modifications. Acetone is used instead of butanone and a teflon cuvette holder is fitted to the top of the jet to avoid scratching cuvettes. Cuvettes and pipettes are

---

[13] C. D. Thurmond, *J. Polym. Sci.* **8**, 607 (1952).

allowed to dry in the still, not exposed to room air. On removal from the still, the cuvette is capped immediately and stored under an inverted beaker to prevent the accumulation of dust that might subsequently be introduced into the sample.

Centrifugation of Sample and Cuvette

Once the cuvette is loaded and capped, a low power laser can be used to check for residual dust as described above. If large flashes are noticed, or if they are seen after mounting the cuvette in the spectrometer, or if the count rate fluctuations seem to indicate contamination, there is one post-loading technique that will often clean the samples further. The cuvette with the sample can be centrifuged in a swinging-bucket rotor at 500 g. For centrifugation, the stoppered cuvette is placed in a centrifuge tube adapter that has a rubber stopper in the bottom to provide a flat cushion base. With the cuvette in the adapter, water is added to the adapter to a level that approximately matches the level of the solution in the cuvette. This reduces the net forces acting on the cuvette itself, decreasing the likelihood of breakage. This procedure has been used very effectively to clean up protein solutions. Centrifuging at much higher relative centrifugal forces is not recommended due to the danger of breaking the cuvette.

Thermal Bath

Precise PCS measurements of diffusion coefficients require both accurate determination of the scattering angle and temperature regulation to ±0.05° or better. The thermostatted cuvette holder (Fig. 1), achieves this degree of temperature regulation and allows access to the scattered light at all angles from 5 to 150 degrees. Instead of using optical flats fixed at several different scattering angles, this design[14] uses a single optical flat as an entrance window and a nearly hemispherical zone exit window. The hemispherical zone window was manufactured to highest attainable optical tolerances by Muffoletto Optical Co., Baltimore, Maryland. Once aligned, this bath permits photon correlation measurements with no need for refractive index corrections of the scattering angle. The back-reflection of the laser beam from the exit window–air interface is eliminated by using a solid glass Woods horn mounted on an XYZ translator and coupled to the exit window by refractive index matching with a thin film of microscope immersion oil. The back-reflections from the water–glass interfaces of the two surfaces of the exit wall of the cuvette and the exit window of the water bath are not eliminated with this arrangement.

[14] Suggested by Richard C. Haskell in our laboratory.

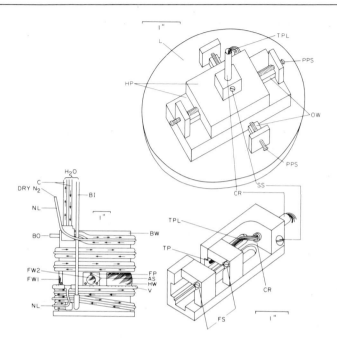

Fig. 1. Thermostatted cuvette holder and bath. The cuvette is immersed in a water bath that is temperature regulated by circulation of water from a Forma-Temp Jr. bath and circulator (Forma Scientific, Marietta, Ohio) through coils (C) surrounding the bath wall (BW). Adjacent coils carry water in opposite directions and are soldered together. Water can also be circulated through the bath itself through an inlet (BI) and overflow (BO). This permits rapid temperature equilibration and filtration of the bath water (3 μm filter) to remove dust. Temperature is monitored on a digital thermometer to ±0.01° with a thermistor probe (TP) located adjacent to the cuvette. A 1-inch-thick styrofoam insulator (not shown) is positioned around the bath. The entrance window for the laser beam is one of two optically flat windows (FW). The choice of the entrance window is made by rotating the bath on its supporting base, which is thermally isolated on three stainless steel feet. One entrance window (FW1) is positioned so that the beam exits through the center of the hemispherical window, and the other (FW2) so that the exit is near one end. Each is attached with silicone rubber adhesive sealant (General Electric, Waterford, New York) to a mounting plate located inside the bath. The plate is pulled by three screws (AS) against an O-ring seal seated on a flat plate (FP) mounted in the wall of the bath. Adjustment of these screws allows optical alignment of the window. The hemispherical equatorial zone window (HW) is glued directly to the brass cylindrical wall of the bath with silicone rubber sealant. The center of curvature of this window is permanently fixed on the axis of rotation of the bath, and optical alignment of all other components is done relative to the hemispherical window. Condensation on the outside surface of the windows is prevented by flowing dry nitrogen, which has been equilibrated to the bath temperature as its input line (NL) passes along the water coils, through vents (V) at the base of each window. The cuvette is held in the cuvette holder by two phosphor-bronze flat springs (FS) pressing one corner of the cuvette into the opposite corner of the cuvette holder. The holder is attached to a translator on the bath lid (L) by a hollow plexiglass connecting rod (CR) through which the leads (TPL) of the temperature

These back-reflections give rise to no errors in the measurement of the diffusion coefficient at a scattering angle of 90 degrees. For all other scattering angles the back-reflections give rise to a systematic error, the magnitude of which depends on the scattering angle and the scatterer's form factor. Errors due to the back-reflections can be eliminated or greatly reduced by placing a beam stop within the sample cuvette.[15] Alternatively, the cuvette can be tilted so that the back-reflected beam does not pass through the scattering volume and eliminating the back-reflection from the exit window with a beam stop placed in the water bath.

Data Collection and "Editing"

Removing dust prior to data collection is the most desirable way to deal with the problem, but it is not always possible to do so. By using appropriate data collection methods, the data can be "edited" to eliminate some of the effects of dust. This is done by taking advantage of the fact that dust, as defined in this paper, is large in comparison to the particles being studied. Therefore, when a particle of dust enters the scattering volume its contribution to the scattered intensity is large.

When the amount of dust is small, an occasional particle may drift through the scattering volume, corresponding to case 4 under Sample Preparation. If data are collected in a sequence of experiments of short duration, a histogram of total counts in each experiment makes possible the identification of experiments with inordinately high counts. These experiments may be presumed to be contaminated by dust and can be eliminated prior to further analysis. Data can be collected easily in this way with a digital correlator or spectrometer operating on line to a computer. A point of caution with this procedure: each short experiment, if treated independently of the others (e.g., normalized by the background determined from that short experiment), rather than summed with the exclusion of selected experiments, must have a duration long enough to get a valid estimator of averages for that short experiment (see Oliver[16] for details of appropriate values of experimental parameters). An added ben-

---

[15] T. J. Racey, R. Hallett, and B. Nickel, *Biophys. J.* **35**, 557 (1981).
[16] C. J. Oliver, *Adv. Phys.* **27**, 387 (1978).

---

probe pass. The translator positions the cuvette in the bath with two sets of push-push screws (PPS) pushing on horizontally translatable plates (HP) that move along two orthogonal ways (OW) and by vertical adjustment of the connecting rod (CR). All metal parts within the bath were treated with Black Magic anodizing solution (Mitchell-Bradford Chemical Co., Milford, Connecticut) to reduce stray reflections.

efit of treating each short experiment separately is that it will yield an experimental standard deviation for each data point.

Neiuwenhuysen[17] has described a method for analyzing clipped photon autocorrelation data obtained from samples containing dust. His method determines the diffusion constant in the presence of dust by either fitting $g^{(2)}(\tau) - 1$ to a single exponential plus a constant or by extrapolating the first cumulant in a cumulant fit (Koppel[18]) of $\ln(g^{(2)}(\tau) - 1)$ to zero quality factor, Q, where $Q = 4X/r^2$, X = second cumulant, and r = first cumulant.

In our experience, time and effort spent removing dust from the sample during its preparation and care in avoiding subsequent contamination by dust are the most effective ways of dealing with this problem. In any case, procedures for editing or correcting for dust are useful only when the amount of contamination by dust is small.

[17] P. Nieuwenhuysen, *Macromolecules* **11**, 832 (1978).
[18] D. E. Koppel, *J. Chem. Phys.* **57**, 4814 (1972).

[50] Hydrodynamic Approaches to the Study of High Molecular Weight Proteins and Large Macromolecular Assemblies

By WALTER F. STAFFORD, III

Several hydrodynamic techniques have been exploited or developed for the study of high molecular weight proteins and large macromolecular assemblies. Successful application of these techniques generally requires highly purified, chemically homogeneous preparations. Removal of dust, aggregates, and low molecular weight components is mandatory. Purification by transport methods that can yield individual thermodynamic components is often necessary if meaningful conclusions are to be drawn. Gel filtration and density gradient sedimentation under native conditions are the most suitable techniques for resolving components.

Sedimentation Analysis

Sedimentation velocity analysis offers several new approaches to the investigation of the physical properties of large proteins. A technique employing the ultraviolet (UV) photoelectric scanner of the Beckman

Instruments Model E ultracentrifuge at low protein concentrations developed by Van Holde and Weischet[1] is useful for obtaining sedimentation coefficient distributions for solutions of large macromolecules. Extrapolations to infinite time eliminate the effect of diffusion, making it possible to observe polydispersity. The sedimenting boundary is transformed at each concentration level into an apparent integral sedimentation coefficient distribution using the relation

$$S_w^* = \ln(r/r_0)/\omega^2 t$$

where S_w^* is the apparent sedimentation coefficient, r is the radius at the point chosen in the boundary, r_0 is the radius of the meniscus, ω is the angular velocity of the rotor, and t is the time.

The values of S_w^* at each concentration are plotted as a function of $t^{-1/2}$ and extrapolated to infinite time to yield the sedimentation coefficient corresponding to each point in the boundary. For a monodisperse system all graphs will extrapolate to a single value. The method is most useful, however, for detecting polydispersity in systems exhibiting either a single boundary or overlapping boundaries. It appears to be the only ultracentrifuge method that will give information concerning the individual species in several overlapping boundaries. In favorable cases, the diffusion coefficient of a monodisperse preparation may be determined with an accuracy of about 10–20%.

Online Sedimentation Analysis

Another technique was developed by Laue and Yphantis.[2] Their online analysis of sedimentation velocity experiments gives sedimentation coefficient distributions for solutions of large macromolecules. The technique employs a microprocessor-controlled Reticon line scanner located at the film plane of the Rayleigh camera. Fringe deflection at a single radial position is followed as a function of time while the boundary sediments. The time dependence of the fringe displacements is analyzed to give the sedimentation coefficient distribution of the sample. The method is especially useful for studies of molecules for which diffusion may be ignored on the time scale of the sedimentation experiments. The technique has been successfully applied to a study of a microtubule-neurofilament complex by Runge et al.[3]

[1] K. E. Van Holde and W. O. Weischet, *Biopolymers* **17**, 1387 (1978).
[2] T. M. Laue and D. A. Yphantis, *Fed. Proc., Fed. Am. Soc. Exp. Biol.* **39**, 1602 (1980).
[3] M. S. Runge, T. M. Laue, D. A. Yphantis, M. R. Lifsics, A. Saito, M. Altin, K. Reinke, and R. C. Williams, Jr., *Proc. Natl. Acad. Sci. U.S.A.* **78**, 1431 (1981).

Turbidimetric Ultracentrifugation

Turbidimetric ultracentrifugation allows one to measure sedimentation coefficient distributions of very large particles by employing the UV photoelectric scanner system of the Model E ultracentrifuge at nonabsorbing wavelengths to measure the turbidity of the sedimenting sample. The method has the advantage over other light-scattering techniques that dust and large aggregates are removed during centrifugation. This technique was first employed by Ma *et al.*[4] to study the sedimentation coefficient distribution for very low density lipoproteins of human serum. Since turbidity is a function of the product of the protein concentration and molecular weight, its measurement is relatively insensitive to the presence of low molecular weight contaminants. The utility of this technique has been expanded by Berkowitz and Day[5] to include the measurement of mass per unit length of filamentous viruses. By performing turbidimetric measurements with the scanner at various wavelengths and combining them with concentration determinations using the Rayleigh optical system, one can estimate the mass per unit length for large asymmetric particles. Determination of the wavelength dependence of turbidity provides the same type of information as the measurements of the angular dependence of the Rayleigh ratio in conventional light-scattering techniques.

Non-Newtonian Viscosity

Non-Newtonian viscosity measurements can provide potentially useful information about the shape and aggregation behavior of large asymmetric proteins. This technique has not yet been exploited extensively to study muscle proteins. The potential and limitations of the method have been discussed in a review by Yang,[6] and the general details of procedures for high precision viscometry have been reviewed in this series.[7]

Falling-Ball Viscometry

Falling-ball viscometry has been employed by MacLean-Fletcher and Pollard[8,9] to study *Acanthamoeba* actin filament crosslinking and cell extract gelation. The falling-ball viscometer consists of a capillary tube containing both the protein sample and a stainless steel ball. The ap-

[4] S. K. Ma, V. N. Schumaker, and C. M. Knobler, *J. Biol. Chem.* **252**, 1728 (1977).
[5] S. A. Berkowitz and L. A. Day, *Biochemistry* **199**, 2696 (1980).
[6] J. T. Yang, *Adv. Protein Chem.* **16**, 323 (1961).
[7] J. E. McKie and J. F. Brandts, this series, Vol. 26, p. 257.
[8] S. D. MacLean-Fletcher and T. D. Pollard, *J. Cell Biol.* **85**, 414 (1980).
[9] T. D. Pollard, *J. Biol. Chem.* **256**, 7666 (1981).

paratus allows one to measure both the apparent viscosity and the yield point of the sample. A complete description of this technique is presented in this volume [20].

Yield point determinations have also been made by Brotschi et al.[10] with a simple device that allows one to determine the hydrostatic pressure required to disrupt a gelled solution.

Quasi-Elastic Light Scattering

Quasi-elastic light-scattering (QELS) techniques either alone or combined with various transport methods provide a useful class of methods to study mixtures of macromolecules. Alone these methods are used to determine the diffusion coefficient: this technique is reviewed in this volume.[11] It was first applied to myosin by Herbert and Carlson[12] to study its self-association in high salt.

Sedimentation and Quasi-Elastic Light Scattering

In combination with sedimentation in gradients of different density, QELS methods yield molecular weight distributions of polydisperse preparations.[13] Similar techniques offer potential for investigation of contractile proteins.

Preparation of Proteins

Preparative procedures for contractile proteins have been amply described in the literature and by the other contributors to this volume; they will not be repeated here. Special precautions for viscosity measurements must be taken to remove dust and aggregates either by Millipore filtration or centrifugation. The tendency for many of these proteins to aggregate when diluted requires that samples be centrifuged after dilutions are performed. Removal of dust or aggregates for turbidimetric measurements during analytical ultracentrifugation occurs automatically and, therefore, is not a problem.

[10] E. A. Brotschi, J. H. Hartwig, and T. P. Stossel, J. Biol. Chem. 253, (1978).
[11] C. Montague and F. D. Carlson, this volume [49].
[12] T. J. Herbert and F. D. Carlson, Biopolymers 10, 2231 (1971).
[13] S. T. Kunitake, E. Loh, V. N. Schumaker, S. K. Ma, C. M. Knobler, J. P. Kane, and R. L. Hamilton, Biochemistry 10, 1936 (1978).

[51] Fluorescence as a Probe of Contractile Systems

By ROGER COOKE

Fluorescence spectroscopy has provided a powerful tool for the study of biological systems. The excitation and emission spectra of intrinsic, extrinsic, and prosthetic chromophores can give information on protein conformation, ligand binding, and the accessibility of the chromophore to quenchers. The polarization of the fluorescence can be used to measure the rate of brownian rotations and to define the orientation of chromophores in systems whose structure is organized on a macroscopic scale. The transfer of energy between chromophores has been used as a "molecular meterstick" to measure the distance between them. Contractile systems have provided a particularly suitable subject for fluorescence studies, and all of the above techniques have been applied to them; one can investigate the proteins isolated and in solution as well as in the organized filament array of the muscle fiber. In the field of nonmuscle contractility, fluorescent dyes on both the contractile proteins and antibodies against the contractile proteins have produced a visually dramatic picture of the cytoskeleton. This chapter will first give a brief overview of the parameters of fluorescence that are commonly measured and the information that can be obtained from them. Each technique will then be discussed, along with a description of the methodology, the type of data obtained, the interpretation of these data, and a few representative applications. In general, the apparatus employed is complex, and a detailed description of it would go beyond the scope of this chapter. Thus, I have briefly described some equipment that has been used, indicated where more detailed accounts can be obtained, and attempted to outline the major experimental pitfalls that can plague the investigations. The basic concepts of fluorescence have been reviewed in previous volumes in this series[1,2] and in monographs[3,4]; thus, the discussion given below is very brief.

Absorption of electromagnetic radiation causes the transition of an electron to a molecular orbital of higher energy, represented by the transition from S_0 to S_2 in Fig. 1. This process occurs rather rapidly compared to other processes, taking approximately 10^{-15} sec, and the positions of the nuclei of the molecule do not change during this time. Because the cloud

[1] J. Yguerabide, this series, Vol. 26, p. 498.

[2] G. Weber, see this series, Vol. 16, p. 380.

[3] A. J. Pesce, C. Rosen, and T. L. Pasby, "Fluorescence Spectroscopy: An Introduction for Biology and Medicine." Dekker, New York, 1971.

[4] G. G. Guilbault, "Practical Fluorescence." Dekker, New York, 1973.

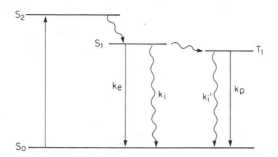

FIG. 1. Energy level diagram for a fluorophore. S_2 and S_1 represent singlet excited electronic states, and S_0 is the ground state. T_1 is an excited triplet state. Straight and wavy lines represent radiative and nonradiative processes, respectively.

of electron density is different in the excited state, the nuclei of the excited molecule as well as charged groups in the surroundings are no longer in their equilibrium positions. These groups relax to their new equilibrium positions in a radiationless transition, S_2 to S_1, that requires approximately 10^{-12} sec to complete. Several processes now compete for the transition from the lowest excited state back to the ground state. The excited state can decay with emission of a photon, i.e., a fluorescent transition, or its energy can be dissipated into molecular vibrations in a nonradiative transition known as internal conversion. Alternatively, the spin of an electron can flip, giving rise to a triplet excited state that can decay via emission of a photon, i.e., phosphorescence, or by internal conversion. Under appropriate conditions, discussed more fully below, the energy of the excited state can be transferred to another chromophore in a process known as Förster energy transfer. The processes outlined above and the information obtained on contractile systems will be discussed in more detail in the appropriate sections below.

Probes of Conformation

The environment that surrounds an excited chromophore can influence its fluorescence in several ways. Reorientation of charged groups partly determines the amount of energy dissipated in the transition from S_2 to S_1, and thus determines the energy of the photon emitted in a radiative transition back to the ground state. Charged groups also provide interactions that enhance the rate of internal conversion and, therefore, influence the amount of fluorescence emitted. Both of these processes rely on the fact that the excited state lifetime is long compared to the period of a molecular vibration. Because absorption occurs more rapidly than a molecular vibration, the fluorescence spectrum of a molecule is more sensitive to its surroundings than is its absorption spectrum.

Most of the fluorescent molecules that have been used to study contractile systems undergo a $\pi \rightarrow \pi^*$ transition upon absorption of a photon. The π^* excited state has a larger dipole moment than the ground state. For these molecules the reorientation of charged groups in the vicinity of the excited molecule results in a decrease in the energy of S_1 and a shift of the emitted fluorescent transition to longer wavelengths. In practice, the effect of the polarity of the environment upon the fluorescent spectrum is ascertained experimentally by observing the fluorescence emitted in media of different polarities.

The amount of fluorescence emitted is known as the quantum yield; defined as the number of photons emitted divided by the number absorbed. It can also be defined by the rate constants given in Fig. 1.

$$Q = k_e/(k_e + k_i) \tag{1}$$

Owing to technical problems it is difficult to measure accurately the number of photons emitted. Thus, changes in quantum yield are often measured by changes in lifetime. The decay of a fluorescent state follows first-order kinetics

$$I(t) = I(0) \exp(t/\tau) \tag{2}$$

where I is the intensity and τ is the lifetime, $\tau^{-1} = k_e + k_i$. Equations (1) and (2) can be combined to give a relation between changes in Q and changes in τ.

$$Q/Q' = \tau/\tau' \tag{3}$$

where the primed quantities refer to the parameters of a probe in a different environment. Equation (3) describes those situations where k_e, the rate of emission of fluorescence, is the same in the two environments, an assumption that is generally valid.

Changes in the intensity and energy of the emitted fluorescence are often measured by observing the fluorescence of samples that are excited by a steady illumination. A representative apparatus is shown schematically in Fig. 2. Fluorometers of this type can be assembled fairly inexpensively from the separate components or can be purchased from several manufacturers.[5,6]

The major problems that arise in the attempt to measure the intensity of emitted fluorescence are due to the fact that none of the components outlined in Fig. 2 work perfectly. Monochromators based on gratings will also pass higher-order reflections. These are often reduced by the use of filters in tandem with the monochromator. Because the systems used to

[5] C. A. Parker, "Photoluminescence of Solutions." Elsevier, Amsterdam, 1965.
[6] J. M. Fitzgerald, in "Modern Fluorescence Spectroscopy" (E. L. Wehry, ed.). Plenum, New York, 1976.

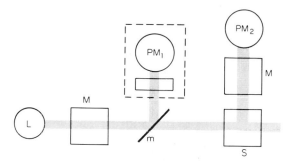

FIG. 2. Schematic of a fluorometer designed to measure steady-state fluorescence. Light from a source, L_1, passes through, M, which represents monochromators and/or filters and polarizers used to define the characterististics of the exciting beam. A half-silvered mirror, m, reflects a portion of the beam into a quantum counter and photomultiplier, PM_1, which is used as a reference signal to correct for fluctuations in lamp intensity. The fluorescence from the sample S is detected by a second photomultiplier, PM_2, after passing through a set of monochromators, etc.

determine the wavelength of the exciting and emitted light are not absolute, some small fraction of the exciting light can be reflected into the sensing photomultiplier by turbidity in the sample. Often the exciting beam is intense and the fluorescence signal is weak so that only a very small fraction of the exciting light has to reach the photomultiplier in order to cause artifacts. The problem is more acute when working with turbid samples, such as those encountered when working with polymers of actin or myosin. For instance, changes in turbidity, such as that which occurs on the binding of subfragment-1 to actin, can result in artifactual changes in the observed fluorescence. The effect of scattered light in a fluorometer should be ascertained by observing the shifts in the photomultiplier output that occur when the turbidity of the sample is altered via addition of a turbid substance, such as polystyrene spheres. In some situations the amount of scattered light can be minimized by using an excitation beam that is polarized along the direction of observation. The observed fluorescence can also be influenced by the optical density of the sample, since the optical density can attenuate both the exciting and emitted light. All of the above factors can cause artifactual changes in the measured intensity of the fluorescence. Another problem is introduced by the fact that the efficiency of many elements in the system (filters, photomultipliers, etc.) are wavelength dependent; however, commercial instruments that can correct for this problem are now available. Thus, the measurement of Q by determining the number of emitted photons is possible but subject to experimental difficulties. In general, however, the measurement of changes in observed intensity are sufficient to give useful information about the system, and uncorrected spectra are often reported.

In spite of these technical difficulties, steady-state fluorescence spectra do provide information on the environment of the fluorophore, and this approach has been used to study a number of contractile proteins. The investigations have used both the intrinsic fluorescence of tryptophan and the fluorescence of attached dyes. A few selected examples, which illustrate the type of information that can be obtained, are discussed below.

Werber et al.[7] showed that the binding of nucleotides to myosin causes an increase in the intensity of its tryptophan fluorescence and a shift of the maximum intensity to shorter wavelengths. During the steady-state hydrolysis of ATP the intensity of fluorescence increases by 17%. This effect does not require hydrolysis as evidenced by the fact that the binding of ADP causes an increase in fluorescence of 6%. The binding of PP_i produces very little change in the intensity, indicating that the base is required to alter the fluorescence. All the spectral changes described above are consistent with one or more tryptophans shifting from a polar to a nonpolar environment upon binding of nucleotide. The hypothesis that tryptophans are buried within the hydrophobic interior of the protein was tested by the use of quenching agents, described in more detail below. These studies indicated that the binding of a nucleotide caused two tryptophans to move from a polar environment, which was accessible to the solvent, to a nonpolar environment, which was inaccessible to the solvent.[7]

The changes in tryptophan fluorescence described above have been used to define some kinetic intermediates of the myosin ATPase reaction.[8,9] These studies have shown that the binding of nucleotides to myosin is at least a two-step process involving first a collision intermediate followed by a transition to a second state that exhibits increased tryptophan fluorescence. Measurements of the rapid changes in fluorescence that arise following the mixing of protein and nucleotide were made in a stopped-flow apparatus. The adaptation of the apparatus shown in Fig. 2 for use in a stopped-flow apparatus is relatively straightforward. Often the results obtained in the rapid-mixing apparatus are compared to those obtained in steady-state machines. This comparison must be performed with care, as discussed by Johnson and Taylor,[9] because uncorrected spectra from one fluorometer cannot necessarily be compared quantitatively with that from another.

Extrinsic fluorophores enjoy some advantages over intrinsic fluorophores, but their use also raises some problems. The probes can be

[7] M. M. Werber, A. G. Szent-Györgyi, and G. D. Fasman, *Biochemistry* **11**, 2872 (1972).
[8] C. R. Bagshaw, J. F. Eccleston, F. Eckstein, R. Goody, H. Gutfreund, and D. R. Trentham, *Biochem. J.* **141**, 352 (1974).
[9] K. A. Johnson and E. W. Taylor, *Biochemistry* **17**, 3432 (1978).

chosen to have desirable characteristics, such as a long lifetime or a large response to changes in polarity. They provide the most useful information when attached specifically to a single site, and such specificity has been obtained with a variety of probes having reactivities for sulfhydryl or amino groups. The attached probes must be sensitive to the effect that one desires to monitor and should not greatly perturb the function of the proteins to which they are attached. When these requirements can be met, the signals obtained from the attached probes can be more informative than those from intrinsic fluorophores, since the abundance of common intrinsic fluorophores such as tryptophan complicates the interpretation of their spectra.

The binding of calcium to troponin C results in conformational changes in the protein that have been monitored by both extrinsic probes and intrinsic tyrosine. The single cysteine could be labeled selectively with S-mercuric-N-dansylcysteine, and Potter and co-workers[10] found that the fluorescence intensity of this probe increased upon binding of calcium to either the high- or low-affinity sites. This effect was used to monitor the binding of calcium to troponin C. The fluorescence of tyrosine was also found to increase when calcium bound to these two sites.[11] The reaction of dansylaziridine with methionine-25 of troponin C resulted in a derivative whose fluorescence increased only when calcium bound to the low-affinity calcium-specific sites.[12] These probes have been used to investigate the rates of calcium binding to, and dissociation from, troponin. They are useful tools for probing the function of troponin, and the fluorescence of dansylaziridine-labeled troponin C can provide a method for measuring calcium concentrations.

The investigations addressed above illustrate the type of information that can be obtained on protein conformation. This information is obtained mainly from the emission spectrum of the fluorophore. The spectrum is determined by mechanisms, usually involving adjacent charged groups, which alter the energy of S_1 and the rate, k_i, of internal conversion. These mechanisms are complex and interpretations of the fluorescence spectra are often confined to statements concerning the polarity of the environment of the fluorophore. Although more detailed information on protein conformations cannot be obtained, the sensitivity of these probes to their environment nonetheless makes them valuable probes of changes that occur in protein conformation. Rather simple steady-state fluorometers have been used to measure changes induced by the binding of ligands, denaturation, etc. The sensitivity of the instrumentation allows

[10] J. D. Potter, J. L. Seidel, P. Laevis, S. S. Lehrer, and J. Gergely, *J. Biol. Chem.* **251**, 7551 (1976).
[11] J. D. Johnson and J. D. Potter, *J. Biol. Chem.* **253**, 3775 (1978).
[12] J. D. Johnson and A. Schwartz, *J. Biol. Chem.* **253**, 5243 (1978).

one to monitor the fluorescence of very dilute probes (<1 μM). In conjunction with a rapid mixing apparatus, the kinetics of these conformational changes can be monitored.

Probes of Accessibility

A number of small molecules can quench the fluorescence of an excited fluorophore. A collision between these molecules and the fluorophore causes an increase in the rate of nonradiative transitions, k_i in Fig. 1. Commonly used quenchers include molecules that have polarizable charge distributions, such as I^-, or unpaired electron spins, such as O_2. When the quenching process involves a diffusion controlled collision between the quencher and the fluorophore, the dependence of the fluorescence yield, F, on the concentration of the quencher, [Q], is given by the Stern–Volmer equation[13] [Eq. (4)].

$$F_0/F = 1 + k_+\tau[Q] \tag{4}$$

where F_0 is the fluorescence observed in the absence of Q, τ is the lifetime of the dye in the absence of Q and k_+ is the bimolecular quenching constant. The constant k_+ describes the rate of collisions between quencher and dye and is given by:

$$k_+ = 4\pi aDN \tag{5}$$

where D is the sum of the diffusion constants for the two molecules, a is the sum of their radii, and N is the number of molecules per millimole (when Q is expressed in millimoles per liter). In a number of studies it has been found that the Stern–Volmer equation is obeyed; i.e., plots of $1/F$ vs [Q] give a straight line with a slope of $k_+\tau$.

The ability of hydrophilic quenching agents to decrease the fluorescence of fluorophores that are incorporated in some protein structure has been used to define the accessibility of the fluorophores to the solvent. The work of Werber et al.[7] suggested that some tryptophans of myosin were transferred to a hydrophobic environment upon binding of nucleotides to the myosin. They checked this possibility by observing the ability of KI to quench the fluorescence of the tryptophans. Owing to the presence of several classes of tryptophans, some accessible and others inaccessible to the I^-, Stern–Volmer plots were not linear and a modified plot, due to Lehrer,[14] was employed.

$$F_0/\Delta F = \frac{1}{[Q]f_a}\, k_+\tau + \frac{1}{f_a} \tag{6}$$

---

[13] O. Stern and M. Volmer, *Phys. Z.* **20**, 183 (1919).
[14] S. S. Lehrer, *Biochemistry* **10**, 3254 (1971).

where f_a is the fluorescence of the class of accessible fluorophores. An analysis of these data in the presence and the absence of nucleotides indicated that the interaction of the myosin head with ATP results in two tryptophans becoming inaccessible to the quenchers.

The electric potential surrounding a fluorophore bound to the reactive sulfhydryl on the myosin head was measured by contrasting the quenching of the fluorescence by a negatively charged ion (iodine), a positively charged ion (thallium), and a neutral molecule (acrylamide).[15] When the quencher carries a charge, the rate constant k_+ is influenced by the interaction of the charge and the electric potential surrounding the fluorophore. In practice the ionic strength of the medium is changed and its influence on the rate of quenching is measured. For instance, it was found that increasing the ionic strength, which decreases electric potentials by shielding charged groups, increased the effectiveness of I^-, decreased the effectiveness of thallium, and did not greatly influence the effectiveness of acrylamide as quenchers. All these effects are those expected if the fluorophore is in an electronegative environment.

Probes of Orientation

In the classical view of the interaction of light with a molecule, the light is absorbed by and emitted from the molecule via electric dipole oscillators. These oscillators have certain orientations relative to the molecule. The probability that light is absorbed by a molecule is proportional to $\cos^2\theta$, where θ is the angle between the absorption dipole and the electric vector of the exciting light. Thus, those molecules whose absorption dipoles are aligned along the direction of the polarization of the exciting light are preferentially excited. The properties of the emitted light are governed by the emission dipole. The light can be emitted at any angle, with a probability that is proportional to $\sin^2\phi$, where ϕ is the angle between the emission dipole and the direction of propagation of the emitted light. The electric vector of the emitted light is constrained to lie in a plane defined by the emission dipole and the direction of propagation. These properties allow one to determine the orientation of fluorophores that are rigidly attached to macroscopically oriented systems such as a muscle fiber. When the fluorophores are attached to molecules that are rotating in solution, one can use measurements of fluorescence polarization to determine their rate of rotation as discussed in the next section.

Several experimental configurations have been used to study the polarization of fluorescence originating from muscle fibers, and a representative apparatus is outlined in Fig. 3.[16] The fiber is illuminated

[15] T. Ando, H. Fujisaki, and H. Asai, *J. Biochem.* **88**, 265 (1980).
[16] C. G. Dos Remedios, R. G. C. Millikan, and M. F. Morales, *J. Gen. Physiol.* **59**, 103 (1972).

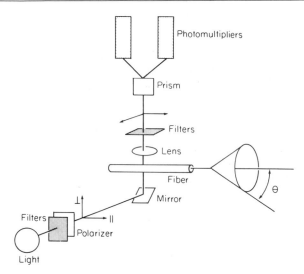

FIG. 3. A fluorometer designed to measure the components of the polarization of the fluorescence emitted from a muscle fiber.

through a lens system by light that is polarized either parallel or perpendicular to the fiber axis. The fluorescent light is collected through a second lens system and passed through a Wollaston prism that resolves the beam into two components with orthogonal polarization. This arrangement of the optics allows the same system of lenses to collect both components of the emitted fluorescence. Each component is detected by a photomultiplier whose output is usually fed into a minicomputer that stores and manipulates the data. The Wollaston prism is arranged to select two light beams whose electric vectors are parallel and perpendicular, respectively, to that of the exciting light. Taking the intensities of these two beams as $I_{\parallel}$ and $I_{\perp}$, the polarization is defined by

$$P = (I_{\parallel} - I_{\perp})/(I_{\parallel} + I_{\perp}) \tag{7}$$

When the incident light is polarized perpendicular to the fiber axis the above function is termed $P_{\perp}$ and when the light is polarized parallel to the fiber axis it is termed $P_{\parallel}$. The wavelengths of the emitted and exciting beams are defined by filters. In the configuration pictured in Fig. 3, the excitation beam is focused on the fiber along a line that is coincident with the direction in which the fluorescence is viewed. Other arrangements have also been used. The excitation beam can encounter the fiber at 90 degrees to the direction in which the fluorescence is viewed, thereby making it easier to separate the fluorescence from the exciting light.[17]

[17] J. Borejdo and S. Putnam, *Biochim. Biophys. Acta* **459**, 578 (1977).

Alternatively, the exciting light can be introduced into the optical path above the fiber and reflected down by a dichroic mirror through the same optical system that is used to collect the fluorescence.[18,19]

The apparatus shown in Fig. 3 is capable of making continuous measurement of the polarization of single muscle fibers. Many of the problems that arise are due to the fact that the components tend to alter the polarization of light that passes through them. Lenses will depolarize light, particularly when short focal points are used. Dichroic mirrors will partially polarize the light that they reflect. As with other fluorometers the possibility of the incident beam reaching the photomultiplier must be considered.

The filament array of the muscle fiber provides a macroscopically oriented sample in which most of the proteins, particularly those of the filaments, are in a helical array aligned along the fiber axis. If a helical array of fluorophores is excited by light polarized either parallel or perpendicular to the axis of the helix then a certain subset of fluorophores will be selectively excited. This set of fluorophores will have a highly asymmetric distribution of emission dipoles, and thus, its fluorescence will be polarized. Four different intensities can be measured in the experimental apparatus outlined in Fig. 3; i.e., $I_\parallel$ and $I_\perp$ can be measured for the two cases when the exciting light is polarized either parallel or perpendicular to the fiber axis. For some well defined cases, the components of the fluorescence can determine the orientation of the fluorophore array in the fiber. The polarization expected from the interaction of a helix of fluorophores with a plane-polarized excitation beam has been calculated by Tregear and Mendelson.[20] The cylindrical symmetry of the fiber requires that all dipoles that make a specific angle with the fiber axis lie on the surface of a cone; such a cone is illustrated in Fig. 3. Using the appropriate relations governing the absorption and emission of light, discussed very briefly above, and averaging over a cylindrically symmetric ensemble of probes, the fluorescent intensities expected for various probe angles were calculated. The polarization expected for probes with random orientations was also calculated. However, because the information is limited, i.e., there are only three independently measurable intensities, the measurements can only uniquely define the angular distribution of the fluorophores in the fiber in the case where the angular distribution is a simple one. This case has been realized for some extrinsic probes, as discussed below.

The first study of the polarization of the fluorescence of muscle fibers was carried out by Aronson and Morales,[21] who found that the polariza-

[18] K. Guth, *Biophys. Struct. Mech.* **6**, 81 (1980).
[19] J. Borejdo, S. Putman, and M. F. Morales, *Proc. Natl. Acad. Sci. U.S.A.* **76**, 6346 (1979).
[20] R. Tregear and R. A. Mendelson, *Biophys. J.* **15**, 455 (1975).
[21] J. F. Aronson and M. F. Morales, *Biochemistry* **8**, 4517 (1969).

tion of the fluorescence of tryptophan depended on the state of the muscle. The changes in $P_\perp$ were larger than those in $P_\parallel$. The value of $P_\perp$ increased when the fiber went from the rigor to the relaxed states and was intermediate between these two during contraction. These results were extended by Dos Remedios et al.,[16] who found values for $P_\perp$ of 0.09, rigor; 0.126, relaxation; and 0.110, contraction (all ±0.001). Qualitatively similar results were found for living frog muscle.[16] Guth has claimed that the polarization observed for contracting muscle is identical to that of relaxed muscle when the concentration of ATP bathing the glycerinated fibers is sufficiently high.[18] Although this discrepancy has not been fully resolved, the observation of a difference in polarization between these two states in living frog muscle, a system with an adequate supply of ATP, argues that the orientation of some tryptophans do change upon contraction.

The use of extrinsic probes to measure myosin cross-bridge orientation in muscle fibers introduces some experimental difficulties, but the data obtained are more informative than data obtained from tryptophan fluorescence. A quantitative treatment of the data requires that the extrinsic fluorophore be attached selectively and rigidly to a single site on the myosin head. It has indeed been fortunate for these studies that myosin has one sulfhydryl that is more reactive than any others. This sulfhydryl has provided a probe site for a number of studies involving myosin in solution and in fibers. The fluorophore 1,5-IAEDANS has been found to attach to the reactive sulfhydryl and to be rigidly attached to the myosin head (discussed more fully in the next section).[22,23] To be useful the probes should not affect the function of the myosin. The reaction of fibers with 1,5-IAEDANS did not decrease their ability to exert rigor tension; however, the ability to produce active tension should be explored further.[17] It has been shown that the attachment of a variety of ligands to the reactive sulfhydryl can decrease the ATPase activity of acto S-1, indicating that the probes may also affect some cross-bridge functions in the fiber.[24]

The polarization of 1,5-IAEDANS attached to SH_1 in glycerinated fibers was found to change when the state of the fiber changed.[22] Qualitatively the changes were similar to those observed with tryptophan; i.e., $P_\perp$ was greater in relaxation, less in rigor, and intermediate during contraction, whereas $P_\parallel$ showed much smaller changes. The specificity of this effect was checked by using a fluorophore that reacted predominantly with actin rather than with myosin. Fibers labeled with this fluorophore showed no change of fluorescence polarization upon change of state.[22]

[22] T. Nihei, R. A. Mendelson, and J. Botts, Biophys. J. 14, 236 (1974).

[23] R. A. Mendelson, M. F. Morales, and J. Botts, Biochemistry 12, 2250 (1973).

[24] S. Mulhern and E. Eisenberg, Biochemistry 17, 4419 (1978).

Calculations showed that 1,5 IAEDANS attached to the rigor fiber did not give the polarization expected for a helix in which all dipoles made a single angle with the fiber axis. A fit could be obtained only if a class of fluorophores with random distributions was included, suggesting that the probe was not completely specific for SH_1.[20]

The fluorescence of 1,5-IAEDANS attached to fibers was reinvestigated by Borejdo and Putnam, using a new method for labeling the fibers.[17] If the reaction between dye and fiber is carried out at 0° in relaxed fibers, the label reacts more specifically with SH_1.[25] Scans of the fluorescence of polyacrylamide gels run in the presence of SDS showed that more than 80% of the dye was attached to myosin heavy chains. A comparison between the intensities of the fluorescence observed from these fibers and the theoretical results indicated that most of the fluorophores in the rigor fiber had the same orientation with respect to the fiber axis. When the fiber was relaxed the orientation of the fluorophore became more random, but was still distinguishable from a completely random distribution.

The above studies show that the polarization of a fluorophore can be used to monitor changes in the orientation of myosin heads in muscle fibers. This property has been exploited to measure the dynamics of cross-bridge orientation during contraction.[19] Using an apparatus similar to that of Fig. 3, the fluctuations in $P_\perp$ were monitored as a function of time. To reduce the problems associated with the photobleaching of the probe a rhodamine label was employed. The observation of fluctuations requires that the number of active fluorophores be low, and the fluorescent intensity due to approximately 2×10^6 dyes was measured. Fluctuations in polarization were observed during contraction, but not during relaxation or rigor. This result indicates that cross-bridges can assume several orientations during contraction.

The polarization of the fluorescence of an ATP analog, ϵ-ATP, incorporated into actin filaments in myosin-free fibers was measured by Yanagida and Oosawa.[26] They used the components of the polarized fluorescence to calculate the average angles of the emission and absorption dipoles of the ϵ-ADP as well as the mean square deviation about these angles. The mean square deviation was interpreted as arising from fluctuations in the orientation of the actin filaments and from it they derived a value for the elastic modulus for the bending of the actin. This elastic modulus was shown to decrease upon binding of heavy meromyosin to actin or upon binding of calcium to troponin.

[25] J. Duke, R. Takashi, K. Ue, and M. F. Morales, *Proc. Natl. Acad. Sci. U.S.A.* **71**, 274 (1974).
[26] T. Yanagida and F. Oosawa, *J. Mol. Biol.* **126**, 507 (1978).

The studies cited above show that the measurement of the polarization of the fluorescence that originates from fluorophores in muscle fibers can be a powerful method for investigating the orientation of the fluorophores. The spectra can be obtained from single fibers, and intensities can be recorded continuously with great accuracy. The major experimental difficulties lie not in the construction of the apparatus, but in achieving labeling of specific sites. This problem has been largely solved in some cases. The major difficulty associated with the interpretation of the data is that one obtains only three independent numbers, and these contain insufficient data to define a complex distribution of angles. Thus, the introduction of a class of fluorophores at nonspecific sites severely limits quantitative conclusions. More recently, the spectra of electron paramagnetic probes also attached to the myosin SH_1 in muscle fibers has been observed.[27] The spectra of these probes can define their orientation, and they have corroborated the results found with the fluorescent probes; i.e., the probes have similar orientations in rigor muscle and random orientations in relaxed muscle. The paramagnetic probes can provide more information on angular distributions than can fluorescent probes; however, the signals are not sufficiently strong to allow spectra to be recorded from single fibers. Thus, although these two techniques have considerable similarities, they have different strengths and can provide complementary information.

Detection of Molecular Rotations

In the preceding section it was shown that the polarization of the fluorescence emitted by a static array of fluorophores could be used to measure their orientation. If, during the lifetime of the excited state, the fluorophores change their orientation, the observed polarization will decrease. This effect has been used to measure the Brownian rotation of fluorophores attached to proteins. The basis for the depolarization of fluorescence by rotation of the fluorophores is well understood. One is usually studying protein molecules in solution, so that there is an isotropic distribution of probe orientations. Because the probability of absorption is greater for probes aligned along the electric vector of the exciting light, a plane-polarized beam excites a subset of fluorophores whose distribution of orientations is anisotropic. If the orientation of the fluorophores does not change, the fluorescence observed will also be polarized. Rotation of the fluorophores during the lifetime of the excited state will decrease this polarization.

Several experimental configurations have been used to study the effects of brownian rotations on fluorescence polarization. In early experi-

[27] D. Thomas and R. Cooke, *Biophys. J.* **32**, 891 (1980).

ments the sample was excited with a steady illumination and the steady-state polarization was observed in an apparatus similar to that shown in Fig. 2. For these experiments, a more effective experimental design is to have two photomultipliers, one to monitor each component of the fluorescence. The effects of rotations on the polarization for this case was analyzed by Perrin.[28] The polarization, P, is given by

$$(1/P - \tfrac{1}{3}) = (1/P_0 - \tfrac{1}{3})(1 + 3\tau/\rho)$$

where P_0 is the polarization in the absence of rotations; τ is the lifetime of the dye, and ρ is the rotational relaxation time. This equation was used to analyze the rotations of probes attached to F- and G-actin.[29] The use of a steady illumination has the advantage that measurements can be made quickly, accurately, and easily. Thus, it could be used to monitor changes in rotations in a variety of situations, such as that following the rapid mixing of reactants in a stopped-flow apparatus. There are several disadvantages to steady-state measurements. The measurement of P under one set of conditions does not contain sufficient information to define both P_0 and ρ. More information can be obtained by varying ρ, via changes in viscosity, but this method introduces new uncertainties since the changes in temperature or in solvent composition used to vary viscosity may alter the system under investigation. Thus, more recent investigations have employed the measurement of the time course of polarization following a transient excitation.

Weber and co-workers have developed a method for measuring the rate of decay of the polarization of fluorescence using an exciting light whose intensity is modulated sinusoidly at high frequencies, 1–20 MHz.[30] When the frequency of the modulation is of the same order as the inverse lifetime of the excited state, the observed fluorescence will also be modulated. Owing to the lifetime of the excited state, the maximum intensity of the fluorescence will lag behind the maximum in excitation. By measuring the phase difference between the parallel and perpendicular components of the fluorescence, the rate of decay of the polarization can be measured. This method has the advantage that measurements can be made quickly, and that the time resolution is limited only by the modulation frequency and the ability to determine phase differences. Thus, sub-nanosecond decays can be measured. The disadvantage of this method is that a measurement at a single modulation frequency will not allow one to resolve different decay rates in samples in which the polarization decays with multiple rates. The use of multiple modulation frequencies could

[28] F. Perrin, *J. Phys. Radium* **1**, 39 (1926).
[29] H. C. Cheung, R. Cooke, and L. Smith, *Arch. Biochem. Biophys.* **142**, 333 (1971).
[30] G. Weber and G. W. Mitchell, *Excited States Biol. Mol., Proc. Int. Conf. 1974* (1974).

resolve multiple decay processes; however, it is difficult to build instruments that have multiple modulation frequencies.

The most direct method of observing the effect of rotations on fluorescence polarization is to observe its decay with time following excitation by a nanosecond pulse of light. This method has been used to investigate the contractile proteins and is described in more detail below.

The methodology for producing nanosecond light pulses and measuring the time course of fluorescent decay has been reviewed by Yguerabide,[1] and a representative instrument is described by Mendelson et al.[23] The source of illumination is a flash lamp producing a train of light pulses with a width of a few nanoseconds. These flashes are plane polarized by a Glan–Thompson prism, and the fluorescence is observed at right angles to both the incident beam and its direction of polarization. A Wollaston prism splits the fluorescence into two orthogonally polarized beams that are detected by fast photomultipliers. The optical paths are thus a modification of those shown in Fig. 2. In order to measure accurately the time course of the components of fluorescence, the time delay between the pulse of exciting light and the arrival of a photon at one of the photomultipliers is measured using a time-to-amplitude converter. To avoid distortion of the time course, the observed intensities should be such that the probability of detecting more than one photon per flash is low. This is achieved by using low illumination intensities or dilute samples, or both. Typical flash rates are about 100 kHz and a photon is detected in about 10% of the flashes. The results of multiple flashes are stored in a multichannel analyzer or minicomputer. The time course of the decay of the polarized fluorescent intensities can be measured directly from several nanoseconds up to several hundred nanoseconds using fluorophores with lifetimes of around 20 nsec. The decay of multiple components can thus be resolved if they differ greatly in their lifetimes. Brownian rotations with rotational relaxation times ranging from 10 to 1000 nsec have been studied.

One problem that has hindered the use of fluorescence polarization to measure rotational relaxation rates in contractile proteins is that the rotational relaxation rates are slow, $\rho \sim 200–1000$ nsec compared to the lifetime of conventional fluorophores ~ 20 nsec. Thus, the fluorescence decays more rapidly than the polarization. In this situation the single-photon method described above has several advantages. Because the photomultiplier is used only to measure the time of arrival of a photon, nonlinearities in the sensitivity of the system do not distort the curve of polarization vs time. Thus, the decay of the fluorescence components can be recorded over several decades of intensity, allowing one to measure the rather slow rates of the decay of polarization. Rotational correlation times of up to 1000 nsec have been measured by this technique, but this represents an upper limit with current technology and fluorophores. Unfortunately, many rotations, such as those of the myosin heads in muscle fibers, appear to have correlation times of 1000 nsec or slower. These slow relaxation times could be measured using fluorophores with longer lifetimes. Since the decay of phosphorescence is often slow, the development of probes that decay from a triplet state (Fig. 1) could allow present instruments to measure the rates of slower processes. Alternatively, the development of new faster electronics may drastically improve the ability to measure fast decay processes following a modulated light source with a continuously variable frequency.

The flexibility of myosin was studied by measuring the Brownian rotations of 1,5-IAEDANS attached to the reactive sulfhydryl on its heads.[23] The polarization correlation time ϕ was found for dyes attached to subfragment-1, heavy meromyosin, and intact myosin to be 220, 400, and 450 nsec, respectively. The fact that ϕ measured for dyes attached to myosin is only twofold greater than that measured for subfragment-1 shows that the heads of the intact myosin have considerable rotational flexibility relative to the entire molecule. A theoretical analysis concluded that the heads were attached to the rod by a "universal joint" that did not greatly restrict their ability to rotate.

Measurements of fluorescence depolarization were used by Highsmith, Mendelson, and Morales to determine the binding constant for the acto subfragment-1 complex.[31,32] The difference in the rate of Brownian rotations, measured as described above, between subfragment-1 free in solution and bound to actin allows the determination of the fraction bound. Because the fluorescence intensities can be very low, in fact have to be low, submicromolar concentrations of subfragment-1 can be studied, and the tight binding of subfragment-1 to actin, with binding constants around $10^6 M^{-1}$, could be determined. Although the measurements were lengthy, the methodology is straightforward, and the effects of salts, temperature, and ligands on the association constant were readily measured.[33] The interaction energy between the nucleotide and actin binding sites was also quantitated.

The studies cited above show that measurements of fluorescence polarization can be a powerful tool for investigating molecular rotations. As in the experiments on probe orientation, the results obtained on the rotation of fluorescent probes can be contrasted with those obtained using electron paramagnetic probes. The rates of rotation of myosin and its subfragments were almost identical when measured by the two tech-

[31] S. Highsmith, R. A. Mendelson, and M. F. Morales, *Proc. Natl. Acad. Sci. U.S.A.* **73**, 133 (1976).
[32] S. Highsmith, *J. Biol. Chem.* **251**, 6170 (1976).
[33] S. Highsmith, *Arch. Biochem. Biophys.* **180**, 404 (1977).

niques.[34] The paramagnetic probes can be used to measure rotational correlation times over a wider range 10^{-11} to 10^{-4} sec than is easily accessible to fluorescent probes; however, their utility is limited by their signal strength, requiring several orders of magnitude greater concentrations of probes.

Probes of Distance

In addition to the mechanisms discussed above, an electronically excited molecule can also lose its energy by transferring it to an adjacent molecule. This process is the result of a resonance between the two molecules, and it does not involve the emission and reabsorption of a photon, a process known as trivial reabsorption. Classically, the resonance can be thought of as the interaction of the oscillating electric field of the excited donor molecule with the electric dipole moment of the acceptor molecule. Quantum mechanically it is the result of the transfer of a virtual photon between donor and acceptor. Fluorescence energy transfer and its application to biological systems has been reviewed in detail.[35,36]

There are a number of conditions that determine whether resonant energy transfer can occur and, if so, the amount of energy transferred. The absorption spectrum of the acceptor must overlap the emission spectrum of the donor. The spectral overlap integral, J, can be calculated from the known spectra. The rate of transfer is proportional to the inverse sixth power of the distance between donor and acceptor. It is this dependence on distance that allows the effect to be used as a molecular yardstick. The rate of transfer also depends on κ^2, where κ is a function involving the relative orientation of the emission dipole of the donor and the absorption dipole of the acceptor. An elementary formula for the interaction energy between two dipoles yields

$$\kappa = \cos \theta_{AD} - 3 \cos \theta_A \cos \theta_D \tag{8}$$

where θ_A and θ_D are the angles between the individual oscillators and the line joining them, and θ_{AD} is the angle between the two dipoles. The efficiency of fluorescent energy transfer is then given by a formula first derived by Förster.[37]

$$\text{Efficiency} = [1 + (R/R_0)^6]^{-1} \tag{9}$$

R_0 is given by

[34] D. P. Thomas, S. Ishiwata, J. C. Seidel, and J. Gergely, *Biophys. J.* **32**, 893 (1980).
[35] L. Stryer, *Annu. Rev. Biochem.* **47**, 819 (1978).
[36] V. Blumenfield, *Q. Rev. Biophys.* **11**, 251 (1978).
[37] T. Förster, *Discuss. Faraday Soc.* **27**, 7 (1959).

$$R_0 = 9.7 \times 10^3 \times (J\kappa^2 Q n^{-4}) \qquad (10)$$

where n is the index of refraction of the medium between the two probes and Q is the quantum yield of the donor in the absence of transfer.

The above formula can be used to calculate the distance between two fluorophores from the measured efficiency of energy transfer between them. The overlap integral can be calculated accurately, the quantum yield Q can be measured, and the index of refraction can be estimated. The major apparent problem in the application of Eq. (9) is the orientation factor κ^2. This factor, which can vary from 0 to 4, is usually not known. As a first approximation the value for an isotropic distribution, $\kappa^2 = \frac{2}{3}$, can be assumed. Although at first sight this may appear to be a poor approximation for a function that can vary arbitrarily from 0 to 4, in practice it appears to work reasonably well. Several factors operate to constrain the possible range of κ^2. If one of the fluorophores has some rotational freedom relative to the other, the isotropic approximation is a good one. The presence of rotational motion can be checked by measuring the polarization of the fluorescence of the two probes as discussed in the preceding section. However, even in the case where the probes are rigidly attached to the protein structure, the dynamics of that structure ensures that some distribution of angles is sampled. As discussed by Stryer,[35] the presence of even a small amount of flexibility will provide rather severe constraints on the values of κ^2. Experimentally, the problems introduced by the angular factors can be attacked by measuring the distance between two sites using several different sets of probes. When this has been done, the distances calculated from the different sets are generally in agreement, indicating that the isotropic approximation for κ is reasonable.[35,36] A variation of this method is to measure the efficiency of energy transfer with probe A at site 1 and probe B at site 2 and then to switch probes and remeasure the transfer. This approach eliminates other uncertainties, and agreement between the two results indicates that angular distributions are probably not a problem. When distances have been measured in proteins whose structures are known from X-ray diffraction, they have generally been found to be in agreement with the X-ray results.[35,36] Thus, in spite of some uncertainties in the method, the measurement of molecular distances by fluorescent energy transfer appears to be valid. The range of distances that are assessible to this method depends on the magnitude of the factors that determine R_0. Generally, the commonly used fluorophores can measure distances between 10 and 80 Å.

Experimentally one can measure the efficiency of energy transfer in one of several ways. First, transfer decreases the quantum yield and the lifetime of the donor. Second, excitation of the donor will result in an

Donor site	Acceptor site	Distance (nm)	References
ϵ-ATP in actin	DDPM at actin Cys^{373}	3.0	39
1,5-IAEDANS at actin Cys^{373}	5IAF at myosin SH_1	6.0	44
1,5-IAEDANS at myosin SH_1	5IAF at actin Cys^{373}	6.0	44
1,5-IAEDANS at Cys^{177} of myosin light chain 1	5IAF at myosin SH_1	3.9	38
1,5-IAEDANS at myosin SH_1	5IAF at Cys^{177} of myosin light chain 1	4.0	38
1,5-IAEDANS at myosin SH_1	TNP at a reactive Lys on myosin	2.6	40
1,5-IAEDANS at Cys^{177} of myosin light chain 1	TNP-ADP on myosin	5.5	41
ϵ-ADP on myosin	Crosslinker between myosin SH_1 and SH_2	1.5–2.5	42
1,5-IAEDANS at myosin SH_1	TNP-ADP on myosin	3.8	43
ϵ-ADP on myosin	TNP at a reactive Lys on myosin	4.2	43

increase in the fluorescence of the acceptor. A steady-state fluorometer such as that shown in Fig. 1, can be used to measure corrected excitation and emission spectra in the presence and in the absence of the acceptor, and the efficiency of transfer can be calculated from the change in quantum yield. A better method is to measure the lifetime of the excited state of the donor in the absence and in the presence of the acceptor. The lifetime can be measured very accurately using pulsed light sources as described above, and its interpretation is less open to ambiguity.

In addition to the theoretical uncertainties discussed above, several experimental uncertainties must also be dealt with in the calculation of distances between probes. If stoichiometric labeling is not achieved, the presence of the different populations must be taken into account. This involves the determination of the stoichiometries and a summation over all possible species as outlined by Marsh and Lowey.[38] A second problem in interpretation is introduced by nonspecific labeling. Because the efficiency of transfer depends on the inverse sixth power of the distance between probes, acceptors that are very close to the donor will be very efficient at promoting transfer, making the calculated distance too small. Both of the above problems can be approached by the measurement of the time-resolved decay of the fluorescence of the donor. If several populations of donor acceptor pairs exist, the decay of fluorescence will be biphasic, and under some conditions the different components can be resolved from one another. Thus, the presence of different species can be detected, and the change in lifetime for each decay of each component can be used to calculate the transfer efficiency for its configuration of probes.

A number of sites including reactive sulfhydryls, lysines, and nucle-

[38] D. J. Marsh and S. Lowey, *Biochemistry* **19**, 774 (1980).

otide binding sites can be specifically labeled both on the myosin head and on actin. Distances have been measured between a number of these sites,[38-44] and these are listed in the table. As can be seen in several instances where a distance has been determined by two different pairs of probes, the results agree. In the absence of a three-dimensional structure of these proteins, the method of energy transfer provides the only reasonable method to determine the distance between sites. In addition to determining the distance between sites, the measurement of energy transfer provides a very sensitive probe of conformational changes in protein structure. Because the efficiency of transfer depends on the inverse sixth power of the distance, small changes in this distance, of a few ångstroms, will produce measurable changes in the fluorescence. Thus, the observation of Marsh and Lowey[38] that the binding of nucleotides and actin produced no change in the distance between a site on the light chain of myosin and SH_1 indicates that these ligands do not induce global conformational changes in the portion of the myosin that lies between these two sites.

The above examples have illustrated the use of fluorescence spectroscopy as a tool for investigating some diverse aspects of the contractile proteins both in solution and in the organized array of the fiber. The apparatus required for these studies ranges from simple steady-state fluorometers that can be purchased relatively cheaply and require little theoretical or experimental expertise for their operation, to rather complex machines measuring the arrival times of single photons, whose successful operation requires considerably more knowledge of optics, electronics, and the theory of the decay processes being measured. The introduction of microprocessors to collect and to manipulate data, the advances in electronics and optics, and especially the development of new fluorophores, will extend the sphere of future experiments. Fluorescence will most probably be used even more extensively in further studies of contractile systems.

Acknowledgments

I would like to thank Drs. Stefan Highsmith and Robert Mendelson for useful discussions. This work was supported by grants HL-16683 and AM 00479 from USPHS.

[39] M. Miki and K. Mihashi, *Biochim. Biophys. Acta* **533**, 163 (1978).
[40] R. Takashi, A. Muhlrad, T. Hozumi, and J. Botts, *Fed. Proc., Fed. Am. Soc. Exp. Biol.* **39**, 1936 (1980).
[41] D. J. Moss and D. R. Trentham, *Fed. Proc., Fed. Am. Soc. Exp. Biol.* **39**, 1935 (1980).
[42] J. Perkins, J. A. Wells, and R. G. Yount, *Fed. Proc., Fed. Am. Soc. Exp. Biol.* **39**, 1937 (1980).
[43] R. Takashi, *U.S./Jpn. Seminar, Supporo, Japan* (1980).
[44] R. Takashi, *Biochemistry* **18**, 5164 (1979).

[52] Electron Paramagnetic Resonance of Contractile Systems

By JOHN C. SEIDEL

The application of electron paramagnetic resonance (EPR) to studies of the proteins of the myofibril have centered largely on the use of nitroxide spin labels covalently bound to purified proteins and more recently to labels selectively bound to myosin heads of intact myofibrils and glycerinated muscle fibers. In addition to their use as reporter groups when bound to proteins[1,2] nitroxides have also found use in evaluating changes in environmental polarity,[3] flexibility and rotational motion of proteins or segments of proteins,[4,5] and distances between two paramagnetic centers[6,7] or a paramagnetic center and a nuclear spin.[8] Nitroxides have been extensively employed in studies on membranes and lipid bilayers where they have been used to study fluidity, structure, and protein mobility.[9]

The application of spin labels to the study of the contractile system of muscle has developed in parallel with the use of fluorescence spectroscopy; the two have often provided similar information and generally have supplemented each other in providing a more detailed understanding of biochemical and biophysical processes underlying contraction. Conventional EPR spectroscopy using nitroxide labels closely parallels steady-state fluorescence spectroscopy, providing a means of investigating local changes in protein conformation, while saturation-transfer EPR is analogous in many respects to time-resolved fluorescence depolarization[10] permitting measurement of rotational motions of individual proteins and of myosin heads. Saturation transfer, although it developed later, has

[1] L. J. Berliner, this series, Vol. 49 [18]; J. D. Morrisett, *in* Berliner,[2] p. 273.

[2] L. J. Berliner, ed. "Spin Labeling: Theory and Applications." Academic Press, New York, 1976.

[3] P. C. Jost and O. H. Griffith, this series, Vol. 49 [17], p. 369; O. H. Griffith and P. C. Jost, *in* Berliner,[2] p. 453.

[4] J. S. Hyde, this series, Vol. 49 [19], p. 480; J. S. Hyde and L. R. Dalton, *in* Berliner,[5] p. 1.

[5] L. J. Berliner, ed. "Spin Labeling II" Academic Press, New York, 1979.

[6] J. S. Hyde, H. M. Swartz, and W. E. Antholine, *in* Berliner,[5] p. 71; S. S. Eaton and G. R. Eaton, *Coord. Chem. Rev.* **26**, 207 (1978).

[7] G. I. Likhtenshtein, "Spin Labeling Methods in Molecular Biology," Chapter 4, p. 66. Wiley, New York, 1976.

[8] T. R. Krugh, *in* Berliner,[2] p. 339.

[9] H. M. McConnell, *in* Berliner,[2] p. 525; O. H. Griffith and P. C. Jost, *in* Berliner,[2] p. 453; I. C. P. Smith and K. W. Butler, *in* Berliner,[2] p. 411.

[10] D. D. Thomas, *Biophys. J.* **24**, 439 (1978).

extended the time scale accessible to spectroscopic measurements to encompass the microsecond range and permitted measurements of rotation of myosin heads bound to actin and internal rotations of F-actin.

A second area of application of EPR to muscle proteins lies in the use of transition metals, primarily Mn^{2+}, and more recently lanthanides, as probes of the metal binding sites in actin, myosin, and troponin C.[6,11] The use of paramagnetic metal ions has been combined with the use of spin labels, giving rise to a technique that has been referred to as the spin probe–spin label method in which the effects of a perturbing spin, usually the metal ion, on an observed spin, usually the nitroxide, are investigated.[6,7] In the study of proteins, the method provides a means of estimating distances between the two magnetic centers and can provide a more direct and perhaps more precise approach to changes in protein conformation. Again the analogy to fluorescence arises, in this case to energy transfer measurements.

A sometimes overlooked application of the method is in measurement of protein–protein interactions and ligand binding. The interaction between actin and myosin has been investigated by conventional EPR and by saturation transfer using labels bound either to actin or to myosin. A promising potential application lies in the quantitative estimation of the number of myosin heads bound to actin in myofibrils and muscle fibers.

Nitroxides are useful as probes primarily because of the anisotropy in their magnetic properties. This provides sensitivity not only to rotational motion but also gives information about the orientation of the radical relative to the applied magnetic field, the latter providing a method of detecting the orientation of F-actin filaments and of myosin heads in glycerinated muscle fibers. The spin labeling technique itself will be described here only briefly. Relevant aspects including saturation transfer and the spin probe–spin label method have been described elsewhere; in particular, reference should be made to earlier volumes in this series[1,3,4,11] and to the two volumes of "Spin Labeling: Theory and Application."[1–6,8,9] The emphasis in this chapter is on the application of present techniques in EPR to proteins of the myofibril with reference to newer aspects of EPR that may in the future be applied to contractile systems.

Labeling Procedures

The success or failure of using nitroxide labels to study a given protein depends to a large extent on how and where the label can be attached. In devising a strategy for covalent labeling of a protein, several criteria can be used as a guide. The label must be attached at a site at which its

---

[11] G. Palmer, this series, Vol. 19, p. 594; J. A. Fee, this series, Vol. 49 [20], p. 512.

spectrum will reflect the molecular changes to be studied, and its binding to the protein should produce minimal changes in enzymic activity, binding of ligands, or other relevant activities. To obtain the maximal amount of useful information, it is desirable that a single site be labeled. Although numerous labels have been synthesized,[12] virtually all successful selective labeling of myofibrillar proteins has been at reactive sulfhydryl groups, employing N-(1-oxyl-2,2,6,6-tetramethyl-4-piperidinyl)iodoacetamide (IASL), N-(1-oxyl-2,2,6,6-tetramethyl-4-piperidinyl)maleimide (MSL), or their pyrrolidine analogs.

Selectivity and Stoichiometry

Knowledge of the stoichiometry of labeling is an important aid in interpreting spectra and spectral changes and in some cases is essential, e.g., when attempting to localize spectral changes to a particular region of the protein and in assessing spin–spin interactions. The situation is straightforward when the molecule to be labeled has only a single reactive site, but more complex when many potential sites are present. In some cases the extent and rate of labeling can be determined by mixing the protein and label in the EPR cell and following the time course of the decrease in amplitude of the peak corresponding to free label. If the label becomes strongly immobilized upon binding, the appearance of a new downfield peak can be monitored. If the label becomes only weakly immobilized no new peaks appear, but the peaks will broaden with a significant decrease in amplitude, the largest decrease occurring in the upfield peak ($m_I = -1$). Use of reducing agents to terminate the reaction should usually be avoided, since they will often destroy the EPR signal. It is more convenient to vary the concentration of label and allow the reaction to go to completion before removing unreacted label, although it may be possible to follow the time course of slow reactions by precipitating the protein and bound label with ammonium sulfate. The concentration of free label can then be determined directly from the amplitude of any one of the spectral peaks. The concentration of weakly immobilized labels can be estimated from the product of the peak amplitude and the square of the line width,[13] whereas the concentration of more slowly tumbling labels is best estimated by double integration of the spectrum as described in detail in this series.[3]

[12] B. J. Gaffney, in Berliner,[2] p. 184; J. W. F. Keana, in Berliner,[5] p. 115; E. G. Rozantsev, "Free Nitroxyl Radicals." Plenum, New York, 1970.

[13] H. M. Swartz, in "Biological Applications of Electron Spin Resonance" (H. M. Swartz, J. R. Bolton, and D. C. Borg, eds.), Chapter 11. Wiley, New York, 1972.

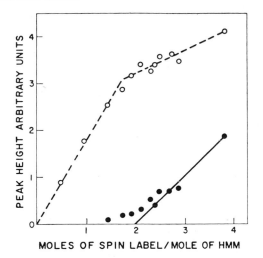

FIG. 1. Dependence of peak heights in electron spin resonance (EPR) spectra of labeled heavy meromyosin (HMM) on concentration of added label. Reaction of HMM (37 mg/ml) was carried out in 0.1 M Tris (pH 8.0) with varying concentrations of IASL for 1 hr at 0°, and EPR spectra were immediately recorded. ○, Height of strongly immobilized peak; ●, height of weakly immobilized peak. Reproduced, with permission, from Seidel et al.[14]

If two spectral components corresponding to weakly and strongly immobilized labels appear at different rates, the amplitudes of their respective peaks can be used to assess the stoichiometry of labeling. When myosin is labeled with IASL, the labels bound to SH-1 groups become strongly immobilized whereas those bound to less reactive groups become weakly immobilized; the dependence of the height of either spectral peak on the concentration of added label is described by two linear regions that intersect at 2 mol of label per mole of myosin (Fig. 1).[14]

When the labeling reaction leads to a change in enzymic activity, the extent of the change can be compared with the concentration of bound label. The labeling of SH-1 groups of myosin with IASL is accompanied by loss of the K^+(EDTA)-ATPase activity that, plotted against the concentration of bound label, can be fitted with a straight line extrapolating to a value of 2 mol of IASL per mole of myosin, one label per enzymic site (Fig. 2).[14] In this case the labeling of one reactive site inactivates one enzymic site, and when the labeling is limited to less than 75–80% inactivation an insignificant number of labels is bound to other sites. A general equation for chemical modification of proteins has been derived[15] that

[14] J. C. Seidel, M. Chopek, and J. Gergely, *Biochemistry* **9**, 3265 (1970).
[15] E. Stevens and R. F. Coleman, *Bull. Math. Biol.* **42**, 239 (1980).

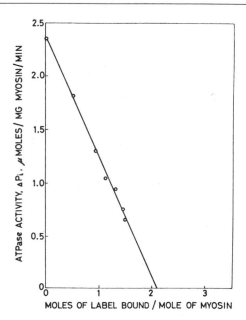

FIG. 2. Loss of K^+-activated ATPase activity and binding of spin label to myosin. Reaction of myosin (10 mg/ml) with IASL was carried out in a solution containing 0.5 M KCl, 0.02 M Tris, and 5 × 10^{-5} M IASL at pH 7.0 and 0°. Samples were taken for determination of ATPase activity and electron spin resonance spectrum. The last sample was taken 90 min after initiation of the reaction. Reproduced, with permission, from Seidel *et al.*[14]

relates the fraction of enzymic activity remaining at a given time to the number of sites modified. This equation may be useful in quantitatively evaluating the stoichiometry and selectivity of labeling when changes in activity occur.

Modification of Spectrum after Labeling

Nitroxides, although relatively stable radicals, can be reduced or oxidized to diamagnetic species. Such chemical reactions have been used to probe the accessibility of labels to solutes; ascorbate has been employed to destroy labels selectively on the outer surface of lipid vesicles,[16] and $Fe(CN)_6^{3-}$ has been used to restore labels that have been reduced after incorporation into intact cells, the restoration being restricted to the outer surface membrane.[17] This chemical lability of spin labels has been exploited selectively to remove unwanted components of the EPR spec-

[16] R. D. Kornberg and H. M. McConnell *Biochemistry* **10**, 1111 (1971).
[17] J. Kaplan, P. G. Canonica, and W. S. Caspary, *Proc. Natl. Acad. Sci. U.S.A.* **70**, 66 (1973).

trum when selective labeling could not be obtained.[18] Labeling of myosin with MSL results in both rigidly immobilized and weakly immobilized spectral components, the latter interfering with the estimation of rotational correlation times by saturation transfer. This weakly immobilized component can be selectively removed upon treatment of the labeled myosin with 25–50 mM $K_3Fe(CN)_6$ for periods of 24–48 hr at 0°. The loss of signal is the result of a chemical reaction involving a cysteine residue near the bound nitroxide, which also results in modification of the cysteine residue. Addition of cysteine alone to MSL-labeled myosin also selectively removes the weakly immobilized component of the spectrum, presumably by reduction of the nitroxide to the hydroxylamine, but the loss is slower and less complete than that obtained with $Fe(CN)_6^{3-}$.[18]

Saturation Transfer

The range of rotational motion accessible to EPR measurements has been greatly expanded by the development of saturation transfer spectroscopy. While the conventional EPR methods are sensitive to motions having rotational correlation times between 10^{-12} and 10^{-7} sec, saturation transfer can extend this to 10^{-3} sec.[19] Saturation transfer depends on diffusion of saturation brought about by rotational motion on a time scale comparable to the spin lattice relaxation time, for nitroxides approximately 10 μsec. Rotation must be sufficiently rapid to occur before saturation can decay by other relaxation mechanisms. Detailed descriptions of the technique can be found in a number of reviews,[4,10,20–22] including several that cover the application of the technique to the myofibrillar proteins.[4,10,21,22] The paper by Thomas et al.[19] is of particular importance.

Saturation transfer measurements can be carried out by several methods including (a) continuous-wave saturation using conventional detection methods (V_1)[23]; (b) second harmonic absorption detected with the phase-sensitive detector 90 degrees out of phase with respect to the magnetic field modulation (V_2'); (c) first harmonic dispersion detected 90 degrees out of phase (U_1'); and (d) electron double resonance (ELDOR).[24]

[18] P. Graceffa and J. C. Seidel, *Biochemistry* **19**, 33 (1980).
[19] D. D. Thomas, L. R. Dalton, and J. S. Hyde, *J. Chem. Phys.* **65**, 3006 (1976).
[20] L. R. Dalton, B. H. Robinson, L. A. Dalton, and P. Coffey, *Adv. Magn. Reson.* **8**, 149 (1976).
[21] J. S. Hyde and D. D. Thomas, *Annu. Rev. Phys. Chem.* **31**, 293 (1980).
[22] J. C. Seidel, *Appl. Spectrosc.* **34**, 280 (1980).
[23] The *V* refers to absorption and *U* to dispersion measurements, subscripts 1 and 2 to first and second harmonics, and the primes to detection 90 degrees out of phase.
[24] M. D. Smigel, L. R. Dalton, J. S. Hyde, and L. A. Dalton, *Proc. Natl. Acad. Sci. U.S.A.* **71**, 1925 (1974).

Experiments have commonly employed a microwave frequency of 9.5 GHz (X band), although measurements at 35 GHz (Q band) have indicated advantages in analysis of anisotropic rotation and sensitivity to slower motion.[25]

Of the four methods of detection, continuous wave saturation can be carried out without modifications to existing spectrometers, but it is relatively insensitive to motion having correlation times longer than 10^{-6} sec and is strongly influenced by spin–lattice and spin–spin relaxation processes. On the other hand, conventional detection methods are sensitive to substantially lower spin concentrations and allow more rapid accumulation of data than is possible by V_2' detection. If the distribution of labels is restricted to two states widely separated in terms of τ_2, such as the free and actin-bound states of MSL-heavy meromyosin (HMM),[26,27] then continuous-wave saturation should be faster and more sensitive than the V_2' method. The intensities of U_1' spectra are as much as 10 times greater than those of V_2' spectra but have approximately a 10-fold lower signal-to-noise ratio due largely to klystron frequency modulation noise.[19] The problems of detecting dispersion signals may be resolved with a newly developed bimodal TM-110 cavity.[28] With this cavity, signal-to-noise ratios for U_1' spectra of MSL-labeled bovine serum albumin are comparable to those obtained for V_1 and V_2' spectra and the amplitudes of U_1' spectra are 3–5 times those of V_2' and equal to those of V_1 spectra. At present the most commonly used method is second harmonic absorption detected with the phase-sensitive detector operated at 100 kHz and 90 degrees out of phase with respect to the 50 kHz magnetic field modulation.[19]

In recording V_2' spectra precise setting of the detector phase is essential in order to minimize the in-phase signal, which may distort the spectrum giving rise to spurious values of the motional parameters. In determining the correct phase it is desirable to use the sample to be measured, i.e., the self-null method, since the phase null depends on the nature of the sample.[19] The null is determined by interpolation with instrument settings to be used in recording the spectra except for the use of low microwave power, ~ 1 mW, by carefully recording the amplitude of the signal on both sides of the null, approximately 0.1 degree away. Further details of this method have been described.[21] The shape of the first and second harmonic

[25] M. E. Johnson and J. S. Hyde, *Biochemistry* **20**, 2875 (1981).

[26] J. C. Seidel, *Arch. Biochem. Biophys.* **157**, 588 (1973).

[27] D. D. Thomas, J. C. Seidel, J. S. Hyde, and J. Gergely, *Proc. Natl. Acad. Sci. U.S.A.* **72**, 1929 (1975); D. D. Thomas, J. C. Seidel, J. Gergely, and J. S. Hyde, *J. Supramol. Struct.* **3**, 376 (1975).

[28] C. Mailer, H. Thomann, B. H. Robinson, and L. R. Dalton, *Rev. Sci. Instrum.* **51**, 1714 (1980).

absorption spectra of a freely tumbling nitroxide has been shown to be sensitive to phase angle, and this sensitivity has been proposed as a means of determining phase nulls.[29] Samples to be studied by saturation transfer generally will not be freely tumbling, thus effects of rotational frequency on line shape must be considered. A more thorough investigation of the dependence of the line shape on detector phase including evaluation of effects of varying τ_2 and careful comparison with nulls obtained by the self-null method seem warranted before employing this method. An approach to the problem of phasing has been to bypass the phase-sensitive detector and to digitize and process the signal by Fourier transformation, the phase being determined by simultaneously processing the modulation signal.[30]

In addition to phasing, the two important instrumental parameters that influence both the amplitude and shape of the V'_2 spectrum are the modulation amplitude, H_m, and the microwave power, P_0. Because the signal-to-noise ratio is usually limiting with biological samples, H_m and P_0 are chosen primarily to maximize the signal amplitude and secondarily to minimize spectral distortion.[19] The chosen value of P_0 determines the microwave field intensity, H_1. The value of H_1 used to record V'_2 spectra must be the same as that used to record or simulate the reference spectra if valid estimates of τ_2 are to be obtained. The relationship between P_0 and H_1 is

$$P_0^{1/2} = CH_1 \tag{1}$$

the proportionality constant, C, depending on characteristics of the cavity and sample cell and on the presence or the absence of the Dewar used for temperature control. Prior to measurement of V'_2 spectra, H_m and H_1 should be calibrated, H_1 for the particular conformation of the cavity to be used. This can be done using 0.9 mM peroxylamine disulfonate (Fremy's salt) in 10 mM K_2CO_3 saturated with nitrogen[19] (available from Thiokol, Alfa Products, Danvers, Massachusetts). H_1 is calibrated by determining the value of $P_0^{1/2}$ at half saturation and evaluating C. The signal amplitude increases linearly with P_0 in the absence of saturation but deviates as saturation is reached, half-saturation being the point where the amplitude is half that expected in the absence of saturation (Fig. 3). The value of H_1 at half-saturation was determined by McCalley[31] to be 0.1067 ± 0.0007 gauss.

[29] A. H. Beth, R. Wilder, L. S. Wilkerson, R. C. Perkins, B. P. Meriwether, L. R. Dalton, C. R. Park, and J. H. Park, *J. Chem. Phys.* **71**, 2074 (1979).

[30] T. Watanabe, T. Sasaki, K. Sawatari, and S. Fujiwara, *Appl. Spectrosc.* **34**, 456 (1980); T. Sasaki, Y. Kanaoka, T. Watanabe, and S. Fujiwara, *J. Magn. Reson.* **38**, 385 (1980).

[31] R. W. McCalley, quoted *in* Thomas *et al.*[19]

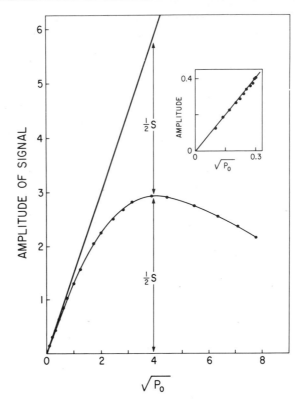

FIG. 3. Calibration of microwave field strength, H_1, from power saturation. A fresh 0.9 mM solution of peroxylamine disulfonate in 0.05 M K_2CO_3 is deoxygenated with nitrogen for several minutes. The slope of the linear region (inset) can be obtained between 0.01 and 0.1 mW and the line extrapolated to estimate the amplitude expected in the absence of saturation at any value of P_0. Then $H_1/(P_0)^{1/2} = 0.1067/(P_{0(1/2)})^{1/2}$ (see Thomas *et al.*[19]).

Using high microwave powers, cooling becomes important; this can be accomplished by insulating the cavity with styrofoam and passing N_2 gas through the radiation slits of the cavity. The use of a Dewar in the cavity necessitates use of a smaller sample cell decreasing the signal-to-noise ratio and necessitating use of a different microwave power to obtain the same value of H_1.

Interpretation of V_2' Spectra. Motional parameters, L''/L, C'/C, and H''/H, have been chosen from the low-field, central, and high-field regions of the spectra, and estimates of rotational correlation times, τ_2, can be obtained by comparison with sets of reference spectra that have been obtained experimentally using MSL-hemoglobin (Hb) or IASL-sub-fragment-1 (S-1)[19,27] or by computer simulation using several theoreti-

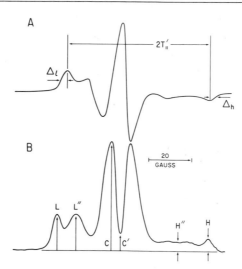

FIG. 4. Experimental parameters of the V_1 spectrum, A, and the V'_2 spectrum, B, for the slow and very slow tumbling time range.[19,68]

cal approaches.[19,32] The motional parameters for V_1 and for V'_2 spectra are illustrated in Fig. 4. Empirical reference spectra were obtained with MSL-Hb at several temperatures and viscosities. Then correlation times were calculated from the expression

$$\tau_2 = (4\pi\eta R^3)/(3kT) \qquad (2)$$

assuming Hb to be a sphere with a radius of 29 Å.[27,33] R is the molecular radius and η is the measured viscosity of the solvent. Curves relating the three motional parameters to τ_2 have been constructed for experimental as well as for computer-simulated spectra (Fig. 5).[19] For experimental results, virtually identical curves are obtained using IASL-S-1 provided that values of τ_2 for S-1 obtained from Eq. (2) are multiplied by 3.4 to correct for the ellipsoidal shape of S-1.[27] Good agreement between empirical and simulated reference spectra is obtained with L''/L and H''/H, but the agreement with C'/C is less satisfactory, owing at least in part to the assumption of axial symmetry in the simulations (Fig. 5).[19]

The three motional parameters differ in the minimal concentration of label required and in the accessible time scale. For a given spectrum the signal intensity and the signal-to-noise ratio will be greatest in the central and least in the high-field region, therefore a 5- to 10-fold higher concentration may be needed when using the high-field region to obtain a com-

[32] D. D. Thomas and H. M. McConnell, *Chem. Phys. Lett.* **25**, 470 (1974).
[33] J. S. Hyde and D. D. Thomas, *Ann. N. Y. Acad. Sci.* **222**, 680 (1973).

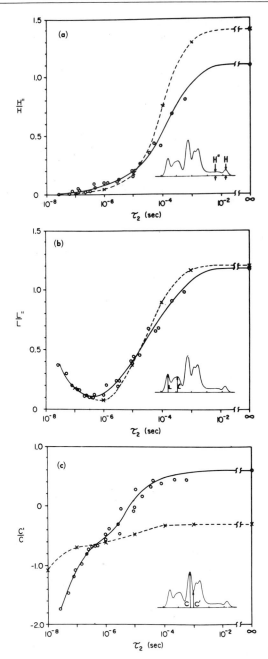

FIG. 5. Dependence on τ_2 of parameters derived from V_2' spectra. O——O, From MSL-Hb experiments; ×---×, from computer-simulated spectra. Panel (a) H''/H; (b) L''/L; (c) C'/C. Reproduced, with permission from Thomas *et al.*[19]

parable signal-to-noise ratio. With spin-labeled myosin or F-actin it is difficult to obtain sufficiently high concentrations of label (10^{-4} M) to permit accurate determination of H''/H. In addition to the greater sensitivity to low concentrations, the central region has greater sensitivity to rotations having correlation times between 10^{-6} and 10^{-8} sec, making C'/C the parameter of choice for motion in this range.[19] On the other hand C'/C shows greater sensitivity to the presence of weakly immobilized labels and to anisotropic rotation. In the presence of in-phase spectral components arising from weakly immobilized labels neither L''/L nor C'/C give accurate values of τ_2, although L''/L is less perturbed than C'/C. If the concentration of labels is high enough, H''/H should be used; L''/L is the usual compromise at low concentrations.

In addition to rotational motion of the protein, changes in T_1 or T_2 can also influence V'_2 spectra, as can changes in local rotation of the nitroxide relative to the protein itself. Direct measurements of T_1 can be made by pulsed methods such as saturation recovery,[34] but the necessary instrumentation is not widely available. An alternative means to assess contributions from T_1 and T_2 is continuous-wave saturation, which can be expected to be relatively more sensitive to relaxation processes other than rotational motion than is the V'_2 method. If changes in V'_2 spectra are not accompanied by changes in saturation behavior assessed by plotting of signal amplitude against $P_0^{1/2}$, it is unlikely that they can be attributed to changes in T_1 or T_2. It is prudent as a routine procedure to determine the dependence of V_1 spectral amplitude on microwave power as a control accompanying V'_2 measurements.

Although the V_1 method is not sensitive to very slow rotation the V'_2 method does detect motion having values of τ_2 shorter than 10^{-7} sec. To eliminate the possibility that changes in V'_2 spectra might reflect changes in fast rotation of a small fraction of labels, both V'_2 and V_1 spectra should be recorded. If both the V_1 and V'_2 spectra change, then the change in V'_2 cannot with certainty be ascribed to a change in microseconds of rotational motion.

Anisotropic Rotation in the Slow and Very Slow Tumbling Regions

Anisotropic rotation in the slow tumbling region and on the microsecond time scale has been discussed in several reviews covering conventional detection,[3,35] and saturation transfer.[4,21] Differences in sensitivity of V'_2 motional parameters to anisotropic motion are illustrated by the effect of temperature on a fatty acid label in dipalmitoyl phosphatidyl-

[34] P. W. Percival and J. S. Hyde, *J. Magn. Reson.* **23**, 249 (1976).
[35] J. Seelig, *in* Berliner,[2] p. 373.

choline vesicles.[36] H''/H and L''/L give the same values of τ_2, while C'/C gives somewhat shorter times diverging sharply from H''/H and L''/L as the temperature is increased. Similarly, evidence for anisotropic motion was obtained by Kusumi et al.,[37] who obtained correlation times of 20, 9.6, and 13 μsec from L''/L, C'/C, and H''/H, using maleimide spin-labeled rhodopsin in rod outer-segment membranes. A spin-labeled cholestane as a thiourea adduct has been used as an experimental model for anisotropic rotation about the y axis of the nitroxide.[38] As the temperature was varied throughout the range of sensitivity, L''/L determined from V'_2 spectra varied between 2.0 and 0.8 but did not decrease below 0.8 even though it reached zero for isotropic rotation. Similar results were obtained with computed spectra for rotation about the y axis.[39] Spectral simulation has provided some insight into the effects of anisotropic rotation on V'_2 and U'_1 spectra,[39,40] but a complete systematic approach to the analysis of anisotropic rotation has not yet been developed.

Rigidity of Attachment of the Label

Because of its sensitivity to rotational motion on the microsecond scale, one of the first applications of saturation transfer has been to rotations of large protein molecules. To interpret V'_2 spectra in terms of rotation of a protein or a segment of it, it is necessary to show that rotation of the label relative to the protein does not make a significant contribution to the overall motion; that is, such local rotation is slow compared to that of the protein itself.[4,22,41] One approach is to vary τ_2 systematically by changing the temperature and viscosity of the protein solution. Local motion having apparent correlation times in the nanosecond range can be detected by conventional (V_1) EPR,[41] the correlation time τ_2 being a linear function of the motional parameters Δ_1 and A'_2 which equals $2T'_{\parallel}$ (see Fig. 4). Plotting the line width, Δ_1, against $(T/n)^{b'_m}$, where b'_m is a parameter obtained by spectral simulation, and extrapolating to the ordinate gives a value for the rigid limit. By varying the viscosity at several different temperatures it has been possible to show that the rigid limit linewidth for MSL-Hb is greater at 45° than at 25° or 3°, suggesting local rotation of the label at the higher temperature.[41] This method provides a systematic approach but is limited to V_1 detection and thus cannot assess local rotation in the microsecond range.

[36] D. Marsh, *Biochemistry* **19**, 1632 (1980).

[37] A. Kusumi, S. Ohnishi, T. Ito, and T. Yoshizawa, *Biochim. Biophys. Acta* **507**, 539 (1978).

[38] B. J. Gaffney, *J. Phys. Chem.* **83**, 3345 (1979).

[39] B. H. Robinson and L. R. Dalton, *Chem. Phys.* in press.

[40] B. H. Robinson and L. R. Dalton, *J. Chem. Phys.* **72**, 1312 (1980).

[41] M. E. Johnson, *Biochemistry* **17**, 1223 (1978); M. E. Johnson, Biochemistry **18**, 378 (1979).

On the microsecond time scale, evidence that IASL remains rigidly bound to S-1 has been obtained by estimating τ_2 over a range of viscosities and temperatures and comparing them with those of MSL-Hb, where strong evidence for rigid binding of the label has been obtained. Rigid binding is implied by the fact that $\tau_2(S\text{-}1)/\tau_2(Hb)$ remains constant as τ_2 is varied from 10^{-7} to 10^{-4} sec.[27] A comparison of the measured τ_2 with that calculated for a sphere of equivalent size should also be made; a measured value less than the calculated value strongly suggests local rotation. Other methods of testing for rigid immobilization that attempt to reduce rotation of the protein without changing local rotation include precipitation or crystallization of the protein, covalent attachment to a rigid support such as Sepharose 4B or glass beads, and binding to another macromolecule, e.g., the binding of MSL-labeled myosin to actin. Since any of the methods of immobilization of the protein may also suppress local motion the use of several methods will increase the degree of certainty that local motion does not contribute significantly to the overall observed motion. The greatest success in rigid labeling has been obtained with maleimide labels, which have been used to label rigidly hemoglobin, myosin, actin, the ATPase of the sarcoplasmic reticulum, cytochrome oxidase, and rhodopsin.[42] Whether this immobilization is a general property of the maleimide label is unclear, as is the mechanism by which immobilization occurs.

Labeling of Specific Proteins

Myosin. Myosin from rabbit skeletal muscle, although it contains over 40 cysteine residues per mole, can be selectively labeled at a highly reactive pair, the SH-1 groups, provided that the conditions of labeling are carefully chosen. Selective labeling is favored by using stoichiometric concentrations of label, pH 7.0, and concentrations of monovalent salt below 100 mM, where the myosin is assembled into filaments.[43] When labeled with IASL under these conditions, myosin exhibits a spectrum indicating the presence of only strongly immobilized labels, while a weakly immobilized component appears upon labeling at pH 8.0 and 0.5 M KCl.[14] With myosin the selectivity also depends on the choice of label, the maleimide label being less selective than the iodoacetamide label. Comparison of the loss of K^+(EDTA)-activated activity with the amount of label bound indicates that more MSL needs to be bound to produce a given decrease in ATPase activity.[14]

Myosin can also be selectively labeled in glycerinated myofibrils in

---

[42] See discussions in reviews.[4,21,22]

[43] D. D. Thomas, S. Ishiwata, J. C. Seidel, and J. Gergely, *Biophys. J.* **32**, 873 (1980).

suspensions (10 mg of myofibrillar protein per milliliter) containing 5 mM MgCl$_2$, 1–10 mM potassium pyrophosphate (PP$_i$), 1 mM EGTA, and 50 mM KCl at pH 7.0, labeling being carried out for 10 min at 0° with 2 mol of IASL or 4 mol of MSL per mole of myosin assuming 50% of the myofibrillar protein is myosin.[43] ATP is as effective as PP$_i$ in promoting selective labeling of myosin, but the use of PP$_i$ prevents contraction of the myofibrils that occurs upon removal of ATP.

Selective labeling with MSL requires prelabeling with Ma1NEt (0.5–1.0 mol per mole of myosin), which blocks sites on thin-filament proteins. Spectra obtained by this method contain a small "weakly immobilized" component that can be removed by subsequent treatment for 40 hr at 0° with 25 mM K$_3$Fe(CN)$_6$. Measurements of K$^+$(EDTA)-ATPase suggest that IASL labels mainly SH-1 groups, MSL also labeling other groups, but both labels are confined to the myosin head as shown by the fact that subfragment-1 prepared from myofibrils labeled with MSL or IASL contains 90–95% of the bound label. The actin-activated ATPase activity of myofibrils or myosin prepared from labeled myofibrils is essentially unchanged by labeling and treatment with Fe(CN)$_6^{3-}$, and sensitivity of the myofibrillar ATPase to micromolar concentrations of Ca^{2+} is maintained.[43] Similar procedures using longer labeling time have selectively labeled myosin heads in intact glycerinated rabbit psoas fibers.[44]

The labeling of SH-2 groups of myosin represents a case where selectivity can be enhanced by taking advantage of effects of small molecules or ions on reactivity of specific amino acid residues. Reactivity of SH-2 groups of myosin is greatly accelerated by MgADP or MgATP, and if SH-1 groups are preblocked with a diamagnetic reagent, the SH-2 groups can be selectively labeled in the presence of MgADP.[45]

Purified LC$_2$ light chain has been spin-labeled with MSL and subsequently incorporated into myosin by exchange of labeled LC$_2$ with the unlabeled LC$_2$ of intact myosin in 0.6 M KSCN.[46] The spectrum of this selectively labeled myosin responds to enzymic phosphorylation of the labeled light chain and to Ca^{2+} binding to it. This technique has also been used to incorporate a fluorescent labeled light chain into myosin from chicken gizzard[47] and spin-labeled light chains into molluscan myosin or myofibrils.[48]

Actin. When F-actin is labeled with MSL, modification of a single residue, Cys-373, is reflected in the dependence of the spectral amplitude on concentration of added label, the spectra being recorded after the

[44] D. D. Thomas and R. Cooke, *Biophys. J.* **32**, 891 (1980).
[45] J. C. Seidel, *Arch. Biochem. Biophys.* **152**, 839 (1972).
[46] Y. Okamoto and K. Yagi, *J. Biochem. (Tokyo)* **82**, 835 (1977).
[47] Y. Okamoto and T. Sekine, *J. Biochem. (Tokyo)* **87**, 167 (1980).
[48] P. D. Chantler and A. G. Szënt-Gyorgyi, *Biochemistry* **17**, 5440 (1978).

reaction has been allowed to go to completion and the unreacted label has been removed.[49] The amplitude is linearly related to concentration up to 1 mol of MSL per mole of actin monomer, and no further change occurs upon adding up to 10 mol of label per mole of monomer.[49] To be selective, labeling must be carried out with actin in the F form.[50] Actin prepared with 0.1 mM dithiothreitol is labeled in a solution containing 0.1 M KCl, 1 mM MgCl$_2$ at pH 7.5 with sufficient MSL to provide a molar ratio of label to actin monomer between 1 and 2, in excess of the dithiothreitol present. Labeling can be carried out for 8–12 hr at 0°[51] but 1 hr at 4° may be adequate.[49] Removal of unreacted label by dialysis or by repeated depolymerization–polymerization cycles is often accompanied by partial denaturation. Removal of unreacted label by repeated sedimentation without depolymerization with 10^{-4} M ATP added to the F-actin buffer, followed by a single depolymerization–polymerization is recommended.

Tropomyosin. The SH groups of tropomyosin are relatively unreactive compared to Cys-373 of actin or the SH-1 groups of myosin; however, the stability of tropomyosin allows reversible denaturation and renaturation so that labeling can be carried out in 5 M guanidine followed by reassembly of dissociated chains by exhaustive dialysis against a solution containing 1 M KCl.[52] Under these conditions labeling was carried out for 8 hr at 25° after reduction with dithiothreitol.

Troponin C (TnC). The ideal situation for labeling in which the protein has a single reactive site, as encountered with Cys-98 of TnC, makes selective labeling relatively straightforward. In this case, selective labeling is indicated by the fact that 0.8–1.2 mol of label are bound per mole and labeling is accompanied by loss of the single sulfhydryl group.[53,54] The labeling of Cys-98 has been accomplished with the iodoacetamide[53] or maleimide[54] derivatives of 1-oxyl-2,2,6,6-tetramethylpiperidine. The labeling is strongly depressed by low concentrations of Ca^{2+} and should be carried out in the presence of chelating agents. A methionine residue of TnC has selectively labeled in the presence of Ca^{2+} with the fluorescence probe, dansylaziridine,[55] suggesting that similar labeling with nitroxylaziridine may be feasible.

[49] R. W. Burley, J. C. Seidel, and J. Gergely, *Arch. Biochem. Biophys.* **146**, 597 (1971).
[50] D. B. Stone, S. C. Prevost, and J. Botts, *Biochemistry* **9**, 3937 (1970).
[51] D. D. Thomas, J. C. Seidel, and J. Gergely, *J. Mol. Biol.* **132**, 257 (1979).
[52] Y. Y. H. Chao and A. Holtzer, *Biochemistry* **14**, 2164 (1975).
[53] J. D. Potter, J. Seidel, P. Leavis, S. S. Lehrer, and J. Gergely, *in* "Calcium Binding Proteins" (W. Drabikowski, H. Strzelecka-Golaszewska, and E. Carafoli, eds.), pp. 129. Elsevier, Amsterdam, 1974; J. D. Potter, J. C. Seidel, P. Leavis, S. S. Lehrer, and J. Gergely, *J. Biol. Chem.* **251**, 7551 (1976).
[54] S. Ebashi, S. Ohnishi, S. Abe, and K. Maruyama, *J. Biochem. (Tokyo)* **75**, 211 (1974); S. Ohnishi, K. Maruyama, and S. Ebashi, *J. Biochem. (Tokyo)* **78**, 73 (1975).
[55] J. D. Johnson, S. C. Charlton, and J. D. Potter, *J. Biol. Chem.* **254**, 3497 (1979).

Protein Conformational Changes

Nitroxide spin labels were first applied as a class of reporter groups to detect localized changes in protein structure accompanying binding of small molecules.[56] Their usefulness is based largely on their sensitivity to rotational motion, which arises from the anisotropy of their magnetic properties, the g factor, and the nuclear hyperfine coupling. The approach has been to attach the label covalently in such a way that a change in protein conformation changes the rotational motion of the label, which in turn alters the EPR spectrum. Since spectral changes may also accompany change in rotational motion of part or all of the protein, additional evidence obtained by an independent technique will strengthen the argument for a conformational change.

Myosin. In early studies it was found that the iodoacetamide label bound to SH-1 groups becomes more mobile upon addition of MgATP (Fig. 6). The more mobile spectrum persists during the steady state of ATP hydrolysis; when hydrolysis is complete, it decays to a spectrum indicating a mobility between that observed in the absence of nucleotide and that seen during ATP hydrolysis and being identical to that produced by ADP.[57] Similar changes in the tryptophan fluorescence of heavy meromyosin were independently observed during the steady state of ATP hydrolysis, after completion of hydrolysis, and upon adding ADP.[58] These spectroscopic results, together with the finding that the hydrolysis of ATP on the enzyme was rapid and was followed by a rate-limiting step involving product dissociation,[59] led to the following scheme for enzymic hydrolysis of ATP.

$$M + ATP \rightleftharpoons M^*ATP \rightleftharpoons M^{**}ADP - P_i \rightleftharpoons M^*ADP + P_i \rightleftharpoons M + ADP + P_i$$

The occurrence of a conformational change associated with enzymic hydrolysis of ATP was suggested by the fact that nonhydrolyzable analogs of ATP give rise to (a) an EPR spectrum indicating less mobility than is observed in the steady-state product complex; and (b) a lower tryptophan fluorescence than the steady-state complex. That the conformational transition from the M to the M** species occurs in two stages is also indicated by kinetic studies showing that part of the change in fluorescence occurred faster than the hydrolytic step at pH 6.9 and 3°.[60] Using

[56] T. J. Stone, T. Buckman, P. L. Nordio, and H. M. McConnell, *Proc. Natl. Acad. Sci. U.S.A.* **54**, 1010 (1965).
[57] J. C. Seidel and J. Gergely, *Biochem. Biophys. Res. Commun.* **44**, 826 (1971); J. C. Seidel and J. Gergely, *Arch. Biochem. Biophys.* **158**, 853 (1973).
[58] M. Werber, A. G. Szent-Györgyi, and G. Fasman, *Biochemistry* **11**, 2872 (1972).
[59] R. W. Lymn and E. W. Taylor, *Biochemistry* **9**, 2975 (1970).
[60] E. W. Taylor, *Biochemistry* **16**, 732 (1977).

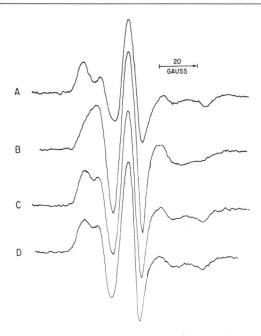

FIG. 6. Effect of MgATP and MgADP on electron spin resonance EPR spectra of spin-labeled myosin. Selective spin labeling at the SH-1 thiol groups with IASL. Spectra of solutions containing labeled myosin, 14 mg/ml, 0.4 M KCl, and 0.04 M Tris, pH 7.5, were recorded at room temperature. A, no further addition; B, 5 mM MgCl$_2$ and 5 mM ATP, recorded 2 min after addition of ATP; C, as B, recorded 10 min after addition of ATP; D, 5 mM MgCl$_2$ and 5 mM ADP. Reproduced, with permission, from Seidel and Gergely.[57]

myofibrils in which the myosin head has been selectively labeled with IASL, it has been possible to observe the increased spin label mobility characteristic of the steady-state of ATP hydrolysis in a system in which the geometry of the contractile apparatus has been preserved.[61]

Troponin C. Troponin was one of the first myofibrillar proteins investigated by the spin-labeling method by Tonomura *et al.*,[62] who prior to the isolation and characterization of the troponin components separated the thin-filament proteins actin, tropomyosin, and "relaxing protein" (troponin), attached spin labels to each, and studied the effects of Ca^{2+} on the individual proteins and reconstituted systems. They observed that Ca^{2+} altered the EPR spectrum of spin-labeled tropomyosin, but only in the presence of unlabeled troponin, and also altered that of spin-labeled actin, but only in the presence of unlabeled tropomyosin and unlabeled troponin. On the basis of these results they proposed a mechanism in which a

[61] D. D. Thomas, J. C. Seidel, and J. Gergely, unpublished observations.
[62] Y. Tonomura, S. Watanabe, and M. Morales, *Biochemistry* **8**, 2171 (1969).

Ca^{2+}-induced conformational change in troponin is transmitted to actin through tropomyosin.

More recently, TnC has been labeled with several nitroxides including IASL,[53] MSL,[54] and a tyrosine-directed analog of N-acetylimidazole.[54] The reaction with IASL is strongly inhibited by micromolar concentrations of Ca^{2+}, thus the labeling reaction itself can be used to monitor the binding of Ca^{2+}.[53] The spectra of each of the three labels bound to TnC change upon addition of Ca^{2+} or upon addition of either troponin T (TnT) or troponin I (TnI), the spectral changes indicating reduced rotational mobility of the label. That the Ca^{2+}-induced change represents a conformational change is supported by changes in fluorescence[53] and in circular dichroism[63] occurring at similar Ca^{2+} concentrations. The dependence of the spectral change on the concentration of Ca^{2+} was used to determine whether the spectral change arises from binding to the high- or low-affinity Ca^{2+} binding sites of the IASL-labeled protein. Approximately $5 \times 10^{-8} M$ Ca^{2+} is required to produce a half-maximal spectral change and a 5- to 10-fold higher concentration is required in the presence of 2 mM $MgCl_2$, indicating that the high affinity Ca^{2+}-Mg^{2+} sites are involved.[53] These correspond to sites III and IV in the model of Kretsinger and Barry[64] and therefore may not reflect a conformational change directly related to regulation of the actin–myosin interaction, which appears to involve binding to sites I and II.

The labeled TnC also exhibits slower rotational motion upon binding to TnI or TnT and can be used to monitor the binding of these proteins to TnC. Sharp isosbestic points are observed upon addition of TnT suggesting the presence of only two spectral states,[54] and a one-to-one stoichiometry is found for both TnT and TnI.[53] It is not clear whether these spectral changes reflect (a) slower local rotation of the label due perhaps to a conformational change or (b) slower rotation of the entire protein associated with formation of binary complexes. The spectra indicate correlation times of the order of 10 nsec or longer, approximately that expected for the entire protein; thus the observed rotation may contain components from local rotation and from rotation of the protein. If these components were of approximately equal magnitude, changes in the spectrum could reflect changes in either one.

Saturation Transfer Studies on Myosin and Actin

Rigid attachment of MSL bound to either the head region of the myosin molecule or to Cys-373 of F-actin have permitted the investigation

[63] J. P. van Eerd and Y. Kawasaki, *Biochem. Biophys. Res. Commun.* **47**, 859 (1972); Y. Kawasaki and J. P. van Eerd, *Biochem. Biophys. Res. Commun.* **49**, 898 (1972).
[64] R. H. Kretsinger and C. D. Barry, *Biochim. Biophys. Acta* **405**, 40 (1975).

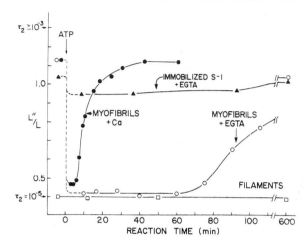

FIG. 7. L''/L, from saturation transfer electron paramagnetic resonance spectra (V'_2), plotted as a function of time after adding ATP to myosin filaments ($\square$), myofibrils ($\bigcirc$,$\bullet$), and immobilized subfragment-1 (S-1) ($\blacktriangle$). The reaction mixtures contained 20–30 mg of protein per milliliter in buffer containing 50 mM creatine phosphate, 0.5 mg of creatine phosphokinase, 5 mM ATP, and 0.1 mM EGTA, except for the filled circles ($\bullet$), where 0.1 mM CaCl$_2$ was substituted for EGTA. Reproduced, with permission, from Thomas et al.[43]

of rotational motion of both on the microsecond time scale. A comparison of the correlation times obtained for myosin, HMM, and S-1[27] supported the evidence obtained by fluorescence depolarization measurements[65] indicating that the head is joined to the myosin rod through a flexible region that allows rotation of the head relative to the remainder of the molecule. In addition, ST-EPR measurements indicate that the myosin head is bound rigidly to the actin filament and that in complexes of F-actin with myosin, HMM, or S-1 the rotation of the head appears to be determined by an internal mode of rotation of the actin filament.[27] HMM and S-1 induce cooperative changes in this internal mode of rotation of F-actin.[51] Selective labeling of myosin heads in glycerinated myofibrils[43] and muscle fibers[44] have provided a means of investigating the rotational motion of myosin heads in experimental systems having the geometric organization of the contractile apparatus of the intact muscle (Fig. 7).

Interaction of Myosin and Actin

Conventional and saturation-transfer EPR can be used to monitor the interaction between myosin and actin. Actin added to IASL-labeled myosin increases $2T'_\parallel$, the separation of the outer spectral peaks of the V_1

[65] R. A. Mendelson, M. F. Morales, and J. Botts, Biochemistry 12, 2250 (1973).

spectrum, suggesting a decrease in label mobility.[66] With saturating microwave power a 50% decrease in spectral amplitude occurs when actin combines with MSL-labeled HMM, reflecting a decrease in the microwave power at which saturation occurs, an effect that can be readily used to quantitate the fraction of HMM bound.[26] Three possible physical interpretations were considered: (a) a conformational change in HMM upon binding to actin, further immobilizing the label; (b) a direct interaction of actin with the spin-labeled site; or (c) a decrease in the rotational motion of the myosin head on interaction with actin. Although the first interpretation was favored originally, more recent studies using saturation transfer EPR indicate that the third interpretation is correct.[27] This follows from the observations that the iodoacetamide label attached to SH-1 groups is rigidly bound to the head, and the head itself becomes immobilized on interaction with actin; thus a decrease in rotational motion could only arise from a decrease in the rotational motion of the head.

Myosin labeled with MSL at the SH-2 groups also exhibits changes in saturation transfer spectra when it binds to F-actin. In this case the nitroxide moiety is not rigidly attached to the protein and undergoes considerable local rotational motion with respect to the protein, which is slowed approximately 60-fold upon interaction with actin as compared to about a 1000-fold change for myosin labeled at the SH-1 sulfhydryls.[27]

The use of spin labels bound to myosin heads to reflect their interaction with actin has been extended to systems of greater complexity, such as myofibrils[43] and glycerinated muscle fibers.[44] The decrease in rotational mobility of the heads upon binding to actin can be monitored in MSL-labeled myofibrils by saturation transfer, and the ordering of bound heads imparted by the actin filament is reflected in conventional EPR spectra of oriented labeled fibers as discussed below.

The interaction can also be monitored using labeled actin and unlabeled myosin, HMM, or S-1. The initial approach made use of the spectral change associated with polymerization of actin and the fact that binding of HMM induces polymerization. The spectrum of G-actin labeled at Cys-373 with MSL contains components corresponding to both rapidly and slowly tumbling labels that change on polymerization, the amplitude of the peak arising from slowly tumbling labels shifting to lower field strength suggesting a slowing of rotation.[50] This spectral change has been used to estimate the stoichiometric relationships involved in the polymerization induced by HMM.[67] The combined use of V_1 and V_2' spectra of MSL-labeled F-actin has revealed two distinct effects of HMM, both reflected in the spectral properties of MSL bound to Cys-373 (Fig.

[66] T. Tokiwa, *Biochem. Biophys. Res. Commun.* **44**, 471 (1971).
[67] R. Cooke and M. F. Morales, *J. Mol. Biol.* **60**, 249 (1971).

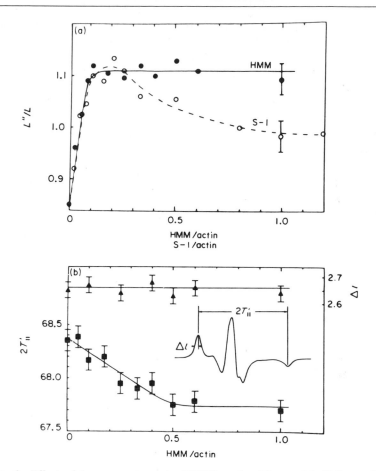

FIG. 8. Effect of heavy meromyosin (HMM) and subfragment-1 (S-1) on electron paramagnetic resonance (EPR) spectral parameters of maleimide spin-labeled F-actin. Final F-actin concentration was 2.8 mg/ml at 20°. (a) L''/L, from saturation transfer EPR spectra, plotted against the molar ratio of HMM (●——●) or S-1 (○---○) to actin monomers. The estimated uncertainty for each point is indicated by the error bars at the right. (b) $2T'_\parallel$ (■——■), scale at left) and Δ_1 (▲——▲, scale at right), from conventional EPR spectra, plotted against the molar ratio of HMM to actin monomers. Reproduced, with permission, from Thomas *et al.*[51]

8).[51] The V'_2 spectral reveal an increase in the motional parameter L''/L as HMM is added to MSL–F-actin saturating at 1 mol of HMM per 10 actin monomers. On the other hand, $2T'_\parallel$, the separation of the outer spectral peaks, observed in V_1 spectra decreased with added HMM saturating at a ratio of one head per actin monomer. The half-width of the downfield peak, Δ_1, which is sensitive to rotation but not to changes in polarity of the environment,[68] does not change on adding HMM, suggesting that the change in $2T'_\parallel$ reflects a change in polarity. Thus on addition of HMM the rotational motion of the label changes in a cooperative manner while the polarity changes in direct proportion to the fraction of heads bound.

Anisotropic Rotation and Sample Orientation

When the nitroxide is ordered in a crystalline lattice, the spectra will be anisotropic and information about the orientation with respect to the lattice can be obtained by orienting the sample in the magnetic field.[3] If the label is incorporated into a lipid bilayer or bound to an asymmetric protein in such a way that its rotation is restricted, orientation of the bilayer or protein in the magnetic field will provide similar information. For labels undergoing rapid anisotropic rotation about one of the three axes of the nitroxide or rapid rotation of restricted amplitude, characteristic spectra can be obtained without orienting the sample. These cases have been extensively explored in connection with studies on lipid bilayers and membranes.[3,35] For immobilized labels or for slowly tumbling labels undergoing rotation of restricted amplitude, the separation of the outer spectral peaks will be maximal when the z axis of the nitroxide is parallel to the field and minimal when it is perpendicular. In favorable situations the angle between the principal axis and an axis of the protein as well as the deviation from this angle can be obtained. Oriented films of F-actin can be produced by streaking an F-actin gel between the outer flat surface of the EPR sample cell and a glass coverslip in a direction perpendicular to the long axis of the cell.[49]

By rotating the cell in the cavity, the actin film can be aligned either parallel or perpendicular to the magnetic field. The separation of the outer peaks is maximal when the cell surface is parallel to the field, indicating an orientation of the principal axis of the label approximately parallel to the axis of the actin filament (Fig. 9). Virtually identical spectra are obtained whether the actin is streaked in a direction parallel or perpendicular to the long axis of the cell,[69] suggesting that the filaments are randomly oriented within planes parallel to the surface of the cell. When the orientation of the cell is perpendicular to the field, virtually all filaments would be perpendicular, but in the parallel orientation only a fraction of the filaments would be parallel or nearly parallel and the spectrum would preferentially sample this fraction. Anisotropy was observed in the spectrum of Mn^{2+} bound to F-actin oriented in this way when the sample was rotated about an axis perpendicular to the direction of streak.[70] This unexpected result

[68] R. P. Mason and J. H. Freed, *J. Phys. Chem.* **78**, 1321 (1974).
[69] H. Stedman, unpublished observations.
[70] J. Loscalso and G. H. Reed, *Biochemistry* **15**, 5407 (1976).

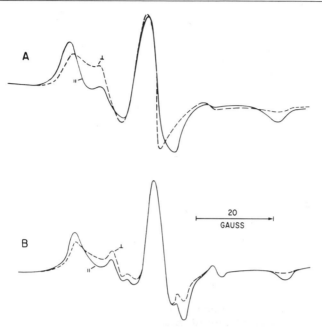

FIG. 9. The electron paramagnetic resonance spectra of oriented films of F-actin labeled with maleimide spin labels. Actin was labeled in the F form with the piperidinyl maleimide (A), or the pyrrolidinyl maleimide (B). In (B) some unreacted label was present. ----, Film oriented perpendicular to the magnetic field; ———, film oriented parallel to the magnetic field. Reproduced, with permission, from Burley et al.[49]

would be explained by random orientation of filaments within planes parallel to the cell surface. Filaments of MSL-labeled tropomyosin have also been prepared and their spectra recorded with the fiber axis at various angles to the field; the principal axis of the label was estimated to be at an angle of 50° to the fiber axis.

Thomas and Cooke[44] have utilized V_1 EPR spectra of oriented glycerinated muscle fibers in which the myosin heads have been selectively labeled with either MSL or IASL to estimate the angle between the fiber axis and the principal axis of the nitroxide (Fig. 10). Three angles are considered in their treatment: θ, the angle between the z axis and the direction of the applied magnetic field (H_0); θ', that between the fiber axis and the z axis; and ψ, that between the fiber axis and H_0. If the fiber axis is coincident with the direction of the magnetic field ($\psi = 0$), the angle between the z axis and the fiber becomes equal to that between the z axis and the field ($\theta' = \theta$) and only a single angle, θ_0, need be determined. The distribution of labels in fibers in rigor can then be described by a gaussian function and on comparing experimental and simulated spectra, the center of the gaussian distribution, θ_0, and the width at half-maximal height, $\Delta\theta$,

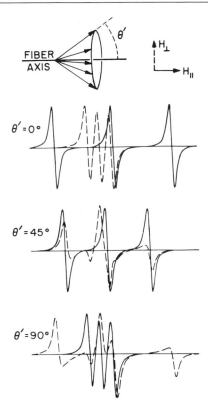

FIG. 10. Simulated electron paramagnetic resonance spectra for spin labels uniformly oriented relative to the axes of uniformly oriented muscle fibers. Arrows indicate the orientations of the principal axes of the spin label, which describe the surface of a cone about the fiber axis. The unique angle between the fiber axis and the principal axis of the spin label is θ', and the angle between the magnetic field and the fiber axis is $\psi = 0°$ ($H_\parallel$, solid curves) or $\psi = 90°$ ($H_\perp$, dashed curves). These spectra were produced by summing single-θ spectra over values of ϕ' (the azimuthal angle of the label axis) from $0°$ to $90°$. Reproduced, with permission, from Thomas and Cooke.[44]

can be estimated. For MSL-labeled fibers in rigor these were estimated to be 68 and 17 degrees, respectively, while for IASL-labeled fibers they were 82 and 15 degrees. The same narrow distribution of labels is observed with spin-labeled S-1 or HMM diffused into unlabeled fibers indicating that the order arises from myosin heads bound to the thin filaments.

Upon addition of a relaxing solution containing MgATP and EGTA, the separation of the outer spectral peaks increases, giving rise to a spectrum characteristic of an isotropic distribution of labels immobilized in the time frame accessible to V_1 spectra. The spectral peaks arising from oriented and isotropic components are sufficiently well separated that they can be readily distinguished in spectra containing both, such as those of stretched fibers or fibers containing PP_i or adenylyl imidodiphosphate (AMPPNP).

The isotropic spectra could represent labels in either a static state of disorder or in a dynamic state in which they undergo large-amplitude rotational diffusion at a frequency too slow to be detected by V_1 EPR. Saturation transfer measurements on MSL-labeled myofibrils in the presence of ATP, AMPPNP, or PP_i suggest that the dynamic state more accurately describes the ensemble of heads giving rise to the isotropic component.[43,71] This component presumably reflects the rotation of myosin heads not bound to actin characterized by a correlation time of about 10 μsec.

Restriction in Amplitude of Rotation

An additional type of deviation from isotropic Brownian rotation is illustrated by the restricted random-walk model, which has been used to describe rotation of an asymmetric lipid in biological membranes.[3] The molecule or spin label undergoes random rotation or wobbles within a cone described by an angle that is less than 90 degrees. Spin labels bound to myosin heads and to F-actin may very well undergo similar motion on a slower time scale, and this must be taken into consideration in evaluating correlation times for these proteins deduced from saturation-transfer measurements. In conventional EPR spectra such a change in amplitude of rotation cannot be readily distinguished from a change in frequency of isotropic rotation.[3] If the amplitude of rotation is restricted, correlation times based on an isotropic model may be too long and changes in amplitude may be mistaken for changes in τ_2.

Mn^{2+} as a Probe for Divalent Cation Binding Sites

The binding of Mn^{2+} to myosin and to other proteins can be measured directly by determining the amplitude of the EPR spectra of Mn^{2+} in solutions containing varying amounts of added Mn^{2+}.[72] When Mn^{2+} binds to most proteins the amplitude of the EPR signal is substantially reduced and the decrease in signal amplitude can be used to estimate the amount bound. Measurement of Mn^{2+} binding to myosin indicated the presence of two sites with an affinity constant of $10^6 \ M^{-1}$. Competition experiments indicate that binding of Ca^{2+} and Mg^{2+} to these sites is characterized by constants of 3.2×10^3 and $9.4 \times 10^2 \ M^{-1}$, respectively.[73]

[71] S. Ishiwata, J. C. Seidel, and J. Gergely, *Biophys. J.* **25**, 19a (1979).

[72] M. Cohn and J. Townsend, *Nature (London)* **173**, 1090 (1954).

[73] M. C. Beinfeld, D. A. Bryce, D. Kochavy, and A. Martonosi, *J. Biol. Chem.* **250**, 6282 (1975).

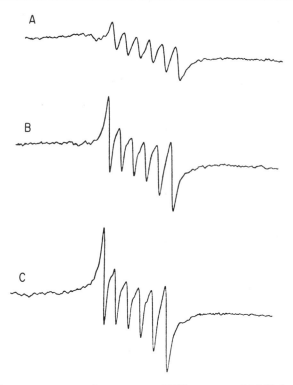

Fig. 11. Electron paramagnetic resonance (EPR) spectra of Mn^{2+} bound to hybrid myofibrils. Scallop myofibrils (*Pecten*) were desensitized with 10 mM EDTA, and hybrids were formed by adding regulatory light chains in the presence of 1 mM Mg^{2+}. Gel electrophoresis confirmed that the hybrids had taken up about 1 mol of foreign light chain. The myofibril preparations were equilibrated against 4 μM Mn^{2+}, and the EPR spectra were obtained. (A) Desensitized scallop myofibrils; (B) hybrid scallop myofibrils containing a rabbit DTNB light chain; (C) hybrid scallop myofibrils containing a clam (*Mercenaria*) regulatory light chain. The amplitude of the difference spectrum (b) − (a), corresponds to a concentration of 50 μM Mn^{2+} bound. Hence, the DTNB light chain in the hybrid myofibrils is nearly saturated with Mn^{2+}, indicating the $K_{Mn}^{app} < 4 \mu M$. Reproduced, with permission, from Bagshaw and Kendrick-Jones.[74]

Mn^{2+} bound to the nonspecific sites that are located on the DTNB light chain (LC$_2$) of rabbit skeletal muscle myosin give rise to a well-resolved EPR spectrum compared to a partially resolved spectrum seen with myosin from scallop or clam adductor (Fig. 11).[74] This resolved spectrum has been used to identify the nonspecific binding sites in light chains from a number of myosins from other sources and can be used to monitor the

[74] C. R. Bagshaw and J. Kendrick-Jones, *J. Mol. Biol.* **130**, 317 (1979); C. R. Bagshaw, *Biochemistry* **16**, 59 (1977); C. R. Bagshaw and G. H. Reed, *FEBS Lett.* **81**, 386 (1977).

formation of hybrids of purified light chains with "desensitized" scallop myofibrils from which regulatory light chains have been partially removed. The resolved spectral pattern originating from the purified light chain dominates the observed spectrum and can be readily identified even in the presence of residual scallop light chains.

Spin Probe– Spin Label Methods

Taylor *et al.*[75] observed a dipolar interaction between Mn^{2+} and a nitroxide spin label bound to creatine kinase, in which the addition of Mn^{2+} altered the amplitude of the nitroxide spectrum without significant change in the shape, although the shape might be expected to show significant broadening as a result of the proximity of a paramagnetic metal ion. A theory to account for this behavior was put forth by Leigh[76] in which the linewidth is expressed as

$$H = C(1 - 3 \cos^2 \theta'_R)^2 + \delta H_0 \tag{3}$$

with

$$C = g\beta\mu^2\tau/r^6\hbar \tag{4}$$

where θ'_R is the angle between the direction of the magnetic field and the line connecting the two magnetic centers, and δH_0 is the linewidth in the absence of the perturbing paramagnetic species. If the labeled molecule is randomly oriented with respect to the field, then for a certain fraction of labels $C(1 - 3 \cos^2\theta'_R)^2$ will be small compared to δH_0, and their signal will appear unperturbed; for others $C(1 - 3 \cos^2\theta'_R)^2$ will be much greater than δH_0, and their spectrum will be broadened sufficiently to be undetectable, the net result being a decrease in amplitude with little or no apparent change in shape. The value of C exhibits an inverse sixth-power dependence on r, the distance between magnetic centers. Provided the relative geometry of the two spins is essentially unchanged during the lifetime of the spin states, the correlation time for the dipolar interaction may be taken to equal the spin-lattice relaxation time, T_1, of the perturbing metal.[76] When the nitroxide spectrum indicates rapid rotation of the label, the addition of glycerol has been employed to reduce the likelihood of a significant change in the spin geometry occurring during the lifetime of the dipolar interaction, which is approximately 10^{-9} sec.[77] Knowledge of T_1 together with a set of computer-simulated spectra (Fig. 12) for various values of C allows calculation of r from the extent of the decrease in

[75] J. S. Taylor, J. S. Leigh, Jr., and M. Cohn, *Proc. Natl. Acad. Sci. U.S.A.* **64**, 219 (1969).
[76] J. S. Leigh, Jr., *J. Chem. Phys.* **52**, 2608 (1970).
[77] C. R. Bagshaw and J. Kendrick-Jones, *J. Mol. Biol.* **140**, 411 (1980).

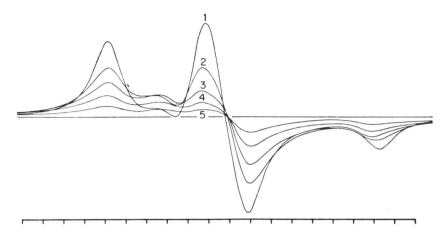

FIG. 12. Computed "nitroxide" electron paramagnetic resonance spectra; the parameters used were $g_x = 2.0089$, $g_y = 2.0061$, $g_z = 2.0027$, $A_x = 1.25 \times 10^8$ rad/sec, $A_y = 0.986 \times 10^8$ rad/sec, $A_z = 5.63 \times 10^8$ rad/sec, and $\delta H_0 = 4.5\ g$, $\theta_R = 0.2\pi$, $\phi_R = \pi/2$. The dipolar interaction coefficient $C = 0$, 3, 10, 30, and 100 in curves 1–5, respectively. The magnetic field scale begins at 3200 g, and the marks show 5-g increments. Reproduced, with permission, from Leigh.[76]

spectral amplitude.[75,76] For creatine kinase r was estimated to be between 7 and 10 Å. The lower limit was calculated using a value of 1.2×10^{-9} sec for τ, which is the value of T_1 measured for Mn^{2+} in the nucleotide–metal–enzyme complex, and the upper limit using 1×10^{-8} sec, T_1 of free $Mn(H_2O)_6^{2+}$. Hyde and co-workers[6] have pointed out that spin exchange must also be considered in evaluating interactions between two paramagnetic centers. They have studied the interaction of a number of paramagnetic cations including transition metals and lanthanides on free radicals in biological systems and present a detailed discussion of the theoretical basis and practical applications of the method.

A decrease occurs in the amplitude of the nitroxide signal of MSL-labeled actin upon incorporation of Mn^{2+} into the metal-binding site.[78] The decrease is proportional to the concentration of bound Mn^{2+} reaching a maximal value of 50% with one Mn^{2+} per actin monomer, and is reversed either by replacing Mn^{2+} with Ca^{2+} or Mg^{2+} or by denaturation of the protein. The fact that quantitatively the same decrease in amplitude occurs with G- or F-actin indicates that the dipolar interaction occurs between spin label and Mn^{2+} on the same actin monomer and precludes significant intersubunit interaction. From the magnitude of the decrease in signal amplitude the distance between the bound Mn^{2+} and the nitroxide

[78] R. W. Burley, J. C. Seidel, and J. Gergely, *Arch. Biochem. Biophys.* **150**, 792 (1972).

was calculated to be between 16 and 23 Å using the values of T_1 for Mn^{2+} bound to creatine kinase or that of the free ion, respectively.[78] The nitroxide was shown to be bound to Cys-373 penultimate to the C-terminal residue when actin is labeled under the usual conditions, but when actin is stored without nucleotide prior to labeling it can be labeled at the cysteine residue in the N-terminal peptide region of the amino acid sequence.[79] The effects of Mn^{2+} on the spectral amplitude are the same for actin labeled in either of the two positions, suggesting that in the tertiary structure the Mn^{2+} binding site is about equidistant from the cysteine residues in the C- and N-terminal regions. In contrast to the effect on actin, addition of Mn^{2+} to myosin spin-labeled at the SH-1 groups had no effect on the nitroxide spectrum; however, an interaction between spin labels bound to SH-1 and those bound to SH-2 groups was reported, and the distance between the two paramagnetic centers was estimated to be less than 17Å.[80]

Based on homology between the amino acid sequence of LC_2 from myosin of rabbit skeletal muscle and the Ca^{2+} binding proteins parvalbumin and troponin C, it has been suggested that the Ca^{2+} binding sites may also have homologous structures.[64] Of the four binding domains of the proteins in the calcium-binding family, only the first appears to be a likely candidate as the binding site of LC_2 based on the amino acid sequence. To test this hypothesis, Bagshaw and Kendrick-Jones[77] have made use of the dipolar interaction between Mn^{2+} and nitroxide spin labels bound to the metal binding light chains of myosin from the clam adductor muscle and from rabbit skeletal muscle. Spin labeling was carried out at the single cysteine near the first binding domain of the clam light chain and at the two cysteine residues of the rabbit light chain, which are in domains III and IV. A dipolar interaction would be expected in the clam light chain if the Mn^{2+} binding site were in domain I and in the rabbit light chain if it were in domains III or IV. That the binding site is in domain I was indicated by the 25% decrease in signal amplitude on binding of Mn^{2+} to the clam light chain, with less than 5% change with the rabbit light chain.

An apparent dipolar interaction of the type described by Taylor et al.[75] between the maleimide spin label bound to Cys-98 of TnC and Gd^{3+} bound to the Ca^{2+} binding sites has been observed by Wang et al.[81] Digestion of the labeled TnC molecule with trypsin in the presence of either Ca^{2+} or EDTA gives rise to cleavage on the N- or C-terminal side of Cys-98,

[79] R. W. Sleigh and R. W. Burley, Arch. Biochem. Biophys. 159, 792 (1973).

[80] G. I. Likhtenshtein, "Spin Labeling Methods in Molecular Biology," pp. 62, 113. Wiley, New York, 1976; Y. B. Grebenshchikov, G. G. Charkviani, N. I. Gachechiladze, Y. V. Kokhanov, and G. I. Likhtenshtein, Biofizika 17, 794 (1972).

[81] C. L. A. Wang, P. L. Leavis, J. C. Seidel, and J. Gergely, Biophys. J. 33, 237a (1981) (abstract).

respectively, giving rise to fragments in which the label reports binding either to sites I and II or to sites III and IV. Ca^{2+} and Gd^{3+} alter the spectra of both fragments. In addition, titration of labeled and unlabeled TnC with metal ions reveals that labeling with maleimides, but not with iodoacetamide derivatives alters the affinity of one of the two high-affinity sites.

Acknowledgments

I thank Dr. Larry R. Dalton, Dr. James S. Hyde, and Dr. David D. Thomas for making results of their work available prior to publication and Dr. John Gergely and Dr. Philip Graceffa for helpful comments on the manuscript. Supported by grants from NIH (HL-15391 and HL-23249) and from the Muscular Dystrophy Association.

[53] Phosphorus-31 Nuclear Magnetic Resonance of Contractile Systems

By MICHAEL BÁRÁNY and THOMAS GLONEK

Since the first phosphorus-31 nuclear magnetic resonance (^{31}P NMR) spectrum of intact skeletal muscles was recorded,[1,2] NMR has been widely applied to the study of various muscles. This trend was natural, since muscle is extremely rich in phosphorus compounds, and these compounds play major roles in normal muscle function and in the diseased state. Thus, it was possible to measure the breakdown of phosphocreatine (PCr) during muscle contraction,[3] to follow changes in the concentration of several phosphates during the anaerobic metabolism of skeletal muscle,[4-7] and to study the phosphate profiles of normoxic and ischemic hearts.[8,9]

[1] D. I. Hoult, S. J. W. Busby, D. G. Gadian, G. K. Radda, R. E. Richards, and P. J. Seeley, *Nature (London)* **252**, 285 (1974).

[2] M. Bárány, K. Bárány, C. T. Burt, T. Glonek, and T. C. Myers, *J. Supramol. Struct.* **3**, 125 (1975).

[3] M. J. Dawson, D. G. Gadian, and D. R. Wilkie, *J. Physiol. (London)* **267**, 703 (1977).

[4] C. T. Burt, T. Glonek, and M. Bárány, *J. Biol. Chem.* **251**, 2584 (1976).

[5] C. T. Burt, T. Glonek and M. Bárány, *Science* **195**, 145 (1977).

[6] S. J. W. Busby, D. G. Gadian, G. K. Radda, R. E. Richards, and P. J. Seeley, *Biochem. J.* **170**, 103 (1978).

[7] K. Yoshizaki, *J. Biochem. (Tokyo)* **84**, 11 (1978).

[8] D. G. Gadian, D. I. Hoult, G. K. Radda, P. J. Seeley, B. Chance, and C. Barlow, *Proc. Natl. Acad. Sci. U.S.A.* **73**, 4446 (1976).

[9] W. E. Jacobus, G. J. Taylor, IV, D. P. Hollis, and R. L. Nunnally, *Nature (London)* **265**, 756 (1977).

This chapter focuses on techniques and procedures for recording and evaluating the ³¹P NMR spectra of intact muscles and of perchloric acid extracts of muscle. The chapter will close with a brief discussion of applications to selected nonmuscle systems.

Basic Principles

NMR Parameters

Phosphorus NMR. The spectral sensitivity of ³¹P relative to that of other nuclides is reasonably high. For example, considering equal populations of nuclei in the same magnetic field, ³¹P is 4.2 times as sensitive as ¹³C, 6.9 times as sensitive as ²H, and 63.7 times as sensitive as ¹⁵N. (It is, however, only 1/15th as sensitive as the proton, ¹H.) Moreover, since all naturally occurring phosphorus exists as the ³¹P isotope, the relative sensitivity of ³¹P spectroscopy is high in comparison, for example, to ¹³C, which is only 1% naturally abundant, and the need to incorporate enriched compounds into the tissues under study is precluded. Further, the NMR window for ³¹P is not cluttered by the resonances of any other nuclide; only resonance signals from phosphorus are detected in the usual high-resolution ³¹P spectrum. In addition, ³¹P is present in relatively few compounds whose cellular concentrations are in excess of 0.5 mM; however, these compounds (e.g., inorganic orthophosphate and ATP), are among the most important tissue metabolites.

In considering biological systems, the chemical forms of phosphorus ordinarily encountered are members of the families of tetraconnected phosphorus oxyacids, the phosphates and their esters, phosphonates and their esters, and certain nitrogen and sulfur phosphoroamidate and thiophosphate derivatives. Most of these occur as anions, and as a result, the resonance signals are frequently sensitive indicators of the nature of the solvent. For example, weak acid phosphates yield chemical shift titration curves that can be used to determine intracellular pH.[4,10,11] All phosphates coordinate metallic countercations to a greater or lesser degree, and this is reflected in their ³¹P NMR parameters so that the phosphates can be effective reporter groups monitoring the ionic makeup of the solvents.[12-14] The phosphorus oxyacids hydrogen-bond to water and are in-

[10] M. M. Crutchfield, C. H. Dungan, J. H. Letcher, V. Mark, and J. R. Van Wazer, *Top. Phosphorus Chem.* **5**, 1 (1967).

[11] R. B. Moon and J. H. Richards, *J. Biol. Chem.* **248**, 7276 (1973).

[12] T. Glonek, J. R. Van Wazer, M. Mudgett, and T. C. Myers, *Inorg. Chem.* **11**, 567 (1972).

[13] T. Glonek, R. A. Kleps, E. J. Griffith, and T. C. Myers, *Phosphorus* **5**, 157 (1975).

[14] T. Glonek, R. A. Kleps, E. J. Griffith, and T. C. Myers, *Phosphorus* **5**, 165 (1975).

corporated into the water "structure." As a result their ^{31}P parameters are sensitive to the relative hydrophobicity of the medium and may be used in the future to determine precisely the physical state of the intracellular water.

All NMR analysis is physiologically innocuous because of the low energies involved; thus, its nondestructive property permits intact tissues to be resampled and allows for the determination of *in vivo* kinetic parameters or, at the clinical level, the repeated imaging of human anatomy in hospital patients.[15]

At the present time, high-resolution NMR is probably the best general-purpose analytical tool available for studying the chemistry of phosphorus and its compounds. With a ^{31}P NMR probe in a high-resolution NMR spectrometer, the analyst can determine easily, and usually within a few minutes, whether phosphorus is present at concentrations greater than 100 μM in fluid samples, characterize each type of phosphorus-containing group present, and obtain a quantitative analysis of the various species present. For a spectrometer in which a magnetic field of 23 kiloGauss is employed so that the resonance position of protons is 100 MHz, the ^{31}P resonance will be found at 40.5 MHz; in a magnetic field 3.6 times this, the resonance will be found at 145.8 MHz.

Conventional-mode, high-resolution NMR spectra, including signal-averaged Fourier-transform spectra, are ordinarily analyzed in terms of four kinds of parameters. One of these is the chemical shift, δ, which is referenced to some arbitrarily chosen standard compound and describes the relative magnetic shielding of the particular nucleus being observed. Another is the spin–spin coupling constant, J, which describes the interaction between pairs of magnetically active nuclei that are present in the same compound. The third is the NMR peak area which, when the spectra are properly obtained, is proportional to the relative amount of the active nuclei, i.e., ^{31}P, giving rise to that particular peak. And the fourth is the resonance linewidth at half height. Broadening of a resonance is attributable to relatively rapid substituent-exchange processes and/or to the presence of nuclei having quadrupole moments. Both of these effects have been put to considerable use in chemical investigations of phosphorus-containing molecules.

Chemical Shift and Referencing. Nuclear magnetic resonance chemical shifts are usually reported in terms that are independent of the laboratory magnetic field value. Thus, the chemical shift, δ, is given as the relative change in the magnetic field, as measured in parts per million (ppm), between the resonance of the chosen magnetically active nucleus in the

[15] B. Chance, J. S. Leigh, Jr., S. Eleff, and C. D'Ambrosio, "Nuclear Magnetic Resonance Imaging Symposium Abstracts," *J. Computer Assisted Tomography* **5**, 295 (1981).

compound under study and that observed for the reference compound

$$\delta = \frac{\nu_{ref} - \nu_{obs}}{\nu_{inst}} \times 10^6 \tag{1}$$

where the frequencies ν are measured in Hertz, and ν_{inst} is the operating frequency of the instrument. With Eq. (1) one can directly compare the chemical shifts that were determined by various spectrometers operating at different frequencies. The range of phosphorus chemical shifts is very large and extends from ca. 250 ppm to ca. −500 ppm relative to inorganic orthophosphoric acid.[10,16]

For ^{31}P the usual reference standard is 85% orthophosphoric acid (H_3PO_4), although phosphorus trioxide (P_4O_6), trimethylphosphate [$(CH_3O)_3PO$], pyrophosphate ($Na_4P_2O_7$), white phosphorus (P_4), and other phosphorus-containing compounds have been advocated as phosphorus shift references from time to time. There is no entirely satisfactory phosphorus shift reference material as there is, for example, in the use of tetramethylsilane as a shift reference for hydrogen, carbon, and silicon NMR. The phosphorus reference material must be isolated from the bulk sample to prevent alterations through chemical reactions, this usually being accomplished by sealing the substance in a glass capillary tube 1 mm or less in diameter and coaxially mounting the capillary in the NMR sample tube along with the sample under study. Phosphorus shifts are also quite temperature sensitive, and so the temperature at which experiments are performed should always be stated. In addition, the shifts may also be altered by several parts per million by changing the countercation in solution,[4,13,14] the salt concentration[16] (in organic as well as inorganic media), or the solvent. Liquid–crystalline, intact tissue, or solid-state determinations will depend entirely on the availability of a suitable natural resonance in the material under study.

With modern field-frequency-stabilized spectrometers, it is also possible to reference the phosphorus magnetic resonance spectrum to the resonance signal of another nuclide, e.g., the deuterons of HDO. This technique circumvents many of the problems inherent in the use of phosphorus references; however, it requires the use of sophisticated, and often expensive, NMR hardware, as well as a knowledge of the spectroscopic properties of two nuclides.

For those fortunate enough to possess a cryospectrometer system, an internal reference may be unnecessary for most applications. Such systems possess stability on the order of 1 or 2 Hz per week. Thus, once the location of 0 ppm is established for a given set of experimental conditions,

---

[16] J. R. Van Wazer, in "Determination of Organic Structures by Physical Methods" (F. C. Nachod and J. J. Zuckerman, eds.), Vol. 4, p. 323. Academic Press, New York, 1971.

which may be accomplished through the use of a test sample approximating the materials to be analyzed, repeated analyses may be performed without fear that the accuracy of the chemical shift scale will be compromised.

INTERNAL CHEMICAL SHIFT REFERENCE SIGNALS USEFUL FOR CALIBRATING INTACT TISSUE AND TISSUE EXTRACT SPECTRA. Unfortunately for biological ^{31}P NMR studies, the 85% inorganic orthophosphoric acid signal frequently falls in a region of the phosphorus spectrum crowded with other resonance signals so that these latter may be masked by the reference. To get around this, we use a 1.0 M solution of methylenediphosphonic acid [(HO)$_2$OPCH$_2$PO(OH)$_2$], in D$_2$O (pD = 9.5 with sodium as the countercation) contained in a sealed and coaxially mounted 1 mm capillary tube. The resonance position of this compound so prepared is 16.1 ppm at 25°, which places it in a region of the spectrum ordinarily free of naturally occurring phosphorus-containing molecular resonances,[17] and the preparation is chemically and spectroscopically stable. Stabilization of the magnetic field may be achieved through the deuterium signal from the capillary, and the area of the phosphonic acid resonance, after suitable calibration procedures,[18] may serve as the reference for relative area measurements. This sample is, however, an external NMR reference, since it is isolated from the substance under analysis by the glass capillary tube, and this fact gives rise to small unpredictable errors that render its use unsuitable for precise chemical shift measurements as, for example, may be desired in determining a change in intracellular pH.

For precise chemical shift determinations, an internal reference, i.e., one that is incorporated into the chemical matrix of the substance under analysis, is necessary. For intact tissue analysis, the signal from PCr has been used for this purpose,[3] and it is probably the best internal reference for tissues rich in this substance as, for example, is muscle. The dissociable secondary proton is a relatively strong acid (pK_a = 4.7); thus, its phosphorus chemical shift is stable at tissue pH values of about 6 and higher. PCr is rapidly degraded by intact tissues, however, and it is readily hydrolyzed during preparative procedures.

There are, perhaps, two additional phosphorus resonance signals that may serve as chemical shift references for both intact tissue and liquid sample analysis; these are sn-glycerol 3-phosphorylcholine (GPC) and the pyrophosphate resonance of nicotine adenine dinucleotide (NAD). Of these, GPC is by far the best choice. It is a component of all tissues thus

---

[17] Resonances of the phosphonic acids, unfortunately, are in the region of 16–20 ppm. For samples containing such compounds, 85% orthophosphate can usually be used as the reference.

[18] T. Glonek and S. F. Marotta, *Horm. Metab. Res.* **10**, 365 (1978).

far examined, and its resonance position, at -0.13 ppm, lies in an uncluttered region of the spectrum.[19] Further, the proton-decoupled linewidth is the narrowest in the intact tissue spectrum, and even proton coupled, the signal is easily resolved from its neighbors in the phosphodiester spectral region. The molecule contains no weakly acidic or basic groups, and so its chemical shift is pH independent to pH values as low as 3. It does not appear to form aqueous solution complexes with any of the common metal ions, and the salt dependence of the phosphorus shift is only 2–3 Hz over salt concentrations ranging up to 6 M. Moreover, the molecule is soluble in a variety of organic solvents as well as water. It has a disadvantage as a shift reference, in that its concentration in most tissues is less than 60 μM; i.e., the signal area ordinarily accounts for about 4% of the total tissue phosphate profile.

The NAD signal lies at -11.34 ppm. NAD is also a minor phosphate profile component; its signal is often covered by the α-group doublets of the adenosine polyphosphates. In addition, its phosphorus shift may be metal-ion dependent, and it is not ordinarily soluble in nonaqueous media.

Regardless of which method or substance is used to determine the phosphorus chemical shift, the data should always be reported as positive downfield, in accord with the recommendation of the International Union of Pure and Applied Chemistry and in which the American Chemical Society is in agreement. Although there is no official position on this next point, it is also strongly recommended that shifts be reported relative to 85% *inorganic orthophosphoric acid* (0 ppm). Most published phosphorus NMR data are referenced to 85% phosphoric acid.

Coupling Constants

A magnetically active nucleus associated with a given resonance can interact through the electronic structure with other magnetically active nuclei in the molecule. The effect of this is to split the resonance into smaller multiplet peaks that sum to give the same total intensity. This phenomenon is called spin–spin splitting, and the strength of the coupling between the magnetically active nuclei is measured in terms of the spin–spin coupling constant, J, which is invariant to the field strength when expressed in Hertz. Spectra exhibiting spin–spin splitting can be divided into two fundamental categories: simple or first-order spectra, including pseudo-first-order spectra, and second and higher-order spectra.

As regards phosphatic molecules of biological origin, only first-order spectra are ordinarily observed. The commonest of these are true first-order splittings of the type POCH between phosphorus and protons in organophos-

---

[19] C. T. Burt, T. Glonek, and M. Bárány, *Biochemistry* 15, 4850 (1976).

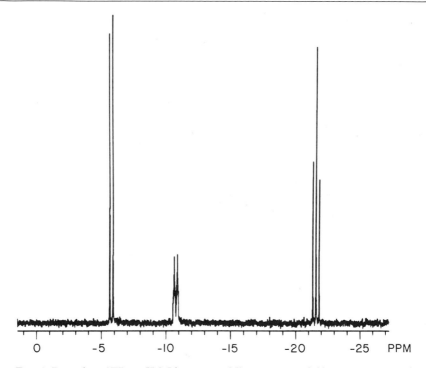

FIG. 1. Potassium ATP at pH 9.5 in water at 25°: α group, -10.92 ppm; β group, -21.45 ppm; γ group, -5.80 ppm. All three groups show the large POP spin couplings, and, in addition, the α group shows the effect of the much smaller POCH coupling. Also see Crutchfield et al.[10] for an early but detailed spectrum of ATP.

phate esters, e.g., α-glycerophosphate [$(OH)_2P(O)OCH_2CH(OH)CH_2OH$]. In this example, the phosphorus atom spin-couples to the two neighboring α-CH_2 protons to produce a $1:2:1$ triplet in the phosphorus spectrum exhibiting a J value, or spacing between the arms of the multiplet, of 6.68 Hz. Most phosphorus signals of biological origin will exhibit POCH couplings; the important exceptions are inorganic orthophosphate, phosphocreatine, and the ionized end-group phosphates of the nucleoside polyphosphates which are not proton spin-coupled.

    Adenosine triphosphate (ATP) is an example of a molecule that gives rise to a pseudo-first-order spectrum (Fig. 1). In this example the α and the γ end group phosphates each spin-couple to the β middle group to produce the characteristic doublet, doublet, triplet pattern of ATP exhibiting, in aqueous solution, two identical POP J values of 19.44 Hz (the spacing between the arms of the end-group doublets). As the ATP spectrum is ordinarily displayed, the multiplets will not appear as perfect $1:1$

doublets and a 1 : 2 : 1 triplet, but, instead, will be skewed toward the center of the overall pattern. This is a property of pseudo-first-order spectra that arises from the fact that the spin-coupled multiplets are relatively close together. Only the inorganic- and nucleoside-polyphosphates give rise to pseudo-first-order splitting patterns in spectra from biological sources. All the remaining phosphorus-containing molecules exhibit only first-order splittings, which, under the influence of broad-band proton decoupling, appear as single resonance lines.

Line Widths

SINGLE LINES. Although individual phosphorus signals, such as one of the 10 from neat trimethylphosphate or the single resonance from proton broad band-decoupled GPC, have been resolved (signal width at half height) to less than 0.07 Hz, the usual ^{31}P NMR signals exhibit linewidths greater than this. Line broadening of ^{31}P NMR signals can be attributed to the nature of the solvent or the sample, or it can be an inherent property of the compound under investigation or its organization within the matrix of the sample. In aqueous solutions prepared from deionized water and either containing ethylenedinitrilotetraacetic acid (EDTA) or having been treated with a chelating resin, signals from the low molecular weight phosphates, such as glucose 1-phosphate, can be resolved to values under 1 Hz. Signals from molecules such as ATP or a phosphorylated decapeptide show about 2 Hz resolution, whereas signals from some compounds, such as phosphonoalanine (1-amino-2-phosphonopropionic acid) may exhibit linewidths of 6–10 Hz in the same solvent. Moreover, when phosphatic residues, such as those mentioned above, are incorporated into macromolecules or into macromolecular arrays, such as phospholipid residues in biomembranes, the single resonance lines arising from these units can become so broad as to be undetectable by the usual methods employed in high-resolution NMR. In such cases, some chemical operation, such as acid hydrolysis, is usually required to effect identification of the structural fragment by phosphorus NMR. In general, phosphorus NMR linewidths can be said to be similar to proton NMR linewidths obtained on similar compounds. The very much larger chemical shift range of ^{31}P over ^{1}H, however, permits larger numbers of individual compounds to be simultaneously and separately determined by phosphorus NMR than by proton NMR spectroscopy.

ENVELOPES. There is at least one other important aspect of signal width measurement worthy of note, an aspect that is well illustrated by the middle-group phosphate signal of the amorphous glassy polyphosphates, shown in Fig. 2, which is also observed in the usual polynucleotide resonance spectrum. In such cases, each individual phosphorus

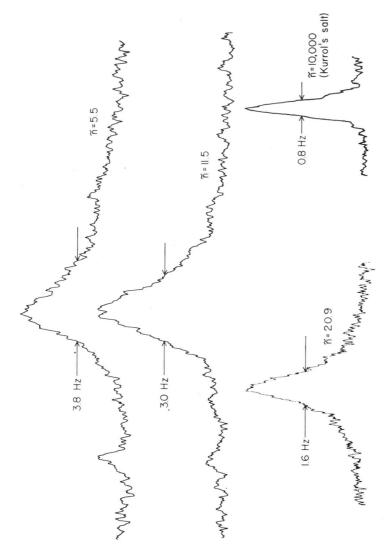

FIG. 2. The ^{31}P NMR spectrum of signals arising from the nearly identical middle-group phosphates of the linear chain inorganic polyphosphates. Samples of various chain lengths, $\bar{n}$, were dissolved in deionized water at pH 7 containing a small amount of EDTA (sodium countercation). The signal width at half height is indicated for each sample. The shift position of each resonance is about −20 ppm.

atom in a molecule experiences a chemical shift somewhat different from that of its neighbors, which depends on its proximity to the end of the chain.[19a] This difference diminishes as one moves from the chain end, until, near the center of the chain, the environments, and hence the chemical shifts, are identical. Each individual signal in such a case is so close to its neighbor that the spectrometer cannot resolve the difference between them, and what is observed is the envelope of the various absorption signals. For example (Fig. 2), in a mixture of small-chain phosphates, where the ratio of end-phosphate groups to middle-phosphates groups is high, the absorption signals of the middles close to the chain end are relatively strong, while the signals from .the interior middles are relatively weak. Thus, the overall spectrum of the middle-group phosphates is a broad line. When the chains are very long, however, this ratio is reversed and the signals from the essentially identical interior middles account for most of the area of the overall resonance band. The result is a rather narrow signal, which, nonetheless, is an envelope of the various absorption signals.

A similar situation arises when a single phosphorus functional group is present, as a building unit, in a family of similar molecules, such as the phosphocholine residue in a preparation of egg yolk lecithin. If there are a sufficient number of analogs of a given substance present in the sample, as would be the case here where alterations in the fatty acid side-chain residues generate a variety of closely related molecular species, the resultant closely spaced overlap of the phosphorus multiplets cannot be resolved by the spectrometer, and, again, only the envelope of the entire absorption band is observed.

With respect to examining intact tissues and their extracts by ^{31}P NMR, the following observations have been made. The signals from intact tissues arise from the low molecular weight phosphorus-containing molecules of intermediate metabolism; their signal widths are on the order of 10–50 Hz. The phospholipid signals of the membrane phospholipids are a component of the base line in high-resolution phosphorus spectra and are ordinarily not observed. They can be detected in experiments designed for the determination of wide lines, and under these conditions they exhibit a single convoluted signal with a line width on the order of 6000 Hz (40–50 pm). Likewise, polynucleotide and phosphoprotein signals from intact tissues are not ordinarily observed, although a phosphoprotein signal has been detected in the intact adrenal gland[18] and in demineralized bone and human dentin. In the case of intact collagen, the organic phosphate resonance is a relatively narrow 20 Hz.

[19a] A. J. R. Costello, T. Glonek, and J. R. Van Wazer, *Inorg. Chem.* **15**, 972 (1976).

When phosphates interact with alkaline earth and transition metal ions, their signals are frequently extensively broadened. Hence, perchloric acid tissue extracts must be treated with EDTA or a chelating resin before high-resolution spectra may be obtained. Without such treatment, the phosphorus signals are actually 4–6 times wider than they are when obtained from the intact tissue.[2] Complex formation with macromolecular substances may also broaden phosphorus NMR signals. For example, adenosine diphosphate, in simple aqueous solutions, exhibits a signal width of about 2 Hz; when combined with the protein actin, the signal is broadened to the point of being undetectable under high resolution NMR conditions.[2] The hemoglobin–2,3-diphosphoglycerate complex, however, gives rise to a highly resolved phosphorus NMR spectrum, both in simple aqueous solutions and in the intact erythrocyte.[20]

It is difficult to extrapolate NMR line width data. The only advice that can be offered is never to assume that usable ^{31}P cannot be obtained from any given system. It is usually worth the effort to test the system in the spectrometer.

Relaxation Times. Because of the large number of phosphorus-containing compounds and the relative high sensitivity of the ^{31}P nuclide, it should be expected that the phosphorus NMR relaxation time parameters, T_1 and T_2, ought to have the same general value to phosphorus chemistry as, for example, ^{13}C relaxation has had for carbon chemistry. This has not been the case, thus far, primarily because of the complexity of the phosphorus relaxation process and the inability in most systems, of establishing a set of simplifying assumptions. First considering small, nearly spherical nonionizable molecules, generally examined as their neat liquids, the relaxation mechanisms that have been observed to dominate the relaxation process consist of the interaction of the nuclear spin with the molecular rotation (spin rotation), the dipole–dipole interaction between pairs of phosphorus nuclei or between phosphorus and hydrogen, and relaxation due to the tumbling of residues exhibiting large chemical shift anisotropy. In the case of the phosphorus oxyacids, scalar relaxation due to exchange of dissociable protons between the oxyacid and the solvent water molecules must also be considered.

A detailed relaxation time analysis[21] of concentrated inorganic orthophosphoric acid revealed the following: 88% of the total ^{31}P-T_1 relaxation occurred through the scalar mechanism; about 11% occurred through the intermolecular dipole–dipole mechanism, the rest being attributable to the spin-rotation mechanism. For this molecule, the chemical-shift

[20] T. O. Henderson, A. J. R. Costello, and A. Omachi, *Proc. Natl. Acad. Sci. U.S.A.* **71**, 2487 (1974).
[21] W. E. Morgan and J. R. Van Wazer, *J. Am. Chem. Soc.* **97**, 6347 (1975).

anisotropy contribution was neglected. Considering the symmetrical PO_4^{3-} this surely is a reasonable assumption, and it is in accord with Gillen's argument.[22] It was also estimated that about 90% of the observed phosphorus line broadening was attributable to fluctuations in the local magnetic shielding associated with exchange of the acidic proton. A similar interpretation was applied to the analysis of a series of low molecular weight, water-soluble phosphate esters.[23]

T_2 relaxation data obtained since that report confirmed the earlier assumption that the phosphorus line width is an accurate reflection of the bulk T_2 relaxation. In water the ^{31}P-T_2 of the family of orthophosphates at most pH values and with most countercations is very much shorter than T_1 (100 msec or less).

Analysis of inorganic and nucleoside polyphosphates incorporates all of the above mechanisms, and, in addition, the intramolecular ^{31}P–^{31}P dipole–dipole mechanism operates and must be considered.[24] The simultaneous contributions of all these mechanisms considerably complicate interpretations of ^{31}P relaxation data, generally weakening most subsequent theoretical developments of molecular dynamics. The data, however, are quite useful as experimental determinants, and they are essential for the optimization of spectroscopic scan conditions and the gathering of good quantitative data.

Considering macromolecules and macromolecular arrays, the most definitive work has involved the phospholipid components of the biomembrane. T_1 values of the human circulating lipoprotein phospholipids[25] have been measured, as have the T_1 values of these same residues in spherical vesicles,[26,27] lamellar arrays,[28,29] various membrane preparations, and viruses.[30] The values obtained are on the order of tenths of a second. T_2 values have also been determined for phospholipid dispersions,[31] and these are on the order of 200 μsec. The linewidths, which are related to T_2, range from as narrow as 4 Hz, in human circulating high-

[22] K. T. Gillen, *J. Chem. Phys.* **56**, 1573 (1972).

[23] T. Glonek and J. R. Van Wazer, *J. Phys. Chem.* **80**, 639 (1976).

[24] T. Glonek, P. J. Wang, and J. R. Van Wazer, *J. Am. Chem. Soc.* **98**, 7968 (1976).

[25] G. Assmann, D. A. Sokoloski, and H. B. Brewer, Jr., *Proc. Natl. Acad. Sci. U.S.A.* **71**, 549 (1974).

[26] P. L. Yeagle, R. G. Langdon, and R. B. Martin, *Biochemistry* **16**, 3487 (1977).

[27] A. C. McLaughlin, P. R. Cullis, J. A. Berden, and R. E. Richards, *J. Magn. Reson.* **20**, 146 (1975).

[28] S. J. Kohler and M. P. Klein, *Biochemistry* **16**, 519 (1977).

[29] D. G. Davis, *Biochem. Biophys. Res. Commun.* **49**, 1492 (1972).

[30] N. F. Moore, E. J. Patzer, R. R. Wagner, P. L. Yeagle, W. C. Hutton, and R. B. Martin, *Biochim. Biophys. Acta* **464**, 234 (1977).

[31] R. W. Barker, J. D. Bell, G. K. Radda, and R. E. Richards, *Biochim. Biophys. Acta* **260**, 161 (1972).

density lipoproteins to about 5000 Hz in *Acholeplasma laidlawii* cell membranes derived from cells grown on elaidic acid.[32] Ordinarily nuclear relaxation of the phospholipids in bilayer arrays takes place through the chemical shift anisotropy mechanism.

Instrumentation

The Spectrometer. All modern high-resolution NMR spectrometers operate at relatively high magnetic fields, 1.8–8.3 T (32–146 MHz for ^{31}P), where the detection of phosphorus is usually a routine operation. In addition, they incorporate an internal proton or deuterium lock, various modules for multiple resonance experiments, a pulse programmer, amplifiers, and a sophisticated instrument computer and software package.

Current instruments employ a master oscillator to control all the radio frequencies needed in the NMR experiment and, thereby, to prevent their relative drifts. The lock, or field-frequency stabilization unit, serves to fix the value of these frequencies to the strength of the polarizing magnetic field, allowing each and every point in a spectrum to be precisely reproduced and permitting time averaging of the data for the purpose of improving the signal-to-noise ratio. This is essential for most experiments employing biomedical applications of phosphorus NMR, where the signal intensity on a single scan can be expected to lie beneath the spectral noise level. For a spectrum of a muscle biopsy extract from as little as 70 mg of muscle, 150,000 individual scans, each composed of 16,384 separate data points, are summed over a 24-hr period to produce the final phosphorus spectrum. A drift rate of no more than 0.3 Hz or 3.7 parts in 10^9 per 24 hr is necessary for the successful completion of such an experiment.

The latest generation of spectrometers are set up so that both homonuclear and heteronuclear decoupling may be carried out in a single-frequency or noise-modulated mode. Examples of proton decouplings, typical of those that are routinely applied to determinations on muscle extracts, are presented in Fig. 3.

Noise-decoupling is particularly advantageous if, for example, one wishes to suppress all of the spin–spin splitting due to, say ^{1}H, in a phosphorus spectrum. Such a spectroscopic manipulation, however, introduces an additional complexity where the spectrum intensities may not be proportional to the concentration of the phosphate. This phenomenon is called the nuclear Overhauser enhancement,[33] and it acts in proton-decoupled phosphorus spectra to generate a peak of greater intensity than would ordinarily be obtained.

[32] B. DeKruijff, P. R. Cullis, G. K. Radda, and R. E. Richards, *Biochim. Biophys. Acta* **419**, 411 (1976).
[33] J. H. Noggle and R. E. Schirmer, "The Nuclear Overhauser Effect: Chemical Applications." Academic Press, New York, 1971.

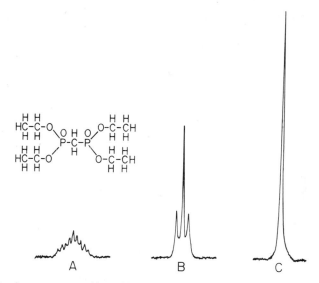

FIG. 3. Continuous-wave and broad-band (noise) decoupling of protons while observing a ^{31}P NMR signal. Spectrum A shows the regular high-resolution spectrum of neat tetraethyl-methylenediphosphonate. The spectrum consists of a complex pattern due to the coupling of the equivalent phosphorus atoms with the CH_2 protons of the methylenediphosphonate moiety and of the four equivalent ethyl groups. Spectrum B shows the signal obtained during continuous-wave decoupling of only the methylene protons of the ethyl groups. The triplet pattern due to the remaining coupling of the phosphorus atoms with the methylene phosphonate protons, an example of a first-order PCH coupling, is now clearly visible. Spectrum C shows the signal obtained during broad-band decoupling of the entire proton spectrum. The single resonance line of the two equivalent phosphorus atoms is obtained. The three spectra were recorded using the same gain settings; the increased peak amplitude and area of spectra B and C over that of A are the result of reduced signal multiplicity coupled with a small Overhauser enhancement.

Decoupling experiments are also useful in removing line broadening by coupled nuclides having spins greater than $\frac{1}{2}$. An example of such broadening is found in phosphorus spectra of compounds containing phosphorus–nitrogen bonds, where the effect is attributable to the phosphorus bonded ^{14}N atoms of spin 1. In muscle extract spectra, line broadening due to nitrogen-coupled phosphorus is observed in the phosphocreatine resonance.

Tuning the Magnet. The most critical procedure in the operation of the NMR spectrometer is proper tuning of the magnetic field. Under high-resolution conditions, i.e., where the natural NMR line widths may be less than 0.1 Hz, the homogeneity of the laboratory field determines, in large measure, the amount of information that can be gleaned from a single analysis. In the perfect magnetic field, each and every signal would have a line width characterized by the value of its T_2 relaxation time. For low molecular weight phosphates in ideal solutions, particularly at high magnetic field strengths, these T_2 values are long and, hence, the natural linewidths are narrow. Such NMR signals in the ideal magnetic field would be detected as very sharp resonances that, even in a complex mixture, could be resolved one from the other. Flaws in the magnetic field act to broaden these resonance signals artificially so that closely spaced neighboring components coalesce. The result is a loss of information.

Most notable among other spectral artifacts that can be generated by a poorly tuned magnetic field are the spinning side bands, which are harmonics of a central peak generated by the fact that the sample is rotating in a magnetic field that is not uniform. In the NMR spectrum, these side bands appear as small peaks lying to either side of a central resonance. The easiest of these to detect are the first-order side bands, which lie symmetrically displaced about the central resonance at a distance equal to the sample's rotational frequency, or multiple of this frequency, expressed in Hertz (Fig. 4). Usually, only one or two pairs of first-order side bands are detected in the modern high resolution instrument, and in phosphorus spectra, where the spectral noise is often considerable, only spinning side bands of the very largest peaks will ever be detected. These peaks can be assigned with certainty by the simple expedient of changing the sample spinning rate. With such an alteration in the experimental

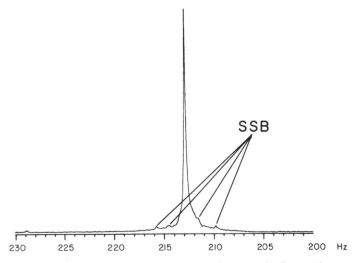

FIG. 4. ^{31}P NMR spectrum from orthophosphate in a marginally tuned magnetic field showing first-order spinning side bands (SSB).

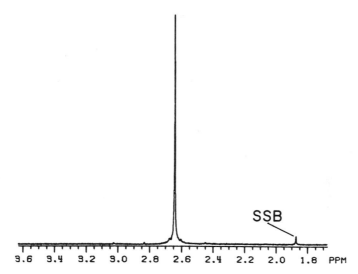

FIG. 5. Orthophosphate spectrum showing a higher-order spinning side band (SSB).

design, the spinning side band signals will change position by an appropriate amount whereas genuine magnetic resonance signals will remain fixed.

Far more difficult artifacts to detect, particularly in complicated extract spectra, are higher-order side bands. Again, these are generated by the rotation of the sample in the magnetic field; however, they do not necessarily come in symmetrical pairs and may appear in the spectrum a considerable distance from the parent peak. As can be seen in the example given in Fig. 5, the artifact lies almost 1 ppm upfield of the central resonance. If this peak were a component of a complicated muscle perchloric acid extract, it could easily be mistaken for the resonance of some other phosphate, but not, as it is, a spectral artifact from the orthophosphate signal.

Figure 6 shows the inorganic orthophosphate resonance obtained in our instrument with a properly tuned magnetic field. The peak is symmetrical, narrow, and does not show any evidence of either first-order or higher-order spinning side bands. An NMR field thus tuned can be trusted to give accurate as well as precise signal area measurements. These signals should be compared to that shown in Fig. 4. Note that while the peak of Fig. 4 is still fairly sharp, the base of the peak has been broadened. An integral of this peak must include not only the area of the central resonance but the area of the skirt, including the area of the spinning side bands. In this example, the skirt extends for at least 14 Hz.

When high-resolution NMR experiments are conducted on older spec-

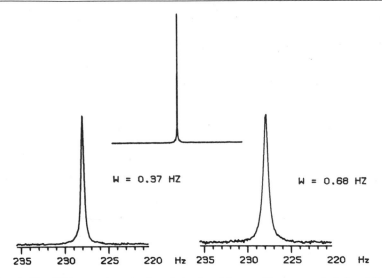

FIG. 6. The ^{31}P inorganic orthophosphate signal from a 12-mm sample tube at 25° in a properly tuned magnetic field. The aqueous sample was 0.01 M in phosphorus, adjusted to pH 10.0 with 10 N KOH after treatment with potassium Chelex-100. (It is important to use 10 N KOH, as the strong alkali precipitates heavy metal and transition element hydroxides.) The center spectrum was taken while spinning the sample; the total sweep width was 500 Hz. The two lower spectra show the same signal on an expanded scale. The narrower signal was obtained while spinning the sample; the wider signal was obtained without spinning. The signal widths at half-height W are indicated for each expanded spectrum.

trometers (up to 1975), it is always necessary to spin the sample in order to obtain highly resolved spectra. The modern instruments, however, with their highly refined basic magnetic fields, can generate refined phosphorus spectra without spinning the samples. This is illustrated in Fig. 6, which shows the inorganic orthophosphate signal obtained on our instrument without spinning the sample. As can be seen by comparing this spectrum to that obtained by spinning the sample, the peak is broadened and does exhibit a slightly different line shape; however, the skirt of the peak is still reasonably narrow and the loss in signal-to-noise ratio (a consequence of containing the same signal area in a resonance of greater width) is tolerable for most NMR measurements. The generation of such a highly resolved NMR resonance in the absence of sample spinning has two advantages. First, there are no spinning side-band artifacts in the spectrum; and second, the spectrometer can be used to gather data on intact tissue samples that are either too delicate to be spun (i.e., lamellar membrane preparations or intact mammalian lenses) or are mounted in the NMR tube so as to preclude spinning (such as would be the case in the analysis of a perfused working mammalian heart).

State-of-the-art instruments have homogeneous magnetic fields over large volumes so that analysis of quite large samples (20–25 mm in diameter) can be undertaken without any real loss in the information content of the spectrum. For example, our instrument is capable of determinations using NMR tubes 20 mm in diameter, a volume large enough to contain a perfused rabbit heart. The consequence of the use of these large tubes, besides the fact that larger samples can be accommodated, is that there is now considerable space in the probe for other equipment, for example, tension-measuring devices with mounted muscles. In addition, since the sensitivity of the NMR experiment is proportional to the volume of material undergoing analysis, the large tube permits the use of very dilute samples.

A typical experiment, conveniently performed in a large tube, is determination of a phosphorylated protein, where the molecular weight of the macromolecule precludes the preparation of a concentrated sample. The spinning and nonspinning signals obtained on our instrument using the 20-mm in diameter sample tube is shown in Fig. 7. Note that in comparison to the narrow tube sample exhibited in Fig. 6, the peaks are broader, have considerably larger skirts, and always when spun exhibit first-order side bands of ~1%. The base of the peak in the nonspinning example is also asymmetric, and, in this particular case, the asymmetry results from a higher-order inhomogeneity in the magnetic field over this large volume. The shape of this signal is an artifact of the applied field; in other instruments in other laboratories, the shape will not be the same.

The ultimate application of field homogeneity technology in the biomedical fields is the production of limb and whole-body scanning devices, particularly devices with a high-resolution phosphorus detection

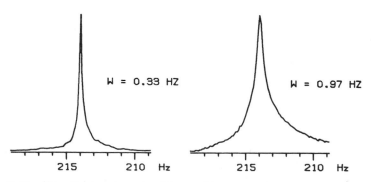

FIG. 7. The ³¹P orthophosphate signal from a 20-mm sample tube at 25° is shown. The narrower signal was obtained while spinning the sample; the wider signal was obtained without spinning. The signal widths at half height, W, are indicated for each resonance.

capability. As of this writing there is at least one commercial phosphorus detection device available for the determination of phosphate profiles in human limbs. It is a cryogenic device featuring a solenoid magnet with a usable working area 20 cm in diameter and a resolving power of 0.3 ppm (3 Hz for phosphorus at the laboratory field employed). In addition, we are aware of a number of development programs directed toward the production of whole-body scanning devices based on phosphorus high-resolution detection operating at fields approaching 2 T. Technological development is moving swiftly, and it will not be long before commercial units will be available for use in a hospital setting. The potential of such devices in the diagnosis and treatment of human diseases is enormous.

Tuning the Probe. The most common limiting factor in the use of phosphorus NMR for biochemical analysis is the sensitivity of the method. Thus, it is essential that the radiofrequency (RF) component of the NMR experiment be as well tuned as the magnetic field. Changing the sample always changes the precise tune of the RF transmittor system, and this effect becomes more pronounced as the RF increases with the magnetic field strength in modern high-powered instruments or as the sample volume increases. We have observed changes in RF tune as great as 90% upon changing from an organic sample in a cylindrical configuration to an aqueous sample contained in a microcell. Obviously, such a dramatic change in the balance of the RF system would have a profound effect on the signal-to-noise ratio eventually obtained upon completion of the analysis. Thus, failure to "peak" the RF system may result in loss of information due to decreased signal-to-noise ratio or loss of instrument time necessitated by the need for prolonged signal-averaging to improve the signal-to-noise ratio. Further, a badly tuned RF system can actually create spurious signals in the spectrum as a result of the complex harmonic interactions among the components of the receiver network. It seems trivial to mention such an obvious component of NMR analysis; however, in visiting many laboratories over the years, we have become aware that proper adjustment of the RF system is all too frequently overlooked.

Scan Conditions. Rarely does analysis of biological substances permit the analyst to establish scan conditions that are optimal from the NMR point of view, i.e., large numbers of data points per spectrum, long delay times, 90-degree flip angles. Most of the time, conditions are dictated by the need to optimize the signal-to-noise ratio per unit time under circumstances where extensive signal-averaging will be performed. Thus, in our laboratory, the following conditions have been found to be optimal for the analysis of intact tissue: sweepwidth of ±2500 Hz, incorporating 8192 data points in the pulse free-induction decay; acquisition time, 819 msec;

pulse length, 8 μsec, corresponding to a 45-degree flip angle. In addition, we incorporate quadrature phase detection with RF filtering corresponding to the 2500 sweepwidth employed and 8- or 16-bit digitizer resolution. Proton broad-band decoupling may or may not be used, depending upon the experimental design. In any event, the RF power levels employed for proton decoupling when used are approximately an order of magnitude less than is needed for a corresponding decoupled carbon spectrum, and heating of the sample does not occur to any significant degree. When analyzing extract samples, where the signals are much narrower, it is necessary to increase the number of data points per scan to 16,384, thereby increasing the acquisition time proportionately.

With such scan conditions, there is some RF saturation of the NMR peaks. However, this saturation is essentially the same for all the biological phosphates in either intact tissue or extract spectra run at the alkaline pH values customarily used, the only major exceptions being a relatively greater saturation of the inorganic orthophosphate resonance or, in blood analysis, of the 2,3-diphosphoglycerate peaks. Thus, area measurements are relatively unaffected by the need to scan the data rapidly. Nonetheless, we calibrate our scan conditions with samples of known constitution so that integration errors may be appropriately corrected. In fact, such a procedure is recommended for any system for which calibration procedures can be devised, since phosphorus relaxation behavior, which determines relative saturation behavior in any given sample, is difficult, if not impossible, to predict.

Computerized Processing of Data. State-of-the-art instruments are equipped with sophisticated computer technology that permits refined reduction of raw NMR data. Routine operations involve various types of filtering applications performed on pulse free-induction-decay (FID) data, Fourier transformation of the time-domain FID data to the frequency domain intensity versus frequency plots customarily read by the analyst, peak position calculations and integrations. With more sophisticated programming and appropriate NMR hardware, T_1 and T_2 experiments can be performed. Such experiments require the additional computer capability of disk or some other form of high-density storage, normalization programs so that one spectrum in a set can be equated with another, and, usually, some sort of peak-picking routine. Regardless of the type of data being analyzed, the experiment usually benefits from the application of filtering and smoothing operations, which in addition to improving resolution or signal-to-noise ratio also generate plots that can be easily interpreted. The value of a good display should never be overlooked by the analyst when there is a need to explain the data to another scientist unfamiliar with the nuances of NMR technology.

Preparation of Samples

Intact Muscle. The preparation of intact muscle samples may be as simple as placing the dissected muscles into NMR tubes,[4] or as involved as the mounting of four frog sartorii in a specially designed NMR sample tube that is perfused with oxygenated Ringer solution and possesses arrangements for stimulating the muscles electrically and recording their tension.[3]

In our laboratories, cold-blooded animals are chilled in ice water, before sacrifice, and the muscles are dissected on a watch glass kept on ice. Warm-blooded animals are killed with Nembutal overdose to keep the muscles relaxed; the dissected muscles are quickly covered by ice. Other muscle samples, e.g., those of human biopsies, are transported on ice to the NMR facility. Intact muscles or muscle biopsies are inserted directly into ice-chilled NMR tubes that contain an open 1-mm capillary. The muscles are worked to the bottom of the tube with a large glass plunger, the entrapped air escaping through the capillary. After the muscle is positioned this capillary is withdrawn, and the muscle is covered with a tight-fitting Teflon plug. (For determination of the concentration of phosphate metabolites, an NMR reference capillary may be inserted through the Teflon plug and through the muscle to the bottom of the tube). The NMR tube is then capped and placed in the spectrometer probe for analysis. During the preparation of the sample tube (3–5 min) the small amount of oxygen left in the dissected muscle is consumed; thus, during recording of the muscle spectrum *anaerobic* conditions prevail.

For routine analysis the muscle is placed in the NMR tube without any solution.[4] Thus, the total volume of the tube is occupied by muscle, which maximizes the signal-to-noise ratio. However, when the muscle is stimulated under anaerobic conditions, it is bathed in Ringer solution that contains 2 mM NaCN and has been additionally gassed with nitrogen.[34]

Phosphocreatine (PCr) is the predominant resonance in the spectrum of a freshly dissected intact chicken muscle placed in the NMR tube in ice (Fig. 8). Only a small peak is observed in the resonance position of P_i, and no peak is detected around 7 ppm, the position of sugar phosphates (SP) in this spectrum. Similar spectra were recorded with resting frog muscles at 4°.[3,35] In contrast, a muscle that has been aged or has been injured during dissection shows a reduced PCr resonance and increased P_i and SP peaks. ATP, however, remains constant, provided the muscle has been handled with reasonable care.

[34] M. J. Dawson, D. G. Gadian, and D. R. Wilkie, *Nature (London)* **274**, 861 (1978).
[35] T. Glonek, C. T. Burt, R. J. Labotka, M. J. Danon, M. D. Vuolo, and M. Bárány, *in* "Benjamin Goldberg Symposium" (J. Marbarger, ed.), p. 214. Univ. Illinois Med. Ctr., Chicago, 1977.

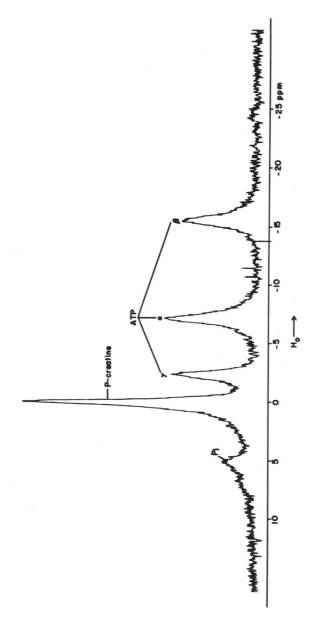

FIG. 8. Proton-coupled ³¹P NMR spectrum of intact chicken pectoralis muscle, initially at 0°. A Bruker CXP-180 spectrometer operating at 72.88 MHz in quadrature detection mode was used. The muscle was placed into a 20-mm tube that was spun at 24 Hz at 24°. Spectrometer conditions: sweepwidth and filter bandwidth 10,000 Hz; dwell time, 50 μsec; 8 K data point; cycling time 1 sec; pulsewidth, 5 μsec; line broadening, 2 Hz. The spectrum shown is the signal average of 600 scans. Chemical shifts are relative to the phosphocreatine resonance in the muscle. Unpublished data of D. D. Doyle and M. Bárány, 1980.

For physiological experiments the muscles must be perfused with O_2-saturated Ringer's solution. The sartorius muscles from small frogs are thin enough for the center of the muscle to be reached by diffusion of oxygen from the solution. Virtually no other muscle can be adequately oxygenated for studying changes in phosphate metabolites during contraction and recovery. Details of recirculating perfusion systems for such studies were described.[3] The idea of skeletal muscle perfusion through the muscle's own circulatory system was put forward some time ago,[36] and Kushmerick et al.[37] perfused the biceps brachii muscle of cat in vitro through its arterial branches with an oxygenated suspension of fluorocarbon micelles in a physiological saline at 28°. This preparation gave stable[31] ^{31}P NMR spectra for at least 12 hr.

In contrast to skeletal muscles, which require special conditions to be kept alive, perfused hearts of almost any size can be kept functioning for many hours.[38] Specially designed heart perfusion systems have been described.[39,40] The pressure developed by the heart can be monitored in a continuous manner,[40] and changes in the mechanical performance of the heart can be correlated with changes in the ^{31}P profile. NMR sample tubes ranging from 8 mm to 12 mm inner diameter have been used for rat hearts,[9,41] whereas 25-mm NMR tubes have been used for rabbit hearts weighing 6–8 g.[42]

Although ^{31}P NMR spectra recorded from excised and perfused muscle may be considered to reflect the physiological phosphate metabolism of muscle, technical advances have made possible the detection of phosphate compounds in muscles of living animals.[43,44] The rat was secured in

[36] J. Dawson, D. G. Gadian, and D. R. Wilkie, in "NMR in Biology" (R. A. Dwek, I. D. Campbell, R. E. Richards, and R. J. P. Williams, eds.), p. 289. Academic Press, New York, 1977.

[37] M. J. Kushmerick, T. Brown, and M. Crow, Fed. Proc., Fed. Am. Soc. Exp. Biol. 39, 1934 (1980).

[38] D. G. Gadian, G. K. Radda, R. E. Richards, and P. J. Seeley, in "Biological Applications of Magnetic Resonance" (R. G. Shulman, ed.), p. 463. Academic Press, New York, 1979.

[39] P. J. Seeley, P. A. Sehr, D. G. Gadian, P. B. Garlick, and G. K. Radda, in "NMR in Biology" (R. A. Dwek, I. D. Campbell, R. E. Richards, and R. J. P. Williams, eds.), p. 247. Academic Press, New York, 1977.

[40] D. P. Hollis, R. L. Nunnally, G. J. Taylor, IV, M. L. Weisfeldt, and W. E. Jacobus, J. Magn. Reson. 29, 319 (1978).

[41] P. B. Garlick, G. K. Radda, P. J. Seeley, and B. Chance, Biochem. Biophys. Res. Commun. 74, 1256 (1977).

[42] D. P. Hollis, R. L. Nunnally, W. E. Jacobus, and G. J. Taylor, IV, Biochem. Biophys. Res. Commun. 75, 1086 (1977).

[43] T. H. Grove, J. J. H. Ackerman, G. K. Radda, and P. J. Bore, Proc. Natl. Acad. Sci. U.S.A. 77, 299 (1980).

[44] J. J. H. Ackerman, T. H. Grove, G. G. Wong, D. G. Gadian, and G. K. Radda, Nature (London) 283, 167 (1980).

the NMR probe in a vertical position, a four-turn receiver coil was introduced into the thorax and placed around the heart[43] or the coil was placed against the leg of an anesthetized rat,[44] and ³¹P NMR spectra were recorded in 8 and 25 min, respecitvely. These studies are the frontrunners of a new level of biological ³¹P NMR, which already includes the measurement of changes in PCr and P_i concentrations during the exercise of human limbs[45] and will eventually be incorporated into whole-body scanners.

Perchloric Acid Extracts. Frozen tissue, preferably tissue frozen with liquid nitrogen-chilled Wollenberger clamps, is pulverized with a liquid nitrogen-chilled stainless steel mortar and pestle to a fine powder. The mortar is maintained in a liquid-nitrogen bath throughout the pulverization procedure. After pulverization, the tissue is transferred to a centrifuge tube containing 0.1 v/w 60% perchloric acid (PCA), prefrozen with liquid nitrogen or Dry-Ice. The powder is stirred with a glass rod until the temperature rises to approximately $-20°$, when a liquid phase can be first discerned.[18] After thorough mixing at below freezing temperatures to provide optimal extraction, the contents are centrifuged at 43,000 g for 15 min in the high-speed centrifuge at $-4°$. The supernatant solution is transferred immediately into a volume of 10 N KOH, equal to the volume of PCA initially used, and the pH is immediately adjusted to values between 9.5 and 10.0 to prevent acid hydrolysis of labile phosphates and to precipitate the potassium perchlorate. (The protein pellets are dissolved in 100 ml of 0.1 N NaOH for analysis by the biuret method.) This suspension is centrifuged at $-4°$ for 10 min at 43,000 g to remove the potassium perchlorate. The final supernatant solution is washed through a potassium Chelex-100 (Bio-Rad, 200–400 mesh) column to remove polyvalent cations. The pH of the sample is then adjusted to 10.0 and lyophilized. The lyophilizate from tissue samples under 1 g is dissolved in 0.4 ml of 20% D_2O and filtered through washed glass wool, to remove any sediment ($KClO_4$), into a NMR microcell for spectroscopic analysis. No more than 2% of the sample's PCr will be lost to hydrolysis when this procedure is properly carried out.

The use of PCA muscle extracts has the advantage that the extract spectrum is more highly resolved than that from the corresponding fresh muscle. This is illustrated in Fig. 9, which compares the ³¹P NMR spectrum of frog muscle with that of its extract after treatment with Chelex resin to remove the alkaline earth metals. In the spectrum of the PCA muscle extract (lower part of Fig. 9), the characteristic multiple patterns of ATP and the sugar phosphate resonances are clearly seen. On the other

---

[45] S. Eleff, J. S. Leigh, and G. G. McDonald, *Fed. Proc., Fed. Am. Soc. Exp. Biol.* **39**, 1807 (1980).

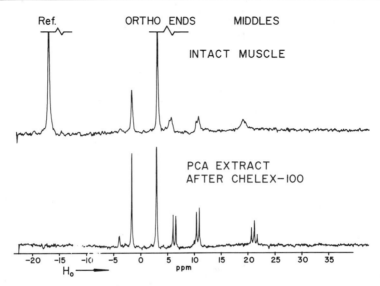

FIG. 9. Proton-decoupled ^{31}P NMR spectra of frog leg muscle (upper spectrum) and a perchloric acid extract of this muscle (lower spectrum) taken from the other leg of the same frog. A Bruker HFX-5 spectrometer with ^{2}H stabilization operating at 36.43 MHz for ^{31}P and containing modules for broad-band and continuous-wave heteronuclear ^{1}H decoupling was used. The muscle and the extract were placed into 10-mm tubes, which were spun at 45 Hz at 31°. Spectrometer conditions: sweepwidth, 2500 Hz (200 μsec/data point; 4096 data points); cycling time, 832 msec; filter bandwidth, 2500 Hz; pulsewidth, 4.5 μsec; 1.2 Hz line broadening. The spectrum shown is the signal average of 2160 scans. "Ortho, Ends and Middles" refer to the characteristic regions of the ^{31}P spectrum. Peak assignments in the muscle spectrum from left to right: the external methylenediphosphonate reference compound (Ref), -16.3 ppm; SP, -3.7 ppm; P_i, -1.7 ppm; phosphocreatine (PCr), 3.2 ppm; the phosphate groups of ATP: γ, 5.6 ppm; α, 10.7 ppm; and β, 19.1 ppm. Chemical shift data are relative to 85% H_3PO_4 and follow the old convention; i.e., upfield shifts are positive. In the spectrum of the PCA extract the resonance positions of the γ-, α-, and β-phosphate groups of ATP are displaced upfield by 0.8, 0.4, and 2.1 ppm, respectively, caused by the lack of complex formation between ATP and Mg^{2+}. Reproduced, with permission, from Burt et al.[4]

hand, the relative area under the P_i peak is larger in the extract than in the intact muscle (upper part of Fig. 9) owing to the breakdown of PCr during the workup of this particular sample.

The PCA extracts are invaluable in the detection of minor differences in the ^{31}P profile between different muscles[4] and in the identification of unknown resonances.[35]

The use of PCA extracts is obligatory when a large number of samples are obtained simultaneously for comparative purposes.[46] Neutralized

[46] J. M. Chalovich, C. T. Burt, M. J. Danon, T. Glonek, and M. Bárány, Ann. N.Y. Acad. Sci. 317, 649 (1979).

PCA extracts are stable in the deep freeze for longer times and may be analyzed conveniently when the spectrometer is available. Finally, muscle samples as little as 70 mg may be effectively analyzed through the use of neutralized PCA extracts (see the section Phosphate Spectra of Perchloric Acid Extracts, below).

Concentration Determinations

The integral of the phosphate signals can be used for the determination of phosphate concentrations in the intact muscle. A 90-degree excitation pulse (tilt angle) yields the maximum signal at the spectrometer's detector, provided the spin system is fully equilibrated beforehand,[47] i.e., the time allowed for equilibration is at least 5 times the greatest ^{31}P T_1 value of the sample. Few measurements of T_1 of ^{31}P nuclei in intact tissue have been reported, and the data available show quite a variation. Thus, T_1 values of the α, β, and γ resonances of ATP were found to be 0.35, 0.31, and 0.28 sec, respectively, in the foot muscle of a marine invertebrate,[48] 0.24 sec for the γ-ATP in *Escherichia coli* cells,[49] and about 1 sec for the β-ATP in rat leg muscle.[44] T_1 values for P$_i$ were found to be 0.38 sec in *E. coli*,[49] 0.6 sec in rat liver mitochondria,[50] approximately 3 sec in either the foot muscle[48] or in rat skeletal muscle,[44] and 3–4 sec in rat heart.[43] T_1 values for PCr were about 3 sec in rat skeletal,[44] 3–4 sec in rat heart,[43] and 4.8 sec in frog muscle at 4°.[51] Although, the T_1 of PCr in muscle was found to be comparable to that of PCr in solution,[51] T_1 values for ATP and P$_i$ in aqueous solution were significantly longer than those in intact tissues.[52]

Assuming 4 sec for the longest relaxation time of a muscle phosphate, about a 20-sec interval would be required between sampling pulses to maximize the signal-to-noise ratio with a 90-degree excitation pulse. To collect 100 scans from an intact muscle 5 min would be needed; this is too few scans and too long a time for accurate integration of the phosphate resonance areas. During the course of 25 min, significant changes in the concentration of most of the phosphate metabolites take place. For that reason optimum signal-to-noise ratios are often achieved most effectively by using 45-degree excitation pulses and shorter repetition times. As mentioned previously, however, such less-than-ideal conditions require calibration of the spectrometer for accurate concentration determination.

[47] C. T. Burt, S. M. Cohen, and M. Bárány, *Annu. Rev. Biophys. Bioeng.* **8**, 1 (1979).
[48] K. D. Barrow, D. D. Jamieson, and R. S. Norton, *Eur. J. Biochem.* **103**, 289 (1980).
[49] T. R. Brown, K. Ugurbil, and R. G. Shulman, *Proc. Natl. Acad. Sci. U.S.A.* **74**, 5551 (1977).
[50] S. Ogawa, H. Rottenberg, T. R. Brown, R. G. Shulman, C. L. Castillo, and P. Glynn, *Proc. Natl. Acad. Sci. U.S.A.* **75**, 1796 (1978).
[51] S. M. Cohen and C. T. Burt, *Proc. Natl. Acad. Sci. U.S.A.* **74**, 4271 (1977).
[52] T. Glonek, *Biochem. Med.* **19**, 246 (1978).

TABLE I

CHANGES IN INTEGRATION UNIT VALUES OF VARIOUS PHOSPHATE RESONANCES DURING
ANAEROBIC INCUBATION OF INTACT DYSTROPHIC CHICKEN PECTORALIS MUSCLE[a]

Time (min)	SP + NMP[b]	P_i	SEP[b]	PCr	γ-ATP	α-ATP	NAD	β-ATP	Total
10–20	131	379	35	618	225	315	15	215	1933
20–30	108	496	29	497	222	244	43	212	1851
30–40	107	598	43	409	223	245	48	218	1891
40–50	81	733	51	328	221	221	71	216	1922
50–60	81	734	48	260	210	196	79	200	1808
60–70	99	859	31	220	208	210	64	195	1886
70–80	83	912	30	168	200	201	60	156	1810
80–90	95	936	46	141	191	180	73	181	1843
90–100	86	992	40	120	192	192	60	175	1857

[a] Spectrometer conditions: One pulse sequence employing a 45 degree (15 μsec) excitation pulse; sweepwidth of ± 2500 Hz digitized to 8192 data points (quadrature detection, 12-bit digitizer resolution); acquisition time 819.4 msec; spectrometer frequency for ^{31}P, 80.987 MHz. The integration units are listed in the same order as they appeared in the printout, corresponding to the increasing magnetic field. The integration units are rounded up to the nearest integer.

[b] SP, sugar phosphates; NMP, nucleoside monophosphates; SEP, serine ethanolamine phosphodiester.

In our laboratory two methods are used for concentration determination. At 35.43 MHz, with a 4.5-μsec pulse width and 832 msec cycling time, a capillary (1 mm in diameter) containing 1.0 M methylenediphosphonic acid is calibrated against 3 ml of 0.02 M Na_2HPO_4. This reference capillary is coaxially mounted in the sample tube containing at least 3 ml of muscle. The concentration of the various muscle phosphates can be determined by comparing the integral of their signals to that of the capillary. Using the capillary method, we have measured changes in the concentrations of PCr, P-arginine, P_i, and SP in frog[4] and barnacle muscles[53] for 2–3 hr and in human muscles[5] for 20–40 min.

In another method, with a 20-mm probe in a Nicolet-200 spectrometer, at 80.99 MHz, using a 15 μsec pulsewidth, and a 0.82 sec recycling time, we followed the peak areas of the phosphates in intact muscle, and determined, in other aliquots of the same muscle, the PCA-extractable total phosphate and the ATP and ADP content of the fresh muscle. Table I shows the computer-determined integration units for the various phosphates as a function of muscle incubation under anaerobic conditions. Two major changes are apparent: the increasing values under

[53] M. Bárány, J. M. Chalovich, C. T. Burt, and T. Glonek, in "Diseases of the Motor Unit" (D. L. Schotland, ed.), p. 697. Wiley, New York.

the P_i peak, and their corresponding decrease under the PCr peak. The sugar phosphate area (which also includes nucleoside monophosphates) remains constant, surprisingly, as does the area of serine ethanolamine phosphodiester (SEP), a compound characteristic of chicken hereditary muscular dystrophy.[54] The γ-, α-, and β-phosphates of ATP remain constant until 50 min and, thereafter, exhibit a slight decline. Between 10 and 20 min, nicotine adenine dinucleotide (NAD) is incompletely resolved from the α-phosphate of ATP (hence the NAD value is low and that of the α-phosphate is high); subsequently the NAD area increases to a constant value (after 40 min). It is important to point out that the total integration unit values remain the same, within the experimental error, during the 10–100-min experiment.

The PCA-extractable total phosphate was quantitated in terms of total inorganic phosphate after ashing an aliquot of extract; 44.87 μmol of P_i per gram of fresh muscle was found. The ATP and ADP content of the muscle was determined by extracting the nucleotides from the muscle with PCA and separating them by ion exchange chromatography; 4.20 μmol of ATP and 0.85 μmol of ADP were found per gram of fresh muscle. (In terms of total phosphate these values are 12.60 μmol and 1.70 μmol, respectively.) ADP in the muscle is bound to actin and, as such, is not detectable by ^{31}P NMR (see Bárány et al.[2]). There is no difference in the integration unit values between γ-ATP and β-ATP (Table I); therefore, ADP phosphate content must be subtracted from the total phosphate in the PCA extract; this leaves 43.17 μmol of total phosphate per gram of fresh muscle. The 12.60 μmol of phosphate due to ATP corresponds to the integration unit value sum under the α, β, and γ phosphate peaks of ATP until 40–50 min; that is, throughout the time during which the ATP concentration does not decrease. Subtracting both the ADP- and ATP-phosphate from the total phosphate yields 30.57 μmol of phosphate per gram of fresh muscle; this corresponds to the phosphate concentrations for SP and NMP, P_i, SEP, PCr, and NAD. Since the relaxation times of P_i and PCr (which together comprise 85% of the peak areas) do not differ,[6] a good estimate for the concentration of phosphate metabolites may be obtained by multiplying 30.57 by the fractional peak areas (Table I).

These data can be further used to obtain the phosphate concentrations from peak areas as a function of time. From the constant value of 215 for the β-phosphate of ATP up to 50 min, 645 total integration units appear for the α, β, and γ phosphates of ATP. Substituting 645 uniformly for the ATP phosphates for time periods of 10–20, 20–30, 30–40, and 40–50 min, and calculating for each time period the micromoles of phosphate per gram of

[54] J. M. Chalovich, C. T. Burt, S. M. Cohen, T. Glonek, and M. Bárány, Arch. Biochem. Biophys. **182**, 683 (1977).

muscle for SP + NMP, P_i, SEP, PCr, and NAD, one can show that 12.60 μmol of phosphate in ATP per gram of muscle corresponds to 497 integration units under the α, β, and γ peaks, or 166 units under the β-peak. In other words, the correction factor for the β-ATP peak area is 166/215 = 0.77, reflecting the faster relaxation rate of ATP as compared to the other phosphates under our conditions. By multiplying the ATP peak areas by 0.77, the concentration of all phosphates may be calculated from the fractional peak areas and the 43.17 μmol of total phosphate per gram of muscle.

For calculation of phosphate concentrations in fresh muscle the values for P_i and PCr are extrapolated to zero time, and the values for SP + NMP, SEP, NAD are averaged. The intact dystrophic chicken pectoralis muscle (Table I) contains (μmol/g): 2.42 SP + NMP, 7.08 P_i, 0.95 SEP, 22.39 PCr, 0.71 NAD, and 4.20 ATP. The concentration of the phosphate metabolites in the muscle water, which is actually measured by ^{31}P NMR, may be calculated from the 76.8% water content (which was determined for the muscle shown in Table I): 3.15 mM SP + NMP, 9.23 mM P_i, 1.24 mM SEP, 29.15 mM PCr, 0.92 mM NAD, and 5.47 mM ATP.

Identification of Resonances

Peak assignments in intact muscle are essentially based on accurate measurement of the chemical shift of a resonance. Within the same laboratory this is usually reproducible with ± 0.1 ppm, and among different laboratories with ± 0.5 ppm. Table II lists the chemical shifts of phosphates known to occur in muscle, measured both in ^1H-coupled and -decoupled spectra. The chemical shifts of other phosphates that have been observed in nonmuscle tissues are given in Table III.

The resonances of the sugar phosphates, P_i, GPC or SEP, PCr, the γ-, α-, and β-phosphates of ATP are well separated in muscle, and thus the identification of these metabolites presents no problem. Within the sugar phosphate group only glucose 1-phosphate can be resolved from the other hexosephosphates and from the triosephosphates. A better resolution of the sugar phosphates requires 145.8 MHz for ^{31}P and proton decoupling. In muscle, it is very difficult to separate the sugar phosphates from AMP or IMP, and it is difficult to separate AMP and IMP from each other (a weak point of ^{31}P NMR). Differentiation of arginine phosphate from creatine phosphate is relatively easy, and NADH–NAD$^+$ appear as an individual peak at 72.88 MHz, although they are incorporated in the α-phosphate peak of ATP at 36.43 MHz. The chemical shift caused by complexing of the β-P of ATP with Mg^{2+} is easily determined at 36.43 MHz.

TABLE II

CHEMICAL SHIFTS OF MUSCLE PHOSPHATES AT ABOUT pH 7[a]

Phosphate	Chemical shift[b] (ppm)
Dihydroxyacetone phosphate	4.5–5.1
Glucose 6-phosphate	4.2–4.7
Glycerol 1-phosphate	4.4
Glyceraldehyde 3-phosphate	4.2–4.4
Fructose 1,6-diphosphate, 1-P	4.0–4.4
3-Phosphoglycerate	4.2
Glycerol 2 phosphate	4.1
Pyridoxal phosphate	4.0
Fructose 1,6-diphosphate, 6-P	3.8–4.0
AMP, IMP	3.7–4.0
Fructose 6-phosphate	2.6–4.1
2-Phosphoglycerate	3.5
Glucose 1-phosphate	2.6
P_i	1.7–2.3
sn-Glycerol 3-phosphorylserine and 3-ethanolamine	0.4
Phosphoenolpyruvate	0.3–(−0.9)
sn-Glycerol 3-phosphorylcholine	−0.1–(−0.13)
Serine ethanolamine phosphate	−0.4
Phosphocreatine	−2.3–(−3.2)
PArg	−3.5
γ-P of ATP (Mg bound)	−4.8
β-P of ADP (Mg bound)	−5.4
γ-P of ATP	−5.7–(−6.7)
β-P of ADP	−6.1–(−7.4)
Inorganic pyrophosphate	−7.0–(−7.5)
α-P of ADP (Mg bound)	−9.4
α-P of ATP (Mg bound)	−9.9
α-P of ADP	−10.3–(−10.4)
α-P of ATP	−10.4–(−10.8)
NADH and NAD+	−10.5–(−11.3)
β-P of ATP (Mg bound)	−18.5–(−19.5)
β-P of ATP	−21.2–(−21.4)

[a] Values were taken from Burt et al.[47]; Gadian et al.[38]; and unpublished results from our laboratories.

[b] Relative to 85% H_3PO_4; downfield shifts are positive.

If doubt exists concerning the identity of a resonance, following the chemical shift with aging of the muscle may be helpful. The acidification of the internal millieu of muscle may result in a characteristic chemical shift. If the questionable resonance survives the muscle aging, it should be extracted with PCA and its identity further investigated at the level of the extract.

TABLE III
CHEMICAL SHIFTS OF OTHER TISSUE PHOSPHATES[a]

Phosphate[b]	Chemical shift (ppm)
Phosphonate in snail eggs	~22
Phosphorylethanolamine	3.4
3-P of DPG	3.3
Phosphorylcholine	3.1
2-P of DPG	2.6
Phosphoproteins	1–4
Nucleic acids	1–(−1)
Phospholipids	0
Sphingomyelin and phosphatidylethanolamine	−0.3
Phosphatidylcholine	−0.9
Nucleohistones	−1.2
γ-P of NTP	−4.7–(−5.8)
β-P of NDP	−5.1–(−5.3)
α-P of NDP	−9.8–(−10.0)
α-P of NTP	−10.9
CDP-choline	−10.6–(−11.5)
β-P of NTP	−20.3–(21.3)
Polyphosphates (yeast)	>−22

[a] Values taken from Burt et al.[47]; and R. C. Carroll, E. B. Edelheit, and P. G. Schmidt, *Biochemistry* **19**, 3861 (1980).
[b] DPG, diphosphoglycerate; NTP and NDP, nucleoside tri- and diphosphate.

Phosphates can be identified in PCA extracts by simply adding a known quantity of a pure compound to the sample and observing the position of resonance. Coincidence of the resonance of the added material with that of the signal of the sample is usually sufficient to identify an unknown resonance. Figures 10 and 11 illustrate such examples. The assignment can be reinforced by selectively shifting the resonance of interest with respect to those from the other compound in the sample by adjustments of pH, ionic strength, or both. In the Appendix of their paper, Gadian et al.[38] describe the pH and ionic strength dependence of the chemical shifts of several muscle phosphate compounds.

If a resonance cannot be assigned to a known phosphate compound, its purification may be attempted. The selective solubility of organic phosphates in barium salts and ethanol, and their strong affinity to anion exchange columns, renders such an endeavor feasible,[19,54] although it may be time consuming.

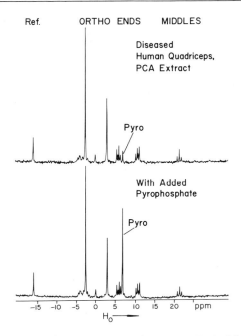

Ref.   ORTHO ENDS   MIDDLES

Diseased
Human Quadriceps,
PCA Extract

Pyro

With Added
Pyrophosphate

Pyro

-15 -10 -5   0   5   10 15 20   ppm

H$_o$

FIG. 10. Proton-decoupled ^{31}P phosphate profile (36.43 MHz) of limb-girdle dystrophy. The top spectrum was obtained from a PCA extract of the diseased human muscle. The resonance from pyrophosphate occurs at 7.0 ppm. The bottom spectrum shows the spectroscopic identification of pyrophosphate. The addition of known pyrophosphate to the sample caused the resonance at 7.0 ppm to be uniformly enhanced. For spectrometer conditions and peak assignments, see the legend to Fig. 9. Reproduced, with permission, from Glonek et al.[35]

Muscle Sample Spectral Profiles

Phosphate Spectra of Intact Muscles

Normal Muscle. ^{31}P spectra of normal muscles may exhibit considerable variation. Figure 12 compares the spectra of various intact muscles. The gastrocnemius muscle of the North American winter frog shows the presence of GPC and SEP (0.1 and 0.4 ppm, respectively). Toad gastrocnemius contains these compounds in a much larger quantity. GPC but not SEP is noticeable in human gastrocnemius, whereas no phosphodiester is visible in the abalone mantle muscle. PCr or PArg (3.2 or 3.5 ppm) is a major peak in all muscles with the exception of the human, which was taken from a gangrenous leg and shows P$_i$ (around -2 ppm) as its predominant peak.

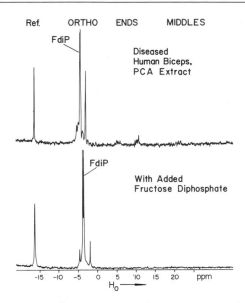

FIG. 11. Proton-decoupled ^{31}P phosphate profile (36.43 MHz) of a type II atrophy of undetermined origin. The top spectrum was obtained from a PCA extract of the diseased human muscle. The pronounced signals at -3.9 and -4.0 ppm arise from fructose 1,6-diphosphate (FdiP). The bottom spectrum shows the identification of the FdiP resonances through the addition of known FdiP. For spectrometer conditions and peak assignments, see the legend to Fig. 9. Reproduced, with permission, from Glonek et al.[35]

Splitting of the P_i resonance into two components as a function of time is demonstrated in Fig. 13 with a human pectoralis muscle sample. The same phenomenon was reported by Seeley et al.[55] with rabbit hind-leg white muscle, and the femoral biceps muscle of the bullfrog by Yoshizaki et al.[56] Since the frequency of the P_i signal is sensitive to pH, the data have been interpreted[55] to indicate that there are two phosphate compartments in muscle with pH values of 6.8 and 6.4. Subsequent work[6] indicated that in high-energy phosphate-depleted muscle, the intracellular pH is not uniform within the muscle volume, and that the P_i is distributed among different compartments in the muscle cell. Indeed after disruption of the transverse tubules with glycerol, only one P_i resonance was found.[56] Furthermore, in isolated rat liver cells, mitochondrial and cytosolic phosphate pools with different pH values were described.[57] The pH gradient hypoth-

[55] P. J. Seeley, S. J. W. Busby, D. G. Gadian, G. K. Radda, and R. E. Richards, Biochem. Soc. Trans. 4, 62 (1976).
[56] K. Yoshizaki, H. Nishikawa, S. Yamada, T. Morimoto, and H. Watari, Jpn. J. Physiol. 29, 211 (1979).
[57] S. M. Cohen, S. Ogawa, H. Rottenberg, H. Glynn, P. Yamane, T. Brown, R. G. Shulman, and J. R. Williamson, Nature (London) 273, 554 (1978).

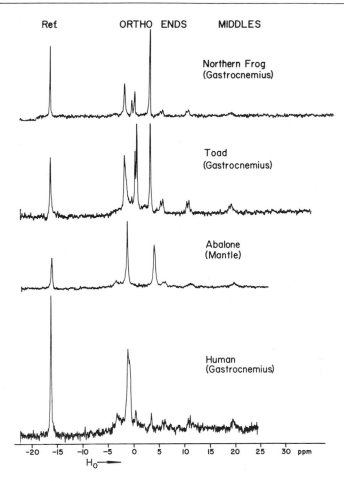

Ref	ORTHO ENDS	MIDDLES

Northern Frog
(Gastrocnemius)

Toad
(Gastrocnemius)

Abalone
(Mantle)

Human
(Gastrocnemius)

$H_0 \longrightarrow$

FIG. 12. Proton-decoupled ^{31}P NMR spectra (36.43 MHz) of various intact muscles. The spectra shown are the signal average from 1440 to 3240 scans. For the spectrometer conditions and peak assignments, see the legend to Fig. 9. Reproduced, with permission, from Burt *et al.*[4]

esis of Seeley *et al.*[55] is further supported by their finding that the P_i resonance split does not occur in rabbit hind-leg red muscle, which, unlike the white muscle, does not accumulate lactic acid.

In all muscle spectra, the phosphate groups of ATP are displaced downfield from their resonance positions in simple aqueous solutions, the displacement being 2.1 ppm for the β, 0.8 ppm for the γ, and 0.4 ppm for the α group. For other phosphates, however, such as PCr or SP, similar relative shift changes between those in simple aqueous solutions and those in muscle are not observed. The shift changes for ATP have been inter-

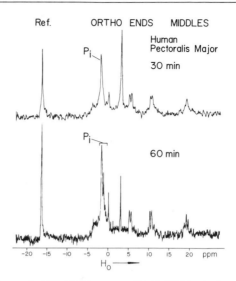

FIG. 13. Proton-decoupled ^{31}P NMR spectra (36.43 MHz) of a human pectoralis muscle showing appearance of multiple signals in the inorganic orthophosphate peak: *Top:* Signal-averaged spectrum for 5 to 30 min after biopsy; the P_i resonance shows a single, fairly broad (20 Hz) signal. *Bottom:* The same muscle during the period from 5 to 60 min; the P_i resonance is clearly split into two resonance signals indicating at least two discrete pools of P_i in the muscle. The width of the P_i resonance in both spectra, however, suggests the presence of still more phosphate pools. For the spectrometer conditions and peak assignments, other than P_i, see the legend to Fig. 9. Reproduced, with permission, from Glonek *et al.*[35]

preted to arise as the result of complex formation between ATP and Mg^{2+} ions.[1-3] Several lines of evidence indicated that ATP forms a complex with Mg^{2+} in muscle.[4]

Initial spectra of fresh vertebrate skeletal muscle are characteristic in that they show PCr as the major phosphate present, little P_i, and even less SP^3 (see Fig. 8). Indeed ^{31}P NMR is more suitable for PCr and P_i determinations than conventional chemical procedures. Moreover, the ATP, PCr, and P_i content of frog and chicken muscles measured by ^{31}P spectroscopy agreed well with literature data wherein a variety of chemical methods were employed.[3,4] As mentioned above under Basic Principles, subsection on concentration determinations, the NMR technique ordinarily detects only signals from compounds in solution, therefore, the phosphate concentrations are expressed in terms of the total free muscle water rather than in unit weight of the muscle.

The nondestructive nature of NMR, and its ability to measure the concentration of phosphate metabolites simultaneously and repeatedly, makes it an attractive analytical tool. Because of the low sensitivity of ^{31}P

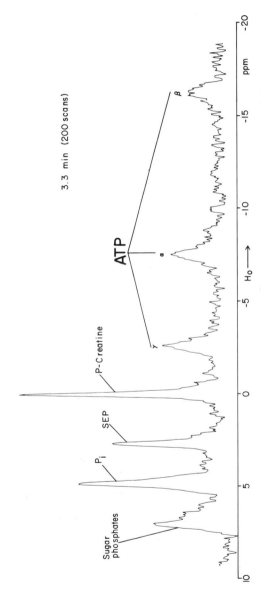

FIG. 14. Proton-coupled ³¹P NMR spectrum (72.88 MHz) of intact turtle muscle. For spectrometer conditions, see the legend to Fig. 8. This spectrum was obtained on a muscle 5 hr after dissection from the animal, which explains the high level of P_i and sugar phosphates. Serine ethanolamine phosphodiester (SEP) is present in high concentration in turtle skeletal muscle. Unpublished data of D. D. Doyle and M. Bárány, 1980.

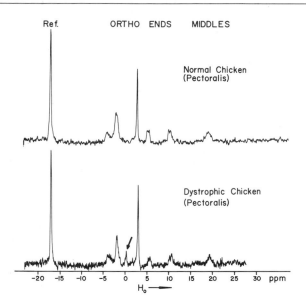

FIG. 15. Comparison of proton-decoupled ^{31}P NMR spectra (36.43 MHz) of normal and dystrophic chicken breast muscle. The arrow indicates the presence of SEP in the dystrophic but not in normal chicken. For the spectrometer conditions and peak assignments see the legend to Fig. 9. Reproduced, with permission, from Burt et al.[4]

NMR, however, signal-averaging processes are required to obtain a usable muscle spectrum (i.e., several minutes are needed for a muscle phosphate analysis). With 20-mm sample tubes and at 72.88 MHz, however, useful data can be obtained within about 3 min (Fig. 14). Dawson et al.[58] obtained a spectrum from 1.5 g of frog gastrocnemii in a *single* scan, and Kushmerick et al.[37] achieved the same with cat muscle.

Diseased Muscle. ^{31}P NMR studies on intact muscles have revealed the presence of several phosphates that are not commonly encountered.[4,19,55] Two of these phosphates were proved to be invaluable markers for muscle diseases: SEP for hereditary chicken dystrophy and GPC for human diseases.

Figure 15 compares the ^{31}P NMR spectra of intact normal and dystrophic chicken breast muscle. There is no significant difference in the concentration of the common phosphates from normal and dystrophic muscles; however, a significant extra signal, SEP at 0.4 ppm, appears in dystrophic muscle.

Figure 16 compares the spectra of human quadriceps muscles from a healthy patient and from a patient afflicted with nemaline rod disease. The

[58] M. J. Dawson, D. G. Gadian, and D. R. Wilkie, *Philos. Trans. R. Soc. London Ser.* B **289**, 445 (1980).

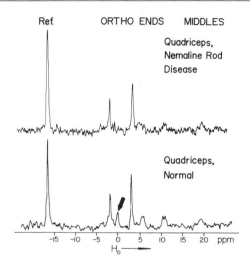

FIG. 16. Comparison of proton-decoupled ^{31}P NMR spectra (36.43 MHz) of normal and nemaline rod-diseased human quadriceps. The arrow indicates the presence of GPC in the normal muscle, but not in the diseased muscle. For the spectrometer conditions and peak assignments, see the legend to Fig. 9. Reproduced, with permission, from Glonek et al.[35]

GPC resonance, at 0.1 ppm in the normal muscle, is missing in the diseased spectrum. In addition, in the diseased muscle a slight reduction in the concentrations of all other phosphates is apparent.

The differences that exist between normal and diseased muscle are better analyzed in PCA extracts.

Phosphate Spectra of Perchloric Acid Extracts

The all-or-nothing existence of a compound in diseased muscle is best analyzed through the use of PCA extracts that permit a 10- to 20-fold concentration of the muscle phosphates.

Perchloric acid extracts are also useful for analysis of phosphate concentrations when only a small amount of muscle is available.[46,59] Addition of carrier-free ^{32}P (about 10^6 cpm) to the extract, before it is separated from the residue, ensures quantitation of the recovered phosphates. The amount of muscle used for the analysis is quantitated by dissolving the PCA-insoluble residue in 0.1 N NaOH, clarifying the sample with high speed centrifugation. Total noncollagenous proteins are determined in the supernatant. One gram of human skeletal muscle contains 173 mg of noncollagenous proteins, whereas for other skeletal muscles the range is 170–

[59] M. Bárány, C. T. Burt, R. J. Labotka, M. J. Danon, T. Glonek, and B. H. Huncke, *in* "Pathogenesis of Human Muscular Dystrophies" (L. P. Rowland, ed.), p. 337. Excerpta Med. Found., Amsterdam, 1977.

200 mg. In the calculation of the P_i and PCr content of the muscle (μmol/g), it is considered that, during PCA extraction, PCr is partially hydrolyzed to P_i. Any excess P_i over 1 μmol per gram of normal muscle is likely to originate from PCr.

Dynamic Changes in Phosphate Profiles

Changes in Resting Muscle

Kinetics of Phosphocreatine Breakdown. The ability of NMR to follow changes in the concentration of phosphate metabolites allows kinetic measurements in intact muscle. Of the various phosphates, PCr is the phosphate to be monitored, because (*a*) it appears as a sharp resonance and, therefore, changes in its peak area can be measured accurately; (*b*) its concentration in the resting skeletal muscle is above 25 mM, more than the sum of the other phosphates, and, therefore, changes in PCr concentration are usually large; (*c*) it is in rapid equilibrium with ATP, and, therefore, PCr participates in all energy-requiring reactions of the muscle cell.

Changes in PCr concentration are most easily followed under anaerobic conditions. Figure 17 compares the ^{31}P spectra of a chicken dystrophic muscle at time intervals of 20–30, 40–50, and 60–70 min after killing the animal. The area of PCr decreases and that of P_i increases as a function of time, whereas the areas of the ATP phosphates and the sugar phosphates remain the same.

Figure 18 compares the logarithm of the percentage of PCr peak area as compared to the sum of peak areas of all the phosphates, as a function of incubation time of normal and dystrophic chicken pectoralis muscles under anaerobic conditions. In both cases, the plots are straight lines, indicating that under these conditions the breakdown of PCr is first order. From the slopes, the rate constants were calculated to be 0.011 and 0.024 min^{-1} for normal and dystrophic muscles, respectively, thus showing about a twofold greater PCr utilization for the dystrophic muscle relative to the normal muscle. Similar experiments may be carried out with other muscle types. There is no need to know the absolute PCr concentration in muscle for such a plot; only the relative peak areas are required.

Myosin ATPase Activity. Under anaerobic conditions the rate of breakdown of PCr may be used to estimate the ATPase activity of myosin in muscle. This is based on Eqs. (2)–(4).

$$\text{ATP} \xrightarrow[\text{myosin}]{} \text{ADP} + \text{P}_i \tag{2}$$

$$\text{ADP} + \text{PCr} \xrightarrow[\text{creatine phosphotransferase}]{} \text{ATP} + \text{Cr} \tag{3}$$

$$\overline{\text{PCr} \qquad\qquad = \qquad\qquad \text{Cr} + \text{P}_i} \tag{4}$$

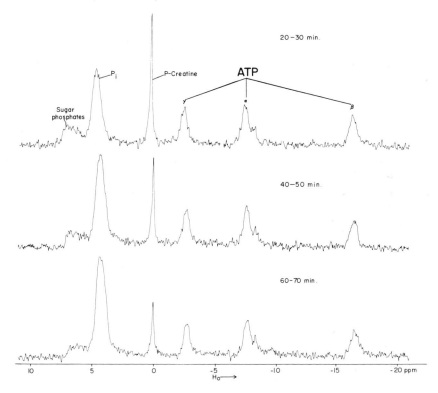

FIG. 17. A series of proton-coupled ^{31}P NMR spectra (72.88 MHz) of intact dystrophic chicken pectoralis muscle. For spectrometer conditions, see the legend to Fig. 8. Unpublished data of D. D. Doyle and M. Bárány, 1980.

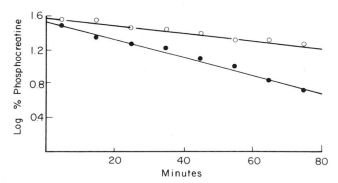

FIG. 18. ^{31}P NMR time course for the utilization of phosphocreatine by intact normal (O——O) and dystrophic (●——●) chicken pectoralis muscle. The ordinate represents the log of the percentage of phosphorus present as phosphocreatine. Proton-coupled ^{31}P spectroscopy (72.88 MHz) was carried out under conditions as described in the legend of Fig. 8. Unpublished data of D. D. Doyle and M. Bárány, 1980.

Since the equilibrium in reaction (3) is shifted far to the right[60] under anaerobic conditions and before the onset of glycolysis (which uses P_i for the formation of glucose 1-phosphate from glycogen), the initial breakdown of PCr is equal to the breakdown of ATP. In agreement with the above scheme, we found a decrease in PCr and a stoichiometric increase in P_i while ATP remained constant.[61]

In muscle, in addition to myosin ATPase, the Na^+- and K^+-ATPase and the sarcoplasmic reticulum (SR) ATPase may contribute to the hydrolysis of PCr. In resting muscle, the contribution from the Na^+- and K^+-ATPase is negligible, since no difference in PCr breakdown was observed between ouabain-treated and untreated muscles.[61] In skeletal muscle, the amount of SR protein (about 2 mg per gram of muscle) is much lower than that of myosin (about 70 mg per gram of muscle). Furthermore, in resting muscle (at $\sim 10^{-7}$ M Ca^{2+} in the myoplasm) the activity of SR-ATPase is low. Thus, in resting skeletal muscle the rate of PCr hydrolysis is mainly due to the myosin ATPase activity.

In order to quantitate the myosin ATPase, the absolute concentration of PCr in resting muscle must be known. From the values of the millimoles of PCr and the precentage area of PCr péak (as compared to the total phosphate peak area) at zero time, the millimoles of PCr can be calculated from the percentage area of the PCr peak at any given time. Such data will allow the calculation of the steady-state rate constant for myosin ATPase in muscle, taking the concentration of myosin subfragment-1 as 0.28 μmol per gram of muscle. This is approximately 0.01 sec^{-1} at 30°.[53] With a temperature control unit in the NMR probe, the temperature dependence of myosin ATPase in muscle may be determined.

Creatine Phosphotransferase Activity. Brown *et al.*[62] have applied saturation transfer NMR to the measurement of creatine phosphotransferase activities in skeletal and heart muscles. Figures 19 and 20 illustrate the principle of saturation NMR when applied to creatine phosphotransferase activity in heart muscle. Figure 19A shows the heart spectrum when the saturation pulse is applied at a different frequency from that of either the γ-P of ATP or PCr ("control irradiation"). In spectrum B, the γ-P of ATP is irradiated; this eliminates the γ-ATP peak and also reduces the intensity of the PCr peak by 75%. These differences are shown in spectrum C (Fig. 19). Figure 20 shows the reverse experiment: irradiation of PCr results in a reduction of 30% in the γ-ATP resonance.

[60] F. D. Carlson and D. R. Wilkie, "Muscle Physiology," p. 90. Prentice-Hall, New York, 1974.
[61] M. Bárány and C. T. Burt, *Fed. Proc., Fed. Am. Soc. Exp. Biol.* **38**, 338 (1979).
[62] T. R. Brown, D. G. Gadian, P. B. Garlick, G. K. Radda, P. J. Seeley, and P. Styles, *in* "Frontiers of Biological Energetics" (P. L. Dutton, J. S. Leigh, and A. Scarpa, eds.), Vol. 2, p. 1341. Academic Press, New York, 1978.

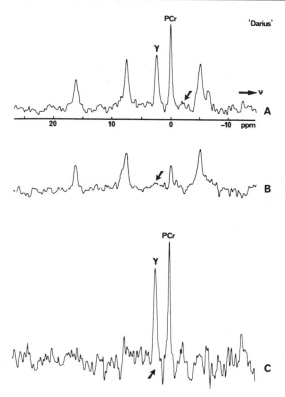

FIG. 19. Illustration of the principle of saturation transfer ^{31}P NMR as applied to creatine phosphotransferase activity in perfused rat heart (Darius). Spectrum A: 64, 90-degree radiofrequency pulses applied at 12 sec intervals (control irradiation); B: 64, 90-degree pulses applied at 12-sec intervals (γ-ATP irradiation); C: difference spectrum (A − B). Spectra are interleaved in blocks of 32 scans. The intensity of the β-ATP peak (around 16 ppm) is artificially reduced by filtering. Reproduced, with permission, from Brown *et al.*[62]

These changes in the magnetization form the basis for determination of the rate constants in the creatine phosphotransferase-catalyzed reaction (5).

$$\text{ATP} + \text{Cr} \underset{k_2}{\overset{k_1}{\rightleftharpoons}} \text{PCr} + \text{ADP} + \text{H}^+ \tag{5}$$

An additional parameter that has to be known for the calculation of rate constants is the "inherent longitudinal relaxation time" of the γP-ATP and PCr nuclei magnetized with the PCr and γP-ATP spins, respectively, under radiofrequency saturation.

Brown *et al.*[62] reported the following values for the rate constants: skeletal muscle, $k_1 = 0.22$ sec^{-1}, $k_2 = 0.037$ sec^{-1}; heart muscle, $k_1 = 0.1$–0.3 sec^{-1}, $k_2 = 0.35$ sec^{-1}.

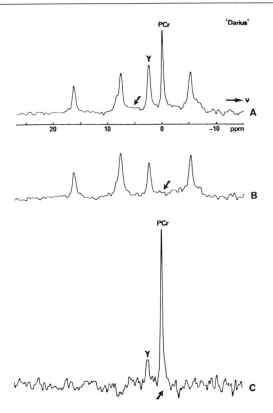

FIG. 20. Further illustration of the principle of saturation transfer ^{31}P NMR as applied to creatine phosphotransferase activity in perfused rat heart (Darius). Spectrum A: 256, 90-degree radiofrequency (RF) pulses applied at 8-sec intervals (control irradiation); B: 256, 90-degree RF pulses applied at intervals of 8 sec (PCr irradiation); C: difference spectrum (A − B). Reproduced, with permission, from Brown *et al.*[62]

The ability of saturation transfer NMR to measure enzymic rates in intact muscle with resolution times less than 1 sec opens new avenues in muscle research. Gadian *et al.*[63] and Kushmerick *et al.*[37] applied this technique to the comparison of creatine phosphotransferase activity between resting and contracting muscles.

Changes in Stimulated Muscle

The use of NMR to measure phosphate changes in physiologically contracting muscle requires a special setup[3] (Fig. 21). Figure 22 shows the effect of a 35-sec tetanus on muscle spectra in oxygenated Ringer's solu-

[63] D. G. Gadian, G. K. Radda, T. R. Brown, E. M. Chance, M. J. Dawson, and D. R. Wilkie, *Biochem. J.* **194,** 215 (1981).

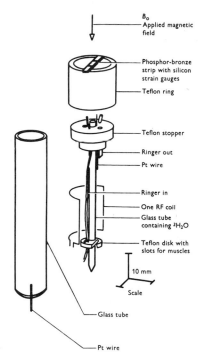

FIG. 21. Design of an NMR sample tube (muscle chamber) that accommodates stimulating electrodes and a force transducer without interfering with the magnetic field and radiofrequency. Reproduced, with permission, from Dawson et al.[3]

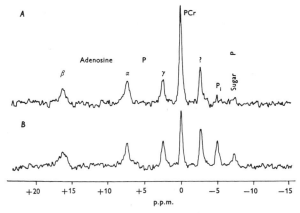

FIG. 22. The effect of a 35-sec tetanus on the ³¹P spectrum (129.2 MHz) of toad gastrocnemius. Spectrum A: Resting muscle in oxygenated Ringer's solution at 4°; B: stimulated muscle under the same conditions. Each spectrum was obtained over a period of 7 min and represents an accumulation of 200 scans at 2-sec intervals. Reproduced, with permission, from Dawson et al.[3]

tion at 4°. PCr decreases to about half of its resting value, P_i increases greatly, SP are increasing slightly, and ATP remains constant.

To follow tetani of 1 sec and 25 sec duration and to follow recovery from tetani, Dawson *et al.*[3] used "gated" NMR. A computer program was written to synchronize the accumulation of NMR data with the electrical stimulation of the muscle. The spectra are successively sampled by the NMR spectrometer over a series of short time increments, the data from each time increment being stored in separate bins of the spectrometer's computer. Perfusion permits the muscle to be maintained in a constant physiological state over the long term, while storage of the separate NMR spectra in separate computer bins allows for the signal-averaging of spectra corresponding to small increments of time.

Because the spectra corresponding to separate increments of time are separately stored, they may be subtracted, one from the other, to yield difference spectra that show only those resonance signals that actually changed in the selected time interval. The results show that about 1% of PCr (about 0.25 mM) was broken down per second of contraction and that this was approximately equal to the increase in P_i concentration. Both PCr and P_i recovered with a half-time of ~10 min. The level of sugar phosphates remained high throughout the recovery period following long contractions.

Muscular fatigue may also be investigated by ^{31}P NMR.[34,64] As stimulation continues, under anaerobic conditions, the muscle uses up its stores of PCr faster than it can be replenished by glycolysis; it also shows a decrease in pH as a result of lactic acid formation. The pH change can be determined by NMR from the chemical shift of the P_i peak, and, if the buffering capacity of the muscles is known, it is possible to estimate the amount of lactic acid formed (see section Intracellular pH). It is thereby possible to determine the extent to which glycolysis proceeds under a variety of conditions, and in this way to assess what factors control the rate of glycolysis in muscle.[38]

In muscles poisoned with 2,4-dinitro-1-fluorobenzene and iodoacetate, ATP disappears while PCr remains, temporarily, as shown by the ^{31}P NMR studies of Yoshizaki.[7] Iodoacetate alone, under anaerobic conditions, does not change the spectrum markedly; however, after tetanic contraction, the PCr and ATP signals disappear with concomitant increases in P_i and SP resonances.[7]

In addition to electrical stimulation, muscles also may be stimulated by chemical agents, notably by caffeine.[5] There is no major requirement to carry out this type of experiment in the NMR tube. Very small glass hooks can be fused to the bottoms of the usual thin-wall NMR tubes, and a

[64] M. J. Dawson, D. G. Gadian, and D. R. Wilkie, *J. Physiol.* (*London*) **299**, 465 (1980).

threaded Teflon rod with a hook on the end can be mounted on a tapped Teflon NMR tube cap, so that a muscle can be fixed between these two hooks and stretched to any desired length for isometric contracture. The modified NMR sample tube can be spun in the NMR spectrometer probe to provide enhanced resolution of the ^{31}P signals.[5]

Upon 20 mM caffeine treatment,[5] frog muscle rapidly hydrolyzes all its PCr and ATP, and accumulates P$_i$, SP, and presumably IMP. With less caffeine added, the rate of PCr breakdown can be followed and the log of the percentage of phosphorus present as phosphocreatine may be plotted against time. Caffeine is suitable for comparative experiments with mammalian muscle; with some improvement of the setup described above, one may correlate changes in ^{31}P profile with quantitative tension measurements.

Perfused Heart

^{31}P NMR of heart has a great potential in physiological chemistry as well as in clinical medicine. At 72.9 MHz it takes only 5 min to accumulate a good spectrum from a perfused rabbit heart,[65] (Fig. 23, spectrum A), and usable spectra have been obtained in as little as 30 sec.[40] A notable feature of the heart spectrum is the much lower ratio of the PCr resonance area to the β-P ATP resonance area: 1.5–1.9 in heart relative to 6–6.7 in skeletal muscle.[38,62] Further, the area of the γ-P of ATP in heart includes the signal from the β-P of ADP, and the α-P of ATP contains in addition to the α-P of ADP resonances from NAD and FAD. The free ADP and dinucleotide concentrations in heart are much higher than those in skeletal muscle.

The isolated, perfused heart is extremely sensitive to alterations in oxygen supply and to changes in flow through the coronary arteries.[66] During ischemia, a rapid decrease in PCr, a slower decrease in ATP, and a marked increase in the P$_i$ resonance takes place[40,41,67] (see Fig. 24). There is also an upfield shift of the P$_i$ signal. This phenomenon is also shown in spectrum B of Fig. 23. Upon reperfusion, there is a rapid resynthesis of PCr and a reverse in the shift of P$_i$ signal, whereas the recovery of ATP and reduction of P$_i$ are relatively low.[9,40,68]

^{31}P NMR analysis of rat heart in the live animal has now been reported.[43] Respiratory arrest led to the classic metabolic pattern in the rat:

[65] D. P. Hollis, R. L. Nunnally, W. E. Jacobus, and G. J. Taylor, IV, *Biochem. Biophys. Res. Commun.* **75**, 1086 (1977).
[66] D. P. Hollis, *Bull. Magn. Reson.* **1**, 27 (1979).
[67] J. M. Salhany, G. M. Pieper, S. Wu, G. L. Todd, F. C. Clayton, and R. S. Eliot, *J. Mol. Cell. Cardiol.* **11**, 601 (1979).
[68] P. B. Garlick, G. K. Radda, and P. J. Seeley, *Biochem. J.* **184**, 547 (1979).

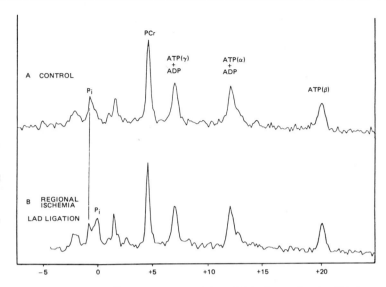

FIG. 23. Proton-coupled ^{31}P NMR spectra (72.9 MHz) of a perfused rabbit heart. Each spectrum is the signal average of 150 scans with 2-sec cycling time. Chemical shifts (ppm) are relative to a solution of 0.2 M H_3PO_4 in 15% $HClO_4$ contained in a capillary tube 1 mm in diameter. Spectrum A: Fully perfused heart; B: the same heart after ligation of the left anterior descending coronary artery. The bar line calibrates the location of the control phosphate peak. Reproduced, with permission, from Hollis et al.[65]

PCr, then ATP, decreased while P_i increased in concentration. The effects of various pathophysiological perturbations on the metabolism of heart in living animals are now accessible to measurement.

Intracellular pH

In muscle, the chemical shifts of ATP, P_i, and SP vary with the physiological pH. From the ^{31}P NMR titration of muscle phosphates as a function of pH, in a salt milieu of muscle, it was concluded that the resonance of P_i is the most suitable to estimate the pH in muscle.[4]

The chemical shift of P_i as a function of pH under physiological conditions and ionic strength follows the Henderson–Hasselbach equation[3,56,66]

$$pH = pK + \log [(\delta - \delta_1)/(\delta_2 - \delta)] \tag{6}$$

where δ is the observed chemical shift and δ_1 and δ_2 are the shifts of $H_2PO_4^{1-}$ and HPO_4^{2-}, respectively. pK values of 6.88 and 6.90, δ_1 values of -3.35 and -3.29, and δ_2 values of -5.60 and -5.81, were found in the laboratories of Wilkie[3] and Hollis,[66] respectively.

It is important to calibrate the chemical shift of the P_i resonance versus

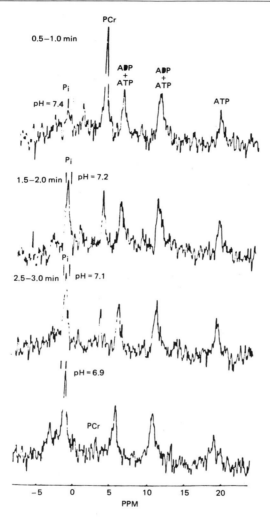

FIG. 24. Proton-coupled ³¹P NMR spectrum (72.9 MHz) of a perfused beating rabbit heart during the onset of total ischemia. Sequential 30-sec spectra were collected after the aortic perfusate line was clamped. The lower spectrum is the 5.5–6.0 min scan. pH values were estimated from the chemical shift of a P_i standard solution. Reproduced, with permission, from Hollis *et al.*[40]

pH under the ionic composition of a particular muscle. For instance, marine invertebrate muscles have significantly higher ionic strength in their cytoplasm than vertebrate muscles, and in relevant pH titration experiments a marked effect of ionic composition of the medium on the titration curve was found.[48]

By comparing the chemical shift of the P_i resonance in the muscle to that of the standard, a value may be assigned to the intracellular pH of muscle. The pH values of unstimulated muscles are in the range of 7.1–7.3 at room temperature.[2-4,48,56] Similarly for perfused guinea pig, rat, and dog hearts, pH 7.1–7.2 was reported at 37°.[66-68] At 4°, the pH of the frog muscle was found to be 7.5.[3] After a 25-sec tetanus, the pH became acid by a few tenths of a pH unit and then returned to its prestimulation value before the end of the recovery period.[3] After a 1-sec contraction of toad gastrocnemius, the pH became slightly more alkaline for the first few seconds.[3]

For accurate pH determination in muscle, the spectrometer's magnet should be shimmed maximally in order to narrow the linewidth of P_i resonance in the muscle. If the pH standard was calibrated with a solution containing D_2O, a capillary tube containing D_2O should be inserted into the NMR tube containing the muscle. A good test of the spectrometer's sensitivity for internal pH determination is to place the muscle in physiological salt solution containing phosphate buffer of pH 7.1, 7.2, 7.3, 7.4, and determine at what extracellular pH the internal and external phosphate signals are resolved (the external P_i does not penetrate into muscle to a significant extent).

The simplest demonstration of pH changes in muscle is to follow the chemical shift of the P_i resonance as a function of time under anaerobic conditions.[4] Acidification in anaerobic frog muscle is caused by glycolysis, and the quantity of lactic acid (LA) produced can be estimated from the pH changes and the buffering capacity of frog muscle, as shown by Dawson et al.[34]

$$\Delta[LA^-] = \Delta\left(\frac{36 \times 10^{-3}[H^+]}{3.162 \times 10^{-7} + [H^+]} + \frac{14 \times 10^{-3}[H^+]}{5.011 \times 10^{-8} + [H^+]} + \frac{[P_i][H^+]}{1.995 \times 10^{-7} + [H^+]} \right) \quad (7)$$

where $\Delta[LA^-]$ is the increase in the concentration of dissociated lactic acid, which, practically, corresponds to all lactic acid because at physiological pH the concentration of undissociated LA ($pK = 3.9$) is negligible. The numbers in the numerators of the first and second terms correspond to the concentrations of protein histidine and carnosine, respectively, and the numbers in the denominators correspond to their dissociation constants, respectively. The number in the denominator of the third term corresponds to the second dissociation constant of phosphoric acid. The concentration of hydrogen ions is determined from the position of the P_i peak, and the concentration of P_i is determined from the peak area.

Dawson et al.[34] also showed that from the hydrogen ion concentration, determined by NMR, one can calculate the free ADP concentration in skeletal muscle. This is based on the rearrangement of the creatine phosphotransferase-catalyzed equilibrium reaction [see Eq. (5)].

$$[ADP] = [ATP] \times [Cr]/K\ [PCr] \times [H^+] \qquad (8)$$

where K is the equilibrium constant. [ATP] and [PCr] are determined from their peak areas, and [Cr] is calculated from the ratio [PCr]/([PCr] + [Cr]), which is between 0.85 and 0.88 in resting muscle.[34] A free ADP concentration of 29 μM is calculated from Eq. (8), which is about 10% of the analytical [ADP] in resting frog muscle.[64]

The ability of ³¹P NMR to measure the pH noninvasively was applied to ischemic heart, and a decline in pH with ischemia was demonstrated.[8,9,40,67,68] Figure 24 shows the shift in the position of P_i peak (intracellular pH 7.4) upfield (pH 6.9) during the onset of global ischemia. After reperfusion, the cellular pH returns to its value prior to ischemia.[38,40] The shift of the P_i resonance when combined with NMR imaging methods could provide a diagnostic tool for evaluation of myocardial tissue damage.[65]

Intracellular Free Mg²⁺ Concentration

³¹P NMR is also used for the estimation of free Mg²⁺ concentration in intact muscle. Cohen and Burt[51] determined the spin–spin relaxation time (T_2) for PCr in intact frog muscle and in model solutions at 4°. T_2 was greatly affected by the free Mg²⁺ concentration. With a calibration curve of $1/T_2$ versus the fraction of PCr bound to Mg²⁺ (F_B), and with an apparent stability constant (K_{app}) of 20 M^{-1}, F_B in muscle was estimated to be 0.08. Substituting these values into Eq. (9)

$$\text{m}M \text{ free } [Mg^{2+}] = \frac{10^3 F_B}{K_{app}(1 - F_B)} \qquad (9)$$

gives 4.4 mM as the intracellular free Mg²⁺ concentration. With a more recent K_{app} of 40 M^{-1} for Mg binding by PCr, however, the estimated F_B was 0.109 in muscle, and the free Mg²⁺ concentration in frog gastrocnemius became 3.0 mM.[51]

In the procedure of Gupta and Moore[69] the separation of the α-P and β-P resonances of ATP in the NMR spectrum of intact frog skeletal muscle was compared with the separation of α-P and β-P resonances in solutions of free ATP and MgATP, determined under physiological conditions. From these chemical shifts measurements one can calculate the fraction of the total ATP that is not complexed to Mg²⁺ in the muscle,

$$\phi = [ATP]_f/[ATP]_T \qquad (10)$$

[69] R. K. Gupta and R. D. Moore, *J. Biol. Chem.* **255**, 3987 (1980).

where f and T subscripts refer to free and total concentrations, respectively.

The free Mg^{2+} concentration in the muscle water is calculated from the equation:

$$[Mg]_f = K_D^{MgATP} (\phi^{-1} - 1) \tag{11}$$

where $K_D^{MgATP} = 45 \ \mu M$.

From an average ϕ value of 0.07, the free Mg^{2+} concentration in frog sartorius and gastrocnemius muscles was calculated to be 0.6 mM at 25°. Clearly, the estimated value of free Mg^{2+} ions in the muscle is directly proportional to the value of the dissociation constant, determined *in vitro*.

^{31}P NMR Phosphate Profiles from Nonmuscle Systems

Figure 25 shows ^{31}P spectra obtained from intact rabbit erythrocytes, rabbit reticulocytes, and sea urchin eggs, and an extract from human platelets. A number of the resonances observed in these spectra are similar to those seen in the spectra of intact muscles. For example, the SP resonance band at 3.7 ppm is observed in each of the spectra, as well as the signal from P_i; however, the amounts of these, relative to the total phosphate profile, vary considerably. In erythrocytes and reticulocytes, for example, the P_i signal is relatively weak and appears as a small shoulder on the upfield side of the highfield signal from 2,3-diphosphoglycerate. In the spectra from the sea urchin eggs and platelets, however, the P_i signal is considerably enhanced and is readily distinguished from that of the other phosphates.

ATP is readily observed in the erythrocyte, reticulocyte, and platelet spectra, but is absent from the sea urchin spectrum. The sea urchin spectrum, however, shows the resonance from PCr (-3.2 ppm), which is not present in the spectra from the three blood components. This spectrum also shows a large signal due to stored phospholipids. The main peak is from lecithin, while the downfield shoulder is principally sphingomyelin. The upfield group of resonances, at about -11 ppm, arises from the symmetrically esterified pyrophosphate groups of dinucleotide and related cofactors.

The spectra from both reticulocytes and platelets show the resonances from unidentified phosphate diesters at 0 ppm, and the available spectroscopic evidence, along with some thin-layer chromatographic data, suggests that the compounds giving rise to these resonances are the same as those in the muscle. The reticulocyte and platelet compounds, however, have yet to be isolated in pure form and rigorously characterized.

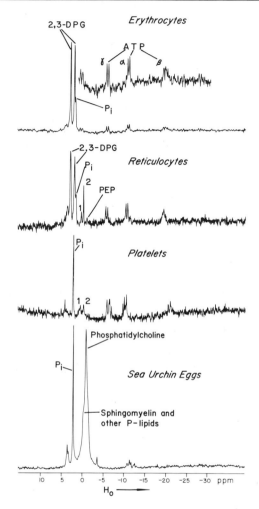

FIG. 25. ³¹P NMR spectra (36.43 MHz) from other cellular systems: rabbit erythrocytes, rabbit reticulocytes (treated with 10 mM sodium ferricyanide, which serves to maintain the reticulocytes in their native state), a neutralized PCA extract of human platelets obtained from 0.5 unit whole blood, and sea urchin eggs. Because of the relatively reduced concentration of phosphates in these samples, signal-averaging times up to 14 hr were used in some instances to obtain these spectra. The usual signal-averaging times, however, were between 0.5 and 2 hr. In the figure, the resonances from 2,3-diphosphoglycerate (2,3-DPG) and P$_i$ are designated. Resonances from phosphodiesters are denoted by the numerals 1 and 2. Unpublished data of T. Glonek, 1978.

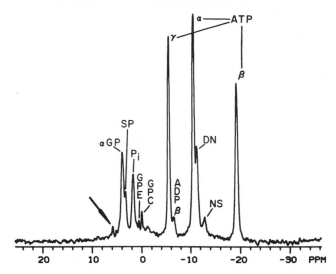

FIG. 26. ^{31}P NMR spectrum (80.99 MHz) of three rabbit lenses accumulated over a period of 6 hr to reveal the fine structure of the spectrum. During the experiment, the lenses were maintained by Earle's buffer (with glucose) at pH 7.4 at 37°. The symbols are as follows: the arrow indicates a new phosphorus metabolite at 6 ppm; αGP, α-glycerophosphate; SP, sugar phosphates; P_i, inorganic orthophosphate; GPE and GPC, respectively, glycerol 3-phosphorylethanolamine and -choline; ADP and ATP, adenosine di- and triphosphates; DN, the dinucleotides; and NS, nucleoside diphosphosugars and related species. Unpublished data of J. V. Greiner, S. J. Kopp, D. R. Sanders, and T. Glonek, 1981.

The spectrum from the reticulocytes also shows a small but distinct resonance from phosphoenolpyruvate at −0.9 ppm. This is the only tissue thus far examined that shows phosphoenolpyruvate resonance.

Figure 26 shows the phosphorus spectrum obtained from intact rabbit lenses incubated in Earle's buffer at 37°. This is the most highly resolved spectrum we have yet obtained from intact tissue, and it shows, clearly, a number of the minor phosphate metabolites of the lens, including a resonance from an unidentified phosphatic substance. The prominent triose phosphate in this tissue is α-glycerophosphate. Its chemical shift as well as that from P_i have been used in this case as indexes of the intralens pH. The lens pH is 6.9, and the values obtained using both phosphate metabolites agreed within 0.01 pH unit.

Acknowledgments

This work was supported by the Muscular Dystrophy Association, by Grant NS-12172 from the United States National Institutes of Health, and by an intramural grant from the Chicago College of Osteopathic Medicine.

[54] Optical Activity Measurements for Elucidating Structure–Function Relationships in Muscle Protein Systems

By WILLIAM D. MCCUBBIN and CYRIL M. KAY

The aim of this chapter is to present circular dichroism (CD) and occasional optical rotatory dispersion (ORD) data obtained from several representative muscle protein systems. The proteins considered will be restricted to the structural or regulatory ones that make up the myofibril; any of the soluble enzymes that have muscle as their source will not be discussed. An attempt will be made to outline some of the more novel ways ORD and CD methodologies have been applied to solving various problems in the muscle protein area. It will be assumed in this presentation that the reader is familiar with the basic theories of optical activity as applied to proteins, and only a very limited treatment will be given, outlining the units of measurement and the parameters that are measured. For a more in-depth study of the relevant theory and methodological practices, earlier chapters in this series by Fasman[1] and Adler *et al.*[2] as well as an article by Gratzer[3] are recommended.

Just as with other proteins, the elucidation of the complete three-dimensional positioning of the polypeptide backbone as well as all the side chains in muscle proteins can only come from X-ray diffraction analysis. However, for a variety of reasons, chief among which is a lack of usable crystals, these data are very limited and only in one case is the complete three-dimensional structure of a muscle protein known. The protein in question is the calcium-binding protein parvalbumin, isolated from carp muscle.[4] To fill this gap, the technique of ORD, now almost exclusively replaced by the more powerful sister technique of CD, may be profitably employed. Small quantities suffice for a rapid, and usually fairly accurate, estimate of the amounts of the various conformational forms present in dilute solution. Perhaps of more interest and utility than the establishment of the amount of these conformers present in any particular protein is an understanding of any changes in these values when, for example, the protein interacts with a metal, another protein, a variety of solvent systems, a substrate or inhibitor (if the protein displays any enzymic activ-

---

[1] G. D. Fasman, see this series, Vol. 6 [126].

[2] A. J. Adler, N. J. Greenfield, and G. D. Fasman, see this series, Vol. 27 [27].

[3] W. B. Gratzer, *in* "Techniques in the Life Sciences B108" (H. L. Kornberg *et al.*, eds.), pp. 1–43. Elsevier/North-Holland, Amsterdam, 1978.

[4] R. H. Kretsinger and C. E. Nockolds, *J. Biol. Chem.* **248**, 3313 (1973).

METHODS IN ENZYMOLOGY, VOL. 85

ity), or when the protein undergoes changes in temperature. It is this sensitivity of ORD and CD to conformational changes that has led to its widespread satisfactory use.

Units and Calculations

The specific rotation is a measure of the optical activity of a compound. It may be defined by Eq. (1).

$$[\alpha]_\lambda^T = \frac{\alpha_{\lambda,\text{obs}}^T \times 100}{l \times c} \tag{1}$$

where $\alpha_{\lambda,\text{obs}}$ is the observed rotation of the plane of polarized light in angular degrees at temperature T and wavelength λ, l is path length of cell in decimeters, and c is concentration of the compound in grams per 100 ml.

The molar rotation is defined by

$$[m]_\lambda = \frac{MW}{100} [\alpha]_\lambda \tag{2}$$

where MW is the molecular weight of the compound. In the case of biopolymers, composed of repeating units, the data are usually reported in terms of the mean residue rotation, defined by

$$[m] = \frac{MRW}{100} [\alpha]_\lambda \tag{3}$$

where MRW is the mean residue weight. For proteins, in many cases this value may be taken as 115. Circular dichroism is the difference in molar (or mean residue) absorptivities, $(\epsilon_L - \epsilon_R)$, between the left- and right-handed circularly polarized components of a linearly polarized light. As the emergent light is elliptically polarized, CD may be expressed in terms of molar (or mean residue) ellipticity, $[\theta]$, defined by the following relationship.

$$[\theta]_\lambda = \frac{\theta_{\lambda,\text{obs}} \times MW \text{ (or MRW)}}{10 \times d \times c} \tag{4}$$

where λ = wavelength, θ_{obs} = observed ellipticity in degrees, MW = molecular weight, MRW = mean residue weight, c = concentration in grams per milliliter, d = path length in centimeters.

The units of $[\theta]$ are deg cm^2 decimole^{-1}. CD and ellipticity are related by Eq. (5).

$$[\theta] = 3300(\epsilon_L - \epsilon_R) \tag{5}$$

Data Handling

Applications of ORD and CD for the study of protein conformation attempt to fit the observed spectra to those constructed from mixtures of α-helix, β-structure, and aperiodic forms. From the outset the primary assumptions are made that only three conformations exist in proteins and that their separate optical activities are additive. This is illustrated by Eq. (6).

$$X = f_H X_H + f_\beta X_\beta + f_R X_R \tag{6}$$

at any wavelength λ. Here X represents either $[\theta]$ or $[m]$. The X_H, X_β, and X_R terms are the reference values of the α-helix (H), β-structure (β), and aperiodic (R) forms. The f's are the fractional amounts of the three forms in a protein molecule ($\Sigma f = 1$).

There are two approaches currently in vogue for the choice of reference values. One group uses data from synthetic polypeptides as the model compounds,[5,6] whereas the other prefers to compute the reference values from the CD or ORD spectra of proteins of known structure.[7] As an extension of the latter approach, consideration has also been given to the chain-length dependence of the optical activity of helices.[8]

Tables of the percentages of the three forms may be constructed for muscle proteins under a variety of conditions, or, alternatively, one may simply present plots of $[\theta]$ versus wavelength. From both approaches it is a simple matter to follow conformational changes. It can be seen from results of this type that muscle proteins are no different from proteins isolated from other sources. High, low, and intermediate amounts of helix are found.

In the next section of this chapter some specific examples have been selected to demonstrate the versatility of the technique. Ellipticity values or the amounts of the three forms will be presented, but, in addition, an attempt will be made to show how the study has contributed to a better understanding of the structure–function relationship for the particular muscle protein under discussion.

Use of Optical Activity Measurements in Establishing Structure– Function Relationships in Some Muscle Protein Systems

Myosin and Its Subfragments: Interactions with Substrates

Myosin plays a key role in muscle contraction. The central process in contraction is the interaction of ATP with the ATPase sites on myosin,

[5] N. J. Greenfield, B. Davidson, and G. D. Fasman, *Biochemistry* 6, 1630 (1967).
[6] N. J. Greenfield and G. D. Fasman, *Biochemistry* 8, 4108 (1969).

accompanied by a specific interaction between myosin and actin. When a muscle contracts, ATP hydrolysis occurs in the globular myosin heads; the resulting energy released from this reaction somehow enables the heads to move along the actin filaments.[9] How chemical energy is transformed into movement is not understood. A variety of techniques including electron spin resonance,[10] fluorometry,[11] and absorption spectrophotometry[12] have detected nucleoside triphosphate hydrolysis-dependent conformational changes in the myosin subfragment-1 (S-1) globular heads. Since ORD[13] and CD[14] have detected essentially no change in α-helical content, a controversy has arisen as to the extent and position of these changes.

Although the far-UV ORD or CD spectra indicate no change in helicity upon interaction with substrate, examination can be made of the near-UV region, i.e., 250–320 nm, since these spectra arise from perturbations in the environments around aromatic amino acid residues or disulfide bridges,[15] and these may be sensitive to binding of nucleotide. Murphy[16] has examined the near-UV CD spectra of myosin, heavy meromyosin (HMM), and S-1 when various nucleotides were added under different conditions. Figure 1 shows CD spectra for myosin, HMM, and S-1 in the spectral range 250–320 nm. These spectra are characterized by four positive bands with extrema near 299, 272, 265, and 259 nm and by two negative bands centered near 290 and 283 nm. It can be seen that the three spectra are qualitatively very similar, implying a similar environment of the aromatic amino acids. The binding of adenosine 5'-diphosphate (ADP) to HMM produced little alteration of the CD spectrum, small differences being noted only in the 250–280 nm region, where the bands became less intense. The binding of the analog nucleotide, adenylyl imidodiphosphate, was also studied; the spectral differences were even less than was noted for the HMM · ADP complex.

When an adenosine 5'-triphosphate (ATP) generating system was employed, the CD spectrum of the steady-state complex HMM · $\overset{*}{A}$DP · P_i could be measured. A 50% increase in the magnitude of the bands at 283

[7] I. P. Saxena and D. B. Wetlaufer, *Proc. Natl. Acad. Sci. U.S.A.* **68**, 969 (1971).

[8] Y.-H. Chen, J. T. Yang, and K. H. Chau, *Biochemistry* **13**, 3350 (1974).

[9] H. E. Huxley, *Cold Spring Harbor Symp. Quant. Biol.* **37**, 698 (1972).

[10] J. C. Seidel and J. Gergely, *Biochem. Biophys. Res. Commun.* **44**, 826 (1971).

[11] M. M. Werber, A. G. Szent-Györgyi, and G. D. Fasman, *Biochemistry* **11**, 1872 (1972).

[12] F. Morita, *J. Biol. Chem.* **242**, 4501 (1967).

[13] W. B. Gratzer and S. Lowey, *J. Biol. Chem.* **244**, 22 (1969).

[14] J. Y. Cassim and T.-I. Lin, *J. Supramol. Struct.* **3**, 510 (1975).

[15] M. Goodman, G. W. Davis, and E. Benedetti, *Acc. Chem. Res.* **1**, 275 (1968).

[16] A. J. Murphy, *Arch. Biochem. Biophys.* **163**, 290 (1974).

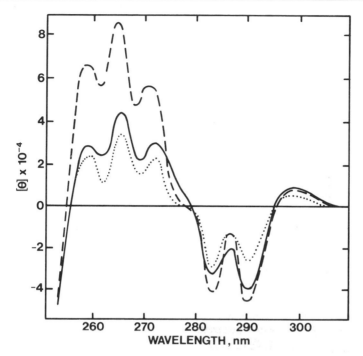

FIG. 1. Circular dichroism spectra of myosin, heavy meromyosin (HMM), and subfragment-1 (S-1) in the near-UV region: myosin (---), HMM (——), S-1 (· · · · ·). Solutions contained 1–7 mg of protein per milliliter, 0.1 M morpholinopropanesulfonic acid, 0.5 M KCl (myosin) or 0.1 M KCl (HMM and S-1) and were at pH 7.0. From Murphy.[16]

and 259 nm and a 50% decrease in the one at 272 nm occurred (Fig. 2). The bands at 299 and 290 nm are unchanged. It was concluded from these data that an aromatic residue, most probably tryptophan, in light of additional results from fluorometry,[11] moves near the purine of the nucleotide upon hydrolysis to HMM · $\overset{*}{A}DP$ · P_i and moves away during the conversion to the post-steady-state complex, HMM · ADP · P_i. This movement of an amino acid side chain and nucleotide at the active site may be the start of the mechanism that converts the chemical energy of hydrolysis to the mechanical energy of movement.

Cassim and Lin[14] have coined the terms "delocalized" and "localized" conformational changes and have used CD to look at the myosin–ATP system for evidence of either phenomenon. They define a delocalized conformational change as a change in the geometry of the protein that results in the transmission of conformational distortion over a considerable distance from the region of origin. On the other hand, a localized change is one where the effect does not proceed from the region

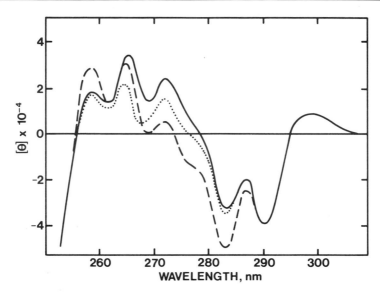

Fɪɢ. 2. Circular dichroism (CD) spectra of the ATP steady-state and post-steady-state complexes of heavy meromyosin (HMM). ——, Algebraic sum of the CD spectra of HMM and ATP, 2 mol of nucleotide per 3.4×10^5 g of HMM. ----, HMM · ADP complex (ADP generated *in situ* by the addition of ATP), with the addition of pyruvate kinase to 2 μg/ml. · · · ·, Same solution 2.5–3 hr after the addition of pyruvate kinase. All solutions contained 0.1 M morpholinopropanesulfonic acid, 0.1 M KCl, pH 7.0. From Murphy.[16]

of origin. CD studies were carried out both in the far and near-UV spectral regions under a variety of conditions. It was found that no change could be detected in the ellipticity at 222 nm whether ATP hydrolysis was in steady state or was complete, or whether either ADP or pyrophosphate were added. In separate experiments the same conclusions were reached for HMM.

Measurements in the 250–300-nm region showed that the optically active substrate ATP produced a perturbation in the CD spectrum only at wavelengths below 280 nm, which coincided with the regions of its optical activity. The authors' claim that this is a very important point and present an argument as follows. If the conformational changes resulting from nucleotide interaction affect only a few side chains, then this might not act as a sufficient perturbant to enough aromatic/cystine residues to produce any significant change in the total CD spectrum (a summation of contributions from all such residues). Alternatively, if the conformational change spreads to a large region of the polypeptide chain, then this perturbation would affect more aromatic/cystine residues and one might expect to see dramatic changes in the near-UV CD, both above and below 280 nm.

Assuming that the aromatic/cystine residues are randomly situated in the myosin head, one would then anticipate a significant CD change in the near-UV, both above and below 280 nm, if the conformational change resulting from the action of ATP at the active site was transmitted over a major portion of the head. Since no changes in CD were seen above 280 nm, this possibility appears to be precluded. This argument also is strengthened by the fact that pyrophosphate, itself optically inactive and a competitive inhibitor of ATP, caused no CD change and that ATP and ADP produced CD changes only at the wavelengths of their own optical activity. This is in accord with the earlier study.[16]

Chantler and Szent-Györgyi have utilized CD in a study of invertebrate myosin.[17] The far-UV CD spectrum of pure scallop myosin was found to be unchanged when the following alterations were made in the buffer: NaCl replacing KCl; Mg^{2+} added to 1 mM and Ca^{2+} to 0.2 mM; addition of Mg^{2+} to 1 mM and ATP to 50 μM. The spectrum of desensitized myosin (treatment with EDTA to remove a specific light chain) was also identical under these varying conditions. Near-UV CD spectra of scallop myosin and desensitized scallop myosin were similar and showed a minimum at 285 nm and a maximum at 265 nm. The curves were unchanged, within experimental error, upon addition of MgATP and/or calcium. Thus, with this myosin, CD studies do not indicate a conformational change accompanying the nucleotide interaction.

Another interesting study has utilized CD in a slightly different manner to look at myosin–nucleotide interaction.[18] It had already been demonstrated that there was one functional amino group, located at or near the ATPase active site of myosin, since this residue reacted very much faster with trinitrobenzene sulfonate than with the rest of the amino groups, with a resulting dramatic change in the ATPase activity of the protein.[19] The effect of nucleotides on the CD spectrum of specifically trinitrophenylated myosin was studied.

Although the trinitrophenyl group is itself optically inactive, its interaction with asymmetric centers on a protein molecule can generate extrinsic Cotton effects in the near-UV and visible region. This was found to be the case with trinitrophenylated S-1, where the CD spectrum showed two maxima at 295 and 360 nm, and two minima at 320 and 420 nm, respectively. Upon addition of MgATP to the system the CD spectrum was altered dramatically. The two maxima and the trough at 420 nm decreased, whereas the trough at 320 nm became considerably deeper. Addition of the nucleotide analog adenosine 5'-(β,γ-imino)triphosphate

[17] P. D. Chantler and A. G. Szent-Györgyi, *Biochemistry* **17**, 5440 (1978).
[18] A. Muhlrad, *Biochim. Biophys. Acta* **493**, 154 (1977).
[19] A. Muhlrad, R. Lamed, and A. Oplatka, *J. Biol. Chem.* **250**, 175 (1975).

produced an even greater effect than MgATP, and addition of MgADP produced smaller changes. The presence of Mg^{2+} was not essential for these nucleotide-induced CD changes; however, the spectra in the absence of Mg^{2+} were not the same as those observed in the presence of the cation. This also demonstrates that the conformation of S-1 is sensitive to Mg^{2+}. Derivatized myosin and HMM behaved like modified S-1 and the perturbing effect of nucleotides operated similarly. It was concluded from these results, along with difference absorption spectral work, that ATP and its analogs induce a conformational change in the myosin head that extends to the environment of the functional lysine residue and makes it less polar.

Another novel application of CD to the myosin subfragment systems is its use to follow titrations. The nucleotide-induced change in $[\theta]_{282\,nm}$ has been used to follow the binding of ADP to HMM and S-1.[20] Typical titrations (Fig. 3) are consistent with binding data obtained by others employing different methodology. The basic conclusion of this study was that the active sites on the two heads of myosin are equivalent and independent. A more recent investigation extended this approach to cardiac myosin and demonstrated similar changes upon nucleotide interaction, although these were smaller in overall magnitude.[21]

Actin

When the ionic strength of a solution of globular, or G-actin is raised, the protein polymerizes to the fibrous form (F-form). It is of obvious interest to ascertain if any changes in secondary and/or tertiary structure accompany this important transition.

In active muscle, F-actin interacts with myosin heads to allow activation of the MgATPase and facilitation of contraction. Upon activation of muscle, conformational changes might be expected to occur in the actin filaments. Such a possibility was, in fact, demonstrated in an UV-CD study,[22] which indicated that F-actin in the thin filament undergoes some conformational change with the binding of Ca^{2+} to troponin, and the change is amplified by the interaction of myosin with actin.

Circular dichroism has been employed in a limited fashion to explore possible conformational changes associated with the transformation of G- to F-actin. A near- and far-UV CD study of actin under a variety of conditions[23] showed that the far-UV CD spectrum was essentially unchanged in going from G- to F-actin; thus, no alteration in secondary

[20] D. J. Marsh, A. d'Albis, and W. Gratzer, *Eur. J. Biochem.* **82**, 219 (1978).
[21] S. J. Smith, *FEBS Lett.* **105**, 197 (1979).
[22] T. Yanagida, M. Taniguchi, and F. Oosawa, *J. Mol. Biol.* **90**, 509 (1974).
[23] A. J. Murphy, *Biochemistry* **10**, 3723 (1971).

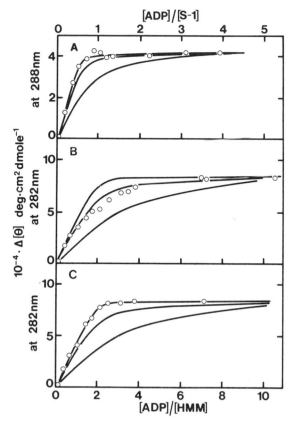

FIG. 3. Typical circular dichroism titrations with ADP. (A) Subfragment-1 (S-1) at 20° (4.2 mg/ml); (B) myosin at 20°; (C) myosin at 7° (both 3.1 mg/ml). The curves are calculated for identical sites with association constants of 10^7, 10^6, and $10^5 M^{-1}$. HMM, heavy meromyosin. From March et al.[20]

structure accompanies this transformation. On the other hand, perturbations were observed in the CD signal in the region 250–300 nm, suggesting that the environment of at least some of the aromatic amino acids was dramatically changed upon polymerization; in particular an aromatic residue interacts with the nucleotide purine in F-actin but not in G-actin.

There are many questions still to be answered, and CD could clearly play a key role in their evaluation. For example, the conformational change induced in F-actin by Ca^{2+} binding to troponin needs to be explored in greater detail to establish whether or not the actual binding of troponin itself to the actin filament produces conformational changes. Some preliminary observations from this laboratory indicate that this, in fact, could be the case.

TABLE I

EXTREMA OF THE CIRCULAR DICHROISM (CD) AND OPTICAL ROTATORY DISPERSION (ORD) SPECTRA OF MUSCLE PROTEINS BELOW 250 nm AT $25°$[a-c]

Protein[d]	CD						ORD			
	λ(nm)	[θ]	λ(nm)	[θ]	λ(nm)	[θ]	λ(nm)	[m]	λ(nm)	[m]
G-actin	219	−12,900	210	−14,000	192	20,800	230	−5,800	200	28,000
Myosin	222	−26,200	210	−25,100	192	56,200	233	−12,900	199	60,100
HMM	222	−22,000	210	−21,200	192	44,800	233	−10,500	199	51,100
S-1	222	−15,200	210	−15,300	192	29,500	233	−7,300	199	33,300
LMM Fr.1	222	−39,800	210	−38,200	192	87,100	233	−19,400	199	100,000
Tropomyosin	222	−38,800	210	−37,100	192	87,400	233	−18,500	199	85,000
Troponins										
TN-C	222	−12,800	207	−16,100	192	25,300	233	−6,400	199	34,100
TN-I	221	−9,000	205	−15,500	190	10,700	232	−5,600	199	20,400
TN-T	222	−14,900	207	−19,700	191	23,200	232	−7,600	199	36,300
Light chains										
LC$_1$	220	−8,200	207	−13,000	190	12,900	233	−4,700	199	21,000
LC$_2$	221	−7,500	207	−11,800	191	9,800	233	−4,600	199	20,200
LC$_3$	221	−10,100	208	−13,200	192	20,000	233	−5,300	199	23,000

[a] From Wu and Yang.[24]

[b] Solvent: 2 mM sodium phosphate (pH 7.0) for G-actin; 40 mM sodium pyrophosphate (pH 7.5) for myosin, heavy meromyosin (HMM), light meromyosin (LMM) Fr. I, and tropomyosin; 40 mM sodium pyrophosphate (pH 7.0) for subfragment-1 (S-1) and light chains.; 40 mM sodium pyrophosphate and 0.5 mM EGTA (pH 7.5) for troponins; 10 mM sodium phosphate (pH 7.0) for subfragment-1 (S-1) and light chains.

[c] Dimension of [θ] and [m]: deg cm^2 dmol^{-1}.

[d] M_0 used for CD and ORD calculations (in the descending order): 112, 115, 115, 115, 115, 115, 113, 117, 118, 109, 110, 111.

TABLE II

Estimated f_H, $\bar{n}$, i, and f_β of Muscle Proteins from Optical Activity[a]

Proteins	f_H	$\bar{n}$ [b]	i [b]	f_β
G-actin	0.45	7	24	0.27
	(0.37)	(8)	(17)	(0.25)
Myosin	0.78	23	154	0
HMM	0.70	13	147	0.07
S-1	0.60	8	85	0.16
LMM Fr.1	0.94	505	2[c]	0
Tropomyosin	0.96	284	2[c]	0
	(0.98)			(0)
Troponins				
TN-C	0.51	8	11	0.13
	(0.64)	(11)	(9)	(0.11)
TN-I	0.29	12	5	0.20
	(0.53)	(11)	(9)	(0.14)
TN-T	0.38			0.14
Light chains of myosin				
LC$_1$	0.37	7	11	0.23
	(0.53)	(14)	(7)	(0.21)
LC$_2$	0.33	7	8	0.22
	(0.41)	(9)	(8)	(0.28)
LC$_3$	0.48	6	11	0.19
	(0.56)	(14)	(6)	(0.26)

[a] The values in parentheses are based on the predictive method from amino acid sequence [P. Y. Chou and G. D. Fasman, *Biochemistry* 13, 224 (1974)]; predicted β turns are not listed.

[b] $\bar{n}$ is related to the number of helical segments, i, by $i = f_H N/\bar{n}$ (N = total number of amino acid residues).

[c] Assume $i = 2$. Modified from Wu and Yang.[24]

The Troponin System

The paper by Wu and Yang[24] gives an excellent compilation of ORD and CD spectra of several muscle proteins including myosin and its subfragments along with the subunits of the troponin complex and tropomyosin, the latter two protein systems constituting the relaxing system of muscle, located on the thin filaments. Table I lists the ellipticity values for these proteins, and Table II documents the amounts of α-helix, β structure, and random coil computed from these ellipticities for the various systems. For the purposes of this chapter it is advantageous to extend this approach and to present some examples of how CD studies

[24] C.-S. C. Wu and J. T. Yang, *Biochemistry* 15, 3007 (1976).

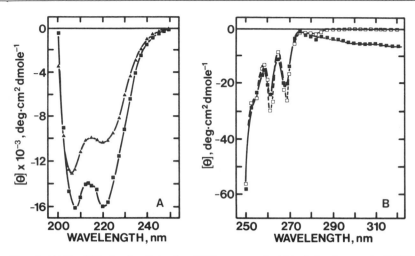

FIG. 4. (A) Far-UV circular dichroism (CD) spectra of rabbit skeletal troponin C (TN-C) in 50 mM Tris-HCl, pH 7.6, 1 mM EGTA (▲), and in 50 mM Tris-HCl, pH 7.6, 5 × 10^{-4} M CaCl$_2$ (■). (B) Near-UV CD spectra of rabbit skeletal TN-C in 50 mM Tris-HCl, pH 7.6, 1 mM EGTA (■), and in 50 mM Tris-HCl, pH 7.6, 5 × 10^{-4} M CaCl$_2$ (□). (A) and (B) from Murray and Kay.[26]

have contributed to the overall knowledge of the structure–function relationships of these important proteins.

TROPONIN C (TN-C)

This protein avidly binds Ca^{2+} ions and, in conjunction with tropomyosin and the other subunits of the troponin complex, is responsible for coupling excitation with muscle contraction. A Ca^{2+}-dependent conformational change in TN-C is believed to be transmitted through protein interaction to actomyosin. A key feature of this hypothesis is a Ca^{2+}-induced conformational change in one protein that induces a series of events ultimately leading to muscle contraction. For an in-depth discussion of this system the reader is referred to a review article.[25] A great deal of evidence supports the occurrence of such a conformational change in TN-C, and CD has played a major role in its study.

Initial observations on the far-UV CD spectrum of TN-C revealed ellipticity change of ~60% in $[\theta]_{222\,nm}$ when the protein bound Ca^{2+}.[26] In addition, the near-UV CD spectrum, which is dominated by vibronic fine structure from the 10 phenylalanine residues in this protein, undergoes a considerable sharpening and increase in magnitude when the cation is bound (Fig. 4).

[25] W. D. McCubbin and C. M. Kay, *Acc. Chem. Res.* **13**, 185 (1980).
[26] A. C. Murray and C. M. Kay, *Biochemistry* **11**, 2622 (1972).

It was determined from equilibrium dialysis studies that skeletal muscle TN-C could bind 4 mol of Ca^{2+},[27] whereas the cardiac analog could only bind 3 mol.[28] These binding sites were arranged in two classes: the high-affinity or Ca^{2+}–Mg^{2+} sites and the low-affinity or Ca^{2+} specific site(s). Circular dichroism studies have contributed to the determination of which sites are which and to establishment of the biological implication of each type of binding.

Circular dichroism Ca^{2+} titrations were performed by monitoring changes in $[\theta]_{222}$ for skeletal and cardiac muscle TN-C as functions of metal ion concentration. By application of computer analysis to the resulting titration curves, the resolution of the Ca^{2+}-binding parameters for the different classes of binding sites was achieved for both proteins.[29] For skeletal muscle TN-C, it was found that although the major percentage of the Ca^{2+}-induced change resulted from interaction with the high-affinity sites, a substantial contribution, about 30%, arose from the low-affinity binding.

Similar conclusions were reached by Johnson and Potter,[30] who established that approximately 62% of the total α-helical change occurred with Ca^{2+} binding to a class of sites with a $K_{Ca^{2+}}$ of approximately 2.7×10^7 M^{-1} (high affinity). The remainder of the spectral change arose from binding of Ca^{2+} to a class of sites with a $K_{Ca^{2+}}$ of $3.1 \times 10^5 M^{-1}$ (low affinity). Because the Ca^{2+}–Mg^{2+} sites are always occupied by cations while the Ca^{2+}-specific sites with a lower affinity for Ca^{2+} can be occupied by Ca^{2+} during contraction only when the available Ca^{2+} is elevated, these Ca^{2+}-specific sites must be involved in the regulation of muscle contraction.

Each of the Ca^{2+} binding sites in TN-C, by analogy with the known three-dimensional structure of parvalbumin,[4] is made up of a Ca^{2+}-binding loop with an α-helix segment on either side.[31] These regions have been located in the primary structure, but several points are not yet clear. First, how are the high-affinity Ca^{2+}–Mg^{2+} sites and the low-affinity Ca^{2+}-specific sites arranged in the polypeptide chain? Second, how is the overall structural integrity of TN-C stabilized by Ca^{2+} binding to each class of site? Finally, how is the conformational change induced in the TN-C relayed to the actomyosin regulatory process?

Enzymic and chemical cleavage of TN-C generated fragments containing one or more of the four binding regions (I to IV as outlined in Fig. 5).[32]

[27] J. P. Potter and J. Gergely, *J. Biol. Chem.* **250**, 4628 (1975).

[28] L. D. Burtnick and C. M. Kay, *FEBS Lett.* **75**, 105 (1977).

[29] M. T. Hincke, W. D. McCubbin, and C. M. Kay, *Can. J. Biochem.* **56**, 384 (1978).

[30] J. D. Johnson and J. D. Potter, *J. Biol. Chem.* **253**, 3775 (1978).

[31] J. H. Collins, *Biochem. Biophys. Res. Commun.* **58**, 301 (1974).

[32] P. C. Leavis, S. S. Rosenfeld, J. Gergely, Z. Graberek, and W. Drabikowski, *J. Biol. Chem.* **253**, 5452 (1978).

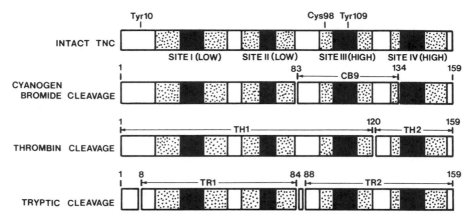

Fig. 5. Schematic diagram of troponin C (TN-C) (top) and the fragments isolated for study after digestion with trypsin, thrombin, and cyanogen bromide. Isolated fragments are indicated by arrows, the numbers indicating the included residues. Stippled area, calcium-binding loops proper; shaded areas, putative flanking α helices. The four regions are indicated as sites I–IV. From Leavis et al.[32]

These fragments were then studied by CD, fluorescence spectroscopy, and their ability to bind Ca^{2+}, as measured by an ion-specific electrode. The largest change in the CD spectra arises from fragments containing sites III and IV, and site III yields the major contribution to the spectral change; therefore, these sites are the high-affinity ones. Fragments with sites I and II showed no change in CD spectra upon Ca^{2+} binding and are, therefore, the low-affinity sites. The CNBr fragment of TN-C (CB9) shows a large change in both far and near-UV CD spectra upon binding Ca^{2+}; this confirms that binding site III is a high-affinity one.[33]

More recently the technique of CD has been utilized to study the effect of Ca^{2+} binding on a completely synthetic peptide analog of Ca^{2+} binding site III in skeletal muscle TN-C, which lacks the helix on the NH_2-terminal side of the loop; i.e., it consists of a loop–helix rather than the complete helix–loop–helix in the native protein.[34] The ellipticity changes noted when the peptide bound Ca^{2+} indicated that the complete helix–loop–helix assembly is not required for Ca^{2+} binding within the loop.

Of relevance to TN-C are CD studies carried out on three highly homologous protein systems: calmodulin, parvalbumin, and the light chains of myosin.

Calmodulin is a ubiquitous, acidic Ca^{2+}-binding protein that apparently links the intracellular second messengers—calcium and the cyclic

[33] B. Nagy, J. D. Potter, and J. Gergely, J. Biol. Chem. 253, 5971 (1978).
[34] R. E. Reid, D. M. Clare, and R. S. Hodges, J. Biol. Chem. 255, 3642 (1980).

nucleotides—through its ability to regulate numerous enzymic activities.[35] The amino acid sequence shows a marked homology to TN-C and, like TN-C, four potential Ca^{2+}-binding regions in the sequence can be identified.[36] When calmodulin binds Ca^{2+}, the molecule undergoes a conformational change, essential for its biological activity. The far-UV CD spectra of calmodulin have been measured in the presence and in the absence of Ca^{2+}. These data suggested 49% α-helix in the presence of Ca^{2+} compared with 40% in the absence of the cation. Ionic strength changes do not affect the CD pattern. The near-UV CD spectrum of the protein, dominated by signals from phenylalanine chromophores, is relatively insensitive to Ca^{2+} binding, merely undergoing a slight sharpening effect below 270 nm, along with the appearance of a shallow trough (tyrosine contribution) above 270 nm.[38] These changes are comparable to those already observed for TN-C.[26]

Parvalbumins, closely related to TN-C but of smaller molecular weight (11,500 as opposed to 18,000), have been extensively studied by CD. These molecules bind two Ca^{2+} ions with an affinity constant of $10^9 M^{-1}$. ORD[39] and CD[40] studies have shown that when Ca^{2+} is bound, the helical content of the protein increases from 39–47% of the amino acid residues.

In this same context, one should mention the light chains of myosin. The amino acid sequences of the so-called alkali I,[41] DTNB,[42] and scallop EDTA[43] light chains from myosin show an approximately 30% homology with skeletal muscle TN-C and carp parvalbumin. In addition to this homology, conservative amino acid substitution occurs among the nonidentical sequences, and this enables models to be constructed for these molecules based on X-ray data from carp parvalbumin.[4] In attempts to extend these observations of homology, CD studies have been undertaken to establish whether the conformations of these proteins show any Ca^{2+} sensitivity. Initial observations indicated that the far-UV CD spectrum of scallop EDTA light chain was unresponsive to Ca^{2+} concentrations $\leq 10^{-4} M$[44]; however, when these measurements were extended to higher Ca^{2+} concentrations, a fairly large CD change was found,[17] ($[\theta]_{222}$ in

[35] A. R. Means and J. R. Dedman, *Nature (London)* **285**, 73 (1980).

[36] T. C. Vanaman, F. Sharief, and D. M. Watterson, *in* "Calcium Binding Proteins and Calcium Functions (R. H. Wasserman *et al.*, eds.), p. 107. Elsevier, Amsterdam, 1977.

[37] C. B. Klee, *Biochemistry* **16**, 1017 (1977).

[38] M. Walsh, F. C. Stevens, K. Oikawa, and C. M. Kay, *Can. J. Biochem.* **57**, 267 (1979).

[39] A. Cave, M. Pages, P. Morin, and C. M. Dobson, *Biochimie* **61**, 607 (1979).

[40] H. Donato, Jr. and R. B. Martin, *Biochemistry* **13**, 4575 (1974).

[41] G. Frank and A. G. Weeds, *Eur. J. Biochem.* **44**, 317 (1974).

[42] J. H. Collins, *Nature (London)* **259**, 699 (1976).

[43] J. Kendrick-Jones and R. Jakes, *Myocard. Failure (Int. Symp. 1976,* p. 28 (1977).

[44] W. F. Stafford and A. G. Szent-Györgyi, *Biochemistry* **17**, 607 (1978).

the absence of Ca^{2+} $\simeq$ $-11,900°$; in the presence of the cation $[\theta]_{222}$ $\simeq$ $-13,800°$). Similar results were obtained for the rabbit DTNB light chain. It has still not been possible, however, to relate this low-affinity binding ($pK_{Ca^{2+}}$ = 3.7 for EDTA light chain and $pK_{Ca^{2+}}$ = 5.1 for DTNB light chains) with the high-affinity binding present when the light chain is in intact myosin.

TROPONIN SUBUNIT INTERACTIONS

The troponin molecule is a complex of the three subunits: TN-C, the Ca^{2+}-binding protein, TN-I, the myofibrillar ATPase inhibitory protein, and TN-T, which serves to anchor the complex to the protein tropomyosin. In an attempt to understand in greater detail the biological function of this complex, *in vitro* formation of various complexes has been carried out and CD has been very powerful in establishing whether conformational rearrangements accompany these interactions.

Troponin I–Troponin C (TN-IC). The far-UV CD spectrum of a 1 : 1 molar ratio mixture of TN-I and TN-C has ellipticities in the absence and, particularly, in the presence of Ca^{2+} considerably greater than the theoretical ellipticities calculated by summing separate contributions from TN-I and TN-C[45] (Fig. 6). A net gain in secondary structure accompanies the interprotein interaction. Sedimentation velocity studies verify that complex formation occurs. Furthermore, the Ca^{2+}-induced conformational change in TN-C[26] still occurs in the complex, but the CD results suggest that Ca^{2+} binds more strongly in the complex.

Another interesting use of CD was made in this study, whereby the ellipticity value at 221 nm was followed as a function of temperature. As a result, a CD melting profile was established for the protein. These experiments demonstrated that Ca^{2+} stabilizes the structure of the protein against thermal denaturation. This methodology has been applied to a comparative study of skeletal and cardiac TN-C, with essentially similar conclusions being reached.[46]

Troponin T–Troponin C (TN-CT) and Reconstituted Troponin (TN-ICT). Circular dichroism measurements indicate that the Ca^{2+}-induced conformational change in TN-C occurs both in TN-CT and in TN-ICT, and in both instances the binding of Ca^{2+} is enhanced. The CD melting experiments reveal that for both complexes the structures produced in the presence of Ca^{2+} are more resistant to thermal denaturation than those produced in its absence.[47]

[45] W. D. McCubbin, R. S. Mani, and C. M. Kay, *Biochemistry* **13**, 2689 (1974).

[46] W. D. McCubbin, M. T. Hincke, and C. M. Kay, *Can. J. Biochem.* **58**, 683 (1980).

[47] R. S. Mani, W. D. McCubbin, and C. M. Kay, *Biochemistry* **13**, 5003 (1974).

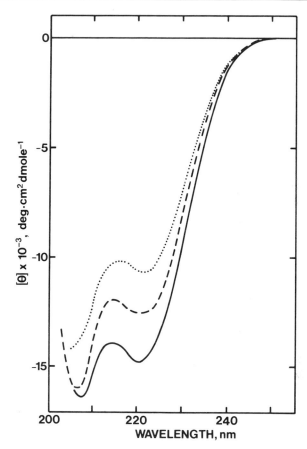

FIG. 6. Far-UV circular dichroism (CD) spectra of troponin I–troponin C (TN-IC) protein complex in 0.5 M KCl, 50 mM Tris-HCl, 1 mM EGTA at pH 8.0 (---) and in 0.5 M KCl, 50 mM Tris-HCl, 1 mM EGTA, $5 \times 10^{-4} M$ free Ca^{2+} at pH 8.0 (——). Theoretical CD spectrum for TN-IC in 0.5 M KCl, 50 mM Tris-HCl, 1 mM EGTA at pH 8.0 ($\cdots\cdots$). From McCubbin et al.[45]

Troponin I–Troponin T (TN-IT). Interaction between these subunits has been difficult to demonstrate, in large part, because of the many difficulties in handling these proteins (e.g., insolubility at low ionic strength, ease of aggregation, and oxidation in solution). After a study employing crosslinking reagents revealed that these subunits lie very close to each other, within 0.6 nm or so in the *in vivo* complex,[48] fresh efforts were devoted to trying to establish unequivocally whether these two subunits interacted.

[48] S. E. Hitchcock, *Biochemistry* **14,** 5162 (1975).

Horwitz et al.[49] used gel filtration and monitored the near-UV CD spectra of the proteins, employing a computer enhancement technique to establish clearly that if proper care was taken to maintain the protein in a reduced state, interaction does indeed occur and conformational changes accompany this interaction. Studies from this laboratory[50] extend these observations to TN-I and TN-T isolated from bovine heart muscle with essentially similar conclusions. In addition, this work showed that the far-UV CD spectrum of the TN-IT complex is significantly different ($\Delta[\theta]_{220\,nm} = 1000°$) from the calculated sum of the individual components and provides extra evidence for interprotein interaction.

Troponin T–Tropomyosin Complexes (TN-T-TM). Similar studies were extended to the complex of troponin T with tropomyosin (TN-T-TM) and this complex with TN-C (TN-CT-TM).[51] The observed $[\theta]_{221}$ for TN-T-TM is 2500° less than the theoretical value, reflecting a fairly substantial conformational alteration, perhaps at the level of the TM moiety. In the complex, TN-CT-TM, the far-UV CD signal is sensitive to Ca^{2+}, a rather important finding because it adds credence to the idea that a conformational change originating in one protein, TN-C, could be relayed through the whole troponin complex and thence to tropomyosin, resulting in the alteration in the interaction of myosin heads with actin.

Cardiac Troponin–Tropomyosin Interactions. This work was extended to a study of the interactions between the analogous proteins from bovine cardiac muscle.[52] In a similar fashion, interactions occur between TN-T and tropomyosin, TN-I and TN-C, and TN-C with TN-T. Complexes that contain TN-C can undergo a Ca^{2+}-induced conformational change, and in the case of reconstituted troponin, there appears to be an amplification of the effect of binding Ca^{2+} to TN-C.

In a later study, hybrid combinations of the components of the regulatory systems of bovine cardiac and rabbit skeletal muscles were examined by CD and biological activity techniques.[53] The results indicate that functional hybrids can be prepared from the troponin subunits. When cardiac TN-C was present in any complex, a potentiation of the Ca^{2+}-induced conformational change was detected, whereas no such potentiation could be discerned for skeletal muscle TN-C.

Tropomyosin

Tropomyosin is a long rodlike molecule consisting of two polypeptide chains arranged as α-helices in a coiled-coil structure. *In vivo* head-to-tail

[49] J. Horwitz, B. Bullard, and D. Mercola, *J. Biol. Chem.* **254**, 350 (1979).
[50] M. T. Hincke, W. D. McCubbin, and C. M. Kay, *Can. J. Biochem.* **57**, 768 (1979).
[51] R. S. Mani, W. D. McCubbin, and C. M. Kay, *FEBS Lett.* **52**, 127 (1975).
[52] L. D. Burtnick and C. M. Kay, *FEBS Lett.* **65**, 234 (1976).
[53] M. T. Hincke, W. D. McCubbin, and C. M. Kay, *FEBS Lett.* **83**, 131 (1977).

aggregates form extended filaments that lie in the two grooves of the F-actin helical structure. Binding of troponin to tropomyosin is essential for the regulation of muscle contraction induced by Ca^{2+} ions.[54] ORD[55] as well as CD studies[56] indicated that the molecule exists in solution as a structure with more than 95% α-helix.

The question immediately arises: Are the constituent chains parallel and in register? Model building studies[57] have shown that the most stereochemically favored structure is a coiled-coil of variable radius, with the chains in register and symmetrical. The fact that an intramolecular disulfide bridge can be readily formed confirms that the chains are in register.[58]

A study of the near-UV CD of tropomyosin was successful in shedding light on this question.[59] The CD spectrum in the region 250–320 nm in this protein arises from tyrosine and disulfide bonds, and the contribution of each may be distinguished by oxidation and reduction of the protein. The CD due to tyrosine is much more intense than expected (e.g., the highly helical coiled-coil protein, light meromyosin, which has the same number of tyrosine residues per mole as tropomyosin, has a tyrosyl CD of less than one-fifth the value for tropomyosin). This intensity is very likely due to tyrosine–tyrosine interactions at distances of less than 8 Å.

If the coiled-coil model is used as the reference to analyze the spectra, the conclusion is that tyrosines of one helical subunit interact with those of the other. This also implies that the chains must be in register, since only then will five tyrosines of one chain be opposite those of the other. In addition, the very large CD noted for the disulfide bond of oxidized tropomyosin indicates the presence of intramolecular disulfide bonds between chains, which can only be present in the nonstaggered model for tropomyosin.

M-Band Proteins

Circular dichroism studies, in conjunction with gel filtration chromatography and sedimentation equilibrium molecular weight measurements, have been carried out to examine possible interactions between several of the M-band systems (see this volume [15]). Both creatine kinase (CPK) and the 165,000 molecular weight components are characterized by the presence of a low amount of α-helix, 15–25%. By compari-

[54] S. Ebashi, M. Endo, and I. Ohtsuki, *Q. Rev. Biophys.* **2**, 351 (1969).
[55] W. D. McCubbin, R. F. Kouba, and C. M. Kay, *Biochemistry* **6**, 2417 (1967).
[56] K. Oikawa, C. M. Kay, and W. D. McCubbin, *Biochim. Biophys. Acta* **168**, 164 (1968).
[57] A. D. McLachlan and M. Stewart, *J. Mol. Biol.* **98**, 298 (1975).
[58] P. Johnson and L. B. Smillie, *Biochem. Biophys. Res. Commun.* **64**, 1316 (1975).
[59] B. Bullard, D. A. Mercola, and W. F. H. M. Mommaerts, *Biochim. Biophys. Acta* **434**, 90 (1976).

son of the observed far-UV CD spectra with the computed ones, it was shown that conformational changes are induced when CPK interacts with myosin, HMM, and S-1, the extent of interaction being strongest with the intact myosin molecule.[60,61] In like fasion the 165,000 molecular weight component was found to interact with both myosin and heavy meromyosin S-2.[62]

This use of CD nicely complements both the sedimentation equilibrium and gel filtration studies in establishing that interactions occur between these various protein members and that conformational changes accompany the interactions.

Other Myofibrillar Protein Components

Very little optical activity work has been done on other myofibrillar proteins, such as the actinins. On occasion, far-UV CD spectra have been measured and, by using the equations of Chen,[8] estimates have been made of the amount of α-helix, β structure, and random coil. There have been no examples of the use of ORD and CD to examine any interactive phenomena in any of these systems. Two systems will be briefly described as points of interest.

α-*Actinin.* The far-UV CD spectrum shows two negative extremes at 208 and 222 nm, the values of which suggest an α-helix content of 74%.[63] A point of interest about this spectrum is that $[\theta]_{208\,nm}$ is slightly larger than $[\theta]_{222\,nm}$. This may be correlated with the known tendency of the protein to aggregate[64] and also suggests little or no β structure.[6]

β-*Actinin.*[65] From the $[\theta]_{208}$ and $[\theta]_{222}$ of the CD spectrum[66] a helical content of 13–15% was calculated; this agrees closely with the value calculated from the ORD[66] data. No optical activity work has been done on the interaction of this protein with F-actin.

Particular Problems in Measuring Optical Activity of Muscle Proteins

Gratzer[3] gives an excellent description of the pitfalls, precautions, and care required in the measurement of ORD and CD of protein solutions in

[60] R. S. Mani and C. M. Kay. *Biochim. Biophys. Acta* **453**, 391 (1976).
[61] O. S. Herasymowych, R. S. Mani, and C. M. Kay, *Biochim. Biophys. Acta* **534**, 38 (1978).
[62] R. S. Mani and C. M. Kay, *Biochim. Biophys. Acta* **536**, 134 (1978).
[63] A. Suzuki, D. E. Goll, I. Singh, R. E. Allen, R. M. Robson, and M. H. Stomer, *J. Biol. Chem.* **251**, 6860 (1976).
[64] D. E. Goll, W. F. H. M. Mommaerts, M. K. Reedy, and K. Seraydarian, *Biochim. Biophys. Acta* **175**, 174 (1969).
[65] K. Maruyama, S. Kimura, T. Ishi, M. Kuroda, K. Ohashi, and S. Muramatsu, *J. Biochem.* (*Tokyo*) **81**, 215 (1977).
[66] K. Maruyama, *J. Biochem.* (*Tokyo*) **69**, 369 (1971).

general. Needless to say, all those suggestions should be considered when handling muscle proteins, but, in addition, a few extra pecularities of these materials must be considered as well.

Many of the muscle proteins perform a structural role, and their resulting fibrous nature often makes them very difficult to handle. They are certainly, in most cases, quite different from the highly soluble, usually crystalline, proteins common to most biochemists. Several of them, e.g., myosin and troponin T, are soluble at neutral pH only above a certain ionic strength, and, even then, concentrations of only a few milligrams per milliliter can be prepared. The very asymmetric protein tropomyosin, although soluble in water, exists in such an aggregated form that its viscous gel-like solutions can hardly be introduced into the optical cells.

The methods of isolation and purification of many of the muscle proteins are usually tedious and time consuming, and the same protein isolated in different laboratories by slightly different approaches may show subtle differences in optical and other properties. Often these differences may be attributed to different degrees of denaturation. Of paramount importance when trying to compare optical properties of proteins prepared by different laboratories is the necessity to establish the degree of sample purity. It is essential to describe in detail the criteria used in establishing purity: e.g., Did the protein migrate as a single band on polyacrylamide gel electrophoresis? Was a single symmetrical peak detected by sedimentation velocity in the analytical ultracentrifuge? If the protein is an enzyme, what was the value of the specific activity?

In addition, agreement among the various optical parameters will depend to a great extent on the way in which the protein concentration is measured. It really does not matter which established procedure one decides to use, provided explicit details are given. For example, if the concentration is established from ultraviolet absorption, then it is imperative to quote not only the particular extinction coefficient at the wavelength used, but also how this figure was established and whether due correction was made for light scattering. For the large muscle proteins, this latter correction is very important.

When G-actin is transformed to F-actin, there is an enormous increase in viscosity. This is a difficult process to follow by ORD or CD in view of the inhomogeneity in solution that results. It is essential to employ low protein concentrations, $0.1-0.5$ mg/ml in cells of short path length, such as 0.05 cm, and to mix the sample carefully and then allow sufficient relaxation time to eliminate artifacts prior to recording the spectra. Similar problems arise in handling myosin, but, additionally, this protein has a pronounced tendency to adhere to optical cell windows, which necessitates scrupulous cleaning with enzyme detergent or chromic acid, followed by extensive water washing and careful drying between separate runs.

In conclusion if one is aware of the pitfalls and possible shortcomings inherent in these methods, ORD and, in particular, CD are powerful techniques to study the conformations and, moreover, conformational changes of muscle proteins in solution.

[55] Special Instrumentation and Techniques for Kinetic Studies of Contractile Systems

By HOWARD D. WHITE

Steady-State Rate Measurements

The pH Stat Method

Owing to the rather slow rate constant for MgATP hydrolysis by myosin and its subfragments (0.05 sec^{-1}), the pH stat must be able to make deliveries of ca 1–100 μl/min. A suitable commercial instrument is produced by Radiometer Corporation equipped with a 0.25-ml autoburette. The author has built according to a published design[1] a pH stat-titrator that performs as well as the commercial instrument and is approximately one-tenth the cost.

The following experimental conditions are convenient for measuring ATP hydrolysis by skeletal actomyosin subfragment-1 (S-1) primarily because the hydrolysis rates are ideal for the pH stat method (40 mM KCl, 5 mM MgCl$_2$, pH 8.0, 20°).

Solutions

A: 10 mM ATP, pH 8.0 (stable frozen at $-20°$ for up to 6 months)
B: 0.4 M KCl, 50 mM MgCl$_2$
C: 1–10 mg of myosin S-1 per milliliter in 1.0 mM Tris-HCl, pH 7.9
D: 1.0–10 mg of F-actin per milliliter in 40 mM KCl, 5 mM MgCl$_2$, 1 mM Tris pH 7.9
E: 10^{-2} N NaOH, prepared in CO$_2$-free glass-distilled water
F: CO$_2$-free glass-distilled water

[1] B. D. Warner, G. Boehme, M. S. Urdea, K. Pool, and J. I. Legg, *Anal. Biochem.* **106,** 175 (1980).

Experimental Procedures

1. Burette reservoir is filled with solution E and fitted with a soda-lime trap to prevent the entry of CO_2. The NaOH concentration should be monitored frequently by titrating 2.5 μmol of a standard such as oxalic acid.
2. Solutions A (1.0 ml), B (1.0 ml), and F (7.0 ml) are allowed to equilibrate for 5 min at 20° beneath a slow stream of nitrogen (1 cm³/sec). The pH should be kept constant and may require a small addition of NaOH (less than 1.0 μl/min).
3. The hydrolysis is begun by the addition of 1 mg of myosin S-1, solution C. The initial rapid consumption of NaOH (to bring the pH to 8.0) should cease within a minute and be followed by a steady state of typically 2.5 μl/min-mg of myosin S-1 in the absence of actin.

Addition of actin, solution D, will increase the steady-state rate of myosin S-1 ATP hydrolysis approximately 40-fold per milligram of actin added under these conditions. Rates of ATP hydrolysis should be within 5% of linearity for the first 0.25 ml of NaOH consumed.

Problems and Limitations of the Method. The maximum rate that can be measured by the pH stat method ultimately depends upon the time response of the instrument and electrodes, mixing efficiency, and rate of titrant delivery.

A frequent problem is nonuniform delivery of NaOH, which results in an erratic record. This problem is usually attributable to excess buffering capacity, viscous solution, or poor electrode response time. Excess buffering capacity (greater than 1 mM at the pH of the reaction) reduces the pH excursion from the set point for a given amount of $[H^+]$ or $[OH^-]$, which will produce an inadequate instrumental time response and a pattern of undershoot and overshoot. At high concentrations of actin (>2 mg/ml) the viscosity may reduce mixing efficiency sufficiently to give similar symptoms. More efficient stirring will alleviate viscosity-related problems. Separate glass and calomel electrodes tend to have better response times than combined electrodes.

The rate of ATP hydrolysis is not equal to the rate of NaOH consumed, but must be corrected for the incomplete dissociation of $H_2PO_4^-$ to $HPO_4^{2-} + H^+$ [Eq. (1)].

$$[ATP] = [NaOH] \frac{1 + \text{antilog} [pH - pK_a]}{\text{antilog} [pH - pK_a]} \tag{1}$$

The pK_a of $H_2PO_4^-$ is 7.2 under the conditions given here. While this correction (1.15) is small at pH 8.0, it increases to 2.6 at pH 7.0. Hence, the use of the pH stat method for measurement of rates of ATP hydrolysis is not very useful at pH much lower than 7.

Separation of the reference electrode from the protein solution by a salt bridge containing the same salt concentration as the reaction mixture will prolong useful life of the electrode. A suitable salt bridge can be easily constructed by pulling out a 1/2-inch piece of glass tubing to a fine point. The electrode is inserted through a tightly fitting hole through a rubber stopper, which is inserted into the drawn-out glass tube.

Regulation of the temperature to ±0.1° during the pH stat measurement is required. Temperature variations can result in significant absorption or release of protons by temperature-dependent buffer. The rate of actomyosin S-1 ATP hydrolysis is also extremely temperature sensitive; a temperature change of 1° may result in up to a 30% change in the rate of ATP hydrolysis. The large dependence of the rate upon ionic strength (reducing the ionic strength from 0.1 to 0.04 results in a 20-fold increase in actin-activated ATP hydrolysis) requires that considerable care be taken to ensure that constant ionic strength is maintained for a proper comparisons among different sets of data.

Colorimetric Measurements of ATP Hydrolysis

If a pH stat is not conveniently available, steady-state rates of ATP hydrolysis can be measured by colorimetric phosphate determination. Measurements of ATP hydrolysis by colorimetric determination of phosphate release by the method of Tausky[2] or subsequent modification have been used to determine the steady-state kinetics of actomyosin ATP hydrolysis. A weakness in these methods has been continual phosphate release due to instability of ATP in acid after quenching the enzymic reaction. The following modification by R. Siemankowski offers considerable improvement in the method.

Solutions

A: Myosin (with or without actin), incubated with ATP for a sufficient length of time to produce 25–600 nmol of phosphate in a total volume of 0.75 ml

B: 13.3% sodium dodecyl sulfate, 0.12 M EDTA, pH 7.0

C: 0.5% (w/v) ferrous sulfate, 0.5% (w/v) ammonium molybdate, 0.5 M H_2SO_4, made fresh daily from a solution of 10% ammonium molybdate, 10 M H_2SO_4 (stable up to 6 months at room temperature), solid ferrous sulfate, and distilled water

Procedure. The hydrolysis reaction, solution A, is terminated by the addition of 0.25 ml of solution B. Color development is initiated by addition of 2.0 ml of solution C; it is complete after 15 min and stable for at

[2] H. H. Tausky and E. Shorr, *J. Biol. Chem.* **202**, 675 (1963).

least 1 hr. The absorbance of 1.03×10^3 OD-cm/nmol at 550 nm is linear up to 600 nmol of phosphate. The relatively low background and color stability enable as little as 25 nmol of phosphate to be measured with a precision of approximately $\pm 10\%$. If the initial reaction mixture (A) contains a high concentration of potassium ions (greater than ca 0.3 M), the final solution is briefly warmed to 37° to dissolve the $K^+ \cdot$ SDS precipitate, prior to measurement of the absorbance.

Problems and Limitations. The same precautions should be taken with respect to pH and ionic strength using colorimetric as are required for pH stat rate measurements. Owing to the nature of the colorimetric rate measurements, good practice suggests measurement of at least two different times to ensure linearity of ATP hydrolysis with time. An enzyme blank (reagent B added to the enzyme moiety of reagent A prior to addition of ATP) should accompany the determinations, in order to correct for P_i contamination of any reagents or any nonenzymic hydrolysis of ATP, which occurs slowly after addition of reagent C.

Transient Kinetic Measurements

Perhaps the most surprising aspect of the application of transient kinetic studies to myosin and actomyosin ATP hydrolysis has been how well the standard techniques have worked despite the size and complexity of the protein molecules involved. The table summarizes the interaction of nucleotides with myosin S-1, HMM, intact myosin, and actomyosin measured by enhancement of the tryptophan fluorescence, proton release, and ^{32}P release from $[\gamma\text{-}^{32}P]ATP$. The proteolytic subfragments of myosin have been studied in more detail than intact myosin, primarily because they are soluble at physiological ionic strengths and appear to be good models for myosin, at least with respect to actin binding and ATP hydrolysis activities.

Basic Equations for a Two-Step Binding Mechanism

Nucleotide binding to myosin S-1 and actomyosin S-1 is at least a two-step process as shown in Eqs. (2) and (3).

$$M + N \underset{k_{-1}}{\overset{k_1}{\rightleftarrows}} M \cdot N \underset{k_{-2}}{\overset{k_2}{\rightleftarrows}} M^* \cdot N \tag{2}$$

$$AM + N \underset{k_{-1a}}{\overset{k_{1a}}{\rightleftarrows}} AM \cdot N \underset{k_{-2a}}{\overset{k_{2a}}{\rightleftarrows}} A + M \cdot N \tag{3}$$

The general strategy of transient kinetics is to measure the observed rate of formation, k_{obs}, of a state with altered physical properties ($M^* \cdot N$ from M or $M \cdot N$ from $A \cdot M$) at a series of ligand (N) concentrations. In

BASIC TRANSIENT KINETIC EXPERIMENTS WITH MYOSIN (S-1 OR HMM) AND ACTOMYOSIN

Reaction[a]	Method[b]	Physical parameter	Measurement
i M + A →	SF	Light scattering	Rate of myosin binding to actin
ii AM + ATP →			
1. [ATP] ≫ [AM]	SF	Light scattering	Rate of actomyosin dissociated by ATP
	SF	Fluorescence	Rate of $M \cdot ATP \rightarrow M \cdot ADP \cdot P_i$
	QF	^{32}P release	Rate of $M \cdot ATP \rightarrow M \cdot ADP \cdot P_i$
2. [ATP] ~ [AM]	SF	Light scattering	Dissociation and recombination $AM \rightarrow A + M \cdot ATP \rightleftarrows M \cdot ADP \cdot P_i + A \rightarrow AM + ADP + P_i$
iii ($M \cdot ADP \cdot P_i$ / M + ATP, A)	SF	Light scattering	Double mixing measurement of $M \cdot ADP \cdot P_i$ binding to actin
iv M + N →			
1. [N] ≫ Myosin	SF	Fluorescence; H+ release	Rate of nucleotide binding to myosin, and (for $-^{32}P$-ATP) $M \cdot ATP \rightarrow M \cdot ADP \cdot P_i$
2. [ATP] ~ myosin	QF	^{32}P release	Single turnover of ATP hydrolysis
		Fluorescence	Active site stoichiometry, rate and equilibria of
		^{32}P release	$M \cdot ATP \rightleftarrows M \cdot ADP \cdot P_i$
v AM · ADP + ATP →	SF	Light scattering	Rate and equilibria of $AM \rightleftarrows AM \cdot ADP$

[a] A / M ⟶ indicates the mixing of actin and myosin in a rapid-mixing device. Actin and myosin-S1 concentrations 0.05 mg/ml to 1.0 mg/ml give satisfactory results, although concentrations of actin up to 5 mg/ml or S-1 or HMM up to 50 mg/ml can be used.
[b] SF, stopped flow; QF, quenched flow.

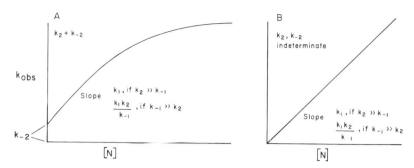

FIG. 1. The dependence of k_{obs} upon [N]. (A) $k_{+2} + k_{-2} \lessgtr 1000$ sec^{-1}. (B) $k_{+2} + k_2 \gtrless 1000$ sec^{-1}.

the most favorable cases, one can dissect out the equilibrium constants for each of the steps and values for two of the rate constants. For example if $k_{-1} \gg k_2$,

$$\Delta = \Delta_0 e^{-k_{obs}t} \tag{4}$$

the rate of change in the physical property (Δ) will always fit a single exponential [Eq. (4)].

$$k_{obs} = k_{-2} + \frac{k_2}{1 + (k_{-1}/k_1[N])} \tag{5}$$

The dependance of k_{obs} upon [N] predicted by Eq. (5) is illustrated in Fig. 1A. Values for $k_{+2} + k_{-2}$ can be obtained from the k_{obs} at saturating [N] and k_{-2} from the intercept as [N] approaches 0. The equilibrium constant for the first step is obtained from k_2 and the slope at low [N].

However, under less favorable conditions, k_2 being very large or k_{-2} very small, the dependence of k_{obs} upon [N] may not plateau at high [N] or have an intercept at low [N] making k_2 and k_{-2} indeterminate. Therefore only $(k_1/k_{-1})k_2$ can be measured. Figure 1B, in the absence of any additional information, is also consistent with an irreversible, one-step binding mechanism.

A second general case of two-step binding occurs if $k_2 \gg k_{-1}$. The observed rate of change will follow Eq. (4) at high and low concentrations of [N], but at intermediate concentrations a lag phase may occur that will require two exponentials to fit the data correctly [Eq. (6)].

$$\Delta = \Delta_1 e^{-k_1 \text{obs} t} + \Delta_2 e^{-k_2 \text{obs} t} \tag{6}$$

The dependence of k_{obs} upon [N] is similar to Fig. 1 except that at intermediate values of [N] two k_{obs} are measured.

At low [N] the slope equals k_1. At high [N], k_{obs} may reach a maximum rate of $k_2 + k_{-2}$. The appearance of a lag phase indicates $k_{-1} \ll k_2$; single exponential changes over all concentrations of [N] suggest $k_{-1} \gg k_2$.

Stopped-Flow Method—Instrumentation

The design of stopped-flow instrumentation has been comprehensively treated in this series.[3] Therefore the discussion of instrumentation will be limited to the choice of light source and cell geometry. In addition, several stopped-flow fluorometers are now available commercially, although generally at a cost at least 5 times that of laboratory-built equipment.

Light Sources. The light output of tungsten filament or quartz iodide–tungsten lamps is insufficient in the wavelength region of 280–295 nm for use in measuring changes in tryptophan fluorescence of proteins. Xenon and mercury arc lamps provide both the required near-ultraviolet light output and greater intensity than tungsten lamps. Problems involving arc wander can be overcome with stabilized power supplies in the constant current mode. A suitable power supply is available from Olis Inc., Silver Spring, Georgia. Although designed for 150 W xenon arc, satisfactory results can also be used with 100 W mercury arc (Osram HBO W/2) or xenon 75 W arc (Osram XB 75 W/2) by reducing the current to 4–5 amp.

The smaller mercury and xenon arc lamps have the brightest arc and are more efficient for illuminating the small surface areas (1–4 nm^2) of stopped-flow cells. It is possible to obtain short-term (1 msec–10 sec) stability of 0.1% in light intensity with these components. High-pressure mercury arc lamps have approximately 10-fold greater light intensity in the region of tryptophan excitation than xenon arc lamps but suffer from having an intense line at 340 nm near the tryptophan emission maximum.

Monochromatic light can be produced either by use of a monochrometer or interference filters. Monochrometers have the advantage of being more flexible with respect to wavelength and bandwidth; interference filters have the advantage of making the optical system very simple. Interference filters are more suitable for the line spectral output of mercury arc whereas the continuous output of the xenon arc is more appropriately used with a monochrometer. Interference filters with 20% transmittance at 294 nm and with a 10-nm bandwidth are available from several optical supply firms.

Although a complete quantitative comparison has not been made, the author has found that both systems give signal-to-noise ratios of less than 5% upon mixing myosin S-1 (at concentrations as low as 0.1 mg/ml) with MgATP and a filter time constant of 1 msec.

Wavelength selection of the emitted light can be obtained by using a

[3] Q. H. Gibson, this series, Vol. 16, p. 187.

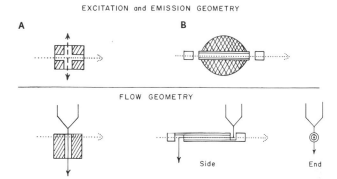

FIG. 2. Flow and optical geometries of square cross section (A) and tubular cross section (B) flow cells. Exciting light path (dotted arrow); emission light path (dashed arrow); black quartz (hatched); Mirror surface (crosshatched); flow path (→).

sharp-cutoff filter, band-pass filter, or interference filters (if a narrower bandwidth of emission spectra is required). Cutoff and band-pass filters allow more light to be transmitted and are usually adequate for kinetic studies of myosin S-1.

Cell Geometry. The design of the stopped-flow cell provides a choice of either optimizing the signal-to-noise ratio or the fastest measurable maximum first-order rate constant. Two types of cells that have been used in the study of contractile proteins are illustrated in Fig. 2.

The square cross-sectional cell (Fig. 2A) features a straight-through flow path between mixer and cell, and observation of only the top 2–3 mm of the cell can be obtained by masking. This achieves a deadtime of ~1 msec enabling measurement of rate constants of up to 1000 sec^{-1} with a 67% signal loss. Disadvantages are poor light collection efficiency and high background of light scattered from the internal cell walls. These contribute to a fairly low signal-to-noise ratio. The amount of light scattered can be reduced by constructing the cell with black quartz corners. The short path length (1–2 mm) makes the measurement of absorbance changes impractical except for species with very high extinction coefficients. The two perpendicular emission ports permit simultaneous measurements of light scattering and of fluorescence or fluorescence emission at two wavelengths. The latter has been found to be quite useful for simultaneous measurement of the rates of dissociation of acto-S-1 by MgATP and changes of fluorescence of myosin S-1.

The tubular flow cell (Fig. 2B) has the advantage of greater than 10-fold increased signal-to-noise ratio, due primarily to improved light collection geometry and masking of the cell surfaces to reduce internally scattered light. The improved signal-to-noise ratio is, however, offset somewhat by a longer deadtime of ~4 msec, which reduces the maximum measurable first-order rate constant to ~250 sec^{-1}. Placing a spherical mirror surface behind the cell[4] gives an approximately 2-fold further increase in the light collection efficiency. The tubular cell is of particular advantage in studying small changes in fluorescence or weakly absorbing and fluorescing proteins (such as troponin C and calmodulin, which contain tyrosine but not tryptophan) or where large quantities of proteins cannot be easily obtained. In addition, reactions having very large second-order rate constants can be brought within measure by dilution of protein until a sufficiently slow rate is achieved. The long path length (1.0–2.0 cm) is ideal for absorption measurements, but care must be taken to ensure that absorption of the exciting light does not interfere with fluorescence measurements of high protein (or ligand) concentrations. Although simultaneous observations of two emission wavelengths is not easily made, absorption (or turbidity) and fluorescence emission measurements can be made simultaneously.

Basic Transient Kinetic Techniques for Studying Actomyosin ATP Hydrolysis

Light Scattering. Although actin filaments may be sheared to some extent by the turbulence in a stopped-flow mixer, filament lengths of up to 0.5 μm can be observed a few seconds after mixing.[5] In addition, light scattered by actin, acto-S-1, and acto-HMM remains constant after mixing, indicating that no major changes in physical state of the proteins has occurred. Figure 3 illustrates the increase in the amount of light scattered at 90 degrees by actin in the presence of HMM. At actin monomer concentrations of less than 10 μM, there is a very nearly linear increase in the amount of light scattered at 90 degrees upon the addition of S-1 or HMM up to concentrations where myosin heads equal the number of actin subunits. At actin monomer concentrations greater than 10 μM, the relationship between light scattered and [S-1] or [HMM] is pronouncedly S shaped and the light scattering is no longer proportional to actin concentration.

The stoichiometric relationship between amount of S-1 (or HMM) bound to actin and amount of light scattering provides an empirically observable physical parameter for measuring S-1 (and HMM) binding to actin and was used to make the transient kinetic measurements in the table.

Light-scattering measurements of actomyosin could be made at any

---

[4] A spherical mirror surface can be easily placed behind the observation tube by milling a cavity with a ball end mill. A good reflective surface can be achieved with "Semichrome" polish if the cell body is constructed from aluminum or by polishing followed by vacuum deposition of aluminum if the cell body is stainless steel.

[5] Unpublished observations of Howard White and Jonathan Seymour.

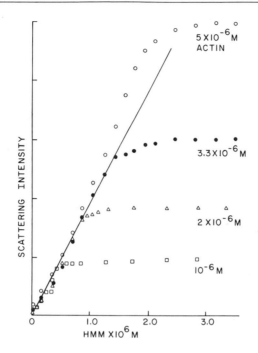

FIG. 3. Light scattering at 340 nm of F-actin with the addition of heavy meromyosin (HMM). Conditions: 0.1 M KCl, 10 mM Tris, 5 mM MgCl$_2$, pH 8.

convenient wavelength where absorption and fluorescence do not occur. The best signal-to-noise ratio is obtained by using an interference filter to isolate the 337-nm Hg line.

Simultaneous light-scattering and fluorescence measurements of the dissociation of acto-S-1 by ATP are best performed using a 290-nm interference filter for excitation and light scattering and a 340-nm interference filter to isolate the fluorescence. Light scattering causes some interference in the fluorescence measurement at relatively low [ATP], but at [ATP] > 100 μM the dissociation is essentially complete before the beginning of the fluorescence signal. The square-cell geometry is particularly well suited for simultaneous scattering and fluorescence measurements, although turbidity and fluorescence can be measured simultaneously using the cylindrical-cell geometry.

Double Mixing and Single-Turnover Experiments. The kinetics of the recombination of M · ADP · P$_i$ to actin, can be measured either by a single-turnover experiment [Eq. (7a)] or by double mixing [Eq. (7b)].

$$\text{AM} + \text{ATP} \xrightarrow{k_{\text{diss}}} \text{M} \cdot \text{ATP} \rightleftarrows \text{M} \cdot \text{ADP} \cdot \text{P}_i + \text{actin} \xrightarrow{k_{\text{rec}}} \text{AM} + \text{ADP} + \text{P}_i \qquad (7a)$$

$$\text{M} + \text{ATP} \xrightarrow{k_{\text{a}}} \text{M} \cdot \text{ATP} \rightleftarrows \text{M} \cdot \text{ADP} \cdot \text{P}_i + \text{actin} \xrightarrow{k_{\text{rec}}} \text{AM} + \text{ADP} + \text{P}_i \qquad (7b)$$

In Eq. (7a) [MgATP] equal to the concentration of myosin active sites is mixed with the actomyosin. A rapid dissociation is observed from the decrease in light scattering occurring upon the formation of the $M \cdot ATP$ and $M \cdot ADP \cdot P_i$ intermediate and is then followed by a slower recombination to re-form AM. As long as k_{diss} is much faster (100-fold or more) than k_{rec}, the observed rate of recombination will be near k_{rec}. If, however, the two processes occur with rate constants of less than 100-fold difference, the observed rate of recombination will be slower than k_{rec} (by a factor of ~2 if $k_{diss} = 10 k_{rec}$).

In the case of extremely good quality data (with high signal-to-noise ratio and stable end point), the two rate constants can, however, be dissected out by fitting the change in light scattering data with Eq. (6).

Alternatively, the myosin may first be mixed with stoichiometric ATP for sufficient time to ensure maximum ATP binding and then mixed with actin to observe k_{rec}. Considerable advantage is obtained by double mixing, as the length of aging can be adjusted so that very little [MgATP] remains free in solution. In practice the maximum amount of binding is limited by the rate of $M \cdot ADP \cdot P_i$ decomposition to M in the absence of actin. The optimum time of delay, T_d (maximum $M \cdot ADP \cdot P_i$ with minimum myosin and free ATP), can be estimated using Eq. (8).

$$T_d^{-1} = (k_a k_d [M])^{1/2} \qquad (8)$$

and is ca 1 sec using the rate constants $k_a = 10^6\,M^{-1}\,sec^{-1}$, $k_d = 0.05\,sec^{-1}$, $[M] = 10^{-5}$. Under these conditions less than 5% of M and ATP are free.

H⁺ Release. Proton-release measurements are the most demanding of the stopped-flow measurements. A pH indicator such as chlorophenol is included with the protein to measure proton release. If the following conditions are met, reasonably good data can be obtained.

1. Exclusion of carbon dioxide is necessary. All solutions should be made from boiled distilled water and stored under nitrogen or a soda lime trap.
2. The pH of the protein and ATP should be adjusted to be as nearly identical as possible just before the experiment.
3. It should be demonstrated that the indicator is free in solution and does not bind to any of the components of the reaction.
4. No components of the stopped-flow apparatus can produce or absorb protons. Cells containing stainless steel or aluminum components may contribute to artifactual changes and should be avoided.

[56] Interaction of Actin and Myosin in the Presence and the Absence of ATP

By Lois E. Greene and Evan Eisenberg

The fundamental process that drives muscle contraction as well as many forms of cell motility is the interaction of actin, myosin, and ATP. Two major problems arise in studying this interaction *in vitro*. First, under conditions where actin and myosin interact *in vivo,* both proteins occur as filaments, which makes it difficult to study their interaction quantitatively *in vitro*. This has led to the widespread use of the soluble fragments of myosin that are produced by proteolytic digestion of the myosin molecule. Two types of fragments have been used: heavy meromyosin (HMM), the two-headed fragment of myosin, and subfragment-one (S-1), the single-headed fragment of myosin.[1] Both HMM and S-1 retain the ATP and actin-binding properties of the parent myosin molecule. At the same time, both are soluble at low ionic strength, so that their interaction with actin and ATP can be quantitatively studied *in vitro*.

Studies in a number of laboratories using HMM and S-1 have shown that these fragments bind to actin both in the presence and in the absence of ATP.[2] This leads to the second major problem that arises in studying the interaction of actin, myosin, and ATP: the binding constant of myosin to actin varies over four orders of magnitude depending on the state of the nucleotide bound to the active site of myosin. Under conditions where the binding constant of S-1 to actin is $>10^8 \, M^{-1}$, the binding constant of S-1 · ATP is only about $10^4 \, M^{-1}$.[2,3] In addition, studies in the presence of ATP must take into consideration that ATP hydrolysis is occurring as the measurement is taken. Thus quite different techniques are used to study the interaction of actin and myosin in the presence and in the absence of ATP.

We will first discuss the methods used to study the interaction of actin and myosin in the absence of ATP hydrolysis, i.e., under equilibrium conditions. Second, we will discuss the methods used to study the much weaker interaction of actin and myosin that occurs during steady-state ATP hydrolysis.

---

[1] S. Lowey, *in* "Fibrous Proteins: Scientific, Industrial and Medical Aspects" (D. A. D. Parry and L. K. Creamer, eds.), Vol. 1, p. 1. Academic Press, New York, 1979.

[2] R. S. Adelstein and E. Eisenberg, *Annu. Rev. Biochem.* **49,** 921 (1980).

[3] L. A. Stein, R. P. Schwarz, P. B. Chock, and E. Eisenberg, *Biochemistry* **18,** 3895 (1979).

METHODS IN ENZYMOLOGY, VOL. 85   ISBN 0-12-181985-X

Equilibrium Binding of S-1 to F-actin

The binding of S-1, the single-headed soluble fragment of myosin, to actin has been examined under equilibrium conditions both in the presence and in the absence of ATP analogs (nonhydrolyzable nucleotides), e.g., AMP-PNP and ADP. As shown by Scheme 1, the effect of ATP analogs is to weaken the binding of S-1 to actin,[4,5]

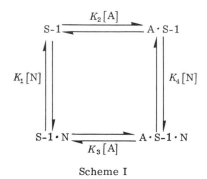

Scheme I

where A = actin, M = S-1, N = ATP analog, and K = association constants. Experimentally, this means that, in the presence of ATP analogs, higher concentrations of S-1 and actin are needed to measure the association constant of S-1 to F-actin than in the absence of analogs. As with any binding study, it is important to work over a range of ligand concentration (either actin or S-1 can be considered the ligand) where the extent of S-1 binding to F-actin can be easily detected, yet at the same time, not all of the added ligand is bound. It is difficult to measure directly a relatively strong ($>10^6\,M^{-1}$) actin–S-1 association constant because there is a lower workable limit for the concentration of both free S-1 and F-actin due to denaturation of S-1 and depolymerization of F-actin. It is also difficult to measure a relatively weak ($<10^4\,M^{-1}$) actin–S-1 association constant because there is an upper workable limit for the concentration of S-1 and actin due to the technical problem of concentrating the S-1 and actin. In addition, since S-1 is a proteolytic fragment and may therefore be slightly heterogeneous, it is particularly important to examine a representative portion of the entire S-1 population. The same considerations also apply in measuring the binding of HMM to F-actin, which will be discussed below.

Increasing ionic strength decreases the actin–S-1 association constants both in the presence[6] and in the absence of ATP analogs.[7,8] Therefore,

[4] S. Highsmith, *J. Biol. Chem.* **251**, 6170 (1976).
[5] L. E. Greene and E. Eisenberg, *J. Biol. Chem.* **255**, 543 (1980).
[6] L. E. Greene and E. Eisenberg, *Proc. Natl. Acad. Sci. U.S.A.* **71**, 54 (1978).
[7] S. Marston and A. Weber, *Biochemistry* **14**, 3868 (1975).
[8] S. Highsmith, *Arch. Biochem. Biophys.* **180**, 404 (1977).

ionic strength can be adjusted so that binding of S-1 to actin can be measured without encountering the technical problems described above.

Centrifugation Method

The most common method of investigating the equilibrium binding of S-1 to F-actin is by ultracentrifugation. This method requires no special equipment; analytical,[6,9] preparative,[7,10] and airfuge[11] centrifuges have all been used in these binding studies. The centrifugation method takes advantage of the property of F-actin to sediment readily. Thus, both free actin and acto · S-1 cosediment upon centrifugation, while the unbound S-1 remains in the supernatant. This method can be used to measure binding because, as is shown by Eq. (1),

$$K_2 = \frac{[A \cdot S\text{-}1]}{[A]_{\text{free}}[S\text{-}1]_{\text{free}}} \qquad (1)$$

neither the ratio of $[A \cdot S\text{-}1]/[A]_{\text{free}}$ nor the concentration of $[S\text{-}1]_{\text{free}}$ changes during centrifugation; the $[A]_{\text{free}}$ and $[A \cdot S\text{-}1]$ must sediment at the same rate, since they occur on the same filaments.

In using the centrifugation method, controls should be conducted to check that most (>90%) of the S-1 remains in the supernatant in the absence of actin, but binds to actin in the absence of nucleotide and at low ionic strength. In addition, in the absence of S-1, most (>90%) of the F-actin should sediment upon centrifugation. Although it is generally not considered to be necessary, the equilibrium constants obtained using this method can be corrected for the effect of hydrostatic pressure.[12] To determine whether such a correction is necessary, the equilibrium constants should be measured by centrifuging the actin–S-1 solution at two different speeds.

After the actin-bound S-1 is sedimented, there are many different methods used to measure the concentration of free S-1. The most direct method is to measure the absorbance of the supernatant during centrifugation by using an analytical ultracentrifuge equipped with absorbance optics.[6,9] The advantage of this method is that it requires less centrifugation time since a tight pellet of actin does not have to be formed. Hence, virtually no sedimentation of the S-1 occurs. On the other hand, when a preparative ultracentrifuge or airfuge is used, a tight pellet is required so that the supernatant can easily be removed after centrifugation. In this case, the concentration of unbound S-1 in the supernatant can be deter-

[9] S. S. Margossian and S. Lowey, *Biochemistry* **17**, 5431 (1978).
[10] L. E. Greene, *Biochemistry* **20**, 2120 (1981).
[11] A. Inoue and Y. Tonomura, *J. Biochem. (Tokyo)* **88**, 1643 (1980).
[12] S. B. Marston, *J. Muscle Res. Cell Motil.* **1**, 305 (1980).

mined by protein determination,[13] by ATPase assay,[11] or, most commonly, by modifying S-1 with a radioactive label that enables the concentration of unbound S-1 to be determined from the radioactivity in the supernatant.[7,10] This is the most sensitive method of measuring the concentration of unbound S-1, but with this method, it is necessary to determine the effect of modification on the binding of S-1 to actin. Generally, a radioactive label has been incorporated into S-1 by blocking the SH_1 group with iodoacetamide[7,10,14] or N-ethylmaleimide.[15] These modifications have been shown to have only a slight weakening effect on the binding constant (less than twofold).

Analysis of Binding Data

In analyzing the binding of S-1 to F-actin, several points should be considered. First, S-1 binds independently along the F-actin filament both in the presence and in the absence of ATP analogs,[5,6,9] which allows the data to be plotted according to the Scatchard equation.[16] Second, actin and S-1 bind in a one-to-one complex both in the presence and in the absence of analogs.[6,17,18] Since this stoichiometry is so well established, in some binding studies the actin–S-1 association constant has been calculated by assuming this stoichiometry.[7] Third, in measuring the binding of S-1 to F-actin in the presence of ATP analogs, this experiment can be conducted at an analog concentration where all of the S-1 and acto · S-1 are saturated with analog so as to measure directly the binding constant of S-1 · analog to actin (K_3). K_3 can also be determined by measuring the binding of S-1 to actin at varying concentrations of analog and then extrapolated to infinite analog concentration using Eq. (2).[6,19]

$$K_{app} = \frac{K_3}{K_4[N]} + K_3 \tag{2}$$

The data are then plotted as K_{app} vs $1/[N]$ so that the ordinate intercept equals K_3 and the abscissa intercept equals $-1/K_4$ (K_4 = binding constant of ATP analog to acto · S-1). Having determined values for K_3 and K_4, the binding constant of analog to S-1 alone (K_1) can be measured under the same conditions, e.g., by equilibrium dialysis. The actin–S-1 association constant (K_2) can then be determined from the following equation derived from Scheme 1.

[13] K. Takeuchi and Y. Tonomura, *J. Biochem. (Tokyo)* **70,** 1011 (1971).
[14] L. E. Greene and E. Eisenberg, *J. Biol. Chem.* **255,** 549 (1980).
[15] J. M. Murray, A. Weber, and M. Knox, *Biochemistry* **20,** 641 (1981).
[16] G. Scatchard, *Ann. N.Y. Acad. Sci.* **51,** 660 (1949).
[17] E. Eisenberg, L. Dobkin, and W. Kielley, *Biochemistry* **11,** 4657 (1972).
[18] S. S. Margossian and S. Lowey, *J. Mol. Biol.* **74,** 313 (1973).
[19] W. Hofmann and R. S. Goody, *FEBS Lett.* **89,** 169 (1978).

$$K_2 = (K_1 K_3)/K_4 \qquad (3)$$

Determining a value for K_2 in this manner avoids the problems often associated with directly measuring the relatively strong actin–S-1 association constant ($K_2 > 10^6 \, M^{-1}$).[5,7–9]

The centrifugation method has also been used in competition studies. In this type of study, two types of myosin fragments compete for sites on actin (e.g., unmodified S-1 and modified S-1, or S-1 and HMM).[7,14] The competition method does not measure a binding constant directly, but instead measures the relative affinity of the two types of fragments for actin. Thus, it can be used under conditions where the binding constant of each of the fragments is too strong to be measured directly. An essential control in a competition experiment is to check that changing the order of addition of the two types of fragments has no effect on the results.[14]

After centrifugation, the fraction of each type of fragment bound to actin must be determined. Therefore, with this method it is necessary to be able to distinguish between the two types of fragments. Radioactivity or sodium dodecyl sulfate (SDS) gels can be used for this purpose.

Other Methods

Although the centrifugation method is by far the most popular method used in equilibrium binding studies of S-1 to actin, several other methods have also been used. Measuring the turbidity or light scattering of the acto·S-1 complex has frequently been used in binding studies.[11,19,20] In general, turbidity has been shown to be a reliable measure of the fraction of S-1 bound. However, under certain conditions problems may arise. For example, as shown by Marston,[12] the turbidity did not change proportionately with the fraction of S-1 · AMP-PNP bound to dansylaziridine-modified actin. (However, this problem did not occur with unmodified actin.) Turbidity has also been used to measure the rate of S-1 binding to actin in kinetic studies where the rates of S-1 association with and dissociation from actin are measured. Using these rate constants, White and Taylor[21] have calculated the actin–S-1 association constant. However, the time course of S-1 binding to actin cannot always be fitted by a single exponential,[20] which limits the usefulness of this method.

Another technique used to measure the binding of S-1 to actin is fluorescence depolarization, which was developed by Mendelson and co-workers.[22,23] This technique uses the difference in rotational mobility

[20] K. Trybus and E. W. Taylor, *Proc. Natl. Acad. Sci. U.S.A.* **77**, 7209 (1980).

[21] H. D. White and E. W. Taylor, *Biochemistry* **15**, 5818 (1976).

[22] R. Mendelson, M. Morales, and J. Botts, *Biochemistry* **12**, 2250 (1973).

[23] S. Highsmith, R. Mendelson, and M. Morales, *Proc. Natl. Acad. Sci. U.S.A.* **76**, 133 (1976).

between free and bound S-1 to determine the actin–S-1 association constant. Values of the actin–S-1 association constant obtained by this method in the absence of ATP analogs are in agreement with the values obtained by other laboratories.[5,7,9,23] However, in the presence of AMP-PNP, this method[4] gave a much weaker actin–S-1 association constant compared to values obtained by other methods.[5,19]

Equilibrium Binding of HMM to F-actin

The binding of HMM, the two-headed fragment of myosin, to actin has also been studied primarily by the centrifugation method, but the binding of HMM is not as well understood as the binding of S-1. In the absence of ATP analogs, the two heads of HMM mainly appear to bind to adjacent F-actin monomers (one HMM molecule per two F-actin monomers). The mathematical analysis of this binding (see Hill[24] for equations) is rather complex because of the "parking problem"; i.e., the number of points where two adjacent actin sites are available for binding decreases more rapidly than the total number of free actin sites. Because of the parking problem, it is incorrect to analyze the data using the Scatchard equation; a somewhat more complex equation[24] should be used.

To determine the role of each of the HMM heads in the binding of HMM to actin, this binding can be analyzed as a two-step process.[14,25,26] The first head binds in a second-order reaction with a binding constant similar to that of S-1. This is followed by the binding of the second head of HMM in a first-order reaction at an adjacent F-actin monomer. Therefore, the ratio of the actin–HMM association constant to the actin–S-1 association constant provides a measure of the strength of binding of the second head of HMM to F-actin.[10] The strength of binding of the second head depends not only on its binding energy, but also on the mobility of the second HMM head before it binds to actin, on the overall mobility of the HMM molecule after both heads bind, and on any distortion that may occur after both heads bind.[26] If there was no distortion when the two HMM heads bound to F-actin (i.e., "independent" binding of the heads occurred), the actin-HMM association constant could not be equal to the square of the actin–S-1 association constant since the units would be different.[14]

The strong binding of HMM to F-actin that occurs in the absence of ATP analogs ($K \geq 10^7 \ M^{-1}$) has been measured using several different methods. Margossian and Lowey[9] have measured this binding directly by

[24] T. L. Hill, *Nature* (*London*) **274**, 285 (1978).
[25] S. Highsmith, *Biochemistry* **17**, 22 (1978).
[26] T. L. Hill and E. Eisenberg, *Biophys. Chem.* **11**, 271 (1980).

determining the unbound HMM after sedimenting the acto·HMM complex. They determined that the actin–HMM association constant ($\sim 10^7$ M^{-1}) was only 10-fold stronger than the actin–S-1 association constant. A similar result was obtained by Highsmith[25] using fluorescence depolarization. In contrast, we found that HMM binds 600-fold stronger to F-actin than does S-1.[14] We obtained this measurement by having HMM and S-1 compete for sites on F-actin. In these experiments, varying concentrations of S-1 were added to a fixed concentration of HMM and F-actin, and the amounts of bound HMM and S-1 were determined after centrifugation. This method gives the ratio of the actin-HMM to the actin–S-1 association constant. The actin–S-1 association constant was determined under the same conditions, enabling the actin–HMM association constant to be calculated ($K \sim 10^9 M^{-1}$). It is not yet clear why the results of the direct method and competition methods for measuring the binding of HMM to actin differ so greatly.

In the presence of ATP analogs, the binding of the HMM · analog complex to actin is relatively weak, enabling the actin–HMM association constant to be easily measured. Under these conditions, agreement between various methods is better. Centrifugation is again the most widely used method,[9–11] although light scattering has also been used to measure the binding of HMM to actin in the presence of AMP-PNP.[11] There are, however, problems in measuring HMM binding to actin by turbidity. Absorbance changes with HMM are not as linear as with S-1.[21] In addition, there are problems with standardizing turbidity measurements under conditions where HMM may bind to actin with only one head. One-headed binding of HMM to F-actin should predominate over two-headed binding when the actin–HMM association constant approximately equals the actin–S-1 association constant (i.e., at high ionic strength in the presence of AMP-PNP).[10] In this case, HMM should bind with a stoichiometry of one HMM molecule per F-actin monomer unless the unbound head interferes with the binding of another HMM molecule at an adjacent F-actin site. Whether this, in fact, occurs has not yet been determined. Labeling of the HMM heads with paramagnetic probes that detect mobility may help to resolve this question.[27]

Equilibrium Binding of S-1 to Regulated Actin

The equilibrium binding of S-1 to the troponin–tropomyosin–actin complex (regulated actin) shows positive cooperativity in the presence and in the absence of nucleotides.[15,20,28] At low levels of saturation of the actin with S-1, S-1 binds more weakly to actin than it does at high levels of

[27] D. D. Thomas, S. Ishiwata, J. Seidel, and J. Gergely, *Biophys. J.* **32**, 873 (1980).
[28] L. E. Greene and E. Eisenberg, *Proc. Natl. Acad. Sci. U.S.A.* **77**, 2616 (1980).

saturation. Therefore, to measure the cooperativity of S-1 binding to regulated actin, it is necessary to work at a much lower level of saturation of actin with S-1 ($\leq 10\%$) than is usually employed when determining the actin–S-1 association constant with unregulated actin. It is important to verify that the reduction in S-1 binding to regulated actin observed at low levels of saturation is due to cooperativity, not to denaturation of the S-1 at the very low S-1 concentrations being used. This can be done by measuring the binding of S-1 to unregulated actin under the same conditions (and in particular at the same low free S-1 concentrations) used in the binding studies with regulated actin. If no artifact is occurring, the binding of S-1 to unregulated actin will be much greater than to regulated actin.[28] In addition, it is important to establish that the actin is saturated with the troponin–tropomyosin complex under the condition of the binding study by varying the troponin–tropomyosin concentration.[28] Likewise, in binding studies conducted in the presence of ATP analogs, saturation of the acto · S-1 complex with analog should be checked both in the weak and strong binding regions by establishing that doubling the analog concentration does not change the binding of S-1.[28] Calcium concentration affects the cooperativity and therefore must be carefully controlled in these experiments. Finally, diadenosine pentaphosphate should be added to inhibit myokinase activity.[29] This is especially important when the binding studies are conducted in the presence of ADP.

The same methods used in measuring the binding of S-1 to unregulated actin have been used to study the binding of S-1 to regulated actin. In addition, Trybus and Taylor[20] have used fluorescently labeled troponin to study the binding of S-1 to regulated actin. Rather than measure actual binding, this probe appears to reflect movement of the tropomyosin molecule when S-1 binds to regulated actin.

The data for the binding of S-1 to regulated actin must be analyzed using a cooperative binding model. A theoretical model for the cooperative binding of S-1 to regulated actin has been developed by Hill et al.[30] In this model, actin is postulated to exist in two forms, a form that binds S-1 weakly (weak form) and a form that binds S-1 strongly (strong form). The cooperative response stems from two sources: the seven actin sites act as a single unit when they change from the weak to the strong form and there are interactions between adjacent tropomyosin molecules. On the other hand, at very low and very high levels of saturation of the actin with S-1, S-1 should bind noncooperatively to actin[28] and the data can be treated accordingly. In these regions, Scheme 1, as well as the equations based on this scheme, can now be applied, assuming that in

[29] G. E. Lienhard and I. I. Secemski, *J. Biol. Chem.* **248**, 1121 (1973).
[30] T. L. Hill, E. Eisenberg, and L. E. Greene, *Proc. Natl. Acad. Sci. U.S.A.* **77**, 3186 (1980).

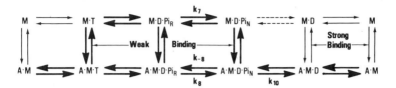

FIG. 1. The kinetic scheme of Stein *et al.*[3] The predominant pathway for the actomyosin ATPase cycle is shown by the heavy solid lines. The rate constants, k_7 and k_8, determine the rate-limiting steps in the actomyosin ATPase cycle. The dashed arrow indicates the rate-limiting step in the myosin ATPase cycle in the absence of actin. M = S-1; A = an actin monomer in an F-actin filament; T = ATP; D = ADP; and R and N subscripts indicate the refractory and nonrefractory states, respectively.

these regions the regulated actin is totally either in the weak or the strong form. Although there is no difficulty in working in the region where the actin is completely in the strong form,[28] care must be taken to ensure that, in the region of weak binding, the actin is indeed totally in the weak form. The latter point can be checked by working over a range of S-1 concentrations and determining that the association constant of S-1 to regulated actin does not change significantly.

Steady-State Binding of S-1 to Actin

Measurement of K_{ATPase}

The binding of S-1 to F-actin in the presence of ATP can be studied using many of the methods mentioned above, but, in addition, kinetic methods can be used. Figure 1 shows the kinetic model for the actomyosin ATPase proposed by Stein *et al.*[3] This model includes all the major steps proposed for the actomyosin ATPase cycle[2] and thus serves as a useful basis for discussion. One of the key properties of the myosin ATPase is that, in the absence of actin, the rate of P_i release (dashed arrow in Fig. 1), and hence the overall ATPase rate, is very slow. Actin activates the MgATPase activity of myosin by greatly increasing the rate of P_i release. Thus, one of the most common methods for investigating the interaction of actin and myosin in the presence of ATP is to study the dependence of the myosin MgATPase on actin concentration. In the absence of actin, the MgATPase activity of S-1 is very low, whereas in the presence of actin, it can be more than 200-fold increased in activity. This dependence of the S-1 ATPase activity on actin concentration is hyperbolic, and a double-reciprocal plot of ATPase activity per mole of S-1 vs actin concentration is linear.[31] When the soluble proteolytic fragments of myosin, HMM, or S-1

[31] E. Eisenberg and C. Moos, *J. Biol. Chem.* **245**, 2451 (1970).

are used, the double-reciprocal plot is linear over a wide range of conditions; i.e., the actin-activated ATPase activity follows simple Michaelis–Menten kinetics.[31] The ordinate and abscissa intercepts of this double-reciprocal plot give values for V_{max} (the maximum ATPase rate) and K_{ATPase} (the actin concentration where the ATPase activity is half of V_{max}), respectively.

In using the double-reciprocal plot several precautions should be taken. First, and most important, the plot is meaningful only if the actin concentration plotted on the abscissa is the free actin concentration.[31] Unfortunately, in experiments where myosin filaments rather than HMM or S-1 are employed, this point is sometimes ignored. Under some conditions linear double-reciprocal plots are obtained with myosin filaments, but these plots generally have lower values for both V_{max} and K_{ATPase}.[32] The exact reasons for these differences remain unclear. Reisler[32] has developed a method for making uniform preparations of small myosin filaments known as myosin minifilaments. Using these myosin minifilaments, he obtained a double-reciprocal plot identical to that obtained with HMM. Therefore, myosin minifilaments may prove to be useful in studying the actin-activated ATPase activity of myosin in a quantitative manner.

A second precaution that should be taken when using the double-reciprocal plot of ATPase activity vs actin concentration is to work over at least a 10-fold range of actin concentration that brackets K_{ATPase}. This enables accurate determination of both V_{max} and K_{ATPase}. K_{ATPase} greatly increases as the ionic strength is increased.[33] Therefore, to obtain accurate double-reciprocal plots at physiological ionic strength would require extremely high actin concentrations that may be technically unobtainable. In work at high actin concentration, solutions are very viscous, and therefore adequate mixing must be employed. For this reason, it is best to assay the ATPase activity directly by measuring $[\gamma\text{-}^{32}P]P_i$ production at high actin concentration, rather than by using the pH-stat method.[34] In general, studies of the actin-activated HMM or S-1 ATPase activity have been conducted at very low ionic strength, where there is no need for high actin concentrations ($K_{ATPase} \simeq 10 \ \mu M$).[3,34]

Although determination of the double-reciprocal plot of ATPase activity vs actin concentration is the most commonly employed method of studying the interaction of actin, myosin, and ATP, other kinetic methods have also been used. A simple variant of the double-reciprocal plot is obtained by holding the actin concentration fixed at a low concentration (1–5 μM) and then varying the S-1 concentration.[18,35] In this case, after

[32] E. Reisler, *J. Biol. Chem.* **255**, 9541 (1980).
[33] A. A. Rizzino, W. W. Barouch, E. Eisenberg, and C. Moos, *Biochemistry* **9**, 2402 (1970).
[34] L. Stein, P. B. Chock, and E. Eisenberg, *Proc. Natl. Acad. Sci. U.S.A.* **78**, 1346 (1981).

correcting for the ATPase activity of the S-1 alone, a linear double-reciprocal plot of ATPase rate per actin monomer vs free S-1 concentration is obtained. In general, the ordinate and abscissa intercepts in this plot are not identical to the intercepts in the usual double-reciprocal plot of ATPase rate (per S-1) vs actin concentration.[18,35,36] Based on the simplest kinetic models, these intercepts would be expected to be identical.[18] As described below, differences between the intercepts of the two types of plots can be explained by the kinetic model shown in Fig. 1. Nevertheless, because of difficulties in interpretation, the double-reciprocal plot of ATPase rate per actin monomer vs S-1 concentration is not used nearly as much as the double-reciprocal plot of ATPase rate per S-1 vs actin concentration.

A third kinetic method that has been used to study the interaction of actin and myosin is based on the ability of actin to *inhibit* the K^+-ATPase activity of myosin in the absence of Mg^{2+}.[37-39] Since this inhibition occurs only in the complete absence of Mg^{2+}, EDTA must be present in the assay mixture. Furthermore, since K^+ is the activating ion, the assay is generally carried out at high KCl concentration (e.g., $0.3\ M$ KCl). The ability of actin to inhibt the K^+-ATPase activity of myosin at $0.5\ M$ KCl has been used to study the interaction of actin and myosin under conditions where the myosin exists as monomers.[39] These studies showed that skeletal and smooth muscle myosin bind quite differently to actin. It must be noted, however, that assays conducted in the absence of Mg^{2+} are probably much less reflective of physiological behavior than those done in the presence of Mg^{2+}. Furthermore, like the analysis of the actin-activated myosin ATPase activity, the analysis of actin inhibition requires a kinetic model. Only in this case, much less work has been done in developing a complete kinetic model, so interpretation of the results are somewhat more ambiguous.

Measurement of $K_{binding}$

New methods have been developed for directly studying the interaction of actin and myosin in the presence of ATP. In contrast to the kinetic measurements, the binding constant ($K_{binding}$) obtained by direct measurement of the binding of actin and myosin in the presence of ATP is not as dependent on a kinetic scheme for interpretation of the results. One

[35] P. D. Wagner and A. G. Weeds, *Biochemistry* **18**, 2260 (1979).
[36] E. Eisenberg and W. W. Kielley, *Cold Spring Harbor Symp. Quant. Biol.* **37**, 145 (1972).
[37] S. Barron, E. Eisenberg, and C. Moos, *Science* **151**, 1541 (1966).
[38] R. Cooke and K. Franks, *Biochemistry* **19**, 2265 (1980).
[39] J. M. Krisanda and R. A. Murphy, *J. Biol. Chem.* **255**, 10771 (1980).

method that has been used to obtain K_{binding} is to measure the turbidity of the actin–S-1–ATP mixture during steady-state ATP hydrolysis. Since the ATPase rate is very high, except at low temperature, this measurement is most easily made on a stopped-flow apparatus by measuring percentage transmission.[3] The stopped-flow apparatus is standardized by setting the voltage reading with buffer to zero absorbance and the voltage reading with the shutter closed to 0% transmission. In this way, absolute values of absorbance are obtained. The relationship between absorbance and the amount of S-1 bound to actin is then empirically standardized by using the values of absorbance obtained when S-1 binds to actin in the absence of ATP. The major assumption of this technique is that the binding of a given amount of S-1 to actin yields the same change in absorbance whether it occurs in the presence or in the absence of ATP. This assumption has been verified by measuring the binding of S-1 to actin in the presence of ATP using an airfuge.[40,41]

M · T and M · D · P_{i_R} (see Fig. 1) are in very rapid equilibrium with A · M · T and A · M · D · P_{i_R}, respectively[3]; equilibrium is reached in less than 1 msec. Since the half-life of the transition between M · T and M · D · P_{i_R} is more than 20 msec at 15°,[42,43] a direct determination of the binding constants of both M · T and M · D · P_{i_R} to actin can be obtained by following the absorbance changes that occur either when S-1 is directly mixed with actin and ATP or S-1 is premixed with ATP to form S-1 · ADP · P_{i_R} and then mixed with actin.[3] The results of such experiments show that the binding constants of M · T and M · D · P_{i_R} to actin are similar; both are about 100-fold weaker than the binding constant of S-1 to actin in the presence of AMP-PNP,[3] which in turn is about 500-fold weaker than in the absence of nucleotide.[5]

The use of stopped-flow absorbance measurements to determine the binding constants of M · T and M · D · P_{i_R} to actin have been confirmed by direct measurement of binding in the presence of ATP using the preparative ultracentrifuge or airfuge. In these experiments, the free concentration of S-1 in the supernatant is assayed either by SDS gel electrophoresis[35] or by measuring the NH_4^+-ATPase activity in the presence of EDTA.[40,41] In these experiments the S-1 concentration must be kept low enough so that ATP is not completely hydrolyzed during the centrifugation. In the case of the airfuge it takes about 20 min to sediment the actin completely.

The binding constant of S-1 · ATP to actin has also been determined using a kinetic method involving the ability of actin to cause exchange of

[40] A. Inoue, M. Ikebe, and Y. Tonomura, J. Biochem. (Tokyo) 88, 1633 (1980).
[41] J. Chalovich and E. Eisenberg, J. Biol. Chem. 257, 2432 (1982).
[42] K. A. Johnson and E. W. Taylor, Biochemistry 17, 3432 (1978).
[43] S. P. Chock, P. B. Chock, and E. Eisenberg, J. Biol. Chem. 254, 3236 (1979).

S-1-bound ATP with free ATP.[44] All of these methods give similar values for the binding constants of S-1 · ATP and S-1 · ADP · P_{i_R} to actin. The method of choice will depend on the conditions of the particular experiment involved. In most studies, these methods have been employed with S-1 only in the presence of ATP. However, in a few studies, HMM has also been employed.[40] Particularly in experiments involving absorbance measurements, results with HMM must be interpreted with caution, as discussed above. Both HMM heads bind to actin in the absence of ATP, whereas only one head may bind to actin in the presence of ATP.[10] This makes standardization of the absorbance changes more complex than with S-1.

Analysis of K_{ATPase} and $K_{binding}$

The interaction of actin and myosin in the presence of ATP is described by the following two constants: K_{ATPase}, a kinetic constant obtained from the double-reciprocal plot of S-1 ATPase vs actin concentration and $K_{binding}$, a true binding constant obtained in direct binding studies of S-1 to actin in the presence of ATP. With the simplest kinetic models, K_{ATPase} would be expected to equal $K_{binding}$. However, with skeletal muscle S-1 at low ionic strength and temperatures below 15°, K_{ATPase} is found to be 5- to 10-fold stronger than $K_{binding}$; i.e., the actin activated ATPase activity reaches its half maximal value at a considerably lower actin concentration than is required for half-maximal binding of S-1 to actin in the presence of ATP.[35,36] Stein et al.[3] have suggested that this phenomenon is a variant of a common kinetic observation: Michaelis constants obtained from kinetic experiments often differ in value from true binding constants. In the model in Fig. 1, this effect occurs because of the presence of a rate-limiting conformational change that occurs at the same rate whether the S-1 is attached to or detached from actin (M · D · P_{i_R} → M · D · P_{i_N} and A · M · D · P_{i_R} → A · M · D · P_{i_N}) and which precedes the fast release of P_i when S-1 is bound to actin (A · M · D · P_{i_N} → A · M · D). In this model, the differences between K_{ATPase} and $K_{binding}$ mainly depend on the ratio of k_{-8} and k_{10}; changes in this ratio could explain why the difference between K_{ATPase} and $K_{binding}$ depends on experimental conditions. This model also gives a simple meaning to V_{max}; it is equal to k_7 or k_8, which are assumed to be equal. Finally this model can account for the difference in the intercepts observed when actin is varied at fixed S-1 concentration or S-1 is varied at fixed actin concentration (see above). The difference in these intercepts are proportional to the difference observed between K_{ATPase} and $K_{binding}$.[45]

[44] J. A. Sleep and R. L. Hutton, *Biochemistry* **17**, 5423 (1978).
[45] L. Stein, unpublished results.

Taylor[46] has suggested possible alternative models for the difference observed between K_{ATPase} and $K_{binding}$ that involve differences in the binding constants of S-1 · ATP and S-1 · ADP · P_{i_R} to actin, and work is presently underway in several laboratories to test the kinetic model shown in Fig. 1. In any event, whatever the final model turns out to be, it is clear that care must be taken not to identify constants obtained from kinetic measurements with true equilibrium constants or single-rate constants.

In regard to interpretation of the binding constants of S-1 to actin obtained *in vitro*, another problem should be noted. When S-1 or S-1 · ATP interact with actin *in vitro*, a second-order equilibrium constant is obtained; i.e., the amount of bound S-1 depends on the actin concentration. This is not the case *in vivo*, where the binding of the cross-bridge to actin is a first-order process.[26] Because the actin and myosin filaments exist in a highly structured array *in vivo*, the cross-bridge does not "see" an actin concentration, it sees only one or two binding sites along the actin filament. Therefore, the values of the equilibrium constants for actin binding obtained *in vitro* cannot be directly applied to the *in vivo* situation. Such an application requires an estimate of the first-order rate of attachment of the cross-bridge to actin *in vivo*.[26] This necessitates an estimate of the "effective" actin concentration, which is the actin concentration required *in vitro* to mimic the *in vivo* situation. Similar problems in estimating the binding constant of the cross-bridges to actin occur when actin and myosin filaments interact *in vitro* or in myofibrils.

Steady-State Binding of S-1 to Regulated Actin

The interaction of S-1 with regulated actin in the presence of ATP is a considerably more complex kinetic problem than this interaction in the absence of the regulatory proteins, troponin–tropomyosin. The difficulty occurs in the kinetic studies rather than in the binding studies, which appear to be quite straightforward. The same methods have been used to study the binding of S-1 to regulated actin in the presence of ATP as were used in the absence of the regulatory proteins, and the results are quite similar. Unlike the large cooperative effect observed when S-1 or S-1 · ADP binds to regulated actin,[15,20,28] troponin–tropomyosin has almost no effect on the binding of S-1 · ATP or S-1 · ADP · P_{i_R} to actin.[41,47] Therefore, in measuring the direct binding of S-1 to regulated actin in the presence of ATP, the same methods (absorbance and ultracentrifugation) can be used with and without the regulatory proteins.

[46] E. W. Taylor, *CRC Crit. Rev. Biochem.* **6**, 103 (1979).
[47] J. M. Chalovich, P. B. Chock, and E. Eisenberg, *J. Biol. Chem.* **256**, 575 (1981).

In contrast, kinetic studies examining the effect of tropomyosin or troponin–tropomyosin on the actin-activated ATPase rates are much more complex. Cooperative effects occur both at relatively high ratios of S-1 to actin and at low ATP concentration.[48,49] Interpretation of these effects requires a kinetic model similar to that of Fig. 1, but with the major addition that the cooperative units of the actin filament themselves can exist in at least two forms and cooperative transitions occur between these two forms. Obviously, such a kinetic model is quite complex, but such a model must be employed to analyze quantitatively the cooperative effects that occur in the presence of tropomyosin or troponin–tropomyosin. A preliminary form of such a kinetic model has been described.[50]

Potential Methods for Studying Actomyosin Interaction

In addition to the techniques thus far described, several new techniques provide a possibility for studying the interaction of actin and myosin both under equilibrium conditions and in the presence of ATP. One of the areas that is least well understood is the interaction of the actin and myosin filaments. Here, the use of the myosin minifilaments described by Reisler[32] should be useful since their kinetic interaction with actin and ATP appears to be similar to the kinetic interaction observed with HMM. Another technique that may prove to be useful in studying the interaction of actin and myosin filaments is ^{18}O exchange.[51–53] It is clear that actin has a major effect on ^{18}O exchange, reducing the magnitude of the ^{18}O exchange from four to one ^{18}O per P_i.[52,53] With S-1, one is the minimum number possible during ATP hydrolysis. In contrast, with actomyosin the magnitude of the ^{18}O exchange decreases to two ^{18}O per P_i at very low ratios of actin to myosin and then remains constant at this level although the ATPase activity is still increasing as the actin concentration is increased.[54,55] Although a complete explanation of this effect is not yet available, ^{18}O exchange may prove to be a useful method for studying the effect of actin in structured systems such as actomyosin, myofibrils, and

[48] R. D. Bremel and A. Weber, *Nature* (*London*) **97**, 238 (1972).
[49] R. D. Bremel, J. M. Murray, and A. Weber, *Cold Spring Harbor Symp. Quant. Biol.* **37**, 267 (1972).
[50] T. L. Hill, E. Eisenberg, and J. M. Cahlovich, *Biophys. J.* **35**, 99 (1981).
[51] C. R. Bagshaw, D. R. Trentham, R. G. Wolcott, and P. D. Boyer, *Proc. Natl. Acad. Sci. U.S.A.* **72**, 2592 (1975).
[52] J. A. Sleep and P. D. Boyer, *Biochemistry* **17**, 5417 (1978).
[53] K. K. Shukla and H. M. Levy, *Biochemistry* **16**, 132 (1977).
[54] J. A. Sleep, D. Hackney, and P. D. Boyer, *J. Biol. Chem.* **255**, 4094 (1980).
[55] K. K. Shulka, H. M. Levy, F. Ramirez, J. Marecek, S. Meyerson, and E. Kuhn, *J. Biol. Chem.* **255**, 11344 (1980).

skinned muscle fibers. Two other methods that may also be useful in the future are the use of actin labeled with fluorescent derivatives that are sensitive to S-1 binding[56] and the effect of actin on the nuclear magnetic resonance properties of S-1.[57]

[56] T. Kouyama and K. Mihashi, *Eur. J. Biochem.* **114**, 33 (1981).

[57] S. Highsmith, K. Akasaka, M. Konrad, R. Goody, K. Holmes, N. N. Wade-Jardetzky, and O. Jardetzky, *Biochemistry* **18**, 4238 (1979).

Author Index

Numbers in parentheses are reference numbers and indicate that an author's work is referred to although the name is not cited in the text.

Fuchs, S., 522
Fujisaki, H., 581
Fujiwara, K., 181, 202, 536
Fukada, E., 165, 211, 212(14), 213(14), 217(14), 224(14), 226(14), 231, 232, 481
Fusenig, N. E., 489

G

Gabbiani, G., 490, 554
Gaber, B. P., 113
Gachechiladze, N. I., 623
Gadasi, H., 72, 341, 342(25), 346(25), 347(25), 359, 360, 361(1)
Gadd, J. O., 192, 193(22)
Gadian, D. G., 624, 628(3), 644(3), 646(3), 647(44), 649(44), 651(6), 653, 654, 656(6), 657(55), 658(3), 660(55), 664, 665(62), 666(3, 62), 667(3), 668(3, 34, 38), 669(38, 62), 670(3), 672(3, 34), 673(8, 34, 38, 64)
Gaffney, B. J., 596, 606
Galardy, R. E., 285
Gallagher, M., 99
Gallo, M., 322, 323(2)
Ganapathi, M., 309
Gard, D. L., 489, 490(7, 9), 491, 492(9), 494(9), 505, 506
Garlick, P. B., 646, 664, 665(62), 666(62), 669(41, 62), 672(68), 673(68)
Gaskin, F., 376, 386, 405, 417, 434, 435, 436, 438(7)
Gaylinn, B. D., 160
Gazith, J., 75, 76(25)
Gealt, M. A., 391
Geiger, B., 534, 536, 551(60)
Geils, R. H., 217
Gekko, K., 405
Gergely, J., 53, 54(60), 55(60), 85, 89, 99, 165, 168(11), 177, 242, 579, 590, 597, 598(14), 600, 602(27), 603(27), 607(14, 27), 608(43), 609, 610, 611, 612(53), 613(27, 43), 614(27, 43), 615(51), 616(49), 617(49), 619(43), 622, 623(78), 680, 689, 690(32), 715
Gershman, L. C., 72
Gethner, J. S., 435, 438(7)
Gibbons, B. H., 452, 454, 457, 459, 464(32), 465, 469(54), 472, 473, 474

Gibbons, I. R., 389, 451, 452, 454(1, 8), 457(22), 459(22), 460(22), 462(10), 464, 465(22), 466(2, 22), 467(2), 468(1, 2, 31, 41), 469(8, 22, 54), 472(8, 20, 22), 473(10, 22), 474(22, 31)
Gibson, Q. H., 704
Gibson, W., 128
Giglio, M., 562(6), 563
Gillen, K. T., 635
Gitler, C., 206
Glascoe, P. K., 20
Glenney, J. R., 191
Glonek, T., 624, 625(4), 627(4, 13, 14), 628, 629, 633(18), 634(2), 635, 644(4), 647(18), 648(4, 35), 649, 650(4, 5), 651(2), 654(19), 655, 656, 657(4), 658(4), 660(4, 19), 661(46), 664(53), 668(5), 669(5), 670(4), 672(2, 4)
Glover, J. S., 500
Glynn, P., 649, 656
Goad, W. B., 408
Godfrey, J. E., 135
Goldman, A. E., 490, 494(13)
Goldman, M., 515, 520(6), 521(6)
Goldman, R., 181, 332(27), 343(27), 344, 345(27), 356(27)
Goldman, R. D., 364, 490, 494(13), 515, 557
Goldstein, G., 503
Goldstein, L., 67
Goldstein, M., 194
Goll, D. E., 316, 696
Gomer, R. H., 507
Gominak, S., 393
Goodman, M., 680
Goodno, C. C., 116, 117, 118(6), 119(6), 120, 122(6)
Goody, R. S., 56, 373, 712, 713(19), 714(19)
Gorbunoff, M. H., 438
Gordon, D. J., 181, 337, 339, 373
Gordon, J., 502, 521
Gordon, W. G., 521
Gorecka, A., 308, 364, 365
Gottesman, M. M., 490
Gouvea, M. A., 177
Graberek, Z., 689, 690(32)
Graceffa, P., 599
Graessmann, A., 553, 554, 556(77, 91)
Graessmann, M., 553, 554, 557(77, 91)
Grandmont-Leblanc, A., 343, 351
Grange, N., 285

Z

Subject Index

A

A segment, preparation, from fresh
  muscle, 19
Acanthamoeba castellanii
  myosin, 5, 357–363
    aggregation studies, 38
    actomyosin Mg^{2+}-ATPase activity
      assay, 359
    assay, 357–359
      general considerations, 357
    Ca^{2+}-ATPase activity, 358
    ion exchange chromatography,
      337–339
    K^+(EDTA)-ATPase activity assay,
      357, 358
    single-headed, 72
  myosin I
    affinity chromatography, antibody
      column, 343, 352, 353
    agarose adsorption
      chromatography, 341, 342
    purification, 340
    Sepharose-adipic acid hydrazide-
      ADP affinity chromatography,
      346, 347
  myosin IA, 357
    properties, 360, 361
    purification, 359–361
  myosin IB, 357
    properties, 360, 361
    purification, 359–361
  myosin II, 356, 357
    affinity chromatography, antibody
      column, 343, 352, 353
    ATPase activity, 124
    properties, 363
    purification, 361–363
    storage, 354
Acetone muscle powder
  actin extraction from, 165, 166–176
  actin proteolysis, 169, 173–176
  paramyosin extraction from, 153–155

preparation, 153, 166, 167
  from cardiac muscle, 236
  from rabbit skeletal muscle, 236
tropomyosin purification from, 236,
  237
Acetone treatment, cold, for cell fixation,
  517
Acid-paramyosin, 162, 163
Actin, 1
  assembly, 165
  capillary viscometry assay, 183,
    185–190
    effect of frequency on actin
      polymerization, 188–190
  circular dichroism study, 684, 685
  contamination, myosin, 170, 173
  critical concentration, 184, 185
    and degree of flow birefringence,
      195
    in DNase I inhibition assay, 205,
      206
    effect of actin-binding proteins,
      210
    measurement, 209, 210
  crosslinking, by filamin, 322
  cytoplasmic
    assay, 371
    preparation, 371–373
    properties, 372, 373
    purification, 371, 372
  depolymerization, 170, 171
  difference spectra assay, 183, 190, 191
  displacement of vanadate from
    myosin active site, 118–120
  distances between sites in,
    measurement, by fluorescent
    energy transfer, 592, 593
  DNase I inhibition assay, 183,
    204–207
    fluorescent, 206
    modifications, 206
  dual-labeling, 536

G